GLIAL-NEURONAL INTERACTION

ANNALS OF THE NEW YORK ACADEMY OF SCIENCES
Volume 633

GLIAL-NEURONAL INTERACTION

Edited by N. J. Abbott

The New York Academy of Sciences
New York, New York
1991

Library of Congress Cataloging-in-Publication Data

Glial-neuronal interaction/edited by N.J. Abbott.
p. cm. — (Annals of the New York Academy of Sciences, ISSN 0077-8923; v. 633)
Result of a conference sponsored by the New York Academy of Sciences and held on Sept. 4–7, 1990 at the University of Cambridge, Cambridge, England.
Includes bibliographical references and index.
ISBN 0-89766-679-8 (cloth: alk. paper). — ISBN 0-89766-680-1 (paper: alk. paper)
1. Neuroglia—Congresses. 2. Neurons—Congresses. 3. Cell interaction—Congresses. I. Abbott, N. J. (N. Joan) II. New York Academy of Sciences. III. Series.
[DNLM: 1. Cell Communication—congresses. 2. Nerve Regeneration––congresses. 3. Neuroglia—congresses. W1 AN626YL v. 633 / WL 102 G55875 1990]
Q11.N5 vol. 633
[QP363.2]
500 s—dc20
[591.1'88]
DNLM/DLC
for Library of Congress

91-41038
CIP

SP
Printed in the United States of America
ISBN 0-89766-679-8 (cloth)
ISBN 0-89766-680-1 (paper)
ISSN 0077-8923

ANNALS OF THE NEW YORK ACADEMY OF SCIENCES

Volume 633
December 15, 1991

GLIAL-NEURONAL INTERACTION[a]

Editor
N. J. ABBOTT

Conference Organizers
N. J. ABBOTT, E. M. LIEBERMAN, AND M. C. RAFF

CONTENTS

[a]This volume is the result of a conference entitled Glial-Neuronal Interaction, which was sponsored by the New York Academy of Sciences and held on September 4–7, 1990 at the University of Cambridge, Cambridge, England.

Part III. Myelin and Demyelination

Part IV. Glial-Neuronal Interaction in Regeneration

Part V. Glial Electrophysiology and Transport

Part VI. Glia and the Blood-Brain Barrier

Part VII. Glial Receptors, Signaling, and Second Messengers

Poster Papers

Financial assistance was received from:

Supporter

- NATIONAL SCIENCE FOUNDATION

Contributors

- ABBOTT LABORATORIES
- BAYER UK plc
- BRISTOL-MEYERS SQUIBB (USA)
- ELSEVIER PUBLISHERS (TRENDS IN NEUROSCIENCES)
- FISONS PHARMACEUTICALS (USA)
- FISONS plc (UK)
- GENENTECH, INC. (USA)

- GLAXO RESEARCH LABORATORIES (USA)
- HEFFERS BOOKSHOP, CAMBRIDGE
- HOFFMANN-LA ROCHE INC. (USA)
- ICI AMERICAS INC.
- LILLY RESEARCH CENTRE (UK)
- MERCK SHARPE & DOHME plc (UK)
- MULTIPLE SCLEROSIS SOCIETY (UK)
- NATIONAL MULTIPLE SCLEROSIS SOCIETY (USA)
- PFIZER CENTRAL RESEARCH (UK)
- SMITH KLINE & BEECHAM (UK)
- THE COMPANY OF BIOLOGISTS (UK)
- THE WELLCOME TRUST (UK)

Preface

N. JOAN ABBOTT[a] AND MARTIN C. RAFF[b]

[a]*Physiology Group*
Biomedical Sciences Division
King's College
London WC2R 2LS, UK

[b]*Department of Biology*
Medawar Building
University College
Gower Street
London WC1E 6BT, UK

Most cells in the nervous system are either neurons or glial cells. The major classes of glia in the vertebrate central nervous system (CNS) are astrocytes and oligodendrocytes, whereas Schwann cells associated with axons are the predominant glial type in the peripheral nervous system (PNS). The close anatomical associations between glia and neurons and between glia and blood vessels were recognized by the earliest neurohistologists, including Golgi and Cajal. Physiological understanding is much more recent, dating from pioneering work by S. W. Kuffler and J. G. Nicholls in the 1960s using electrophysiological and cell biological techniques. Invertebrate glia contributed significantly to these studies, and where comparison was possible, vertebrate and invertebrate glia appeared to share key properties.

There are few areas in biology that have not been revolutionized in recent years by the advent of powerful new technologies, especially by monoclonal antibody, recombinant DNA, and patch clamp technology. This is certainly the case for glial cells, which are now recognized to play crucial roles in neurogenesis, neural regeneration, and neural function. Glia have been shown to interact with neurons in complex ways that are essential for the normal health of the nervous system. The interactions are either direct, via diffusive and cell-surface signals, or indirect, via effects on the interstitial medium and the blood-brain barrier. This book presents the proceedings of an international symposium held on September 4–7, 1990, in Cambridge, England. The meeting was sponsored by the New York Academy of Sciences and was hosted by the Department of Zoology, University of Cambridge. It aimed to review recent advances in the nature of glial-neuronal interactions at the cell and molecular level, which have come from a variety of approaches: tissue culture, developmental studies *in situ,* molecular biology and genetics, electrophysiology, and study of membrane receptors and transport. No single preparation can answer all the key questions in this field, and the meeting was designed to bring together the best work on *in vitro* and *in situ* models from both vertebrate and invertebrate nervous systems.

The book follows the format of the meeting, with the addition of introductions to each section by the chairmen of sessions, and summary overviews by discussants dealing with key areas of excitement and controversy. The paper by J. Z. Young was given as the John Treherne Memorial Lecture; a tribute to John Treherne, to whom this volume is dedicated, follows the Preface. John Nicholls provides a personal epilogue, describing the early days of glial studies with Stephen Kuffler when some of the major questions of current interest were identified.

Vertebrate Glial Subtypes, Lineage, and Morphology. The emphasis of the volume is on glial-neuronal interactions, but the first section deals with glial subclasses and glial cell lineage, which must be the starting point for any discussion of what glial cells do and how they interact with neurons. Studies are most advanced for the mammalian nervous system, where a range of antibodies recognizing particular marker antigens has made possible rigorous experimental work (Ransom). There remains considerable controversy, generated by the differences between findings in culture (Fulton/Raff, Noble, Jessen, Kettenmann) and *in vivo* (Skoff, Price, Butt). The powerful new method of using retroviral gene transfer to trace glial cell lineage *in vivo* is reviewed by Price.

Glial-Neuronal Interaction in Development. Glial cells are now known to play a crucial role in neuronal development, influencing neuronal shape, both neuroblast and growth cone migration, and neuronal survival and differentiation (Rakic, Scholes). This section considers what has been learned about the molecules and processes that mediate glial-neuronal interactions during development (Tomaselli, Schachner, Stitt/Hatten, Laywell/Steindler, Pringle/Richardson). The most direct evidence for the role of glial cells in patterning neuronal connections has come from ablation experiments in insects, where the combined power of classical and molecular genetics can hopefully be harnessed to work out the molecular basis for the interactions. This approach is discussed by Klämbt.

Myelin and Demyelination. The interactions between glial and neuronal cells operate in both directions: glia influence neuronal development and neurons in turn regulate glial cell development. The most dramatic example of the latter type of interaction is seen during myelination. Myelin structure is summarized by Davison, and the molecular genetics of central and peripheral myelin are presented by Campagnoni, Lemke, and Brophy. Oligodendrocyte damage in inflammatory demyelination of the CNS is reviewed by Compston.

Glial-Neuronal Interaction in Regeneration. If the peripheral nervous system is damaged, a certain amount of regrowth and reconnection can occur. Key questions in this field are, why is the same not true of the mammalian CNS, and can any of the differences be attributed to differences in the glial properties (Raisman, Holder). The role of glial cells in promoting or inhibiting axonal regrowth after injury is presented by Bray, Bunge, and Cadelli.

Glial Electrophysiology and Transport. The earliest electrophysiological studies of CNS glial cells showed them to be inexcitable, that is, incapable of generating action potentials, but with a high membrane selectivity to K^+. This led to the idea that K^+ might be a signal linking neuronal and glial function; the K^+ permeability would also enable glia to buffer neuronally released K^+ (Orkand). The modern story is more interesting still, with glia showing a range of ion channels including voltage-operated ones (Barres, Ritchie), regional differences in channel distribution over the cell surface (Minturn/Waxman), and evidence that neurons may control glial channel expression. Key factors influencing glial K^+ regulation are discussed by Reichenbach. The membrane transport processes of glia also contribute to the homeostasis of the interstitial microenvironment (Schlue, Kempski), a delicate balance disturbed in conditions like hypoxia and ischemia. The transfer of large molecules between neurons and glia is discussed by Grossfeld.

Glia and the Blood-Brain Barrier. All vertebrates and some invertebrates have a blood-brain barrier, but the invertebrate barriers are formed by glia (Lane), whereas in all vertebrates except elasmobranchs, the barrier is endothelial (Brightman). In evolution and ontogeny the glial barrier apparently came first, which leads to insights into the adult mammalian system (Abbott). Some of the glial barriers show interesting modulation by pH, [Ca^{2+}], and humoral agents (Butt). In higher vertebrates, perivascular glia provide at least some of the factors that induce barrier properties in the endothelial layer (Nagy); the inductive agents are now being studied at the level of cell and molecular biology (Risau, Rubin). It may soon prove possible to generate an *in vitro* blood-brain barrier that shows all the *in vivo* properties (Greenwood).

Glial Receptors, Signaling, and Second Messengers. As glia and neurons interact in several ways during development, adult neural function, and pathological states, it would be advantageous for glia to be able to communicate with other cells of the nervous system in order to coordinate these activities. Signaling from neurons to glia via K^+ is one possibility, but neurotransmitter signaling is potentially more specific and precise (Pearce, Mudge). Pioneering studies by J. Villegas 20 years ago showed that squid Schwann cells respond to neurotransmitters; transmitter receptors have now also been found on vertebrate CNS glia. The squid system can be studied while maintaining the normal neuron-glial geometry (Evans/Villegas, Lieberman), while patch clamp recording is giving new information about mammalian glial receptors (Cull-Candy). Glial peptide receptors show interesting regional differences within the CNS, and linkage to intracellular second messenger systems (Wilkin). Glia may in turn release neuroactive agents including taurine (Dutton).

Selected posters from the meeting are presented as short papers in the final section.

This volume concentrates on aspects of basic neuroscience, a first step towards understanding the mechanisms at work in normal physiology, and the ways in which these processes are disturbed in pathology.

We wish to express our gratitude to the many individuals in London, Cambridge, and in the offices of the New York Academy of Sciences, who assisted with the symposium, in particular Charles Nicholson, Ed Lieberman, Gabriel Horn, Nancy Lane, Tricia Revest, Margaret Clements, Reg Clements, Rebecca Treherne, Maria Simpson, and Sherryl Greenberg. We are grateful to Angela Fink for editorial work on the volume.

JOHN TREHERNE

John Treherne (1929–1989): An Appreciation

Reader in Invertebrate Physiology University of Cambridge

Director, AFRC Unit of Insect Neurophysiology and Pharmacology

Fellow of Downing College

Writer

N. JOAN ABBOTT
Physiology Group
Biomedical Sciences Division
King's College
Strand, London WC2R 2LS, UK

In 1981, John Treherne organized a very successful meeting on Glial-Neuronal Interaction at Titisee in the Black Forest. He deliberately included work on both invertebrate and vertebrate preparations, as he believed, with August Krogh and Stephen Kuffler, that once a biological problem is identified, a search should be made for the most appropriate preparation for its investigation. In glial studies, as in many other areas of neuroscience, no one species holds all the clues to function. The nervous system of mammals, amphibia, annelids, insects, crustacea, and cephalopods had already yielded advantageous glial preparations that were helping to resolve important questions concerning the physiology of glial cells and their interactions with neurons. The meeting was published as a Review Volume in the *Journal of Experimental Biology*.

Since 1981 the field has expanded dramatically, and a whole range of new molecular and electrophysiological techniques has been added to the repertoire of tools available to the experimental scientist. When we began to plan the Cambridge meeting in late 1988, John was an enthusiastic adviser and was very much in favor of bringing together the novel work on vertebrate and invertebrate preparations. He died in September 1989; there is no doubt that had he been with us at the meeting, he would have been the life and soul of the party. With so many of his friends, colleagues, past students, and his family gathered in Cambridge in September 1990, and such frequent references to his work, many of us saw the meeting as a splendid chance to celebrate John's contribution to glial physiology and to the advancement of a transphyletic approach to neuroscience. This volume is therefore dedicated to the memory of John Treherne, a much loved colleague and friend.

John Edwin Treherne was born on May 15, 1929 in Wiltshire and educated at Headlands School, Swindon. His childhood friends included Desmond Morris and

Diana Dors, two flamboyant extroverts whose companionship did nothing to discourage John's own ebullience. His interest in natural history, especially bird life, was developed in the Wiltshire countryside. He was an outstanding zoology student at Bristol University, where he was both an undergraduate and a PhD student. It was a summer vacation on a studentship at Uppsala University, Sweden, that determined his interest in insect physiology.

By great good fortune, he was drafted for his National Service into the same section of the Army (the Royal Army Medical Corps) as Trevor Shaw, then a young physiologist from Cambridge; their experiences in the RAMC led to a lifetime's friendship. John used to tell with mock shame how the four men on guard duty were discussing evolution, one howling January night on Salisbury Plain. The officer in charge, a fundamentalist, decreed that the penalty for believing in Darwin was to patrol in the snow for 2 hours. Trevor Shaw staunchly defended Darwin and was sent out into the gale for his pains. John agreed with the officer that fossils were probably planted by the devil to mislead innocent scientists and stayed by the stove. John and Trevor later used this episode to debate the role of pragmatism and opportunism in the survival of the fittest.

After demobilization, John wrote to Vincent Wigglesworth in Cambridge, the foremost insect physiologist of the time, and in 1955 was offered a position at the Agricultural Research Council's Unit of Insect Physiology to study insect digestion, using the cockroach *Periplaneta americana* as an experimental model. He made rapid progress at a time when radioisotopic studies were in their infancy, and showed that glucose uptake in the gut was linked to dimerization to the insect sugar trehalose. He soon become interested in the insect nervous system and the blood-brain barrier; a paradox appeared to exist in that neural activity was relatively resistant to the addition of certain neurotoxic compounds to the bathing medium, yet radioisotopic studies showed rapid flux of compounds between bathing medium and tissue. In 1963 he published with David Smith an influential review on the structure and function of the insect nervous system, which clearly showed that the outer glial layer, the perineurium, had the specialized anatomy of a transporting epithelium. When Wigglesworth's Unit closed in 1967, John was appointed to a University Lectureship in the Department of Zoology and continued his studies on the nervous system in collaboration with Simon Maddrell, Yves Pichon, and Roger Moreton. He was Vice-President of the Royal Entomological Society (1967–1968) and in 1968 was awarded the Scientific Medal of the Zoological Society.

In 1971 John became Reader in Experimental Biology and was appointed Director of the new ARC Unit of Invertebrate Chemistry and Physiology, later renamed, under the AFRC, the Unit of Insect Neurophysiology and Pharmacology. The Unit was to become one of John's most successful creations. He assembled a group of gifted experimentalists, including Nancy Lane, Simon Maddrell, Mike Berridge, David Sattelle, Peter Evans, Philip Schofield, and Helen Skaer, with close associations within the Department of Zoology (Michael Bate and Malcolm Burrows). From the group has come not only definitive work on the physiology of the insect blood-brain barrier and its location in the glial perineurium, but also seminal observations on insect neuropharmacology and fluid secretion and on the role of phosphoinositides as intracellular messengers. John nourished the talents of his group and sustained the family atmosphere by a real personal concern and by regular social gatherings, with birthday celebrations and fancy dress parties the highpoints of the year.

Having established the principles of normal blood-brain barrier function to his satisfaction, John characteristically looked for a new topic and soon recognized that the repair of damage to the nervous system was a key issue, particularly the question of what happened to the blood-brain barrier once breached and the nature of the restoration of barrier function. A series of papers with Peter Smith, Elizabeth Howes, and Colin Leech showed, somewhat surprisingly, that blood-derived cells (hemocytes) were responsible for temporary repair of the barrier and that these cells either themselves transformed into functional glial cells or induced proliferation of endogenous glia. Analogous inductive interactions between macrophages and Schwann cells were recently described in repair of the vertebrate nervous system. The very recent finding that outer but not inner perineurial cells are mesodermal derivatives in insects (Edwards, this volume) may help to resolve the transformation/induction question.

In addition to being an excellent laboratory scientist, John was an enthusiastic field worker and, together with William Foster, first at Scolt Head Island in Norfolk, later in the Galapagos Islands, studied the population ecology of marine insects. He brought together his love of history and ecology in coining the term "the Trafalgar effect" to describe how rafts of floating insects (*Halobates*) relayed signals to each other about approaching predators, in the same way that Nelson's signals revealed the movements of the French fleet.

John wrote with fluency and elegance and had a sharp editorial eye. As Editor of the *Journal of Experimental Biology* from 1974, he rapidly established an innovative style and an international standing, so that the *Journal* is now a key publication in neurobiology and comparative physiology. Manuscripts from foreign authors received meticulous attention and patient help with rewriting. With Simon Maddrell he was responsible for turning the *Journal* into a profitable venture, the proceeds helping to fund advanced workshops and scientific meetings (including the one at Titisee).

After he reached 50, John found he could manage with very little sleep (a legendary 3–5 hours per night), so he began to get up early and write for pleasure. He began by applying the techniques of scientific research to the investigation of intriguing unsolved mysteries of human behavior and crime (*The Galapagos Affair,* 1983; *The Strange History of Bonnie and Clyde,* 1984; *Dangerous Precincts,* 1987). His novels include the lyrical evocation of a Wiltshire boyhood (*The Trap*) and a humorous account of academic rivalries and College High Table preoccupations (*Mangrove Chronicle*). *The Walk to Acorn Bridge* and *The Canning Enigma* were published in 1989. P. D. James, an admirer of his writing, noted his sensitivity to the natural world, his compassion, and his acute awareness of the transitoriness of life and happiness. Dick Orkand recounts the pleasure he received from being sent copies of John's books instead of the statutory £15 cheque for reviewing for *JEB*. John himself enjoyed the world of US chat shows opened up by his writing, claiming that he was treated with much more respect and interest at American parties if he described himself as "novelist" rather than "professor."

Most of his colleagues knew John as a man brimming with scientific ideas, and a charismatic leader and coordinator of scientific effort. But we all also have happy individual memories of John Treherne the man—a wonderfully erudite and entertaining companion on expeditions, a caring and inspiring teacher and colleague, with a knack for sensing when someone needed help and support and when they were ready to fly independently. His interest and care extended far beyond his research group, to students and staff of Downing College, to visiting Fellows, to his friends and their students in Japan, South America, and China. He quickly established rapport with everyone around him, across every kind of social and language barrier. He is fondly remembered in Downing College kitchen as the Fellow who showed the most personal interest in and appreciation of the cooking; he had only to whisper "jam roly-poly" in the ear of the chef for his favorite pudding to appear on the next day's dinner table. I remember a hilarious conversation in the tiny fishing village of Mochima, Venezuela, where John was discussing the menu (barracuda steaks and unleavened bread) with the fisherman/chef. Perfect communication was established using only John's pidgin French, the chef's Mochima Spanish, and judicious hand waving. The morning after one of the neuroscience seminars for advanced students organized by Jorge Villegas in Caracas, John was even more famous for his performance on the drums at 2 AM than for the brilliant talks he had given on (a) the insect blood-brain barrier and (b) the way insects had changed the course of history.

In 1955 John married June Vivienne Freeman, who shared his passion for history and nurtured their parallel interests in antiques. John took great pleasure and pride in the achievements of his family and was delighted at the completeness of the circle, when his son Mark went to work with John Nicholls. John Treherne had himself been inspired by the work of Stephen Kuffler, with whom John Nicholls had made pioneering studies on the glial cells of the leech. John would have been very touched to see the expert way in which his daughter Rebecca and personal assistant Margaret Clements worked together to ensure the smooth running of this meeting.

We felt John's absence acutely during the days of the Cambridge meeting, but take heart from the fact that he would have applauded its aims and been delighted at the richness of ideas, techniques, and findings that were presented. We are happy to record here our debt to his inspiring teaching and example and to the spirit of enthusiasm and enquiry that he embodied. We know that John Treherne would have enjoyed this meeting, and we dedicate this volume to him.

ACKNOWLEDGMENTS

I am grateful to June Treherne, Margaret Clements, John Nicholls, and Lord Butterfield for help in the preparation of this appreciation.

SELECTED BIBLIOGRAPHY

TREHERNE, J. E. 1961. Sodium and potassium fluxes in the abdominal nerve cord of the cockroach, *Periplaneta americana*. J. Exp. Biol. **38:** 315–322.

TREHERNE, J. E. 1962. The distribution and exchange of some ions and molecules in the central nervous system of *Periplaneta americana* L. J. Exp. Biol. **39:** 193–217.

TREHERNE, J. E. 1962. Some effects of the ionic composition of the extracellular fluid on the electrical activity of the cockroach abdominal nerve cord. J. Exp. Biol. **39:** 631–641.

SMITH, D. S. & J. E. TREHERNE. 1963. Functional aspects of the organization of the insect nervous system. *In* Advances in Insect Physiology. J. W. L. Beament, J. E. Treherne & V. B. Wigglesworth, eds. Vol. 1: 401–484. Academic Press. New York & London.

TREHERNE, J. E. 1965a. Some preliminary observations on the effects of cations on conduction processes in the abdominal nerve cord of the stick insect, *Carausius morosus*. J. Exp. Biol. **42:** 1–6.

TREHERNE, J. E. 1965b. The distribution and exchange of inorganic ions in the central nervous system of the stick insect, *Carausius morosus*. J. Exp. Biol. **42:** 7–27.

TREHERNE, J. E. 1966. The Neurochemistry of Arthropods. Cambridge University Press. Cambridge, UK.

TREHERNE, J. E. & S. H. P. MADDRELL. 1967a. Membrane potentials in the central nervous system of a phytogagous insect (*Carausius morosus*). J. Exp. Biol. **47:** 235–247.

TREHERNE, J. E. & S. H. P. MADDRELL. 1967b. Axonal function and ionic regulation in the central nervous system of a phytophagous insect (*Carausius morosus*). J. Exp. Biol. **47:** 235–247.

MADDRELL, S. H. P. & J. E. TREHERNE. 1967. The ultrastructure of the perineurium of two insect species, *Carausius morosus* and *Periplaneta americana*. J. Cell Sci. **2:** 119–128.

TREHERNE, J. E., N. J. LANE, R. B. MORETON & Y. PICHON. 1970. A quantitative study of potassium movements in the central nervous system of *Periplaneta americana*. J. Exp. Biol. **53:** 109–136.

PICHON, Y. & J. E. TREHERNE. 1970. Extraneuronal potentials and potassium depolarization in cockroach giant axons. J. Exp. Biol. **53:** 485–493.

LANE, N. J. & J. E. TREHERNE. 1970. Uptake of peroxidase by the cockroach central nervous system. Tissue & Cell **2:** 413–425.

TREHERNE, J. E. & R. B. MORETON. 1970. The environment and function of invertebrate nerve cells. Int. Rev. Cytol. **28:** 45–88.

PICHON, Y., R. B. MORETON & J. E. TREHERNE. 1971. A quantitative study of the ionic basis of extraneuronal potential changes in the central nervous system of the cockroach (*Periplaneta americana* L). J. Exp. Biol. **54:** 757–777.

LANE, N. J. & J. E. TREHERNE. 1972. Studies on perineurial junctional complexes and the sites of uptake of microperoxidase and lanthanum in the cockroach central nervous system. Tissue & Cell **4:** 427–436.

TREHERNE, J. E. & Y. PICHON. 1972. The insect blood-brain barrier. *In* Advances in Insect Physiology. J. E. Treherne, M. J. Berridge & V. B. Wigglesworth, eds. Vol 9: 257–308. Academic Press. London & New York.

PICHON, Y. & J. E. TREHERNE. 1973. An electrophysiological study of sodium and potassium permeabilities of insect peripheral nerves. J. Exp. Biol. **59:** 447–461.

LANE, N. J. & J. E. TREHERNE. 1973. The ultrastructural organization of peripheral nerves in two insect species (*Periplaneta americana* and *Schistocerca gregaria*). Tissue & Cell **5:** 703–714.

TREHERNE, J. E., P. K. SCHOFIELD, & N. J. LANE. 1973. Experimental disruption of the blood-brain barrier system in an insect (*Periplaneta americana*). J. Exp. Biol. **59:** 711–723.

TREHERNE, J. E. 1974. The environment and function of insect nerve cells. *In* Insect Neurobiology, Frontiers of Biology. Vol 35: 187–244. J. E. Treherne, ed. North Holland Publishing. Amsterdam.

BENNETT, R. R., P. B. BUCHAN & J. E. TREHERNE. 1975. Sodium and lithium movements and axonal function in cockroach nerve cords. J. Exp. Biol. **62:** 231–241.

SCHOFIELD, P. K. & J. E. TREHERNE. 1975. Sodium and lithium movements across the insect blood-brain barrier. Nature (Lond.) **255:** 723–725.

THOMAS, M. V. & J. E. TREHERNE. 1975. An electrophysiological analysis of extra-axodal sodium and potassium concentrations in the central nervous system of the cockroach (*Periplaneta americana* L.). J. Exp. Biol. **63:** 801–811.

LETTAU, J., W. A. FOSTER, J. E. HARKER & J. E. TREHERNE. 1977. Diel changes in potassium activity in the haemolymph of the cockroach *Leucophaea maderae*. J. Exp. Biol. **71:** 171–186.

ABBOTT, N. J. & J. E. TREHERNE. 1977. Homeostasis of the brain microenvironment: A comparative account. *In* Transport of Ions and Water in Animals. B. L. Gupta, R. B. Moreton, J. L. Oschman & B. J. Wall, eds.: 481–510. Academic Press. London & New York.

TREHERNE, J. E. & P. K. SCHOFIELD. 1978. A model for extracellular sodium regulation in the central nervous system of an insect (*Periplaneta americana*). J. Exp. Biol. **77:** 251–254.

LANE, N. J. & J. E. TREHERNE. 1980. Junctional organization of arthropod neuroglia. *In* Insect Biology in the Future, 'VBW 80', M. Locke & D. S. Smith, eds.: 765–795. Academic Press. London and New York.

TREHERNE, J. E. & SCHOFIELD, P. K. 1981. Mechanisms of ionic homeostasis in the central nervous system of an insect. J. Exp. Biol. **95:** 61–73.

TREHERNE, J. E., P. K. SCHOFIELD & N. J. LANE. 1982. Physiological and ultrastructural evidence for an extracellular anion matrix in the central nervous system of an insect (*Periplaneta americana*). Brain Res. **247:** 255–267.

SCHOFIELD, P. K., L. S. SWALES & J. E. TREHERNE. 1984a. Potentials associated with the blood-brain barrier of an insect: Recordings from identified neuroglia. J. Exp. Biol. **109:** 307–318.

SCHOFIELD, P. K. & J. E. TREHERNE. 1984. Localization of the blood-brain barrier of an insect: Electrical model and analysis. J. Exp. Biol. **109:** 319–332.

SCHOFIELD, P. K., L. S. SWALES & J. E. TREHERNE. 1984b. Quantitative analysis of cellular and

paracellular effects involved in disruption of the blood-brain barrier of an insect: Electrical model and analysis. J. Exp. Biol. **1090:** 333–340.
SMITH, P. J. S., C. A. LEECH & J. E. TREHERNE. 1984. Glial repair in an insect nervous system: Effects of selective glial disruption. J. Neurosci. **4:** 2698–2711.
SCHOFIELD, P.K. & J. E. TREHERNE. 1985. Octopamine reduces the potassium permeability of the glia that form the insect blood-brain barrier. Brain Res. **360:** 344–348.
TREHERNE, J. E. 1985. Blood-brain barrier. *In* Comprehensive Insect Physiology, Biochemistry and Pharmacology. Vol 5. Nervous System: Structure and Motor Function, G. A. Kerkut & L. I. Gilbert, eds.: 115–137. Pergamon Press. Oxford and New York.
SCHOFIELD, P. K. & J. E. TREHERNE. 1986. Octopamine sensitivity of the blood-brain barrier of an insect. J. Exp. Biol. **123:** 423–439.
SMITH, P. J. S., E. A. HOWES & J. E. TREHERNE. 1987. Mechanisms of glial regeneration in an insect central nervous system. J. Exp. Biol. **132:** 59–78.
TREHERNE, J. E., E. A. HOWES & P. J. S. SMITH. 1987. Glial repair in an insect. J. Physiol. Paris **82:** 239–247.
SMITH, P. J. S., E. A. HOWES & J. E. TREHERNE. 1988. Reactive cells and their role in glial regeneration in an insect CNS. Am. Zool. **28:** 1145–1153.
TREHERNE, J. E., P. J. S. SMITH & E. A. HOWES. 1988. Neural repair and regeneration in insects. Adv. Insect Physiol. **21:** 35–84.
SMITH, P. J. S., E. A. HOWES & J. E. TREHERNE. 1990. Cell proliferation in the repairing adult insect central nervous system: Incorporation of the thymidine analogue 5-bromo-2-deoxyuridine *in vivo.* J. Cell Sci. **95:** 599–604.
TREHERNE, J. E., E. A. HOWES, P. K. SCHOFIELD & P. J. S. SMITH. 1990. Modulation and repair of the insect blood-brain barrier. *In* Insect Neurochemistry and Neurophysiology. 1989. A. B. Borkovec & E. P. Masler, eds. Humana Press.

SPECIAL LECTURE

The Concept of Neuroglia

J. Z. YOUNG

Department of Experimental Psychology
University of Oxford
South Parks Road
Oxford OX1 3UD, UK

In organizing this lecture I began with the idea of simply describing the views of some early workers on glia but soon realized that it would be much more interesting to see how the sequence of discoveries led to current views especially about functions. When it was decided to make this lecture a tribute to John Treherne, it became particularly interesting to follow the history of ideas about the glia as a barrier to permeability in both the brain and peripheral nerve. It has been difficult to carry out such a program fully because the spread of time and of techniques is so great. The first vague ideas about the neuroglia in the brain were formulated by Virchow in the 1840s. It was 130 years later that Treherne and his colleagues in the 1970s proved its efficacy as a barrier in the CNS and peripheral nerves of insects. I cannot possibly give a coherent account of all that went on in between in the study of the glia in brain and nerves. It is characteristic of the progress of biological science that some of the key advances in understanding have come from comparative study, and it is typical of John Treherne that he had the vision to see the special opportunities presented by the great variations in the fluids around insect nerves. It is also characteristic of him to have carried the study to its ecological aspects, including the insects of salt marshes. So to be true to his memory, let us first look at the early history, mainly of the glia in mammalian brains, and then end up with some recent comparative studies in insects and cephalopods, which last contain the only types of glia that I personally know first hand, and which also provided the material for research on the blood brain barrier by our convenor, Joan Abbott.

WHAT IS NEUROGLIA?

Interest in the history of any intellectual concept is not in the date at which a certain word was used but in the new ideas about the subject that were first identified. By examining the origins of a new point of view we may be led to recognize that today we see things only from one particular aspect and that this will inevitably change.

The word neuroglia, or nerve glue, was first used by the German pathologist Virchow in 1856. What new thoughts did he bring with the word? To answer that question we have to solve the much more difficult one of how scientists had thought about nervous tissue before that date. This also takes us to a metaquestion, a question about such questions. If I ask for an opinion about the nature of neuroglia, what sort of answer do I expect? If I am talking to a surgeon, he will tell me about the soft, almost fluid nature of most brain substance. Neurophysiologists or biochemists will have quite different descriptions—of electrical signals or chemical substances. A light microscopist will talk about star-shaped cells with long processes, whereas for the electron microscopist it is all a question of membranes and junctions, mitochondria, filaments, and other organelles.

The metaquestion is whether it is possible to have a concept that covers all these ideas at once? The answer may be that there is no such thing as one concept or one view of the nature of neuroglia (or indeed anything else). It all depends on what we call one's "point of view," which in a scientific context usually means the instrumentation with which to deal with the material in question. As Cajal puts it in his last book on the "Neuron Theory" (1932), "It is a regrettable truth that almost no-one can be entirely free from the tradition and fashions of his day." He died in 1934.

In following the history of the concept of neuroglia it may be useful to consider how the various techniques applied have influenced opinion on certain questions that now seem to us to be important:

1. What are the various types of glia?
2. What is the form of the glia? Is it a sort of connective tissue? Does it consist of fibers or sheets?
3. What is the relation of glial cells to nerve cells and fibers. Do the glial processes act as wrappings?
4. What is the relation of glial cells to blood vessels and the intercellular spaces?
5. What active actions do glial cells perform and what part do their actions play in nervous functioning?
6. What is their role in pathologic nervous conditions? Can they divide and can they become phagocytic?

In 1991 we can give answers to some of these questions, and it will be interesting to see how these answers emerge with new techniques. I shall pursue the answers given with the light microscope, but can follow the use of the electron microscope only to some primitive observations that we made in 1954.

DISCOVERY OF THE GLIA

Virchow

In following the history of the concept of neuroglia we are fortunate to be able to begin from the very beginning, as no comparable concept existed or could exist in the eighteenth century. The material of the brain and other organs was then described by using the analogy of "tissues," woven fabrics or pieces of cloth, exemplified in the lectures of Xavier Bichat, given in postrevolutionary France in the 1790s but published only after his death in 1802 at the age of 31. He hardly deals with the tissues of the brain as we know them but only with its membranes and their diseases. His ideas were based on physical consistency as revealed by teasing, with little use of the microscope.

The notion of the glia came gradually to Virchow in the late 1840s and 1850s as he used the microscope and applied Schwann's cell theory of 1839 to the brain. His discovery of the glia was not sudden. A paper of 1846 is usually given as his first report, but he does not clearly describe the glia or mention the name. His only concern then was to show that some sort of connective tissue, *Bindegewebe,* lies within the ependyma. The name neuroglia came out only 10 years later when he republished the 1846 paper with a new long footnote (Virchow, 1856) stating that, "the ependyma is no isolated layer in the narrow sense of the word, but only the surface layer of the *Zwischen-Bindesubstanz* of the brain matter. This binding substance makes in the brain . . . a sort of *Kitt* (neuroglia) in which the nerve elements are embedded." The German word *Kitt* literally means putty; *neuroglia* was taken from the Greek and translated as "glue," but the German word for glue is *Leim* or *Klebstoff* (as pointed out by Somjen in 1988). Virchow's first figure of the glia was reproduced later still in a textbook in 1858 in which he shows a set of spheres (FIG. 1). He does not say how the preparation was made. The idea of the *Nervenkitt* represents an important advance. In this way Virchow gives the idea of a substance or stuff, not just a layer, a membrane, or a set of connective tissue fibers. This idea of a glue was in some ways a better conception than that given by the branching star-shaped cells that were shown later by Golgi preparations.

In relation to our questions, therefore, Virchow recognized a set of cells in the brain, besides the neurons, that were related to the ependyma. He showed no details of relations to nerve cells or blood vessels and had no clear idea as to function. He suspected his *glia* might play a part in pathologic processes: "This very interstitial tissue of the brain and spinal marrow is one of the most frequent seats of morbid change, as for example of fatty degeneration" (1858). He later described the structure of gliomas.

The first pictures of branched glial cells were given by Deiters in 1865 (FIG. 2). He isolated them by teasing and called them "connective tissue cells." It is not really clear if what he saw were glia or neurons, although he himself was quite sure of this. His figure is only in a small corner among his wonderful first pictures of neurons with their axons and dendrites. In subsequent decades the glial cells were often called Deiters' cells. Although Deiters showed the glia as fibrous, he also referred to them as "*schwammigpörose*"; spongy-porous perhaps gives as good an idea as any we have today.

Golgi

Camillo Golgi was the first to give really good descriptions of glia in a paper of 1870 entitled, "*Sulla sostanza connettiva del cervello.*" This was one of his earliest papers (it is placed first in his Opera Omnia [1903] and he was 27 at the time). He mentions Virchow at once and uses the name neuroglia, calling it the "*stroma connettiva.*" His first preparations were thin sections made after fixation in ½% osmic acid. He describes round cells "from which run various quite long very fine processes in different directions — and many are directed around the blood vessels." He could also show the glial cells by teasing. There are no figures in this first paper, but Golgi gives several in the next in 1871. They show rather thin feeble fibers, but are more convincing than Deiters' and can be considered the first reliable pictures of neuroglia.

After the discovery of his silver-chromate staining method in 1873 Golgi provided much better pictures of glia (1885). He held that the absence of an axon was the only certain way of distinguishing between glia and nerve cells. He believed that their processes join each other and form a continuum with those of the neurons. He described the various types of glia, including fibrous and protoplasmic, but these names were, in fact, given later by Andriezen (1893). Golgi (like Nissl) seems to have believed that there were transitions between glia and nerve cells.

Some of the different representations of the neuroglia were collected by Nakai (1963) and are reproduced in FIGURES 3–7. They give an idea of the improvement through the years. The figure taken from Golgi (1894) is clearly overly simple (FIG. 4f) and some of his efforts were better than this.

Golgi thought that the glia serves for transport between the blood vessels and the nerve cells, but he did not produce any special evidence for this. He continued to regard the glia as a form of connective tissue and made few references to it after 1900. It is not mentioned in his Nobel Lecture of 1906.

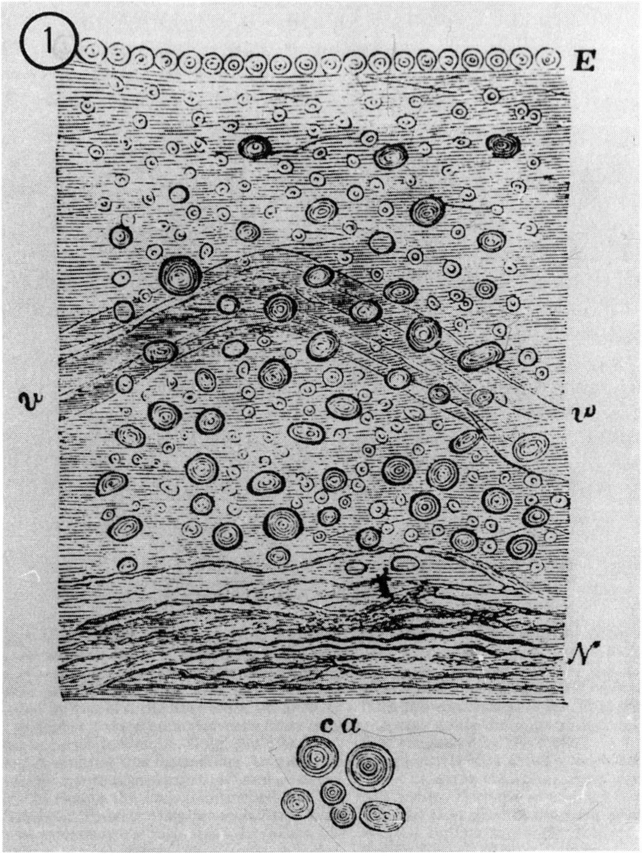

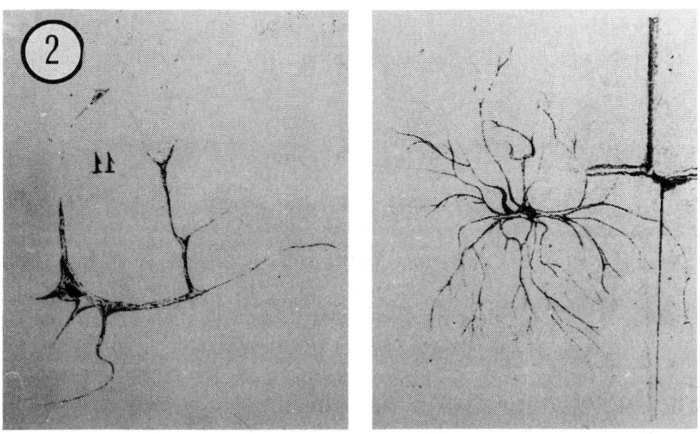

FIGURE 1. The first figure of neuroglia. Virchow, 1858.

FIGURE 2. The figures of glia given by Deiters, 1865.

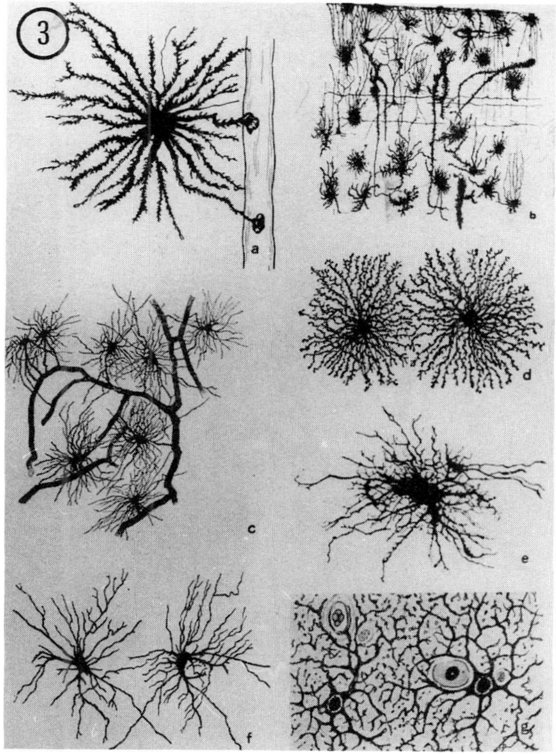

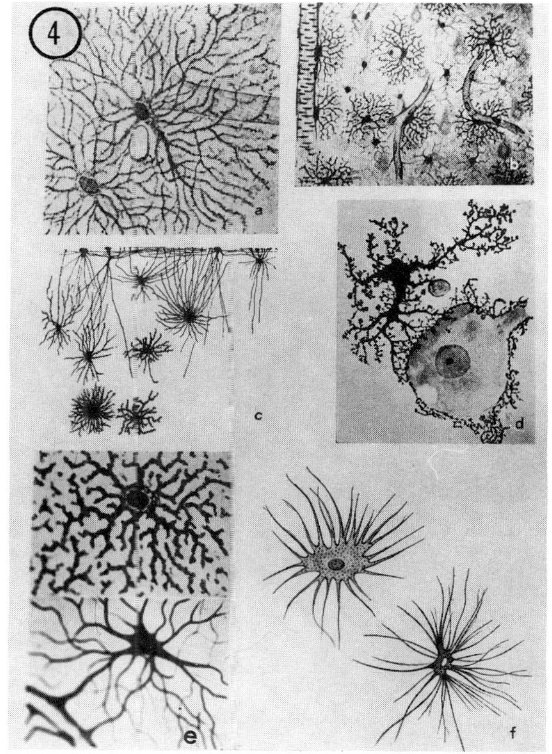

FIGURE 3. Protoplasmic astrocytes, all seen by the Golgi method except **g** which is the gold sublimate method. (**a**) Andreizen (1893), human cortex layer IX; (**b**) Retzius (1894), human fetus 45 mm, frontal lobe; (**c**) Kölliker (1896), rabbit cortex; (**d**) Cajal (1911); (**e**) Schaffer (1922), 9-month-old infant, spinal cord; (**f**) Kölliker (1896), infant, white matter; (**g**) Cajal (1933), human cortex.

FIGURE 4. Protoplasmic astrocytes, mainly Golgi method: (**a** and **b**) Hortega, 1933; (**c**) Retzius, 1894; 1-year-old infant, frontal cortex; (**d**) De Castro, 1951; (**e**) Hortega, 1920; fibrous and protoplasmic astrocytes; (**f**) Golgi, 1894; spinal cord, protoplasmic (*left*) and fibrous (*right*) astrocytes.

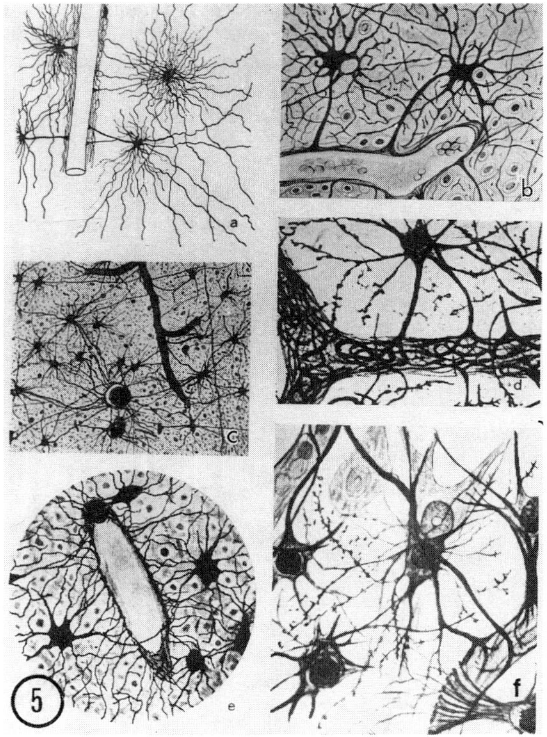

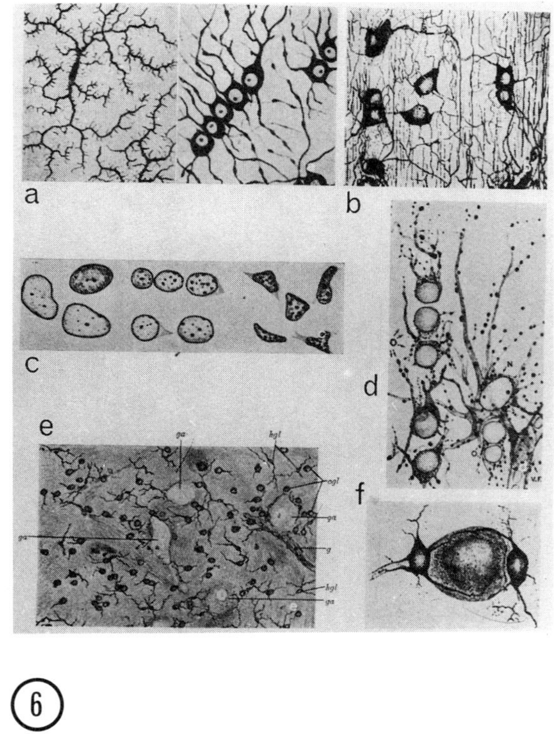

FIGURE 5. Vascular feet of fibrous astrocytes: (**a**) Cajal (1911), Golgi method; (**b**, **d**, & **f**) Cajal (1931), gold sublimate method; (**c**) Bauer (1953), human cerebral white matter; (**e**) Glees (1955), *Macacca,* occipital cortex, Hortega method.

FIGURE 6. Oligodendroglia. (**a**) Hortega (1920), oligodendroglia (*right*) and microglia (*left*); (**b**) Penfield (1924), cat cerebellar white matter, silver carbonate; (**c**) Penfield (1924), cat cerebellum, Nissl stain to show nuclei; from the left, astrocytes, oligodendroglia, and microglia. (**d**) Penfield (1924), rabbit cerebral white matter stained for gliosomes; N = astrocyte; O = oligodendroglia; (**e**) Jacob (1928), monkey, nucleus dentatus, Hortega, microglia (hgl), and oligodendroglia (ogl) (**f**) Penfield (1924), cerebral cortex, oligodendroglia around a neuron.

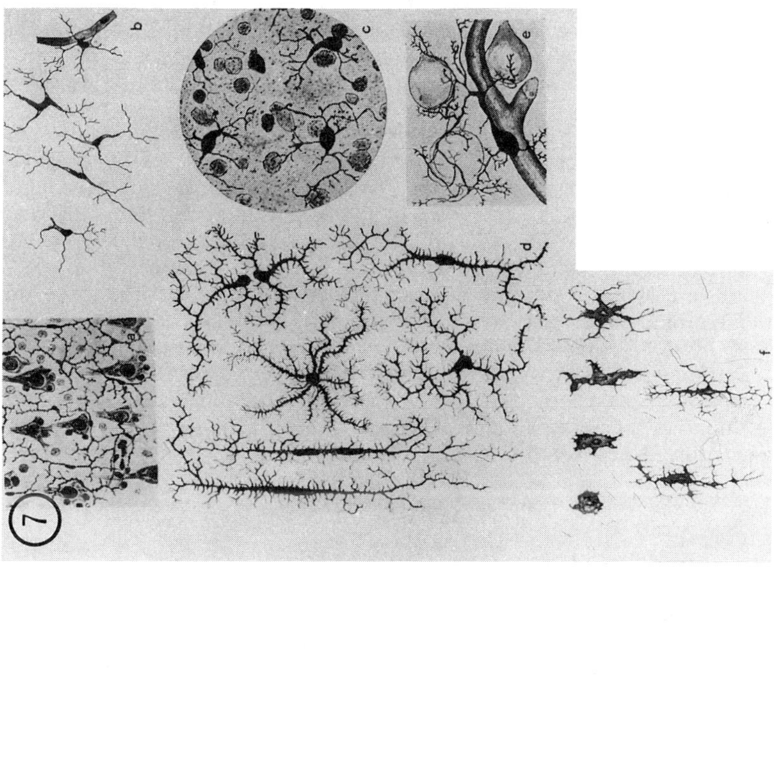

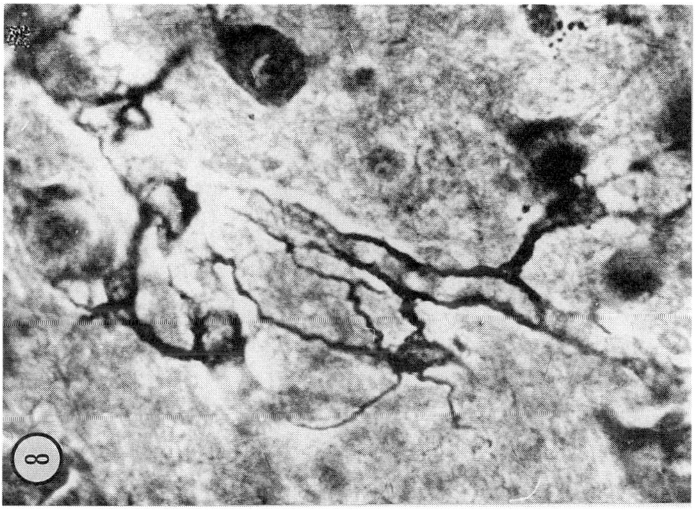

FIGURE 7. Microglia. (**a**) Cajal (1933), cerebral cortex, Achúcarro's method; (**b**) Schaffer (1926), silver carbonate method; (**c**) Glees (1955), rabbit, cortex; (**d**) Hortega (1920); (**e**) Penfield (1924), microglia, perivascular satellite; (**f**) Kirshman (1939), stages in differentiation of microglia.

FIGURE 8. Vascular end feet of fibrous astrocytes from rabbit rhinencephalon. Gold sublimate stain. Magnification given is 40 × 10 (Nakai, 1963).

Golgi therefore was the first to supply answers to some of our questions. He distinguished two types of glia and the ependyma. He thought that it was related to connective tissue and to blood vessels and that it served for transport to the nerve cells. He was also first to show that neoplasms in the brain consist of branched cells and that gliomas were thus readily distinguished from sarcomas and not liable to spread by metastasis (Golgi, *Opera Omnia,* 1903).

Cajal

Cajal's contribution to the history of the discovery of the neuroglia is both complex and controversial. It is best to consider it in two parts: first, his early work up to about 1900, and secondly, the studies done with his new method of 1913. His earlier views can be taken from the 1909 French edition of *Histology,* which was amended by Cajal from the original Spanish of 1899 entitled, *Textura del Sistema Nervioso del Hombre y de los Vertebrados.* The French 1909 edition was republished in 1952 in celebration of the centenary of Cajal's birth. In the introduction to that edition, J. S. Ibañez comments on Cajal's methods of work. Cajal was not content simply to give histologic details; he had to find the functional significance of the structures. For him as Ibañez says, "the form is not rigid or static, but quite the contrary, unstable, mobile, subject to continuous changes of material expression and cellular activity." In Cajal's own words, "in the form you meet the whole past and present functioning."

In the treatment discussed in his book of 1909, Cajal calls the glia "spider cells" or "Deiters' corpuscles" and makes very little reference to Golgi. He says that the cells form "a warp (*chaine*) of the nervous tissue. It serves both as a framework (*charpente*) and as an insulator (*isolant*)." "In short, as a framework with a thousand girders (*travées*) the neuroglia cells act as a material support and protection of the nerve cells and blood vessels. As an insulator their many branches interpose themselves among the dendritic expansions and nerve fibres which for whatever reason must not be in contact."

These are very interesting thoughts, but the descriptions that he gives of the glia at this time are not really much in advance of those of Golgi and other authors. He recognizes only three main types: (1) fibrillar glia in the white matter, (2) protoplasmic glia of the gray matter, and (3) epithelial glia. Note that at this time (1909) he does not mention the "third element," which he recognized later (see below). In the white matter he describes the glia as producing "an interstitial fibrillar web in which the nerve fibres are plunged (*plongées*)," separating them from each other. The glia also covers large and small vessels with a "neuroglial adventitia" (*adventice névroglique*). He therefore clearly recognized the presence of sheets of glia and states specifically that they are distinct from the collagenous adventitia of the vessels.

In the gray matter he recognizes six types of glia, partly on the basis of the form of the vascular foot. For the present analysis it is interesting that he stresses not only the large number of branches but also the fact that they often make sheets, "*véritables lamelles.*"

Cajal's Specific Stain for Glia

The second period of Cajal's interest in the neuroglia began with his discovery of a method that specifically shows these cells, using chlorides of gold and mercury (Cajal, 1913a) (FIGS. 5 & 8). This method shows beautiful dark glial fibers against a purple background. With this method he gave full descriptions of the two sorts of astrocytes (Cajal, 1913b). The *protoplasmic* astrocytes in the gray matter have frequent branching and wide expansions; the *fibrous* astrocytes in the white matter have fewer branches, which are straight and long. Both sorts have end feet on blood vessels. He emphasized that each cell is a distinct unit and that it is not connected to form a syncytial network. Cajal also now described a third element of small nonnervous cells. These he at least partly identified with mesodermal cells recognized by Robertson (1900) and called mesoglia.

Cajal's final description of astrocytes was far in advance of that of any of his predecessors (FIGS. 3 and 5). Most of the figures here are drawings made from specimens stained with the sublimate-gold chloride method.

Cajal and Function of the Glia

His long discussion in 1909 of function begins by deploring how difficult the study is because we cannot find ways to make the glia act, as we can with nerve cells. All we can do is to make conjectures "by analogy with the hazardous methods of either the physico-chemical or biological sciences. We can according to our taste (*esprit*) use analogies from the industrial techniques used to insulate electric currents or with the function of connective tissue in other organs." He rejects this second course which leads to the nutritive theory of Golgi. He also dismisses the theory of Weigert (1896) that the glia serves to fill up spaces between neurons. He comments that this is a *petitio principi:* given spaces, glia fill them. Perhaps the theory is not as absurd as he supposed. He would have been intrigued to know that the brains of larger animals have fewer neurons and more glia than do those of smaller animals (FIG. 9) (Tower, 1978).

For his own theory Cajal prefers the analogy with electrical devices, which he says was first thought of by his brother, P. Ramón Cajal. He holds that the glial processes "provide a medium resistant to the passage of nerve waves (*ondes nerveuses*). They are always arranged in such a way as to prevent contacts either between fibres without myelin or between dendritic expansions, or finally between the one and the other, but only at the points where these different processes of the neurons must not be in immediate contiguity." In this long sentence he outlines his theory of the "*rôle isolant*" of the glia. He supports this theory with the claim that glial expansions are abundant when there are many dendritic expansions, such as the molecular layer of the cerebellum, the hippocampus, or the retina. Conversely they are rare when there are pericellular networks, as around Purkinje cells, cells of the fascia dentata, or pyramidal cells of the cerebral cortex. He believes that in these places the cells all conduct together (*ont la mème conduction*) and do not need to be insulated. The glia are abundant where there is "fear of psychic confusion," and there must be no "*passage du flux nerveux*" between neighbors. He suggests that the glial cells have an active function in regions of complex connections where "contacts need to be regulated and in some way induced (*amenés*) by the neuroglia so as to put each dendritic appendage in close relation only with a particular set of nerve terminals."

Cajal discusses this theory of insulation in relation to many parts of the brain, emphasizing how many places there are where there is no myelin to act as insulation. He is at a loss only to explain the presence of glia in the white matter. "We have to recognise that this hypothesis loses much of its plausibility if we consider the abundance of neuroglia in the white matter." He must have been pleased by Hortega's later discovery of the relationship of oligoglia to myelin and would have been even more thrilled by the Geren-Robertson discovery that myelin is indeed the surface of the glia and Schwann cells. In fact, in a sense the condition in the white matter is *not an obstacle but a confirmation of his theory of the glia as an insulator.* As we learn more of the interchanges between glia and neuronal tissue, we may realize that we have still not thought hard enough about the extent to which it regulates the *flux nerveux.* But for Cajal in 1909 the glia of the white matter was a stumbling block. He humbly ends his chapter with the thought that it must have "some other functions in the nervous system of which we are completely ignorant." We should still say "amen" to that.

Cajal's early contribution to understanding the glia was thus great in many ways, but incomplete. He recognized only three sorts, besides the ependyma, but he greatly improved our understanding of the nature of astrocytes. His various references to them such as "a *duvet* of the neuronal processes" and "flocculent expansions" show

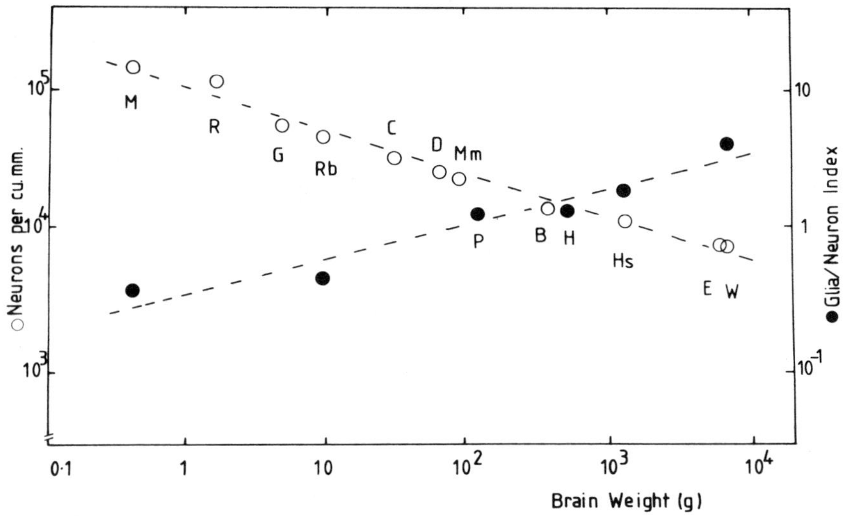

FIGURE 9. Plot showing the cortical neuron density (*open circles, left ordinate*) and the glia/neuron index (*filled circles, right ordinate*) as functions of species brain weight (log scales). M = mouse; Mn = monkey; P = pig; B = beef; H = man; Hs = horse; E = elephant; W = whale. (After Tower, 1978.)

that he understood how widely they penetrate the nervous tissues. He did not believe that they brought nourishment to the neurons, and his theory of insulation might be regarded as the first hint of the functions that we now accept. In addition to all this, Cajal's contribution to the pathology of the glia is to be found in his study entitled, "Degeneration and Regeneration of the Nervous System" (1928).

Hortega and the Oligodendroglia and Mesoglia

The last major step in the revelation of the various types of glia by light microscope was made by Cajal's student P. del Rio Hortega. In the years 1919–1921 he discovered another specific stain for neuroglia, using silver carbonate (Hortega, 1932). With this he showed that Cajal's "third element" includes two quite distinct types of neuroglia (FIGS. 6 and 7). The *oligodendroglia* are of ectodermal origin, like the astrocytes, and are found mainly in white matter, accompanying nerve fibers. The *microglia* is a set of branched cells abundant mainly in the gray matter. These originate from mesodermal tissue in the meninges and are in fact histiocytes, part of the reticuloendothelial system, and able to produce phagocytosis under pathologic conditions. In view of their origin these cells are appropriately called *mesoglia* but are commonly called *microglia.* Hortega realized that the oligodendrocytes were related to medullated nerve fibers, and of course the electron microscope finally revealed that myelin consists of the overlapped surfaces of the enwrapping oligocytes (Geren, 1955; Robertson, 1956).

Hortega also greatly advanced knowledge of the pathologic changes in glia, already begun by Golgi and others. Working with Wilder Penfield, he described how astrocytes undergo gliosis, then become swollen and may divide, perhaps by amitosis (Penfield, 1932). Microglia become phagocytic and divide mitotically. They described later stages of gliosis depending on whether the wound was a clean cut or left damaged tissue. It is of the greatest interest that Wilder Penfield's great success as an innovative brain surgeon was preceded by personal studies of the glia in normal tissues and after experimental injury. Consequently, he was able to have a total view or experience of the nature of the brain tissue upon which he operated and its subsequent behavior.

ANSWERS TO OUR QUESTIONS

Hortega thus provided the last step in our knowledge of the major types of glia, leaving us with five sorts, two types of astrocyte, protoplasmic and fibrous, oligodendroglia, microglia, and ependyma. This gives a satisfactory answer to the first of our initial questions, although many subtypes still have to be considered.

Light microscopy never fully described the extent of the neuroglia as we now know it from electron microscopy. The specific stains showing many processes have tended to obscure the way that the glia penetrates throughout the tissues. Cajal perhaps came closest with the analogy of a sponge and of wrapping as by a duvet. Yet the light microscopists have given a better idea of the whole cell and especially of the vascular feet than is possible from thin sections and electron microscopy. The large diameter of these trunks leading to the blood vessels is very well seen in the best light micrographs (FIGS. 5 and 8). The dimensions of the processes are presumably important for the propagation of electrical effects, perhaps for active material transport. Is it significant that the processes leading to blood vessels are nearly always larger than are any others leaving the cell body? For answers to such questions we shall have to wait until we find further ways to make the glia "play," as Cajal regretted he could not do in his time. I cannot now relate the history of the discovery of the electrical activities of glial cells and of the ionic interchanges by which they may regulate potassium and other ions in the intercellular spaces. Evidence of their intervention in synaptic functioning is still not clear, but Cajal's suggestion of an insulating function is certainly true for oligocytes. The function of the microglia as phagocytes is not in doubt.

EARLY ELECTRON MICROSCOPY OF THE NEUROGLIA

With the advent of the electron microscope interest soon changed from the study of star-shaped cells to membranes and intercellular spaces. I cannot trace the whole

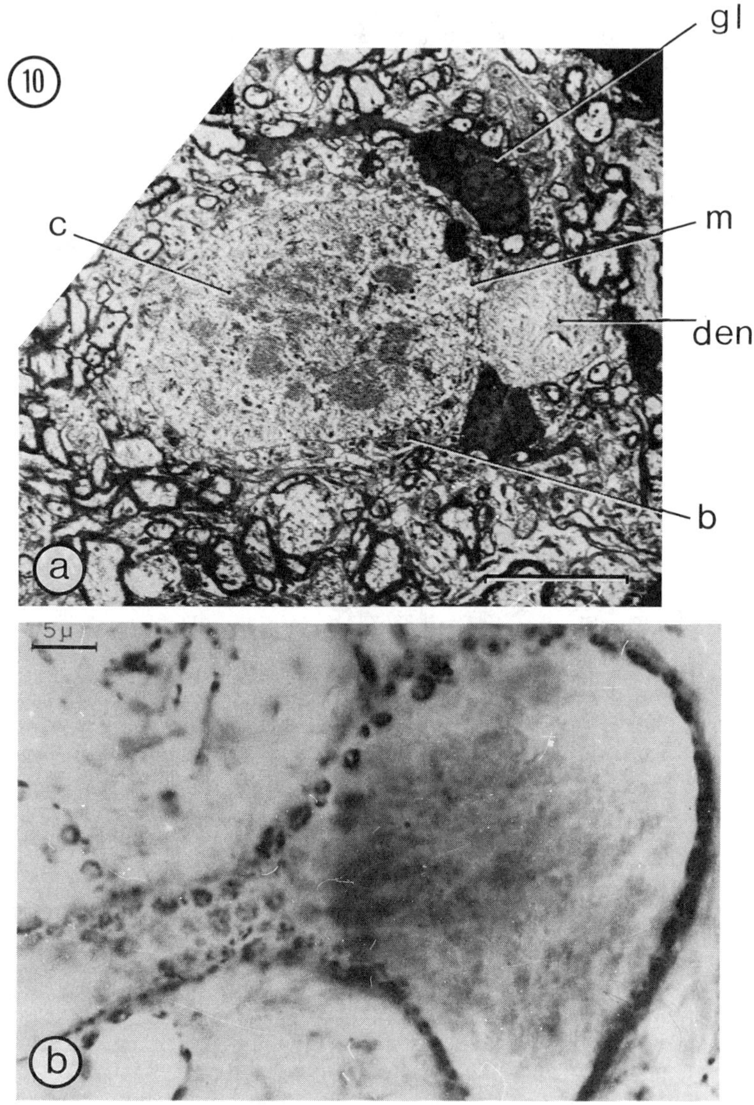

FIGURE 10. (**a**) An early electronmicrograph of a section of a neuron and glial cell in the spinal cord of a rabbit. b = synaptic end feet; c = nerve cell body; den. = dendrite; gl. = glial cell; m = cell membrane. Scale bar, 5 μm. (Wyckoff and Young, 1955). (**b**) Ventral horn cell from the spinal cord of a cat, showing the synaptic end feet preserved after fixation with formalin, postchroming, and staining with Weigert's method. Scale bar, 5 μm. (Wyckoff and Young, 1955).

history, but a curious episode involved the United States Embassy in London. The Science Attaché there in 1954–1955 was R. W. G. Wyckoff, who had played a large part in developing the instrument. He insisted on having his own Phillips M100 microscope in the Embassy, but had little material to study with it. He was a Foreign Member of the Royal Society and I met him at one of the dinners of the Royal Society Club. Hearing of his dilemma I suggested that we look at neural structure. So we fixed some rabbit spinal cord by perfusion with an 0.5% osmic acid, embedded it in methacrylate, and sectioned it on a Minot microtome with a glass knife. We wanted to see the relationships of the cells rather than fine details, and for this the

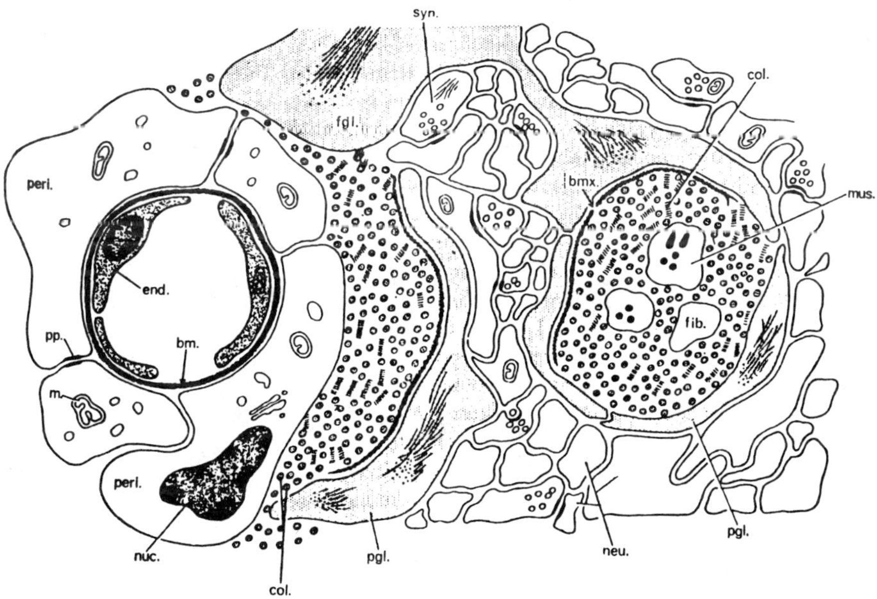

FIGURE 11. Diagrammatic section of gliovascular relationships as seen with the electron microscope in sections of subesophageal ganglia of *Octopus.* The section shows a capillary on the left and a gliovascular tunnel on the right (Gray, 1968). art. = artery; bm. = basement membrane of blood vessel; bmx.. = basement membrane of extracellular zones; col. = collagen; end. = blood vessel endothelium; fib. = fibrocyte; fgl. = fibrous glia; lu. = lumen of blood vessel; ly. = lymph channels; mus. = muscle cell or fiber; neur. = neuropil; nuc. = nucleus; peri. = blood vessel pericyte; pgl. = protoplasmic glia; syn. = synapse; ve. = vein.

methacrylate was removed and the section stained with palladium (Wyckoff and Young, 1956). The results were poor by modern standards but enabled us to study large areas and so to realize the great extent of the glia around the nerve cells, dendrites, and synapses (FIG. 10a). We showed some of the earliest pictures of synapses, although they were little better than the best we could do with the light microscope (FIG. 10b). We were surprised to see so little intercellular space; at that time there was still talk of "ground substance" in the central nervous system. We wrote that, "The glial protoplasm appears to fill most, perhaps all, of the tissue that is not occupied by nervous elements. There are no large spaces or intercellular matrix."

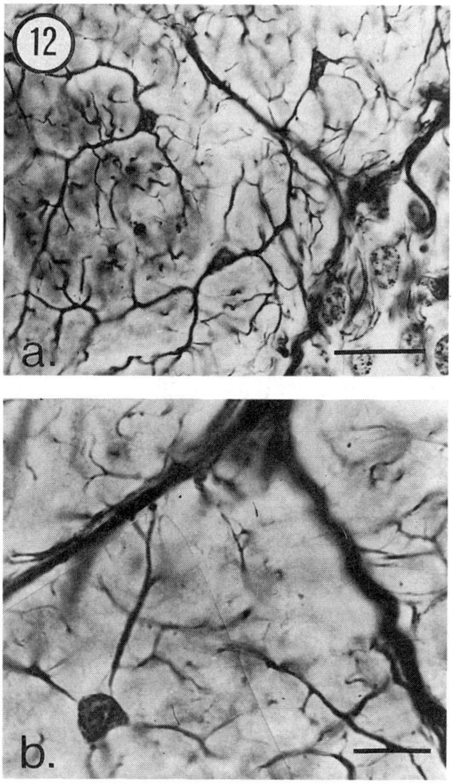

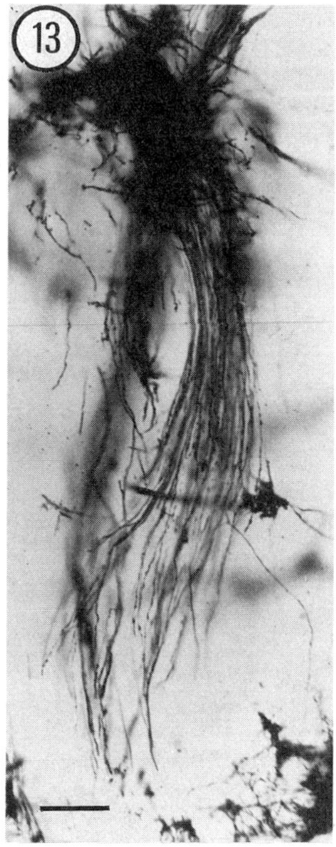

FIGURE 12. Protoplasmic glial cells from the subesophageal lobe of *Octopus.* Hortega method. Scale bars: **a**, 25 μm; **b**, 10 μm. (Stephens and Young, 1969.)

FIGURE 13. Fibrous: glia cell from the subvertical lobe of *Octopus.* Hortega method. Scale bar, 50 μm. (Stephens and Young, 1969.)

The pictures were all at low resolution and do not show membranes properly. The three-layered structure of glial unit membranes was shown by Robertson (1955).

COMPARATIVE STUDY OF NEUROGLIA

In a lecture honoring the memory of John Treherne it is appropriate to consider the comparative aspects of glia to which he greatly contributed. Indeed, his work on the neuroglia of insects is a good example of how the differences among animals can give clues to basic functions. In insects the glia forms an essential blood-brain barrier. Treherne and his colleagues used knowledge of the detailed anatomy and pharmacology to destroy the glia in a neuronal pathway of the cockroach, leaving axons and other tissues intact. As a result, potassium and other ions were able to penetrate, and function was impaired. The authors were then able to follow the whole process of repair and recovery of function. This was a striking contribution to the demonstration that the blood-brain barrier in insects is glial, whereas in vertebrates, except elasmobranchs, the barrier is mainly endothelial (Abbott *et al.,* 1985a). However, in cephalopods as in insects, it is also probably glial. In these animals the endothelium is incomplete; nevertheless, there is restriction of permeability to albumin in the brain of *Sepia* (Abbott *et al.,* 1985b). The barrier is presumed to be the complete layer of glia that covers the vessels (FIG. 11) (Gray, 1969). Small nonelectrolytes also penetrate the brain slowly and reach an equilibrium at about half their plasma concentration. Therefore evidence exists for continuous turnover of interstitial fluid. This is especially interesting because these animals have no cerebral ventricles; therefore, some source of interstitial fluid other than the choroidal vessels of vertebrates must exist.

The glia of cephalopods is, in some ways, remarkably like that of vertebrates (Stephens and Young, 1969). It stains readily with Hortega's method, and protoplasmic and fibrous types can easily be identified (FIGS. 12 and 13). In addition, the tissue is penetrated by a web of fibers that have been called "gliovascular tissue" (Bogoraze and Cazal, 1944), which are channels attached to the blood vessels. They contain collagen and muscle fibers and are surrounded by glia (FIGS. 14 and 15). In addition, there are descriptions of a system of "lymph spaces," opening to veins, that are filled by arterial injections (FIG. 16). The presence of muscle in the gliovascular network suggests that this may be part of a system for ensuring interstitial fluid flow, perhaps also using the "lymph channels." A difficulty is that the gliovascular tissue has been described mostly in the subesophageal lobes of *Octopus,* whereas experiments on the interstitial fluid have been made on the vertical and optic lobes of *Sepia,* where

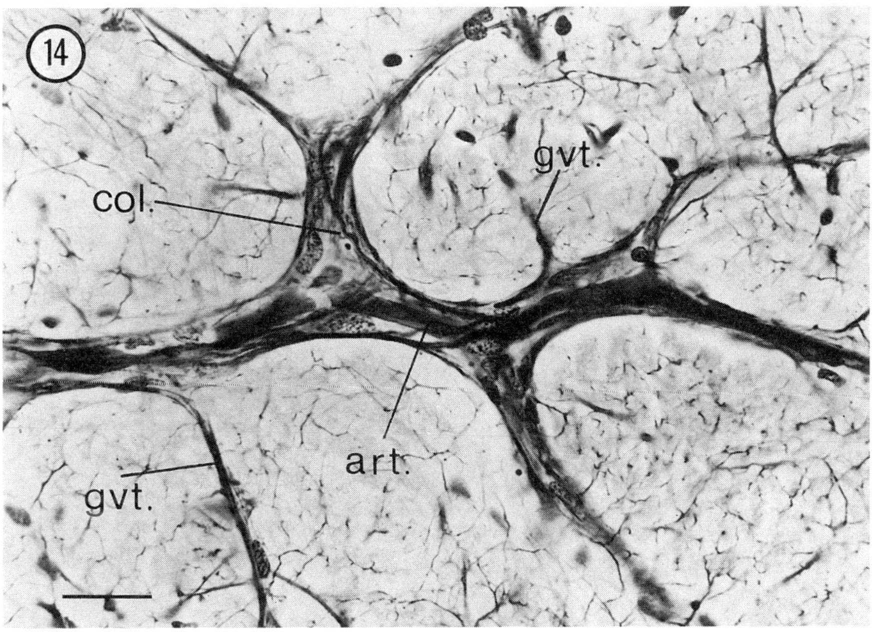

FIGURE 14. Small branching blood vessels and attached gliovascular strands in the neuropil of the pedal lobe of *Octopus;* Hortega method. art. = arterial vessel; col. = collagen; gvt. = gliovascular channel. Scale bar, 25 μm. (Stephens and Young, 1969.)

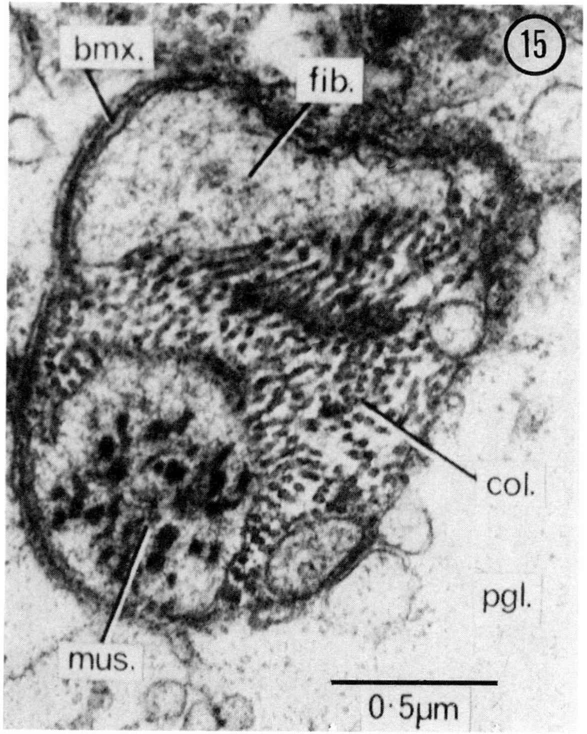

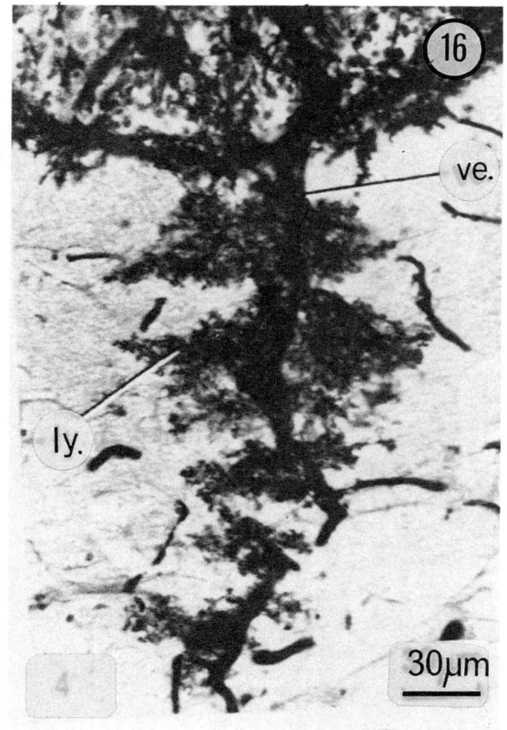

FIGURE 15. Elecron micrograph of a section of a gliovascular tunnel in the optic lobe of *Octopus* (Gray, 1969). Lettering as in FIGURE 11.

FIGURE 16. Section of the pedal lobe of an *Octopus* after injection of india ink through the aorta. The veins (ve) and lymph channels (ly) have been filled. (Stephens and Young, 1969.)

further anatomic study is needed. Already, sufficient is known to make it clear that studies of the glia of invertebrates can provide a valuable complement to that of vertebrates. Physiologic and anatomic studies of the various species and lobes of the brain should provide further information about the functions of the glia.

CONCLUSION

In the 144 years since the discovery of the glia we have advanced a very long way. We know that these cells are a universal, integral, and essential part of nervous systems. They are as important as the neurons, which could not function without them. They wrap around and enclose the axons, dendrites, synapses, and blood vessels. They probably control the composition of the interstitial medium, which is fundamental to the operation of a system of computation and transmission that is dependent upon membrane currents. Even this is partly speculative, and we can only begin to glimpse how they do it. I feel sure that you will find out much about this in the future, by work on many different animal species, as John Treherne would have advised.

REFERENCES

ABBOTT, N. J., M. BUNDGAARD & H. F. CSERR. 1985a. Comparative physiology of the blood-brain barrier. *In* The Blood-Brain in Health & Disease. A. J. Suckling, M. G. Rumsby & M. W. B. Bradbury, eds. Chichester. Ellis Horwood.

ABBOTT, N. J., M. BUNDGAARD & H. F. CSERR. 1985b. Tightness of the blood-brain barrier and evidence for brain interstitial fluid flow in the cuttlefish, *Sepia officinalis.* J. Physiol. **368:** 213–226.

ANDRIEZEN, W. 1893. The neuroglia elements in the human brain. Brit. Med. J. **2:** 227.

BICHAT, M. F. XAVIER. 1802. Traité des Membranes en Général et des Diverses Membranes en Particulier. Paris.

BICHAT, M. F. XAVIER. 1805. Recherches Physiologiques sur la Vie et la Mort. 3rd ed. Paris.

BICHAT, M. F. XAVIER. 1826. Anatomie pathologique. *Dernier Cours.* P. A. Béclard, ed. Paris. (From notes of 1905.)

BOGORAZE, D. & P. CAZAL. 1944. Recherches histologiques sur le système nerveux du Poulpe. Les neurones, le tissu interstitial et les éléments neurocrines. Arch. Zool. Exp. Gen. **83:** 413–444.

CAJAL, S. R. 1909. Histologie du système nerveux de l'homme et des vertébrés. Maleme. Paris. (Reprinted in 1952, Madrid).

CAJAL, S. R. 1913a. Sobre un nuevo proceder de impregnación de la neuroglia y sus resultados en los centros nerviosos del hombre y animales. Trab. Lab. Invest. Biol. Univ. Madrid. **11:** 219–238.

CAJAL, S. R. 1913b. Contribución al conociemento de la neuroglia del cerebro humano. Trab. Lab. Invest. Biol. Univ. Madrid. **11:** 255–315.

CAJAL, S. R. 1916. El proceder del oro para la coloración de la neuroglia. Trab. Lab. Invest. Biol. Univ. Madrid. **14:** 155–162.

CAJAL, S. R. 1917. Contribución al conociemento de la retina y centros opticos de los cefalópodos Trab. Lab. Invest. Biol. Univ. Madrid. **15:** 1–82.

CAJAL, S. R. 1920. Algunas consideraciones sobre la mesoglia de Robertson y Rio Hortega. Trab. Lab. Invest. Biol. Univ. Madrid **17:**109–128.

CAJAL, S. R. 1925. Contribution à la connaissance de la nevroglia cérébrale et cérebeleuse dans la paralysie générale progressive. Trab. Lab. Invest. Biol. Univ. Madrid **23:** 157–216.

CAJAL, S. R. 1928. Degeneration and Regeneration of the Nervous System. R. M. May, transl. Oxford University Press.

CAJAL, S. R. 1932. Neuron theory or reticular theory. M. W. Purkiss & C. Fox, transl. Consejo superior de investigacions cièntificas Instituto "Ramon y Cajal", Madrid, 1954.
CAJAL, S. R. & D. SANCHEZ. 1915. Contribución al conocimiento de los centros nerviosos de los insectos. Trab. Lab. Invest. Biol. Univ. Madrid **13:** 1–168.
DEITERS, O. 1865. Untersuchungen über Gehirn u. Rückenmark des Menschen u. der Saugetiere. Braunschweig.
GEREN, B. B. 1954. The formation from the Schwann cell surface of myelin in the peripheral nerves of chick embryos. Exp. Cell Res. **7:** 558.
GLEES, P. 1955. Neuroglia. Morphology & Function. Blackwell. Oxford.
GOLGI, C. 1870. Sulla sostanza connettiva del cervello. Gaz. Med. Ital. **19:**
GOLGI, C. 1871. Contribuzione alla fina anatomia degli organi centrali del sistema nervosa. Riv. Clin., Novembre.
GOLGI, C. 1873. Sulla sostanza grigia del cervello. Gaz. Med. Ital.
GOLGI, C. 1885. Sulla fina anatomia del la sistema nervosa. Riv. Sper. Freniatr. 8.
GOLGI, C. 1903. Opera omnia. Milan.
GRAY, E. G. 1969. Electron microscopy of the gliovascular organisation of the brain of the Octopus. Phil. Trans. Roy. Soc. **B. 255:** 13–32.
HELD, H. 1909. Ueber die Neuroglia marginalis der menschlichen Grosshirnrinde. Monatschr. Psychol. Neurol. **26:** 360–416.
HORTEGA, P. DE RIO. 1920. La microglia y sa trasformación en celulas en bastoncito e cuerpas gránulo-adiposos. Trab. Lab. Invest. Biol. Univ. Madrid. **18:** 37–82.
HORTEGA, P. DEL RIO. 1932. Microglia in Cytology and Cellular Pathology of the Nervous System. W. Penfield, ed. Hoeber. New York.
KÖLLIKER, R. A. VON. 1896. Handbuch der Geweblehre der Menschen. Leipzig.
NAKAI, J. 1963. Morphology of neuroglia. Igaku Shoin. Osaka.
PENFIELD, W. 1924. Oligodendroglia and its relation to classical neuroglia. Brain **47:** 430–452.
ROBERTSON, J. D. 1955. The ultrastructure of adult vertebrate peripheral myelinated nerve fibres in relation to myelinogenesis. J. Biophys. Biochem. Cytol. **1:** 271.
ROBERTSON, W. F. 1900. A Textbook of Pathology in Relation to Mental Diseases. Clay. Edinburgh.
SOMJEN, C. G. 1988. Nervenkitt. Notes on the history of the concept of neuroglia. Glia **1:** 2–9.
STEPHENS, P. R. & J. Z. YOUNG. 1969. The gliovascular system of cephalopods. Phil. Trans. Roy Soc. B. **255:** 1–12.
TOWER, D. B. 1978. *In* Dynamic Properties of Glia Cells. E. Schoffeniels *et al.,* eds. Pergamon Press. Oxford.
TREHERNE, J. E., P. J. S. SMITH & E. A. HOWES. 1988. Neural repair and regeneration in insects. Adv. Insect Physiol. **21:** 35–84.
VIRCHOW, R. 1946. Ueber das granulierte Ansehen der Wandungen der Gehirnventrikel. Allg. Zeitschchr. Psychiatr. **3:** 242–250.
VIRCHOW, R. 1856. Gesammelte Abhandlungen zur Wissenschaftlichen Medicin. Hamm. Frankfurt.
VIRCHOW, R. 1858. Cellular pathology as based upon physiological and pathological histology. Translated by F. Chance from 2nd ed., 1859. Dover. New York, 1971.
WEIGERT, C. 1896. Kenntnis der normalen menschlichen Neuroglia. Abh. Senckenberg, Naturforsch. Gessel. **19:** 65–216.
WYCKOFF, R. W. G. & J. Z. YOUNG. 1956. The motoneuron surface. Proc. Roy. Soc. B. **144:** 440–450.
YOUNG, J. Z. 1971. The Anatomy of the Nervous System of *Octopus vulgaris.* Clarendon Press. Oxford.

Vertebrate Glial Classification, Lineage, and Heterogeneity[a]

BRUCE R. RANSOM

Yale University School of Medicine
Department of Neurology, LCI 702
333 Cedar Street
New Haven, Connecticut 06510

The topics addressed in the first section of this volume, namely, vertebrate glial subtypes, lineage, and morphology, are among the oldest in the field of glial research. Impressive progress has been made in understanding these fundamental aspects of glial biology, but many important questions remain. As so often happens, modern techniques have raised fascinating new questions and provided evidence that challenges complacent attitudes about long-held theories. All of this is clear in the articles that follow. After a few historical comments about neuroglia, I will raise the disquieting idea that we appear to be encountering biochemical, physiological, and functional differences among neuroglial cells that are not anticipated by our current classification schemes.

THE PROBLEM AND THE TOOLS

In 1932, Penfield[1] stated, "knowledge of the form and function of neuroglia is the result of progressive improvement in histological technique in the hands of careful workers. . . . Insight into the principles involved has been made difficult by numerous and uncritical publications of men who, although doubtless sincere, are opportunists rather than cytologists." This enunciates clearly the historical approach to glial investigation and alludes to the controversies, sometimes bitter, that arose among these early workers. The connection between glial subtypes, morphology, and lineage was perfectly obvious during this period; cellular morphology alone defined both the subtypes of glial cells and their histogenetic origin or lineage. Today, morphology is only one of the features that serves to define subtypes of neuroglia and their lineage (to be discussed).

Virchow is usually acknowledged as the first to recognize neuroglia (in 1846), but the primitive microscopic and tissue-staining techniques of his day imposed severe limitations, and what he mainly saw was a kind of ground substance or "connective tissue" that filled all the space around neurons. He found what he was searching for, however; as a pathologist, he had observed inflammatory changes in the brain and felt strongly that some form of connective tissue was necessary for such changes to occur, as that was the prevailing theory. He termed this material "nervenkitt" or neuroglia, which has been translated as "nerve-glue" or "nerve-cement".[2,3] A further

[a] This work was supported by NIH grant 15589.

implication of his thinking was that neuroglial cells must derive from mesoderm, and this notion held sway for many years.

In his autobiography, Cajal talks about the technical advance that furthered his studies on glia: "my repeated inquiries upon the technique of coloring the neuroglia selectively, . . . led, in 1913, to my discovering the method of gold sublimate, a most simple procedure which allows one to impregnate specifically with a purple violet color the two types of neuroglia of the cerebral cortex, and especially the protoplasmic form."[4] A clear histologic picture of fibrous and protoplasmic glia (the "two types of neuroglia" referred to by Cajal), found respectively in white or gray matter, emerged with the work of Andriezen and Cajal. Cajal attached the name astrocyte to both of these cells and correctly insisted that they derived from ectoderm (see ref. 3). Still there were other non-nervous cells; these were poorly stained by available methods, and Cajal referred to them as the "third element." Del Rio-Hortega[5] introduced the silver carbonate staining method and demonstrated that Cajal's "third element" consisted of oligodendroglia and microglia (Del Rio-Hortega's terms), completing the cast of cellular players in the vertebrate brain (Robertson may also have demonstrated these cells, calling them mesoglia; see ref. 1). He asserted that microglia originated from mesoderm and were the true third element of Cajal, to be distinguished from the first element, neurons, and the second element, the astrocytes.[5]

Classically then, glial cells were defined on the basis of morphology and staining characteristics.[1] Astrocytes tend to be stellate in shape and are well stained by Cajal's gold chloride-sublimate method. Protoplasmic astrocytes have numerous radial processes that pass in all directions and branch profusely; fibrous astrocytes give rise to processes that are straighter, less numerous, and longer. Some of the processes of both cell types terminate on blood vessels or on the pial surface. Oligodendrocytes, compared to astrocytes, have fewer and shorter processes (essentially the meaning of the term "oligodendroglia"), stain well using the silver carbonate method of Del Rio-Hortega, and do not give rise to perivascular or subpial endfeet.

Early lineage studies employed a very straightforward method, and to some extent this method has validity today. Lineage was simply determined by recognizing morphologic similarities between certain cells in developing tissue and adult glial subtypes; alike cells constituted the same lineage. Cajal favored the notion that astrocytes derived from radial glial cells (called "supportive spongioblasts" below) and supported this argument by demonstrating what he viewed as transitional forms midway between embryonic radial glia and adult astrocytes.[6] The Spanish savant's perceptions have received support in modern studies[7] (see also ref. 8). Overall, however, simple histology alone was unable to provide a reliable scheme of neuroglial lineage. Histologic techniques were imprecise and capricious, especially in fully revealing the appearance of oligodendrocytes (FIG. 1; ref. 9; see also Butt, this volume). Moreover, there was no reliable way to know if glial cells arrived at their adult form by multiple cell divisions, with increasing differentiation, or by continuous transformation of individual immature cells. According to the lineage scheme that emerged solely on the basis of histologic investigation (FIG. 2; ref. 1), neuroepithelium gave rise to two classes of primitive precursor cells: primitive spongioblasts, the source of glia, and germinal cells that begot neurons. Schaper[1] believed that germinal cells could give rise to both neurons and glia, but the question mark next to these steps indicates that this was not regarded as likely. Three types of spongioblasts were recognized depending on the attachment of their processes to the pial surface (supportive), a blood vessel (polar), or neither (migratory). Astrocytes and oligodendrocytes were believed to have a common precursor in adult animals, the migratory spongioblast.

THE SAME PROBLEM AND NEW TOOLS

The articles that follow introduce some of the powerful newer methods that have been brought to bear on questions relating to glial subtypes and lineage. Glial cells can now be categorized on the basis of biochemical features (e.g., carbonic anhydrase, oligodendrocytes; glutamine synthetase, astrocytes), expression of surface molecules (e.g., A2B5, astrocytes; galactocerebroside, oligodendrocytes), ultrastructural features (e.g., intermediate filaments, astrocytes; myelin, oligodendrocytes), physiological properties (e.g., strong electrical coupling, astrocytes; weak electrical coupling, oligodendrocytes; see also Kettenmann *et al.,* this volume), and gene expression (e.g., mRNA for myelin basic protein, oligodendrocytes; mRNA for GFAP, astrocytes). Lineage can be assessed by labeling dividing cells and following the label with autoradiographic techniques, as discussed by Skoff and Knapp (this volume). Tissue culture techniques can be creatively employed to study glial cell lineage, and a great deal of information about the development of macroglial cells *in vitro* has been obtained in this manner (Fulton *et al* and Noble *et al,* this volume). The newest approach to lineage analysis is the use of retroviral vectors to label germinal cells genetically and histochemically (Price, this volume). This approach can be used both *in vitro* and *in vivo.*

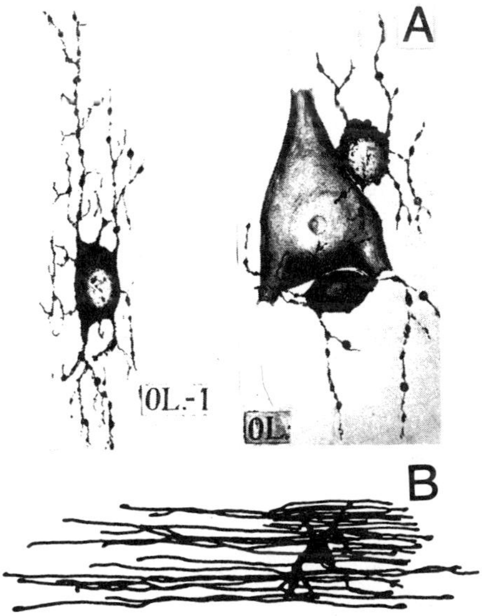

FIGURE 1. Light microscopic images of oligodendrocytes illustrating the degree of morphologic resolution afforded by different staining techniques. **(A)** *Right panel:* Oligodendrocytes (OL) in white matter. *Left panel:* Oligodendrocytes in gray matter. **(B)** Camera lucida drawing of an oligodendrocyte in rat optic nerve stained by intracellular injection of horseradish peroxidase. The intracellular staining technique **(B)**, compared to metal impregnation staining **(A)**, provides clearer and more reliable resolution of oligodendrocyte morphology (see Butt, this volume). Scale bar represents 100 μm. Panel **A** from Penfield (1932) and panel **B** from Butt and Ransom,[9] reproduced with permission.

The view of glial lineage and development that has resulted from the use of such new techniques and from the refinement of classical approaches is, not unexpectedly, far more complex and detailed then previously thought (FIG. 2B; see also ref. 10). Some of the same controversies continue unresolved. For instance, it remains unclear if neurons and glia derive from a single ventricular precursor or from two distinct pools of cells (refs. 8 and 11; Price *et al.,* this volume). Part of the additional complexity is a direct result of the new precision with which glial cells can be defined.

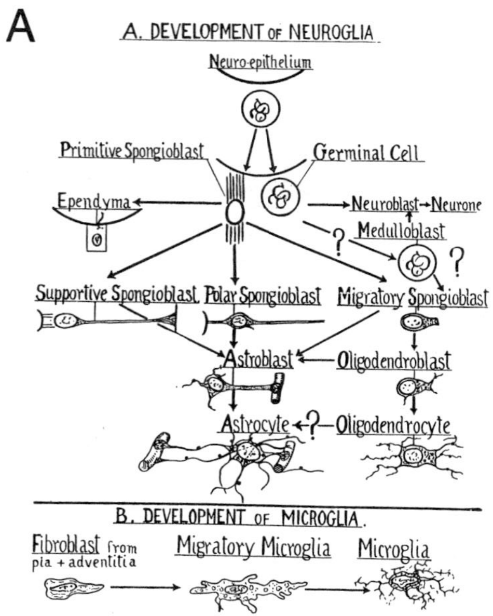

FIGURE 2. Diagrams of glial cell lineage showing the evolution of thinking about this topic. **(A)** This lineage diagram was constructed by Penfield based on knowledge available to him in 1932. **(B)** Modern conception of possible glial cell lineages that occur during the development of the cerebral cortex. The acquisition or loss of identifying antigens is indicated by *arrows* with + or − signs. Cameron and Rakic[8] should be consulted for a more detailed explanation of diagram **B.** Panel **A** is taken from Penfield[1] and panel **B** is from Cameron and Rakic,[8] reproduced with permission.

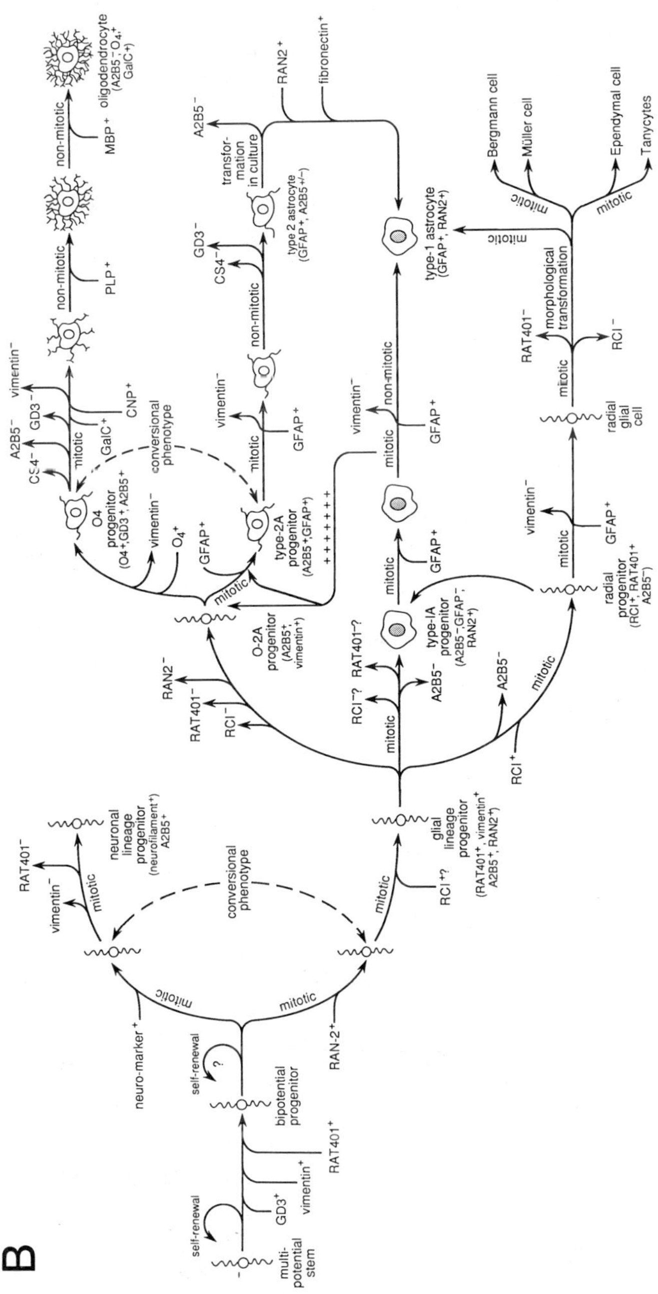

FIGURE 2B. Legend on page 22, above.

Whereas previously an astrocyte was a cell with a particular light microscopic appearance when stained with the Golgi technique, today's astrocyte can be further categorized according to its affinity for a number of different molecular markers (FIG. 2B; see also discussion of Schwann cells by Jessen *et al.,* this volume). As more is learned about glial cells, the apparent number of glial subtypes may enlarge, perhaps greatly, necessitating more revisions in the lineage scheme (to follow).

GLIAL CELL HETEROGENEITY: THE IMPLICATIONS

The nervous system is characterized by tremendous cellular diversity and in this way is unique among the body's organs. This vast diversity has been the central problem for developmental neurobiologists: How do the apparently homogeneous cells in the early neural tube give rise to the awesome variety of cell forms that comprise the adult CNS?[11] The tendency has been to focus on the diversity of neurons within the brain, both because this is so obvious and because it has such important functional implications. Without being explicit about it, work on glia has often proceeded as if the two major subpopulations of macroglial cells were each sufficiently uniform that the results on cells from any brain region could readily be generalized to all like-glial cells.

We must now ask if a similar degree of cellular diversity might not characterize glial cells, as well as neurons, in the highly specialized brains of vertebrates. The importance of framing this question is conceptual: (1) The possible existence of great glial diversity, perhaps by brain region primarily, suggests that glial cells might serve special functions in different areas of the brain. Why bother with such diversity if one "generic" cell type is sufficient everywhere in the brain? (2) The existence of glial diversity would have important implications for developmental neurobiologists who would need to address the issue of how this comes about. Are the different cell types a consequence of genetic programming or the result of unique regional influences on "generic" glial cells? (3) If glial cells are very heterogeneous, there are critical implications for those who analyze basic glial properties. We must recognize that experimental results could vary greatly on the basis of the area of the brain from which the glial cell came and on its developmental history.

In fact, considerable evidence of glial heterogeneity exists.[12] I will briefly, and incompletely, review some of these data that are available for astrocytes. Wilkin *et al.*[12] point out many examples of regional differences in astrocytic expression of proteins (e.g., microtubule-associated protein 2), mRNAs (e.g., somatostatin), transmitter receptors (e.g., adrenergic binding sites), and transmitter uptake capacity (e.g., GABA). In our studies on ion channel expression in astrocytes from rat hippocampus, optic nerve, and spinal cord, we, as well as others,[13] found remarkable regional and developmental variability.[14,15] Spinal cord astrocytes exhibited a much greater density of Na^+ channels than did astrocytes from other regions; after 6 days or more in culture, type 1-like astrocytes (see Fulton *et al.,* this volume) also exhibited the smallest density of K^+ channels. In rat optic nerve type-1 astrocytes, Na^+ channels are only expressed during a "time window" from about 3–10 days *in vitro.* Also noted were regional differences in the nature of the Na^+ channels expressed; the Na^+ channels expressed in spinal cord astrocytes were relatively resistant to blockade by tetrodotoxin compared to the channels expressed by optic nerve and hippocampal astrocytes[14] (Sontheimer, Waxman, Black, and Ransom, unpublished observations). Another electrophysiological feature that showed impressive variability was gap-junctional communication among glial cells.[16] The frequency of dye-coupling (a standard method of demonstrating functional gap-junctions) was variable, depending on the region of origin, and rat optic nerve type-2 astrocytes (see Fulton *et al.,* this volume) never exhibited coupling.[16,17]

I have emphasized differences between glial cells based on derivation from different regions in the brain. Differential expression of cellular properties could occur because of exposure of "generic" (or undifferentiated) glial cells to unique regional influences (e.g., neurotransmitters, neuronal surface molecules, extracellular ion changes, etc.). It follows that at some stage in development these "generic" glia could be removed from the brain before local conditions cause them to commit to some specific phenotype; under these conditions they would uniformly express a set of constitutive cellular features. This could, of course, be tested. Alternatively, the commitment to some specific, region-dependent set of cellular properties might be determined very early in development by genetic commands that are not influenced by local conditions. Phenotypic heterogeneity also occurs within glial populations from the same area of the brain (e.g., refs. 14 and 16), and the last mentioned mechanism would be best suited to account for this observation.

It is daunting to imagine a similar magnitude of variable cell types for glia as we know exist for neurons, and this clearly has not yet been determined. Accumulating data, however, suggest that we are moving in this direction. Most of the work has been done *in vitro,* and *in vivo* confirmation of regional patterns of glial heterogeneity is necessary. The significance of the noted differences among glial cells has also to be addressed. It is by no means certain, for example, that differences in astrocyte Na^+ channel density have functional consequences. The papers that follow in this section are broadly relevant to many of the questions raised in this essay and are representative of the new scientific approaches that are now possible in pursuing answers. The "decade of the brain" may also turn out to be a decade of unprecedented progress in understanding the extent and mechanisms of functional diversity of glia and the consequences of this for interactions between glia and neurons.

ACKNOWLEDGMENTS

I thank Drs. Harry Sontheimer, Steve Waxman, and George Richerson for many helpful suggestions.

REFERENCES

1. PENFIELD, W. 1932. Neuroglia, normal and pathological. *In* Cytology and Cellular Pathology of the Nervous System. W. Penfield, ed. Vol. **2:** 423–479. Paul B. Hoeber. New York, NY.
2. KUFFLER, S. W. & J. G. NICHOLLS. 1966. The physiology of neuroglia cells. Ergeb. Physiol. Biol. Chem. Explt. Pharmakol. **57:** 1–90.
3. SOMJEN, G. G. 1988. Nervenkett: Notes on the history of the concept of neuroglia. Glia **1:** 2–9.
4. RAMON Y CAJAL, S. 1937. Recollections of my Life. The MIT Press. Cambridge, Massachusetts.
5. DEL RIO-HORTEGA, P. 1932. Microglia. *In* Cytology and Cellular Pathology of the Nervous System. W. Penfield, ed. Vol. **2:** 423–479. Paul B. Hoeber. New York, NY.
6. RAMON Y, CAJAL, S. 1929. Studies on Vertebrate Neurogenesis.: 245–249. Charles C Thomas. Springfield, Illinois.
7. VOIGHT, T. 1989. Development of glial cells in the cerebral wall of ferrets: Direct tracing of their transformation from radial glia into astrocytes. J. Comp. Neurol. **289:** 74–88.

8. Cameron, R. S. & P. Rakic. 1991. Glial cell lineage in the cerebral cortex: A review and synthesis. Glia **4:** 124–137.
9. Butt, A. M. & B. R. Ransom. 1989. Visualization of oligodendrocytes and astrocytes in the intact rat optic nerve by intracellular injection of lucifer yellow and horseradish peroxidase. Glia **2:** 470–475.
10. Special Issue: Glial cell lineage. 1991. Glia **4:** 121–243.
11. McConnell, S. K. 1988. Development and decision-making in the mammalian cerebral cortex. Brain Res. Rev. **13:** 1–23.
12. Wilkin, G. P., D. R. Marriott & A. J. Cholewinski. 1990. Astrocyte heterogeneity. TINS **13:** 43–46.
13. Barres, B. A., L. L. Y. Chun & D. P. Corey. 1989. Glial and neuronal forms of the voltage-dependent sodium channel: Characteristics and cell-type distribution. Neuron **2:** 1375–1388.
14. Sontheimer, H., B. R. Ransom, A. H. Cornell-Bell, J. A. Black & S. G. Waxman. 1991a. Na^+-current expression in rat hippocampal astrocytes *in vitro:* Alterations during development. J. Neurophysiol. **65:** 3–19.
15. Sontheimer, H., J. E. Minturn, J. A. Black, B. R. Ransom & S. G. Waxman. 1991b. Two types of Na^+-currents in cultured rat optic nerve astrocytes: Changes with time in culture and with age of culture derivation. J. Neurosci. Res., in press.
16. Sontheimer, H., J. E. Minturn, J. A. Black, S. G. Waxman & B. R. Ransom. 1990. Specificity of cell-cell coupling in rat optic nerve astrocytes *in vitro.* Proc. Natl. Acad. Sci. USA **87:** 9833–9837.
17. Sontheimer, H., S. G. Waxman & B. R. Ransom. 1991c. Relationship between Na^+ current expression and cell-cell coupling in astrocytes cultured from rat hippocampus. J. Neurophysiol. **65:** 989–1002.

Glial Cells in the Rat Optic Nerve

The Search for the Type-2 Astrocyte[a]

BARBARA P. FULTON,[b] JULIA F. BURNE, AND
MARTIN C. RAFF

*MRC Developmental Neurobiology Programme
Department of Biology, Medawar Building, and
[b]Department of Anatomy & Developmental Biology
University College London
London WC1E 6BT, UK*

The development of central nervous system (CNS) macroglial cells has been studied extensively in cell cultures of the developing rat optic nerve, where some of the complex cell-cell interactions that underly macroglial cell differentiation, at least *in vitro*, have been defined.[1,2] Such cultures contain three antigenically distinct types of macroglial cells, namely, oligodendrocytes and two types of astrocytes (type-1 and type-2). These three cell types appear at different stages of development and arise from two separate lineages: type-1 astrocytes develop from one type of precursor cell, whereas oligodendrocytes and type-2 astrocytes develop from a common, bipotential precursor called the O-2A progenitor cell.[3] Whereas O-2A progenitor cells in culture differentiate autonomously into oligodendrocytes,[4] their differentiation into type-2 astrocytes, at least in serum-free cultures, requires two or more inducing signals produced by other cells; these signals include ciliary neurotrophic factor (CNTF), probably produced by type-1 astrocytes, and extracellular matrix-associated signals, probably produced by mesenchymal cells.[5]

The use of cell culture techniques to study developmental processes raises the question of how accurately findings *in vitro* represent what happens *in vivo.* Does the type-2 astrocyte exist *in vivo,* for example, and, if so, what is its morphology, distribution, and relationship with other cell types in the nerve? All of the signals required for the development of type-2 astrocytes *in vitro* are present in the developing rat optic nerve,[5–7] but it has yet to be shown definitively that O-2A cells develop into type-2 astrocytes *in vivo.* Before functional relevance can be ascribed to the properties of type-2 astrocytes defined in culture, it is crucial to know that these cells exist *in vivo* and what some of their characteristics are. At the same time, however, properties determined in culture can provide useful tools to aid in the identification of these cells *in vivo.* Here we shall take stock of the evidence for and against the existence of the type-2 astrocyte *in vivo* and of some previous suggestions about the possible identity and function of this cell.

SOME PROPERTIES OF TYPE-1 AND TYPE-2 ASTROCYTES IN CULTURE

All astrocytes contain a characteristic type of intermediate filament, the glial filament, which is not found in other cell types in the rat CNS. This unique property permits immunohistochemical identification of astrocytes, both in culture and in

[a]This work was supported by the Medical Research Council, U.K..

tissue sections, by the use of antibodies directed against the major component of glial filaments, glial fibrillary acidic protein (GFAP).[8] In cultures of rat optic nerve, $GFAP^+$ cells can be further subdivided into type-1 and type-2 astrocytes by a combination of morphologic and antigenic criteria,[9] neither of which on its own is totally reliable for distinguishing between the two astrocyte types. Whereas type-1 astrocytes are usually flat and fibroblast-like in optic nerve cell cultures, type-2 astrocytes are almost always process-bearing. The two types of astrocyte can also be distinguished immunohistochemically in these cultures, using a number of antibodies to cell-surface antigens. The one most frequently used for this purpose is the monoclonal antibody A2B5,[10] which recognizes specific gangliosides and labels most cells with the morphology of type-2 astrocytes, but few cells with the morphology of type-1 astrocytes.[9] Another monoclonal antibody, 04, which recognizes sulfatide and one or more related glycolipids,[11,12] is more specific for O-2A lineage cells: it labels a proportion of type-2 astrocytes, but not any type-1 astrocytes in culture.[11,13,14] A2B5 also labels O-2A progenitor cells and newly differentiated oligodendrocytes in culture,[3] as does 04, but 04 only labels progenitors beginning several days after birth[15] whereas A2B5 labels them at all stages of development. Antibodies that recognize the NG2 chondroitin-sulfate proteoglycan,[16] stain cells in cultures of rat optic nerve in much the same way as does A2B5,[17] but they also label meningeal cells.

Patch-clamp studies of type-1 and type-2 astrocytes in culture have demonstrated differences in their ion channel phenotypes in terms of both voltage-activated[18–20] and transmitter-activated channels.[21,22] For example, type-1 astrocytes express a slow, glial type of voltage-activated sodium channel and no glutamate-receptor channels, whereas type-2 astrocytes express a fast, neuronal type of voltage-activated sodium channel as well as glutamate-receptor channels of the quisqualate and kainate types.

$GFAP^+$ O-2A LINEAGE CELLS IN CELL SUSPENSIONS

The original evidence that type-2 astrocytes exist *in vivo* and are not simply an artifact of culture was that cells with the antigenic phenotype of type-2 astrocytes are present in suspensions of freshly dissociated optic nerve cells before plating. $A2B5^+$, $GFAP^+$ cells are found in cell suspensions prepared from $\geq$ postnatal day 8 (P8) rat optic nerves, but not from younger nerves[23]; similar results are obtained with cell suspensions of rat brain.[24] However, four difficulties exist in interpreting these findings: (1) A2B5 cannot be used as a definitive marker of type-2 astrocytes as, in culture at least, occasional type-1 astrocytes are $A2B5^+$; (2) morphology cannot be used to help distinguish between different types of astrocytes in cell suspensions, where most of the cells look much the same; (3) only small numbers of $A2B5^+$, $GFAP^+$ cells are found in such cell suspensions and these numbers do not increase much with age (M.R., unpublished observations); and (4) it is not certain whether $GFAP^+$ O-2A lineage cells in the developing CNS are stable type-2 astrocytes or, alternatively, are O-2A progenitor cells that have been transiently induced to express GFAP before going on to become oligodendrocytes, as has been shown to occur under some circumstances in culture.[25]

Two recent findings have overcome two of these four difficulties. First, small numbers of 04^+, $GFAP^+$ cells are present in optic nerve cell suspensions (M.R. and B. Barres, unpublished observations), and thus far 04 has not been seen on type-1 astrocytes or on any other type of non-O-2A lineage cells in cultures of rat optic nerve. Second, if cells dissociated from 2-week-old rat optic nerve are cultured for 2 hours in serum-free medium, to allow them to reform processes and re-express surface antigens that were removed by the enzymatic dissociation procedure, small numbers of process-bearing GFAP$^+$ cells are found that express a number of surface antigens characteristic (but not diagnostic) of type-2 astrocytes, including HNK-1 and NSP-4[26] and NG2[17] (M.R., unpublished observations). These findings suggest that some O-2A lineage cells are induced to express GFAP *in vivo* during normal development.

IMMUNOHISTOCHEMISTRY OF OPTIC NERVE SECTIONS

Of the many strategies that have been adopted to look for cells corresponding to type-2 astrocytes in the intact rat optic nerve, the most direct involves the use of immunohistochemical techniques to identify cells with the antigenic phenotype of type-2 astrocytes in sections of optic nerve. In general, however, this approach has produced ambiguous results and has highlighted a number of problems in comparing antigenic phenotypes in culture and in tissue sections. When A2B5 was used to label semi-thin frozen sections of adult rat optic nerve,[27] large numbers of GFAP$^+$ astrocytes in the interior of the nerve were A2B5$^+$, whereas GFAP$^+$ astrocytes that formed the *glia limitans* were A2B5 . Some radially oriented GFAP$^+$ processes and some GFAP$^+$ processes associated with blood vessels were A2B5$^+$. In developmental studies, A2B5$^+$, GFAP$^+$ cells were not seen in sections until the second week of postnatal development, which is when cells with the antigenic phenotype of type-2 astrocytes are first found in cell suspensions and is equivalent to the time when type-2 astrocytes first appear in cultures of embryonic brain cells[24] or in platelet-derived growth factor (PDGF)-treated, serum-free cultures of newborn optic nerve cells.[25] Initially, the occurrence of A2B5$^+$, GFAP$^+$ cells in sections of the adult nerve, coupled with the timing of their appearance during development, was taken as evidence for the presence of type-2 astrocytes *in vivo,* although this interpretation was complicated by differences in the cellular location of the A2B5 staining in sections compared to that in culture: whereas A2B5 is used as a surface marker of cells in culture, its staining in sections is largely, if not exclusively, intracellular. The nature of the intracellular staining and of the subpopulation of astrocytes that show it is still unknown, but it now seems unlikely that the cells are type-2 astrocytes.

The first clue that the astrocytes labeled in optic nerve sections by A2B5 might not be type-2 astrocytes came from studies using another set of antibodies, all of which mainly recognize the adhesion glycoproteins called J1 160 and J1 180 in Western blots of adult optic nerve extracts.[28] These include monoclonal antibodies HNK-1,[29] L2,[30] and anti-NSP-4[31] and polyclonal antibodies against J1.[32] In semi-thin frozen sections of adult nerve, these antibodies label small GFAP$^+$ astrocyte processes around nodes of Ranvier, as well as some fine processes running longitudinally within the nerve.[26,28] In culture, they label O-2A progenitor cells, newly formed oligodendrocytes, and the majority of type-2 astrocytes, but few type-1 astrocytes,[26] much like A2B5. These results suggested that, *in vivo,* type-2 astrocytes might give rise to most or all of the perinodal astrocyte processes, previously described in CNS white matter,[33,34] and that the processes of type-2 astrocytes might run predominantly longitudinally within the nerve. In view of the lineage relationship between type-2 astrocytes and oligodendrocytes, the idea that they might collaborate in ensheathing axons is appealing. This is not the only possible explanation of these observations, however. It is possible, for example, that oligodendrocytes, which also express J1 160 and J1 180 *in vitro,*[35] are the source of these proteins at nodes *in vivo* (especially as these proteins are thought to be located mainly in the extracellular matrix[36]), or that a subset of type-1 astrocytes might express J1 160 and J1 180 and extend processes to nodes.

Immunoelectron microscopic studies of adult rat optic nerve, using the antibodies against purified rat brain sodium channels,[37] have demonstrated intense staining in the axon membrane at nodes of Ranvier and also in perinodal and longitudinally oriented astrocyte processes.[38,39] Intense cytoplasmic immunostaining was also seen in the cell bodies and major processes of astrocytes, whereas relatively little staining was seen in perivascular astrocyte processes and in astrocyte processes contributing to the *glia limitans,* and no staining was seen in oligodendrocytes. It is not clear whether the pattern of staining observed in optic nerve astrocytes reflects differences in sodium channel expression in different types of astrocytes or in different regions of a single type of astrocyte, or both.

MORPHOLOGICAL STUDIES OF INDIVIDUAL DEVELOPING AND MATURE OPTIC NERVE GLIA *IN SITU*

To search for cells *in vivo* that might correspond to type-2 astrocytes *in vitro,* the morphology and organization of glial cells in the rat optic nerve at different stages of development were examined using two approaches[40] (B.F., J.B., and M.R., in preparation). In the first, the Golgi impregnation technique was used to examine the general organization of glial cells in the developing and adult nerve.[40] In sections of Golgi-stained adult nerve, two distinct types of astrocyte-like cells could be seen, one with a predominantly radial orientation, the other with a predominantly longitudinal orientation. The radially oriented cells were present at birth, but the longitudinally oriented cells were not seen until about 2 weeks later. In the second approach, individual glial cells were microinjected with horseradish peroxidase (HRP) or Lucifer yellow at different stages of development[40] (B.F., J.B., and M.R., in preparation). At all developmental stages studied, a number of distinct morphological cell types could be seen, including astrocyte-like cells. During early postnatal development the astrocyte-like cells had a largely radial orientation, but from about P14 onward a few had a mainly longitudinal orientation. Although these "longitudinal astrocytes" were encountered more frequently after P14, they usually constituted less than 10% of injected cells. In a similar dye-injection study by Butt and Ransom,[41] cells of this morphological type were not described.

When HRP-filled astrocytes in adult optic nerve were examined with the electron microscope, those with a predominantly radial orientation were found to send processes to blood vessels and the pial surface, whereas those with a predominantly longitudinal orientation did not appear to contact either blood vessels or pia, but instead seemed to contact myelinated axons at structures that we interpreted to be nodes of Ranvier (although, with hindsight, the relatively poor preservation of the tissue makes this interpretation insecure).

The results of these experiments, taken together with those of the earlier immunohistochemical studies,[26,28] led us to propose that the cells with mainly radial processes correspond to type 1 astrocytes in cultures of perinatal rat optic nerve, whereas those with mainly longitudinal processes correspond to type-2 astrocytes in culture. We further suggested that there might be a division of labor between these two cell types as they seemed to interact with different structures in the nerve. Subsequent electron microscopic studies of serial thin sections of adult rat optic nerve,[42] however, showed that some astrocytes associated with the pial surface or with blood vessels, also made contact with nodes of Ranvier, making it unlikely that a strict division of labor exists among astrocytes or that all perinodal processes derive from longitudinally oriented astrocytes.

The suggestion that longitudinally oriented astrocytes might correspond to type-2 astrocytes[40] was based partly on the developmental stage at which they are first seen in the nerve and partly on the staining of longitudinal processes with the antibodies that recognize the J1 160 and J1 180 glycoproteins.[28] There is, however, an alternative explanation for the late appearance of astrocytes with a longitudinal orientation: these astrocytes encounter a structurally more complex environment, especially after the onset of myelination, compared to astrocytes that develop earlier, and so their processes might be physically constrained to extend longitudinally rather than radially. In such a scheme, type-2 astrocytes, because of their late development, would be likely to adopt a predominantly longitudinal orientation, but any late-developing type-1 astrocytes would also do so.

Taking all of the immunohistochemical and morphological observations into consideration, it is still uncertain whether the *in vivo* counterpart of the type-2 astrocyte is a cell that gives rise to processes that run longitudinally and associate mainly with nodes of Ranvier.

FAILURES TO FIND CANDIDATES FOR TYPE-2 ASTROCYTES *IN VIVO*

Two reported observations raise serious doubts about the existence of type-2 astrocytes *in vivo.* In a combined autoradiographic, immunocytochemical, and electron microscopic study, Skoff[43] failed to find substantial numbers of astrocytes in the rat optic nerve that were born after the second postnatal week. If dividing O-2A progenitor cells give rise to type-2 astrocytes beginning during the second postnatal week, as we have proposed,[23] Skoff should have found such astrocytes in his study, unless they are not recognizable as astrocytes, or continue to divide rapidly after they develop, or are produced in only very small numbers. In an immunohistochemical and patch-clamp analysis of freshly dissociated glial cells in tissue prints of rat optic nerve in the second postnatal week, Barres and colleagues[20] failed to find any cells with an astrocyte-like morphology that were $A2B5^+$, and only one of hundreds of astrocyte-like cells examined expressed fast neuronal-type Na^+ channels characteristic of type-2 astrocytes. This finding suggests that if type-2 astrocytes exist in the optic nerve at this age, they must either be difficult to isolate with this method, not have a recognizable astrocyte morphology, or be present in very small numbers.

CONCLUSIONS

The case for the existence of type-2 astrocytes *in vivo* remains weak. Although it has been shown that all of the signals required to induce type-2 astrocyte development *in vitro* are present in the developing optic nerve, and that small numbers of O-2A progenitors are induced to express GFAP during normal *in vivo* development, it is still uncertain whether any of these $GFAP^+$ O-2A lineage cells go on to become functional astrocytes and, if they do, what they look like and what they do.

It seems likely that the $NG2^+$ glial cells that have been described in the adult rat cerebellar and cerebral cortex[13,44] are O-2A lineage cells[17]; although they were called smooth protoplasmic astrocytes, they do not contain detectable GFAP or intermediate filaments, and might be similar to the O-2A progenitor cells that are found in the adult optic nerve,[15,45] which have a number of properties that distinguish them from

O-2A progenitor cells in the developing optic nerve,[46] but, nevertheless, can differentiate into either oligodendrocytes or type 2 astrocytes in culture. It is uncertain whether adult O-2A progenitor cells, which seem to develop from perinatal O-2A progenitor cells,[47] and the cells described by Levine and colleagues[44] serve only as a source of new differentiated glial cells or have some other function in the adult CNS.

Experiments in culture have demonstrated that O-2A progenitor cells have the potential to develop into astrocytes as well as into oligodendrocytes *in vitro.* More sophisticated experiments will be required to determine whether these progenitor cells ever express this potential *in vivo.*

REFERENCES

1. RAFF, M. C. 1989. Glial cell diversification in the rat optic nerve. Science **243:** 1450–1455.
2. LILLIEN, L. E. & M. C. RAFF. 1990. Differentiation signals in the CNS: Type-2 astrocyte development *in vitro* as a model system. Neuron **5:** 111–119.
3. RAFF, M. C., R. H. MILLER & M. NOBLE. 1983. A glial progenitor cell that develops *in vitro* into an astrocyte or an oligodendrocyte depending on the culture medium. Nature **303:** 390–396.
4. TEMPLE, S. & M. C. RAFF. 1985. Differentiation of a bipotential glial progenitor cell in single cell microculture. Nature **313:** 223–225.
5. LILLIEN, L. E., M. SENDTNER & M. C. RAFF. 1990. Extracellular-matrix-associated molecules collaborate with ciliary neurotrophic factor to induce type-2 astrocyte development. J. Cell Biol. **111:** 635–644.
6. HUGHES, S. & M. C. RAFF. 1987. An inducer protein may control the timing of fate switching in a bipotential glial progenitor cell in the rat optic nerve. Development **101:** 157–167.
7. HUGHES, S., L. E. LILLIEN, M. C. RAFF, H. ROHRER & M. SENDTNER. 1988. Ciliary neurotrophic factor (CNTF) induces type-2 astrocyte differentiation in culture. Nature **335:** 70–73.
8. BIGNAMI, A., L. F. ENG, D. DAHL & C. T. UYEDA. 1972. Localization of the glial fibrillary acidic protein in astrocytes by immunofluorescence. Brain Res. **43:** 429–435.
9. RAFF, M. C., E. R. ABNEY, J. COHEN, R. LINDSAY & M. NOBLE. 1983. Two types of astrocytes in cultures of developing rat white matter: Differences in morphology, surface gangliosides and growth characteristics. J. Neurosci. **3:** 1289–1300.
10. EISENBARTH, G. S., F. S. WALSH & M. NIRENBERG. 1979. Monoclonal antibody to a plasma membrane antigen of neurons. Proc. Natl. Acad. Sci. USA **76:** 4913–4917.
11. SOMMER, I. & M. SCHACHNER. 1981. Monoclonal antibodies (01–04) for oligodendrocyte cell surfaces: An immunological study in the central nervous system. Dev. Biol. **83:** 311–327.
12. BANSAL, R., A. E. WARRINGTON, A. L. GARD, B. RANSCHT & S. E. PFEIFFER. 1989. Multiple and novel specificities of monoclonal antibodies 01, 04, and R-mAb used in the analysis of oligodendrocyte development. J. Neurosci. Res. **24:** 548–557.
13. LEVINE, J. M. & W. B. STALLCUP. 1987. Plasticity of developing cerebellar cells *in vitro* studied with antibodies against the NG2 antigen. J. Neurosci. **7:** 2721–2731.
14. LEVI, G., F. ALOISI & G. P. WILKIN. 1987. Differentiation of cerebellar bipotential glial precursors into oligodendrocytes in primary culture: Developmental profile of surface antigens and mitotic activity. J. Neurosci. Res. **18:** 407–417.
15. WOLSWIJK, G. & M. NOBLE. 1989. Identification of an adult-specific glial progenitor cell. Development **105:** 387–400.
16. WILSON, S.-S., E. E. BAETGE & W. B. STALLCUP. 1980. Antisera specific for cell lines with mixed neuronal and glial properties. Dev. Biol. **83:** 146–153.
17. STALLCUP, W. B. & L. BEASLEY. 1987. Bipotential glial precursor cells of the optic nerve express the NG2 proteoglycan. J. Neurosci. **7:** 2737–2744.
18. BARRES, B. A., L. L. Y. CHUN & D. P. COREY. 1988. Ion channel expression by white matter glia: I. Type-2 astrocytes and oligodendrocytes. Glia **1:** 10–30.

19. BARRES, B. A., L. L. Y. CHUN & D. P. COREY. 1989. Glial and neuronal forms of the voltage-dependent sodium channel: Characteristics and cell-type distribution. Neuron **2:** 1375–1388.
20. BARRES, B. A., W. J. KOROSHETZ, L. L. Y. CHUN & D. P. COREY. 1990. Ion channel expression by white matter glia. The type-1 astrocyte. Neuron **5:** 527–544.
21. CULL-CANDY, S. G., A. MATHIE, C. J. SYMONDS & D. J. A. WYLLIE. 1989. Distribution of quisqualate and kainate receptors in rat type-2 astrocytes and their progenitor cells in culture. J. Physiol. (Lond.) **418:** 195P.
22. WYLLIE, D. J. A., A. MATHIE, C. J. SYMONDS & S. G. CULL-CANDY. 1991. Activation of glutamate-receptors and glutamate-uptake in identified macroglial cells of the rat. J. Physiol. (Lond.) **432:** 235–258.
23. MILLER, R. H., S. DAVID, R. PATEL, E. R. ABNEY & M. C. RAFF. 1985. A quantitative immunohistochemical study of macroglial cell development in the rat optic nerve: *In vivo* evidence for two distinct astrocyte lineages. Dev. Biol. **111:** 35–41.
24. WILLIAMS, B. P., E. R. ABNEY & M. C. RAFF. 1985. Macroglial cell development in embryonic rat brain: Studies using monoclonal antibodies, fluorescence activated cell sorting and cell culture. Dev. Biol. **112:** 126–134.
25. LILLIEN, L. E. & M. C. RAFF. 1990. Analysis of the cell-cell interactions that control type-2 astrocyte development. Neuron **4:** 525–534.
26. FFRENCH-CONSTANT, C. & M. C. RAFF. 1986. The oligodendrocyte-type-2 astrocyte cell lineage is specialized for myelination. Nature **323:** 335–338.
27. MILLER, R. H. & M. C. RAFF. 1984. Fibrous and protoplasmic astrocytes are biochemically and developmentally distinct. J. Neurosci. **4:** 585–592.
28. FFRENCH-CONSTANT, C., R. H. MILLER, J. KRUSE, M. SCHACHNER & M. C. RAFF. 1986. Molecular specialization of astrocyte processes at nodes of Ranvier in rat optic nerve. J. Cell Biol. **102:** 844–852.
29. ABO, T. & C. M. BALCH. 1981. A differentiation antigen of human NK and K cells identified by a monoclonal antibody (HNK-1). J. Immunol. **127:** 1024–1029.
30. KRUSE, J., R. MAILHAMMER, H. WERNECKE, A. FAISSNER, I. SOMMER, C. GORIDIS & M. SCHACHNER. 1984. Neural cell adhesion molecules and myelin associated glycoprotein share a common carbohydrate moiety recognised by monoclonal antibodies L2 and HNK-1. Nature **311:** 153–155.
31. ROUGON, G., M. R. HIRSCH, M. HIRN, J. L. GUENET & C. GORIDIS. 1983. Monoclonal antibody to neural cell surface protein: Identification of a glycoprotein family of restricted cellular localization. Neuroscience **10:** 511–520.
32. KRUSE, J., G. KEILHAUER, A. FAISSNER, R. TIMPL & M. SCHACHNER. 1985. The J1 glycoprotein—a novel nervous system cell adhesion molecule of the L2/HNK-1 family. Nature **316:** 146–148.
33. HILDEBRAND, C. 1971. Ultrastructural and light-microscopic studies of the nodal region in large myelinated fibers of the adult feline spinal cord white matter. Acta Physiol. Scand. Suppl. **364:** 43–71.
34. HILDEBRAND, C. & S. G. WAXMAN. 1984. Postnatal differentiation of rat optic nerve fibres: Electron microscope observations on the development of nodes of Ranvier and axoglial relations. J. Comp. Neurol. **224:** 25–37.
35. PESHEVA, P., E. SPEISS & M. SCHACHNER. 1989. J1-160 and J1-180 are oligodendrocyte-secreted nonpermissive substrates for cell adhesion. J. Cell Biol. **109:** 1765–1778.
36. FAISSNER, A., J. KRUSE, K. KÜHN & M. SCHACHNER. 1990. Binding of the J1 adhesion molecules to extracellular matrix constituents. J. Neurochem. **54:** 1004–1015.
37. ELMER, L. W. 1988. Mammalian sodium channel physicochemical characterization and immunocytochemical localization and interaction with the neuronal cytoskeleton as a mechanism of restricted distribution. Ph. D. Dissertation. University Microfilms.
38. BLACK, J. A., B. FRIEDMAN, S. G. WAXMAN, L. W. ELMER & K. J. ANGELIDES. 1989. Immuno-ultrastructural localization of sodium channels at nodes of Ranvier and perinodal astrocytes in rat optic nerve. Proc. R. Soc. Lond. B **238:** 39–51.
39. BLACK, J. A., S. G. WAXMAN, B. FRIEDMAN, L. W. ELMER & K. J. ANGELIDES. 1989. Sodium channels in astrocytes of rat optic nerve in situ: Immuno-electron microscopic studies. Glia **2:** 353–369.

40. MILLER, R. H., B. P. FULTON & M. C. RAFF. 1989. A novel type of glial cell associated with nodes of Ranvier in rat optic nerve. Eur. J. Neurosci. **1:** 172–180.
41. BUTT, A. M. & RANSOM, B. R. 1989. Visualization of oligodendrocytes and astrocytes in the intact rat optic nerve by intracellular injection of Lucifer Yellow and horseradish peroxidase. Glia **2:** 470–475.
42. SUÁREZ, I. & M. C. RAFF. 1989. Subpial and perivascular astrocytes associated with nodes of Ranvier in the rat optic nerve. J. Neurocytol. **18:** 577–582.
43. SKOFF, R. P. 1990. Gliogenesis in rat optic nerve: Astrocytes are generated in a single wave before oligodendrocytes. Dev. Biol. **139:** 149–168.
44. LEVINE, J. M. & J. P. CARD. 1987. Light and electron microscopic localization of a cell surface antigen (NG2) in the rat cerebellum: Association with smooth protoplasmic astrocytes. J. Neurosci. **7:** 2711–2720.
45. FFRENCH-CONSTANT, C. & M. C. RAFF. 1986. Proliferating bipotential glial progenitor cells in adult rat optic nerve. Nature **319:** 499–502.
46. WOLSWIJK, G., P. N. RIDDLE & M. NOBLE. 1990. Coexistence of perinatal and adult forms of a glial progenitor cell during development of the rat optic nerve. Development **109:** 691–698.
47. NOBLE, M., G. WOLSWIJK & D. WREN. 1989. The complex relationship between cell division and differentiation in oligodendrocyte-type-2 astrocyte progenitor cells isolated from perinatal and adult rat optic nerve. Prog. Growth Factor Res. **1:** 179–194.

Development, Regeneration, and Neoplasia of Glial Cells in the Central Nervous System

MARK NOBLE,[a] PARIS ATALIOTIS,[a]
SUSAN C. BARNETT,[a] KAREN BEVAN,[a]
OLIVER BÖGLER,[a] ANDREW GROVES,[a] PARMJIT JAT,[a]
GUUS WOLSWIJK,[a] AND DAMIAN WREN[b]

[a]*Ludwig Institute for Cancer Research*
London W1P 8BT, UK

[b]*Department of Clinical Neurology*
Institute of Neurology
London WC1N 3BG, UK

Our studies on glial cell biology are concerned with three related goals: (1) the elucidation of basic principles involved in the control of glial cell division and differentiation; (2) the development of cellular and molecular biological approaches to understanding diseases of the human central nervous system (CNS) that involve glial cells; and (3) the development of precursor cell transplantation as an approach to the repair of damaged tissues. In this review we summarize our present knowledge about glial precursors, oligodendrocytic differentiation, and the relationship between glial cells and human gliomas. We also discuss a new approach to the isolation of precursors and novel cell lines that we believe will dramatically enhance our abilities to identify new precursors and study their properties.

In several of the studies that we will discuss, interactions with the laboratories of Bill Blakemore, Alastair Compston, Martin Raff, and Bill Richardson have been of considerable importance. A reading of the contributions to the present volume written by these authors, and their colleagues, will give a more complete picture of the body of knowledge relevant to the considerations presented in our own chapter.

BACKGROUND

Our studies on glial development have focused, for the most part, on the optic nerve. By studying this simple white matter tract, we have been able to analyze glial development in a relatively homogeneous tissue. Studies of this tissue are further simplified by the fact that the optic nerve contains no intrinsic neurons; thus, neuron-free preparations of glial cells can be obtained by simple dissection.

In our initial studies on the glia of the optic nerve, we found that cultures derived from this simple tissue contained two antigenically distinct families of astrocytes, which we termed type-1 and type-2 astrocytes.[1] This observation was surprising, as previous anatomic investigations had not detected different types of astrocytes in the optic nerve.[2] As little was known about the mechanisms of diversification among astrocyte populations, we initiated studies on the relationship between these two cell types. We found that type-1 and type-2 astrocytes could not be interconverted and seemed to represent independently derived populations.[1]

The discovery of two groups of astrocytes in the optic nerve raised questions about the developmental relationship between these two cell types. Investigation of this relationship led us to the discovery of the oligodendrocyte-type-2 astrocyte (O-2A) progenitor cell,[3] a cell that could be induced by simple tissue culture manipulations to proceed along either of two distinct differentiation pathways. In cultures derived from the optic nerves of 7-day-old rats and grown in chemically defined medium, differentiation of all O-2A progenitors into oligodendrocytes occurred within 3 days. In contrast, the generation of type-2 astrocytes required the presence of appropriate inducing factors, at least one of which is found in fetal sera of a number of different species.[1,3,4] These initial results suggested that the oligodendrocytic pathway was the preferred (or default) differentiation pathway for O-2A progenitors, whereas the astrocytic pathway was an induced pathway. (For a detailed description of type-2 astrocyte differentiation, the reader is referred to the contribution from Fulton, Burne, and Raff in this volume; our own studies on the O-2A progenitor have more frequently been concerned with oligodendrocytic differentiation.) Although the optic nerve-derived cultures also contained type-1 astrocytes, such cells were apparently derived from a precursor cell different from the O-2A progenitor.[5]

Two paradoxes emerged from our early studies on O-2A progenitors. The first paradox was that although this population of cells was expanding rapidly *in vivo* at the time of dissection,[6,7] cells grown in tissue culture did not divide. The second paradox was that all O-2A progenitors grown in chemically defined medium differentiated into oligodendrocytes within 3 days, even though the generation of new oligodendrocytes *in vivo* continued for several weeks after the age at which optic nerves were dissected,[6] and O-2A progenitors could be isolated from optic nerves of animals of several different ages.

A likely explanation for the failure of O-2A progenitors to behave similarly *in vivo* and *in vitro* was that the cell-cell interactions that supported normal development were disrupted when cells were placed in culture. To identify the particular cell-cell interaction(s) responsible for the promotion of normal development, we initiated reconstitution experiments in tissue culture. Be ause such an experimental approach had proven valuable in our earlier studies on neurite outgrowth and the adhesive properties of glial surfaces[8] and in studies by other investigators, it also seemed likely to prove of value in studying cell-cell interactions that might play a role in the control of division and differentiation.

We found that the growth of O-2A progenitor cells in the presence of type-1 astrocytes, the first glial cell type to appear during optic nerve development,[9] promoted O-2A progenitor division and also the continued generation of oligodendrocytes for extended periods *in vitro.*[10] At about the same time, experiments by Raff and colleagues[11] indicated that O-2A progenitor cells continued to divide *in vivo* in the absence of axons from retinal ganglion neurons. Together, these results indicated that O-2A progenitor division *in vivo* might well be promoted by type-1 astrocytes, and they suggested a new role for astrocytes as a potential source of mitogen(s) during CNS development. Thus, these studies raised the possibility that glial-glial interactions during CNS development might be as interesting and important as glia-neuron interactions.

The mechanism by which the optic nerve comes to contain two distinct glial lineages was addressed next. The possibility that the O-2A lineage contained migratory cells had been suggested by earlier studies indicating that oligodendrocytes, and/or cells that gave rise to oligodendrocytes, were able to migrate over significant distances in the mouse CNS.[12] Our ability to now study O-2A progenitors undergoing apparently normal patterns of division and differentiation *in vitro* allowed us to determine that such cells were indeed extremely motile when grown in the presence of type-1 astrocytes.[13] Moreover, our studies indicated that O-2A progenitors were the only migratory members of this lineage. Motility *in vitro* was complemented by a pattern of *in vivo* development consistent with the view that O-2A progenitors populated the optic nerve by migration from a germinal zone outside of the nerve itself. Along with implicating O-2A progenitor migration as a potentially important mechanism in the formation of the myelinated tracts of the CNS, our studies also suggested that the cells that formed the optic stalk (the embryonic anlage of the optic nerve) were specialized to produce only type-1 astrocytes.[13]

The discovery that type-1 astrocytes had a range of profound effects on the *in vitro* development of O-2A progenitors[10,13–15] was also followed by investigations into the molecular signals responsible for these effects. We found that the activity of type-1 astrocytes was apparently mediated through secretion of platelet-derived growth factor (PDGF). For a fuller description of O-2A lineage-related research on PDGF, the reader is referred to the chapter by Pringle *et al.* in this volume. Briefly, however, we found that O-2A progenitors exposed to PDGF expressed all of the traits seen when these cells were grown in the presence of type-1 astrocytes, and anti-PDGF antibodies were capable of inhibiting the effects of type-1 astrocytes on O-2A progenitors.[16–18] Moreover, further studies by Richardson *et al.*[19,20] offer considerable support for the view that PDGF derived from type-1 astrocytes plays an important role in controlling division and differentiation in the O-2A lineage.

COMPLEX REGULATION OF OLIGODENDROCYTE DIFFERENTIATION

One interesting aspect of our studies with PDGF was the finding that exposure of O-2A progenitors to this single mitogen was associated with the expression of a multifaceted program of behavior. O-2A progenitors exposed to PDGF were bipolar cells that were highly motile, had an average cell-cycle length of 18 hours, and were induced to differentiate into oligodendrocytes *in vitro* with a time course similar to that seen *in vivo.*[16,17] In addition, both type-1 astrocytes and purified PDGF caused O-2A progenitors to undergo symmetric division and differentiation, such that clonally related progenitor cells differentiated into oligodendrocytes synchronously within a limited number of cell divisions.[15,17]

The ability of a single mitogen to elicit a multifaceted program of division and differentiation from O-2A progenitors made it of considerable interest to examine the connection between the induction of division and the expression of the complex phenotype displayed by these cells when they were exposed to PDGF. We had already noted that basic fibroblast growth factor (bFGF) was also able to stimulate O-2A progenitor division, but it did not reproduce the effects of type-1 astrocytes.[16] In more detailed investigations, we found that O-2A progenitors exposed to bFGF were multipolar, were not motile, had average cell-cycle times of 45 hours, and differentiated prematurely into oligodendrocytes.[21] These results demonstrated clearly that the induction of division was not in and of itself sufficient to generate a particular complex pattern of behavior from O-2A progenitors and that the specific identity of the mitogen was crucial in determining the behavioral phenotype of the stimulated cells. One important inference from these results is that the signals transduced as a result of exposure to PDGF and bFGF must differ from each other.

The most intriguing result of our studies with PDGF and bFGF was the discovery that O-2A progenitors exposed simultaneously to these two mitogens were induced to undergo continuous self-renewal in the absence of differentiation.[21] The discovery that cooperation between growth factors can cause prolonged self-renewal of precursors revealed a previously unknown means of regulating self-renewal processes in a precursor population. The extent to which oligodendrocytic differentiation was suppressed is striking; for example, we have been able to grow O-2A progenitors continuously for over a year simply by growing cells in the presence of PDGF and bFGF.[22] However, the "immortalization" induced by mitogen cooperativity is fully conditional, and cells can be readily induced to differentiate into oligodendrocytes or type-2 astrocytes simply by altering the conditions of tissue culture. These findings are intriguing to us not only in the context of the control of precursor differentiation, but also in suggesting that cooperative interactions between growth factors may be as effective at immortalizing cells as are activated nuclear oncogenes. It is also important to note that it is necessary to have cooperation between specific growth factors to achieve indefinite self-renewal; simply exposing O-2A progenitors to any two randomly selected mitogens does not have these effects.

The ability to generate large populations of O-2A progenitors that are conditionally prevented from undergoing oligodendrocytic differentiation by the action of cooperating growth factors now offers us a valuable tool for analyzing the properties of the biological clock that causes clonal families of O-2A progenitors to undergo synchronous and symmetric differentiation if cells are grown in the presence of type-1 astrocytes or PDGF.[15,21] As discussed in more detail elsewhere in this volume (Bögler and Noble), experiments in progress have yielded the surprising result that growth of populations of O-2A progenitors in the presence of PDGF and bFGF for several weeks is associated with a dramatically increased rapidity of differentiation of the entire population of cells following removal of bFGF from the cultures. One interpretation of such results is that the mechanism that would normally limit the number of divisions undergone by an O-2A progenitor grown in the presence of PDGF continues to measure elapsed time, but it is prevented from exercising its inductive (or permissive?) effects on oligodendrocytic differentiation. In other words, it appears to be possible to biologically distinguish the mechanism that measures elapsed time from the effects of such a mechanism on the initiation of differentiation.

Strikingly, it now appears that the observation that growth factor cooperation can promote precursor division will generalize to other lineages. For example, although striatal neuroblasts are not stimulated to undergo division *in vitro* when exposed to either bFGF or nerve growth factor (NGF) as individual compounds, these neuroblasts will divide *in vitro* if exposed to both of these factors.[23]

Our studies with bFGF and PDGF indicate that the relationship between the induction of cell division in O-2A progenitors and the promotion of appropriately timed oligodendrocytic differentiation is more complex than was previously envisaged. The results of studies conducted on O-2A progenitors grown in the presence of type-1 astrocytes or PDGF suggested that the induction of division in these cells is associated with the promotion of activity of a biological clock, which limits clonally related families to a small number of cell divisions before withdrawal from the cell cycle occurs in association with commencement of oligodendrocytic differentiation.[7,14,16] In contrast, our more recent studies (ref. 21; Bögler and Noble, this volume) demonstrate that dividing O-2A progenitor cells can differentiate prematurely into oligodendrocytes when exposed to bFGF (and can continue to divide as oligodendrocytes) or can continue to divide repeatedly in the absence of differentiation when exposed to PDGF and bFGF. Thus, there is no necessary relationship between the induction of cell division and the control of oligodendrocytic differentiation.

It is important to note that it is necessary to reconcile our results with claims that bFGF is widely present in the CNS during development.[24,25] Thus, one possibility is that bFGF availability is not as widespread as has been suggested, for cooperation of this factor with PDGF in the CNS[18] would be expected to inhibit the differentiation of oligodendrocytes *in vivo.* Alternatively, if bFGF were produced in an accessible form *in vivo,* then there must exist an as yet undiscovered means of causing dividing progenitors to generate oligodendrocytes with an appropriate timing, adding a further level of complexity to the control of this developmental pathway.

MYELIN REPAIR WITH CLONAL POPULATIONS OF O-2A PROGENITORS

The ability to generate effectively unlimited numbers of nontransformed O-2A progenitors has now allowed us to achieve repair of demyelinating damage *in vivo* with clonally derived populations of these cells (ref. 22; see also contribution by Crang *et al.* in this volume). To accomplish this goal we generated clonal populations of O-2A progenitors that were genetically modified to express the bacterial β-galactosidase gene (to allow histochemical visualization) and the neomycin resistance gene (to facilitate cloning). These cells were injected into irradiated regions of rat spinal cord that had been demyelinated by prior injection of ethidium bromide. Such lesions are characterized by the presence of demyelinated axons and by an absolute inability of host cells to carry out myelin repair.[26] The cells we injected produced myelin, and their origin could be confirmed by electron microscopic visualization of β-galactosidase reaction product. Thus, with these experiments we have completed the progression from precursor identification and elucidation of cellular and molecular interactions that control precursor differentiation through to the goal of transplanting amplified clonal populations of appropriate precursor cells to obtain repair in a demyelinated CNS lesion.

THE O-2A^{ADULT} PROGENITOR

There were several reasons for undertaking investigations on O-2A progenitors in the adult CNS. An understanding of repair processes in the adult CNS clearly required study of the cells that might be expected to participate in such repair. At the time we initiated these studies, it was not yet clear if precursor cells for oligodendrocytes existed in the adult CNS or if all myelin repair would be carried out by resident oligodendrocytes. Thus, we wished to know if there was a specific progenitor population that might be able to participate in myelin repair. As the physiological environment of the adult CNS is substantially different from that of the developing CNS, we were additionally interested in the possibility that adult progenitor cells might express features that would distinguish them from their perinatal counterparts. Moreover, we wished to determine if the properties of progenitor cells in adult tissue might shed any light on the pathogenesis of the demyelinated lesions that are seen in patients with multiple sclerosis (MS). The origin of these lesions remains unknown despite intensive investigation. (For a more detailed discussion, see the chapter by Compston and Scolding in this volume.) In addition, the failure of MS lesions to become effectively repaired is presently difficult to understand, as demyelinating damage is readily repaired in many experimentally induced lesions.[27–30]

In our initial studies on the O-2A progenitors of the adult optic nerve, we found that these cells differed from their perinatal counterparts in several ways, justifying separate designation as O-2A^{adult} and O-2A^{perinatal} progenitors. O-2A^{adult} progenitors have a unipolar morphology *in vitro,*[31] whereas O-2A^{perinatal} progenitors are usually bipolar.[13,16] In addition, O-2A^{adult} progenitors have a longer average cell-cycle time *in vitro* than do O-2A^{perinatal} progenitors (65 vs 18 hours[13,16,31]), and they take longer to differentiate (5 vs 2 days for 50% differentiation[31]). Furthermore, dividing O-2A^{adult} progenitors are vimentin$^-$O4$^+$, whereas dividing O-2A^{perinatal} progenitors are vimentin$^+$O4$^-$.[31] The long cell cycle times of O-2A^{adult} progenitors seem of particular interest, in that the densely packed environment of the adult CNS seems unlikely to be suited to accommodate cells which, like O-2A^{perinatal} progenitors, would tend to divide rapidly and also to rapidly produce large numbers of oligodendrocytes.

O-2A^{ADULT} PROGENITORS AND A POSSIBLE ORIGIN FOR SELF-RENEWING STEM CELLS

The discovery of the O-2A^{adult} progenitor cell raised two questions. First, what aspect of their biology allows a population of these cells to be maintained in adult rats, while O-2A^{perinatal} progenitors have largely disappeared from the optic nerve by 1 month after birth and cannot be detected in adult rats?[31,32] Second, what is the developmental origin of the O-2A^{adult} progenitor?

In several respects O-2A^{adult} progenitors behave like stem cells, and it seems likely that it is their stem-cell-like properties that account for their maintenance in the adult optic nerve. The evidence for this view is as follows: We have found that the *in vitro* generation of oligodendrocytes by these cells (stimulated to divide by type-1 astrocytes) occurs predominantly through asymmetric division and differentiation,[33] a behavior characteristic of self-renewing stem cells.[34] This behavior is very different from the highly symmetric patterns of differentiation displayed by O-2A^{perinatal} progenitors induced to divide by PDGF or type-1 astrocytes, where clonally related cells tend to synchronously differentiate into oligodendrocytes within a limited number of cell divisions.[15,17] Thus, O-2A^{perinatal} progenitors express a pattern of division and differentiation consistent with the view that these cells represent a self-extinguishing progenitor population. In contrast, O-2A^{adult} progenitors express a pattern of division and differentiation consistent with the view that this population of cells is capable of undergoing prolonged self-renewal in a manner that does not generate large numbers of excess cells. In addition, these cells have the important quality of being maintained in the optic nerve throughout life: the first O-2A^{adult} progenitors appear in the optic nerve during the first week after birth and are still present in the optic nerve for at least 1 year.[31,32]

Surprisingly, O-2A^{adult} progenitors appear to be directly derived from a subpopulation of O-2A^{perinatal} progenitors.[33] Direct time-lapse microcinematographic observation of dividing O-2A^{perinatal} progenitors isolated from optic nerves of 3-week-old rats (an age when these cells are most rapidly being replaced by O-2A^{adult} progenitors *in vivo*) has demonstrated that cells with the bipolar morphology, rapid motility, and short cell-cycle times of O-2A^{perinatal} progenitors can give rise to cells with the unipolar morphology, slow motility, and long cell-cycle times of O-2A^{adult} progenitors. We have also found that serial passaging of O-2A^{perinatal} progenitors is associated with the conversion of the passaged populations from perinatal-like to adult-like, as judged by morphology, the antigenic phenotype of dividing cells, the population doubling times, and the proportion of progenitor cells labeled by a 20-hour pulse with (^{3}H)-thymidine.[33]

As our data thus far support the concept that the O-2A^{adult} progenitor is a stem cell, our studies on the developmental origin of O-2A^{adult} progenitors have revealed a possible mechanism for the generation of stem cells. The generation of O-2A^{adult} progenitors from O-2A^{perinatal} progenitors suggests that precursor cells with properties appropriate for early development give rise not only to terminally differentiated end-stage cells, but also to a second generation of precursors with properties appropriate for later developmental periods. In the O-2A lineage, this second group of precursors represent the self-renewing stem cell.

The striking differences between O-2A^{perinatal} and O-2A^{adult} progenitors make it of interest to consider the molecular mechanisms responsible for these differences. We recently found that the specific properties of O-2A^{adult} progenitors are elicited by PDGF, thus indicating that the differences between these cells and O-2A^{perinatal} progenitors are intrinsic to the cells themselves and are not due to, for example, responsiveness to a different family of growth factors.[40] This view is supported by our observations that O-2A^{adult} and O-2A^{perinatal} progenitors coexist *in vivo* for most of the first month after birth and also express their characteristic properties when grown in the same tissue culture dish.[32]

THE O-2A LINEAGE AND MULTIPLE SCLEROSIS

Our views on MS are obviously speculative, as we are attempting to shed light on a complex human disease by analyzing the properties of rodent cells grown in tissue culture. Nonetheless, a striking congruence exists between the properties of O-2A^{adult} progenitors, oligodendrocytes, and certain features of MS. Space permits presentation of our views only in a highly condensed form in the present chapter, and the reader is referred to references 36–38 and the chapter in this volume by Compston and Scolding for further discussion of the following considerations.

Our particular concerns have been to understand the reasons behind the failure of effective repair in the demyelinated lesions of the patient with MS and the mechanism by which demyelinating damage is first initiated. The failure of effective repair in the MS lesion is puzzling for two reasons. First, it is well established that myelin repair is extensive following most experimental demyelinating lesions in the CNS (eg, refs. 27–30). Second, recent studies have demonstrated that children with optic neuritis, a demyelinating disease of the optic nerve, recover function more effectively than do adults with apparently similar lesions.[39] The mechanism by which demyelinating damage is initiated is mysterious for several reasons. For example, it is intriguing that the vascular inflammation that occurs in the CNS of patients with MS is not restricted to white matter tracts and frequently is seen even in the nonmyelinated retina of many patients.[40] However, the major pathological alterations that occur in these patients seem to be restricted to white matter tracts.[41] In addition, there are many reasons to believe that the damage in MS is mediated by the immune system (see chapter by Compston and refs. 36–38 for a discussion of this evidence), but there has been a complete failure to identify a myelin antigen that can convincingly be shown to be a target for immune attack in patients with MS[42]; this is a very different situation from that seen in the autoimmune disease myasthenia gravis, for example, in which more than 90% of patients have antibodies against acetylcholine receptors.[43] Why is the inflammatory event in MS widespread in its occurrence, whereas the damage is apparently specific to myelinated tracts? In which way might the immune system be involved in demyelination?

Our first indication that fundamental biologic properties of O-2A lineage cells might be relevant to the understanding of myelin repair emerged from attempts to understand the failure of repair in lesions caused by injection of diphtheria toxin into the CNS. Diphtheritic lesions are not remyelinated,[44] a property that distinguishes them from experimental demyelinating lesions induced by spinal barbotage[27] or compression[29] or by injection of lysolecithin[30] or 6-aminonicotinamide.[28] When we examined the sensitivity of the cells of the CNS to diphtheria toxin, we found that oligodendrocytes and type-2 astrocytes were killed by toxin doses that had little or no effect on type-1 astrocytes or CNS neurons.[45] Loss of oligodendrocytes and type-2 astrocytes was irreversible, and these glial populations were not reestablished even in cultures maintained *in vitro* for a month after removal of toxin. Although these studies were carried out before the identification of the O-2A progenitor cell, the failure to obtain replacement of oligodendrocytes and type-2 astrocytes suggests that the O-2A progenitor cell itself is likely to be as sensitive as its differentiated progeny to the effects of diphtheria toxin. Thus, the failure of myelin repair in the diphtheritic lesion may simply be due to a selective elimination of oligodendrocytes and their precursors, those very cells that might most reasonably be expected to be involved in the process of repair. However, support for such a view would first require examination of the diphtheria toxin sensitivity of cells isolated from tissues of adult animals.

The next indication that properties of O-2A lineage cells might be relevant to considering processes of myelin repair came from our investigations of the properties of O-2A^{adult} progenitors. We initially proposed that the slow division rate of O-2A^{adult} progenitors might limit the ability of these cells to participate effectively in lesion repair.[31] As a specific example of our thinking in this regard, we can use the magnetic resonance imaging (MRI) analyses of W. I. McDonald and colleagues to calculate the number of oligodendrocytes needed to repair a demyelinated lesion. MRI analysis indicates that the demyelinated plaques seen in patients with optic neuritis generally involve up to 25% of the fibers in the optic nerve (that is, 250,000 fibers) and extend a distance of 1 cm, that is, about 50 internodes (W.I. McDonald, personal communication). As oligodendrocytes are thought to myelinate an average of 25 axons each,[46] the number of oligodendrocytes needed to repair a lesion can be estimated at 5×10^5. The optimal figures for oligodendrocyte production indicated by current data suggest that 2,000 O-2A^{perinatal} progenitors could produce the required numbers of oligodendrocytes in 8 days, whereas a similar number of O-2A^{adult} progenitors would require 2 months to produce the same number of oligodendrocytes. During these 2 months, the formation of glial scars within the demyelinated plaque might then prevent migration of O-2A progenitors into the lesion site.[48]

A further indication that properties of O-2A lineage cells might be of interest in an analysis of MS came from studies on the sensitivity of cells to lysis by complement in the absence of antibody. We, and Alastair Compston and colleagues, were struck by observations that CNS myelin was able to directly activate complement in the absence of antibody.[48] It was then not surprising to find that oligodendrocytes themselves had the ability to directly bind and activate complement, in the apparent absence of specific anti-oligodendrocyte antibodies, and thus initiate their own destruction.[36,49] However, it was surprising to find that O-2A^{adult} progenitors and O-2A lineage cells that had apparently almost completed the progression from an O-2A^{perinatal} progenitor to a galactocerebroside-expressing oligodendrocyte were the only other CNS cell types that seem to share this unusual trait.[36] Even O-2A^{perinatal} progenitors were resistant to this form of destruction, as were type-1 and type-2 astrocytes, meningeal cells, and even myelin-producing Schwann cells from the PNS.

Thus, the two cells that presently seem most likely to play a role in the repair of myelin damage in the adult, oligodendrocytes and O-2A*adult* progenitors, were exactly those cells that were lysed by complement in an antibody-independent manner.

Could the complement sensitivity of oligodendrocytes and O-2A*adult* progenitors conceivably be relevant to pathogenesis of the MS lesion? It is known that injection of normal serum into the CNS produces demyelination in experimental animals[50] and that perivascular leakage of cells and serum proteins into the CNS occurs in patients with MS,[51] presumably following damage to the blood brain barrier. Moreover, consumption of complement component C9, indicating formation of membrane attack complexes, has been observed in patients with MS[52,53] despite the lack of any convincing demonstrations of specific immune reaction against myelin in this disease.[42] In addition, studies by Compston and colleagues have found characteristic vesicular structures, like those secreted by oligodendrocytes recovering from exposure to sublytic concentrations of complement, in the cerebrospinal fluid of some patients with MS (ref. 54; also, see chapter by Compston and Scolding in this volume). These results are all consistent with the suggestion that leakage of complement through the blood-brain barrier can disrupt oligodendrocytes. In addition, the susceptibility of O-2A*adult* progenitors to complement-induced lysis could also lead to depletion of progenitors from a lesion site, thus precluding oligodendrocyte replacement. All such forms of damage would obviously be enhanced by the production of antibodies that bound to the surfaces of oligodendrocytes or O-2A*adult* progenitors, but the actual antigen that antibodies recognized would be largely irrelevant. In other words, an immune reaction against any of a large variety of proteins and lipids could reasonably be expected to have similar consequences for the exacerbation of damage as disease progressed. These considerations would also apply to an analysis of cell-mediated immunity in MS.

Obviously a great number of other factors must be considered in evaluating the potential relevance of the O-2A lineage cells to understanding the pathogenesis of the MS lesion. These additional considerations are discussed at greater length in reference 37 and in the chapter by Compston and colleagues in this volume, to which the reader is referred.

GLIOMAS

Along with undergoing normal development, glial cells can also become transformed to yield gliomas. Because of our interests in human disease, we therefore initiated a glioma research program. The initial goal of this research program was to develop a lineage-related classification of gliomas that would be rooted in contemporary views about glial development rather than in earlier, and morphologically based, ideas about the origin of these cells. We have initiated this research because of the very poor relationship between the classification of a glioma and the ability to predict responsiveness to particular forms of therapy. This situation is quite different from that in respect to, for example, lymphoid tumors, for which diagnosis is a useful indicator of probable responsiveness to particular forms of therapy.[55–57] Our hope is that the side-by-side analysis of normal glia and of gliomas will yield useful insights applicable to our understanding of this particular human disease, just as analysis of normal lymphoid cells has been of enormous value in the study of lymphoid tumors.

To initiate a lineage analysis of gliomas, we applied the analytic techniques used in our studies on glial development to the study of human gliomas. These studies resulted in the surprising finding that the antigenic expression of most gliomas does not coincide with the morphologic classification given to the cells.[58] We found that in

90% of cases the cells derived from human gliomas did not express any of the antigens that have been useful in the analysis of normal glia. These cells instead expressed cell-surface fibronectin, an extracellular-matrix molecule not generally expressed by glial cells. Fibronectin-positive cultures were derived from all categories of glioma, irrespective of morphologic classification or grade of malignancy. In only 10% of cases did the glioma-derived cultures express glial fibrillary acidic protein (GFAP), a cell-type specific marker of astrocytes and a molecule that is thought by neuropathologists to be much more widely expressed in glial tumors of, for example, the CNS.[59,60] Moreover, we have found that all of the gliomas that were $GFAP^+$ *in vitro* were also labeled by the A4 monoclonal antibody, which is thought to label all cells derived from the neural tube.[61,62] In contrast, none of the fibronectin-positive cultures we have examined are $A4^+$.[63] Thus, our results suggest the existence of two antigenically distinct families of gliomas that are not recognized by histological analysis.

What is the origin of the fibronectin-positive glioma-derived cells? Limited evidence suggests that these cells can be directly derived from $GFAP^+$ cells. For example, in studies by Westphal and colleagues,[64] $GFAP^+$ glioma cells were cloned and analyzed antigenically and karyotypically. It was found that $GFAP^+fibronectin^-$ and $GFAP^-fibronectin^+$ clones could both be derived from one parental cell line. These results raise two provocative possibilities. It may be that the cells that become immortalized in gliomas represent cells for which differentiation along neural and nonneural pathways represents a normal developmental decision. Alternatively, transformation of glial cells might lead to expression of a phenotype that has no apparent relationship with the cell of origin of the tumor. Distinguishing between these alternative explanations is going to require significant expansion in our knowledge about glial precursors and about the effects of expression of activated oncogenes in glial cells.

THE H-2K^b/TSA58 TRANSGENIC MOUSE

The furthering of our work on normal development and our studies on gliomas both require the ability to extend our understanding of the precursor cells of the CNS. Our abilities to analyze both categories of cells would be greatly enhanced by the availability of conditionally immortalized cell lines of different precursor populations. The availability of large homogeneous populations of such cells would facilitate the identification of marker antibodies for different precursors and thus would allow us to more rapidly identify markers of potential use in the study of normal development. Such cell lines would also greatly enhance our ability to study molecular aspects of differentiation control mechanisms in the CNS. Moreover, if such lines were conditionally immortalized, then the ability to turn off the gene conferring immortalization could enable cells to be used in transplantation experimentation.

Although methods do exist for creating conditionally immortalized cell lines through, for example, retroviral- or transfection-mediated insertion of thermolabile nuclear oncogenes, we have felt that these techniques are of limited value in the study of novel precursor cells. There is no capacity for targeting cells of interest, and precursor cells of particular interest may be too rare to allow any opportunity for introducing nuclear oncogenes into them. In addition, every cell line created has a different integration site and a different level of expression of the immortalizing gene used, thus often making it difficult to compare results obtained even with lines of presumed identical cells.

To overcome the problems involved in the creation of novel cell lines, we (together with Dr. Dimitris Kioussis of the National Institute for Medical Research, London) have developed a new approach to this goal which ensures the presence of a conditional immortalizing gene construct with an identical site of insertion in every cell of interest, no matter how rare the cell may be.[65] This approach has involved the generation of transgenic mice that have stably integrated a conditional immortalizing gene under the control of an inducible promoter. These mice undergo normal development, except for showing hyperplastic development of the thymus, but they can be used to generate conditionally immortalized cell lines from a variety of tissues simply by growing cells *in vitro* in appropriate tissue culture conditions. Our transgenic mice thus allow one to aim for cell types present at low frequencies, which could not realistically be obtained in sufficient numbers for retroviral-mediated gene insertion or DNA transfection. In addition, the ability to turn off oncogene function in the conditionally immortalized cells should allow the creation of cell lines of practical use in many fields of cellular biological research, including studies on cell function, cell-cell interactions, and transplantation biology (to name but a few).

The transgenic mice we have created harbor the thermolabile large T antigen encoded by the Simian virus 40 early region mutant tsA58.[66] This large T antigen is functional at the permissive temperature of 33°C, but it is rapidly inactivated at the nonpermissive temperature of 39.5°C. Expression of this gene was regulated by fusing the coding sequences of tsA58 to the 5′-flanking promoter sequences derived from the H-2K^b gene. This promoter is active in a wide variety of tissues at varying levels,[67–70] but it can be superinduced in many cells by exposure to gamma-interferon.[70–72]

Existing studies on the effects of T antigen indicate clearly that this gene expresses its immortalizing function in a very wide variety of cell types.[73–78] Thus, the H-2K^{b}tsA58 transgenic mice we have developed should allow the direct isolation of conditionally immortal cell lines from many tissues. We have thus far readily isolated conditionally immortal lines of fibroblasts and cytokeratin-positive thymic epithelial cells.[65] Moreover, our preliminary experiments in the CNS have suggested that these mice readily yield populations of glial precursor cells that can be induced to undergo astrocytic differentiation in response to differentiation-inducing factors known to be present in the developing brain.

REFERENCES

1. RAFF, M. C., E. R. ABNEY, J. COHEN, R. LINDSAY & M. NOBLE. 1983. J. Neurosci. **3:** 1289–1300.
2. VAUGHN, J. E. & A. PETERS. 1967. Am. J. Anat. **121:** 131–152.
3. RAFF, M. C., R. H. MILLER & M. NOBLE. 1983. Nature **303:** 390–396.
4. RAFF, M. C., R. H. MILLER & M. NOBLE. 1983. Cold Spring Harbor Symp. Quant. Biol. **XLVIII:** 569–572.
5. RAFF, M. C., E. R. ABNEY & R. H. MILLER. 1984. Dev. Biol. **106:** 53–60.
6. SKOFF, R. P., D. L. PRICE & A. STOCKS. 1976. J. Comp. Neurol. **169:** 291–312.
7. SKOFF, R. P., D. L. PRICE & A. STOCKS. 1976. J. Comp. Neurol. **169:** 313–333.
8. NOBLE, M., J. FOK-SEANG & J. COHEN. 1984. J. Neurosci. **4:** 1892–1903.
9. MILLER, R. H., S. DAVID, R. PATEL, E. R. ABNEY & M. RAFF. 1985. Dev. Biol. **111:** 35–41.
10. NOBLE, M. & K. MURRAY. 1984. EMBO J. **3:** 2243–2247.
11. DAVID, S., R. H. MILLER, R. PATEL & M. C. RAFF. 1984. J. Neurocytol. **13:** 961–973.
12. GUMPEL, M., N. BAUMANN, M. RAOUL & C. JACQUE. 1983. Neurosci. Lett. **37:** 307–311.
13. SMALL, R., P. RIDDLE & M. NOBLE. 1987. Nature **328:** 155–157.
14. RAFF, M. C., E. R. ABNEY & J. FOK-SEANG. 1985. Cell **42:** 61–69.

15. Temple, S. & M. C. Raff. 1986. Cell **44:** 773–779.
16. Noble, M., K. Murray, P. Stroobant, M. Waterfield & P. Riddle. 1988. Nature **333:** 560–562.
17. Raff, M. C., L. E. Lillien, W. D. Richardson, J. F. Burne & M. D. Noble. 1988. Nature **333:** 562–565.
18. Richardson, W. D., N. Pringle, M. J. Mosley, B. Westermark & M. Dubois-Dalcq. 1988. Cell **53:** 309–319.
19. Pringle, N., E. J. Collarini, M. J. Mosley, C.-H. Heldin, B. Westermark & W. D. Richardson. 1989. EMBO J. **8:** 1049–1056.
20. Hart, I. K., W. D. Richardson, C.-H. Heldin, B. Westermark & M. C. Raff. 1989. Development **105:** 595–603.
21. Bögler, O., D. Wren, S. C. Barnett, H. Land & M. Noble. 1990. Proc. Natl. Acad. Sci. USA **87:** 6368–6372.
22. Barnett, S. C., W. Blakemore, J. Crang & M. Noble. In preparation.
23. Cattaneo, E. & R. McKay. 1990. Nature **347:** 762–765.
24. Logan, A. & S. D. Logan. 1986. Neurosci. Lett. **69:** 162–165.
25. Gonzalez, A.-M., M. Buscaglia, M. Ong & A. Baird. 1990. J. Cell Biol. **110:** 753–765.
26. Blakemore, W. F. & R. C. Patterson. 1978. Acta Neuropathol. **42:** 105–113.
27. Bunge, M. B., R. P. Bunge, R. P. & H. Ris. 1961. J. Biophys. Biochem. Cytol. **10:** 67–94.
28. Blakemore, W. F. 1972. J. Neurocytol. **1:** 413–426.
29. Harrison, B. M. & W. I. McDonald. 1977. Ann. Neurol. **1:** 542–551.
30. Blakemore, W. F., R. A. Eames, K. J. Smith & W. I. McDonald. 1977. J. Neurol. Sci. **33:** 31–43.
31. Wolswijk, G. & M. Noble. 1989. Development **105:** 387–400.
32. Wolswijk, G., P. N. Riddle & M. Noble. 1990. Development **109:** 691–698.
33. Wren, D., G. Wolswijk & M. Noble. 1991. J. Cell Biol., in press.
34. Hall, P. A. & F. M. Watt. 1989. Development **106:** 619–633.
35. Wolswijk, G., P. N. Riddle & M. Noble. 1991. Glia, in press.
36. Wren, D. & M. Noble. 1989. Proc. Natl. Acad. Sci. USA **86:** 9025–9029.
37. Compston, D. A. S., N. J. Scolding, D. R. Wren & M. Noble. 1990. Trends Neurosci. **14:** 175–182.
38. Compston, D. A. S. 1990. J. Roy. Coll. Physicians. Lond. **24:** 207–218.
39. Kriss, A., D. A. Francis, F. Cuendet, A. M. Halliday, D. S. Taylor, J. Wilson, J. Keast-Butler, J. R. Batchelor & W. I. McDonald. 1988. J. Neurol. Neurosurg. Psychiatry **51:** 1253–1258.
40. Lightman, S., W. I. McDonald, A. C. Bird, D. A. Francis, A. Hoskins, J. R. Batchelor & A. M. Halliday. 1987. Brain **110:** 405–414.
41. Allen, I. V. 1981. Neuropathol. Appl. Neurobiol. **7:** 169–182.
42. Lisak, R. P. 1986. *In:* Multiple Sclerosis. W. I. McDonald & D. H. Silberberg, eds.: 74–98. Butterworth. London.
43. Newson-Davies, J., G. Harcourt & D. Beeson. 1989. Biochem. Soc. Trans. **17:** 635–637.
44. Harrison, B. M., W. I. McDonald & J. Ochoa. 1972. J. Neurol. Sci. **17:** 293–302.
45. Murray, K. & M. Noble. 1985. J. Neurol. Sci. **70:** 283–293.
46. Peters, A. & J. E. Vaughn. 1970. *In:* Myelination. A. N. Davison & A. Peters, eds.: 3–7. Charles C Thomas. Springfield, IL.
47. Raff, M. C., C. ffrench-Constant & R. H. Miller. 1987. J. Exp. Biol. **132:** 35–41.
48. Vanguri, P. & M. L. Shin. 1986. J. Neurochem. **46:** 1535–1541.
49. Scolding, N. J., B. P. Morgan, A. Houston, A. K. Campbell, C. Linington & D. A. S. Compston. 1989. J. Neurol. Sci. **89:** 289–300.
50. Sergott, R. C., M. J. Brown, D. H. Silberberg & R. P. Lisak. 1984. J. Neurol. Sci. **64:** 297–303.
51. Miller, D. H., M. C. Newton, J. C. Van Der Poel, G. H. DuBoulay, A. M. Halliday, B. E. Kendall, G. Johnson, D. G. MacManus, I. F. Moseley & W. I. McDonald. 1988. Neurology **38:** 175–180.
52. Morgan, B. P., A. K. Campbell & D. A. S. Compston. 1984. Lancet **ii:** 251–255.
53. Sanders, M. E., C. L. Koski, D. Robbins, M. L. Shin, M. M. Franf & K. A. Joiner. 1986. J. Immunol. **136:** 456–459.

54. SCOLDING, N. J., B. P. MORGAN, W. A. J. HOUSTON, C. LININGTON, A. K. CAMPBELL & D. A. S. COMPSTON. 1989. Nature **339:** 620–622.
55. GREAVES, M. F. 1981. Cancer Res. **41:** 4752–4766.
56. GREAVES, M. F., G. JANOSSY, J. PETO & H. KAY. 1981. Br. J. Haematol. **48:** 179–197.
57. VAN EYS, J., J. PUYLLEN, D. HEAD, J. BOYETT, W. CRIST, J. FALLETTA, B. HUMPHREY, J. JACKSON, V. RICCARDI & B. BROCK. 1986. Cancer **57:** 1046–1051.
58. KENNEDY, P., B. WATKINS, D. THOMAS & M. NOBLE. 1987. J. Neuropathol. Appl. Neurobiol. **13:** 327–347.
59. CHRONWALL, B. M., P. E. MCKEVER & P. L. KORNBLITH. 1983. Acta Neuropathol. **59:** 283–287.
60. JAGAGHA, V., W. C. HALLIDAY & L. E. BECKER. 1986. Can. J. Neurol. Sci. **13:** 307–311.
61. COHEN, J. & S. Y. SELVENDRAN. 1981. Nature **291:** 421–423.
62. MILLER, R. H., B. P. WILLIAMS, J. COHEN & M. C. RAFF. 1984. J. Neurocytol. **13:** 329–338.
63. BEVAN, K., B. WATKINS & M. NOBLE. Work in progress.
64. WESTPHAL, M., M. HAENSEL, D. MUELLER, R. LAAS, R. KUNZMANN, E. ROHDE, A. KOENIG, F. HOELZEL & H. D. HERRMANN. 1988. Cancer Res. **48:** 731–740.
65. JAT, P. S., M. NOBLE, P. ATALIOTIS, Y. TANAKA, N. YANNOUTSOS, L. LARSEN & D. KIOUSSIS. 1991. Proc. Natl. Acad. Sci. USA **88:** 5096–5100.
66. TEGTMEYER, P. J. 1975. Virology **15:** 613–618.
67. WEISS, E. H., A. MELLOR, L. GOLDEN, K. FAHRNER, E. SIMPSON, J. HURST & R. A. FLAVELL. 1983. Nature **301:** 671–674.
68. KIMURA, A., A. ISRAEL, O. LE BAIL & P. KOURILSKY. 1986. Cell **44:** 261–272.
69. BALDWIN, A. S., JR. & P. A. SHARP. 1987. Mol. Cell Biol. **7:** 305–313.
70. DAVID-WATTINE, B., A. ISRAËL & P. KOURILSKY. 1990. Immunol. Today **11:** 286–292.
71. WALLACH, D., M. FELLOUS & M. REVEL. 1982. Nature **299:** 833–836.
72. ISRAEL, A., A. KIMURA, A. FOURNIER, M. FELLOUS & P. KOURILSKY. 1986. Nature **322:** 743–746.
73. SANTERRE, R. F., R. A. COOK, R. M. D. CRISEL, J. D. SHARP, R. J. SCHMIDT, D. C. WILLIAMS & C. P. WILSON. 1981. Proc. Natl. Acad. Sci. **78:** 4339–4343.
74. JAT, P. S. & P. A. SHARP. 1986. J. Virol. **59:** 746–750.
75. FREDERIKSEN, K., P. S. JAT, N. VALZ, D. LEVY & R. MCKAY. 1988. Neuron **1:** 439–448.
76. WILLIAMS, D. A., M. F. ROSENBLASTT, D. R. BEIER & R. D. CONE. 1988. Mol. Cell Biol. **8:** 3864–3871.
77. BURNS, J. S., L. LEMOINE, N. R. LEMOINE, E. D. WILLIAMS & D. WYNFORD-THOMAS. 1989. Br. J. Cancer **59:** 755–760.
78. LEMOINE, N. R., E. S. MAYALL, T. JONES, D. SHEER, S. MCDERMID, P. KENDALL-TAYLOR & D. WYNFORD-THOMAS. 1989. Br. J. Cancer **60:** 897–903.

Lineage and Differentiation of Oligodendrocytes in the Brain[a]

ROBERT P. SKOFF AND PAMELA E. KNAPP

Department of Anatomy and Cell Biology
Wayne State University School of Medicine
Detroit, Michigan 48201

What is an oligodendrocyte (OL)? One pictures a cell that makes multiple myelin sheaths around axons in the central nervous system (CNS). Although this statement succinctly describes the most characteristic property of a well-differentiated OL, it is not broad enough to encompass OLs in all situations. The following examples illustrate this point. In dissociated cultures of the CNS a certain population of cells is recognized as OLs even though these cells are often grown in axon-free cultures and do not make myelin sheaths. Their characterization as OLs is based on the fact that they express a set of immunocytochemical markers that are also specifically expressed by myelin-forming OLs *in vivo. In situ,* connections to myelin sheaths are very rarely seen, but OLs can be recognized in ultrastructural studies based on certain morphologic criteria such as cytoplasmic density and composition of organelles.[1] The gray matter of the brain also contains many cells that have the morphologic and immunocytochemical properties of OLs but may not make myelin sheaths. In development, there is a population of cells that have the morphologic features of OLs but have not yet begun to myelinate or are only in the initial stages of myelination. OLs can also be identified in degenerated rat optic nerves 1 year after enucleation, long past the time when axons have disappeared.[2,3]

A broader definition of an OL is a cell that has or had the potential to make myelin, but even this definition is not broad enough. OLs can be morphologically and immunocytochemically identified in the hypomyelinated *jimpy* mouse and other X-linked dysmyelinating mutants, but these OLs do not have the potential to make myelin *in vivo* because of a genetic defect.[4,5]

These illustrations highlight the idea that in practice we operationally define an OL on the basis of expression of a certain subset of the characteristic properties of the cell in its normal milieu. Cells that do not have the potential to express all the properties of a fully differentiated OL may still be classified as OLs. For example, OLs in *shiverer* mice will never, unless genetically engineered, synthesize myelin basic protein (MBP). Thus, an OL is operationally defined by expression of one or more properties specific to OLs. The properties of OLs have gradually been described by investigators using techniques introduced over the last century. In the 1970s we identified OLs by their unique ultrastructural features, and in the 1980s we tended to rely more on immunocytochemical markers. It should be stressed that immunocytochemistry does not supplant electron microscopy, but builds upon this technique, because cells identified ultrastructurally as OLs in the 1970s can also be shown to stain for myelin-specific markers. With the development of each new diagnostic tool it becomes possible to define the properties of a cell more precisely, but these new characterizations ought not to invalidate information gained using earlier tech-

[a] This work was supported by NIH grants NS15338 and NS18898.

niques. Rather, a synthesis of new and existing information should be used to paint a more complete portrait of the life history of the cell.

In the 1990s we predict that glia will be routinely identified by the presence or absence of messenger RNA for specific astrocytic and oligodendrocytic proteins. One advantage of looking at message expression is that it may identify OLs earlier in their ontogeny than may immunocytochemistry or morphology. Investigators working *in vitro* have already identified small, round cells without processes that express message for MBP, carbonic anhydrase II, and proteolipid protein (PLP).[6–9] This message is expressed several days before the specific protein can be detected. Whether this time lag between message and protein detection occurs because *in situ* hybridization is more sensitive than immunocytochemistry or because the message is actually present long before the protein is translated is unclear. However, with this method, OLs can be identified much earlier in their differentiation than was previously possible.

It is equally difficult to decide on the definition of an astrocyte (AS). Is it a cell that makes glial fibrillary acidic protein (GFAP)? Clearly ASs do this, but so do normal lens epithelial cells.[10,11] Schwann cells (which have a completely different embryologic origin) will also express GFAP[12] in certain nerves or when insulted. *In vivo,* GFAP does not stain all ASs in the gray matter, leaving us to employ other criteria for their identification. ASs are also difficult to define functionally because their role probably differs in different CNS regions at different times during the life of an animal.[13]

IDENTIFICATION OF IMMATURE OLIGODENDROCYTES

When does a glioblast become an OL or an AS? We operationally define a cell as an OL or an AS *in vivo* when it expresses a marker that is unique to this cell type and that remains expressed by the differentiated cell type. Even if a cell never makes a myelin sheath, the onset of expression of OL-specific markers identifies it as a committed OL. OLs express a large number of myelin glycolipids and proteins, but they are not all expressed at the same time. Expression of early markers can be used to identify OLs and ASs as they begin to differentiate. As just stated, we believe that the use of molecular probes will allow identification of OLs and ASs at much earlier stages than is now possible with our current battery of antibodies. Thus, cells now considered to be glioblasts or undifferentiated glial progenitors because we lack earlier OL- and AS-specific markers may, in the future, be identified as OLs or ASs on the basis of their expression of AS- or OL-specific messages.

The differentiation of an OL is a process that occurs gradually over time and that involves genetic, biochemical, structural, and physiologic changes within the cell. It is now technically possible to correlate these different processes with one another so that, for example, the expression of OL-specific markers can be correlated with changes in the morphology of the cell. A way of establishing a developmental reference point in the life of a cell is needed if this is to be done in a meaningful way. [^{3}H]-thymidine autoradiography may be used to brand individual cells *in situ* at a particular stage in their differentiation. Thymidine is incorporated into cells while they are in the S phase of the cell cycle. The S phase of the glial cell cycle lasts about 8 hours *in vivo,*[14] and if animals are sacrificed within 1–2 hours of an injection, thymidine-labeled cells represent cells about to undergo mitosis. The combination of [^{3}H]-thymidine autoradiography and immunocytochemistry permits one to determine the relative time of expression of the different myelin markers. By sacrificing animals at various time points after an injection of thymidine, the length of time from the injection of thymidine to the expression of the marker can be determined. The technique is also used to determine if any of these myelin markers are expressed by dividing cells.

DIVISION OF OLIGODENDROCYTES

We have used antibodies that cross-react with the myelin markers galactocerebroside (GC; a mouse monoclonal gift of Dr. B. Ranscht),[a] sulfatide (a rat monoclonal gift of Dr. K. Stefansson),[b] 2′,3′-cyclic nucleotide phosphohydrolase (CNPase) (a gift of Dr. T. Sprinkle), and proteolipid protein (PLP).[15] The glycolipids sulfatide and GC are expressed by many dividing cells *in situ*[16] (FIG. 1). The percentages of glia that are thymidine labeled after a 1-hour pulse (mitotic glia) and that are also sulfatide positive or galactocerebroside positive are at least 40% and 10%, respectively, in presumptive white matter tracts. Exact quantification of the percentages of double-labeled cells is difficult, because some cells are lightly immunostained and near threshold. A further impediment to counting immunostained OLs is that these antibodies stain myelin as well as cell bodies, often obscuring the boundaries between the two.[17] The myelin-specific proteins are apparently expressed at later times in OL differentiation than are the myelin-specific glycolipids. Many thymidine-labeled cells are heavily stained for CNPase 24 hours after a pulse (R. P. Skoff, unpublished observations). However, a few glia pulsed with thymidine for 1 hour appear to be lightly stained for CNPase, suggesting that this protein is synthesized around the time of the final division (R. P. Skoff, unpublished observations). MBP is also expressed within 24 hours of an injection of thymidine.[19] We are particularly interested in PLP because the gene coding for PLP is defective in many of the myelin mutants, and also because it is thought to be one of the last myelin proteins to appear. A lag of about 3–4 days occurs between a thymidine injection and the appearance of PLP in the OL cell body.[15] Thus, the time between the expression of the first known oligodendrocytic specific markers (the glycolipids) and a very late myelin marker (PLP) is at least 3–4 days.

The expression of myelin-specific proteins can be correlated with changes in the morphology of OLs (FIG. 2). The sulfatide-positive and GC-positive cells that are about to undergo mitosis often send long, thin, branching processes out into the parenchyma, but they do not yet appear to have made myelin sheaths that are fully compacted (FIG. 1).[16] This conclusion is difficult to justify simply from light microscopy, but it is supported by our previous electron microscopic autoradiographic studies of gliogenesis.[20] In those studies, the proliferating OLs (oligodendroblasts) are in the early stages of differentiation and do not show definitive connections to myelin sheaths. However, the oligodendroblasts often have long processes that extend out to axons and appear to be in the early stages of axon ensheathment. Visualized ultrastructurally, the oligodendroblasts are packed with Golgi apparatus,

[a] The antibody to GC has been characterized by enzyme-linked immunosorbent assay (ELISA).[18] It reacts strongly with GC and less strongly with psychosine, sulfatide, seminolipid, and other lipids. The antibody is OL specific in both *in vivo* and *in vitro* studies.

[b] The antibody that we refer to as antisulfatide has been characterized by ELISA. It reacts strongly with sulfatide, moderately with seminolipid, but not with GC and many other lipids (J. Benjamins, personal communication). Minor reactivity with other sulfatide glycolipids cannot be eliminated. *In vivo* and *in vitro* this antibody is also OL specific (see text). As it is difficult to determine if the sulfatide and GC antibody in the immunocytochemical experiments binds only to sulfatide or GC, we will refer to them as sulfatide and GC antibodies for ease of discussion.

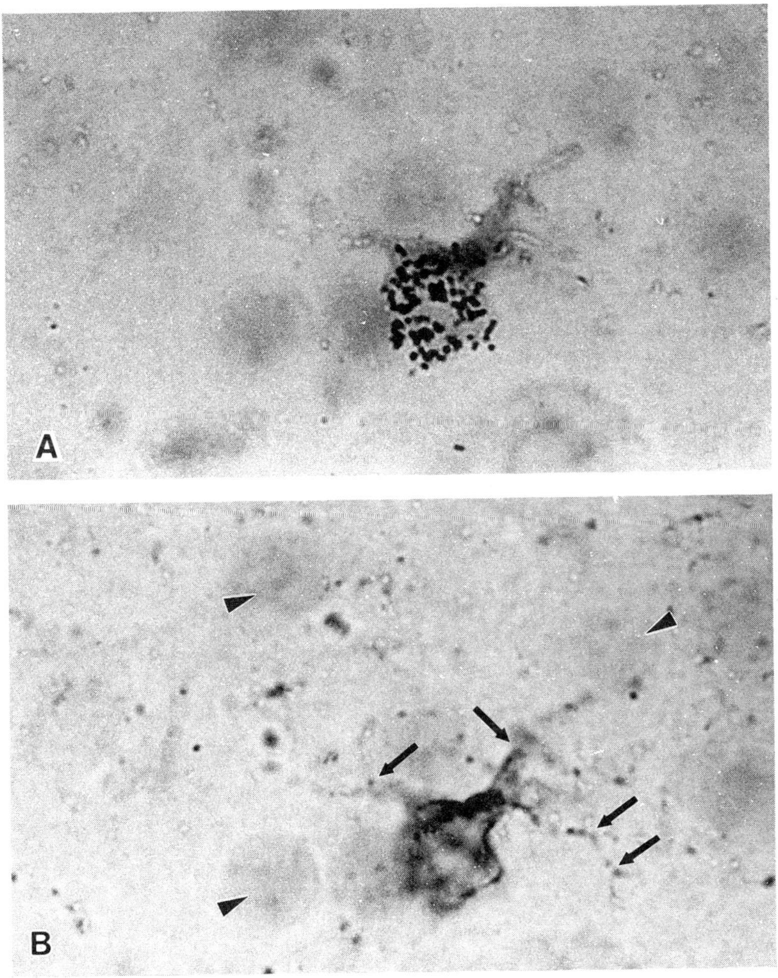

FIGURE 1. Light micrographs from 5-day-old rat brain showing a thymidine-labeled, "sulfatide"-positive cell. The rat was given a single injection of [^{3}H]-thymidine and sacrificed 1 hour afterwards. The brain was "vibratomed," immunostained, embedded in Araldite plastic, sectioned at 1–2 μm, and then processed for autoradiography. (**A**) Focus is on the silver grains; (**B**) focus is on the tissue. The nuclei of several nonimmunostained cells can be discerned throughout the picture (*arrowheads*). The morphology of this cell is typical of many "sulfatide" cells. The nucleus is in an eccentric position, and the cytoplasm gives rise to several fine, branching processes (*arrows*).

numerous free ribosomes and mitochondria, but little rough endoplasmic reticulum. The state of their morphologic differentiation suggests that they may be actively synthesizing glycolipids, but probably not PLP which is synthesized on rough endoplasmic reticulum.[21] We can reasonably conclude that during their period of cell division, the oligodendroblasts have one to several long, thin processes that are closely apposed to axons of relatively large caliber. It is tempting to speculate that the physical proximity of the larger diameter axons to OL processes somehow signals the OL to undergo its final cell division. The time around mitosis clearly is a critical period in the differentiation of the OL, involving the expression of myelin markers and the beginning of myelination.

After the last cell division, OL differentiation proceeds rapidly, for within 24 hours CNPase$^+$ cells with connections to myelin sheaths can be seen. Thus, within a 24-hour period an OL has completed its final cell division and made myelin sheaths of modest thickness. By 72 hours, PLP$^+$ processes emanating from thymidine-labeled

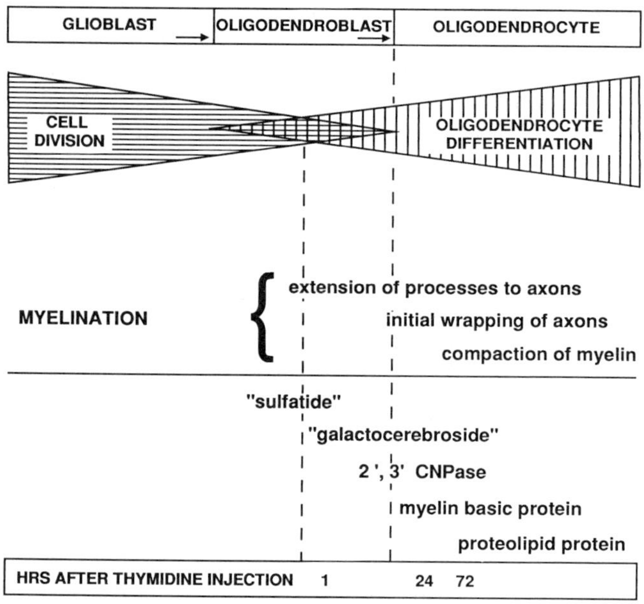

FIGURE 2. A scheme for the development of an oligodendrocyte, relating cell division to changes in the morphology of the cell and to the expression of various myelin markers. The focal point of the figure is the two triangles which show that cell division and differentiation overlap in many cells that have already expressed oligodendrocytic specific markers. The beginning of expression of each myelin-specific marker is correlated with the time after an injection of [^{3}H]-thymidine. The timing for the different events of myelination is based on both electron microscopic autoradiography and immunocytochemistry. They are relative, and more precise information is needed to correlate each event with the different stages of cell division. (See text.)

cells have made many connections to myelin sheaths.[15] PLP is clearly a later appearing myelin marker and is certainly not essential for the early stages of myelination involving ensheathment of axons.

The oligodendroblasts, which are sulfatide positive and/or GC positive, have a very similar morphology, and their transition into mature OLs *in vivo* can be followed with the immunocytochemical preparations. *In vitro,* the sulfatide-positive cells also represent cells in the OL lineage, because when living cells are pulsed with sulfatide antibody and then followed for 4 days, greater than 95% become GC positive but they do not become GFAP positive.[22] These observations indicate that the sulfatide antibody represents an early marker for OLs both *in vivo* and *in vitro.* The dividing sulfatide-positive/GC-positive cells differ considerably in their morphology from GFAP-positive, thymidine-labeled astroblasts. The astroblasts have coarse, stubby processes that are GFAP positive. The nuclei of the astroblasts are often oval or angular shaped and lightly stained, in contrast to the round and densely stained nuclei of the oligodendroblasts. Ultrastructurally, a second population of dividing cells has the characteristic features of immature ASs.[20] The peak proliferation of astroblasts occurs during the first postnatal week, at which time about 40% of thymidine-labeled glioblasts are GFAP positive after a 1-hour pulse.[23] In contrast, the bulk of oligodendroblast proliferation is delayed until the second postnatal week. Not only do the two types of glioblasts have easily discernible differences in ultrastructural morphology and antigen expression, but also their schedules for proliferation are essentially different.

The finding of two morphologically distinct populations of dividing cells, each of which expresses a different constellation of glial-specific markers, argues strongly for separate lineages for the two types of macroglial cells in postnatal development. In drawing this conclusion, we wish to emphasize that there may be distinct subsets of ASs and OLs that cannot be detected using these antibodies. It has been suggested that all OLs and some ASs (type-2 ASs) develop from an uncommitted progenitor cell that expresses neither GFAP nor specific oligodendrocytic markers.[24-26] Such a progenitor cell would need to be present postnatally and divide repeatedly to generate the numbers of OLs and ASs found in the mature CNS. The existence of a bipotential progenitor cell as the sole source of OLs and type-2 ASs seems very unlikely on the basis of classical and recent studies. Using retroviruses to mark individual progenitor cells early in development and to follow their progeny, investigators have found very few clones that give rise to both OLs and ASs. These findings hold for both *in vivo* and *in vitro* studies.[27-29] Our immunocytochemical studies show that the majority of mitotic glia in the postnatal rodent CNS express markers specific for either OLs or ASs. Inasmuch as newly generated OLs and ASs in the brain must be derived from dividing cells, and as a large percentage of dividing cells express the morphology and antigenic profile of committed ASs and OLs, we must conclude that the bulk of glia generated postnatally derive from previously committed cells rather than from uncommitted progenitors.

CONCLUSIONS

The formation of a myelin sheath can be viewed as the last step in OL differentiation. Earlier in its development an OL undergoes many structural and biochemical changes that prepare it for this task. Our ability to recognize OLs in their early stages of differentiation improves as techniques for defining cellular properties become more sophisticated. For example, the identification of OLs is being made earlier in their development through the use of probes for mRNAs of myelin-specific proteins. In time, it may be possible to identify cells committed to the OL lineage even before they migrate away from the ventricular zone. We currently recognize cells as OLs in a variety of situations without demonstrating that they have made myelin sheaths. Thus, cells can be characterized as OLs if they express a subset of properties of OLs *in situ.* Even when cells have lost the capacity to form myelin sheaths because of genetic perturbations, they can be identified as OLs.

In the present study, we operationally define a cell as an OL when it expresses an antigenic marker that is specific for mature OLs *in situ.* We show that cells that express glycolipids specific to mature OLs are still actively proliferating in the neonatal CNS. These proliferating cells are termed oligodendroblasts, and their existence shows that division and differentiation of glial cells are overlapping events. As the vast majority of dividing glia in postnatal rodent brains express markers specific for OLs or ASs and have the morphologic features of immature OLs or ASs, we conclude that most glia generated postnatally must derive from cells already in the OL or AS lineage.

REFERENCES

1. Peters, A., S. Palay & H. deF. Webster. 1991. The Fine Structure of the Nervous System. Oxford University Press. New York.
2. Vaughn, J. & D. Pease. 1970. Electron microscopic studies of Wallerian degeneration in rat optic nerves. II. Astrocytes, oligodendrocytes and adventitial cells. J. Comp. Neurol. **140:** 207–226.
3. Vaughn, J., P. Hinds & R. Skoff. 1970. Electron microscopic studies of Wallerian degeneration in rat optic nerves. I. The multipotential glia. J. Comp. Neurol. **140:** 175–206.
4. Skoff, R. P. & P. E. Knapp. 1990. Expression of the jimpy phenotype in relation to proteolipid protein appearance. Ann. N.Y. Acad. Sci. **605:** 122–134.
5. Skoff, R. P. & P. E. Knapp. 1991. Phenotypic expression of X-linked genetic defects affecting myelination. *In* Myelin: A treatise. Martenson, R., ed. Telford Press. Caldwell, NJ.
6. Holmes, E., G. Hermanson, R. Cole & J. deVellis. 1988. Developmental expression of glial-specific mRNAs in primary cultures of rat brain visualized by *in situ* hybridization. J. Neurosci. Res. **19:** 389–396.
7. De Vitry, F., D. Gomes, P. Rataboul, S. Dumas, J. Hillion, J. Catlon, J.-P. Delaunoy, A. Tixier-Vidal & P. Dupouey. 1989. Expression of carbonic anhydrase II gene in early brain cells as revealed by *in situ* hybridization and immunocytochemistry. J. Neurosci. Res. **22:** 120–129.
8. Amur-Umarjee, S., L. Hall & A. Campagnoni. 1990. Spatial distribution of mRNAs for myelin proteins in primary cultures of mouse brain. Dev. Neurosci. **12:** 263–272.
9. Ghandour, M. S. & R. P. Skoff. 1991. Double-labeling *in situ* hybridization analysis of mRNAs for carbonic anhydrase II and myelin basic protein: Expression in developing cultured glial cells. Glia **4:** 1–10.
10. Hatfield, J., R. Skoff, H. Maisel & L. Eng. 1984. Glial fibrillary acidic protein is localized in the lens epithelium. J. Cell Biol. **98:** 1895–1898.
11. Boyer, S., R. Maunoury, D. Gomes, B. de Nechaud, A. M. Hill & P. Dupouey. 1990. Expression of glial fibrillary acidic protein and vimentin in mouse lens epithelial cells during development *in vivo* and during proliferation and differentiation *in vitro:* Comparison with the developmental appearance of GFAP in the mouse central nervous system. J. Neurosci. Res. **27:** 55–64.
12. Dahl, D., N. Chi, L. Miles, B. Nguyen & A. Bignami. 1982. Glial fibrillary acidic (GFA) proteins in Schwann cells. J. Histochem. Cytochem. **30:** 912–918.
13. Fedoroff, S. & A. Vernadakis. 1986. Astrocytes. Vols. 1–3. Academic Press. Orlando, FL.
14. Korr, H., B. Schultze & W. Maurer. 1975. Autoradiographic investigations of glial proliferation in the brain of adult mice. II. Cycle time and mode of proliferation of neuroglia and endothelial cells. J. Comp. Neurol. **160:** 477–490.
15. Vermeesch, M. K., P. E. Knapp, R. P. Skoff, D. M. Studzinski & J. A. Benjamins. 1990. Death of individual oligodendrocytes in *jimpy* brain precedes expression of proteolipid protein. Dev. Neurosci. **12:** 303–316.
16. Skoff, R. P. & P. E. Knapp. 1991. Division of astroblasts and oligodendroblasts in postnatal rodent brain: Evidence for separate astrocyte and oligodendrocyte lineages. Glia **4:** 165–174.

17. GHANDOUR, M. S. & R. P. SKOFF. 1988. Expression of galactocerebroside in developing normal and *jimpy* oligodendrocytes *in situ*. J. Neurocytol. **17:** 485–498.
18. BANSAL, R., A. E. WARRINGTON, A. L. GARD, B. RANSCHT & S. E. PFEIFFER. 1989. Multiple and novel specificities of monoclonal antibodies 01, 04, and R-mAb used in the analysis of oligodendrocyte development. J. Neurosci. Res. **24:** 548–557.
19. FRIEDRICH, V. & N. STERNBERGER. 1984. Dividing oligodendrocyte precursors do not stain for myelin basic protein. Soc. Neurosci. (Abstr.) **10:** 81.
20. SKOFF, R., D. PRICE & A. STOCKS. 1976. Electron microscopic autoradiographic studies of gliogenesis in rat optic nerve. I. Cell Proliferation. J. Comp. Neurol. **169:** 291–312.
21. COLMAN, D., G. KREIBICH, A. FREY & D. SABATINI. 1982. Synthesis and incorporation of myelin polypeptides into CNS myelin. J. Cell Biol. **95:** 598–608.
22. KNAPP, P. E. 1991. Studies of glial lineage and proliferation *in vitro* using an early marker for committed oligodendrocytes. J. Neurosci. Res., in press.
23. SKOFF, R. 1990. Astrocytes in rat optic nerve are generated in a single wave before oligodendrocytes. Dev. Biol. **139:** 149–168.
24. RAFF, M. C., R. H. MILLER & M. NOBLE. 1983. A glial progenitor that develops *in vitro* into an astrocyte or an oligodendrocyte depending on culture medium. Nature (Lond.) **303:** 390–396.
25. RAFF, M. C., E. R. ABNEY & R. H. MILLER. 1984. Two glial cell lineages diverge prenatally in rat optic nerve. Dev. Biol. **106:** 53–60.
26. RAFF, M., B. WILLIAMS & R. MILLER. 1984. The *in vitro* differentiation of a bipotential glial progenitor cell. EMBO J. **3:** 1857–1864.
27. GALILEO, D., G. GRAY, G. OWENS, J. MAJORS & J. SANES. 1990. Neurons and glia arise from a common progenitor in chicken optic tectum: Demonstration with two retroviruses and cell type specific antibodies. Proc. Natl. Acad. Sci. USA **87:** 458–462.
28. VAYSSE, P. J.-J. & J. GOLDMAN. 1990. A clonal analysis of glial lineages in neonatal forebrain development *in vitro*. Neuron **5:** 227–235.
29. BARFIELD, J., J. PARNAVELAS & M. LUSKIN. 1990. Separate progenitor cells give rise to neurons, astrocytes and oligodendrocytes in the rat. Soc. Neurosci. (Abstr.) **16:** 1272.

Analysis of Cell Lineage in the Rat Cerebral Cortex[a]

JACK PRICE, BRENDA WILLIAMS, ROBERT MOORE, JOANNE READ, AND ELIZABETH GROVE

The National Institute for Medical Research
The Ridgeway
London, NW7 1AA, UK

A characteristic feature of development is the generation of differentiated cells from undifferentiated germinal cells. In the vertebrate nervous system, the original germinal cells are the pseudostratified epithelial cells of the neural tube, whereas the differentiated cells are the various neurons and macroglial cells of the mature central nervous system (CNS). A central question in development is what controls this generation of cellular diversity?

The starting point of our attempt to study this complex process is to consider how the developmental potential of germinal cells changes with developmental time. It is presumably the case, although it has never been proven, that the original germinal cells of the neural plate epithelium each have the potential to generate all the CNS cell types. At some point in development, that potential apparently narrows as each cell takes on its particular fate. The question is then, what developmental pathways are open to cells at any particular point in time, or, put another way, how many germinal cell types are there at any given point in time, and what is their developmental potential and their presumptive fate?

We have approached these issues of genealogy using retroviral lineage marking. These studies began with the demonstration by ourselves and others that retroviral vectors that encode the bacterial lac-Z gene could be used to genetically and histochemically label germinal cells in the embryonic CNS.[1,2] A characteristic of this type of labeling is that all the progeny of the labeled cells inherit an undiluted copy of the vector and consequently are as indelibly labeled as is the parent germinal cell. In recent years, such studies have made substantial inroads into some of those areas of the CNS (the overwhelming majority) where our knowledge of lineage relationships was poor, specifically the retina,[3,4] optic tectum,[5] and the cerebral cortex.[6–8]

With regard to the cerebral cortex, our initial studies suggested a model that explained the lineage relationships of cortical neurons and all the classes of cortical macroglial cells.[6] In this paper we consider how this model has fared in the light of recent data and what important issues remain to be addressed.

CEREBRAL CORTICAL CLONES

We used the BAG retrovirus to label ventricular zone cells of the cerebral cortex in the embryonic day 16 (E16) rat embryo. The virus was administered *in vivo* by injection into a cerebral vesicle. In our initial experiments, a titer of virus was

[a]This work was supported by the Medical Research Council, and the Multiple Sclerosis Society of Great Britain and Northern Ireland.

injected sufficient to infect about 20 ventricular zone cells, and owing to the spread of the virus through all the embryonic brain ventricles, marked clones were found in both cortices and as far afield as the hindbrain. We performed the clonal analysis when the injected animals had reached 14 days of age postpartum.

In these studies we found three different types of clones, but they were found in a relationship that suggested that two distinct lineages existed in the cerebral cortex at this time. One type of clone just contained astrocytes of the grey matter. These cells were identified by their morphology and by their end-feet on blood vessels and the pial surface. These astrocyte clones were large, containing up to 40 cells, and they seemed to express the lac-Z gene product (enzyme β-galactosidase) as detected by histochemical staining throughout their cytoplasm.

A second type of clone just contained neurons. Again these could be identified by morphological criteria, and this type of clone could also be distinguished by size. The largest neuronal clone found was six cells, and the average size was between one and two cells. In addition, the labeled neurons had a particularly distinctive feature. The lac-Z staining was confined to two sites in the cell. One was the area in or around the nucleus; the other was small areas around the base of the principal dendrites, areas otherwise indistinguishable by light microscopy from the rest of the cytoplasm. Only cells in these neuronal clones had this staining pattern. However, some cells in these clones that were clearly neurons did not stain just in these regions, but rather stained throughout the cytoplasm in the manner of astrocytes. This feature is only apparent at this early postnatal time point. If the animals are allowed to reach several months of age, staining in neurons extends throughout the cell in a pattern somewhat reminiscent of a Golgi stain (FIG. 1). We are not the first to see this propensity of the lac-Z gene product to associate with the nucleus.[9] We do not understand its cause, but the most likely explanation is that this prokaryotic protein fortuitously carries a eukaryotic nuclear localization signal that is particularly recognized in neurons. Probably as the cell becomes more mature and the amount of lac-Z production increases, this weak, fortuitous signal is overcome and the enzyme spreads throughout the cell. Whatever its cause, this property of lac-Z is significant, especially if the recent approach of using a retrovirus with a deliberately included nuclear localization signal is considered.[10]

The third type of clone we found was restricted entirely to the white matter. It was made up of cells that were small and morphologically indistinct in most cases. Their main common feature was that they were oriented horizontally along the white matter tract, and for this reason we have called them "horizontal" cells. Some of these cells had processes; often they were bipolar, and sometimes their processes were more elaborate. Despite their lack of distinctive features, we considered it very likely that these cells were white matter glia, mostly because there was little else that they could be. Certainly, the numbers of neurons in this region of the cortex are small. However, these deep neurons are among the earliest cells to be generated in the cortex,[11] and so would have already been postmitotic before the virus was injected. In animals injected at E16 but allowed to survive longer before analysis, we discovered two additional features of these horizontal cell clones. First, after 3 months of postnatal development some of these cells morphologically resemble oligodendrocytes. Second, by this later age some of the horizontal cell clones extend up into the gray matter and so were no longer restricted to the white matter (data not shown).

Other studies using the same approach have indicated that the same three types of clones can be found in the hippocampus (FIG. 2). Many features of hippocampal clones are exactly similar to those found in the cerebral cortex. Clones of hippocampal astrocytes tend to contain many cells, each of which is fully stained (FIG. 2A); the clones of neurons contain only a few cells and have the nuclear staining pattern (FIG. 2B); clones of small, indistinct white matter cells are found, sometimes extending into the fimbria (data not shown).

CORTICAL LINEAGE

Although these studies revealed three different types of clones, we concluded that only two separate lineages were present in the cerebral cortex at this age. The reason was that the neuronal clones and horizontal clones were on the same radial dimension, that is, the labeled horizontal cells were in the white matter exactly

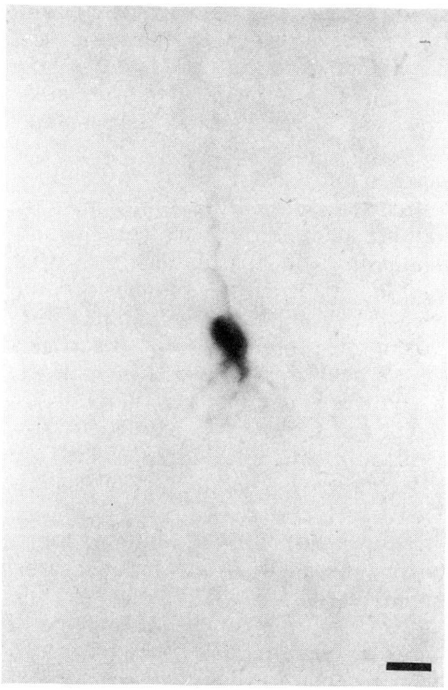

FIGURE 1. A single neuron marked with β-galactosidase. Note how in this older tissue (3 months old) the histochemical reaction product labels the cell body and its processes, whereas in younger tissue (postnatal day 14; see FIG. 2 and ref. 6) the staining has a more restricted distribution. Scale bar = 10 μm.

beneath the gray matter in which a neuronal clone was to be found. In these animals with low density infections, the chance juxtaposition of two clones would be very rare, and in line with this expectation a similar relationship between other types of clone was never found. For example, neuron and astrocyte clones were very rarely found occupying the same cortical space, even though the astrocyte clones were large in cell number and also in terms of the cortical space they occupied. Similarly, astrocyte and horizontal cell clones were never found. The only combination that occurred in this relationship was that of neurons and horizontal cells, and even though they were the least common clonal type, they were still found in 9% of cases (TABLE 1). As all of these cells were originally derived from a ventricular zone cell, the simplest explanation of these data, and the one we favor, is that in these 9% of cases, both the labeled

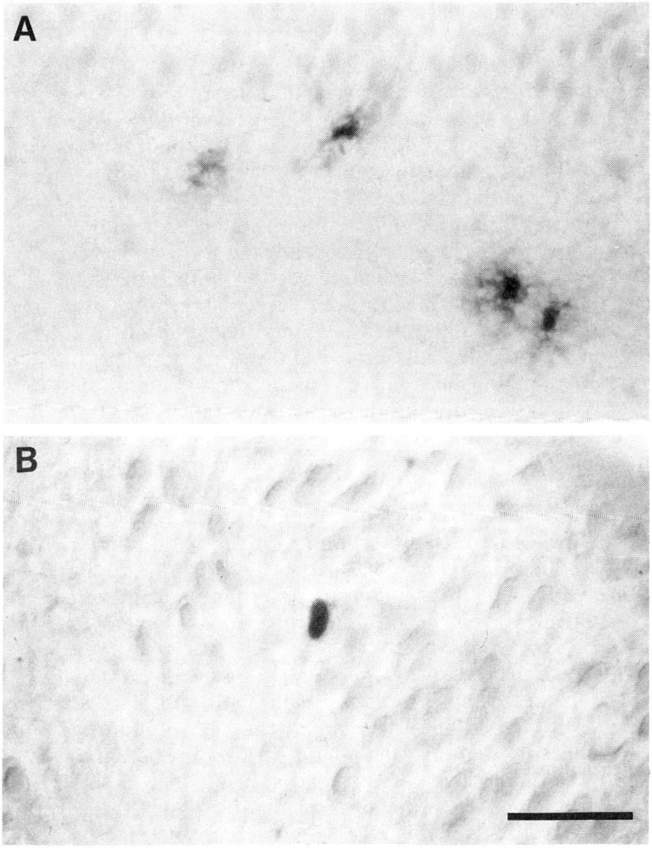

FIGURE 2. Hippocampal clones. (**A**) A clone of astrocytes. Three stained cells are visible as are some processes of a cell whose cell body is out of section. (**B**) A labeled CA3 pyramidal cell. Note that only the kidney-shaped nucleus is stained. Scale bar = 50 μm for both photomicrographs.

TABLE 1. Distribution of Different Clonal Types in the Cerebral Cortices of 14-Day-Old Rats Injected at E16[a]

Clonal Type	Percent
Astrocytes	16
Neurons	43
Horizontal cells	13
Horizontal/neurons	9

[a]Clones not included in this table include the small percentage that could not be identified with certainty and those that fell outside the cerebral cortex.

neurons and the horizontal cells were derived from the same infected ventricular zone cell.

This idea leads us to a model in which there are two cerebral lineages, that is, two types of ventricular zone cells. One gives rise only to gray matter astrocytes; the second is multipotential and generates both neurons and the glia of white matter. The two lineages differ considerably in complexity. Gray matter astrocytes arise from an apparently committed precursor cell. From the work of others, we would guess that this cell is itself a radial glial cell.[12] On the other hand, the multipotential cell seems to have a range of developmental options open to it, and our initial conclusions about the nature of this lineage leave many questions unanswered.

Since these initial studies, we have improved the method by which we analyze this type of experiment. In our early work we analyzed the infected brains by cutting serial 20-μm sections and staining each section histochemically to detect the labeled clones. This step was very labor intensive at the cutting, staining, and analysis steps. To get a reasonable amount of data from such experiments we had to ensure that there were 20–30 clones per brain; otherwise, the return was too small to make the experiments practicable. This number of clones per brain did not generate any problems in clonal detection, as clones at this density are easily differentiated from each other. Nonetheless, we were anxious to obtain the same types of clones even in animals with only two or three clones per brain. We found that we could make such experiments manageable by initially sectioning infected brains into thick sections (ca 200 μm) on a vibratome and then staining the sections. When clones had been identified in these thick-sectioned material, their identity could be confirmed (when necessary) by re-cutting the blocks into thinner sections on a cryostat. Processing was considerably quicker by this method.

In experiments conducted in this manner, we found the same types of clones in very similar proportions to those we had previously found. Particularly noteworthy were the mixed neuron/horizontal cell clones that were found in 6% of clones, a proportion very similar to that recorded in our earlier experiments.

The only unpredicted aspect of these later results was a small proportion of clones that contained neurons, horizontal cells, and gray matter astrocytes. We are still uncertain how to interpret this result, but it might be taken as evidence for a more primitive cell than we had previously identified, that is, a pluripotential cell that can generate both of the lineages that we had previously described. Alternatively, it could be that some of the horizontal cells, the putative white matter glia, may also become astrocytes to the gray matter. As just noted, we have evidence that in older animals the horizontal cell clones do contribute to the gray matter. To interpret these data correctly, we need to know the identity of the horizontal cells. Do they include oligodendrocytes or astrocytes or both? We are currently investigating this issue. It is noteworthy that we have still to find clones that contain neurons and gray matter astrocytes, without horizontal cells. The conclusion that the cortical lineage is primarily divided into the astrocyte and multipotential lineages, therefore, still seems safe.

THE POSITIONAL SPECIFICATION OF CORTICAL CELLS

One of the first observations that followed the application of the retroviral lineage-marking technique to the study of the cerebral cortex was that clones of cells, both astrocytes and neurons, spread horizontally much more than we would have predicted. In our initial studies we estimated that clones of neurons marked on E16 could spread as much as 300 μm in the horizontal direction.[6] The estimate of Luskin *et al.*[8] for mouse cortex was similar. Walsh and Cepko,[7] injecting between E14 and E16 in the rat, found even greater spread, and we concur with the view that clones marked at the earlier time point almost certainly spread more than did those labeled at E16 (Moore and Price, manuscript submitted). Despite the very close correspondence between the data from these different groups, some disagreement has been noted in how the results should be interpreted. A number of investigators have taken these data to support the position that it is the fate of the progeny of a single cell to be oriented radially in the cortex. This has subsequently been taken as support for the contention that the columnar organization of the adult cerebral cortex is the result of the elaboration of a mosaic of germinal units in the cortical ventricular zone.[13,14] However, we are very much of the contrary opinion.

We see a confusion stemming from a failure to differentiate between the pathway of migration of cortical cells and their final positions. This difference can be seen most clearly by labeling cortical clones at embryonic stages (E16, for example) and analyzing the marked clones a short time later (for example, the day of birth). In such an experiment, some clones are seen completing their migration to the cortical plate, some are still migrating, and some have cells in both phases of development (FIG. 3). It is clear from such experiments that the disposition of the migratory and postmigratory cells is quite different. Clonally related migratory cells are predominantly spread along the same radial dimension. The related cells are invariably well separated in the intermediate zone, so a cell's immediate neighbors—the other members of the cohort of migrating cells—are not clonally related (FIG. 2B). Nonetheless, members of a clone can usually be seen to occupy the same or a closely neighboring radial dimension. Clonally related postmigratory cells, however, are almost invariably not on the same radial path. Rather they lie side by side in the cortical plate, often separated by more than 100 μm (FIG. 3C). So even at this early stage, when the cortical gray matter is only a fraction of the size it will become, clonally related cells in the cortical plate have become displaced from each other in the horizontal dimension. Migration, therefore, is predominantly radial, but there is a subsequent displacement of cells in the plane of the cortical plate.

The important conclusion is that although migration is predominantly radial, this fact cannot be taken as evidence supporting the hypothesis that all cells destined for a single cortical column are derived from one, or a small number, of ventricular zone cells. For this hypothesis to be correct, the *final* positions of cells, not just their migration pathways, need to be radial. The question that needs to be addressed is whether the horizontal spread of cells takes them across the boundaries between columnar units in the cortical gray matter. We have examined the spread across the boundaries between barrel columns in the rat somatosensory cortex, and these data will be presented elsewhere (Moore and Price, submitted). Briefly we find that these boundaries are not barriers to clonal spread, that is, clones can contribute to more than one barrel column. We feel, therefore, that the hypothesis just expounded will very unlikely be correct.

This conclusion is consistent with all the studies of cortical cell lineage, both published findings[6–8] and preliminary reports.[15] It is also consistent with chimera studies in the avian forebrain[16] and older studies on cortical development.[17] It also fits well with current concepts of positional plasticity in the rodent cerebral cortex[18] and with the timing of the development of cortical columns and their sensitivity to afferent input.[19] The only surprise is that the cerebral cortex differs so much in this respect from the optic tectum and the retina, the other two regions of the CNS that have been extensively studied in this regard.[3–5] This finding is ironic as the cerebrum is the only region for which a radial clonal distribution had been predicted.[13] Surprising or not, the results seem clear; clones initially migrate radially in the retina,

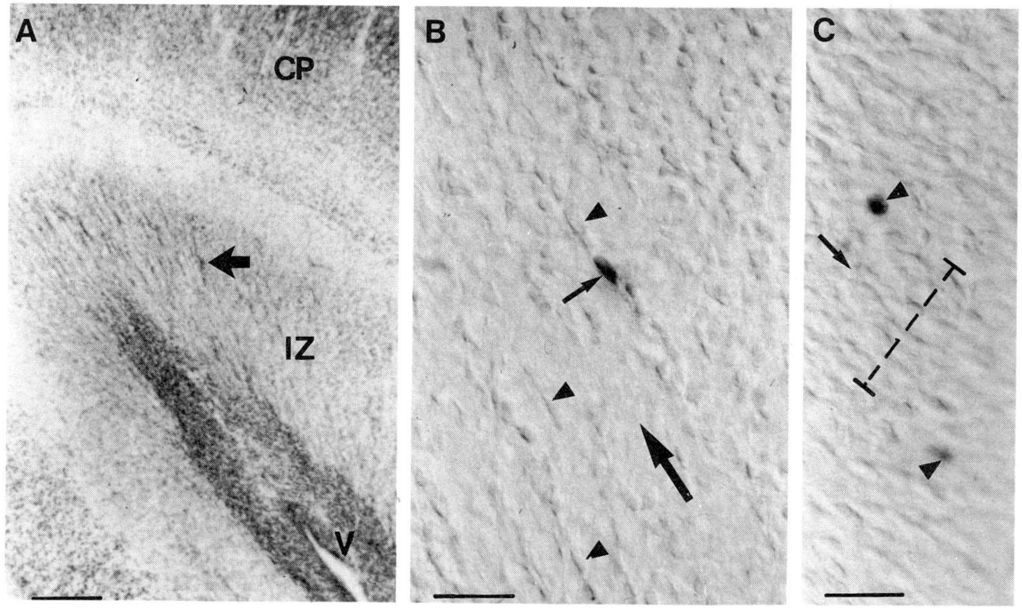

FIGURE 3. Photomicrographs of a labeled clone in sections of newborn rat cerebral cortex. (**A**) Low power photomicrograph of a parasaggital section of presumptive occipital cortex from a rat injected with BAG retrovirus on E16. Filters have been used to enhance the counterstain (orange G) and so highlight the tissue architecture. The ventricular zone (V), intermediate zone (IZ), and cortical plate (CP) can all be seen. The *arrow* points to the position of a single β-galactosidase-stained cell in the IZ. The scale bar is 20 μm. (**B**) Photomicrograph of an adjacent section to that shown in **A.** Again, a single stained cell in this section is indicated by the *small arrow.* It is part of a cohort of cells migrating towards the CP. The direction of migration is indicated by the *large arrow.* Several cohorts are marked with *arrowheads.* The scale bar is 20 μm. (**C**) Photomicrograph of a section from the same series as those shown in **A** and **B.** Two stained cells (indicated by *arrowheads*) are shown in the cortical plate (one lying in a slightly different focal plane). A third cell in this section is not in this field of view. The *arrow* points vertically towards the pial surface. The *dashed line* represents the horizontal distance between the stained cells, a distance of 120 μm. The scale bar represents 100 μm.

optic tectum, and cerebral cortex, but only in the first two of these regions is this radial distribution maintained.

REFERENCES

1. PRICE, J., D. TURNER & C. CEPKO. 1987. Lineage analysis in the vertebrate nervous system by retrovirus-mediated gene transfer. Proc. Natl. Acad. Sci. USA **84:** 156–160.
2. SANES, J. R., J. L. R. RUBENSTEIN & J.-F. NICOLAS. 1986. Use of a recombinant retrovirus to study post-implantation cell lineage in mouse embryos. EMBO J. **5:** 3133–3142.
3. TURNER, D. & C. CEPKO. 1987. Cell lineage in the rat retina: A common progenitor for neurons and glia persists late in development. Nature **328:** 131–136.
4. TURNER, D. L., E. Y. SNYDER & C. L. CEPKO. 1990. Lineage-independent determination of cell type in the embryonic mouse retina. Neuron **4:** 833–845.
5. GRAY, G. E., J. C. GLOVER, J. MAJORS & J. R. SANES. 1988. Radial arrangement of clonally related cells in the chicken optic tectum: Lineage analysis with a recombinant retrovirus. Proc. Natl. Acad. Sci. USA **85:** 7356–7360.
6. PRICE, J. & L. THURLOW. 1988. Cell lineage in the rat cerebral cortex. A study using retroviral-mediated gene transfer. Development **104:** 473–482.
7. WALSH, C. & C. L. CEPKO. 1988. Clonally related cortical cells show several migration patterns. Science **241:** 1342–1345.
8. LUSKIN, M. B., A. L. PEARLMAN & J. R. SANES. 1988. Cell lineage in the cerebral cortex of the mouse studied *in vivo* and *in vitro* with a recombinant retrovirus. Neuron **1:** 635–647.
9. KALDERON, D., B. L. ROBERTS, W. D. RICHARDSON & A. E. SMITH. 1984. A short amino acid sequence able to specify nuclear location. Cell **39:** 499–509.
10. HUGHES, S. M. & H. M. BLAU. 1990. Migration of myoblasts across basal lamina during skeletal muscle development. Nature **345:** 350–353.
11. MARIN-PADILLA, M. 1978. Dual origin of the mammalian neocortex and evolution of the cortical plate. Anat. Embryol. **152:** 109–126.
12. VOIGHT, T. 1989. Development of glial cells in the cerebral wall of ferrets: Direct tracing of their transformation from radial glia into astrocytes. J. Comp. Neurol. **289:** 74–88.
13. RAKIC, P. 1988. Specification of cerebral cortical areas. Science **241:** 170–176.
14. LEISE, E. M. 1990. Modular construction of nervous system: A basic principle of design for invertebrates and vertebrates. Brain Res. Rev. **15:** 1–23.
15. MISSON, J. P., C. AUSTIN, T. TAKAHASHI, C. CEPKO & V. CAVINESS. 1989. Migrating neurons of the murine cerebrum ascend in parallel to radial fibres: Analysis based upon double-labeling of migrating neurons and radial fibres. Soc. Neurosci. Abstr. **15:** 241.7.
16. BALABAN, E., M.-A. TEILLET & N. LE DOUARIN. 1988. Application of the quail-chick chimera system to the study of brain development and behavior. Science **241:** 1339–1345.
17. MELLER, K. & W. TETZLAFF. 1975. Neuronal migration during early development of the cerebral cortex. Cell Tiss. Res. **163:** 313–325.
18. O'LEARY, D. D. M. & B. B. STANFIELD. 1989. Selective elimination of axons extended by developing cortical neurons is dependent on regional locale: Experiments utilizing fetal cortical transplants. J. Neurosci. **9:** 2230–2246.
19. VAN DER LOOS, H. & T. A. WOOLSEY. 1973. Somatosensory cortex: Structural alterations following early injury to sense organs. Science **179:** 395–397.

Physiological Properties of Oligodendrocytes during Development[a]

H. KETTENMANN, G. V. BLANKENFELD,
AND J. TROTTER

Department of Neurobiology
University of Heidelberg
Im Neuenheimer Feld 364
6900-Heidelberg, FRG

Oligodendrocytes are the myelin-forming cells of the central nervous system (CNS). Mature oligodendrocytes and therefore also myelin appear late in the development of the CNS. In the corpus callosum of rodents, for instance, myelination peaks at about 2 weeks after birth.[1] Furthermore, a pool of oligodendrocyte precursor cells is present even in the adult.[2] In contrast, at birth neurons and astrocytes have matured sufficiently to fulfill specialized tasks; neurons are electrically active and have already formed numerous synaptic connections. Oligodendrocytes in the rat optic nerve develop from a bipotential precursor cell which can also give rise to astrocytes.[3] These differentiation rules appear to apply also to other species and other brain regions.[4–6] In this study we have made use of cell culture techniques to isolate oligodendrocytes and their precursor cells. A highly enriched population of these cells, starting from the embryonic brain tissue of mice,[7] can be purified. Under appropriate conditions, the precursor cells in these cultures develop into oligodendrocytes. In addition, some of the precursor cells retain the potential to develop into astrocytes under the influence of fetal calf serum.[6] Different developmental stages of the oligodendrocyte lineage can be identified with the aid of stage-specific and cell-type-specific antibodies.[8–10] Thus, it is possible to characterize the physiologic properties of oligodendrocytes and their precursors during different developmental stages by applying the patch-clamp technique to cells identified on the basis of morphology and marker profile.

Although glial cells classically were regarded as the electrically passive elements of the brain, modern electrophysiologic techniques have demonstrated that glial cells express ionic channels that were previously thought to be exclusive to neurons. Astrocytes, for instance, express Na^+ [11] and Ca^{2+} channels[12] and GABA-receptor-coupled Cl^- channels.[13] In this review we demonstrate that oligodendrocyte precursor cells express a defined set of voltage- and ligand-gated ion channels and that the expression of these channels changes with development.

[a]This research was supported by Bundesministerium für Forschung und Technologie, the Hermann und Lilly Schilling-Stiftung, and the Deutsche Forschungsgemeinschaft (Heisenberg-Stipendium to H.K., SFB 317).

RESULTS

Identification of Oligodendrocytes and Their Precursors

We have characterized membrane properties of identified oligodendrocytes and their precursors in cell cultures established from embryonic mouse brain.[7] After specific lysis of neurons, oligodendrocytes and precursor cells can be shaken free from the underlying astrocytes and cultured further. Many immature precursor cells are bi- or tripolar and express the antigen recognized by the A2B5 antibody (FIG. 1a and b). The subsequent maturation stage expresses the 04 antigen and has a small cell body with three to five processes (FIG. 1c and d). Subsequently the cells also express the 01 antigen and become larger in size (FIG. 1e and f). Finally the 010 antigen is additionally expressed by large cells with large expanses of attached membrane sheets (FIG. 1g and h). Expression of the antigen recognized by the A2B5 antibody is down-regulated with maturation.

Voltage-Gated Channels in Glial Precursor Cells

K^+ Channels

Oligodendrocyte precursor cells from optic nerve express voltage-gated K^+ outward currents with two different inactivation time constants, 20–40 ms and 140–200 ms[14] (FIG. 2a and b). In precursor cells from mouse cortex three different types of K^+ channels could be distinguished: a delayed rectifier, a Ca^{2+}-activated K^+ channel, and an A-type K^+ channel.[15]

The delayed rectifier was activated when the membrane was stepped to potentials more positive than -40 mV. The currents activated within 8 ms; in two thirds of the precursor cells voltage-gated outward currents did not inactivate within the voltage step, whereas in one third of the cells currents inactivated with a mean inactivation time constant of 30 ms. Both types of outward current were equally sensitive to 4-aminopyridine and tetraethylammonium (FIG. 2a and b).

In a subpopulation of precursor cells, an A-type current similar to neuronal A-type K^+ currents could be identified using the appropriate clamp protocol[15] (FIG. 2d). These transient inactivating outward currents decayed fast, and on the basis of inactivation time constants, two populations of cells could be distinguished. The current was blocked by both 4-aminopyridine and tetraethylammonium.

When the Ca^{2+} concentration in the recording pipette was raised to 1 μM, an apamin-sensitive K^+ current was observed[15] (FIG. 2c). This current was voltage-dependent and inactivated within the voltage step. In low internal Ca^{2+} (10 nM) this current was never observed, identifying it as a Ca^{2+}-sensitive K^+ channel. Hyperpolarizing voltage steps did not activate a specific inward current in the precursor cells; thus, the current-to-voltage relation of this cell type shows strong outward rectification[14,15] (FIG. 2e).

Na^+ Channels

Inward currents can be activated by depolarizing voltage steps[14] (FIG. 2f). These currents were identified as Na^+ currents by their dependence on the extracellular $[Na^+]$ and their sensitivity to tetrodotoxin (TTX).[15] The current-to-voltage relation

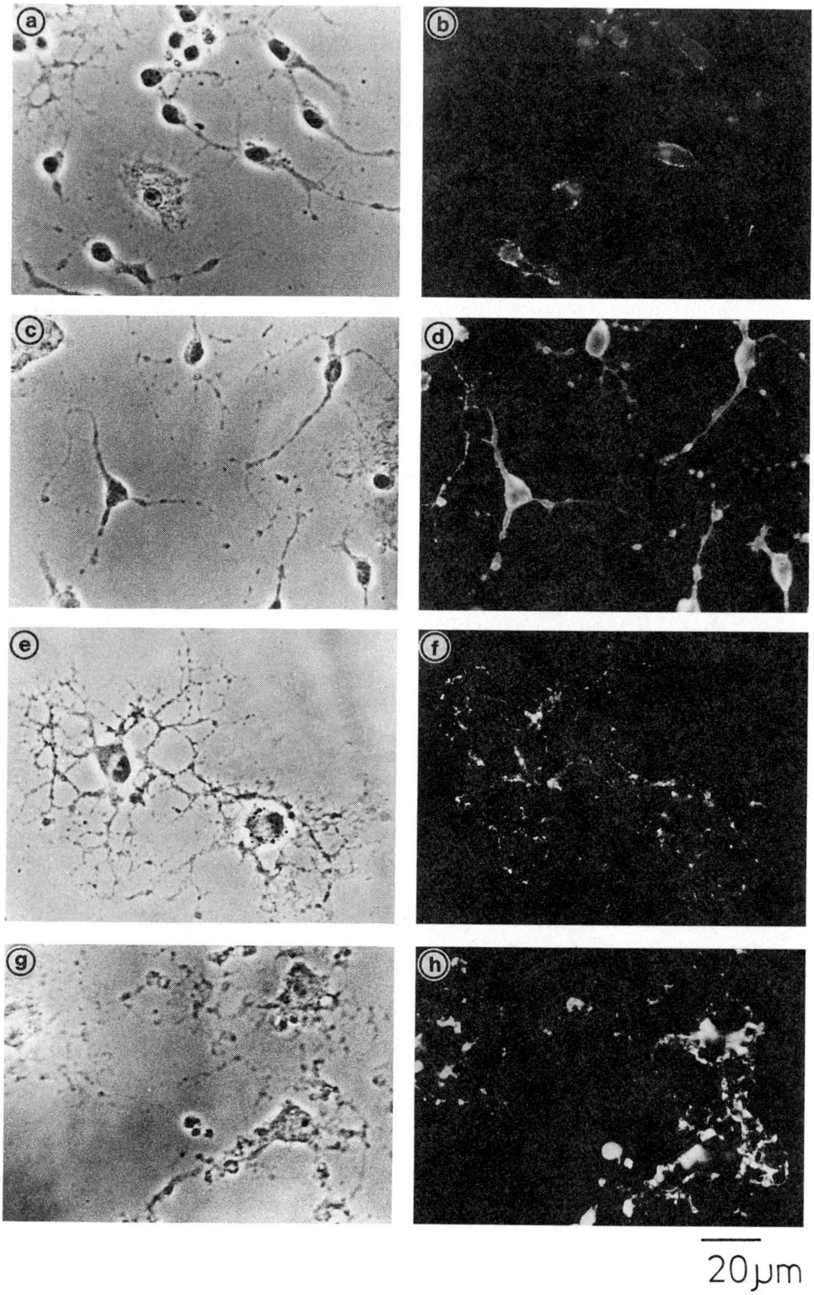

FIGURE 1

indicates a peak of activation at about −10 mV. The steady-state inactivation curve was half-maximal at about −60 mV as with type-2 astrocytes and neurons but not type-1 astrocytes.[16]

Ca^{2+} Channels

In glial precursor cells from mouse cortex, Ca^{2+} currents were identified by recording inward currents in Ba^{2+}-containing solutions[17] (FIG. 3). Two types of Ca^{2+} currents could be distinguished by their peak activation. These currents were recorded using the patch-clamp technique in the whole cell configuration and were found to be similar to the low- and the high-voltage–activated Ca^{2+} channel described in neurons. Although the majority of precursor cells expressed only one of the two types of current, in 30% of the cells both types of current could be identified. In contrast, Barres *et al.*[18] did not observe Ca^{2+} currents in precursor cells from rat optic nerve.

Channels in Oligodendrocytes

Oligodendrocytes commonly possess time-independent, only slightly voltage-dependent K^+ channels[19,20] (FIG. 4). Single-channel conductance varies from 20–200 pS, indicating a population of different K^+ channels with similar gating properties.[19,20] Similar single-channel conductances, namely, 30 and 120 pS, were found in the optic nerve.[21]

In some oligodendrocytes, components of inactivating inward and outward currents can be superimposed on the passive currents.[22,23] Soliven *et al.*[22] described the expression of three different types of potassium current components in cultured bovine oligodendrocytes: a transient outward current, a noninactivating current, and an inwardly rectifying current. Expression of an inwardly rectifying current with a single-channel conductance of about 20 pS has also been described in human oligodendrocytes[24] (FIG. 4c).

FIGURE 1. Morphologic and immunocytochemical characterization of cultures enriched in glial precursor cells and oligodendrocytes. Four cell types can be distinguished, and examples are displayed as phase contrast micrographs in **a**, **c**, **e**, and **g**. Corresponding fields in **b**, **d**, **f**, and **h** show cell type-specific antibody binding visualized by indirect immunofluorescence for A2B5 (**b**), 04 (**d**), 01 (**f**), and 010 (**h**). (**a**, **b**) The most immature cells studied were spindle-shaped bipolar precursor cells with small somata (less than 10 μm), expressing the A2B5 antigen (shown in **b**), but not any of the antigens recognized by the 04, 01, or 010 monoclonal antibody. These cells are referred to as A2B5 antigen-positive precursor cells. (**c**, **d**) In the next recognizable differentiation stage, cells acquire the 04 antigen (**d**), but are not yet 01 or 010 antigen positive. These cells have more than two, but usually less than five processes and a soma diameter of 10–15 μm. Many of these cells have triangular somata that round out as the cells mature. These cells will be referred to as 04 antigen-positive precursor cells. (**e**, **f**) In the subsequent developmental stage, cells retain the 04 antigen expression, gain expression of the 01 antigen (shown in **f**), and are defined as 01 antigen-positive oligodendrocytes. (**g**, **h**) Subsequently, these oligodendrocytes also express the 010 antigen (**h**). Such oligodendrocytes have multiple processes and large somata (larger than 15 μm). In 010 antigen-positive cells the nucleolus is often visible. These cells will be referred to as 010 antigen-positive oligodendrocytes (with modifications from Sontheimer *et al.*[15]).

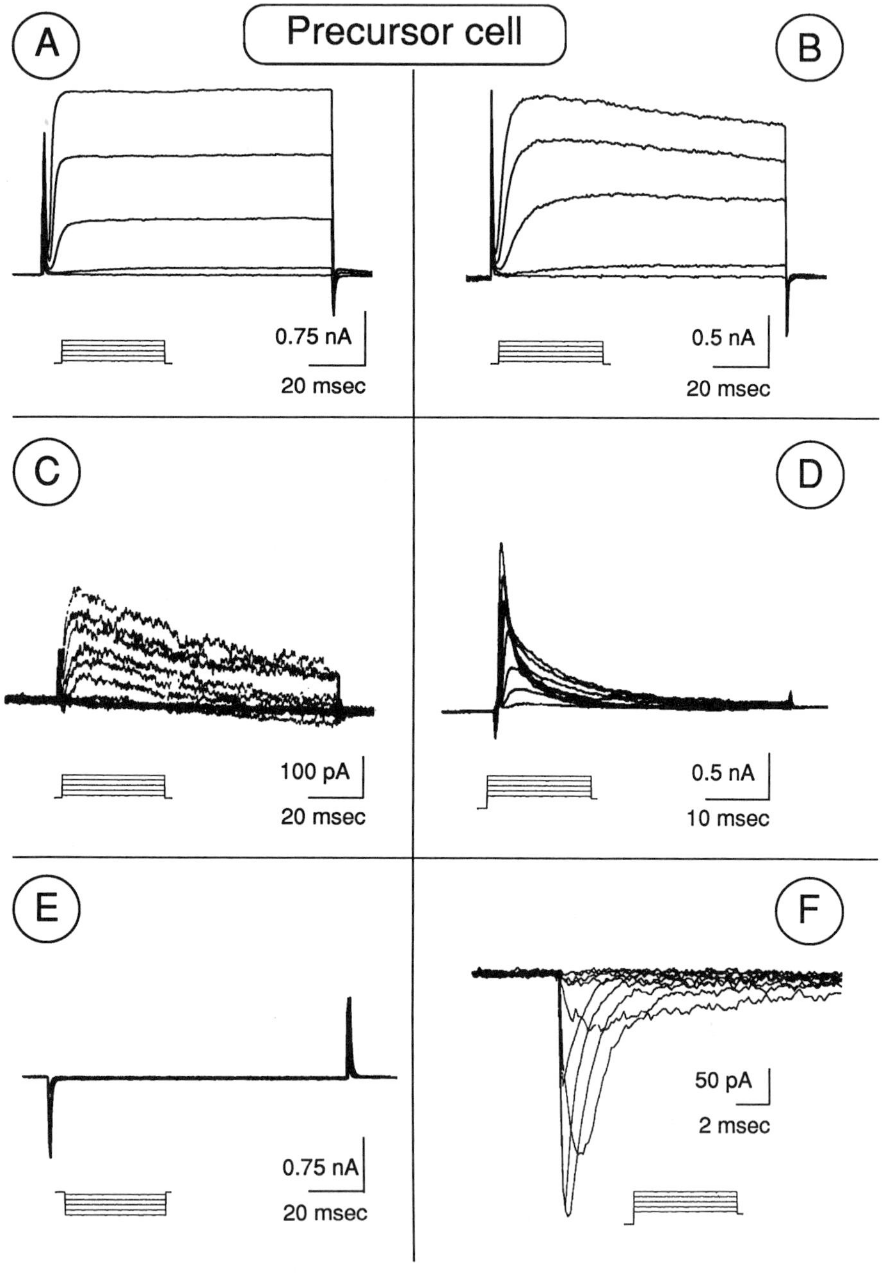
Precursor cell
A
B
0.75 nA
20 msec
0.5 nA
20 msec
C
D
100 pA
20 msec
0.5 nA
10 msec
E
F
50 pA
2 msec
0.75 nA
20 msec

FIGURE 2

pH-Activated Channels in Cells of the Oligodendrocyte Lineage

Oligodendrocytes and their precursors generate large Na^+ currents in response to a step increase of $[H^+]$ ($I_{Na(H)}$)[25] (FIG. 5a). Comparing $I_{Na(H)}$ of oligodendrocytes with $I_{Na(H)}$ of their precursor cells did not reveal a difference in current amplitude, suggesting a higher density of $I_{Na(H)}$ channels on the (smaller) precursor cells. The properties of $I_{Na(H)}$ in glial precursor cells and oligodendrocytes are similar to those previously described in neurons with respect to activation conditions, time course, and the effect of extracellular Ca^{2+} concentration.[25] In contrast to cells of the oligodendrocyte lineage, astrocytes and Schwann cells did not express these currents. The functional role of these channels in both neurons and glial cells is at present unknown.

GABA and Glutamate Receptors in the Oligodendrocyte Lineage

GABA activates a Cl^- conductance in both oligodendrocytes and their precursor cells.[26] The opening of Cl^- channels leads to depolarization of oligodendrocytes[27] because the Cl^- equilibrium potential, in contrast to that in most neurons is more positive than the membrane potential.[28,29] This Cl^- gradient is maintained by the activity of a bumetanide-sensitive, Na^+-dependent Cl^- carrier.[29] The pharmacology of the GABA response is similar to that of neuronal $GABA_A$ receptors in both of these developmental states. These similarities include blockade of the GABA response by the $GABA_A$ receptor antagonists picrotoxin and bicuculline and the modulation by benzodiazepines and barbiturates[26] (FIG. 6). In contrast to astrocytic GABA receptors and in concert with those described in neurons, β-carbolines act as inverse benzodiazepine agonists. However, in precursor cells the dose-response curve shows a slope of less than one, indicating only one binding site for GABA. This finding suggests a molecular diversity in the GABA receptors of oligodendrocytes, neurons, and astrocytes.

FIGURE 2. Voltage-activated membrane currents in precursor cells. (**A**) Outward K^+ currents. Cells were clamped to −70 mV close to the resting membrane potential and depolarized by voltage steps up to 35 mV with 20-mV increments. Outward currents were activated at potentials more positive than −40 mV and did not inactivate within the voltage step. (**B**) Outward currents from a different cell partially inactivated. (**C**) Ca^{2+}-activated K^+ current. A family of outward currents was activated by depolarizing steps (voltage steps indicated in the **inset**), from a holding potential of −70 mV with increased $[Ca^{2+}]$ (1 μM) in the pipette. Outward currents in response to step voltage changes activated upon depolarization beyond −50 mV and inactivated partially within the voltage step (with modifications from Sontheimer *et al.*[15]). (**D**) A-type K^+ current. With the appropriate protocol, A-type K^+ currents were recorded. The resulting family of current traces decayed fast, showed a time-dependent inactivation, and resembled the pattern of neuronal A-type K^+ currents (with modifications from Sontheimer *et al.*[15]). (**E**) Current responses to hyperpolarizing voltage steps. Cells were clamped close to the resting membrane potential and hyperpolarized by voltage steps with 20-mV increments. No currents were activated. (**F**) Currents are displayed corresponding to depolarizing voltage steps from a holding potential of −110 mV to potentials ranging from −70 to 10 mV in 10-mV increments (**lower trace**). Step depolarization to potentials more positive than −40 mV activated fast transient inward currents, which peaked around −10 mV. Recordings were obtained with a pipette containing CsCl and 1 mM Ba^{2+} in the bathing solution to block outward K^+ currents. Leak and capacitive currents were subtracted (with modifications from Sontheimer *et al.*[15]).

Glutamate has been shown to depolarize oligodendrocytes from mouse spinal cord.[27] This depolarization is caused by activation of an inward current (Reichelt and Kettenmann, unpublished observation; FIG. 5b). In precursor cells, Barres *et al.*[18] demonstrated that the glutamate response shares many similarities with a neuronal non-NMDA receptor, as described for astrocytes.[23]

Recent molecular biologic results indicate that Bergmann glia-specific gluta-

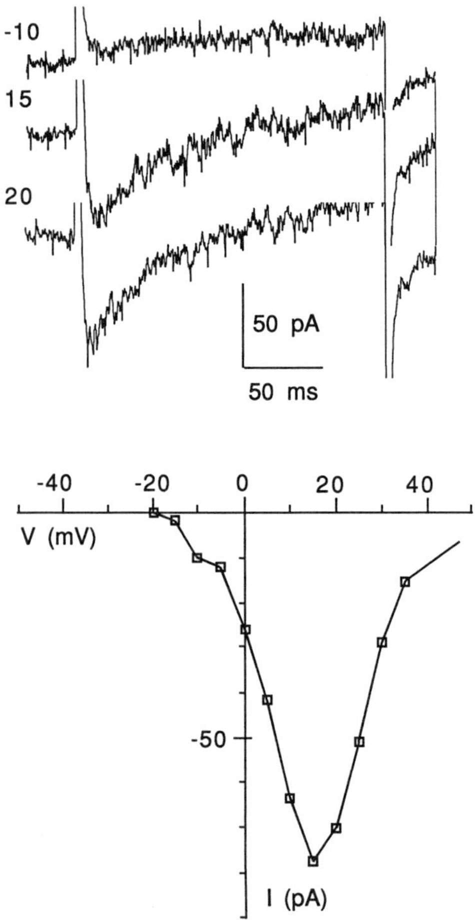

FIGURE 3. Ba^{2+} currents through Ca^{2+} channels. (*Top*) Current traces during a depolarizing voltage step. The holding potential was −75 mV; the potential of the voltage step is indicated above each trace. The electrode contained CsCl to block K^+ currents, the bath 90 mM Ba^{2+} to minimize currents through Na^+ channels and facilitate those via Ca^{2+} channels. (*Bottom*) Peak inward currents, partially shown on *top,* were plotted versus the membrane potential (with modifications from Verkhratsky *et al.*[17]).

mate[30,31] and GABA receptors[32] exist in the brain. The molecular identity and subunit composition of the oligodendrocytic glutamate or GABA receptor are not yet known.

Electrical Coupling

Oligodendrocytes are weakly coupled to one another as determined by electrical and dye-coupling studies.[33,34] Moreover, oligodendrocytes can also form junctions

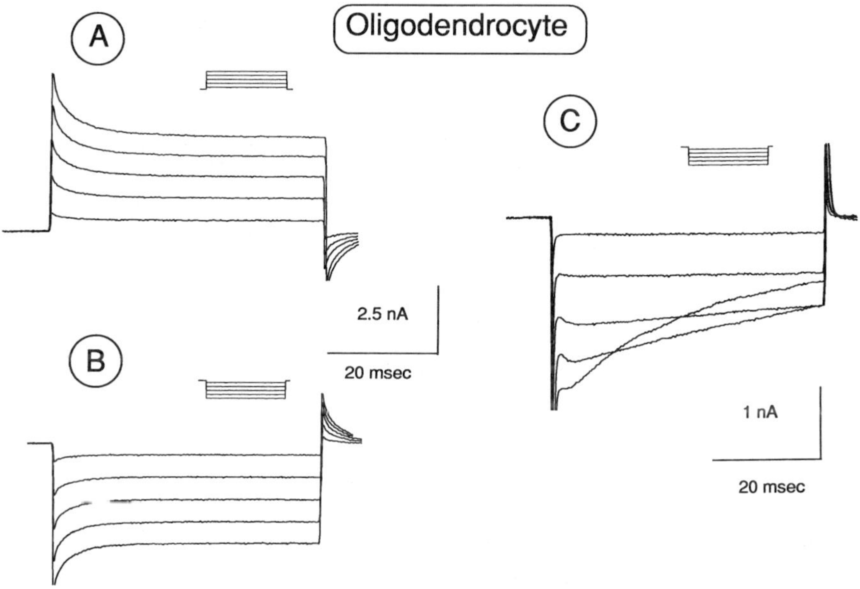

FIGURE 4. Membrane currents in oligodendrocytes. (**A**) Outward K^+ currents. Cells were clamped close to the resting membrane potential and depolarized by voltage steps up to 35 mV in 20-mV increments. (**B**) Inward currents were recorded by clamping the membrane up to −70 mV with 20-mV increments. (**C**) From a different oligodendrocyte, inactivating inward currents were recorded using a similar clamp protocol to that described in **B**.

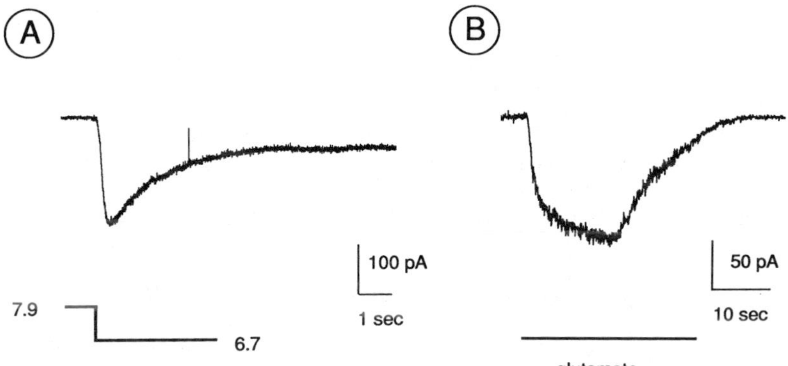

FIGURE 5. pH- and glutamate-activated currents in oligodendrocytes. (**A**) A stepwise change in the external pH from 7.9 to 6.7 (indicated in the bottom traces) induced a large inactivating inward current (with modifications from Sontheimer *et al.*[15]). (**B**) A glutamate-activated inward current was recorded from a different cell. The membrane was clamped at −70 mV, and glutamate was applied at 10^{-4} M.

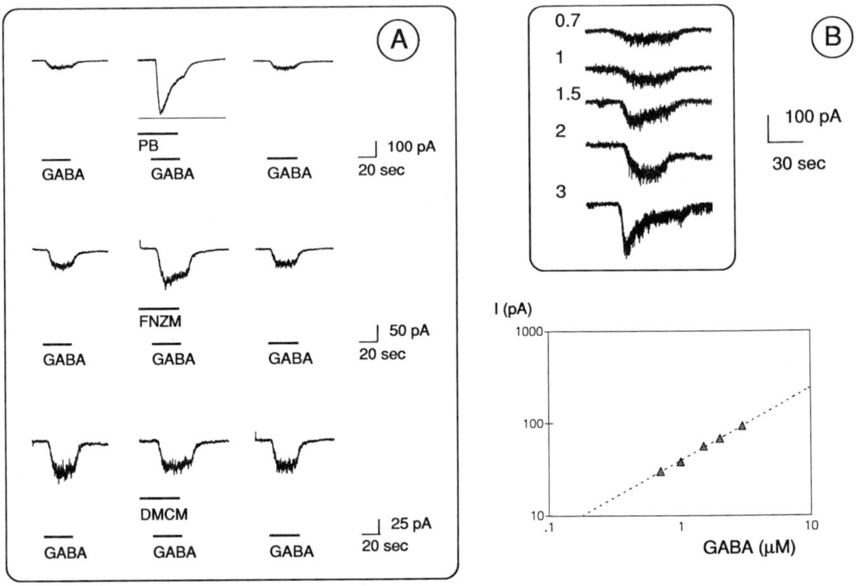

FIGURE 6. Properties of the GABA receptor in glial precursor cells. (**A**) During subsequent recording of membrane currents from a precursor cell clamped at −70 mV, substances were applied as indicated by bars. Whereas pentobarbital (PB, 10^{-4} M) and flunitrazepam (FNZM, 10^{-6} M) resulted in an increase in GABA-induced currents, DMCM (10^{-6} M) reduced the current amplitude. GABA was applied at 10^{-5} M. (**B**) GABA was applied at concentrations indicated at the beginning of each trace (in μM) to a glial precursor cell while recording membrane currents. The membrane potential was clamped at −70 mV. The time elapsing between individual GABA applications was 10 minutes. (*Below*) The log of the peak currents (shown in traces on the *left*) was plotted against the log of the GABA concentration; the resulting linear fit of the dose-response relationship has a slope of less than 1 (with modifications from Blankenfeld *et al.*[26]).

with astrocytes, but do so less frequently.[35,36] The coupling ratio between pairs of oligodendrocytes is much lower than that between pairs of astrocytes, making it very unlikely that these structures serve generally to spread electrical signals between oligodendrocytes.[36] They more likely permit the exchange of small molecules such as second messengers (i.e., cyclic nucleotides) or nutrients such as glycogen.

Precursor cells, in contrast to oligodendrocytes, are electrically isolated, and no coupling between pairs of precursor cells has been observed.[37]

Expression of Voltage- and Ligand-Gated Channels by Glial Cells in Situ

In a recent study Berger *et al.*[38] developed an *in situ* preparation to analyze the membrane properties of oligodendrocytes with the patch-clamp technique. They demonstrated that cells of the oligodendrocyte lineage in corpus callosal brain slices of 6–13-day-old mice show a developmental regulation of ion channels and receptors similar to that observed in cultured oligodendrocytes and their precursor cells. This study also demonstrated that a similar developmental regulation of ion channels occurs in cultured oligodendrocytes and *in vivo.*

Correlation of Channel and Receptor Expression with Cell Differentiation

These results indicate that the expression of ion channels and neurotransmitter receptors by cells of the oligodendrocyte lineage is correlated with the differentiation stage (FIG. 7). The results *in vitro* are also consistent with initial recordings in slices. The youngest cells, including both types of precursor cells that we can identify in

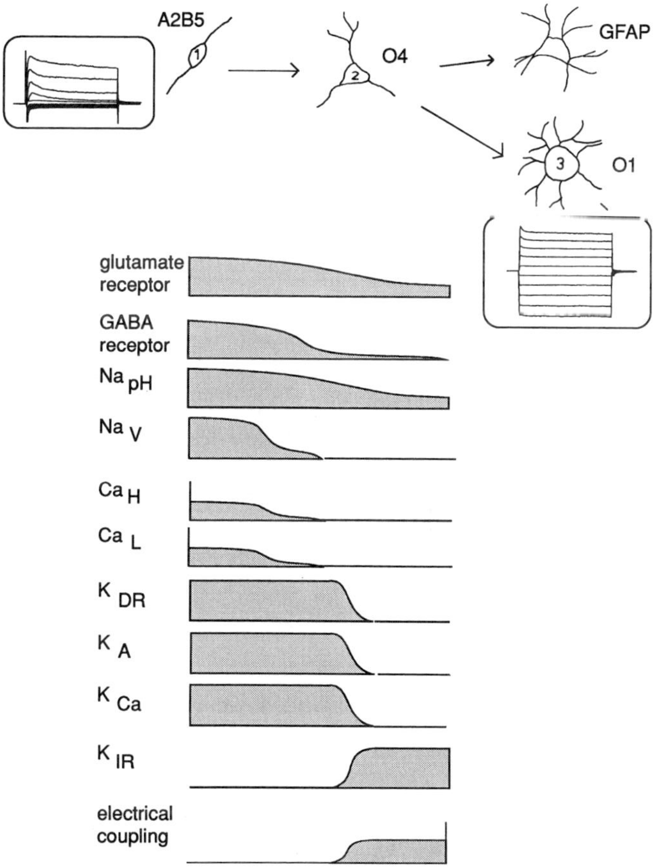

FIGURE 7. Schematic representation of current and receptor expression by cells of the oligodendrocyte lineage. The scheme summarizes the antigen profile and channel expression of glial precursor cells and oligodendrocytes as a function of differentiation stage. A2B5 antigen-positive precursor cells (1) develop into 04 antigen-positive precursors (2). These cells can either give rise to GFAP-positive astrocytes ($GFAP^+$) in the presence of fetal calf serum (FCS) or to 01 antigen-positive oligodendrocytes under the conditions described here (3). Insets on the *left* and *right* show currents in precursor cells and oligodendrocytes, respectively, activated by de- and hyperpolarizing voltage jumps. (*Below*) The density of ion channels and gap junctions during development is schematically displayed. Na_{pH} = pH-activated Na^+ current; Na_V = voltage-activated Na^+ currents; Ca_H = high voltage activated Ca^{2+} current; Ca_L = low voltage activated Ca^{2+} current; K_{DR} = delayed rectifier K^+ current; K_A = A-type K^+ current; K_{Ca} = Ca^{2+}-dependent K^{2+} currents; K_{IR} = time-independent inwardly rectifying K^+ current.

these cultures, express a wide range of ion channels and a high density of receptors for the inhibitory neurotransmitter GABA. The bipolar A2B5 precursor cells that we study are probably analogous to the A2B5-positive O-2A precursors from rat optic nerve and brain. The 04-positive cells in our cultures are also able to give rise to GFAP-positive astrocytes (probably type-2) as well as to oligodendrocytes and seem to represent a later stage of the O-2A lineage. Both precursor cell types are mitotically active in response to appropriate stimuli and exhibit plasticity in their differentiation potential, and the A2B5-positive precursor cells are motile *in vitro* and probably *in vivo.*[39]

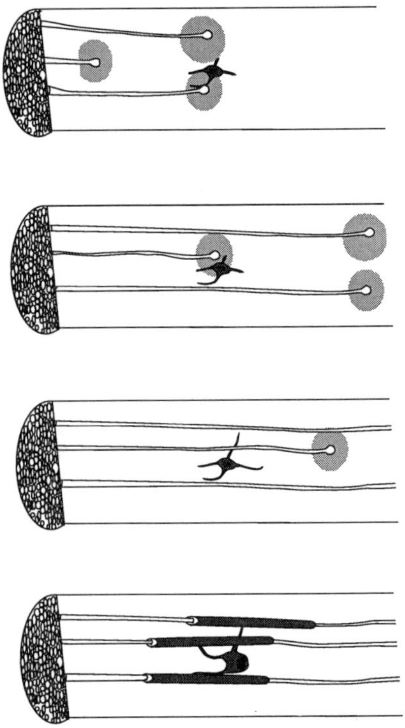

FIGURE 8. Hypothetical model for the role of a glial GABA-activated ion channel during development. For details see text.

In contrast, oligodendrocytes characterized by expression of galactocerebroside or the 01 antigen are nonmitotic under most conditions and no longer mobile. With progression to the 01-positive oligodendrocyte, expression of Na^+ and the delayed rectifier, A, and calcium-dependent K^+ currents is down-regulated and a prominent new current type is observed that corresponds to an inwardly rectifying potassium channel.

Thus, expression of distinct channel types may be linked to properties expressed by the different differentiation stages. In both lymphocytes and glial cells, differentiation stages have been linked to expression of particular K^+ channels.[40,41] K^+ channel blockers inhibited the proliferation of Schwann cells in response to mitotic stimuli. Furthermore, regulation of specific K^+ channels in Schwann cells has been tightly linked with myelinogenesis.[42] Changes in the activity of distinct K^+ channels may also be one of the first indications of adhesion-linked changes in cell behavior.[43]

Possible Function of Transmitter Receptors on Glial Cells

The density of GABA receptors is also greatly reduced with progression along the lineage, and additionally only a subpopulation of 010-positive cells express GABA receptors. It is possible that the release of GABA from neurons may thus regulate glial cell function by binding to precursor cells and oligodendrocytes via specific receptors. For example, neurons may be able to regulate the rate of glial cell differentiation. Neuronal growth cones release transmitter during migration as shown in embryonic amphibian neurons.[44] Bombardment of nearby glial GABA receptors could be the means for oligodendrocyte precursors to detect migrating neurites. When migration ends, bombardment stops, and it is intriguing to speculate that the supply of GABA to precursor GABA receptors may slow their intrinsic developmental clock or even halt these cells in an immature state (FIG. 8). GABA receptors are known to be coupled to Cl^- channels, and it remains to be established if further connections to other intracellular signaling pathways exist that could modulate aspects of glial cell behavior such as maturation.

SUMMARY

The electrical properties of oligodendrocytes during their development in cell culture were analyzed by combining two techniques: cell identification with cell-type and stage-specific antibodies and the patch-clamp technique. The transition from the bipotential precursor cell, which can still develop into astrocytes and oligodendrocytes, into an oligodendrocyte results in a marked change in the ion channel pattern. During this developmental transition, voltage-activated Na^+ and several types of K^+ currents disappear, whereas a comparatively passive, inwardly rectifying K^+ current becomes dominant. $GABA_A$ receptor-mediated Cl^- currents and a pH-activated Na^+ current are down-regulated at this transition but are still present at all developmental stages. In contrast, electrical coupling develops only in oligodendrocytes. This change in the channel repertoire could reflect the transition of a cell in a mobile, mitotic, plastic state (the glial precursor) to a more differentiated specialized state (the oligodendrocyte).

REFERENCES

1. STURROCK, R. R. 1980. Myelination of the mouse corpus callosum. Neuropathol. Appl. Neurobiol. **6:** 415–420.
2. FFRENCH-CONSTANT, C. & M. C. RAFF. 1986. Proliferating bipotential glial progenitor cells in adult rat optic nerve. Nature 499–502.
3. RAFF, M. C., R. H. MILLER & M. NOBLE. 1983. A glial progenitor cell that develops *in vitro* into an astrocyte or an oligodendrocyte depending on the culture medium. Nature **303:** 390–396.
4. RAFF, M. C. 1989. Glial cell diversification in the rat optic nerve. Science **243:** 1450–1455.
5. LEVI, G., F. ALOISI & G. P. WILKIN. 1987. Differentiation of cerebellar bipotential glial precursors into oligodendrocytes in primary culture: Developmental profile of surface antigens and mitotic activity. J. Neurosci. Res. **18:** 407–417.

6. TROTTER, J. & M. SCHACHNER. 1989. Cells positive for the 04 surface antigen isolated by cell sorting are able to differentiate into astrocytes or oligodendrocytes. Dev. Brain Res. **46:** 115–122.
7. TROTTER, J., D. BITTER-SUERMANN & M. SCHACHNER. 1989. Differentiation-regulated loss of the polysialylated embryonic form and expression of the different polypeptides of the neural cell adhesion molecule (N-CAM) by cultured oligodendrocytes and myelin. J. Neurosci. Res. **22:** 369–383.
8. SOMMER, I. & M. SCHACHNER. 1981. Monoclonal antibodies (01 to 04) to oligodendrocyte cell surfaces: An immunocytological study in the central nervous system. Dev. Biol. **83:** 311–323.
9. SOMMER, I. & M. SCHACHNER. 1982. Cells that are 04 antigen-positive and 01 antigen-negative differentiate into O1 antigen-positive oligodendrocytes. Neurosci. Lett. **29:** 183–188.
10. KUHLMANN-KRIEG, S., I. SOMMER & M. SCHACHNER. 1988. Ultrastructural features of cultured oligodendrocytes expressing stage-specific cell-surface antigens. Dev. Brain Res. **39:** 269–280.
11. BEVAN, S., S. Y. CHIU, P. T. A. GRAY & J. M. RITCHIE. 1985. The presence of voltage-gated sodium, potassium and chloride channels in rat cultured astrocytes. Proc. Roy. Soc. Lond. **B225:** 299–313.
12. MACVICAR, B. A. 1984. Voltage dependent Ca^{2+} channels in glial cells. Science **226:** 1345–1347.
13. BORMANN, J. & H. KETTENMANN. 1988. Patch clamp study of GABA receptor Cl^- channels in cultured astrocytes. Proc. Natl. Acad. Sci. USA **85:** 9336–9340.
14. BEVAN, S., R. M. LINDSAY, M. N. PERKINS & M. C. RAFF. 1987. Voltage gated ionic channels in rat cultured astrocytes, reactive astrocytes and an astrocyte-oligodendrocyte progenitor cell. J. Physiol. (Paris) **82:** 327–335.
15. SONTHEIMER, H., J. TROTTER, M. SCHACHNER & H. KETTENMANN. 1989. Developmental regulation of channel expression in cultured oligodendrocytes. Neuron **2:** 1135–1145.
16. BARRES, B. A., L. L. Y. CHUN & D. P. COREY. 1990. Glial and neuronal forms of the voltage-dependent sodium channel: Characteristics and cell-type distribution. Neuron **2:** 1375–1388.
17. VERKHRATSKY, A. N., J. TROTTER & H. KETTENMANN. 1990. Cultured glial precursor cells from mouse cortex express two types of calcium currents. Neurosci. Lett. **112:** 194–198.
18. BARRES, B. A., W. KOROSHETZ, K. J. SCHWARTZ, L. L. Y. CHUN & D. P. COREY. 1990. Ion channel expression by white matter glia: The O-2A progenitor cell. Neuron **4:** 507–524.
19. KETTENMANN, H., R. K. ORKAND, H. D. LUX & M. SCHACHNER. 1982. Single potassium channel currents in cultured mouse oligodendrocytes. Neurosci. Lett. **32:** 41–46.
20. KETTENMANN, H., R. K. ORKAND & H. D. LUX. 1984. Some properties of single potassium channels in cultured oligodendrocytes. Pfluegers Arch. **400:** 215–221.
21. BARRES, B. A., L. L. Y. CHUN & D. P. COREY. 1988. Ion channel expression by white matter glia. I. Type 2 astrocytes and oligodendrocytes. Glia **1:** 10–30.
22. SOLIVEN, B., S. SZUCHET, B. G. W. ARNASON & D. J. NELSON. 1988. Voltage-gated potassium currents in cultured bovine oligodendrocytes. J. Neurosci. **8:** 2131–2141.
23. SONTHEIMER, H. & H. KETTENMANN. 1988. Heterogeneity of potassium currents in cultured oligodendrocytes. Glia **1:** 415–420.
24. MCLARNON, J. G. & S. U. KIM. 1989. Existence of inward potassium currents in adult human oligodendrocytes. Neurosci. Lett. **101:** 107–112.
25. SONTHEIMER, H., M. PEROUANSKY, D. HOPPE, R. GRANTYN, H. D. LUX & H. KETTENMANN. 1989. Expression of pH activated Na^+ channels in glia is restricted to the oligodendrocyte lineage. J. Neurosci. Res. **24:** 496–500.
26. BLANKENFELD, VON G., J. TROTTER & H. KETTENMANN. 1991. Expression of a distinct $GABA_A$ receptor subtype in the oligodendrocyte lineage and its developmental regulation. Eur. J. Neurosci. **3:** 310–316.
27. GILBERT, P., H. KETTENMANN & M. SCHACHNER. 1984. Gamma-aminobutyric acid directly depolarizes cultured oligodendrocytes. J. Neurosci. **4:** 561–569.
28. HOPPE, D. & H. KETTENMANN. 1989. GABA triggers a Cl^- efflux from cultured oligodendrocytes. Neurosci. Lett. **97:** 334–339.

29. Hoppe, D. & H. Kettenmann. 1989. Carrier-mediated Cl^- transport in cultured oligodendrocytes. J. Neurosci. Res. **23:** 467–475.
30. Gregor, P., M. Itzchak, I. Maoz, McKeown & V. I. Teichberg. 1989. Molecular structure of the chick cerebellar kainate-binding subunit of a putative glutamate receptor. Nature **342:** 689–691.
31. Kainunen, K., W. Wisden, B. Sommer, P. Werner, A. Herb, P. A. Verdoorn, B. Sakmann & P. H. Seeburg. 1990. A family of AMPA-selective glutamate receptors. Science **249:** 556–560.
32. Wisden, W., L. A. Naughton, M. G. Darlison, S. P. Hunt & E. A. Barnard. 1989. Differential distribution of $GABA_A$ receptor mRNAs in bovine cerebellum—localization of $alpha_2$ mRNA in Bergmann glia layer. Neurosci. Lett. **106:** 7–12.
33. Kettenmann, H., R. K. Orkand & M. Schachner. 1983. Coupling among identified cells in mammalian nervous system cultures. J. Neurosci. **3:** 506–516.
34. Kettenmann, H. & B. R. Ransom. 1988. Electrical coupling between astrocytes and between oligodendrocytes in mammalian cell cultures. Glia **1:** 64–73.
35. Massa, P. T. & E. Mugnaini. 1982. Cell junctions and intramembrane particles of astrocytes and oligodendrocytes: A freeze fracture study. Neuroscience **7:** 523–538.
36. Ransom, B. R. & H. Kettenmann. 1990. Electrical and dye coupling between astrocytes and oligodendrocytes studied in mammalian cell culture. Glia **3:** 258–266.
37. Blankenfeld, von G., B. R. Ransom & H. Kettenmann. 1991. Electrical and dye coupling during the development of oligodendrocytes. Submitted for publication.
38. Berger, T. J., J. Schnitzer, W. Walz & H. Kettenmann 1990. Developmental changes in voltage and transmitter activated currents of glial cells in the corpus callosum slice. In preparation.
39. Small, K., P. Riddle & M. Noble. 1987. Evidence for migration of oligodendrocyte-type-2 astrocyte progenitor cells into the developing rat optic nerve. Nature **328:** 155–157.
40. Lewis, R. S. & M. D. Cahalan. 1988. The plasticity of ion channels: Parallels between the nervous and immune systems. Trends Neurosci. **11:** 214–218.
41. Chiu, S. Y. & G. F. Wilson. 1989. The role of potassium channels in Schwann cell proliferation in Wallerian degeneration of explant rabbit sciatic nerves. J. Physiol. **408:** 199–222.
42. Wilson, G. F. & S. Y. Chiu. 1990. Potassium channel regulation in Schwann cells during early developmental myelinogenesis. J. Neurosci. **10:** 1615–1625.
43. Sontheimer, H., H. Kettenmann, M. Schachner & J. Trotter. 1991. The neural cell adhesion molecule N-CAM modulates K^+ channels in cultured glial precursor cells. Submitted for publication.
44. Young, S. H. & M. Poo. 1983. Spontaneous release of transmitter from growth cones of embryonic neurones. Nature **305:** 634–637.

Role of Cyclic AMP and Proliferation Controls in Schwann Cell Differentiation[a]

K. R. JESSEN,[b] R. MIRSKY, AND L. MORGAN

Department of Anatomy and Developmental Biology
University College London
London WC1E 6BT, UK

In this paper we describe how antigenic differentiation markers can be used to describe Schwann cell development and to clarify the relationship between myelin-forming and non-myelin-forming Schwann cells. This approach has also made it possible to establish that elevation of cAMP levels has a broad spectrum of effects on Schwann cell differentiation. This includes both myelination-related changes and developmental steps common to both non-myelin-forming and myelin-forming cells and also events characteristic of embryonic Schwann cell development. Lastly, we show that, depending on the proliferative state, cAMP elevation induces different parts of the developmental sequence. In dividing cells, cAMP induces early differentiation only, whereas progression to myelination requires withdrawal from the cell cycle.

MORPHOLOGICAL ASPECTS OF SCHWANN CELL DEVELOPMENT

Some knowledge of the morphologic events that take place during the developmental period is essential for the subsequent discussion. For further information about the morphological aspects, see references 1–4.

In the rat, major peripheral nerves, like the sciatic nerve, can be identified as anatomical structures by embryonic day (E) 14 or 15. Although the sciatic nerve has not yet been studied in detail, work on other systems and our preliminary observations indicate that the nerve at this stage is a rather compact structure consisting of outgrowing axons and proliferating glial cells that form flattened membranous sheets. The extensive penetration of connective tissue and individual ensheathment of axons characteristic of mature nerves is not seen. Significant connective tissue spaces appear in the nerve at approximately E17, and a rapid increase in the number of Schwann cells per axon, a result of ongoing proliferation, gradually leads to a more intimate relationship between individual axons and Schwann cells.

By E21 (birth) many axons of larger diameter are individually surrounded by a single Schwann cell. This attainment of a 1:1 ratio between axons and Schwann cell is followed by cessation of Schwann cell proliferation and it leads, in the rat sciatic nerve, to the formation of myelin sheath in nearly every case. Segregation of large diameter axons, leading to myelination, continues after birth, whereas the axons of smaller diameter segregate only incompletely. They come to lie in troughs running

[a]This work has been supported by grants from the Medical Research Council of Great Britain, Action Research for the Crippled Child, the Wellcome Trust, and the Multiple Sclerosis Society of Great Britain and Northern Ireland.

[b]Address for correspondence: K. R. Jessen, Department of Anatomy and Developmental Biology, University College London, Gower Street, London WC1E 6BT, UK.

along the Schwann cell surface and are most often separated from each other, although several axons share the same Schwann cell. Mature nonmyelin-forming cells of this kind are not seen until the third postnatal week[5]; hence, the phase of myelination largely precedes the phase of maturation of nonmyelin-forming cells in this nerve. Schwann cell proliferation is essentially over by the fifth postnatal week and both nonmyelin-forming and myelin-forming cells are quiescent in unperturbed mature nerves.

A striking positive correlation exists between axon diameter and myelination, because myelin sheaths are essentially restricted to axons above 1 μm in diameter and are found on more or less all axons above that size. This gave rise to the idea that axon diameter *per se* determined whether or not a Schwann cell formed myelin.[6,7] Certain arguments have been put forward against this theory,[7] and in recent years most thought in this area has been influenced by another hypothesis, namely, that myelination is induced by specific molecular signals, "the myelination signal," selectively associated with axons of larger diameter. We will come back to these issues in the context of the regulation of Schwann cell proliferation and the effects of cAMP on the Schwann cell phenotype.

MATURE SCHWANN CELLS CAN BE DISTINGUISHED BY THE USE OF ANTIGENIC MARKERS

Nonmyelin- and myelin-forming Schwann cells can each be identified by a distinct set of antigenic markers, most of which correspond to identified molecules (TABLE 1). The molecular composition of myelin has been under scrutiny for many years, and it is clear that peripheral myelin contains a set of proteins (Group 1, TABLE 1) that are absent from nonmyelin-forming Schwann cells.[8-12] This includes the major protein of peripheral myelin P_0, myelin-associated glycoprotein (MAG), and myelin basic protein (MBP) and possibly also the P_{170k} protein and proteolipid protein (PLP).

More recently, it was found that nonmyelin-forming Schwann cells also express a distinct set of proteins (Group 2, TABLE 1).[13-18] Among these are an intermediate filament protein very similar to glial fibrillary acidic protein (GFAP) in astrocytes and distinct from other intermediate filament proteins, the adhesion molecules N-CAM and L1 (Ng-CAM), nerve growth factor (NGF) receptors, and two surface proteins of unknown function, A5E3 and Ran-2.

Simple experiments can be used to demonstrate that the expression of these two sets of proteins (Groups 1 and 2, TABLE 1) is regulated differently. This involves monitoring protein expression in cells after they have been removed from axonal contact, either by nerve sectioning and axon degeneration *in vivo* or by preparation of neuron-free Schwann cell cultures. In both cases it is observed that expression of myelin proteins and, where investigated, corresponding mRNA declines rapidly.[8,19-21] On the other hand, expression of the proteins of nonmyelin-forming cells is maintained in nonmyelin-forming cells, and these proteins appear in those cells that previously made myelin. From this it can be inferred that expression of the proteins associated with adult nonmyelin-forming cells occurs by default, whereas expression of myelin proteins depends on ongoing signaling from myelinated axons. The effect of axons on Schwann cell development will be discussed further in the following section.

As expected in closely related cells, both types of Schwann cell express several common molecules (Group 3, TABLE 1[13,22-27]). Calcium binding protein S100 is an important Schwann cell marker. Schwann cells are the only nonneuronal cells expressing it within the nerve; it is present in both Schwann cell types and expressed irrespective of whether or not the cell is in contact with axons. No other known Schwann cell antigen fulfills these criteria.

The presence of the lipids galactocerebroside and the 04 antigen (which in Schwann cells is probably sulfatide[26]) in both types of Schwann cell was unexpected because, in central glia, these lipids are restricted to cells in the myelin lineage. The expression of laminin by both Schwann cell types reflects the presence of basal lamina. The Schwann cell basal lamina appears rather early in development, being visible by electron microscopy at E18–19 on some fibers. At this time Schwann cells still envelop axons communally, which implies that the subsequent segregation of axons and the formation of myelinated and unmyelinated fibers must involve Schwann cells and axons breaking out of the early basal lamina tubes. Both types of Schwann cell are surrounded by basal lamina in the mature nerve.

SCHWANN CELL DEVELOPMENT

An outline of major events in Schwann cell development is illustrated in FIGURE 1. In the rat sciatic nerve, Schwann cells, that is, S100-positive cells that survive *in vitro* in routine media, arise from Schwann cell precursors at E15–16, and by E18 over

TABLE 1. Molecular Phenotype of Schwann Cells

Group	Antigen	Myelin-Forming Schwann Cells	Nonmyelin-Forming Schwann Cells	Short-Term Cultured Cells without Neurons	Precursors
1	CNPase	+	nd	+	nd
	P_0	+	−	−	−
	MBP	+	−	−	−
	MAG	+	−	−	−
	P_{170k}	+	nd	−	nd
	PLP	+	nd	−	nd
	P_2	+[a]	−	−	nd
2	GFAP	−	+	+	−
	NGF receptor	−	+	+	+
	N-CAM	−	+	+	nd
	L-1	±[b]	+	+	+
	A5E3	−	+	+	+
	Ran-2	−	+	+	+
3	S100	+	+	+	−
	Vimentin	+	+	+	+
	Laminin	+	+	+	+
	The lipid antigens 04, 08, 09	+	+	−	−
	Galactocerebroside	+	+	−	−
	Seminolipid	+	+	+	−

[a]Present on some myelin-forming Schwann cells only.
[b]Present mainly at nodes of Ranvier, very little elsewhere.
nd = not determined.
The antigens in Group 1 are expressed mainly by myelin-forming cells, those in Group 2 by nonmyelin-forming cells, and those in Group 3 by both. For references, see text.

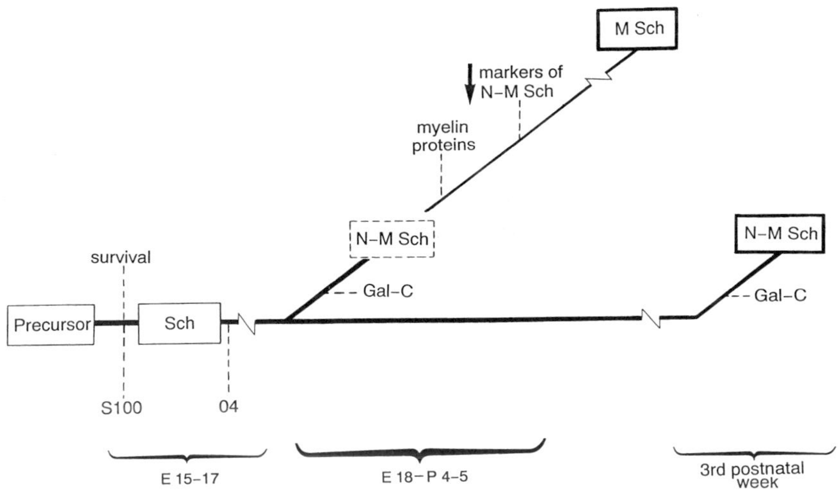

FIGURE 1. Development of the molecular phenotype of myelin-forming and nonmyelin-forming Schwann cells from precursor cells in the rat sciatic nerve. Sch = Schwann cell; N-M Sch in heavy frame = mature nonmyelin-forming Schwann cell; N-M Sch in stippled frame = cells with a molecular phenotype similar to that of mature nonmyelin-forming cells, that will progress to myelination; M Sch = myelin-forming Schwann cell; Gal-C = galactocerebroside. ↓ indicates molecules that are down-regulated during differentiation of the Schwann cell along the myelin pathway. All other Schwann cell molecules or properties indicated are first seen at the time points indicated. Cells acquire the ability to survive alone in culture from embryo day 15 onwards; S100 first appears in Schwann cells between embryo days 15–16 and is present in 95% of Schwann cells at embryo day 18; 04 first appears shortly after S100 at embryo day 16 and is present in 95% of Schwann cells at embryo day 20; and galactocerebroside first appears on cells in the myelin-forming pathway at embryo day 18 and the first P_0-positive cells are seen during the first postnatal day. By postnatal day 5 the first myelin-forming Schwann cells have down-regulated molecules such as GFAP, N-CAM, A5E3, and NGF receptors which are associated with nonmyelin-forming Schwann cells. Cells in the nonmyelin-forming pathway acquire galactocerebroside from the third postnatal week onwards without progressing to myelination. Differentiation of these cells is complete by the fifth to sixth postnatal weeks.

95% of the cells isolated from the nerve are S100-positive Schwann cells. (For discussion of Schwann cell precursors, see ref. 28.) By E15, most of the cells in the nerve already express the proteins that are associated with mature nonmyelin-forming Schwann cells (Group 2, TABLE 1), and all Schwann cells express this group of proteins at birth. The exception is GFAP, which first appears at E18 and is present at low levels in approximately 50% of Schwann cells at birth.[13]

The first step in Schwann cell development is the appearance of the surface lipid antigen 04 (sulfatide) at E16–17. Antigen 04 appears exclusively on S100-positive cells and by E20 essentially all S100-positive Schwann cells also express 04.[26] The related lipid galactocerebroside is first seen on the Schwann cell surface at E18–19.[8,24] The expression of galactocerebroside spreads much more slowly through the Schwann cell population and is not found on all Schwann cells until the fifth postnatal week. During the early part of this period, almost all of the cells that acquire galactocerebroside go on to make myelin. During the third postnatal week, however, cells that acquire galactocerebroside without progressing to myelination first appear.[24] These are the first mature nonmyelin-forming cells, appearing after the major phase of myelination, as discussed in a previous section.

In biochemical terms, myelination involves two sets of changes in protein expression: first, up-regulation in the synthesis of myelin proteins including MAG, P_0, and MBP, and second, down-regulation in the expression of the proteins associated with nonmyelin-forming cells (Group 2, TABLE 1), including GFAP, N-CAM, NGF receptors, and others.[24,29]

If Schwann cells are removed from close apposition or contact with axons by denervation or culturing, they slide backwards along the developmental pathway illustrated in FIGURE 1. As a result, they assume a molecular phenotype similar to that of Schwann cells at their earliest appearance, before 04 acquisition. Unlike these early Schwann cells, however, denervated Schwann cells express low, but unambiguously detectable levels of GFAP. If denervated Schwann cells are allowed to associate with axons again, as during nerve regeneration, the cells are pushed in the reverse direction along their developmental path, resulting in the regeneration of nonmyelin-forming and myelin-forming cells. These observations and many other morphologic studies (eg, refs. 30 and 31) support the notion that Schwann cell development, from the earliest step right to the generation of nonmyelin-forming and myelin-forming cells, is heavily dependent on axon-to-Schwann cell signaling, involving membrane-membrane interactions or short-range soluble factors. The developmental steps that depend on axon-associated signals are the up-regulation of myelin protein expression and the subsequent suppression of protein expression in the myelination pathway, 04 expression in early Schwann cells, and galactocerebroside expression in both nonmyelin- and myelin-forming cells. In addition, GFAP expression in nonmyelin-forming cells apparently is enhanced by contact with unmyelinated axons.[32]

It is likely that intrinsically programmed changes also take part in shaping Schwann cell development, although, as yet, there is only one example of what appears to be an intrinsically programmed change during the developmental period discussed here, that is, the appearance of GFAP. *In vivo* it is first detectable at E18, but it will also appear in cells removed from E15–16 nerves and maintained in culture for a few days, even in the absence of neurons.[13]

EFFECT OF AXON-ASSOCIATED SIGNALS IN SCHWANN CELL DEVELOPMENT ARE MIMICKED BY AGENTS THAT ELEVATE INTRACELLULAR cAMP

The molecular basis of the axon-Schwann cell signaling mechanism(s) is still essentially unknown. It is now possible, however, to mimic different stages of Schwann cell development in purified Schwann cell cultures. This can be done by exposure to agents that elevate intracellular cAMP levels, and manipulation of conditions that govern Schwann cell proliferation.[33] The relevant evidence and the significance of these observations for Schwann cell development *in vivo* will be discussed.

In 1984, Sobue and Pleasure[34] found that agents that elevate intracellular cAMP levels induce reexpression of cell surface galactocerebroside in Schwann cells that had lost galactocerebroside during a period of neuron-free culturing. Since then, similar experiments have established that other phenotypic changes associated with Schwann cell differentiation can also be induced by exposure to cAMP-elevating agents. This includes induction of cell surface expression of the early antigen 04[26] and of several basal lamina components.[35] Cyclic AMP elevation also causes Schwann cells to flatten and form large membranous expansions,[33,36] both of which are morphological changes associated with myelination; it also induces expression of the P_{170k} protein found in myelin,[37] of P_0[33,38] and MBP and their mRNAs, and of the SCIP gene.[38–40]

Thus, in cultured cells cAMP elevation induces events associated with diverse stages in the development of Schwann cells *in vivo,* from their earliest generation to myelination, including developmental steps that occur in both nonmyelin- and myelin-forming cells. It is noteworthy that all of these cAMP-inducible events are considered to be induced *in vivo* by axon-to-Schwann signaling (previous section). Furthermore, some evidence exists that axonal membranes elevate intracellular cAMP levels in Schwann cells, but this is controversial.[41,42] Cyclic AMP affects the expression of Pou type gene regulatory proteins in Schwann cells, and there are precedents for cAMP as the second messenger in the transduction of ligand-receptor interactions at the cell surface to changes in gene activity.[43] All of this taken together suggests that an important part of axon-to-Schwann cell signaling *in vivo* might operate via elevation of Schwann cell cAMP levels. The broad spectrum of phenotypic changes induced by cAMP elevation in turn opens up the interesting possibility that a single axon-associated signaling mechanism, operating via the adenyl cyclase-cyclic AMP pathway, is responsible for driving much of Schwann cell development from E16 to maturity. The axon-associated signal in question should be present as early as E15–16 on all axons and would be responsible not only for myelination, but also for O4 expression and for galactocerebroside expression in both nonmyelin- and myelin-forming cells. These predictions distinguish this idea from an alternative, that is, that cAMP is selectively raised in Schwann cells induced to myelinate, by a signaling mechanism, "the myelination signal," associated only with the larger axons destined to become myelinated.

The idea that a constant input, that is, axon-associated signal-raising Schwann cell cAMP levels, induces a developmental sequence spaced out in time over several days or even weeks raises the question of timing. In the development of myelin-forming cells, for instance, what is the timing mechanism that allows cAMP elevation to trigger antigen O4 expression at E16 but delays P_0 expression in response to the same signal to the first postnatal day, that is, by 4–5 days? In principle, this could be accomplished by an intrinsically programmed alteration in those factors that determine the intracellular translation of the cAMP signal. At present, there is no evidence for such a mechanism.

We do, however, know of an alternative way in which translation of the cAMP signal can be switched to induce early Schwann cell differentiation, that is, 04 expression, on the one hand, or to induce a later developmental stage, that is, P_0 expression, on the other. This can be achieved via regulation of Schwann cell proliferation.[33]

CYCLIC AMP INDUCES O4 EXPRESSION IN PROLIFERATING CELLS, BUT MYELIN-RELATED DIFFERENTIATION IN QUIESCENT ONES

To investigate the relationship between the proliferative state and cAMP-induced Schwann cell differentiation *in vitro,* cAMP levels were elevated using forskolin (2–4 μm) cholera toxin or cAMP analogs (10^{-3} M followed by 10^{-4} M) over a period of 3–5 days under conditions allowing Schwann cell proliferation (ie, in the presence of serum or growth factors) and under essentially quiescent conditions (ie, in serum-free medium or in high-density cultures in the presence of serum).[33] It was found that under conditions allowing DNA synthesis, cAMP elevation effectively

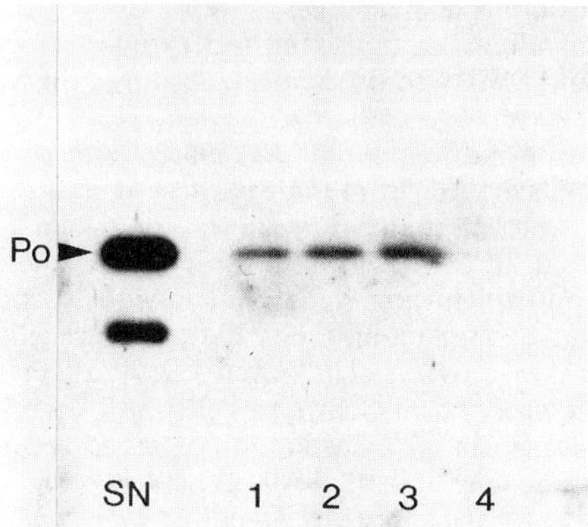

FIGURE 2. Immunoblot of P_0 induction in Schwann cells. (SN) Extract of 5-day-old rat sciatic nerve, 25 μg loaded. *Lanes 1–3:* Extracts from dissociated Schwann cell cultures from 5-day-old rat sciatic nerve treated with agents that elevate or mimic cAMP. *Lane 1:* cAMP analogs; *lane 2:* cholera toxin; *lane 3:* forskolin; *lane 4:* no treatment, 31 μg of extract loaded in each lane. Note that at the P_0 position (*arrow*) an immunoreactive band is visible in the sciatic nerve lane and in each of lanes 1–3, but not in lane 4 (control). In the sciatic nerve extract a lower breakdown product of P_0 is also visible. (Reproduced from ref. 33 with permission.)

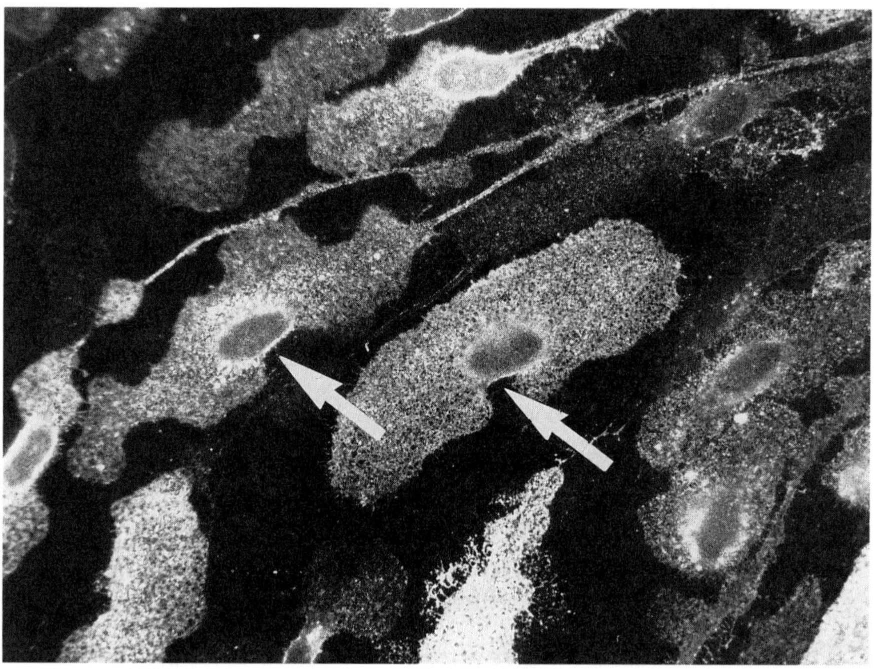

FIGURE 3. Induction of P_0 in Schwann cells treated with cholera toxin. P_0-positive Schwann cells cultured from 5-day-old rat sciatic nerve. Cells were cultured on a laminin substrate, treated with cholera toxin, and labeled with P_0 antibodies after 3 days of treatment. Cells were viewed with fluorescein optics to visualize P_0. Note the flattened shape of the Schwann cells and the P_0 immunofluorescence distributed throughout the whole area of each cell. Nuclei of two intensely P_0-positive cells are indicated by *arrows.* (Reproduced from ref. 33 with permission.)

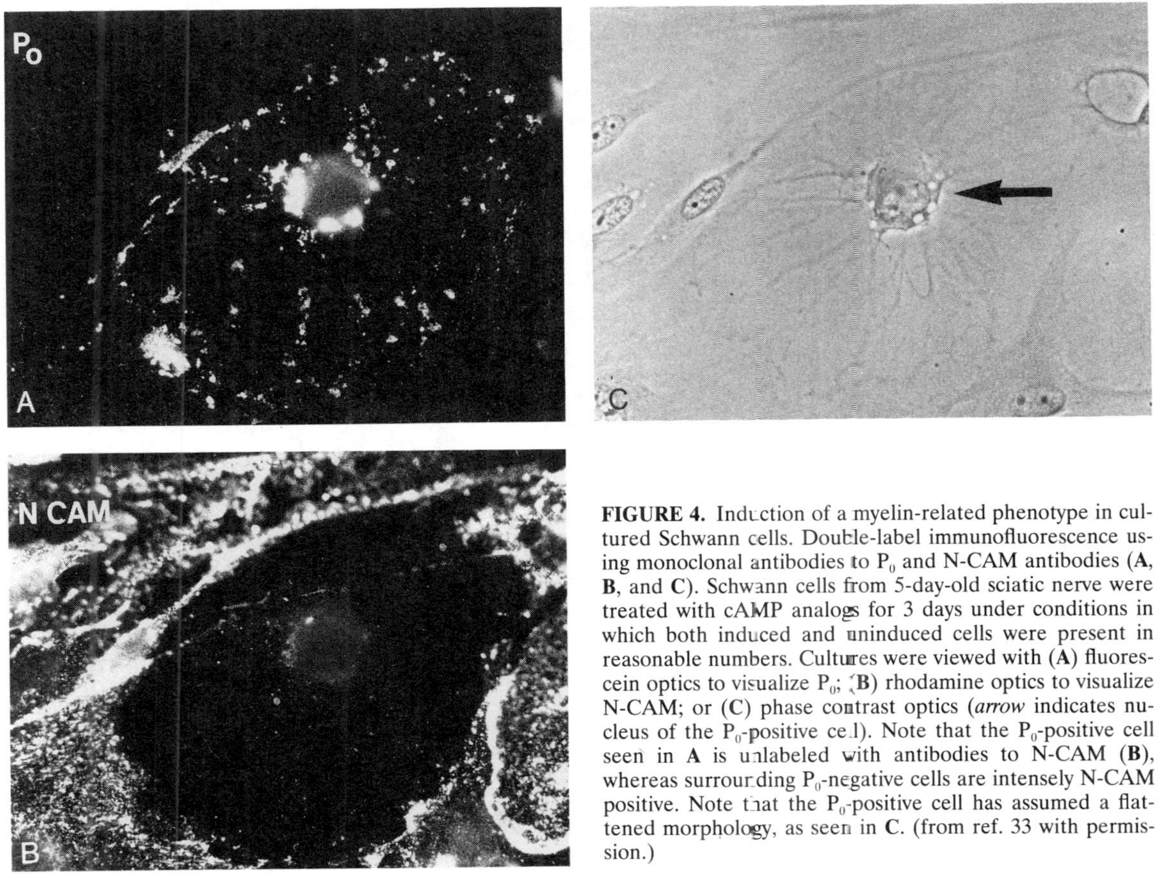

FIGURE 4. Induction of a myelin-related phenotype in cultured Schwann cells. Double-label immunofluorescence using monoclonal antibodies to P_0 and N-CAM antibodies (**A**, **B**, and **C**). Schwann cells from 5-day-old sciatic nerve were treated with cAMP analogs for 3 days under conditions in which both induced and uninduced cells were present in reasonable numbers. Cultures were viewed with (**A**) fluorescein optics to visualize P_0; (**B**) rhodamine optics to visualize N-CAM; or (**C**) phase contrast optics (*arrow* indicates nucleus of the P_0-positive cell). Note that the P_0-positive cell seen in **A** is unlabeled with antibodies to N-CAM (**B**), whereas surrounding P_0-negative cells are intensely N-CAM positive. Note that the P_0-positive cell has assumed a flattened morphology, as seen in **C**. (from ref. 33 with permission.)

induced surface 04 expression and simultaneously enhanced the rate of DNA synthesis. Using the BrdU method to detect DNA synthesis and immunohistochemistry, it was shown that many cells were both BrdU and 04 positive during the third day of cAMP elevation, indicating that induction of 04 expression and DNA synthesis are compatible in individual cells. When the same cultures were examined with P_0 antibodies, however, it became clear that the cells had not progressed to the myelination stage, because essentially no P_0-positive cells were present. If the same experiment was repeated under nonproliferative conditions, or if serum was withdrawn from cells previously exposed to cAMP-elevating agents in serum, a very different response was elicited. Under these conditions cAMP elevation strikingly mimicked the effects of axon-associated signals during myelination, and the effect

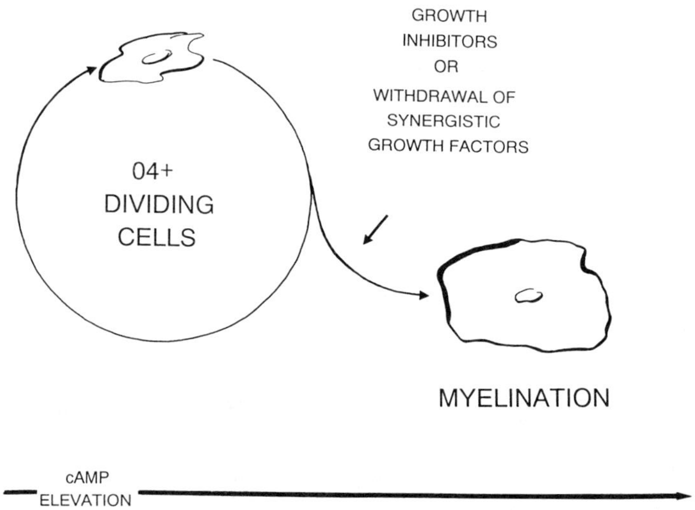

FIGURE 5. The proliferative state may control the response to cAMP *in vivo.* It is envisaged that axon-dependent elevation of cAMP induces early differentiation (04 expression) under conditions in which cAMP simultaneously synergizes with growth factors to promote Schwann cell proliferation. Withdrawal from the cell cycle (brought about as indicated or via other mechanisms) triggers myelination, because in quiescent cells the ongoing cAMP elevation induces myelination.

was seen in the majority of cultured cells. Not only were high levels of P_0 induced (FIGS. 2 and 3), but also the protein down-regulation characteristic of myelination was also effectively obtained, particularly in those cells showing the strongest P_0 induction (FIG. 4). Expression of GFAP, N-CAM, NGF receptors, and the surface protein A5E3 (see Group 2, TABLE 1) were all strikingly suppressed in this way.

These culture experiments show that in cells with raised cAMP, entry of 04-positive proliferating Schwann cells into the myelin pathway is regulated by the proliferative state. Withdrawal from the cell cycle in these cells acts as a trigger for myelin-related differentiation. *In vivo* also, cessation of Schwann cell proliferation precedes the formation of the first myelin wraps on the first day after birth, whereas 04 appears earlier in proliferating cells. Therefore, we suggest that the principle revealed in the culture experiments may also apply *in vivo* (FIG. 5).

Regulation of Schwann cell proliferation *in vivo* is still poorly understood (see ref. 44). Cessation of division might result simply from withdrawal of growth factors or growth factor receptors necessary for driving Schwann cell proliferation. Alternatively, active growth inhibition might be involved, possibly involving collagen type I, interferon-gamma or related molecules, or inhibitory factors secreted by the Schwann cells themselves.

In future work it will be interesting to analyze the intracellular regulatory mechanisms that allow elevation of cAMP to set in motion an extensive phenotype change involving coordinated induction and suppression of a large group of proteins in quiescent cells, but not in proliferating ones. Another central question is whether axon-associated molecules that induce Schwann cells to express O4, galactocerebroside, and myelin proteins can be identified in nerves. If so, do they elevate cAMP levels, and what is their distribution between types of axon and time course of appearance in development? Lastly, a key question not addressed directly in the hypothesis just outlined (FIG. 5) concerns the generation of mature nonmyelin-forming cells. Why do some cells form myelin when withdrawn from the cell cycle, whereas others proceed no further than the galactocerebroside-positive stage? It is possible that the answer lies in different degrees of cAMP elevation. Alternatively, it is possible that the degree of structural polarization of the Schwann cell, between the axonal membrane and the extracellular matrix, is a determining factor. For further discussion of this issue see reference 28.

ACKNOWLEDGMENTS

We would like to thank all the past and present members of our laboratory for their contribution to many of the experiments described here.

REFERENCES

1. FRIEDE, R. L. & T. SAMORAJSKI. 1968. Myelin formation in the sciatic nerve of the rat. J. Neuropathol. Exp. Neurol. **27:** 546–570.
2. WEBSTER, H. DE F., J. R. MARTIN & M. F. O'CONNELL. 1973. The relationships between interphase Schwann cells and axons before myelination: A quantitative electron microscopic study. Dev. Biol. **32:** 401–416.
3. WEBSTER, H. DE F. & J. T. FAVILLA. 1984. Development of peripheral nerve fibres. *In* Peripheral Neuropathy. P. J. Dyck, P. K. Thomas, E. H. Lambert, & R. Bunge, eds. W. B. Saunders. Philadelphia.
4. ZISKIND-CONHAIM, L. 1988. Physiological and morphological changes in developing peripheral nerves of rat embryo. Dev. Brain. Res. **42:** 15–28.
5. DINER, O. 1965. Les cellules de Schwann en mitose et leurs rapports avec les axones au cours du developpement du nerf sciatique chez le rat. C. R. Acad. Sci. **261:** 1731–1734.
6. FRIEDE, R. L. 1972. Control of myelin formation in the sciatic nerve of the rat. J. Neuropathol. Exp. Neurol. **144:** 233–252.
7. GRIFFIN, J. W., J. ROSENFELD, P. N. HOFFMAN, B. G. GOLD & B. TRAPP. 1988. *In* Intrinsic Determinants of Neuronal Form and Function. 403–439. Alan R. Liss Inc. New York.
8. MIRSKY, R., J. WINTER, E. R. ABNEY, R. M. PRUSS, J. GAVRILOVIC & M. C. RAFF. 1980. Myelin-specific proteins and glycolipids in rat Schwann cells and oligodendrocytes in culture. J. Cell Biol. **84:** 483–494.
9. SHUMAN, S., M. HARDY & D. PLEASURE. 1986. Immunochemical characterization of peripheral nervous system myelin 170,000 M_r glycoprotein. J. Neurochem. **35:** 1183–1189.

10. TRAPP, B. D., M. DUBOIS-DALCQ & R. H. QUARLES. 1984. Ultrastructural localization of P_2 protein in actively myelinating Schwann cells. J. Neurochem. **43:** 944–948.
11. TRAPP, B. D., Y. ITOYAMA, N. H. STERNBERGER, R. H. QUARLES & H. DE F. WEBSTER. 1981. Immunocytochemical localization of P_0 protein in Golgi complex membranes and myelin of developing rat Schwann cells. J. Cell Biol. **90:** 1–6.
12. TRAPP, B. D. & R. H. QUARLES. 1984. Immunocytochemical localization of the myelin-associated glycoprotein—fact or artifact. J. Neuroimmunol. **6:** 231–249.
13. JESSEN, K. R., L. MORGAN, H. J. S. STEWART & R. MIRSKY. 1990. Three markers of adult non-myelin-forming Schwann cells, 217c(Ran-1), A5E3 and GFAP: development and regulation by neuron-Schwann cell interactions. Development **109:** 91–103.
14. JESSEN, K. R., R. THORPE & R. MIRSKY. 1984. Molecular identity, distribution and heterogeneity of glial fibrillary acidic protein: An immunoblotting and immunohistochemical study of Schwann cells, satellite cells, enteric glia and astrocytes. J. Neurocytol. **13:** 187–200.
15. JESSEN, K. R. & R. MIRSKY. 1984. Non-myelin-forming Schwann cells coexpress surface proteins and intermediate filaments not found in myelin forming cells: A study of Ran-2, A5E3 antigen and glial fibrillary acidic protein. J. Neurocytol. **13:** 923–934.
16. JOHNSON, E. M., JR., M. TANIUCHI & P. S. DISTEFANO. 1988. Expression and possible function of nerve growth factor receptors on Schwann cells. Trends in Neurosci. **11:** 299–304.
17. MARTINI, R. & M. SCHACHNER. 1986. Immunoelectron microscopic localization of neural cell adhesion molecules (LI, N-CAM and MAG) and their shared carbohydrate epitope and myelin basic protein in developing sciatic nerve. J. Cell Biol. **103:** 2439–2448.
18. YEN, S. & K. L. FIELDS. 1981. Antibodies to neurofilament, glial filament and fibroblast intermediate filament proteins bind to different cell types of the nervous system. J. Cell Biol. **88:** 115–126.
19. GUPTA, S. K., P. F. PODUSLO & C. MEZET. 1988. Temporal changes in P_0 and MBP gene expression after crush-injury of the adult peripheral nerve. Brain Res. **464:** 133–141.
20. LEBLANC, A. C. & J. F. PODUSLO. 1990. Axonal modulation of gene expression in the peripheral nerve. J. Neurosci. Res. **26:** 317–326.
21. TRAPP, B. D., P. HAUER & G. LEMKE. 1988. Axonal regulation of myelin protein mRNA levels in actively myelinating Schwann cells. J. Neurosci. **8:** 3515–3521.
22. ECCLESTON, P. A., R. MIRSKY, K. R. JESSEN, I. SOMMER & M. SCHACHNER. 1987. Postnatal development of rat peripheral nerves: An immunohistochemical study of membrane lipids common to non-myelin forming Schwann cells, myelin forming Schwann cells and oligodendrocytes. Dev. Brain Res. **35:** 249–256.
23. JESSEN, K. R. & R. MIRSKY. 1985. GFA polypeptides in peripheral glia: Molecular weight, heterogeneity and distribution. J. Neuroimmunol. **8:** 377–393.
24. JESSEN, K. R., L. MORGAN, M. BRAMMER & R. MIRSKY. 1985. Galactocerebroside is expressed by non-myelin-forming Schwann cells *in situ.* J. Cell Biol. **101:** 1135–2243.
25. MATA, M., D. ALESSI & D. J. FINK. 1990. S100 is preferentially distributed in myelin-forming Schwann cells. J. Neurocytol. **19:** 432–442.
26. MIRSKY, R., C. DUBOIS, L. MORGAN & K. R. JESSEN. 1990. Prenatal Schwann cell development: Appearance of O4 differentiation antigen in rat embryo sciatic nerve and its regulation by axon-Schwann cell signals. Development **109:** 105–116.
27. STEFANSSON, K., WOLLMAN & B. W. MOORE. 1982. Distribution of S-100 protein outside the central nervous system. Brain Res. **234:** 309–317.
28. JESSEN, K. R. & R. MIRSKY. 1991. Schwann cell precursors and their development. Glia. **4:** 185–194.
29. JESSEN, K. R., L. MORGAN & R. MIRSKY. 1987. Myelinated, but not unmyelinated axons reversibly down-regulate N-CAM in Schwann cells. J. Neurocytol. **16:** 689–700.
30. AGUAYO, A. J., L. CHARRON & G. M. BRAY. 1976. Potential of Schwann cells from unmyelinated nerves to produce myelin: A quantitative ultrastructural and autoradiographic study. J. Neurocytol. **5:** 565–573.
31. WEINBERG, H. & P. S. SPENCER. 1976. Studies on the control of myelinogenesis. II. Evidence for neuronal regulation of myelination. Brain Res. **113:** 363–378.
32. MOKUNO, K., J. KAMHOLZ, T. BEHRMAN, C. BLACK, M. SESSA, D. FEINSTEIN, V. LEE & D.

PLEASURE. 1989. Neuronal modulation of Schwann cell glial fibrillary acidic protein (GFAP). J. Neurosci. Res. **23:** 396–405.
33. MORGAN, L., K. R. JESSEN & R. MIRSKY. 1990. The effects of cyclic AMP on differentiation of cultured Schwann cells: Progression from an early phenotype (O4$^+$) to a myelin phenotype (P_0^+, GFAP$^-$, N-CAM$^-$, NGF-receptor$^-$) depends on growth inhibition. J. Cell Biol. In press.
34. SOBUE, G. & D. PLEASURE. 1984. Schwann cell galactocerebroside induced by derivatives of adenosine 3′, 5′-monophosphate. Science **224:** 72–74.
35. BARON VAN EVERCOOREN, A., A. GANSMULLER, M. GUMPEL, N. BAUMANN & H. K. KLIENMAN. 1986. Schwann cell differentiation *in vitro;* extracellular matrix deposition and interaction. Dev. Neurosci. **8:** 182–196.
36. SOBUE, G., S. SHUMAN & D. PLEASURE. 1986. Schwann cell responses to cyclic AMP: Proliferation, change in shape, and appearance of surface galactocerebroside. Brain Res. **362:** 23–32.
37. SHUMAN, S., M. HARDY & D. PLEASURE. 1988. A cyclic AMP analogue induces synthesis of a myelin-specific glycoprotein by cultured Schwann cells. J. Neurochem. **50:** 190–194.
38. LEMKE, G. & M. CHAO. 1988. Axons regulate Schwann cell expression of the major myelin and NGF receptor genes. Development **102:** 499–504.
39. MONUKI, E. S., G. WEINMASTER, R. KUHN & G. LEMKE. 1989. SCIP: A glial POU domain gene regulated by cyclic AMP. Neuron **3:** 783–793.
40. WEINMASTER, G. & G. LEMKE. 1990. Cell-specific cyclic-AMP-mediated induction of the PDGF receptor. EMBO J. **9:** 915–920.
41. MEADOR-WOODRUFF, J. H., B. L. LEWIS & G. H. DEVRIES. 1984. Cyclic AMP and calcium as potential mediators of stimulation of cultured Schwann cell proliferation by axolemma-enriched and myelin-enriched membrane fractions. Biochem. Biophys. Res. Com. **122:** 373–380.
42. RATNER, N., L. GLASER & R. P. BUNGE. 1989. PC12 cells as a source of neurite-derived cell surface mitogen which stimulates Schwann cell division. J. Cell Biol. **98:** 1150–1155.
43. DUMONT, J. E., J. JAUNIAUX & P. P. ROGER. 1989. The cyclic AMP-mediated stimulation of cell proliferation. Trends Biochem. Sci. **14:** 67–71.
44. MIRSKY, R. & K. R. JESSEN. 1990. Schwann cell development and the regulation of myelination. Semin. Neurosci. **2:** 423–435.

DISCUSSION PAPER

Macroglial Cell Types, Lineage, and Morphology in the CNS

ARTHUR M. BUTT

Sherrington School of Physiology
United Medical and Dental Schools
St. Thomas's Hospital
London SE1 7EH, UK

Classically, from morphological studies using metal impregnation and electron microscopic methods, macroglial cells in the vertebrate central nervous system (CNS) were divided into two types, oligodendrocytes and astrocytes, and the latter were subdivided into protoplasmic or fibrous types, found in the gray matter and white matter, respectively.[1–4] More recently, with immunocytochemical techniques, astrocytes from cultures of the rat optic nerve, a white matter tract, were further subdivided into type 1 and type 2 which derived from separate lineages; type-2 astrocytes developed from a common bipotential precursor with oligodendrocytes, the O-2A progenitor, which persisted as a multipotential precursor in adults.[5–16] Now, because of the lack of unequivocal evidence *in vivo* for type-2 astrocytes or O-2A progenitors in the optic nerve or other CNS tissues (Fulton *et al.*, this volume) there is considerable debate on the extrapolation of findings *in vitro* to the CNS *in vivo* and on the application of this terminology to other CNS tissues in culture or *in vivo.* The descriptions in this volume of an ever-increasing number of glial and precursor cell subtypes in the CNS with distinctive electrophysiologic or immunocytochemical characteristics (e.g. Barres; Kettenman *et al.;* Minturn *et al.;* Price *et al.;* and Sontheimer *et al.*), together with morphologic observations *in vivo* (e.g. Butt and Ransom; Fulton *et al.;* and Skoff and Knapp), have further compounded the issue of glial subtypes and lineage.

EVIDENCE FOR THE O-2A CELL LINE

Astrocytes generally are defined as containing filaments and are immunocytochemically identified by binding antibodies against glial fibrillary acidic protein (GFAP),[17] whereas oligodendrocytes produce myelin sheaths and are labeled by antibodies of galactocerebroside (GC),[18] the major glycolipid in myelin. In culture, rat optic nerve astrocytes were subsequently defined as type 2 when staining positive with monoclonal antibody A2B5, and type 1 when A2B5-[5]; type-2 astrocytes were further characterized as positive to antibodies of cell adhesion glycoproteins J1, NSP-4.[11,13] In such studies, the majority of astrocytes in the optic nerve were found to be type-1 astrocytes (GFAP$^+$, A2B5$^-$) and developed at embryonic day 16, whereas type-2 astrocytes (GFAP$^+$, A2B5$^+$) were first observed 8–10 days postnatal; oligodendrocytes (GC$^+$, GFAP$^-$) developed at birth.[5–7,10] In culture conditions, cells that are GC$^-$, GFAP$^-$, A2B5$^+$ may be induced to develop into type-2 astrocytes (GC$^-$, GFAP$^+$, A2B5$^+$) or oligodendrocytes (GC$^+$, GFAP$^-$, A2B5$^-$), which suggested a common bipotential O-2A progenitor giving rise to both cell types *in vivo.*[5–7,10,12] However,

A2B5 is not diagnostic for type-2 astrocytes *in vitro* (Fulton *et al.,* this volume), and the use of antigenic phenotypes in the rat optic nerve *in vivo* is even more ambiguous because the antibody to A2B5 cannot be used.[15] The large numbers of GFAP+, A2B5+ astrocytes observed in the center of the rat optic nerve and the GFAP+, A2B5− cells associated with the pia[8] do not correspond to type-2 astrocytes and type-1 astrocytes as originally described, and their developmental timetables and functions determined with A2B5 have been questioned. Protoplasmic and fibrous astrocytes cannot be defined by their binding of the A2B5 antibody *in vitro* or *in vivo,*[15] and the association of type-2 astrocytes with the nodes of Ranvier[11,13,16] can no longer be supported because neither A2B5 itself nor the antibodies to adhesion molecules at the nodes of Ranvier are diagnostic for type-2 astrocytes (Fulton *et al.,* this volume). Evidence that glial scars,[9,14] the glial limiting membranes,[8,16] and formation of the blood-brain barrier[15] are functions solely of type-1 astrocytes must also be reconsidered. The use of cobalt to label cells that have glutamate receptor-channels, a characteristic *in vitro* of an O-2A lineage, may help to distinguish these cells *in vivo* (Fulton *et al.,* this volume), but glutamate receptors have also been shown on oligodendrocytes by Kettenman *et al.* (this volume), so this technique may not prove definitive.

DEVELOPMENT OF THE O-2A CELL LINE *IN VIVO*

The foremost evidence for O-2A precursors and the development of type-2 astrocytes *in vivo* is the presence of cells with the A2B5 antigen in cell suspensions and cultures from the developing and adult optic nerve,[6,7,9,12] but whether cells with an O-2A antigenic phenotype develop into both oligodendrocytes and astrocytes *in vivo* is disputed (see Noble and colleagues, this volume). Certainly, in classical electron microscopic studies of gliogenesis in the rat optic nerve there was no evidence for an increase in glial precursors or immature astrocytes in the second week after birth,[3,4] which could correspond to the proposed development of type-2 astrocytes from O-2A progenitors.[5–16] Skoff and Knapp (this volume), using a combination of [^{3}H]-thymidine autoradiography, immunocytochemistry, and electron microscopic analysis in the postnatal rat optic nerve, did not detect a bipotential precursor or observe a second wave of astrocyte production. It was concluded that precursors that fit an O-2A antigenic phenotype (positive to GFAP and sulfatide) are in fact all oligodendroblasts and develop into oligodendrocytes *in vivo.* Undoubtedly, precursor cells that express an O-2A antigenic phenotype are capable of developing into astrocytes and oligodendrocytes in both an *in vitro* and *in vivo* environment (e.g. see this volume for Noble *et al.*), but no direct evidence exists that a bipotential glial precursor develops postnatally into oligodendrocytes and astrocytes *in vivo.*

In short, since astrocytes, oligodendrocytes, and precursors can express a diverse range of immunocytochemical and physiologic characteristics in different culture systems, these traits do not define these cells *in vivo.* The outstanding evidence that type-2 astrocytes or O-2A progenitors are present in the rat optic nerve *in vivo* is that cell suspensions contain cells with the antigenic phenotypes that are characteristic but not diagnostic of these cell types in culture. If type-2 astrocytes are present *in vivo,* they are small in number, they are almost certainly not uniquely responsible for perinodal astrocytic processes, and unequiviocal evidence is needed to prove they develop from a separate astrocytic line.

OLIGODENDROCYTE MORPHOLOGY IN THE INTACT CNS

A picture of glial morphology and development *in vivo* has been obtained in the isolated intact rat optic nerve using intracellular injection of Lucifer yellow or horseradish peroxidase[19,20] (Butt and Ransom, this volume). The two distinguishable cell types observed, interfascicular oligodendrocytes and fibrous astrocytes,[19] were confirmed as such using electron microscopic criteria.[20] Mature oligodendrocytes had a highly stereotypic morphology (FIG. 1), with 15–20 parallel processes running for 200–300 μm exclusively along the long axis of the optic nerve and being connected near their midpoints to the cell body by thin branches 15–30 μm long. The longitudinal processes are the internal and external tongue processes of the myelin sheath, indicating the internodal distance, and they never branched or ended on blood vessels or the pia, but often terminated in fine rings which are the paranodal loops observed in electron microscopic studies.[19,20] In the rat optic nerve, oligodendrocytes may be readily identified by this characteristic morphology, and it is anticipated that oligodendrocytes in other regions of the CNS may be similarly recognizable, but those in regions where axons pass in many directions will have a correspondingly nonuniform orientation of processes.

ASTROCYTE DIVERSITY *IN VIVO*

Dye-filled astrocytes in the rat optic nerve displayed a heterogeneous morphology, possibly indicating a divergence of function and, if anything, suggesting a broad range of astrocytic form rather than two distinct subtypes (FIG. 2). In general, astrocytes had the typical morphology of fibrous astrocytes as defined by metal impregnation studies,[1] with as many as 50–60 fine processes up to 400–600 μm in length extending radially from the cell body. All astrocytes had processes that terminated in end-feet at the pial surface of the nerve and on blood vessels, whereas

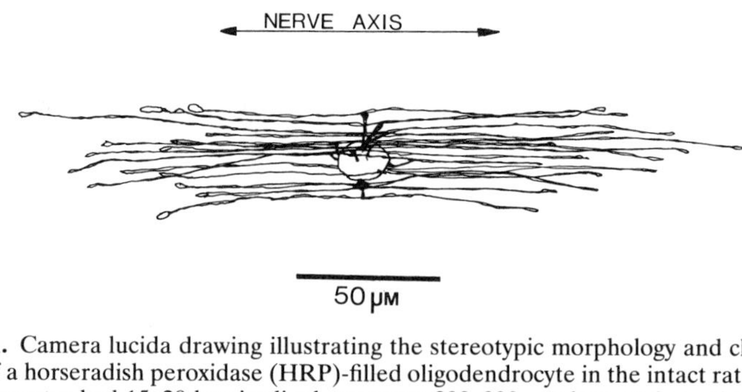

FIGURE 1. Camera lucida drawing illustrating the stereotypic morphology and characteristic features of a horseradish peroxidase (HRP)-filled oligodendrocyte in the intact rat optic nerve. Oligodendrocytes had 15–20 longitudinal processes 200–300 μm long and running exclusively along the nerve axis parallel to the axons. The longitudinal processes never branched or terminated at blood vessels or at the pia, but often ended in fine terminal rings, and were connected to the cell body by fine processes 15–30 μm long. The longitudinal processes were composed of two tightly intertwined processes corresponding to the internal and external tongue processes of the myelin sheaths. Each oligodendrocyte, therefore, provided myelin sheaths with an internodal distance of 200 μm for around 20 axons within 30 μm of the cell body.

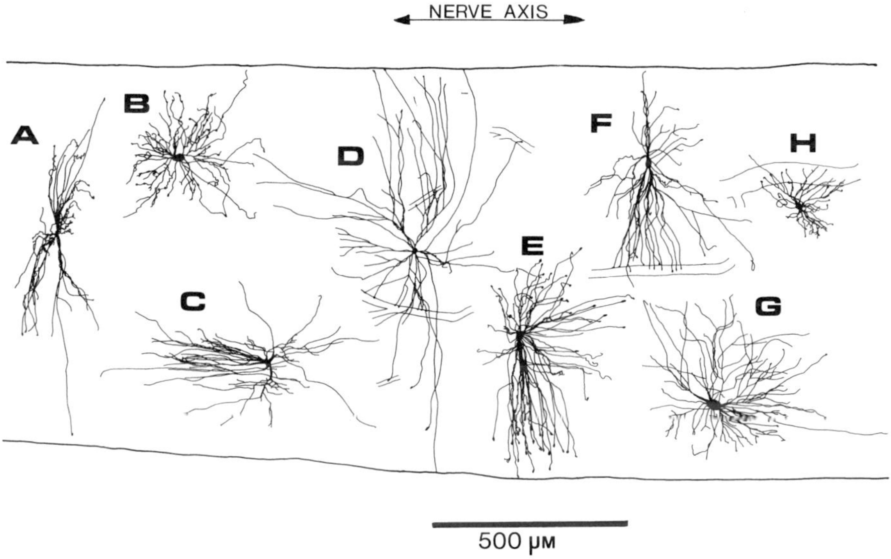

FIGURE 2. Camera lucida drawings illustrating the heterogeneous morphology of HRP-filled astrocytes in the 21–30-day-old rat optic nerve. It is important to recognize that the optic nerve is a cylinder and that the three-dimensional morphology of the cells is necessarily compressed into two dimensions, as viewed from one side through the pial surface of the nerve. Astrocytes could be distinguished as passing processes with a radial (B, G, and H), transverse (A, E, and F), or longitudinal (C) orientation and associated primarily with the pia (A) and blood vessels (F and H) or to passing processes that ended freely within the nerve parenchyma (C and G); the latter may correspond to perinodal processes described in electron microscopic studies. All astrocytes had processes that ended at the pia or on blood vessels, but only radial and longitudinal astrocytes had processes that terminated within the parenchyma. Radial and longitudinal astrocytes were further characterized by branching processes, whereas transverse astrocytes characteristically had processes that did not branch or end freely within the nerve. Astrocytes could also be extremely complex and defy classification (D).

some astrocytes also had processes that ended freely within the nerve parenchyma. It was postulated[19] that these latter processes may correspond to the perinodal processes observed in electron microscopic studies (currently being investigated). The majority of astrocytes in the rat optic nerve had a radial symmetry and passed dichotomously branching processes in all directions, which terminated in end-feet at the pial surface and on blood vessels or ended freely in the nerve (FIG. 2, cells B, G, and H). A second type of astrocyte characteristically passed long straight processes that did not branch and that ended exclusively in end-feet at pia or on blood vessels, and these cells had a distinct transverse orientation when viewed from one side (FIG. 2, cells A, E, and F). A third type of astrocyte passed the majority of processes parallel to the axons. Some of the processes were many hundreds of micra long (Butt and Ransom, unpublished observation), giving the cells a longitudinal appearance, and these processes ended mainly within the parenchyma (FIG. 2, cell C). This last type of cell may correspond to those cells identified with cobalt by Fulton and colleagues (this volume). These observations in the intact nerve suggest that some astrocytes may be particularly suited for structural support (FIG. 2, cell A), others for transport between blood and axon (FIG. 2, cells F and H) or pia and axon (FIG. 2, cells B and E), and still others for providing the perinodal processes (FIG. 2, cells C and G). Whether the variation in form represents different subtypes of astrocyte is not known, and the presence of cells with combinations of these various attributes (FIG. 2, cell D) indicates a less than absolute delineation.

ASTROCYTE DEVELOPMENT *IN VIVO*

Developmental studies did not indicate separate cell lineages during postnatal development; no evidence for a second wave of astrocytic development was noted in the second week, and a continuous sequence of astrocyte and oligodendrocyte differentiation could be traced through this period (Butt and Ransom, this volume). Immature radial and transverse astrocytes were present at birth and throughout the first postnatal weeks and were easily distinguished by their orientation and the presence or absence of branched processes, but the way they proceeded to develop appeared to be related to age and to the position of the cell within the nerve; in this way some radial cells just beneath the surface of the nerve passed processes entirely to the near pial surface (FIG. 2, cell B), whereas others that developed at a later age and deeper within the nerve were more closely associated with blood vessels (FIG. 2, cell H), and some had almost exclusively parenchymal processes (FIG. 2, cell G). The different astrocytic forms and the developmental changes in morphology that were observed in the intact rat optic nerve may signify a large number of separate cell types and lineages, but it is more likely that they represent the growth response of astroblasts to local conditions which themselves change as the nerve develops. Nevertheless, the distinction of radial cells with their branching processes, perivascular and subpial end-feet, and parenchymal processes, from transverse cells that do not have branching or parenchymal processes, together with the fact that both types were present at birth and therefore exposed to the same external growth stimuli, suggests at least two possible subtypes in the rat optic nerve. Cells with a longitudinal orientation were not detectable until the second week, but from morphology alone it is not obvious whether this is a third distinct subtype or the response of radial-like astroblasts to the changing conditions in the nerve.

SUMMARY

The characterization of macroglial antigenic phenotypes has provided considerable information on the classification of glial subtypes and lineage, but the picture *in vivo* appears to more complex than originally surmised from *in vitro* systems. Whether type-2 astrocytes and O-2A progenitors play a significant part *in vivo* remains controversial, but these questions have generated a valuable reassessment of inferences made from *in vitro* studies. It is not known whether disparate regions of the nervous system have different glial cell types with distinct cell lineages, but results in peripheral nerve and the cortex (see Price and colleagues, and Jessen and colleagues, this volume) suggest that this is likely the case. The heterogeneity of astrocytic form and the implied divergence of function *in vivo* parallel but have yet to be equated with the wide range of antigenic and electrophysiologic characteristics that can be induced in glia under certain culture conditions *in vitro.* Whether similar conditions prevail within the developing nervous system is not clear. The continued development of immunocytochemical markers and the use of retroviral vectors, combined with morphologic and electrophysiologic studies *in vivo* and *in vitro,* will undoubtedly answer these and other questions.

REFERENCES

1. Ramón y Cajal, S. 1909–1911. Histologie du Système Nerveux de l'Homme et des Vertébrés. Consejo Superior de Investigaciones Cientificas, Madrid.
2. Del Rio-Hortega, P. 1928. Tercera aportación al conocimiento morfologica e interpretación functional de la oligodendroglia. Mem. Real Soc. Espan. Hist. Nat. **14:** 5–122.
3. Skoff, R., D. Price & A. Stocks. 1976a. Electron microscopic autoradiographic studies of gliogenesis in the rat optic nerve. 1. Cell proliferation. J. Comp. Neurol. **169:** 291–312.
4. Skoff, R., D. Price & A. Stocks. 1976b. Electron microscopic autoradiographic studies of gliogenesis in the rat optic nerve. 2. Time of origin. J. Comp. Neurol. **169:** 313–333.
5. Raff, M. C., E. R. Abney, J. Cohen, R. Lindsay & M. Noble. 1983. Two types of astrocytes in cultures of developing rat white matter: Differences in morphology, surface gangliosides, and growth characteristics. J. Neurosci. **3:** 1289–1300.
6. Raff, M. C., R. H. Miller & M. Noble. 1983. A glial progenitor cell that develops *in vitro* into an astrocyte or an oligodendrocyte depending on culture medium. Nature **303:** 390–396.
7. Raff, M. C., E. R. Abney & R. H. Miller. 1984. Two glial cell lineages diverge prenatally in rat optic nerve. Dev. Biol. **106:** 53–60.
8. Miller, R. H. & M. C. Raff. 1984. Fibrous and protoplasmic astrocytes are biochemically and developmentally distinct. J. Neurosci. **4:** 585–592.
9. David, S., R. H. Miller, R. Patel & M. C. Raff. 1984. Effects of neonatal transection on glial cell development in the rat optic nerve: Evidence that the oligodendrocyte-type-2 astrocyte cell lineage depends on axons for its survival. J. Neurocytol. **13:** 961–974.
10. Miller, R. H., S. David, R. Patel, E. R. Abney & M. C. Raff. 1985. A quantitative immunocytochemical study of macroglial cell development in the rat optic nerve: *In vivo* evidence for two distinct glial cell lineages. Dev. Biol. **111:** 35–41.
11. ffrench-Constant, C., R. H. Miller, J. Kruse, M. Schachner & M. C. Raff. 1986. Molecular specialization of astrocyte processes at nodes of Ranvier in rat optic nerve. J. Cell Biol. **102:** 844–852.
12. ffrench-Constant, C. & M. C. Raff. 1986a. Proliferating bipotential glial progenitor cells in adult rat optic nerve. Nature **319:** 499–502.
13. ffrench-Constant, C. & M. C. Raff. 1986b. The oligodendrocyte-type-2 astrocyte cell lineage is specialized for myelination. Nature **323:** 335–338.
14. Miller, R. H., E. R. Abney, S. David, C. ffrench-Constant, R. Lindsay, R. Patel, J. Stone & M. C. Raff. 1986. Is reactive gliosis a property of a distinct subpopulation of astrocytes? J. Neurosci. **6:** 22–29.
15. Miller, R. H., C. ffrench-Constant & M. C. Raff. 1989. The macroglial cells in the rat optic nerve. Ann. Rev. Neurosci. **12:** 517–534.
16. Miller, R. H., B. P. Fulton & M. C. Raff. 1989. A novel type of astrocyte contributes to the structure of the nodes of Ranvier in rat optic nerve. Eur. J. Neurosci. **1:** 172–180.
17. Bignami, A., L. F. Eng, D. Dahl & C. T. Uyeda. 1972. Localization of the glial fibrillary acidic protein in astrocytes by immunofluorescence. Brain Res. **43:** 429–435.
18. Raff, M. C., R. Mirsky, K. L. Fields, R. P. Lisak, S. H. Dorfman, D. H. Silberberg, N. A. Gregson, S. Liebowitz & M. C. Kennedy. 1978. Galactocerebroside is a specific cell surface antigenic marker for oligodendrocytes in culture. Nature **274:** 813–816.
19. Butt, A. M. & B. R. Ransom. 1989. Visualization of oligodendrocytes and astrocytes in the intact rat optic nerve by intracellular injection of Lucifer yellow and horseradish peroxidase. Glia **2:** 470–475.
20. Ransom, B. R., A. M. Butt & J. Black. 1991. Ultrastructural identification of HRP injected oligodendrocytes in the intact rat optic nerve. Glia **4:** 37–45.

Glial Cells in Development

In Vivo and *In Vitro* Approaches

PASKO RAKIC

Section of Neurobiology
Yale University School of Medicine
New Haven, Connecticut 06510

The term "glia," meaning glue (bindesubstanz), given initially to nonneuronal cells by Rudolph Virchow in the mid-19th century, reflected the widespread notion that these elements are created to subserve the neurons, the "noble" elements of the brain. However, the days when the only role of glial cells was thought to be isolation of individual neurons in the adult central nervous system are long over. Equally invalid proved to be misconceptions ingrained until recently, even in some reputable embryology textbooks, that glial cells emerge only after neurons of a given structure have been generated and have settled in their final positions. In the past two decades it has gradually become established that glial cells play several important roles in the development of both the central and the peripheral nervous system (reviewed in refs. 1 and 2). Furthermore, several lines of evidence indicate that genetically or environmentally induced disturbances of glial cells or neuronal-glial interactions during critical periods lead to various molecular, structural, or functional abnormalities of the brain (e.g., ref. 3). It is also apparent that the astroglial response to brain injury, plasticity, and aging is more multifaceted than initially assumed and considerable progress can be expected in this area of research.[4,5] Because a majority of cells in the mammalian brain are of glial lineages, it is safe to estimate that each neuron is likely to be in direct contact with at least one glial cell during development and probably with several glial cell types during its lifespan. How do these contacts operate, when do the functionally meaningful interactions begin, and what precisely is their its role in the various stages of development?

The first essential step in understanding these issues is to demonstrate that glial cells are indeed generated early enough in embryonic life to be able to affect the proliferation, migration, and differentiation of neurons. Although originally Wilhelm His, on the basis of histologic material made from human embryos, suggested that neurons and glial cells are present side by side from the first trimester of gestation,[6] the hard experimental proof was lacking. The introduction of glial fibrillary acidic protein (GFAP) as the glial-specific cell marker has provided the first direct evidence that a specialized type of glial cell, radial glia, coexists from the onset of neurogenesis in a given structure of the primate brain.[7,8] The application of the same antibody to human postmortem material revealed the coexistence of neurons and glial cells at correspondingly early embryonic ages.[9] Since then, use of numerous cell-specific molecular markers has confirmed the early emergence of glial cell lines and further showed a great diversity of glial cell types during embryonic ages (reviewed in ref. 10). These findings, however, made the task of understanding their role in development even more difficult, because each transient cell phenotype may have a somewhat different function.

Once it was recognized that glial cells are present during the entire course of neural development, the list of possible functions increased rapidly, and I will mention only the most compelling and obvious examples. For instance, it is suggested that contact between neurons and glial cells in the embryo controls the number of various classes of cells and determines their correct proportions. Furthermore, it has been postulated that such contacts may also regulate the appropriate deployment of glial cells, which subsequently may provide boundaries for compartmentalization or play a role in guiding neuronal migration. Similar contacts have been implicated in the induction of specific neural phenotypes, guidance of growth cones, ensheathement of neuronal somas or dendritic spines by astrocytes, or wrapping of their axons by myelin-producing oligodendrocytes. Glial cells are also thought to be essential metabolic intermediaries between endothelial cells of the capillaries and neurons alone or as a participant in the blood-brain barrier. The processes of glial cells are also implicated as intermediary elements in the rearrangement and elimination of synaptic junctions that are an important step in competitive synaptic elimination and neuronal plasticity. Finally, glial membranes are recognized as being actively involved in exchange of nutrients, neurotransmitters, second messengers, ions, and various growth factors, all of which are critical processes of normal brain development.

In contrast to the numerous, relatively well described, developmental, and pathologic phenomena, we are still ignorant about the cellular and molecular mechanisms underlying any of these processes. To start with, we are not even in agreement about the basic classifications of glial cells, phenotypic origin of various glial cell subtypes, and their lineage relationships. The task, however, is not going to be easy because heterogeneity of glial cells and regional differences in their distribution are larger than expected just a few years ago. For example, the findings obtained from the rodent optic nerve do not necessarily apply to chick optic tectum or primate cerebral cortex. Until we learn more about molecular mechanisms that control glial cell proliferation, direct their proper deployment, and regulate their differentiation and death in various brain regions, many generalizations are premature. Simultaneously, we also have to learn about the mechanisms involved in glial response to injury during developmental periods and the normal process of aging.

An essential step in the progress of our understanding of glial place in neurogenesis has been the development of primary astrocyte culture systems. The possibility of separating and maintaining individual glial cell lines *in vitro* has enabled investigators, for the first time, to explore and experimentally manipulate differentiation and the relation of various glial types and neurons as well as their involvement in the pathogenesis of a number of genetic and acquired disorders. However, despite the enormous progress made in the field through the use of cell culture, we should never lose sight of the fact that the final test of any hypothesis of glial function has to be performed within the complexity of the *in vivo* situation. The presentations in this session, however, provide examples of the rapid progress in this field that is currently underway by the application of new methods of cellular and molecular biology in both *in vitro* and *in vivo* situations.

One of the most fruitful areas of research in the last two decades has been based on the assumption that glial cell surface-bound molecules provide substrates that can guide the migration of postmitotic neurons and direct ingrowth of their axons into the proper positions in respect to target cells. Initially, the basic concept of differential cell adhesion was based on the dynamic interpretation of ultrastructural observation of the close association between neuronal and glial cells during development[11,12] (reviewed in refs. 13 and 14). However, in the past decade research has expanded to a new level of inquiry that uses both *in vivo* and *in vitro* approaches. For example, the work presented from Tomaselli and Schacher's labs demonstrates how individual neurons sequentially, or in some instances simultaneously, express multiple receptors, any of which can independently or in combination with others regulate neurite outgrowth along the membrane surface of glial cells (e.g., refs. 15 and 26). These studies provide an instructive example that differentiation of glial and nerve cells can now be characterized by the sequential expression of various molecules that may act as the signals, or by receptors engaged in control of cell proliferation or growth, including the onset and completion of myelinization of axons (e.g., refs. 16 and 17).

Considerable progress has been made in understanding the molecular mechanisms of neuronal migration during normal and genetically altered brain development. The studies of Hatten and her associates show how surface-bound macromolecules become expressed in a specific spatiotemporal pattern constellation in various parts of the mammalian central nervous system. An important advance was made when they began testing *in vitro* the function of various molecules as recognition and adhesion vehicles that might guide cell movement.[18,19] So far, the candidate molecules for this function have been localized on the surface of neural cells, but it is expected that receptor-like complementary elements may be present on the membrane of the glial surface. An understanding of the migratory process can explain certain congenital defects of brain organization, because radial glial fibers apparently not only guide neuronal movement, but also may provide boundaries during compartmentalization of brain modules containing an assembly of neurons that in the mature organism function as separate functional entities.[20] The work of Steindler and his associates is an instructive example of how glycoconjugates associated with radial glial cells may serve as boundaries or barriers to developing neurons and neurites and can eventually lead to their segregation into physiologically definable units in both cerebral cortex and neostriatum.[21]

In recent years it has become apparent that specific growth factors and trophic molecules can influence the multiplication of glial cells and trigger their differentiation into given cell lines (e.g., ref. 22). Richardson and his colleagues in this symposium describe several factors that, at least *in vivo,* can influence the conversion of D-2A progenitor cells into oligodendrocytes, the main myelin-forming cell class in the mammalian brain. On a broader scale, the promising aspect of growth factor research has been inquiring into their possible association with various protooncogenes and tumor suppression genes. Characterizing such cellular events may provide insight into the pathogenesis of malignancy as well as some demyelinization disorders and could lead to suggestions of possible repair mechanisms.

Finally, we are reminded that despite considerable functional, structural, and molecular differences between nonneuronal elements in various species, the basic principles of neural-glial involvement in development may be preserved throughout evolution. Klämbt and his colleagues in Goodman's laboratory provide elegant examples that at least three different classes of glial cells in fruit fly *Drosophila* could prefigure and later ensheathe the major classes of axon tracts, thereby providing a substrate for their formation and possibly cues for positional information.[23] It may not be a coincidence that the similar role of glial cells in the formation of interhemispheric connections has been suggested by Silver and his colleagues who studied the absence of corpus callosum in a mutant mouse.[24] In this respect, the basic concept of the role of nonneuronal cells in insects may not be different from that postulated for mammals, including the highest primates like human. Some morphoregulatory molecules actually are generated by homologous genes in species as diverse as fruit fly and man. However, in many cases the structural details, composition, and nature of the recognition and adhesion molecules may have been altered during evolution.[25] The pattern and propensities of these modifications can provide clues about evolution of the nervous system. Therefore, the study of species at various levels of glial cell elaboration needs to be explored so that from the differences and similarities we may learn about basic biologic principles underlying neural development.

The five chapters that follow exemplify the richness, diversity, and excitement of contemporary research on neuron glial interactions *in vivo* and *in vitro,* and guarantee a bright and promising future for this fast-moving field. However, they also remind us of how much more needs to be done before we understand the complex interrelationship between glial and neural cells during development.

REFERENCES

1. SEARS, T. A. 1981. Neuronal-Glial Cell Interrelationships. Springer.
2. VARON, S. & G. G. SOMJEN. 1979. Neurosci. Res. Prog. Bull. **17:** 9–239.
3. NORENBERG, M. D., L. HERTZ, & A. SCHOUSBOE. 1988. The Biochemical Pathology of Astrocytes. Liss. New York.
4. HATTEN, M. E., R. K. H. LIEM, M. L. SHELANSKI & C. A. MASON. 1991. Glia **4:** 233–243.
5. REISMAN, G. & P. M. FIELD. 1973. Brain Res. **50:** 241–264.
6. HIS, W. 1989. Abhandl. Math. Phys. Konigl. Sach. Ges. Wiss **26:** 313–372.
7. LEVITT, P. & P. RAKIC. 1980. J. Comp. Neurol. **193:** 815–840.
8. LEVITT, P., M. L. COOPER & P. RAKIC. 1991. J. Neurosci. **1:** 27–39.
9. CHOI, B. H. 1986. J. Neuropathol. Exp. Neurol. **45:** 408–418.
10. CAMERON, R. & P. RAKIC. 1991. Gila **4:** 124–137.
11. RAKIC, P. 1971. J. Comp. Neurol. **141:** 283–312.
12. RAKIC, P. 1972. J. Comp. Neurol. **145:** 61–84.
13. RAKIC, P. 1990. Experientia **46:** 882–891.
14. SNOW, D. M., V. LEMMON, D. A. CARRINO, A. I. CAPLESS & J. SILVER. 1990. Exp. Neurol. **109:** 111–130.
15. TOMASELLI, K. J. *et al.* 1988. J. Cell Biol. **107:** 353–362.
16. JESSEL, T. M., P. BOVOLENTA, M. PLACZEK, M. TESSIER-LAVINGE & J. DOOD. 1989. Ciba Foundation Symp. **144:** 255–280.
17. RICHARDSON *et al.* 1990. Sem. Neurosci. In press.
18. EDMONDSON, J. C., R. K. LIEM, J. E. KUSTER & M. E. HATTEN. 1988. J. Cell Biol. **106:** 505–517.
19. HATTEN, M. E. & C. A. MASSON. 1990. Experientia **46:** 907–916.
20. RAKIC, P. 1988. Science **241:** 170–176.
21. STEINDLER, D. A., A. FAISSNER, JR., & M. SCHACHNER. 1989. Comm. Dev. Neurobiol. **1:** 29–60.
22. LIM, R., W. ZHANG, & A. ZAHER. 1990. Cell Regulation **1:** 744–746.
23. KLÄMBT, C., J. R. JACOBS, & C. S. GOODMAN. 1991. Cell. In press.
24. SILVER, J., S. E. LORENZ, D. WAHLSTEN & J. COUGHLIN. 1982. J. Comp. Neurol. **210:** 10–29.
25. KLÄMBT, C. & C. S. GOODMAN. 1991. Glia **4:** 205–213.
26. SEILHEIMER, B. & M. SCHACHNER. 1988. J. Cell Biol. **107:** 431–451.

β_1-Integrin-Mediated Neuronal Responses to Extracellular Matrix Proteins

KEVIN J. TOMASELLI

Athena Neurosciences, Inc.
800F Gateway Boulevard
South San Francisco, California 94080

Patterns of neuronal connectivity are initially established in the embryonic nervous system, when neurons extend their first axons and dendrites. Neuronal process outgrowth is not random, but is shaped by the action of many regulatory factors, both intrinsic and extrinsic, on the growth cone, the motile tip of the elongating nerve fiber. Extrinsic factors include molecules secreted by neurons or nonneuronal cells that are either deposited in intercellular spaces or remain associated with the surfaces of cells. Neurotransmitters, peptide growth factors, and extracellular matrix (ECM) molecules are examples of the former, whereas transmembrane or glycolipid-linked cell-to-cell adhesion molecules provide examples of the latter. Extrinsic factors can exert a range of effects on growth cone motility including inhibition (e.g., the action of some neurotransmitters on certain types of neurons) or stimulation (e.g., the ECM molecule, laminin). Factors that influence the extent and orientation of neuronal process outgrowth are thought to bind to specific receptors on the surfaces of growth cones, initiating a signal cascade that ultimately modifies the growth cone motile apparatus. In this paper, I present our studies aimed at identifying neuronal receptors for ECM molecules.

Several approaches have been taken to studying ECM molecules that regulate neuronal process outgrowth and the neuronal receptors through which they act. Individual approaches usually have their limitations, and conclusions are therefore strengthened by simultaneous use of different approaches. One approach is to culture neurons on substrates coated with purified ECM molecules and to assay the time course or extent of process outgrowth. With this approach many ECM molecules that stimulate neurite outgrowth have been described (reviewed in ref. 1). A second approach, antibody perturbation of neuronal responses to purified ECM molecules, has been particularly useful in identifying possible neuronal receptors that mediate the effects of neurite outgrowth-promoting factors. However, inasmuch as antibodies can exert nonspecific effects, it is necessary to demonstrate biochemically that a proposed interaction between a ligand and a receptor pair actually occurs under physiologic conditions. Antibody perturbation studies were the first to implicate β_1-class integrins as receptors mediating neuronal interactions with the ECM glycoproteins laminin, fibronectin, and type IV collagen. In early experiments, monoclonal antibodies (CSAT and JG22) to the integrin β_1 subunit inhibited neuronal attachment and neurite outgrowth in response to these ECM molecules.[2,3] Interestingly, β_1 subunit MAbs could mimic the ECM ligand and directly stimulate neurite outgrowth when adsorbed to a tissue culture substrate.[4] However, because there are seven characterized $\alpha\beta_1$ integrin heterodimers, some with overlapping ligand specificities (TABLE 1), the inhibitory effects of β_1 subunit MAbs could not be ascribed to a particular $\alpha\beta_1$ integrin. Recently, studies using α subunit-specific antibodies in addition to biochemical studies have led to the identification of specific $\alpha\beta_1$ integrins that function in the responses of neurons and neuronal cell lines to ECM glycoproteins.

The PC12 cell line is a pheochromocytoma-derived line that, upon exposure to nerve growth factor (NGF), differentiates into neuron-like cells. Undifferentiated PC12 cells readily attach to and, in the presence of NGF, extend neurites on substrates coated with purified EHS sarcoma-derived laminin or type IV collagen. Attachment and neurite outgrowth in response to each of these ECM components is nearly completely inhibited by antibodies that recognize the rat integrin β_1 subunit.[5] A mixture of at least two $\alpha\beta_1$ integrin heterodimers, purified from detergent extracts of PC12 cells by antibody affinity chromatography, will bind to laminin and collagen after reconstitution into liposomes.[6] These biochemical studies provide direct evidence that β_1 integrin antibodies block PC12 cell attachment and neurite outgrowth on laminin and collagen by interfering with the functions of molecules that can bind directly to these ECM ligands. Laminin affinity chromatography experiments demonstrated that one of these β_1 integrins, a 140-kD (β)/200-kD (α) dimer, expressed also on the B50 neuroblastoma cell line, binds in a Mg^{2+}-dependent way to laminin.[7] This

TABLE 1. β_1 Integrins and Their Ligands[a]

Integrin	Ligand(s)
$\alpha_1\beta_1$	Laminin, collagen
$\alpha_2\beta_1$	Laminin, collagen
$\alpha_3\beta_1$	Laminin, collagen, fibronectin
$\alpha_4\beta_1$	Fibronectin, VCAM-1
$\alpha_5\beta_1$	Fibronectin
$\alpha_6\beta_1$	Laminin
$\alpha_v\beta_1$	Vitronectin

[a]Seven $\alpha\beta_1$ integrins identified thus far are listed with their known ligands. Additional β_1 integrins will surely be identified in the future as will additional ligands for those integrins listed.

integrin, identified as $\alpha_1\beta_1$ (to be discussed), partly accounts for the previously described dependence on Mg^{2+} of PC12 cell attachment to laminin.[8]

Integrin α subunit-specific antibodies have been used to further dissect PC12 cell integrin expression and ligand recognition. The two major integrins expressed by PC12 cells are $\alpha_1\beta_1$ and $\alpha_3\beta_1$.[9] The $\alpha_1\beta_1$ integrin is recognized by the 3A3 monoclonal antibody, first identified by its ability to inhibit attachment to laminin and collagen.[10,11] Cloning of the cDNA encoding the α subunit recognized by MAb 3A3 demonstrated the greatest sequence similarity to the human α_1 subunit cDNA.[12] $\alpha_1\beta_1$ appears to be the major, if not the only, β_1 integrin receptor for collagen types I and IV on PC12 cells.[9] Although $\alpha_1\beta_1$ is also a laminin receptor on PC12 cells, it is not the only laminin-binding β_1 integrin. Maximal concentrations of MAb 3A3 inhibit PC12 cell neurite outgrowth on laminin by only 60–70% (TABLE 2 and ref. 9). The $\alpha_3\beta_1$ integrin appears to account for some of this residual response to laminin. Alpha$_3$-specific antibodies, although not inhibitory on their own, substantially block neurite outgrowth on laminin when $\alpha_1\beta_1$ function is maximally inhibited (TABLE 2 and ref. 9). Thus, PC12 cells express two β_1 integrins that together mediate responses to laminin.

Previous studies on fragments of laminin, generated by limited proteolysis with elastase, identified two spatially separated neurite outgrowth-promoting domains: one in fragment E1–4, derived from the top of the laminin cross, and one in fragment

TABLE 2. Effects of Integrin α Subunit-Specific Antibodies on PC12 Cell Neurite Outgrowth on Laminin and Collagen IV[a]

Substrate	% Cells with Neurites			
	Control	Anti-$\alpha_1\beta_1$	Anti-$\alpha_3\beta_1$	Anti-$\alpha_1\beta_1$ + Anti-$\alpha_3\beta_1$
Laminin	88 ± 2	35 ± 5	86± 4	7 ± 2
Collagen IV	91 ± 1	3 ± 1	89 ± 3	—

[a]PC12 cell neurite outgrowth on laminin or collagen IV in the presence of MAb 3A3 (anti-$\alpha_1\beta_1$) or polyclonal anti-$\alpha_3\beta_1$ was measured as described by Tomaselli *et al.*[9] Values are the average and range of determinations made on duplicate cultures run in parallel.

E8, derived from the long arm.[13] PC12 cells attach to and extend neurites on either fragment in a β_1 integrin-dependent fashion. Responses to each fragment, however, are mediated by a different β_1 integrin. Anti-$\alpha_1\beta_1$ strongly inhibits responses to fragment E1–4 without noticeably affecting interactions with fragment E8. Conversely, anti-$\alpha_3\beta_1$ strongly inhibits interactions with fragment E8, with little effect on fragment E1–4.[9] These results demonstrate that laminin contains spatially distinct binding sites that are selective for different integrins. These studies are consistent with those on nonneuronal cells that have mapped a binding site for $\alpha_1\beta_1$ to fragment E1–4 and binding sites for $\alpha_3\beta_1$ and $\alpha_6\beta_1$, another laminin-binding integrin, to fragment E8.[14–16]

The $\alpha_1\beta_1$ integrin is most similar to the $\alpha_2\beta_1$ integrin in protein sequence, ligand specificity, and cation dependence.[12] Interestingly, $\alpha_2\beta_1$ ligand specificity appears to be regulated by cell type-specific factors. Thus, $\alpha_2\beta_1$ on some cells recognizes both laminin and collagen, but in others recognizes only collagen.[17,18] Studies on the role of $\alpha_1\beta_1$ in the response of rat sympathetic neurons to laminin and collagen suggest that, like $\alpha_2\beta_1$, the ligand specificity of $\alpha_1\beta_1$ varies in different cell types. When neonatal rat sympathetic neurons are cultured on EHS laminin substrates, anti-$\alpha_1\beta_1$ has no noticeable inhibitory effect on neurite outgrowth. In contrast, anti-$\alpha_3\beta_1$ has a strong, but incomplete, inhibitory effect on laminin (TABLE 3). Two observations demonstrate that the effect of anti-$\alpha_1\beta_1$ is lacking not because these neurons fail to express $\alpha_1\beta_1$. First, substantial amounts of $\alpha_1\beta_1$ can be immunoprecipitated from labeled extracts of >95% pure cultures of sympathetic neurons.[19] Second, anti-$\alpha_1\beta_1$ does block sympathetic neurite outgrowth in response to type IV collagen, demonstrating that these neurons express functional $\alpha_1\beta_1$ integrins on their surface (TABLE 3). Similar observations have been made on rat dorsal root ganglion neurons.[20] Thus, in two primary rat neuronal cell types, $\alpha_1\beta_1$ appears to function as a receptor for

TABLE 3. Effects of Integrin α Subunit-Specific Antibodies on Rat Sympathetic Neurite Outgrowth on Laminin and Collagen IV[a]

Substrate	% Neurons with Neurites		
	Control	Anti-$\alpha_1\beta_1$	Anti-$\alpha_3\beta_1$
Laminin	92 ± 3	94 ± 5	42 ± 3
Collagen IV	79 ± 6	3 ± 1	—

[a]Dissociated neonatal rat superior cervical ganglion neurons were cultured 24 hours in the presence of NGF (50 ng/ml) and antibodies to the $\alpha_1\beta_1$ or $\alpha_3\beta_1$ integrins. The percentage of neurons with a neurite greater than one cell diameter in length was counted. Values are the average and range of determinations made on duplicate cultures.

collagen, but not for laminin, whereas in the rat PC12 cell line, $\alpha_1\beta_1$ clearly functions as a laminin receptor.

SUMMARY

Immunologic and biochemical studies demonstrate that β_1-class integrin heterodimers mediate attachment and neurite outgrowth by primary neurons and neuronal cell lines in response to neurite-promoting ECM molecules. Individual cells are likely to express several $\alpha\beta_1$ integrins, each possessing unique ligand specificities. In some cases, a cell may express two β_1 integrins that recognize different binding domains in a single ECM ligand, as appears to be the case for the $\alpha_3\beta_1$ and $\alpha_1\beta_1$ integrins on PC12 cells. Thus, the responses of a neuron to even a single ligand may be regulated by the combined action of several receptors. Finally, preliminary studies suggest that neurons can regulate the ligand specificity of individual integrin heterodimers (e.g., $\alpha_1\beta_1$). Given the tremendous diversity in the extracellular matrix surrounding them, it is not surprising that neurons possess intrinsic factors that dictate if and how they respond to a particular ECM component.

ACKNOWLEDGMENTS

I would like to acknowledge my collaborators on the studies described in this paper: Louis Reichardt, Caroline Damsky, Deborah Hall, David Turner, Sal Carbonetto, and Kurt Gehlsen. Also, I am grateful to Genine Fong and Alicia Hengst for typing the manuscript.

REFERENCES

1. LANDER, A. D. 1987. Molecules that make axons grow. Mol. Neurobiol. **1:** 213–245.
2. BOZYCZKO, D. & A. F. HORWITZ. 1986. The participation of a putative cell surface receptor for laminin and fibronectin in peripheral neurite extension. J. Neurosci. **6:** 1241–1251.
3. TOMASELLI, K. J., L. F. REICHARDT & J. L. BIXBY. 1986. Distinct molecular interactions mediate neuronal process outgrowth on non-neuronal cell surfaces and extracellular matrices. J. Cell Biol. **103:** 2659–2672.
4. HALL, D. E., K. M. NEUGEBAUER & L. F. REICHARDT. 1987. Embryonic neural retina cell response to extracellular matrix proteins: Development changes and effects of the cell substratum attachment antibody, CSAT. J. Cell Biol. **104:** 623–634.
5. TOMASELLI, K. J., C. H. DAMSKY & L. F. REICHARDT. 1987. Interactions of a neuronal cell line (PC12) with laminin, collagen IV and fibronectin: Identification of integrin-related glycoproteins involved in attachment and process outgrowth. J. Cell Biol. **105:** 2347–2358.
6. TOMASELLI, K. J., C. H. DAMSKY & L. F. REICHARDT. 1988. Purification and characterization of mammalian integrins expressed by a rat neuronal cell line (PC12): Evidence that they function as alpha/beta heterodimeric receptors for collagen IV and laminin. J. Cell Biol. **107:** 1241–1252.
7. IGNATIUS, M. J. & L. F. REICHARDT. 1988. Identification of a neuronal laminin receptor: An Mr 200/120 kD integrin heterodimer that binds laminin in a divalent cation-dependent manner. Neuron **1:** 713–725.
8. TURNER, D. C., L. A. FLIER & S. CARBONETTO. 1987. Magnesium-dependent attachment and neurite outgrowth by PC12 cells on collagen and laminin substrates. Dev. Biol. **121:** 510–525.

9. TOMASELLI, K. J., D. E. HALL, L. A. FLIER, K. R. GEHLSEN, D. C. TURNER, S. CARBONETTO & L. F. REICHARDT. 1990. A neuronal cell line (PC12) expresses two beta1-class integrins—alpha1 beta1 and alpha3 beta1—that recognize different neurite outgrowth-promoting domains in laminin. Neuron **5:** 651–662.
10. TURNER, D. C., L. A. FLIER & S. CARBONETTO. 1989. Identification of a cell-surface protein involved in PC12 cell-substratum adhesion and neurite growth on laminin and collagen. J. Neurosci. **9:** 3287–3296.
11. TAWIL, P., M. HOUDE, R. BLACHER, F. ESCH, L. F. REICHARDT, D. C. TURNER & S. CARBONETTO. 1990. Alpha1 beta1 integrin heterodimer functions as a dual laminin/collagen receptor in neural cells. Biochemistry **29:** 6540–6544.
12. IGNATIUS, M. J., T. H. LARGE, M. HOUDE, J. W. TAWIL, A. BARTON, F. ESCH, S. CARBONETTO & L. F. REICHARDT. 1990. Molecular cloning of the rat integrin alpha1-subunit: A receptor for laminin and collagen. J. Cell Biol. **111:** 709–720.
13. EDGAR, D., R. TIMPL & H. THOENEN. 1984. The heparin binding domain of laminin is responsible for its effects on neurite outgrowth and neuronal survival. EMBO J. **3:** 1463–1468.
14. HALL, D. E., L. F. REICHARDT, E. CROWLEY, B. HOLLEY, H. MOEZZI, A. SONNENBERG & C. H. DAMSKY. 1990. The alpha1/beta1 and alpha6/beta1 integrin heterodimers mediate cell attachment to distinct sites on laminin. J. Cell Biol. **110:** 2175–2184.
15. GEHLSEN, K., K. DICKERSON, W. S. ARGRAVES, E. ENGVALL & E. RUOSLAHTI. 1989. Subunit structure of a laminin-binding integrin and localization of its binding site on laminin. J. Biol. Chem. **264:** 19034–19038.
16. SONNENBERG, A., C. J. T. LINDERS, P. W. MODDERMAN, C. H. DAMSKY, M. AUMAILLEY & R. TIMPL. 1990. Integrin recognition of different cell-binding fragments of laminin (P1, E3, E8) and evidence that alpha6 beta1 but not alpha6 beta4 functions as a major receptor for fragment E8. J. Cell Biol. **110:** 2145–2156.
17. ELICES, M. J. & M. E. HEMLER. 1989. The human integrin VLA-2 is a collagen receptor on some cells and a collagen/laminin receptor on others. Proc. Natl. Acad. Sci. USA **86:** 9906–9910.
18. LANGUINO, R. R., K. R. GEHLSEN, E. WAYNER, W. G. CARTER, E. ENGVALL & E. RUOSLAHTI. 1989. Endothelial cells use alpha2 beta1 integrin as a laminin receptor. J. Cell Biol. **109:** 2455–2462.
19. REICHARDT, L. F., B. BOSSY, S. CARBONETTO, I. DE CURTIS, C. EMMETT, D. E. HALL, M. J. IGNATIUS, F. LEFCORT, E. NAPOLITANO, T. LARGE, K. M. NEUGEBAUER & K. J. TOMASELLI. 1990. Neuronal receptors that regulate axon growth. Cold Spring Harbor Symp. Quant. Biol. **55:** 341–350.
20. LEIN, P. J., D. HIGGINS, D. C. TURNER, L. A. FLIER & V. P. TERRANOVA. 1991. The NC1 domain of type IV collagen promotes axonal growth in sympathetic neurons through interaction with the alpha1 beta1 integrin. J. Cell Biol. **113:** 417–428.

Cell Surface Recognition and Neuron-Glia Interactions[a]

MELITTA SCHACHNER[b]

Departments of Neurobiology
Swiss Federal Institute of Technology
Zürich, Switzerland and
University of Heidelberg
Heidelberg, Germany

Neuron-glia interactions play decisive roles during development, functional maintenance, and regeneration of the nervous system. Glial cells are responsible for the guidance of migrating neuronal cell bodies, promotion of neurite outgrowth, and shaping of neuronal assemblies during development.[1] In the adult, astrocytes and oligodendrocytes contribute to the symbiotic relationship with neurons by controlling the extracellular ion composition and space, neurotransmitter uptake and inactivation, and response to neuronal activity by modification of cellular metabolism.[1] Schwann cells, the glial cells of the peripheral nervous system, are endowed with many of the properties that are characteristic of glial cells of the central nervous system, but, in contrast to these, are able to support regrowth of axons in mammals following trauma.[2]

How the different functional properties of glia arise and how they contribute to the interrelationship with neurons are intriguing problems that require elucidation at the cellular and molecular levels. Molecules that mediate interactions between neurons and glia have been found at the cell surfaces of the interacting partner cells and in the extracellular matrix synthesized and secreted into the extracellular space by neurons or glia. (For reviews, see ref. 3.) Such molecules have been termed adhesion molecules, because antibodies against these molecules or the soluble molecules themselves interfere with adhesion among the two cell types in *in vitro* assay systems. Cell surface or extracellular matrix molecules, however, not only may be adhesive, but also can display antiadhesive properties and therefore should better be termed recognition molecules. Several recognition molecules have been described that, due to their cell type-specific and developmental stage-specific expression, exert their recognition functions in distinct spatiotemporal patterns. On the basis of structural similarities, these molecules fall into families, the members of which are likely to perform related functions.

The immunoglobulin superfamily derives its name from the structural features of immunoglobulin domains originally described in antibodies.[4] These domains appear to confer self-binding, so-called homophilic reactivities of a member of the immunoglobulin superfamily which are enhanced by cooperativity with another member,

[a]This work was supported by Deutsche Forschungsgemeinschaft, Gemeinnützige Hertie-Stiftung, Hermann und Lilly Schilling-Stiftung, Schweizerischer Nationalfonds, Ciba-Geigy Basel, and Kommission zur Förderung der wissenschaftlichen Forschung. A similar review appeared in 1990 in "Neurobiology of Glia," Seminars in the Neurosciences. Saunders Scientific Publications. Baltimore.

[b]Address for correspondence: Department of Neurobiology, Swiss Federal Institute of Technology, Hönggerberg, 8093 Zürich, Switzerland.

probably based on protein-carbohydrate interactions.[5,6] Interactions with non-self, so-called heterophilic partners have also been observed.[7–10] Recognition molecules of the immunoglobulin superfamily are the neural cell adhesion molecules N-CAM,[11,12] L1,[13] the myelin-associated glycoprotein MAG,[14,15] TAG-1,[16] the contactin/F3/F11-group,[17] and the major peripheral nervous system glycoprotein P_0[18–20] in vertebrates. Neuroglian and the fasciclins II and III in flies are also members of the immunoglobulin superfamily (ref. 21 and references therein). The fibronectin type III domain motif, originally described in the extracellular matrix glycoprotein fibronectin, is another structural feature common to some members of the immunoglobulin superfamily, such as N-CAM, L1, TAG-1, and the contactin group, and extracellular matrix molecules such as J1/tenascin/cytotactin.[22–24] Carbohydrate structures common to subsets of recognition molecules[25] form the basis of other family traits and are themselves involved in recognition.[26,27] It is tempting to speculate that these and possibly many other carbohydrate structures contribute, by their immense possibilities of combinations in expression on the protein backbone of a single recognition molecule, to the fine tuning of the recognition process.

The contribution of recognition molecules to neuron-glia interactions will be discussed in the context of neurite outgrowth, demarcation of boundaries between functional neuronal assemblies, and regulation of ionic milieu.

NEURITE OUTGROWTH

Support and guidance of neurite outgrowth appear to be one of the prominent functions of glia during development and, in the adult, during axonal regeneration and formation of new synapses as a consequence of acquisition of newly learned tasks. Neurite outgrowth on astrocytes is promoted by N-CAM, the Ca^{2+}-dependent recognition molecule N-cadherin, L1, and the integrins, which are the cell surface receptors for several extracellular matrix glycoproteins.[9,10,28,29] Schwann cells are generally even better substrates than astrocytes for neurite outgrowth. Neurite outgrowth from dorsal root ganglion neurons placed on a monolayer of Schwann cells is highly L1- and less N-CAM-dependent.[30] Inasmuch as both neurons and Schwann cells express L1, neurite outgrowth may be governed by a homophilic or an assisted homophilic adhesion mechanism. The observation that L1 may not be as dominant in promoting neurite outgrowth on Schwann cells[31] may be due to the use of an assay system in which neurons and Schwann cells are derived from different species and in which the L1 antibodies used were species-specific in their reactivities.

The importance of L1 for neurite outgrowth on Schwann cells is underscored by the observation that L1 expression is increased after nerve lesioning in the distal part of the transected nerve stump,[32] which is the territory into which axons elongate to reinnervate their peripheral targets. During development, both Schwann cells and axons are L1- and N-CAM-positive when they first contact each other. As myelination proceeds, L1 and N-CAM are greatly reduced in expression, whereas nonmyelinating Schwann cells and unmyelinated axons remain L1- and N-CAM-immunoreactive throughout adulthood. After transection of the adult sciatic nerve, both L1 and N-CAM become reexpressed by Schwann cells within only a few days.[33,34]

These observations prompted the investigation of the regulatory mechanisms underlying L1 expression by Schwann cells. L1, but not N-CAM expression is up-regulated by nerve growth factor (NGF) and glia-derived nexin (GDN) previously called GDNPF.[35] Both NGF and GDN are increased in their expression after infliction of a lesion to peripheral nerves of adult mammals.[36,37] In contrast, when a derivative of the central nervous system, the optic nerve of adult mice, is lesioned, no up-regulation of L1 is detectable (Bartsch and Schachner, unpublished observations). In the lesioned optic nerve of fish that display regenerative capacities throughout adulthood,[38] L1 immunoreactivity is significantly increased.[39] It is therefore tempting to speculate that the ability of glial cells to express L1 correlates with their potential to allow regenerative growth of axons.

What are the cues, however, that permit axons to select their pathway so that they can reinnervate their original targets, so that motor neurons show a preference to interact with muscles and sensory neurons with primary sensory organs? As L1 is expressed by all Schwann cells and all axons in the regenerating mouse sciatic nerve, it does not exhibit the signals required for such pathway selection. First hints have now emerged that carbohydrate structures expressed by recognition molecules may be among the determinants conferring such cues. On the basis of the observation that the L2/HNK-1 carbohydrate epitope is prominently expressed in Schwann cell basal laminae and myelin sheaths of ventral, but scarcely of dorsal spinal roots of adult mice,[40] the possibility was investigated that this functionally important carbohydrate structure may be involved in preferential growth of motor axons. From sections of fresh-frozen ventral and dorsal spinal roots as substrates for neurite outgrowth *in vitro*, it was found that motor neurons from the ventral horn of the spinal cord and sensory neurons from dorsal root ganglia showed a particular substrate selection.[41,42] Lengths of neurites from motor neurons were significantly longer on the L2/HNK-1-positive ventral root than on the L2/HNK-1-negative dorsal root, whereas neurite outgrowth of sensory neurons on dorsal or ventral roots was in the same range as that of motor neurons on dorsal roots. Involvement of the L2/HNK-1 epitope in this preferential neurite outgrowth could be shown by specific inhibition of neurite outgrowth by L2 antibodies on ventral, but not on dorsal roots. Furthermore, sensory neurons were not inhibited by the antibodies in their outgrowth on dorsal or ventral roots. To further investigate the involvement of the L2/HNK-1 carbohydrate epitope in neurite outgrowth, the carbohydrate was presented as substrate in the form of glycolipids. Because the glycolipids alone did not show good substrate properties for attachment of neurons, they were offered as substrate in mixture with laminin which is known to be present in Schwann cell basal laminae and to provide a general neurite outgrowth-promoting activity. Outgrowth of neurites from motor neurons was significantly better on the L2 glycolipids than on the control glycolipids, ganglioside GD1b, and sulfatide. Furthermore, motor neurons generally displayed a more elaborate morphology of their neuritic arbors on the L2 glycolipids than on the other two glycolipids.

Thus, motor neurons prefer to extend neurites on the nerve they naturally innervate, that is, the ventral spinal root, in an L2/HNK-1 carbohydrate-dependent manner, implicating this carbohydrate structure in preferential growth of motor neurons. The expression of the L2/HNK-1 carbohydrate in mature myelin sheaths and Schwann cell tubes might be a predisposition of adult peripheral nerves to assure selective axonal regrowth in case of trauma. Specific carbohydrate-based cues could tag particular peripheral nerve structures, such as the neurite outgrowth-promoting interface between basal lamina and Schwann cell surface, with a selective advantage over more general cues that permit Schwann cells to promote neurite outgrowth, such as L1, N-CAM, or P_0. It is not known which protein or glycolipid carries the L2/HNK-1 carbohydrate. How the L2/HNK-1 carbohydrate structure is regulated in its expression and what the nature of the transducing signal in the responding neurites might be will be intriguing to investigate.

GLIAL RECOGNITION MOLECULES DEMARCATING BOUNDARIES

Recognition between interacting partners not only may result in cell adhesion, leading to stabilization of cell contacts over a longer time period, but also may provoke adhesion-inhibitory effects. Functional studies on the J1 family of extracellular matrix glycoproteins that are synthesized by glial cells of the central and peripheral nervous system[30,43] support this view. The J1 family consists of several distinct molecular components with different molecular masses and functions. The members of the high molecular weight group, J1-200 and J1-220, which are immunochemically related to tenascin[44] or cytotactin[45] in the chicken, are expressed by astrocytes and Schwann cells. J1/tenascin is involved in neuron-glia recognition.[43,46] These components most likely also mediate nerve-muscle interactions at the regenerating neuromuscular junction[47] and are locally generated by proliferating fibroblast-like cells.[48] Because, like L1, they are up-regulated after lesioning of peripheral, and less of central nerves in the adult mouse, they are most likely implicated in developmental and regenerative axon growth.[49,50]

The expression of J1/tenascin in the vibrissae-related barrel field boundaries in the somatosensory cortex[51,52] (and see Steindler, this issue) is suggestive of the molecule's ability to demarcate and confine functional neuronal assemblies. The setting up of these boundaries apparently is induced by neuronal activity of thalamic afferents and may be temporally correlated with the arrival of these afferents. Before, J1/tenascin is detectable immunohistochemically in a more diffuse distribution, and it even appears as if the thalamic afferents "hollow out" their individual barrel field boundaries from this evenly distributed background.[53] After establishment of the afferent maps in the somatosensory cortex, J1/tenascin is no longer detectable. These observations implicate neuronal activity in J1/tenascin expression. On the one hand, down-regulation of J1/tenascin during the initial phase of the establishment of the topographic map appears to set up the barrel field boundaries. These boundaries may lead and constrain the later incoming axons into their prospective territory and shape the assembly of cell bodies and associated dendritic target fields of postsynaptic cells. On the other hand, full functional neuronal activity may then shut down J1/tenascin expression altogether.

These observations are worthwhile to consider in the context of the known functional roles of J1/tenascin with regard to neuronal attachment and neurite outgrowth in culture. When offered as a substrate alone or in combination with an adhesive substrate, such as polyornithine, J1/tenascin is inhibitory for cell attachment in situations in which neurons are confronted with the choice to adhere to a more adhesive substrate at high cell-plating densities.[54] When offered in combination with an adhesive substrate without the choice for a more adhesive substrate and at low cell-plating densities, J1/tenascin increases neurite extension when compared with the adhesive substrate alone.[46] As soluble additive to the culture medium, J1/tenascin blocks neurite outgrowth on adhesive substrates.[46] From these observations it is obvious that J1/tenascin may exhibit diverse functional effects depending on its particular cellular and molecular environment. The resultant cell behavior is likely to be mediated by different molecular domains and receptor responses on the neurons that transduce the recognition of J1/tenascin into a cascade of intracellular events underlying the cell's decision on how to interact.

Another set of members of the J1 family, the low molecular weight group of J1/160 and J1/180, becomes detectable late during development.[55] It is expressed by oligodendrocytes and myelin in the central, but not in the peripheral nervous system of mice and appears to play a functional role in the maintenance of the cytoarchitecture at the node of Ranvier in optic nerve myelin.[56] The low molecular weight J1 glycoproteins are distinct from each other and J1/tenascin, but are structurally related.[44,55] Purified J1/160 and J1/180 molecules are nonpermissive substrates for the adhesion and spread of astrocytes, fibroblasts, and small cerebellar neurons. In a mixture with laminin, the J1-160/180 molecules are repulsive for neurons, but not for astrocytes or fibroblasts. In short-term adhesion assays between oligodendrocytes and neurons or between purified J1-160/180 and neurons, these molecules mediate a transitory recognition event between oligodendrocytes and neurons that reverses into inhibition of adhesion as a function of time of interaction between these cells.[57] In the interaction of astrocytes with oligodendrocytes, however, J1-160/180 plays an adhesive role at all times of interaction, ranging from 30 minutes to 24 hours. These observations implicate J1-160/180 in a recognition process between cells that merges into stabilization of cell contacts or adhesion for astrocytes and destabilization of cell interactions or repulsion for neurons. It is likely that the two opposite effects in cell behavior elicited by these J1 molecules result from differential intracellular responses to a cell surface trigger through different cell surface receptors and/or different consequences in intracellular signaling networks. Thus, recognition events mediated at the cell surface are decisive, but not unequivocal determinants of a cell's behavior in that the cell integrates the triggering signals into a constantly changing dynamic interplay of its intracellular affairs.

THE ADHESION MOLECULE ON GLIA (AMOG) AND ITS INFLUENCE ON THE IONIC MILIEU

The link between triggering of cell surface molecules and intracellular events as hypothesized for the J1 glycoproteins and shown to exist for N-CAM and L1[58] gains a new dimension with the adhesion molecule on glia (AMOG) and its structural and functional association with an ion pump, the Na^+/K^+-ATPase.[59] AMOG is expressed by central nervous system glial cells and is involved in neuron-astrocyte, but not astrocyte-astrocyte adhesion. AMOG is expressed during morphogenetically active periods, such as the migration of cerebellar granule cell neurons along Bergmann glial cells.[60,61] Fab fragments of monoclonal AMOG antibodies significantly inhibit this migration.[60] After completion of granule cell migration AMOG remains expressed by astrocytes predominantly in the cerebellum, but also in other brain regions, particularly in those in which the packing density of neuronal cell bodies is high.[61] That AMOG has a direct role in adhesion, thus fitting best the operational definition of a recognition molecule, is indicated by its ability to bind to cells after purification and incorporation into liposomes.[62] Sequence analysis of AMOG reveals a 40% amino acid identity with the ATPase β-subunit whose functional role has remained largely elusive. The 100-kD component tenaciously copurifying with AMOG in immunoaffinity isolation procedures is identical to the α_2 and possibly α_3 isoforms of the Na^+/K^+-ATPase catalytic subunit. A functional link between AMOG and Na^+/K^+-ATPase could be shown by triggering AMOG via monoclonal antibody. The addition of AMOG antibodies to live cultured astrocytes increased the ouabain inhibitable uptake of radioactive Rb^+, indicative of K^+ uptake.[59] The fact that AMOG-mediated adhesion occurs both at 4°C and in the presence of ouabain, an inhibitor of the catalytic activity of Na^+/K^+-ATPase, further underscores that AMOG is indeed a cell recognition molecule. Whether the inhibition of granule cell migration on Bergmann glial cells by monoclonal antibodies to AMOG results from triggering of pump activity in Bergmann glial cells or by modification of the recognition process itself is presently unknown.

Several implications of the association between cell adhesion and ion pump activity are noteworthy in the context of neuron-glia interactions during development and regulation of neuronal activity by astrocytes in the adult. Binding of the putative AMOG receptor on neurons to AMOG on glial cells may modulate Na^+/K^+-ATPase activity directly or indirectly. The activities of both ions, Na^+ and K^+, are known to determine the membrane potential, size of extracellular space, and cell volume. AMOG might modulate these features by influencing ATPase activity. AMOG-mediated morphogenetic events, such as neuronal migration along radial glial processes, may thus result from a feedback loop between recognition and changes in intra- and extracellular ionic balance. In the adult, expression of AMOG at the neuron-glia interface may indicate that cell contacts between neurons and glia remain instrumental in regulating the ionic milieu between neurons and glia. Regulation of the ionic milieu inside and outside the cell may be particularly relevant in the modulation of neuronal excitability and synaptic efficacy.

CONCLUSIONS

Neuron-glia interactions during development and regeneration derive from many molecular events. Glial recognition molecules at the cell surface and in the extracellular matrix display adhesive and antiadhesive properties that reflect their diverse influences on neurons. Glial cells produce neurite outgrowth-promoting recognition molecules, the expression of which correlates with their potential to foster neuronal regeneration in the central and peripheral nervous systems of adult mammals. Glial cells also express carbohydrate structures that appear to influence the pathway selection of neuronal subpopulations in peripheral nerves. Glial recognition molecules form boundaries between units of functional neuronal assemblies during the establishment of topographic maps and display the dual functions of being adhesive and antiadhesive. Barrier functions of antiadhesive glial molecules may contribute to the inability of neurons to regenerate in the central nervous system of adult mammals (see also ref. 63). Finally, glial recognition molecules may influence the intra- and extracellular ionic milieu during development and in the adult. This broad repertoire of functional traits endows glia with hitherto unknown possibilities by which these cells exert their influence on neurons.

ACKNOWLEDGMENTS

The author is grateful to all colleagues who contributed to the work presented in this article.

REFERENCES

1. Neuronal-glial Cell Interrelationships. 1982. Dahlem Workshop Reports. T. A. Sears, ed. :1–427. Springer Verlag. Berlin, Heidelberg, New York.
2. Aguayo, A. J. 1985. *In* Synaptic Plasticity. C. W. Cotman, ed.:457–484. The Guilford Press.
3. Edelman, G. M., B. A. Cunningham & J. P. Thiery (Eds.). 1990. Morphoregulatory Molecules.:1–628. John Wiley & Sons. New York.

4. WILLIAMS, A. F. 1987. Immunol. Today **8:** 298–303.
5. KADMON, G., A. KOWITZ, P. ALTEVOGT & M. SCHACHNER. 1990a. J. Cell Biol. **110:** 193–208.
6. KADMON, G., A. KOWITZ, P. ALTEVOGT & M. SCHACHNER. 1990b. J. Cell Biol. **110:** 209–218.
7. WERZ, W. & M. SCHACHNER. 1988. Dev. Brain Res. **43:** 225–234.
8. GRUMET, M. & G. M. EDELMAN. 1988. J. Cell Biol. **106:** 487–503.
9. DRAZBA, J. & V. LEMMON. 1990. Dev. Biol. **138:** 82–93.
10. SMITH, G. M., U. RUTISHAUSER, J. SILVER & R. H. MILLER. 1990. Dev. Biol. **138:** 377–390.
11. CUNNINGHAM, B. A., J. J. HEMPERLY, B. A. MURRAY, E. A. PREDIGER, R. BRACKENBURY & G. M. EDELMAN. 1987. Science **236:** 799–805.
12. BARTHELS, D., M. J. SANTONI, W. WILLE, C. RUPPERT, J. C. CHAIX, M. R. HIRSCH, J. C. FONTECILLA-CAMPS & C. GORIDIS. 1987. EMBO J. **6:** 907–914.
13. MOOS, M., R. TACKE, H. SCHERER, D. TEPLOW, K. FRÜH & M. SCHACHNER. 1988. Nature **334:** 701–703.
14. ARQUINT, M., J. RODER, L.-S. CHIA, J. DOWN, D. WILKINSON, H. BAYLEY, P. BRAUN & R. DUNN. 1987. Proc. Natl. Acad. Sci. USA **84:** 600–604.
15. POLTORAK, M., R. SADOUL, G. KEILHAUER, C. LANDA, T. FAHRIG & M. SCHACHNER. 1987. J. Cell Biol. **105:** 1893–1899.
16. FURLEY, A. J., S. B. MORTON, D. MANALO, D. KARAGOGEOS, J. DODD & T. M. JESSELL. 1990. Cell **61:** 157–170.
17. GENNARINI, G., G. CIBELLI, G. ROUGON, M.-G. MATTEI & C. GORIDIS. 1989. J. Cell Biol. **109:** 775–788.
18. LEMKE, G. & R. AXEL. 1985. Cell **40:** 501–508.
19. FILBIN, M. T., F. S. WALSH, B. D. TRAPP, J. A. PIZZEY & G. I. TENNEKOON. 1990. Nature **344:** 871–872.
20. SCHNEIDER-SCHAULIES, J., A. VON BRUNN & M. SCHACHNER. 1990. J. Neurosci. Res. **27:** 286–297.
21. BIEBER, A. J., P. M. SNOW, M. HORTSCH, N. H. PATEL, J. R. JACOBS, Z. R. TRAQUINA, J. SCHILLING & C. S. GOODMAN. 1989. Cell **59:** 447–460.
22. SPRING, J., K. BECK & R. CHIQUET-EHRISMANN. 1989. Cell **59:** 325–334.
23. GRUMET, M., S. HOFFMAN, K. L. CROSSIN & G. M. EDELMAN. 1985. Proc. Natl. Acad. Sci. USA **82:** 8075–8079.
24. GULCHER, J. R., D. E. NIES, L. S. MARTON & K. STEFANSSON. 1989. Proc. Natl. Acad. Sci. USA **86:** 1588–1592.
25. KRUSE, J., R. MAILHAMMER, H. WERNECKE, A. FAISSNER, I. SOMMER, C. GORIDIS & M. SCHACHNER. 1984. Nature **311:** 153–155.
26. KÜNEMUND, V., F. B. JUNGALWALA, G. FISCHER, D. K. H. CHOU, G. KEILHAUER & M. SCHACHNER. 1988. J. Cell Biol. **106:** 213–223.
27. FAHRIG, T., B. SCHMITZ, D. WEBER, A. KÜCHERER-EHRET, A. FAISSNER & M. SCHACHNER. 1990. Eur. J. Neurosci. **2:** 153–161.
28. TOMASELLI, K. J. & L. F. REICHARDT. 1988. J. Neurosci. Res. **21:** 275–285.
29. NEUGEBAUER, K. M., K. J. TOMASELLI, J. LILIEN & L. F. REICHARDT. 1988. J. Cell Biol. **107:** 1177–1187.
30. SEILHEIMER, B. & M. SCHACHNER. 1988. J. Cell Biol. **107:** 341–351.
31. BIXBY, J. L., J. LILIEN & L. F. REICHARDT. 1988. J. Cell Biol. **107:** 353–361.
32. NIEKE, J. & M. SCHACHNER. 1985. Differentiation **30:** 141–151.
33. MARTINI, R. & M. SCHACHNER. 1988. J. Cell Biol. **106:** 1735–1746.
34. TACKE, R. & R. MARTINI. 1990. Neurosci. Lett. **120:** 227–230.
35. SEILHEIMER, B. & M. SCHACHNER. 1987. EMBO J. **6:** 1611–1616.
36. MEIER, R., P. SPREYER, R. ORTMANN, A. HAREL & D. MONARD. 1989. Nature **342:** 548–550.
37. HEUMANN, R., D. LINDHOLM, C. BANDTLOW, M. MEYER, M. J. RADEKE, T. P. MISKO, E. SHOOTER & H. THOENEN. 1987. Proc. Natl. Acad. Sci. USA **84:** 8735–8739.
38. HOPKINS, J. M., T. S. FORD-HOLEVINSKY, J. P. MCCOYY & B. W. AGRANOFF. 1985. J. Neurosci. **5:** 3030–3038.

39. Blaugrund, E., U. Bartsch, R. Martini, M. Schachner & M. Schwartz. 1990. Dev. Brain Res.
40. Martini, R., E. Bollensen & M. Schachner. 1988. Dev. Biol. **129:** 330–338.
41. Martini, R., Y. Xin, B. Schmitz & M. Schachner. 1991. Submitted for publication.
42. Schachner, M. 1990. Semin. Neurosci. **2:** 497–507.
43. Kruse, J., G. Keilhauer, A. Faissner, R. Timpl & M. Schachner. 1985. Nature **316:** 146–148.
44. Faissner, A., J. Kruse, R. Chiquet-Ehrismann & E. Mackie. 1988. Differentiation **37:** 104–114.
45. Grumet, M., S. Hoffman, K. L. Crossin & G. M. Edelman. 1985. Proc. Natl. Acad. Sci. USA **82:** 8075–8079.
46. Lochter, A., V. Lloyd, A. Kaplony, A. Prochiantz, M. Schachner & A. Faissner. 1990. J. Cell Biol. **113:** 1159–1171.
47. Sanes, J. R., M. Schachner & J. Covault. 1986. J. Cell Biol. **102:** 420–431.
48. Gatchalian, C. L., M. Schachner & J. R. Sanes. 1989. J. Cell Biol. **108:** 1873–1890.
49. Martini, R., M. Schachner & A. Faissner. 1990. J. Neurocytol. **19:** 601–616.
50. Bartsch, U., S. Bartsch & M. Schachner. 1991. Submitted for publication.
51. Steindler, D. A., N. G. F. Cooper, A. Faissner & M. Schachner. 1989a. Dev. Biol. **131:** 243–260.
52. Steindler, D. A., A. Faissner & M. Schachner. 1989b. Comments Dev. Neurobiol. **1:** 29–60.
53. Steindler, D. A., T. F. O'Brien, E. Laywell, K. Harrington, A. Faissner & M. Schachner. 1990. Exp. Neurol. **109:** 35–56.
54. Faissner, A. & J. Kruse. 1990. Neuron **5:** 627–637.
55. Pesheva, P., E. Spiess & M. Schachner. 1989. J. Cell Biol. **109:** 1765–1778.
56. ffrench-Constant, C., R. H. Miller, J. Kruse, M. Schachner & M. C. Raff. 1986. J. Cell Biol. **102:** 844–852.
57. Morganti, M. C., J. Taylor, P. Pesheva & M. Schachner. 1990. Exp. Neurol. **109:** 98–110.
58. Schuch, U., M. J. Lohse & M. Schachner. 1989. Neuron **3:** 13–20.
59. Gloor, S., H. Antonicek, K. J. Sweadner, S. Pagliusi, R. Frank, M. Moos & M. Schachner. 1990. J. Cell Biol. **110:** 165–174.
60. Antonicek, H., E. Persohn & M. Schachner. 1987. J. Cell Biol. **104:** 1587–1595.
61. Pagliusi, S. R., M. Schachner, P. H. Seeburg & B. D. Shivers. 1990. Eur. J. Neurosci. **2:** 471–480.
62. Antonicek, H. & M. Schachner. 1988. J. Neurosci. **8:** 2961–2966.
63. Schnell, L. & M. E. Schwab. 1990. Nature **343:** 269–272.

Molecular Mechanisms of Glial-Guided Neuronal Migration

TREVOR N. STITT, U. E. GASSER, AND MARY E. HATTEN[a]

Center for Neurobiology and Behavior and Department of Pathology College of Physicians and Surgeons Columbia University New York, New York 10032

The postnatal migration of the cerebellar granule neuron from its displaced proliferative zone, the external granule cell layer, inward to form the internal granule cell layer is perhaps the best documented case of glial-guided neuronal migration in brain. Evidence obtained from both *in vivo* and *in vitro* analyses indicates that the Bergmann fibers of cerebellar radial glia provide the migratory pathway for granule cell migration.[1–6] Moreover, inasmuch as the granule neuron extends its axon, the parallel fiber, tangential to the radial pathway of migration of the cell soma,[7] the granule cell provides an opportunity to determine whether a specific spatial distribution of cell surface molecules occurs on the shaft and growth cones of extending axons as compared with the soma and leading process of migrating neurons. Finally, the small size and great abundance of granule cells, comprising 90% of all cerebellar neurons and numbering millions in the neonatal mouse, have facilitated the purification of granule neurons for *in vitro* analyses of neuronal migration.[5,8]

With purified cerebellar granule neurons and astroglial cells, we have developed an experimental system in which the migration of living, identified neurons along single glial fibers can be observed in real time with video microscopy.[4,7,9,10] A key feature of our *in vitro* system is the small volume (20–50 μl) in which we culture the cells. In microcultures, astroglial cells differentiate into highly elongated radial forms that resemble the glial forms shown to support neuronal migration in brain.[1,4,11] With video-enhanced differential interference contrast (AVEC-DIC) microscopy, migrating neurons are seen to express a highly extended bipolar shape along the glial fiber, forming close apposition with the glial process along the length of the neuronal cell soma[12] and extending a leading process in the direction of migration (FIG. 1).[4,5,9,10] The leading process is a highly motile structure, extending and retracting short filopodia and lamellopodia that enfold the glial fiber. Furthermore, as the neuron moves, organelles flow from the cell soma forward into the leading process.[5]

Our studies on neuronal migration have revealed a number of general patterns of behavior of neurons migrating along astroglial fibers *in vitro*[4,5] (FIG . 2). First, neuronal migration is restricted to radial forms of astroglial cells. Stellate astroglial cells do not provide a substrate for migration.[4] Moreover, the specific cytologies and contact relationships associated with migrating neurons are not seen when the granule cells are cultured on flat tissue culture surfaces. Neuronal migration along radial glial fibers is saltatory, with a motion phase of 4–6 minutes, followed by a resting phase of approximately the same duration. During the motion phase, neurons move at 50–60 μm/h. Migrating neurons move with this periodicity for 30–120

[a] To whom correspondence should be addressed.

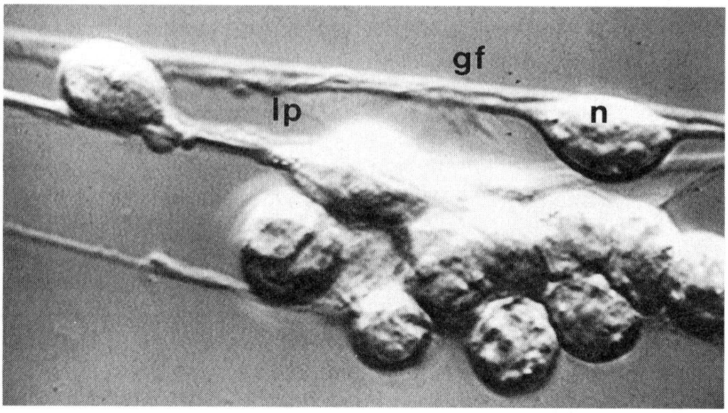

FIGURE 1. Cytology and neuron-glia relationship of a cerebellar granule cell migrating along an astroglial fiber. Cerebellar cells were dissociated from early postnatal mouse cerebellar tissue and cultured in microwells for 24 hours before observation with video-enhanced differential interference contrast microscopy. The granule neuron (n) closely apposes the glial fiber (gf) along the length of the neuronal cell soma, and extends a leading process (lp) in the direction of migration along the glial guide. (Reprinted in modified form from Edmondson and Hatten.[5])

minutes, stop for indeterminate periods, and then move again. Resting neurons have a rounded morphology owing to the retraction of the leading process, and they do not form a migration junction with the glial fiber. Second, with video microscopy, it is evident that migrating neurons can bypass stationary neurons by shoveling their leading process beneath the resting cell and moving underneath it. Third, neurons migrate bidirectionally on glial fibers *in vitro,* generally migrating to the end of the fiber before retracting the leading process, extending it in the opposite direction, and moving in that direction, suggesting that the glial fiber does *not* provide a directionality gradient along its surface. Fourth, migrating neurons shift freely from one glial fiber to another by moving their leading process from one fiber to another. Finally, neuronal migration ceases when the neuron retracts its leading process.

To examine whether the information required for neuronal migration along astroglial fibers is expressed in the neuron or in the glial fiber, we developed methods to purify neurons and glia from cerebellum and hippocampus[9] and analyzed neuronal migration in mosaic cultures. Video microscopy of neurons migrating on glial fibers from the heterotypic region[10] revealed that the neuron-glia relationship, mode of migration, and dynamics of motion of migrating cells are highly conserved in these two brain regions. This suggested that although the glial fibers provide a scaffold for migration, the navigational instructions for migration are encoded in the neuron.[6] Thus, *in vitro* studies on neuronal migration suggest that glial fibers provide a substrate for a specialized form of cell locomotion rather than a specific pathway lined with cues that start and stop migration. The hallmarks of migrating neurons are the formation of an interstitial junction between the neuron and the glial fiber,[12] an oriented flow of cytoplasmic organelles, and the extension of a motile leading process that appears to set the directionality of motion along the glial fiber.

The characteristic neuron-glia relationship of migrating neurons suggests that cell surface adhesion systems provide a molecular mechanism for migration. Antibodies that perturb cell adhesion have been used to define the cell surface molecules that provide adhesive systems, including the calcium-dependent cadherins,[13] and the calcium-independent neural cell adhesion molecules N-CAM,[14] Ng-CAM,[15] and L1.[16] Antibodies that perturb cell substrate and cell matrix interactions have been used to define the receptors for fibronectin and laminin, the integrins.[17–19] Several members of each of these families of adhesion receptor systems are expressed in the cerebellum during the period of granule cell migration.[16,19–22]

To elucidate the molecular basis for granule cell migration, two experimental approaches have been taken. Previously, assays with cerebellar tissue slices suggested a number of mechanisms for granule neuron migration, including a role for proteases,[23] the neural cell adhesion molecules L1[24] and N-CAM,[20,25] cytotactin,[20] and the glial antigen AMOG.[26] To examine the contribution of cell adhesion molecules to granule neuron migration in more detail, in particular to analyze the migration of a single, identified granule cell along a glial fiber, we used microcultures of early postnatal cerebellar cells.[4–6,27] With this culture assay, we did not obtain evidence for a role for L1 and N-CAM in glial-guided neuronal migration. With our *in vitro* assays, however, we identified an immune activity that blocks neuron-glia interactions *in vitro.*[28] We named the activity astrotactin, because by video microscopic analysis, anti-astrotactin antibodies block the establishment of neuronal contacts with glia and the subsequent organization of neuronal positioning by glial processes.[4,28] By immu-

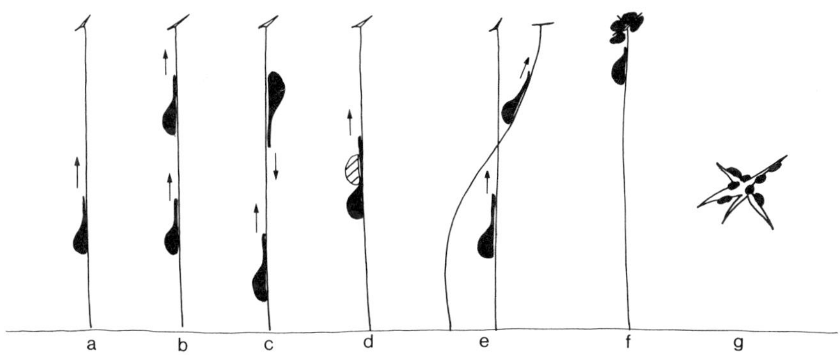

FIGURE 2. Patterns of neuronal migrations along astroglial fibers *in vitro.* Video microscopy of living, migrating granule neurons has revealed that: (a) granule neurons migrate along glial fibers by forming a close apposition with the glial fiber underneath the cell soma and projecting a leading process in the direction of migration (*arrow*), (b) more than one neuron can migrate simultaneously on a single glial fiber, (c) neurons migrate bidirectionally on glial fibers *in vitro,* generally migrating to the end of the fiber before retracting the leading process, extending it in the opposite direction and moving in that direction (d) neurons often pause on the glial fiber for indeterminate periods of time. Resting neurons (*hatched cell*) have a rounded morphology owing to the retraction of the leading process, and do not form a migration junction with the glial fiber. Other, migrating neurons can bypass stationary neurons by shoveling their leading process beneath the resting cell and moving underneath it, (e) migrating neurons can shift from one glial fiber to another by moving their leading process from one fiber to another, and (f) neuronal migration ceases when the neuron retracts its leading process. Migrating neurons generally form neuron-neuron associations with postmigratory neurons clustered around the end-feet of the glial fiber. Migrating neurons are represented as *filled cells,* and the glial fiber as a *solid line. Arrows* indicate the direction of migration of the neuron. (g) Neurons do not migrate on stellate astroglial cells. Rather they form *puncta adherens* junctions with the glial fiber and extend neurites tipped with growth cones. A stellate astroglial cell (*unfilled*) organizes the positioning of granule neurons (*filled cells*). For details, see refs. 4, 5, 9, 10, and 27.

noprecipitation and Western blot analysis, the predominant antigen recognized by the blocking activity of the astrotactin antiserum is a neuronal glycoprotein with an apparent molecular weight of approximately 100 kD.[28,29]

MOLECULAR COMPONENTS OF NEURON-GLIA BINDING

To provide a rapid assay for the molecular components of neuron-glia binding during periods of neuronal migration *in vitro,* we developed an *in vitro* assay using purified granule neurons and astroglia.[30] Such an assay would provide an experimental analysis of the binding of an identified CNS neuron to glia at a defined point in the developmental history of the cell. In this assay, purified granule neurons were metabolically labeled with ^{35}S-methionine, the labeled neurons or neuronal plasma membranes were added to wells containing monolayers of either purified cerebellar astroglia or mouse astrocytoma cell line G26-24, and the kinetics of binding was determined. The kinetics of binding of the granule neurons or membranes was rapid, with maximal binding occurring within 30–60 minutes. Moreover, assays carried out under calcium-free conditions or after calcium-protected proteolysis indicate that granule neuron binding to glia is mediated primarily by calcium-independent receptor mechanisms.[30]

In the presence of the crude astrotactin serum or the IgG fraction of the serum,[30] the binding of radiolabeled granule cell membranes to astroglial cells was reduced 60% (FIG . 3a). Similarly, in the presence of antibodies purified by prior absorption with PC12 cells, the binding of radiolabeled granule neuron membranes to glia was inhibited 60–70%. Antibodies purified by absorption with adult cerebellar membranes also blocked granule cell membrane binding to glia, but to a lesser extent (30%), and microculture assays of the titer of the adult membrane-absorbed antibodies indicated that this purification method significantly reduced the titer of the blocking antibodies.[30] When PC12-absorbed Fab fragments were preincubated with neuronal membranes and removed by washing before the addition of the membranes to the glia, granule cell membrane binding to glia was inhibited by 70% (FIG . 3b). In contrast, when the glial monolayer was preincubated with the Fab fragments, binding of the granule cell membranes remained at control levels. This suggested that the astrotactin Fab fragments blocked neuron-glia binding by binding to the granule cell membrane, not the glial cell. In all cases, Western blot analysis indicates that the antibodies that act to block neuron-glia binding recognize a band of approximately 100 kD in extracts of early postnatal cerebellar cell membranes (FIG . 4).

The membrane binding assay was then used to examine the role of other cell adhesion receptor systems in the adhesion of granule neurons to glia. Antibodies against mouse N-CAM inhibited binding to a small extent (15–20%) but did not exhibit the robust blocking effect seen with astrotactin antibodies, even at high concentrations. The addition of Fab fragments of L1 antibodies actually increased the total amount of neuronal membrane binding to glia, probably by acting to block L1-mediated homotypic binding and subsequent membrane aggregation. We also evaluated antibodies that had been shown to block integrin- and N-cadherin-mediated cell adhesion and found they had no effect on neuron-glia interaction. Experiments using combinations of these antibodies indicated these adhesion systems do not work in concert to promote granule neuron-glia binding.

Aggregation assays of purified granule neurons, carried out in the presence of Fab fragments of astrotactin, N-CAM, and L1 antibodies, indicated that although N-CAM and L1 antibodies inhibited granule neuron aggregation, astrotactin antibodies did not block homotypic granule cell adhesion. Thus, astrotactin does not contribute to homotypic binding of granule neurons. This is consistent with our interpretation that astrotactin functions in heterotypic neuron-glial binding and that N-CAM and L1, in contrast, contribute to homotypic granule neuron binding, but not to heterotypic neuron-glial binding.

These results with a rapid binding assay suggest that astrotactin functions in granule neuron-glia binding. They are consistent with previous studies with *in vitro* assays of the effect of astrotactin antibodies on glial organization of neuronal positioning and the establishment of neuron-glia contacts in microcultures of early

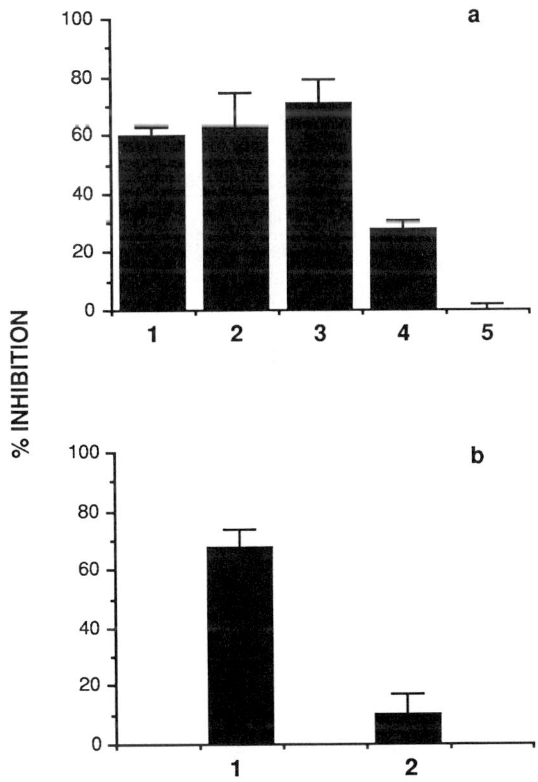

FIGURE 3. Effect of astrotactin antibodies on neuron-glia binding. In **a,** the crude astrotactin serum (1:100) (1) and IgG fractions (1 mg/ml) of the crude serum (2) and PC12-absorbed serum (3) inhibit granule cell membrane binding to glia 60–70%. Inhibition (30%) is also seen with an IgG fraction (1 mg/ml) of antisera absorbed with adult cerebellar tissue (4). No inhibition is seen in the presence of IgG fraction of normal (5) rabbit serum. In **b,** treatment of neuronal membranes (1), but not glial cells (2) with Fab fragments (1 mg/ml) of PC12 cell-absorbed serum inhibits membrane binding to astroglial cells. Binding assays were as described in the text. Data are presented as the mean ± SD of triplicate determinations. (Reprinted with permission from ref. 30.)

postnatal mouse cerebellar cells.[28] At present, the astroglial receptor for the neuronal glycoprotein astrotactin is unknown.

ASTROTACTIN EXPRESSION AND ROLE IN NEURONAL MIGRATION

In vitro labeling experiments indicate that astrotactin is expressed on the surface of cultured granule neurons, but not astroglial cells.[28,29] Together with results of the binding assays, these studies indicate that astrotactin is the neuronal membrane component in neuron-glia binding. To determine the distribution of astrotactin on migrating cells and its contribution to neuronal migration along glia *in vitro,* we immunostained migrating granule neurons and determined whether astrotactin antibodies blocked migration. To accomplish the latter, we developed a new, low-magnification assay of migration in which the behavior of a set of several hundred migrating cells could be followed in a video microscopic assay.[31] These studies showed that astrotactin is expressed along the migration junction between neurons and glial fibers[29] and that astrotactin Fab fragments block migration. Moreover, antibodies against N-CAM, L1, integrin β1, and TAG-1 did not block migration.[31]

Western blot analysis of cerebellar proteins suggests that the expression of astrotactin is developmentally regulated in mouse cerebellum and that astrotactin is transiently expressed by granule neurons in the early postnatal period of cerebellar

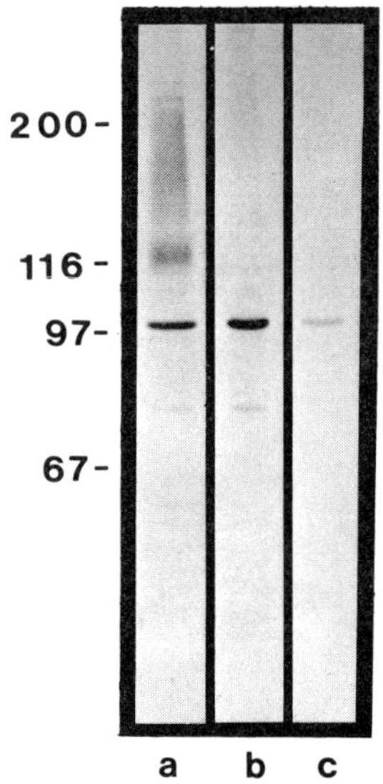

FIGURE 4. Western blot analysis of astrotactin antibodies. Western blot analysis of Triton X-100 extracts of membranes prepared from P8 cerebellar tissue show that although the (1) crude astrotactin antiserum recognizes a number of bands, including a major band of apparent molecular weight of approximately 100 kD, and bands that are likely to represent N-CAM (180 kD) and L1 (140 kD), sera absorbed with (2) PC12 cells or (3) adult cerebellar tissue recognize a band of approximately 100 kD. (Reprinted with permission from ref. 30.)

development. The relative amount of astrotactin protein is low in the late embryonic period, increasing to high levels in the postnatal period and decreasing again to low levels in the adult. The time course of expression follows the timing of granule cell migration and assembly in the internal granule cell layer, which commences just before birth, is most prominent between P2 and P14, and is concluded by the third postnatal week.

Immunocytochemical localization of astrotactin in tissue sections from cerebellum and cerebral cortex, two regions of developing brain with well-studied patterns of glial-guided neuronal migration,[1,11,32] suggest that astrotactin is expressed in zones and at developmental periods, in which neuronal migration along glial fibers and glial-mediated neuronal assembly into cell layers is occurring.[29] Astrotactin is not expressed in proliferative zones, either in the ventricular zone of the cortex or in the external granule layer (EGL) of the cerebellum, or is it expressed on radial glial fibers.

Rather, intensely stained cells appear in migratory zones—the molecular layers of the cerebellum in the postnatal period and the cortical plate of the cortex in embryonic periods.[32] In these areas, particularly in the cerebellum where the characteristics of migrating granule cells have been well studied, labeled cells express the features of migrating neurons.[1,4,5] Moreover, astrotactin immunoreactivity is restricted to the neuronal soma and leading process, with the parallel fibers remaining unstained. In the adult, after the period of neuronal migration has ended,[32] astrotactin expression falls dramatically.

The spatiotemporal pattern of astrotactin expression is unique among adhesion systems described to date, as it is only expressed by cells undergoing glial-guided neuronal migration and assembly into neuronal layers. Moreover, in the cerebellum, a specific spatial distribution of cell surface molecules is seen for the granule neuron. Whereas neural cell adhesion molecules N-CAM[14] and L1[16] are expressed on the parallel fibers, astrotactin expression is restricted to the cell soma and leading process of migrating neurons.[21,29] This suggests that whereas neural cell adhesion molecules and cadherins mediate growth cone locomotion, neuronal migration is mediated by astrotactin.

SUMMARY

The migration of young neurons from their site of origin in proliferative zones out into neuronal layers is a hallmark of cortical development. Neuroanatomic studies show that astroglial fibers provide the primary substrate for neuronal migration. *In vitro* studies on living cells provide evidence that migrating neurons express distinguishing cytologic features including the formation of a specialized junction at the site of neuron-glia contact and the extension of an active leading process in the direction of migration. Our *in vitro* functional assays point to a critical role for astrotactin in neuron-glia binding during the developmental periods of glial-guided cell migration and assembly in brain. Other receptor systems, including neural cell adhesion systems, cadherins, and integrins are expressed by granule cells but do not appear to contribute to neuron-glia binding or to glial-guided neuronal migration. A role for astrotactin in glial-guided migration and assembly is supported by our observation that astrotactin is expressed by neurons and not glial cells[28,29] and by restricted spatiotemporal expression of astrotactin *in vivo,* wherein astrotactin is expressed by migrating neurons and by neurons during periods of assembly into neuronal layers in developing brain.[29] Understanding the regulation of astrotactin expression and its role in migration will provide fundamental insights into the role of glial-guided migration in the histogenesis of the brain.

REFERENCES

1. Rakic, P. 1971. Neuron-glia relationship during granule cell migration in developing cerebellar cortex. A Golgi and electron microscopic study in *Macacus rhesus.* J. Comp. Neurol. **141:** 283–312.
2. Rakic, P. & R. L. Sidman. 1973. *Weaver* mutant mouse cerebellum: Defective neuronal migration secondary to specific abnormality of Bergmann glia Proc. Natl. Acad. Sci. USA **70:** 240–244.
3. Rakic, P. 1978. Neuronal migration and contact guidance in primate telencephalon. Postgrad. Med. J. **54:** 25–40.

4. Hatten, M. E., R. K. H. Liem & C. A. Mason. 1984. Two forms of cerebellar glial cells interact differently with neurons *in vitro.* J. Cell Biol. **98:** 193–204.
5. Edmondson, J. C. & M. E. Hatten. 1987. Glial-guided granule neuron migration *in vitro:* A high-resolution time-lapse video microscopic study. J. Neurosci. **7:** 1928–1934.
6. Hatten, M. E. 1990. Riding the glial monorail: A common mechanism for glial-guided migration in different regions of the developing brain. Trends Neurosci. **13:** 179–184.
7. Ramon y Cajal, S. 1911. Histologie du Systeme Nerveux de l'Homme et des Vertebres, Maloine, Paris. (Reprinted by Consejo Superior de Investigaciones Cientificas, Madrid, 1955.)
8. Hatten, M. E. 1985. Neuronal regulation of astroglial morphology and proliferation *in vitro.* J. Cell Biol. **100:** 384–396.
9. Gasser, U. E. & M. E. Hatten. 1990. Central nervous system neurons migrate on astroglial fibers from heterotypic brain regions *in vitro.* Proc. Natl. Acad. Sci. USA **87:** 4543–4547.
10. Gasser, U. E. & M. E. Hatten. 1990. Neuron-glia interactions of rat hippocampal cells *in vitro:* Glial-guided neuronal migration and neuronal regulation of glial differentiation. J. Neurosci. **10:** 1276–1285.
11. Rakic, P. 1972. Mode of cell migration to the superficial layers of the fetal monkey neocortex. J. Comp. Neurol. **145:** 61–84.
12. Gregory, W. A., J. C. Edmondson, M. E. Hatten & C. A. Mason. 1988. Cytology and neuron-glia apposition of migrating cerebellar granule cells *in vitro.* J. Neurosci. **8:** 1728–1738.
13. Hatta, K., T. S. Okada & M. Takeichi. 1985. A monoclonal antibody disrupting calcium dependent cell-cell adhesion of brain tissue. Possible role of its target antigen in animal pattern formation. Proc. Natl. Acad. Sci. USA **82:** 2789–2793.
14. Thiery, J.-P., R. Brackenbury, U. Rutishauser & G. M. Edelman. 1977. Adhesion among neural cells of the chick embryo. II. Purification and characterization of a cell adhesion molecule from neural retina. J. Biol. Chem. **252:** 6841–6845.
15. Grumet, M. & G. M. Edelman. 1984. Heterotypic binding between neuronal membrane vesicles and glial cells is mediated by a specific cell adhesion molecule. J. Cell Biol. **98:** 1746–1756.
16. Rathjen, F. G. & M. Schachner. 1984. Immunocytochemical and biochemical characterization of a new neuronal cell surface component (L1 antigen) which is involved in cell adhesion. EMBO J. **3:** 1–10.
17. Horwitz, A. F., K. Duggan, R. Greggs, C. Dekker & C. Buck. 1985. The cell substrate attachment (CSAT) antigen has properties of a receptor for laminin and fibronectin. J. Cell Biol. **101:** 2134–2144.
18. Buck, C. A. & A. F. Horwitz. 1987. Integrin, a transmembrane glycoprotein complex mediating cell substratum adhesion. J. Cell Sci. (Suppl) **8:** 231–250.
19. Hynes, R. 1990. Fibronectins. Springer-Verlag. 546 pp.
20. Chuong, C.-M., K. L. Croissin & G. M. Edelman. 1987. Sequential expression and differential function of multiple adhesion molecules during the formation of cerebellar cortical layers. J. Cell Biol. **104:** 331–342.
21. Persohn, E. & M. Schachner. 1987. Immunoelectron microscopic localization of the neural cell adhesion molecules L1 and N-CAM during postnatal development of the cerebellum. J. Cell Biol. **105:** 569–576.
22. Grumet, M., S. Hoffman, K. L. Croissin & G. M. Edelman. 1985. Cytotactin, an extracellular matrix protein of neural and non-neural tissue that mediates glia-neuron interaction. Proc. Natl. Acad. Sci. USA **82:** 8075–8079.
23. Moonen, G., M. P. Grau-Wagemans & I. Selak. 1982. Plasminogen activator-plasmin systems and neuronal migration. Nature (Lond.) **298:** 753–755.
24. Lindner, J., F. G. Rathjen & M. Schachner. 1983. L1 mono- and polyclonal antibodies modify cell migration in early postnatalmouse cerebellum. Nature (Lond.) **305:** 85–90.
25. Lindner, G. Zinser, W. Werz, C. Goridis, B. Bizzini & M. Schachner. 1986. Experimental modification of postnatal cerebellar granule cell migration *in vitro.* Brain Res. **377:** 298–304.
26. Antonicek, H., E. Persohn & M. Schachner. 1987. Biochemical and functional charac-

terization of a novel neuron-glia adhesion molecule that is involved in neuronal migration. J. Cell Biol. **104:** 1587–1595.
27. HATTEN, M. E. & R. K. H. LIEM. 1981. Astroglia provide a template for the positioning of developing cerebellar neurons *in vitro.* J. Cell Biol. **90:** 622–630.
28. EDMONDSON, J. C., R. K. H. LIEM, J. C. KUSTER & M. E. HATTEN. 1988. Astrotactin, a novel cell surface antigen that mediates neuron-glia interactions in cerebellar microcultures. J. Cell Biol. **106:** 505–517.
29. STITT, T. N., C. A. MASON & M. E. HATTEN. Submitted for publication.
30. STITT, T. N. & M. E. HATTEN. 1990. Anti-astrotactin antibodies block granule neuron binding to astroglia. Neuron **5:** 639–649.
31. FISHELL, G. & M. E. HATTEN. 1991. Astrotactin provides a receptor system for CNS neuronal migration. Development, in press.
32. SIDMAN, R. L. & P. RAKIC. 1973. Neuronal migration with special reference to human brain. A review. Brain Res. **62:** 1–35.

Boundaries and Wounds, Glia and Glycoconjugates

Cellular and Molecular Analyses of Developmental Partitions and Adult Brain Lesions[a]

ERIC D. LAYWELL[b] AND DENNIS A. STEINDLER

Department of Anatomy and Neurobiology
University of Tennessee, Memphis
College of Medicine
Memphis, Tennessee 38163

Even though the importance of glial-neuronal interactions during brain development has been known for some time (since the work of Ramon y Cajal), it was not until the landmark studies of Rakic that modern neuroscience has come to appreciate the dependence of young neurons on their glial companions for constructing functional circuitry arrangements. Rakic[1] showed that migrating neurons follow radial glial guideposts in their ascent from the ventricular germinal zone to their adult positions, often great distances away. Glial roles during neuronal migration and neuritic growth and guidance have been well characterized in a variety of invertebrate[2] and vertebrate[3,4] systems. Recent studies of glial-neuronal interactions during development have focused on distinctive molecular expressions that lead to functional cell sorting and neuritic expansion. A substantial number of molecules associated with neurogenesis have been characterized and demonstrated *in situ,* and cellular and molecular interactions that lead to specific circuitries have been elegantly demonstrated in the tissue culture studies of Schachner, Faissner, Hatten, Bunge, Silver, Raper, Kuwada, Raff, Schwab, Bonhoeffer, Prochiantz, and others. (See ref. 5 for review.) Profound advances in cell biology and molecular and genetic neurobiology have led to characterizations of morphogenetic events that even recently were only proposed by students of classical neuroembryology.

This paper is divided into three parts: (1) a history of transient glial and glycoconjugate boundaries since their discovery, in 1986, in the early postnatal mouse brain; (2) an approach toward understanding the cellular constituents of pattern formation boundaries; and (3) an attempt at rekindling the expression of developmentally regulated macromolecules in order to define their possible roles in failed regeneration following lesions created in the mature brain. The term "boundary" implies inhibitory actions of these structures, and to date we can only speculate on their roles during development and in the wounded adult brain. However, the developing and lesioned adult whisker-barrel system presents a unique model in which to resolve these issues inasmuch as form and function are so alike.

[a] This work was supported by grant BNS-8911514 from the National Science Foundation and grant NS 20856 from the National Institutes of Health.

[b] Address for correspondence: Eric D. Laywell, Department of Anatomy and Neurobiology, University of Tennessee, Memphis, College of Medicine, 875 Monroe Ave., Memphis, TN 38163.

TRANSIENT BRAIN BOUNDARIES SURROUND DEVELOPING FUNCTIONAL UNITS

Using the technique of lectin cytochemistry on fixed brain sections developed by other investigators who understood the usefulness of distinct lectin recognition of specific developmentally regulated carbohydrate components of glycoconjugates,[6] Cooper and Steindler[7] found that lectins concanavalin A and peanut agglutinin bound to glycoconjugates in the developing somatosensory cerebral cortex in a

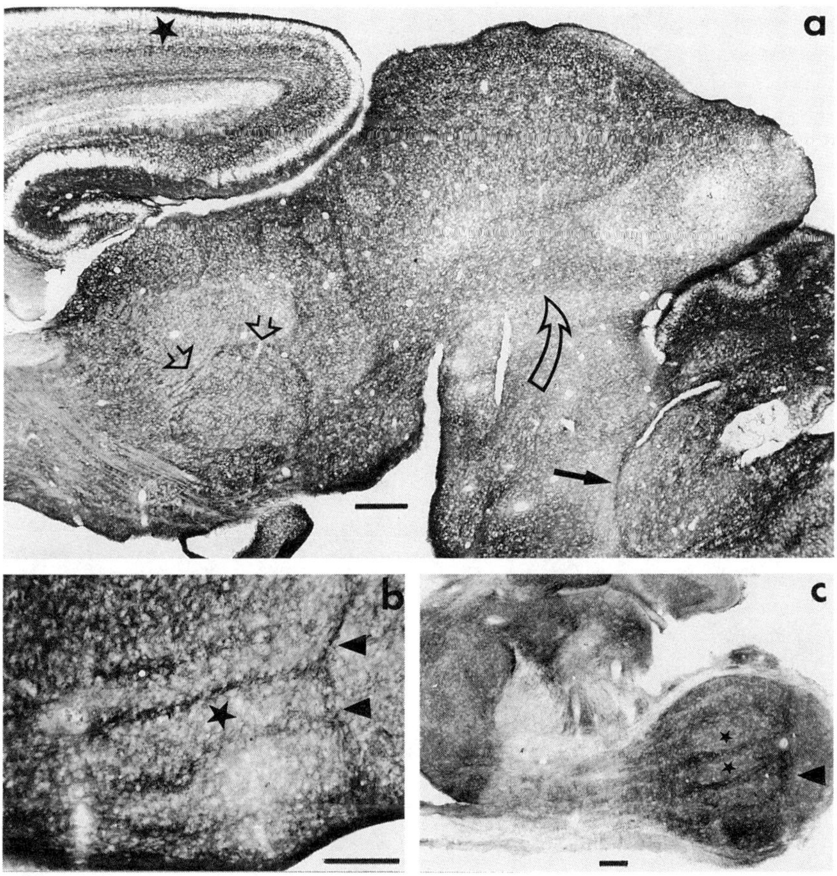

FIGURE 1. (**a**) Postnatal day 2 (P2) mouse brain, parasagittal section (a–c are parasagittal sections with rostral to the left and dorsal up), showing J1/tenascin immunostaining in cerebrocortical layers (*star*), in boundaries around thalamic (*small open arrows*) and brainstem (*filled arrow*) nuclei, and in fiber pathways (*curved arrow*). Scale (in a–c) = 100 μm. (**b**) J1/tenascin boundaries around whisker row representations in the P2 brainstem trigeminal complex (*star* in one row "hollow"; *arrowheads* point out boundary between interpolaris and caudalis subnuclei). (**c**) MAP-2 immunostaining on P18 shows accumulation of trigeminal neuron dendrites within the whisker row boundaries (e.g., between stars) and along the interpolaris-caudalis border (*arrowhead*).

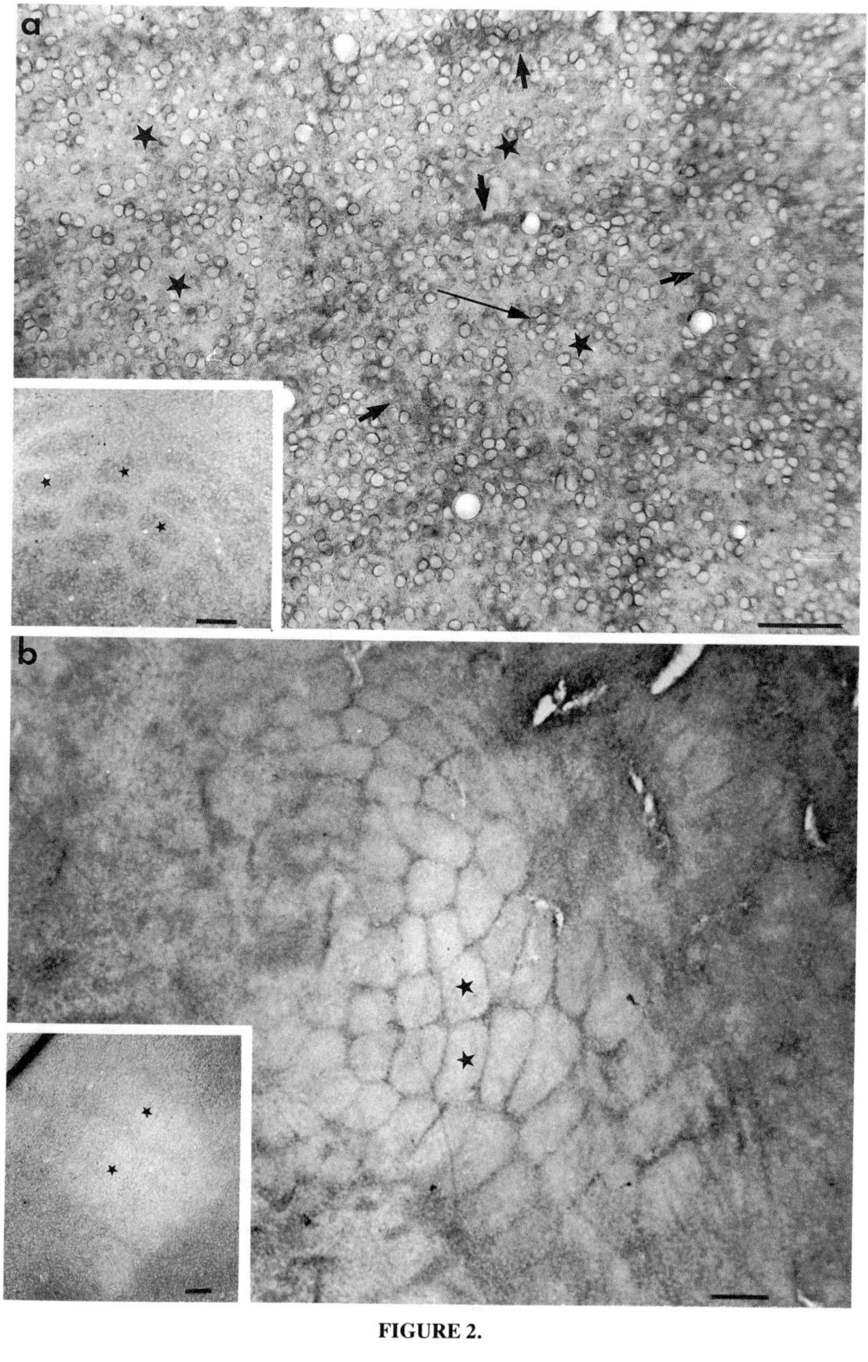

FIGURE 2.

manner that revealed a characteristic functional pattern before the neurons in this region formed such a pattern. From that region of the somatosensory cortex dedicated to the representation of the facial vibrissae,[8] it was found that glycoconjugate boundaries were present between individual vibrissae representations termed "barrels." The lectin-bound glycoconjugates thus revealed a presumptive barrel field early in postnatal life before the final sorting of neurons within the characteristic whisker-barrel field configuration; furthermore, these boundaries, destined to become neuron-sparse "interbarrel septae" in the adult, were transient and disappeared in the second postnatal week. It was also determined that glycoconjugate boundaries were not unique to the developing barrel field; rather, the entire brain revealed boundary plans in accordance with all aspects of cytoarchitectonic parcellation (e.g., nuclei, subnuclei, functionally distinct portions of nuclei, cortical layers, and neostriatal patch-matrix organization, termed striosomes) (FIGS. 1 and 2).

Lectin cytochemistry was informative only to the extent of showing that some molecules, containing certain sugar moieties, were deposited most densely in the interface between functionally different units in the developing central nervous system (CNS). What we did not know was the precise nature of these molecules and which cells were responsible for their expression. Also, because glycoconjugates are present to varying extents throughout the neuropil and fiber tracts of the developing and adult brain, and lectin cytochemistry in addition to autoradiographic demonstrations of tritiated sugar incorporation into glycoconjugates[9] revealed possible spatiotemporal gradients in their expression, more sensitive methods were required to uncover actual changes in the molecules or cells that express them during maturation. Therefore, we used well-characterized antibodies to molecules identified with cell-cell recognition events as well as established markers of cellular phenotype (e.g., the cytoskeletal proteins such as glial fibrillary acidic protein [GFAP]).

Much of this work was recently reviewed,[10–12] but a few highlights from these studies will provide a background for subsequent sections dealing with new findings on the cellular and molecular composition of developmental boundaries and adult brain wounds. FIGURES 1 and 2 show that the 200/220-kD glycoprotein referred to as J1/tenascin (also called glial mesenchymal extracellular matrix molecule, hexabrachion, myotendinous antigen, and cytotactin,[10–12] a putative extracellular matrix (ECM) molecule affiliated with the L2/HNK-1 family of adhesive proteins, defines similar boundary patterns as seen with lectin binding in the late embryonic and early postnatal mouse brain. The appearance of these boundaries coincides in time with neuritic expansion throughout the neuraxis, and synaptogenic events as well as synaptic stabilization. Developing afferent projections as well as dendritic arborizations, defined by immunocytochemical labeling of the synaptic vesicle protein p38 (see FIG. 2a, inset) and the dendritic marker MAP-2 (FIG. 1c), seem to accumulate along the glycoconjugate boundaries. In the developing whisker-barrel system these markers align with boundary zones in the brainstem whisker row delineations of the trigeminal nuclei (FIG. 1b) and in the somatosensory cortical barrel field (FIG. 2a and b). Because the cortical barrel field is typified by individual neuronal units that can

FIGURE 2. (**a**) P3 somatosensory cortex showing lectin (concanavalin A) binding to whisker-barrel boundaries (*short arrows*), with lectin-rimmed neurons (*long arrows*) present within boundaries as well as hollows (*stars*) in no apparent order. Scale = 50 μm. **Inset** shows p38 synaptic vesicle labeling of axon terminal arbors within barrel hollows on P6 (scale = 100 μm). (**b**) P6 barrel field seen with J1/tenascin immunolabeling. *Stars* within some representative hollows. Scale = 200 μm. **Inset** in **b** shows J1/tenascin barely delineating barrels on P16 (*stars* in two hollows). Scale in **b** inset = 100 μm.

be recognized through maturity using conventional cell stains,[8] the boundary-labeling techniques can be superimposed on the cellular whisker-barrel patterns to reveal the earliest to latest boundary-barrel delineations. FIGURE 2 documents the glycoconjugate boundaries around barrel units as early as P2–P3 when young neurons are clearly not yet aligned within the functional barrel patterns (FIG. 2a), as well as the disappearance of barrel boundaries in the second postnatal week (inset in FIG. 2b shows little or no J1/tenascin immunostaining of the P16 barrel field). Therefore, we can conclude that glycoconjugates, including J1/tenascin, delineate the developing whisker barrels in the first postnatal week before the functional alignment of neurons and during the period of neuritic growth and synaptogenesis; furthermore, these boundaries disappear when such events are completed.

Other markers, including well-characterized cytoskeletal proteins such as GFAP, were also investigated to relate the glycoconjugate boundary patterns to particular cellular elements (i.e., astrocytes). We found that GFAP also labeled boundary astrocytes in the first postnatal week and dramatically outlined individual barrels in the P6 barrel field through selective immunostaining of radial glial elements and immature astrocytes that occupy boundary sites[13] (FIG. 3, inset). Although certain differences exist in the developmental regulation of the astrocytic filament protein and cell surface/ECM glycoconjugates,[12] GFAP staining also fails to reveal barrel boundaries after the second postnatal week. FIGURE 3 shows a section through the P11 barrel field where the presence of both GFAP-positive boundary and nonboundary glia can be seen to mar the earlier image of GFAP-delineated developing barrels (compare FIG. 3 and inset). We presume that GFAP-positive radial glia and immature astrocytes are the major, if not exclusive, source of glycoconjugate boundary expression. Electron microscopic immunocytochemistry studies of J1/tenascin in the developing barrel field show the presence of the molecule on the surfaces of glia and neurons and pericellularly in a manner that suggests ECM (FIG. 4). The J1/tenascin glycoprotein has also been localized within subcellular organelles (e.g., Golgi apparatus), linking its synthesis to a population of cells that possess ultrastructural features characteristic of young astrocytes.[10] We have observed other developmentally regulated glycoconjugates that, to varying degrees, reveal boundaries around barrels and other developing functional units. These include a chondroitin sulfate-containing proteoglycan discovered by Andreas Faissner and termed the "473 antigen"[12,14] and the adhesion molecule on glia (AMOG).[15]

The glial and ECM boundaries that we observe during development most likely affect neuritic growth and sorting because they appear "globally" in the late embryonic and early postnatal brain. In fact, their prevalence throughout the developing neuraxis, with some hint of temporal gradients in their expression within different CNS structures, prompted us to refer to the notion of glial/glycoconjugate partitions around developing brain substructure as "cordones."[11] Cordones thus surround developing brain patterns and may have their own genetic program of expression rather than representing an interface zone that is a consequence of two structurally and functionally distinct zones (e.g., barrels and different nuclei) abutting each other.[11,12] As already mentioned, the time frame of cordone, or boundary, expression coincides with neuritic (both dendritic and axonal) growth and exploration throughout the neuraxis, and it disappears when synaptogenesis is complete and synapses are presumed to be anatomically and physiologically stabilized.[12] Glial cells and novel glycoconjugates that they express may act as landmarks, barriers, or both, that guide axonal and dendritic arbors within functional zones (FIG. 1b and c).

The factors that govern axonal or dendritic patterning, including activity, interact with a glial-glycoconjugate boundary network to secure organized circuitry patterns.

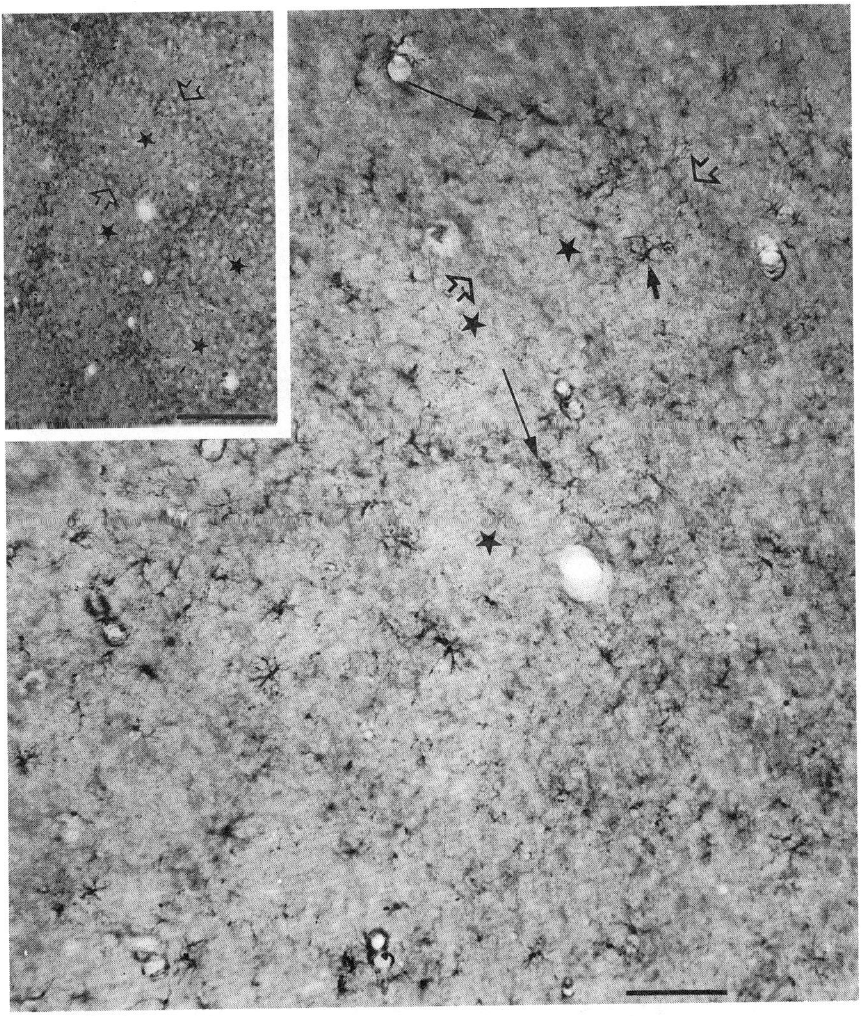

FIGURE 3. On P11, GFAP-immunoreactive astrocytes barely delineate the whisker-barrel field (e.g., a single barrel is visible between the *open arrows; stars* mark some other hollows), with labeled cells present in boundaries (*long arrow*) as well as hollows (*short filled arrow*). Scale = 100 μm. (**Inset**) P6 barrel field is delineated by GFAP immunolabeling of boundary radial glia and young astrocytes (*open arrows*) that surround hollows (*stars*) which contain many less GFAP-positive elements. Scale = 100 μm.

When synaptic arrangements are stabilized, such "tissue" boundaries are no longer visible; instead, glia and ECM molecules now cordon off synaptic complexes on the surfaces of neurons.[16] These more discrete glial and glycoconjugate boundaries seen around synapses in the mature brain have things in common with the ensheathment of nodes of Ranvier by perinodal astrocytes that express novel adhesion and ECM molecules.[17]

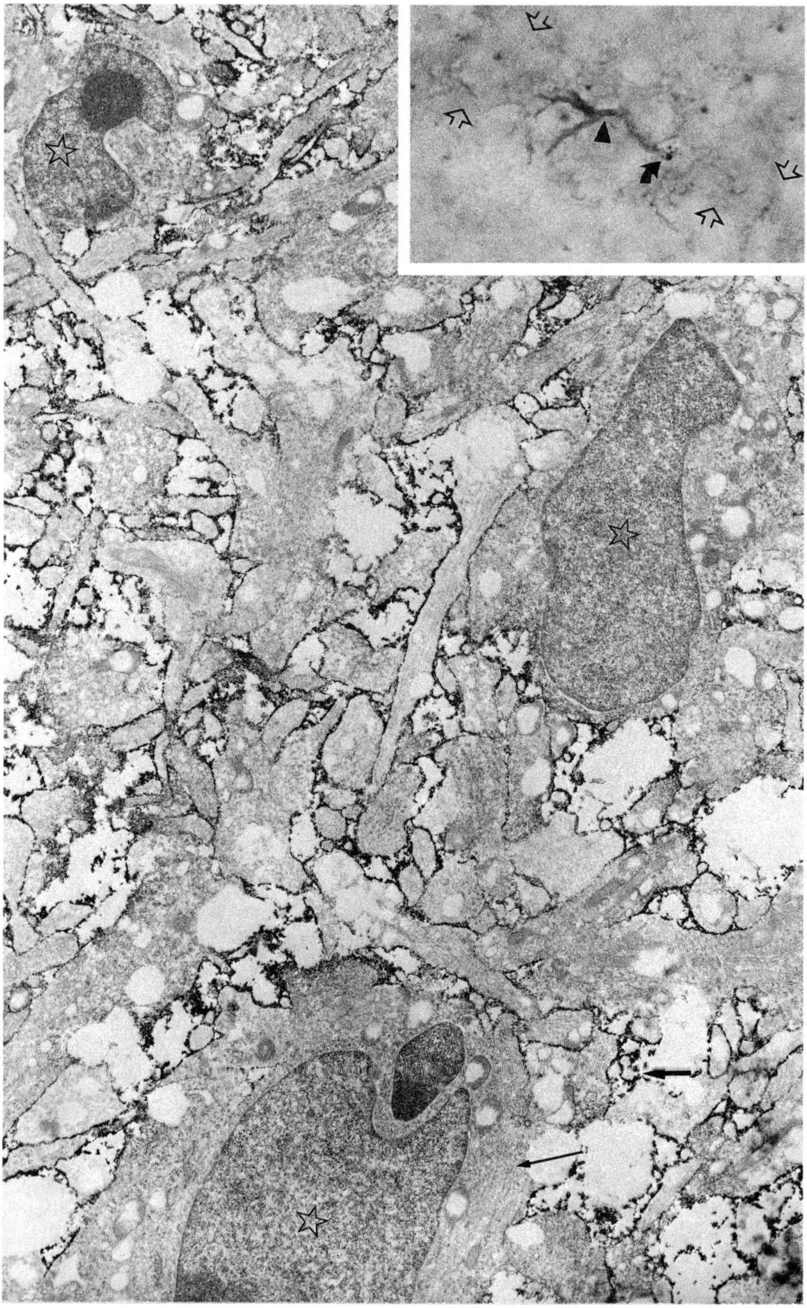

FIGURE 4.

BRAIN LESIONS DURING DEVELOPMENT AND IN THE ADULT REVEAL DISTINCT GLIAL AND GLYCOCONJUGATE DISTRIBUTIONS

Sensory deprivation and deafferentation lesions during development affect glial and glycoconjugate boundaries. Lesions of the whisker pad or activity deprivation during a critical period in the first postnatal week result in a loss of GFAP-stained astrocytes and glycoconjugate-rich boundaries between the deafferented or deprived units.[12,18] After the second postnatal week these peripheral lesions failed to reveal any boundary plans, in accordance with the findings of previous investigators that showed that the critical period of whisker-barrel plasticity is mainly the first few days after birth and certainly within the first postnatal week.[19] We have explored the effects of deafferenting lesions on boundary development in the normal and *reeler* mutant mouse whisker-barrel system[12] and in the developing dopaminergic nigrostriatal system using the glial and glycoconjugate boundaries present around developing patch and matrix components of the neostriatal mosaic.[20] These lesion studies of afferent sources during development all lend credence to the notion that the expression of developmental tissue partitions (boundaries and cordones) is affected by critical numbers and perhaps configurations of afferent axons. The idea was advanced[12] of a threshold whereby numbers of afferent axons within a target above or below a prescribed level would induce alterations in boundary glia and glycoconjugate expression. In the developing brain, partial to near complete deafferentation seemed to consistently produce a down-expression of recognition molecules like J1/tenascin within target structure boundaries. This finding suggested to us that these boundaries may in fact exert a predominantly inhibitory effect on neuritic expansion, and in the young plastic brain, if the number of afferent axons drops below threshold, the boundary is obliterated,[12,20] affording sprouting of other (i.e., intact) neurites within the lesioned target.

Again, during late embryonic and early postnatal development of the mouse somatosensory whisker-barrel system, immature glial elements (radial glia and maturing astrocytes) and developmentally regulated ECM molecules associated with these glia are predominately located in interbarrel boundary regions where they outline a template of the mature barrel field (FIG. 3, inset). After the second postnatal week astrocytes no longer reside exclusively in boundary regions, but rather have an apparently homogeneous distribution throughout barrel hollows as well as interbarrel regions (FIG. 3). Furthermore, the developmentally regulated molecules associated with these astrocytes are no longer detectable after the second postnatal week (except for traces within fiber pathways and around synapses, as just mentioned).

Why is there a temporal difference in the appearance of astrocytes in the boundary versus the hollow? A variety of phenomena could account for this temporal pattern: (1) Astrocytes in the boundary may mature sooner than may their counterparts in the hollows. (2) Boundary astrocytes may move into the hollows after

---

FIGURE 4. Electron microscopic immunocytochemistry of J1/tenascin in the P2 barrel field shows labeling of cell surfaces as well as ECM around processes (*short arrow*). Three cells surrounded by J1/tenascin labeling (*stars* over their nuclei), assumed to be astrocytes based on features such as lobulated nuclei and bundles of filaments (*long arrow*), are presumed to be the same cells GFAP-positive in barrel boundaries. (**Inset** shows a light micrograph of a GFAP-positive astrocyte [*arrowhead*] within a barrel boundary demarcated by *open arrows;* transected labeled radial elements [*curved arrow*] also can be seen.) Magnification × 10,000.

circuitry formation is complete. (3) A quiescent population of astrocytes is present in the barrel hollows that escapes detection early in development.

To address this question we made mechanical lesions (using small needles and survival times of 1 day to 2 weeks) near, and sometimes within, the P6-adult barrel field with the hope that the trauma would accentuate the expression of certain markers (e.g., GFAP) in resident astrocytes. We also looked at the generation of wound-related cells in some of these animals using light microscopic autoradiography and intraperitoneal injections of tritiated thymidine (TdR, as previously de-

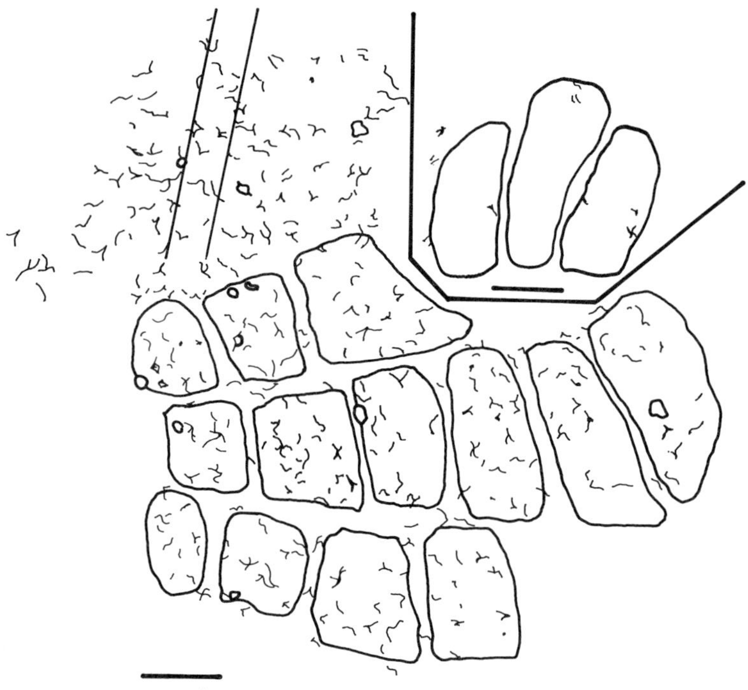

FIGURE 5. Plots from representative sections of GFAP-positive astrocytes within an unlesioned P6 barrel field (**inset**) and the lesion case shown in FIGURE 6. In the unlesioned barrel field on P6, GFAP-labeled immature astrocytes (as well as radial glia; see FIG. 3, inset) reside mostly within boundary sites. In the P6-lesioned barrel field (lesion made on P4), GFAP-positive astrocytes are found within barrel boundaries and hollows. *Smaller open circles* represent blood vessels, and the wound, surrounded by many labeled astrocytes, is marked by *straight lines in the upper left corner.* Scales = 100 μm.

scribed[21,22]). In a normal P6 mouse, GFAP immunocytochemistry sharply reveals developing barrels resulting from the exclusive distribution of labeled astrocytes in boundary regions (FIGS. 3, 4, and 5). However, GFAP immunocytochemistry of lesioned P6 mice (lesion on P4) reveals astrocyte distribution similar to that of a P12 mouse (FIGS. 5 and 6). That is, astrocytes are not confined to boundary areas but are present in hollows as well. These results suggest the presence of at least two populations of astrocytes in the developing mouse barrel field. The first, or boundary, population is present in interbarrel regions very early in development and is

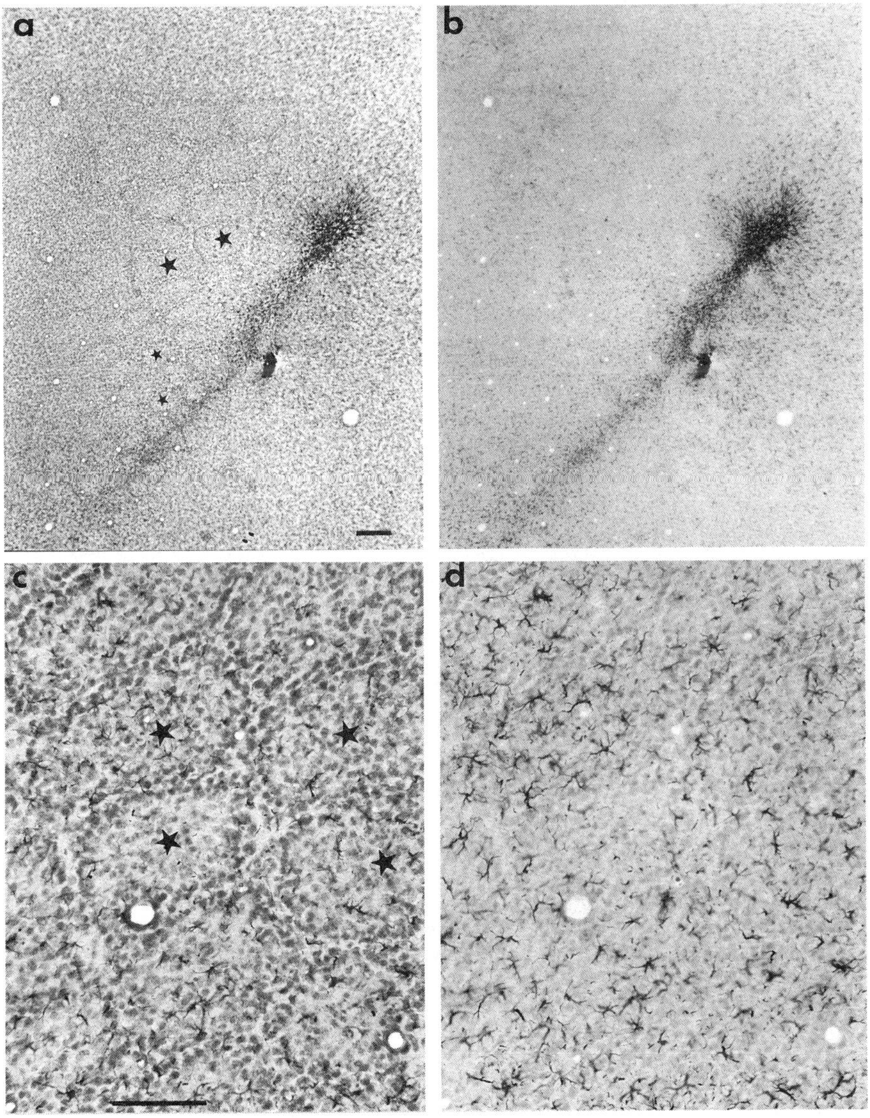

FIGURE 6. (**a** and **b**) P6 barrel field lesion, GFAP and Nissl staining, without (**a**) and with (**b**) a blue filter to subtract Nissl-stained cells. The needle wound involves numerous barrels and seems to miss others (e.g., *large stars* in **a** in representative large barrel hollows; *small stars* in smaller anterolateral barrels). Gliosis is obvious around the needle wound (best seen in **b**). Scale = 200 μm. (**c** and **d**) Higher magnification of barrels in this case show GFAP-positive astrocytes within barrel boundaries and hollows (*stars*). Same field without (**c**) and with (**d**) blue filter. Scale = 100 μm.

responsible for expression and secretion of developmentally regulated ECM constituents (e.g., J1/tenascin, the 473 proteoglycan[10–12]) that may serve to attract afferents, guide neuritic growth, or inhibit dendritic arborization, or a combination of some or all of these. The second population, quiescent until the second postnatal week, resides in barrel hollows and is not significantly associated with the developmentally regulated molecules that are expressed by boundary astrocytes. The second population, present only after afferent segregation is complete, is probably responsible for the duties historically imputed to astrocytes, that is, support and feeding of neurons as well as buffering of the environment around synaptic regions.

The unique geometry of the barrel field and this arrangement of boundary versus nonboundary astrocytes in postnatal developing barrels offer a valuable template within which to evaluate the potential patterning of wound-related astrocytes after intracerebral lesions. FIGURES 7 and 8 document findings from some pilot studies on the distributions of GFAP and TdR birthday identified cells in the P17 and adult barrel field after discrete (needle wound) lesions within the barrel field. (Examples of altered as well as unaffected Nissl-stained barrels could be found in these lesion cases.) Most of the proliferating cells (TdR-identified) reside in and just around the wound (FIG. 7), including the only examples of double-labeled (GFAP + TdR) astrocytes. As previously reported,[22,23] the population of newly generated astrocytes peaked at 3–5 days postlesion, and many examples of densely GFAP-immunopositive "reactive" glia were visible in the vicinity of the wound (FIG. 7). We have found examples of what appear to be GFAP-TdR-labeled dividing cells in these lesions (FIG. 7) as well as single-labeled cells that emanate from the lesion. Only single-labeled (i.e., GFAP or TdR) cells were found in sites away from the lesion within, for example, Nissl-identified barrels (FIG. 8). Plots of P17 and more mature (e.g., 2- to 3-month) cases of barrel field lesion (FIG. 8) show that GFAP- or TdR-labeled cells do not show boundary versus nonboundary preference in their distribution. That is, as in P6 lesions, GFAP- or TdR-labeled cells occupy positions within barrel hollows as well as boundary (interbarrel septae) sites. It also is our impression that many of the TdR-labeled cells are closely associated with blood vessels near the wound as well as in more distant areas. Many of these cells could be vascular-related cells, proliferating in response to the lesion. In addition to blood cells, macrophages, or pericytes, TdR-labeled cells also could be destined to become astrocytes that do not yet express detectable levels of GFAP, and GFAP-positive cells in these regions may owe their proliferation (e.g., they may have been born before or after tritiated thymidine administration) to mitogens released from the lesion site, and their reactive state may be a result of molecular factors (e.g., interleukins and PDGF[24–27]) that are known to alter the biochemical states of astrocytes. It is also possible that reactive glia present in barrels away from the wound site may simply have responded to contortion during mechanical lesioning of the cortex.

At present we do not know for sure what becomes of the boundary astrocytes after the animal has matured. Do these astrocytes maintain their boundary positions throughout the animal's life or is the barrel template lost after development? The GFAP/TdR lesioning studies in the mature barrel cortex suggest that newly born or reactive astrocytes do not show preferential distributions with boundary or nonboundary positions. It is possible that quantitative studies would reveal a subtle preference toward the boundary positions of these cells, especially in P6 and P17 cases. However, we are unable to address completely the issue of whether or not reactive glia generated or already present around the wound might migrate significant distances (as described in some transplant studies[28]) to distribute within seemingly intact barrels. The finding that no double-labeled (GFAP + TdR) cells were observed away from the lesion site, despite the presence of single-labeled cells in these

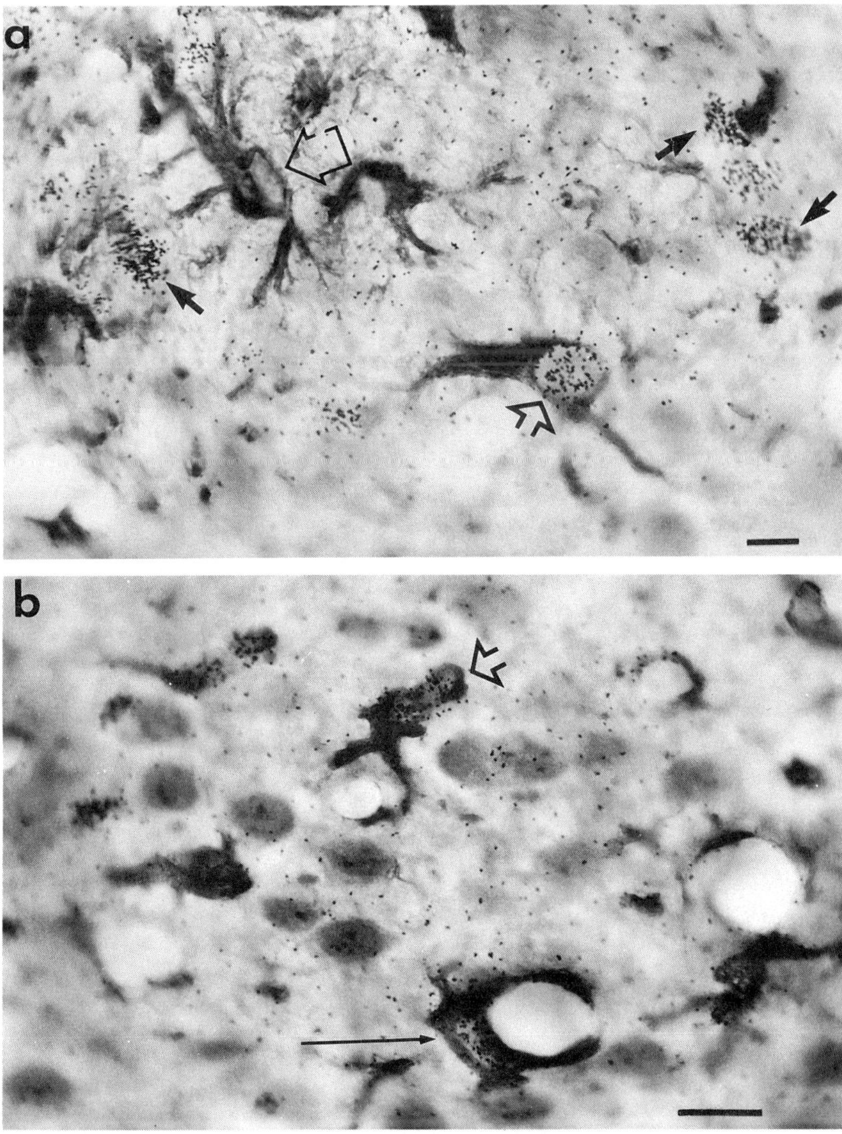

FIGURE 7. (**a**) GFAP (*large open arrow*), tritiated thymidine (*filled arrows*), and double-labeled (*small open arrow*) cells near a needle wound in the barrel cortex of a P17 animal. (**b**) Same case as in **a** shows a double-labeled cell surrounding a blood vessel (*long arrow*) and another double-labeled cell apparently in the process of dividing (*open arrow*). Nissl-counterstained sections. Scales = 10 μm.

FIGURE 8. Plot of a section from the case shown in FIGURE 7 shows GFAP-positive astrocytes (*squares*), tritiated thymidine-labeled cells (*stars*), and two double-labeled cells (*asterisks pointed to by long arrows*) in the P17 barrel cortex 3 days following a needle wound (*dashed lines between open arrows*). *Short filled arrows* point to some cross-sectioned blood vessels, and *curved arrows* point out some representative barrels near the wound (Not all barrel septae are visible within such a multiprocessed section.) Double-labeled cells are found only in the wound, and single-labeled cells distribute away from the wound site within barrel boundaries and hollows. **Inset** shows a section from an older lesion (2- to 3-month-old animal not exposed to tritiated thymidine), showing GFAP-positive astrocytes within barrel boundaries and hollows.

areas, may indicate that such migrations are not common. More studies are needed to disclose the existence of possible biochemically distinct types of boundary and nonboundary glia in the barrel template, and lineage studies may disclose the existence of certain similarities between boundary astrocytes of cortical gray matter and white matter astrocytes (type 2[25,26]) that may limit neuritic expansion during normal development and after lesioning in the mature brain. These types of studies could be carried out *in vitro* so that manipulations for bioassays of cell and molecular roles in neuritic growth could easily be controlled. We have already shown that glial and glycoconjugate boundaries persist in organotypic cultures.[12] Pilot studies indicate that nitrocellulose millipore filters can be implanted within the developing cortex, or they can be acutely blotted against tissue slices and grown in roller tube culture to reveal GFAP-positive astrocytes with varying morphologies (FIGS. 9 and 10). This method (also described by Eng *et al.*[29]) exploits the strongly adhesive nature of boundary glial elements to harvest and study these cells in *in vitro* paradigms. These cells seem to express a variety of ECM molecules, including laminin (FIG. 10), and their morphology and biochemistry can be altered by different controlled environments including the presence of neurons (FIG. 10 and ref. 4). That these cells contain molecules imputed to be conducive to neuritic expansion (e.g., laminin) suggests that boundary astrocytes may express a variety of macromolecules that can deter or facilitate neuritic growth.

WOUNDING THE ADULT BRAIN RESULTS IN A REAPPEARANCE OF DEVELOPMENTALLY REGULATED ECM MOLECULES

J1/tenascin and its associated chondroitin sulfate-containing proteoglycan 473 are developmentally regulated ECM molecules that are correlated with astrocyte distribution in the early postnatal mouse brain. These molecules are present in boundaries around developing subunits (e.g., barrels). When circuit formation is complete, J1/tenascin and 473 tissue boundaries disappear from the brain and are not seen in mature brains except within white matter glia and around synapses. Because of the spatial (boundary) and temporal (circuit formation) distribution of these molecules, it seems fair to impute to them an instructive role in initial neural connectivity. Evidence is mounting which suggests that the instructiveness is manifested in inhibition; that is, these and other molecules (especially proteoglycans and myelin-associated molecules; see ref. 5 for review) may be more or less repulsive and serve to prevent aberrant growth and anomalous connections. Assuming the potential inhibitory nature of these molecules, we wanted to know if lesions of adult brain, which normally has miniscule amounts of J1/tenascin and 473, could induce a reexpression of these molecules. If these molecules are reexpressed after lesioning, then their inhibitory nature may be at least partially responsible for regenerative failure in mature mammalian CNS.

Mechanical brain lesions were made with a hypodermic needle in adult mice to see if such trauma and interruption of existing cytoarchitecture would induce a reexpression of developmentally regulated molecules imputed to have pattern-forming functions (see ref. 30 and FIG. 11). Our lesions in adult mice show, for the first time, that a developmentally regulated molecule that is virtually absent in a normal adult brain can be resurrected after traumatic episodes. Immunohistochemistry of lesions in the cerebral cortex reveals a reexpression of J1/tenascin (FIG. 11b–d) and the 473 proteoglycan as early as 24 hours postlesion, and these molecules are detectable at least 1 week postlesion. In addition to J1/tenascin and the 473 proteoglycan, N-CAM L1 and E-N-CAM controls as well as a variety of other

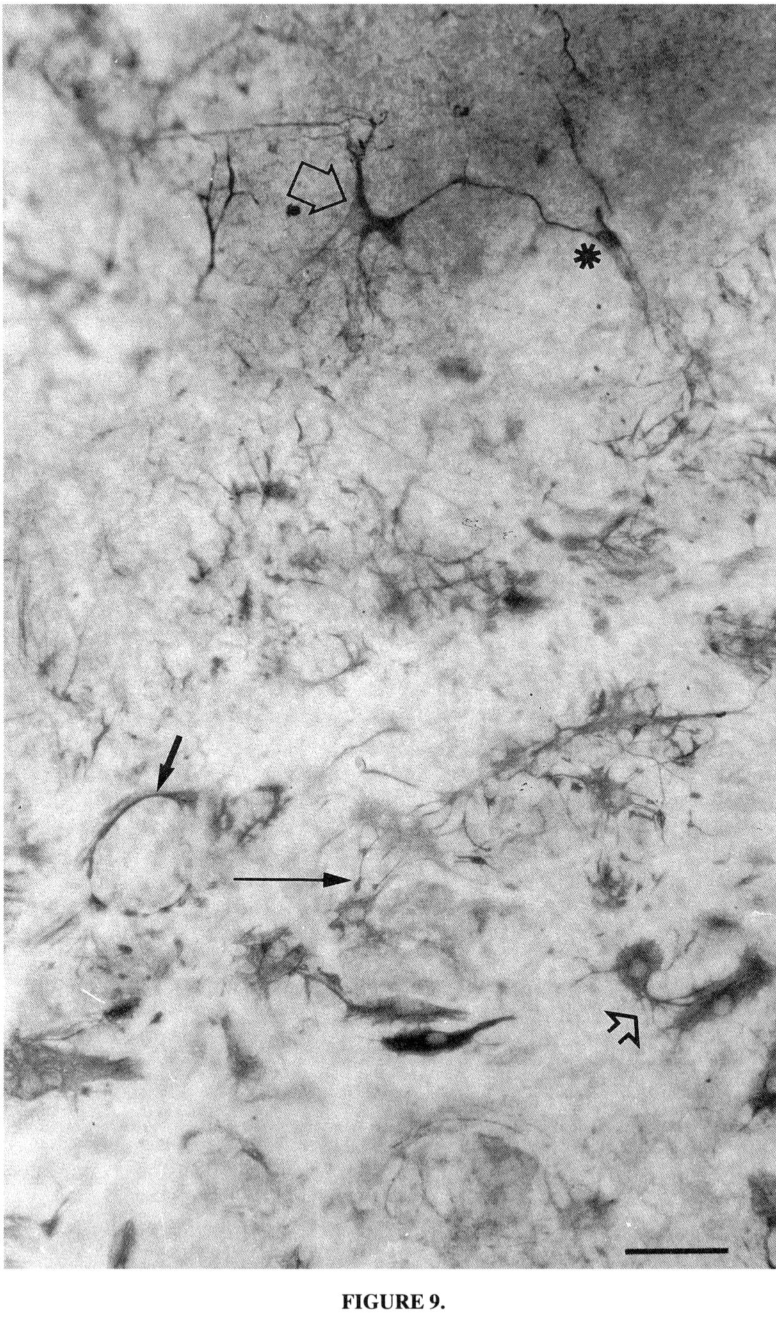

FIGURE 9.

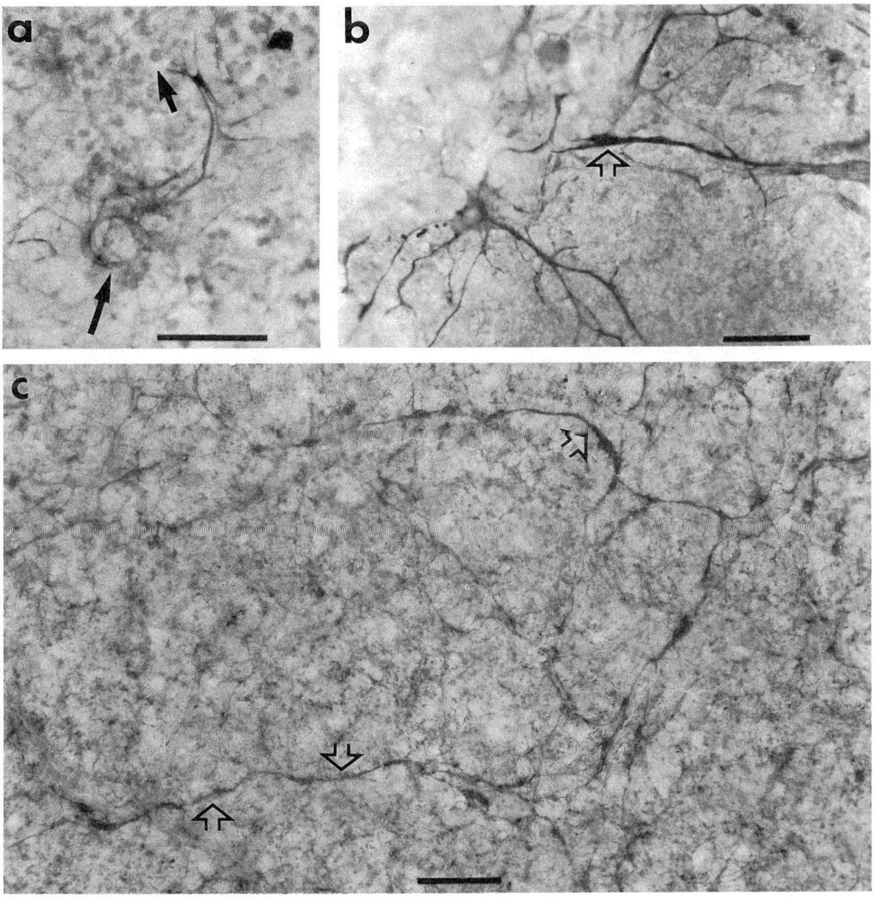

FIGURE 10. Same case as shown in FIGURE 9. (**a**) GFAP-labeled astrocytes (one in close association with what appears to be a blood vessel [*large arrow*]) possess different morphologies when near other cells (presumed neurons that have survived [*small arrow*]). Scale (in a–c) = 50 μm. (**b**) GFAP-positive astrocytes as well as what appear to be radial glia (*open arrow*) survive in these cultures. (**c**) Laminin-immunoreactive processes (*open arrows*) are also present within these nitrocellulose-harvested glial cultures.

FIGURE 9. Nitrocellulose-harvested astrocytes from the neonatal mouse cerebral cortex show a variety of morphologies when grown in roller tube cultures. GFAP immunocytochemistry was used. Cells that appear to be initiating process outgrowth (*small open arrow*) as well as what appear to be more differentiated cells (*large open arrow* marks cell that appears to embrace a blood vessel [*asterisk*]) are present within this field. $GFAP^+$ processes around what appear to be vacular elements (*small filled arrow*) as well as labeled astrocytic "growth cones" (*long arrow*) are also visible. Scale = 50 μm.

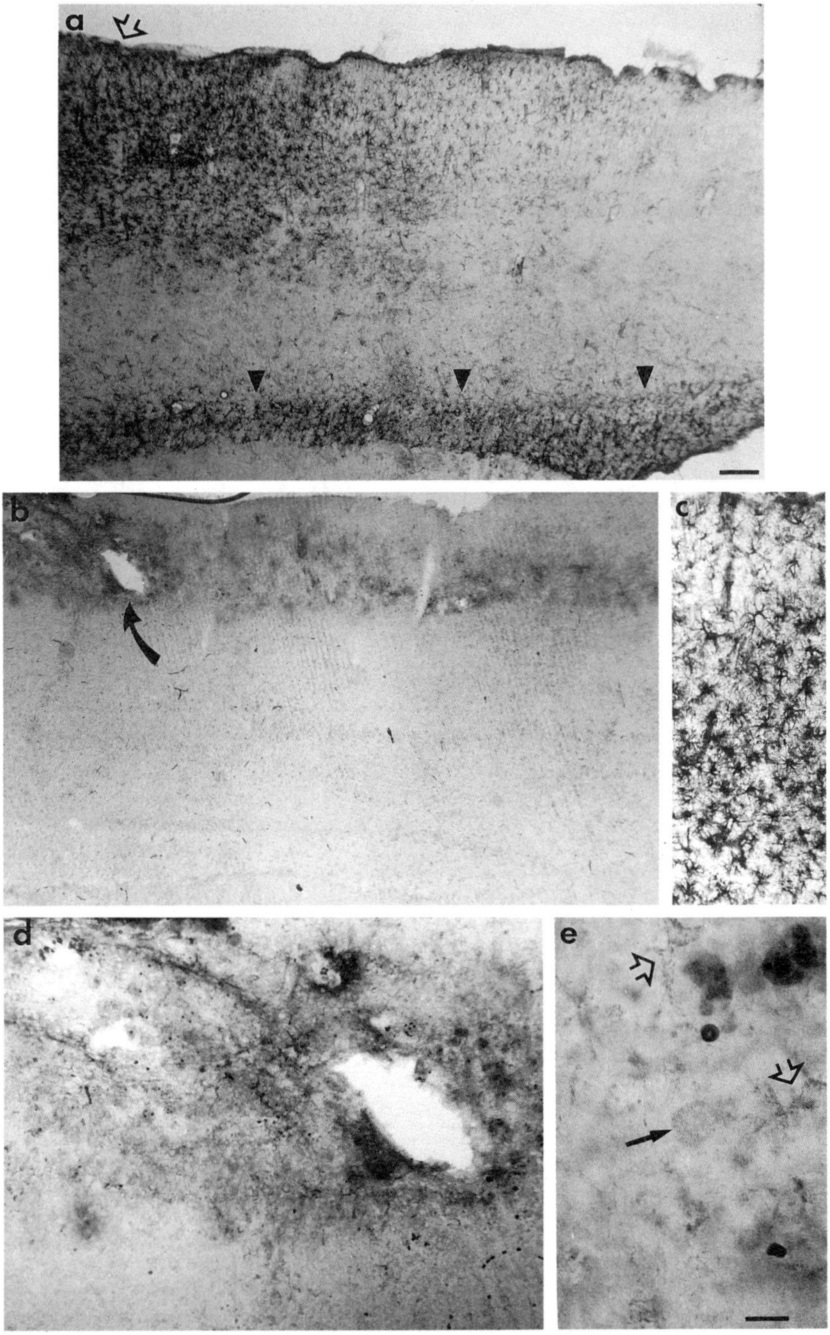

FIGURE 11.

neural antigens fail to be detected in the lesion site (Laywell, Faissner, Schachner, and Steindler, in preparation).

All of the antigens responding to the lesion are associated with astrocytes and are seen in the lesioned cortical gray matter as well as in the subcortical white matter after a particularly large cortical lesion or a lesion in which the needle penetrated the subcortical white matter. (It is our impression that lesions involving the white matter may lead to more widespread staining of J1/tenascin and the 473 proteoglycan.) In cortex, the staining appears neuropilar and is arranged in a gradient fashion around the needle tract. The area of glycoconjugate immunoreactivity is in the midst of a much larger area of GFAP immunoreactivity; indeed, only about 25% of the area expressing GFAP also expresses the glycoconjugates (compare FIGS. 11a and b). The dense neuropilar glycoconjugate staining precludes visualization of cellular elements within the needle tract, but high magnification of the edge of the lesion reveals that a subset of GFAP-immunopositive astrocytes are also immunopositive for the glycoconjugates. We do not yet know if a single astrocyte expresses both J1/tenascin and the 473 proteoglycan. In addition to astrocytes, occasional neurons are surrounded by immunopositivity for these glycoconjugates (FIG. 11e). Neuronal staining of J1/tenascin and the 473 proteoglycan usually follows a large lesion in which bleeding is extensive. In contrast to astrocyte staining, which often densely stains and outlines their processes, neuronal staining is very punctate and appears to be extracellular. Electron microscopic and *in situ* hybridization studies should elucidate these staining patterns. Currently, *in vitro* studies[14,31] suggest that neurons make neither J1/tenascin nor the 473 proteoglycan. The four potential sources of the punctate reaction product around neurons are: (1) immunopositive astrocyte processes that we know normally surround neuronal cell bodies; (2) secreted extracellular supplies of the glycoconjugates bound to neuron membrane; (3) neurons possibly ingesting these molecules; and (4) neurons possibly synthesizing them under particular conditions.

White matter staining of J1/tenascin and the 473 proteoglycan is much different from the staining in gray matter. Whereas gray matter staining is largely neuropilar, white matter staining is extremely discrete and is located between axon fascicles, probably on perinodal astrocytes. After cerebellar lesions, for example, clusters of $GFAP^+$ Bergmann glia and the entire extent of their processes are 473 and J1/tenascin labeled.[30] The area of immunoreactivity in the cerebellum is correlated with the presence of densely neurofilament-positive "swollen" axons as revealed by correlative neurofilament immunostaining.

The notion of developmentally regulated molecules and their reexpression following lesions is interesting, but how can this reexpression be related to the immature mouse who has all of these molecules in a particular pattern during development? Is there a pattern to the reexpression? Studies currently under way are trying to answer this question. Using a reliable anatomic template, the somatosen-

---

FIGURE 11. Stab wound lesions in the adult cerebral cortex show GFAP-immunoreactive astrocytosis (**a**) that extends through numerous cortical laminae (*arrowheads* show heavily labeled cells in the subcortical white matter; *open arrow* marks area near the lesion site). Higher magnification of the wound-related astrocytosis is shown in **c**. (**b**) J1/tenascin immunoreactivity is present around the cortical wound (*curved arrow* points to a portion of the needle track that coursed almost parallel to the cortical surface) and is much more restricted than is that seen with GFAP. (**d**) Higher magnification of the J1/tenascin labeling in the lesion site. (**e**) J1/tenascin immunolabeling is associated with what appear to be astrocytes (*open arrows*) as well as neurons (*filled arrow*) near the wound site. Scales = 100 μm (in **a** and **b**) and 10 μm (in **e**).

sory barrel field, we are trying to determine if there is a pattern of astrocytes that express these molecules.

CONCLUSIONS: BOTH SIMILARITIES AND DIFFERENCES EXIST BETWEEN DEVELOPMENTAL BOUNDARIES AND GLIAL SCARS

We propose that the glial and glycoconjugate partitions that we see around developing functional units in the brain (e.g., whisker barrels, neostriatal striosomes, and nuclear divisions) are boundaries because neurons and their processes seem to accumulate along them. In the absence of true bioassays either *in vivo* or *in vitro,* we can only speculate as to the potential inhibitory nature of these boundaries on neuritic growth. However, the prevalence of boundaries throughout the developing brain, in patterns and time frames in which we know that neurons are functionally segregating, implies such an inhibitory role.

Boundaries are reexpressed after trauma such as the stab lesions used herein. J1/tenascin is but one boundary molecule present both during development and in the injured brain, but it may be a crucial early marker for failed regeneration in the mature mammalian brain. Uncovering the cellular (e.g., nonneural cells recruited in brain injury) and molecular differences between developmental boundaries and wound-induced boundaries could provide therapeutic insights into reversing failed regeneration.

SUMMARY

During brain development, transient partitions of glia and glycoconjugates (glycoproteins, glycolipids, and glyosaminoglycans) surround forming functional units (e.g., nuclear divisions, whisker-related barrels, and neostriatal striosomes). These partitions, which we think of as boundaries, consist of dense aggregates of glial fibrillary acidic protein (GFAP)-positive radial glia, young astrocytes and their processes, and developmentally regulated glycoconjugates (e.g., J1/tenascin and the 473 proteoglycan) that can be thought of as recognition molecules present on membranes or perhaps within the extracellular matrix. When functional patterns have formed and appear to be stabilized, these boundaries are no longer detectable. Lesions of the developing brain show the existence of a more global astrocytic distribution suggestive of biochemically distinct subsets of astrocytes that reside within boundary versus nonboundary positions. Lesions of the adult brain, in addition to showing gliosis, reveal a reexpression of some of the same macromolecules present in transient brain boundaries during development. It is postulated that developmental boundaries and wounds in the adult brain possess some of the same inhibitory and possibly alluring molecular substrates for neuritic expansion.

ACKNOWLEDGMENTS

The authors thank Kristy Harrington for her development of the culture methodologies and Tom O'Brien, Andreas Faissner, and Melitta Schachner for their valuable collaborations.

REFERENCES

1. RAKIC, P. 1972. J. Comp. Neurol. **145:** 61–84.
2. JACOBS, J. R. & C. S. GOODMAN. 1989. J. Neurosci. **9:** 2402–2411.
3. TREVARROW, B., D. L. MARKS & C. B. KIMMEL. 1990. Neuron **4:** 669–679.
4. HATTEN, M. E. & C. A. MASON. 1986. Trends in Neurosci. **9:** 168–174.
5. SILVER, J. & D. A. STEINDLER (Eds.). 1990. Axonal boundaries and inhibitory mechanisms during neural development and regeneration. Exp. Neurol. **109:** 1–139.
6. MARGOLIS, R. U. & R. K. MARGOLIS (Eds.) 1979. Complex Carbohydrates of Nervous Tissue. Plenum. New York.
7. COOPER, N. G. F. & D. A. STEINDLER. 1986a. J. Comp. Neurol. **249:** 157–168.
8. WOOLSEY, T. A. & H. VAN DER LOOS. 1970. Brain Res. **17:** 205–242.
9. STEINDLER, D. A. & N. G. F. COOPER. 1987. Dev. Brain Res. **36:** 27–38.
10. STEINDLER, D. A., N. G. F. COOPER, A. FAISSNER & M. SCHACHNER. 1989. Dev. Biol. **131:** 243–260.
11. STEINDLER, D. A., A. FAISSNER & M. SCHACHNER. 1989. Comm. Dev. Neurobiol. **1:** 29–60.
12. STEINDLER, D. A., T. F. O'BRIEN, E. LAYWELL, K. HARRINGTON, A. FAISSNER & M. SCHACHNER. 1990. Exp. Neurol. **109:** 35–56.
13. COOPER, N. G. F. & D. A. STEINDLER. 1986b. Brain Res. **380:** 341–348.
14. FAISSNER, A. 1988. Neurosci. Abstr. **14:** 920.
15. ANTONICEK, H., E. PERSOHN & M. SCHACHNER. 1987. J. Cell Biol. **104:** 1587–1595.
16. STEINDLER, D. A. & N. G. F. COOPER. 1986. J. Comp. Neurol. **249:** 170–185.
17. FFRENCH-CONSTANT, C., R. H. MILLER, J. KRUSE, M. SCHACHNER & M. RAFF. 1986. J. Cell Biol. **102:** 844–852.
18. COOPER, N. G. F. & D. A. STEINDLER. 1989. Brain Res. **489:** 167–176.
19. VAN DER LOOS, H. & T. A. WOOLSEY. 1973. Science **179:** 395–398.
20. O'BRIEN, T. F., A. FAISSNER, M. SCHACHNER & D. A. STEINDLER. 1990. Neuroscience (Abstr. 16), in press.
21. STEINDLER, D. A. & B. K. TROSKO. 1989. Anat. Embryol. **179:** 423–434.
22. TOPP, K. S., B. T. FADDIS & V. K. VIJAYAN. 1989. Glia **2:** 201–211.
23. ENG, L. F. 1988. *In* The Biochemical Pathology of Astrocytes. M. D. Norenberg, L. Hertz & A. Schousboe, eds.: 79–90. Alan R. Liss, Inc. New York.
24. NIETO-SAMPEDRO, M. 1988. Science **240:** 1784–1786.
25. NOBLE, M., K. MURRAY, P. STROOBANT, M. D. WATERFIELD & P. RIDDLE. 1988. Nature **333:** 560–562.
26. RAFF, M. C., L. E. LILLIEN, W. D. RICHARDSON, J. F. BURNE & M. D. NOBLE. 1988. Nature **333:** 562–565.
27. GUILIAN, D., D. G. YOUNG, J. WOODWARD, D. C. BROWN & L. B. LACHMAN. 1988. J. Neurosci. **8:** 709–714.
28. LINDSAY, R. M., C. EMMETT, G. RAISMAN & P. J. SEELEY. 1987. Ann. N. Y. Acad. Sci. **495:** 35–52.
29. ENG, L. F., E. STOCKLIN, Y.-L. LEE, R. A. SHIURBA, F. CORIA, M. HALKS-MILLER, C. MOZSGAI, G. FUKAYAMA & M. GIBBS. 1986. J. Neurosci. Res. **16:** 239–250.
30. LAYWELL, E. D., A. FAISSNER, M. SCHACHNER & D. A. STEINDLER. 1990. Soc. Neurosci. Abstr. **16:** 163.
31. KRUSE, J., G. KEILHAUER, A. FAISSNER, R. TIMPL & M. SCHACHNER. 1985. Nature **316:** 146–148.

Role of the Midline Glia and Neurons in the Formation of the Axon Commissures in the Central Nervous System of the *Drosophila* Embryo

CHRISTIAN KLÄMBT AND COREY S. GOODMAN

Howard Hughes Medical Institute
Department of Molecular and Cell Biology
519 Life Science Addition
University of California, Berkeley
Berkeley, California 94720

The growth cones of commissural interneurons first head towards the midline and only after crossing it do they dramatically change their behavior as they turn and follow a specific longitudinal pathway. Correlated with this change in behavior is a similar change in their expression of surface glycoproteins; some insect commissural neurons switch from expressing fasciclin I to fasciclin II after crossing the midline,[1] just as some vertebrate commissural neurons switch from expressing TAG-1 to L1.[2]

Not surprisingly, the midline in such diverse organisms as rats[3] and fruit flies[4] is comprised of special cells that appear to play a key role in the guidance of commissural axons. In vertebrate embryos, the cells at the ventral midline of the neural tube, called the floorplate, have a variety of antigenic and physiologic properties that make them distinct from the rest of the neural tube, not the least of which is their ability to chemotropically guide commissural axons towards the midline.[5]

In *Drosophila,* the cells at the midline of the neuroepithelium have such distinct properties that Poulson[6] called them mesectoderm. Initially after gastrulation, these midline cells transiently maintain adhesive contacts with the mesoderm and morphologically look more like the mesodermal cells (that just invaginated) rather than the neuroepithelial cells on either side of them. But these mesectodermal cells ultimately loose their contacts with the mesoderm, and although present in the surface epidermis, they continue to look morphologically different from the neighboring neuroepithelial cells. The midline cells also form a distinct mitotic domain separate from the lateral neuroepithelium.[7] From these midline precursors arise a variety of different neurons and glia including the six midline glia (MGA, MGM, and MGP),[4] the median neuroblast (MNB), the ventral unpaired median neurons (VUMs), and the pair of midline precursor 1 neurons (MP1).

Fortunately, a series of mutants already exists that alters pattern formation at the ventral midline of the *Drosophila* embryo.[8–16] Although selected because of their defects in the ventral cuticle, some of these mutations clearly alter the development of the mesectodermal cells as well. For example, mutations in two different genes, *single-minded* (*sim*)[12,13,17] and *slit,*[14,18] selectively affect the differentiation of all of these midline cells. In both cases, the axon commissures are ultimately missing and all CNS axon pathways collapse together at the midline. Clearly, the midline cells play a major role in organizing the developing CNS, but what signals are involved and from what midline cells they originate remain unknown.

Because of our interest in growth cone guidance, we began a detailed cellular and genetic analysis of the developing midline of the *Drosophila* CNS.[19] Our goal was to uncover the cells and molecules that guide commissural growth cones. During these studies, we discovered that the developing *Drosophila* midline is an ideal model system for the genetic analysis of growth cone guidance, the cell interactions that control the fate of a diverse set of glia and neurons, and the factors that guide the specific migrations of several of these glial cells.

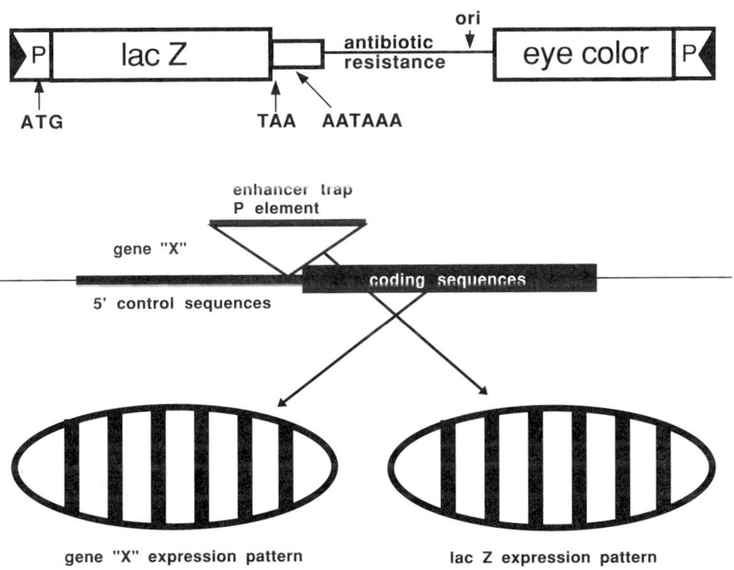

FIGURE 1. The rationale behind an enhancer trap screen. This schematic figure depicts the strategy used in the enhancer trap experiment. A transposon called a P-element is modified (p[*lac*Z]) so that it carries the bacterial *lac*Z gene fused in frame to the translational start site of the P-element transposase gene. This construct provides only the signals for the start of transcription and thus acts as a relatively neutral promoter, showing no or little tissue-specific expression on its own. In addition, the construct contains a *Drosophila* eye color gene (*rosy* or *white*) to identify new insertions of the P[*lac*Z] element. A bacterial antibiotic resistance gene as well as a bacterial origin of replication is included in the construct to allow rapid cloning of a tagged gene. When the P[*lac*Z] element integrates near the 5′ enhancer and promoter control sequences for a normal gene ("X"), the relatively neutral promoter of the P[*lac*Z] element now comes under the control of the surrounding tissue-specific enhancers, so that the *lac*Z expression pattern in many cases mimicks the normal expression pattern of the neighboring gene "X". For details see refs. 22 and 23.

In order to begin a genetic analysis of cell interactions at the midline, we first needed a set of markers that would help to identify the different cell types in normal and mutant embryos. In *Drosophila,* one way to generate a battery of cell-type-specific markers is to use an "enhancer trap" screen with a P-element vector containing the *Escherichia coli lac*Z gene fused to a relatively neutral promoter (FIG. 1).[20–23] By screening over 11,000 lines for embryos displaying spatially restricted patterns of expression of β-galactosidase (β-gal) in subsets of cells at the developing midline, we identified over 40 new cell-type-specific markers for midline cells. In our studies, we used these enhancer trap lines to help elucidate the development of midline cells and axon commissures and then used a series of midline mutants to test the predicted functions of specific midline cells.[19]

EIGHT MIDLINE PRECURSORS IN EACH SEGMENT GENERATE A DISCRETE SET OF SPECIAL GLIA AND NEURONS

At the cellular blastoderm stage, each presumptive segment is about four cells long in the anterior-posterior axis. Along the dorsal-ventral axis, the presumptive pair of bilaterally symmetric neurogenic regions lie ventrolaterally on each side of the embryo, separated by the presumptive mesoderm alone.[24] During gastrulation, the mesoderm invaginates along the ventral midline into the interior of the embryo, forming an anterior-posterior groove called the ventral furrow. As a result of gastrulation, the two neurogenic regions are brought together along the ventral surface of the embryo. But as gastrulation ends, these two lateral neurogenic regions do not actually come in contact with each other at the midline; rather, there is a small group of special cells at the ventral midline, called the mesectoderm,[6] that lies between these two large blocks of neurogenic cells (FIG. 2A).

At the boundary on each side of the cellular blastoderm between the presumptive mesoderm and neurogenic regions sits a single row of mesectodermal cells that will generate these special midline cells. Although not morphologically recognizable until near the end of gastrulation, *in situ* hybridization with a probe to the *single-minded* (*sim*) gene has revealed that the two bilaterally symmetric rows of presumptive midline cells are already distinct by the beginning of gastrulation (FIG. 5B in ref. 11); similar results have been observed with a probe to the *rhomboid* gene.[15] At the end of gastrulation, the mesectodermal cells come together along the ventral midline to form a morphologically discrete group, two cells wide (one from each side) and four cells per segment long, that is distinct from both the mesoderm and the two lateral neurogenic regions (FIG. 2A and B). After gastrulation, during germ band elongation, the relationships and shapes of cells change so that each segment becomes twice as long and half as wide as the whole germ band elongates and doubles over on itself (FIG. 2A). In terms of the midline, the net effect of this morphogenetic process is that the two rows of four cells each per segment interdigitate to become a single row that is eight cells long per segment (FIG. 2B).

As the germ band elongates, neuroblasts (NBs) in the neurogenic region begin to delaminate dorsally from the ventral surface and move to the inside of the epidermis; in total, about one of every four cells in the neurogenic region becomes a NB and the rest remain epidermal cells.[24] In contrast, all of the midline cells delaminate into the same layer as do the NBs (FIG. 2A), leaving a transient gap in the surface epidermis that is quickly filled by cell divisions of the most medial row of epidermal cells.

About this time, each of these eight midline cells divides once to generate 16 midline cells. For some of these cells, as will be described, this cell division is their first and final one. But several of the midline cells, particularly those in the posterior half of each segment, divide several more times. By the beginning of germ band retraction, the initial row of eight midline precursor cells has generated ~26 midline cells that differentiate into both glia and neurons (R. Fetter, unpublished results).

By midline precursor cells, we refer to the initial row of eight cells along the midline of each segment before their first cell divisions. Similar words have been used to refer to a group of midline neuronal precursors in the grasshopper embryo that divide only once to generate two progeny.[25] Most of the grasshopper midline

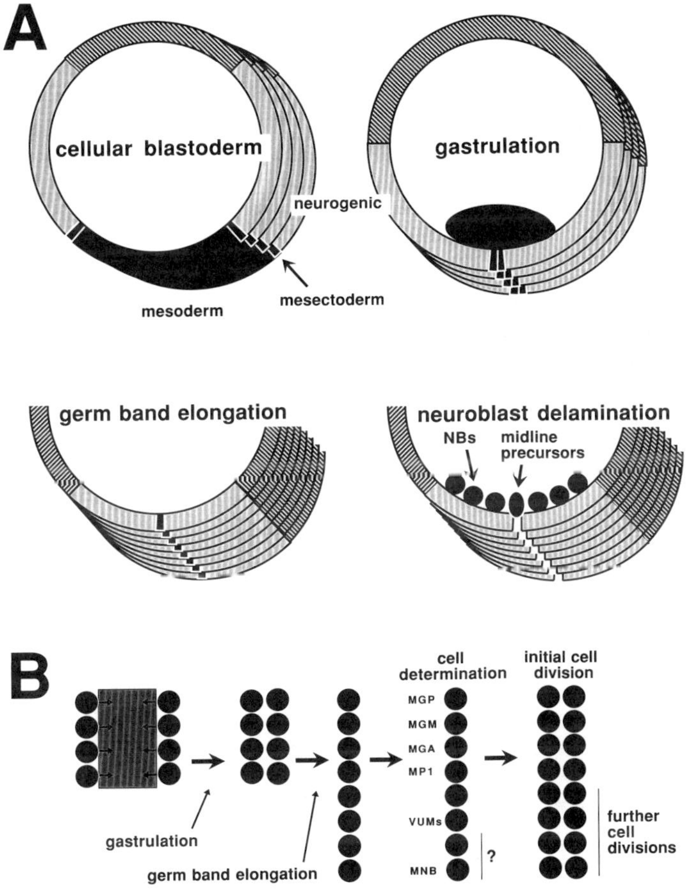

FIGURE 2. Development of the ventral midline of the *Drosophila* embryo. (**A**) Development of the midline is depicted in a series of four schematic cross-sections of embryos from four successive developmental stages (dorsal is up). During the blastoderm stage the mesectodermal or midline precursor cells (*indicated in black*) are located on either side of the mesodermal anlage; they separate the presumptive mesoderm from the presumptive neurogenic ectoderm. As gastrulation proceeds, mesodermal cells invaginate and the two rows of four mesectodermal cells per segment become juxtaposed at the ventral midline. During germ band elongation these cells intermingle to form a row of eight midline precursor cells per segment. Before these precursor cells start to divide to give rise to the neurons and glia of the ventral midline, they all delaminate from the ventral ectoderm into the neuroblast (NB) layer, leaving behind a transient gap, which is quickly filled in by cell divisions of neighboring epidermal cells. (**B**) Schematic drawings of similar stages as in (**A**) but showing a frontal view. During gastrulation the two rows of mesectodermal cells (each row is four cells per segment) initially come together at the midline and then intermingle during germ band elongation to form the row of eight midline precursor cells per segment. As revealed using the enhancer trap lines described in the text, the eight midline precursors become determined to the following fates (from anterior to posterior). The anterior-most three precursors give rise to the midline glia and are: midline glia posterior (MGP), midline glia middle (MGM), and midline glia anterior (MGA). The relative order of these cells changes during development as a result of cell migrations (see FIG. 3). The fourth precursor is MP1 (midline precursor 1, using the grasshopper nomenclature of Bate and Grunewald[25]) and gives rise to the two MP1 neurons. These anterior four midline precursors divide only once. The posterior four midline precursors give rise to the ventral unpaired median (VUM) neurons, the median neuroblast, and their support cells. Two to three precursors give rise to the VUMs (and possibly their support cells), and the fourth precursor gives rise to the MNB (and possibly its support cells). For further details see text.

precursors, with the exception of MP2, have origins along the midline and are likely to be the homologues of the *Drosophila* cells described herein; MP2 arises in the most medial NB row in the lateral neurogenic region.[26] Among the eight *Drosophila* midline precursors described herein, one is the homologue of grasshopper MP1. *Drosophila* does not have an MP3 homologue.[27] It is possible that the six VUMs arise from a trio of midline precursors that are homologues of the MP4, MP5, and MP6 cells in the grasshopper that give rise to a similar set of six unpaired median neurons.

As already indicated herein, some of the glia and neurons that arise from the midline have previously been identified. These cells include three pairs of midline glial cells (anterior, middle, and posterior pair of midline glia, termed MGA, MGM,

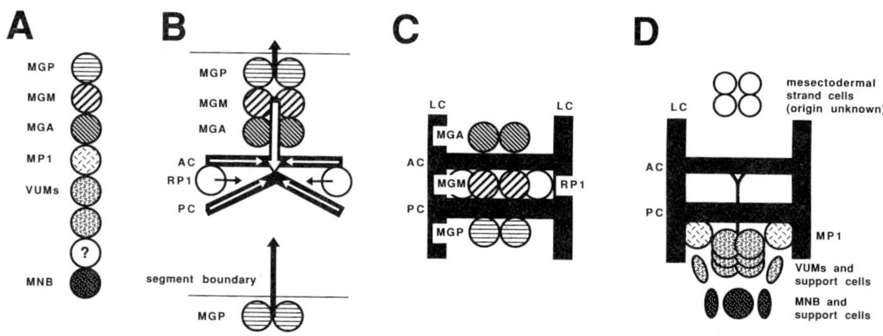

FIGURE 3. Summary of migrations and cell types at the midline. (**A**) Schematic diagram showing the midline precursors and their fates. Different lineages are indicated by different shading throughout this figure and FIGURES 4 and 5. Abbreviations of precursor identities are given in FIGURE 1B. (**B** and **C**) Dorsal, frontal view of the development and migrations of the midline glial cells. In **B** the three pairs of midline glia (MGP, MGM, and MGA) are shown in their original position, with arrows along the midline indicating their directions and extent of migration. The MGP cells migrate into the next anterior segment, whereas the MGM cells migrate posteriorly over the MGA cells to separate the anterior commissure (AC) from the posterior commissure (PC). Note also the RP1 neurons, which are born in the lateral neurogenic region, but migrate medially between the two commissures towards the final location of the MGM glia. (**C**) Migrations of the MGM glia, MGP glia, and RP1 neurons are nearly complete by the time the longitudinal connectives (LC) are formed. **D** is a more ventral, frontal view, showing the relative locations of the other cell types at the midline, including the four mesectodermal strand cells between the segments, the six VUM neurons and their two support cells, the two MP1 neurons, and the median neuroblast (MNB) and its support cells. Note that the VUM axons extend anteriorly along the midline and bifurcate at the posterior edge of the anterior commissure.

and MGP, respectively[4]), a single median neuroblast and its several support cells[27] (which express the *engrailed* gene; see ref. 28), a group of six VUM neurons (which express the antigen recognized by both SOX2 and 22C10 mAbs; eg, ref. 29), and a bilateral pair of MP1 neurons that arise from a single MP1 precursor[25,27] (which both transiently express the *fushi tarazu* gene; see ref. 26). These cells are summarized in FIGURE 3C and D.

In a 12-hour embryo, the nuclei of the three pairs of glia (MGA, MGM, and MGP) lie in front of, between, and just behind the two axon commissures, respectively, and their processes enwrap these axons[4] (FIG. 3C). At the same stage, the MNB and its support cells lie at the posterior end of each segment near the dorsal surface (FIG. 3D). The cell bodies of the six VUM neurons and their two glial support cells (Klämbt *et al.*, unpublished results) come to lie ventrally, just behind the posterior commissure. Their axons extend dorsally and anteriorly to the dorsal surface at the posterior edge of the anterior commissure. Here, in contact with the MGM glia, they bifurcate and extend laterally in the anterior commissure and then turn anteriorly. Most (perhaps all) of their axons join and exit the CNS in the intersegmental nerve of the next anterior segment.[30]

Finally, between each segment lies a group of about four cells that have processes at the midline extending from the ventral epidermis to the dorsal surface of the CNS and that stain with antibodies against most extracellular matrix components.[31] These cells, called the mesectodermal strand[31] (FIG. 3D), are of unknown origin and may or may not arise from the row of eight midline cells (to be described).

ISOLATION OF ENHANCER TRAP LINES AS MARKERS FOR SPECIFIC MIDLINE LINEAGES

Although the midline in each segment starts from only eight midline precursors and generates only ~26 progeny, sorting out exactly who begets whom requires some additional lineage markers. To this end, we generated a battery of cell-type–specific lineage markers by undertaking a series of "enhancer trap" screens using P-element vectors containing *lac*Z fusion genes fused to relatively neutral promoters[10,22,23] (FIG. 1). We first screened about 5,700 P[w^+, *lac*Z] insertions lines (generated using a *lac*Z-*mini white* construct kindly provided by U. Grossniklaus and W. Gehring). We then screened over 5,000 P[ry^+, *lac*Z] insertion lines (generated using a *lac*Z-*rosy* construct kindly provided by Y. Hiromi in our lab). Each of the resultant 11,000 or so lines was examined for β-gal expression during embryogenesis.

In these enhancer trap lines, expression along the midline is a frequent pattern. Of the 11,000 lines screened for β-gal expression, about 5% have some form of β-gal expression along the developing midline. Of these 500 or more lines, we saved 125 (about 1%) for their specific pattern of midline expression. Forty of these have β-gal expression in a discrete subset of midline cells and are thus useful for the studies described here.

Thus far we have identified lineage markers for all of the midline precursor cells and what appears to be all of their progeny. Of all the midline cells, only the mesectodermal strand cells are not stained by any of our enhancer trap lines, which suggests that these cells do not arise from the mesectoderm. One alternative is that they arise from the mesoderm, but proving this awaits a lineage marker for these cells.

Most of the enhancer trap lines isolated in our screen label more than one lineage. All of the cells labeled by these lines were first analyzed as X-gal–stained whole-mount preparations of 12-hour embryos under a dissecting microscope, then under a compound microscope, and finally in light microscopic cryosections or plastic sections. This is a stage for which we already have extensive electron microscopic analysis of the cells at the midline,[30] and therefore one in which we can determine the identity of these different labeled cells. Further conformation of a given cell identity was obtained by double staining the embryos derived from the enhancer trap lines with both X-gal and different antibodies. For example, the anti-engrailed mAb was used to identify the median neuroblast and its support cells, the BP102 mAb was used to stain the developing axon commissures, and the 22C10 mAb was used to stain the VUM neurons and their axons. Once the cell identities were established, we could trace the cells back in successively younger embryos and thus determine the midline precursor cells of origin for several of the midline neurons and glia. The following six enhancer trap lines were used in these studies: T13, AE60, AQ65, AuI, AA142, and X55. (For details of these lines, see ref. 19.)

CELL LINEAGE ANALYSIS OF THE EIGHT MIDLINE PRECURSORS

By following these six enhancer trap lines in double-labeled (X-gal and mAb) embryos, we were able to determine the lineage of most, but not all of the eight midline precursors. This analysis is summarized in FIGURES 2B and 3A. We determined the lineage of five of the eight midline precursors and are confident of the cells generated by the other three, although we do not yet know their exact lineage. From the anterior of the segment, the first three midline precursors divide once each to give rise to the pairs of MGP, MGM, and MGA glia, respectively. These positions were at first surprising, inasmuch as they were exactly opposite to their final relative positions around the mature commissures (ie, initially P-M-A and later A-M-P) (FIG. 3B and C). As will be described in the section on cell migration, the MGP glia migrate anteriorly into the next segment, and the MGM glia migrate posteriorly over the dorsal surface of the MGA glia to separate the two commissures. Both commissures are initially pioneered by growth cones that cross the midline posterior to the four nuclei of the MGM and MGA glia; it is these two pairs of glia that appear to play major roles in the formation of the commissures; the MGP glia appear less important because they arrive at the axons after most of the major events are over.

From anterior to posterior, the fourth midline precursor is MP1. This cell divides once to generate two neuronal progeny (the MP1 progeny), which migrate laterally away from the midline (and a bit ventrally) to come to lie over the two MP2 progeny in the lateral neurogenic region.

The fifth midline precursor initially divides once to give rise to the two most anterior of the VUM neurons. These anterior-most VUM neurons play an important role, and to distinguish them from the rest, we call them V cells. The other VUM neurons, and possibly also their special glial support cells, probably arise not only from the fifth midline precursor, but also from the sixth and seventh midline precursors. The eighth midline precursor probably gives rise to the median neuroblast and its support cells. The reason for the ambiguity in describing the lineage of the sixth through eighth midline precursors is that they divide more than one time. Further lineage analysis will be required to determine their precise number of progeny. Fortunately for the studies described here, we know the most about the lineage of the first through fifth midline precursors, and all of the interesting interactions and guidance cues for the formation of the axon commissures appear to be provided by the progeny of the second through fifth midline precursors (MGM, MGA, MP1, and V).

THE PIONEERING OF THE POSTERIOR COMMISSURE

Formation of the two axon commissures in each segment takes place largely between 8.5 and 10 hours of embryonic development, corresponding to stage 12.[24] As several different discrete and independent events occur in this narrow time window, we have divided this stage into three substages. There is an easy morphologic guide for these stages; between 8 and 10 hours, the embryo undergoes the morphogenetic movement of germ band retraction as the segments that previously had extended around onto the dorsal surface of the embryo (during germ band extension) now retract to the ventral surface. Referring to the number of segments (5, 3, or 0) still on the dorsal side of the embryo, we consider these three substages 12/5, 12/3, and 12/0.

With the use of the BP102 mAb which stains all CNS axons, the first CNS growth cones can be detected at about 8.5 hours of development corresponding to a germ band extended embryo at stage 12/5. The first growth cones arise from neurons about 2–3 cell body diameters lateral to the midline cells and show directed growth straight towards the pair of V cells (the anterior-most of the VUMs) at the midline (see FIG. 4A); these first growth cones, about 2–3 in number from each side, pioneer the posterior commissure.

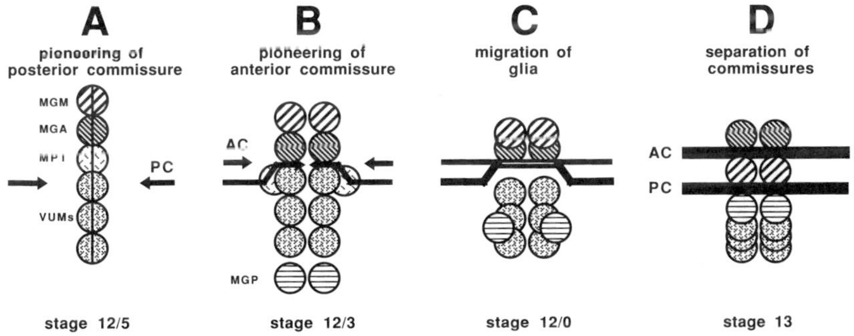

FIGURE 4. Summary of the development of the axon commissures. Development of the two axon commissures takes place in a series of discrete steps: (**A**) pioneering of the posterior commissure (PC), (**B**) pioneering of the anterior commissure (AC), (**C**) migration of the midline glia, and (**D**) the resulting separation of the anterior and posterior commissures. Stages 12/5, 12/3, and 12/0 are further subdivisions from the original stage 12 as described by Campos-Ortega and Hartenstein.[24] (See text for details on staging.) (**A**) The growth cones that pioneer the posterior commissure extend directly towards the anterior-most VUM neurons, called the V-cells. Once these growth cones meet the V-cells, they extend around the anterior end of these cells and cross the midline in the position where the MP1 neurons had just been, as shown in **B**. The growth cones that pioneer the anterior commissure extend directly towards the posterior edge of the MGA glia. (**C** and **D**) Subsequent separation of the two commissures results from the migration of the MGM glia.

As these first growth cones reach the midline, the first cell migrations at the midline can be observed. The pair of MP1 progeny, which are located directly anterior to the V cells and posterior to the MGA and MGM glia, migrate laterally and ventrally away from the midline (FIG. 4B). The pioneering growth cones, having grown straight towards the V cells, now change direction and grow anteriorly around the V cells (in the space where the MP1 cells had just been) (FIG. 4B). After growing around the V cells, they contact and fasciculate with their contralateral homologues at the midline, thus forming the posterior commissure. Because of the migration of the MP1 progeny, this fasciculation point is also just posterior to the MGA glia.

In summary, during the 12/5 stage, the growth cones that pioneer the posterior commissure in each segment of the *Drosophila* CNS first extend straight towards the

V cells (the anterior-most VUM neurons). Once reaching these cells, they extend anteriorly to grow around them in the space formerly occupied by the MP1 progeny. Thus, the MP1 and the anterior-most VUM neurons (the V cells) appear to play an important role in the formation of the posterior commissure.

THE PIONEERING OF THE ANTERIOR COMMISSURE

The lateral and ventral movement of the MP1 progeny brings the MGA and MGM glia in close proximity to the already formed posterior commissure. Beginning in the 12/3 stage, the anterior commissure is pioneered (FIG. 4B and C). Shortly after the pioneering of the anterior commissure, the VUM neurons, their growth cones waiting at the midline, bifurcate and extend laterally in each direction. The anterior commissure initially forms straight across the midline, whereas by this stage, the posterior commissure is "V" shaped, pointing anteriorly towards the midline from both sides (FIGS. 3B and 4C). Both commissures meet at the midline in the space bounded on the posterior side by the V cells and on the anterior side by the MGA glia.

The axons of the two commissures at the 12/3 stage are separated by the midline growth cones of the VUM neurons. At this stage, the cell bodies of the VUM neurons are migrating ventrally, leaving behind their growth cones and medial axons which mark the separation of the anterior and posterior commissures (FIG. 5Aii). The lateral growth cones that pioneer the anterior commissure appear to grow along the surfaces of the MGA glia (to their anterior side) and the VUM growth cones (to their posterior side). Although the anterior commissure develops after the posterior commissure, by 12 hours of development, the anterior commissure contains almost twice as many axon bundles as does the posterior commissure.

MIGRATION OF THE MIDLINE GLIA AND SEPARATION OF THE COMMISSURES

When the two commissures initially form, they are in close proximity at the midline and not yet separated. Separation of the anterior and posterior commissures takes place during stage 12/0; this separation correlates with the posterior migration of the MGM glia (FIGS. 3B, 4D, and 5A) and the medial migration of the RP1 neurons (FIGS. 3B and C). (These neurons arise from the lateral neuroepithelium).[30,32] The pair of MGM glia, and cell bodies of the pair of RP1 neurons, ultimately come to lie between the two commissures (FIG. 3C). The MGM glia already have a cytoplasmic anchor extending over the top of the MGA glia to the point where the two commissures form (FIG. 5Aii). These glial endfeet are in contact with the dorsal surface of the VUM growth cones, and thus these processes separate the initial axons that pioneer the two commissures. The MGM cells now migrate posteriorly over the top of the MGA glia and move along the medial VUM axons in between the two commissures to separate them (FIG. 5Aiii). This migration changes the relative order of these two pairs of midline glia from (starting from most anterior) MGM-MGA to MGA-MGM (FIG. 3B and C).

The posterior pair of glia (MGP) are not directly involved in the process of commissure separation. Rather, the MGP glia undergo a complex migration into the next anterior segment, where they are transiently located either ventral or lateral to

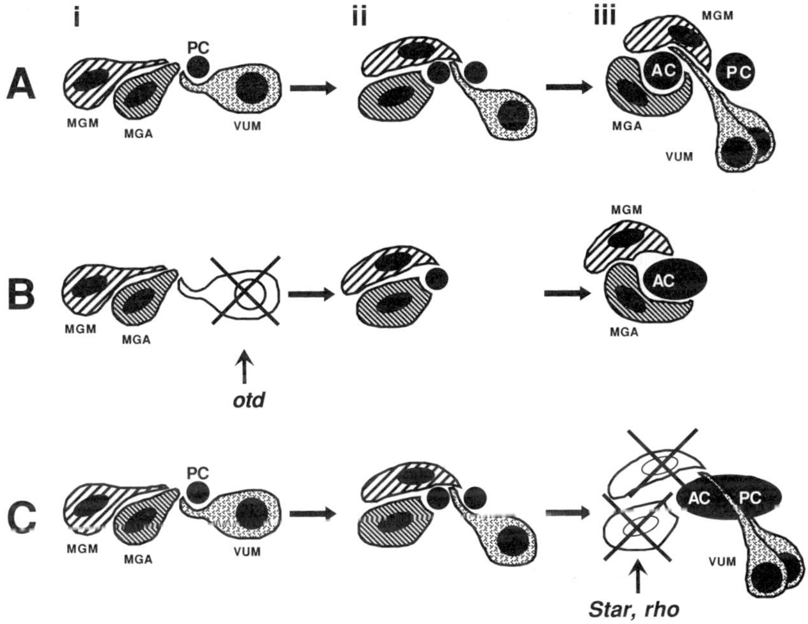

FIGURE 5. Summary of development of midline cells and axon commissures in wild type and mutant embryos. (**A**) Migration of the midline glia in relation to the VUM neurons and the formation of the two commissures in schematic sagittal views of wild-type embryos. (i) The posterior commissure (PC) forms just anterior to the VUM cell bodies on their anterior extensions that are the beginning of their growth cones. The midline glial cells extend endfeet posteriorly which contact the VUM growth cones. (ii) The anterior commissure forms in contact with the MGA and MGM glia on the anterior side of the VUM growth cones (in contrast to the posterior commissure which is now on the posterior side of the VUM growth cones). During this period, the VUM cell bodies begin to drop ventrally and the MGM cells migrate over the top of the MGA cells. (iii) The commissures are separated by the migration of the MGM glia which contact the VUM axons. The VUM growth cones now extend along the MGM glia to the dorsal surface of the anterior commissure where they bifurcate, as their cell bodies continue to drop ventrally. (**B**) The same developmental stages are shown for homozygous *otd* mutant embryos. (i) The VUM neurons fail to differentiate and probably die; as a result, no posterior commissure is formed. (ii) The anterior commissure forms, but (iii) no glial migrations are observed. Rather, in *otd* embryos, the anterior commissure is often thicker than normal, reflecting the fact that many of the axons that normally would have extended in the posterior commissure now extend abnormally in the anterior commissure. (**C**) Development of homozygous *Star* and *rhomboid* (*rho*) mutant embryos. Posterior commissure (i) and anterior commissure (ii) form as they normally do. (iii) However, in later stages of development the MGM cells fail to migrate, resulting in fused commissures. In both *Star* and *rhomboid* the midline glia ultimately die. (See text for further details.)

the median neuroblast and its support cells. After the separation of commissures is complete, the MGP glia finish their anterior migration, arriving at their final position on the dorsal surface of the nervous system just posterior to the posterior commissure. Around this time, the anterior commissure grows rapidly in size, until it contains about twice as many axons as does the posterior commissure.

MODEL FOR THE FORMATION OF THE AXON COMMISSURES

Cellular analysis of the development of the axon commissures in the *Drosophila* embryonic CNS leads us to propose a model of sequential cell interactions (FIGS. 4 and 5A). The model makes a number of predictions about the functions of specific cell types in the formation of the axon commissures. (1) The growth cones that pioneer both commissures appear to be attracted towards the midline from a distance, suggesting a possible role for chemotropism (FIG. 4A and B). (2) The growth cones that pioneer the posterior commissure are guided towards the V cells and MP1 neurons. This commissure appears to form in two steps: initially, the growth cones extend towards the V cells, and once reaching them, they then turn anteriorly over the MP1 neurons to cross the midline anterior to the V cell nuclei (FIG. 4A and B). (3) The growth cones that pioneer the anterior commissure are guided towards the MGA glia and the VUM growth cones (FIG. 4B and C). Thus, different midline cell types, and presumably different signals, appear to be involved in the pioneering of the two commissures. (4) The last step of axon commissure formation requires the separation of the two commissures. This event involves the posterior migration of the MGM glia over the top of the MGA glia and along the medial VUM growth cones between the two commissures (FIGS. 4C and D; 5Aii and iii). Thus, the model predicts that the midline glia play an important role in the formation of the anterior commissure and the key role in the separation of the two commissures, whereas the VUM neurons play the key role in the formation of the posterior commissure and an important role in the formation of the anterior commissure.

GENETIC TEST OF THE MODEL: ANALYSIS OF MIDLINE MUTANTS

The model described herein suggests key roles for both the midline glia (particularly MGM and MGA) and the midline neurons (particularly the VUMs) in the formation of the two commissures in each segment of the *Drosophila* CNS. Disruption of any one of these proposed functions should lead to a predictable phenotype. For example, the absence of the VUM neurons should lead to the absence of the posterior commissure; the absence or loss of function of the MGM and MGA glia should lead to a fused commissure phenotype. The absence of both the VUMs and the midline glia should lead to a lack of commissure phenotype. To test the validity of this model, we undertook a detailed analysis of several midline mutants, taking into account their effect on the absence and/or differentiation of key midline cells and on the development of the axon pattern of the commissures.

Fortunately, a series of mutants that alter pattern formation at the ventral midline already exists.[8–15] One such group of genes was named the spitz group[11] and includes the following zygotically acting genes: *spitz* (*spi*), *Star* (*S*), *single-minded* (*sim*), *pointed* (*pnt*), and *rhomboid* (*rho*). Mutations in all five of these loci cause zygotic lethality, lead to similar cuticular phenotypes with a narrowing of ventral denticle bands, and lead to phenotypes in the developing CNS.

In our studies reviewed here,[19] we consider four of the spitz group genes and two other genes, in the following three groups, according to their final commissural phenotypes: group 1, collapsed longitudinal tracts with no commissures (*sim* and *slit*); group 2, no posterior commissure (*orthodenticle* [*otd*]); and group 3, fused commissures (*spi*, *S*, and *rho*).

The ultimate phenotype in *sim* and *slit* null mutants is the absence of commissures and the collapse of the longitudinal tracts into a single, fused tract at the midline (FIG. 6B). In *sim* embryos, the midline precursors divide at least once, but then they die during germ band retraction (data not shown[17]). In *slit* embryos we do not observe cell death at the midline, but rather the midline cells are displaced ventrally and do not appear to differentiate properly.[18] This death (*sim*) or displacement and lack of further differentiation (*slit*) of the midline cells occur during a developmental stage (12/0) at which the commissures are just being established. Although the commissures are ultimately missing in both mutants and the longitudinal tracts are collapsed on the midline, we wondered about what initially happens to the growth cones that normally pioneer the commissures, and if, in fact, the commissures transiently form. Because recent studies have suggested that the axon scaffold phenotype in *sim* mutants is due to the absence of *slit*,[17] we turned out attention to *slit* mutant embryos, although similar results have been seen in *sim* mutant embryos as well.

In *slit* embryos, the posterior commissure sometimes forms, whereas the anterior commissure is always initially formed. Nevertheless, these commissures appear abnormal from the outset in that they already are narrower and are not as straight and regular in shape. As the midline cells are displaced ventrally, the axon tracts start to fuse and collapse at the midline, and the transient and abnormal appearing commissures disappear. The final phenotype in *slit* mutants gives little indication that the commissural tracts ever formed; rather, the longitudinal tracts are fused at the midline and the CNS is narrower. In summary, analysis of the time course of axon pathway development reveals that in *slit* (and *sim*) mutant embryos, the midline cells differentiate enough to allow one or sometimes both commissures to begin to form (albeit abnormally), and then, with the death (*sim*) or displacement and lack of further differentiation (*slit*) of the midline cells, the nascent commissures collapse and the longitudinal tracts fuse at the midline. The transient appearance of commissures lasts for only about 30 minutes of development before the commissures disappear.

To test the model just proposed, mutants that affect only a subset of lineages at the midline should be more informative about specific cellular functions. The mutation *orthodenticle* (*otd*) has been isolated on the basis of its cuticular phenotype in which structures at the ventral midline are affected.[9] We find that the development of the CNS is also strongly affected by the *otd* mutation: the posterior commissure is missing in every segment, and sometimes in the thoracic segments the formation of the anterior commissure is also affected. As in most mutants that affect the midline lineages, the nervous system is also narrower in *otd* mutant embryos. In analyzing stage 12/5 *otd* embryos, we found that the growth cones that normally pioneer the posterior commissure are extended at the normal time, but do not choose the correct pathway. Rather than growing straight towards the midline and then traversing it, the growth cones in *otd* mutant embryos do not show any affinity for the midline; instead, they grow anteriorly and posteriorly on their own side. In contrast to the aberrant behavior of the posterior commissure pioneers in most segments, the growth cones in *otd* mutant embryos that normally pioneer the anterior commissure grow straight towards the midline.

To analyze the fates of specific midline cells in *otd* mutant embryos, we crossed several enhancer trap lines into the *otd* background. Interestingly, *otd* strongly affects several midline cells: the VUM neurons and probably the MNB degenerate at early stages (12/5 to 12/3). This finding is in agreement with the predictions that emerge from the cellular model, namely, that mutations that delete the VUMs should disrupt the formation of the posterior but not the anterior commissure (FIG. 5B).

In addition, two pieces of evidence suggest that the midline glia may also be altered in *otd* mutants. First, the β-gal expression in the midline glia of the AA142 enhancer trap line is down-regulated in *otd* mutants. Second, the MGM glia fail to migrate posteriorly over the top of the anterior commissure axons. Either or both of these defects might be secondary effects due to loss of the VUM neurons, particularly as the MGM glia contact and migrate along the growth cones of the VUM neurons.

Mutations in any of the genes of this group of three genes lead to the same CNS phenotype: a fusion of the two commissures. The initial development of the axon commissures appears normal through stage 12/0; both commissures develop right on schedule. However, the mutant phenotype appears at the stage at which the MGM glia should normally migrate between the two commissures, resulting in their separation. Instead of separating into distinct anterior and posterior commissures, these axon tracts stay close together, trapping the cell bodies of the RP1 neurons that normally migrate medially to meet the posteriorly migrating MGM glia. Later in development, as more axons cross the midline, the commissures appear to be fused with many axons overgrowing the RP1 neurons. This fusion of the commissures also correlates with a narrowing of the CNS.

To analyze the cellular basis for the fused commissure phenotype shown in all three mutants, we crossed several enhancer trap lines into *spi*, *S*, and *rho* mutant backgrounds. In all three mutants, the glial cell lineage is affected, although at different times of development depending on the mutant background. In *rho* mutant embryos, the midline glial cells appear to be missing from early stages, suggesting a function for this gene in the normal determination or early differentiation of these glial cells (data not shown). Moreover, *rho* acts similarly in the determination of individual cell types in the developing PNS.[15]

In *Star* mutant embryos, the midline glial cells initially develop normally but then the MGM glia fail to migrate. During stage 14, the midline glia finally die. In *spitz* mutant embryos, the midline glial cells behave in a very similar manner to *Star*, in that the MGM glia do not properly migrate in between the two commissures. However, in contrast to *Star,* in which the midline glia ultimately die, in *spitz* these glia come to lie abnormally on the dorsal surface of the CNS where they persist until the end of embryogenesis. In all three cases, the fused commissure phenotype correlates with a defect in the midline glia. In *rho* and *Star,* the midline glia die (FIG. 6D), whereas in *spitz* they behave abnormally. All other midline cells appeared unaffected in *Star* and *spitz.* (We have not yet examined these other lineages in *rho* mutant embryos.)

DISCUSSION

In the central nervous systems of chordates, arthropods, and many other phyla, the two halves of the CNS are interconnected by commissures as many interneurons project across the midline. A common organizational principle in both chordates and arthropods is the observation that the two lateral neurogenic regions are separated at the midline by a special group of cells with properties different from those of their lateral neighbors. *In vitro* experiments in vertebrates[3,5] and genetic ablation experiments in *Drosophila* (this report) show that the midline cells in both rats and fruitflies play a key role in the guidance of commissural axons. In vertebrates this guidance from a distance was shown to be chemotropic; it is only speculation whether a similar mechanism might be at work in *Drosophila,* given the short distances between the midline and the neurons that extend the pioneering growth cones. Two different mutants suggest that some sort of midline signal exists that is attractive to commissural growth cones in *Drosophila;* however, we do not yet know whether the midline signal is secreted or requires cell surface contact. First, with the loss of function of both a neural cell adhesion molecule and a cytoplasmic tyrosine kinase in the *fas I · abl* double mutant, the commissures do not form properly[33]; rather, the growth cones that normally pioneer the posterior commissure do not appear to grow towards the midline as they normally do (Klämbt, unpublished results). Second, in *otd* mutants (as shown here), the posterior commissure does not form because the growth cones that normally pioneer this commissure now extend anteriorly or posteriorly on their own side.

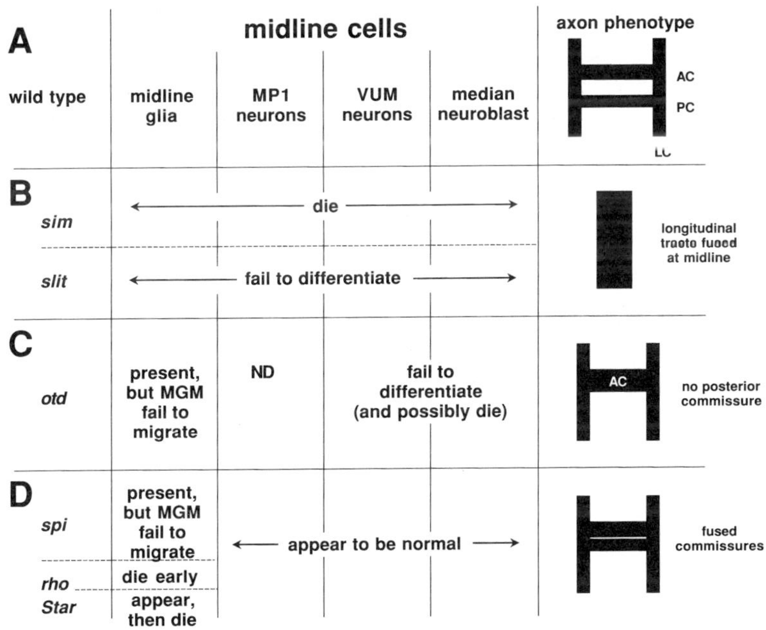

FIGURE 6. Summary of cell fates and axon phenotypes in midline mutants. Schematic figure showing the relation of the final axon phenotype to the differentiation and/or death of specific midline cell lineages in six different midline mutants. ND = not determined. The VUM, MP1, and MNB cell lineages have been analyzed in *rho* mutant embryos by Nomarski optics only. (See text for details.)

In *Drosophila,* the special midline cells are already distinguishable from the onset of gastrulation.[6,12,13,15] Although it seems unlikely that the cells at the ventral midline of the vertebrate neural tube, called the floorplate, and the cells at the midline of the insect neuroepithelium, called the mesectoderm, are true evolutionary homologues, it is nevertheless clear that they both play important roles in the development of commissural axon pathways and may use similar mechanisms to do so.

Given their key role in organizing the pattern of commissural axon pathways in *Drosophila,* mutations that perturb the development of the midline cells not surprisingly lead to profound phenotypic changes in the gross appearance of the commissures. For example, mutations in the *single-minded* (*sim*) gene,[12,13,17] other members of the *spitz* group of genes,[11,15] the *slit* gene,[14,18] the *otd* gene,[16] and the *DER* gene[34,35] lead to severe alterations in the appearance of the axon commissures. Some of these genes which control various aspects of midline development have already been cloned, and they encode a wide range of proteins, including a nuclear protein (*sim*[13]), another putative nuclear protein with a homeobox (*otd*[16]), a transmembrane receptor (*DER*[34]), a putative transmembrane protein (*rho*[15]), and a secreted protein (*slit*[18]). These are some of the features that make the midline such an excellent model system for genetic studies; alterations in specific cells have a dramatic consequence in an easily visualized phenotype—the pattern of commissural axonal pathways, and many mutants that perturb these events are already known.

Shortly after gastrulation, the midline arises from a single row of eight midline precursors that divide during the germ band extended stage to generate ~26 glia and neurons (R. Fetter, unpublished results), over 19 of which have been identified. Three of these eight midline precursors give rise to three pairs of midline glia (MGA, MGM, and MGP), and the other five midline precursors give rise to at least four different types of neurons, including the pair of MP1 neurons, the six VUM neurons, two VUM glia, and the median neuroblast (the MNB, at least two support cells, and its progeny).

Several glia (particularly the MGA and MGM glia) and several neurons (particularly the anterior-most VUMs, called the V-cells, and the MP1 neurons) play important roles in the formation of axon commissures. The growth cones that pioneer the posterior commissure appear to be guided towards the V-cells and MP1 neurons. The growth cones that pioneer the anterior commissure appear to be guided towards the site of contact between the MGA glia and the VUM growth cones. Finally, the last step of axon commissure formation is the separation of the two commissures which involves the posterior migration of the MGM glia over the top of the MGA glia and along the medial VUM growth cones between the two commissures (and the medial migration of the RP1 neurons).[30,32]

These observations give rise to a model that predicts that the midline glia have an important function in the formation of the anterior commissure and the key role in the separation of the two commissures. The VUM neurons play the key role in the formation of the posterior commissure and an important function along with the midline glia in the formation of the anterior commissure. Disruption of any one of these proposed functions should lead to a predictable phenotype. For example, the absence of the VUM neurons should lead to the absence of the posterior commissure; the absence or loss of function of the MGM and MGA glia should lead to a fused commissure phenotype. The absence of both the VUMs and the midline glia should lead to a lack of commissure phenotype. To test the validity of this model, we undertook an analysis of several midline mutants, taking into account their effect on the absence and/or differentiation of key midline cells and on the development of the axon pattern of the commissures.

We studied mutations in six different genes and grouped them according to their commissural phenotypes, as summarized in FIGURES 5 and 6. All of these mutations ultimately lead to gross abnormalities in the pattern of commissural axonal pathways, but on their own these final patterns can be misleading. For example, the absence of commissures in mutants such as *sim* or *slit* might lead one to think that the commissures never form, when in fact the initial commissural pathways do form but shortly thereafter collapse. The key to deciphering these phenotypes has been the use of β-gal expressing enhancer trap lines as cell-specific markers and the examination, using the BP102 mAb and HRP immunocytochemistry, of axon outgrowth and commissure development in young, carefully staged embryos around the time when the commissures are first forming.

In two group 1 mutations, *sim* and *slit,* the cells at the midline do not properly differentiate, and in one case (*sim*) these cells ultimately die. In both cases, although one or sometimes both commissures begin to form (albeit in an abnormal way from the outset), shortly thereafter the midline cells either atrophy or die, leading to the disappearance of the commissures and the collapse and fusion of the longitudinal tracts at the midline. In the single group 2 mutation, *otd,* the VUMs die, and as a result, the posterior commissure does not form; rather, in time, some of the axons that would normally have crossed the midline in the posterior commissure now cross the midline in the anterior commissure. Finally, in three group 3 mutations, *spi*, *S*, and *rho*, both commissures initially form. Yet, by the time the MGM glia should migrate over the MGA glia and separate the two commissures, these glia either die (*S* and *rho*) or fail to properly migrate (*spi*), leading to a fused commissural phenotype.

The midline glia are just one of several distinct classes of glia in the insect embryo.[4,36–40] Some of these embryonic glia appear to establish a prepattern for the development of the scaffold of axonal pathways.[4] Cell ablation experiments have shown that one particular glial cell (the segment boundary cell in the grasshopper) can play an important role in growth cone guidance.[36] Given these results and the final location of the midline glia in front of (MGA), between (MGM), and in back of (MGP) the two commissures,[4] we might have expected that the midline glia would play the key role in guiding growth cones across the commissures. This appears not to be the case, as cell biologic and genetic analyses reveal that the normal development of the two commissures requires the coordinated functions of both the VUM neurons, probably the MP1 neurons, and the midline glia. In fact, the two important pairs of glia, MGA and MGM, start out anterior to the location of the anterior commissure and therefore some distance away from the location of the posterior commissure. As a result, they appear to play no role in the initial development of the posterior commissure and only a shared role in the development of the anterior commissure. The most striking phenotype in their absence (*S* and *rho*) or loss of function (*spi*) is the fusion of the two commissures.

Our initial genetic analysis of six midline mutants has in part confirmed this model; clearly, however, much more work is needed to refine our understanding of the roles played by each of the midline cells. Nevertheless, this initial genetic analysis shows the strength of the system, namely, that genetic lesions in specific midline cells can lead to recognizable commissural phenotypes. Given the existence of a large collection of β-gal expressing enhancer trap lines as cell-specific markers, monoclonal antibodies that stain commissural growth cones and axons, and the knowledge presented herein on midline development, it should now be possible to examine the pattern of axonal commissures in large numbers of mutant embryos in search of genes that control cell fate, cell migration, and growth cone guidance at the midline.

SUMMARY

A row of midline precursor cells separates the two lateral neurogenic regions that give rise to most of the *Drosophila* CNS. From these midline precursors arises a discrete set of special glia and neurons. The growth cones of many CNS neurons initially head straight towards the midline and change their behavior after traversing it, leading to the hypothesis that these midline cells play a key role in the formation of the axon commissures. We have used a variety of cellular and molecular genetic techniques to elucidate the cells and interactions, including specific cell migrations, that are important for the normal formation of the two major commissures in each segment. This cellular analysis has led to a model that proposes a series of sequential cell interactions controlling the three stages in commissure development: (1) formation of the posterior commissure, (2) formation of the anterior commissure, and (3) separation of the two commissures. An initial genetic test of this model has used a number of mutations that, by either eliminating or altering the differentiation of various midline cells, perturb the development of the axon commissures in a predictable manner.

REFERENCES

1. BASTIANI, M. J., A. L. HARRELSON, P. M. SNOW & C. S. GOODMAN. 1987. Cell **48:** 745–755.
2. DODD, J. & T. M. JESSEL. 1988. Science **242:** 692–699.
3. JESSELL, T. M., P. BOVOLENTA, M. PLACZEK, M. TESSIER-LAVINGE & J. DODD. 1989. Ciba Foundation Symp. **144:** 255–280.
4. JACOBS, J. R. & C. S. GOODMAN. 1989. J. Neurosci. **9:** 2402–2411.
5. TESSIER-LAVIGNE, M., M. PLACZEK, A. LUMSDEN, J. DODD & T. JESSELL. 1988. Nature **336:** 775–778.
6. POULSON, D. F. 1950. *In* Biology of *Drosophila.* M. Demerec, ed. :168–274. J. Wiley & Sons, Inc. New York.
7. FOE, V. E. 1989. Development **107:** 1–22.
8. NÜSSLEIN-VOLHARD, C., E. WIESCHAUS & H. KLUDING. 1984. Wilhelms Roux's Arch. **193:** 267–282.
9. WIESCHAUS, E., C. NÜSSLEIN-VOLHARD & G. JÜRGENS. 1984. Wilhelms Roux's Arch. **193:** 283–295.
10. JÜRGENS, G., E. WIESCHAUS, C. NÜSSLEIN-VOLHARD & H. KLUDING. 1984. Wilhelms Roux's Arch. **193:** 283–295.
11. MAYER, U. & C. NÜSSLEIN-VOLHARD. 1988. Genes Dev. **2:** 1496–1511.
12. THOMAS, J. B., S. T. CREWS & C. S. GOODMAN. 1988. Cell **52:** 133–141.
13. CREWS, S. T., J. B. THOMAS & C. S. GOODMAN. 1988. Cell **52:** 143–151.
14. ROTHBERG, J. M., D. A. HARTLEY, Z. WALTHER & S. ARTAVANIS-TSAKONAS. 1988. Cell **55:** 1047–1059.
15. BIER, E., L. Y. JAN & Y. N. JAN. 1990. Genes & Dev. **4:** 190–203.
16. FINKELSTEIN, R., D. SMOUSE, T. M. CAPACI, A. C. SPRADLING & N. PERRIMON. 1990. Genes & Dev. **4:** 1516–1527.
17. NAMBU, J. R., R. G. FRANKS, S. HU & S. T. CREWS. 1990. Cell **63:** 63–75 .
18. ROTHBERG, J. M., J. R. JACOBS, C. S. GOODMAN & S. ARTAVANIS-TSAKONAS. 1990. Genes & Dev. **4:** 2169–2187.
19. KLÄMBT, C., J. R. JACOBS & C. S. GOODMAN. 1990. Cell **64:** 801–815.
20. O'KANE, C., & W. J. GEHRING. 1987. Proc. Natl. Acad. Sci. USA **84:** 9123–9127.
21. GHYSEN, A. & C. O'KANE. 1988. Development **105:** 32–52.
22. BELLEN, H. J., C. J. O'KANE, C. WILSON, U. GROSSNIKLAUS, R. KURTH-PEARSON & W. J. GEHRING. 1989. Genes & Dev. **3:** 1288–1300.
23. BIER, E., H. VÄSSIN, S. SHEPHARD, K. LEE, K. MCCALL, S. BARBEL, L. ACKERMAN, R. CARRETTO, T. UEMARA, E. H. GRELL, L. Y. JAN & Y. N. JAN. 1989. Genes & Dev. **3:** 1273–1287.
24. CAMPOS-ORTEGA, J. A. & V. HARTENSTEIN. 1985. The embryonic development of *Drosophila melanogaster.* Springer-Verlag. Berlin, Heidelberg, New York.
25. BATE, C. M. & E. B. GRUNEWALD. 1981. J. Embryol. Exp. Morphol. **61:** 317–330.
26. DOE, C. Q., Y. HIROMI, W. J. GEHRING & C. S. GOODMAN. 1988. Science **239:** 170–175.
27. THOMAS, J. B., M. J. BASTIANI, C. M. BATE & C. S. GOODMAN. 1984. Nature **310:** 203–207.
28. PATEL, N. H., B. SCHAFER, C. S. GOODMAN & R. HOLMGREN. 1989. Genes & Dev. **3:** 890–904.
29. GOODMAN, C. S., M. J. BASTIANI, C. Q. DOE, S. DU LAC, S. L. HELFAND, J. Y. KUWADA & J. B. THOMAS. 1984. Science **225:** 1271–1279.
30. JACOBS, J. R. & C. S. GOODMAN. 1989. J. Neurosci. **9:** 2412–2422.
31. MONTELL, D. J. & C. S. GOODMAN. 1989. J. Cell Biol. **109:** 2441–2453.
32. PATEL, N. H., P. M. SNOW & C. S. GOODMAN. 1987. Cell **48:** 975–988.

33. Elkins, T., K. Zinn, L. McAllister, F. M. Hoffmann & C. S. Goodman. 1990. Cell **60:** 565–575.
34. Schejter, E. D. & B.-Z. Shilo. 1989. Cell **56:** 1093–1104.
35. Zak, N. B., R. J. Wides, E. D. Schejter, E. Raz & B.-Z. Shilo. 1990. Development **109:** 865–874.
36. Bastiani, M. J. & C. S. Goodman. 1986. J. Neurosci. **6:** 3542–3551.
37. Hoyle, G. 1986. J. Comp. Neurol. **246:** 85–103.
38. Meyer, M. R., G. R. Reddy & J. S. Edwards. 1987. J. Neurosci. **7:** 512–521.
39. Jacobs, J. R., Y. Hiromi, N. H. Patel & C. S. Goodman. 1989. Neuron **2:** 1625–1631.
40. Ferdieu, J. R. & A. P. Mahowald. 1989. Development **106:** 749–757.

Platelet-Derived Growth Factor in Central Nervous System Gliogenesis[a]

N. PRINGLE, E. J. COLLARINI, I. K. HART, M. C. RAFF, AND W. D. RICHARDSON

Department of Biology (Medawar Building)
University College London
London WC1E 6BT, UK

The optic nerve is a typical central nervous system (CNS) white matter tract in that it contains glial cells and axons but no neuronal cell bodies. Because of its simple cellular composition and the fact that it is a relatively accessible part of the CNS, the optic nerve is a popular experimental system for investigating glial cell development. Cultures of dissociated cells from rat optic nerves contain three types of mature glial cells, oligodendrocytes and two types of astrocytes, that belong to two separate cell lineages (see refs. 1 and 2 for reviews). Type-1 astrocytes develop from one kind of precursor cell, whereas oligodendrocytes and type-2 astrocytes develop from a different, bipotential progenitor cell known as the O-2A progenitor.[3] Type-1 astrocytes are flat and fibroblast-like *in vitro,* whereas type-2 astrocytes have a characteristic process-bearing morphology. Because they express some antigenic determinants that are also found around nodes of Ranvier in the intact optic nerve, it has been suggested that type-2 astrocytes *in vivo* may send processes preferentially to nodes.[4,5] Type-1 astrocytes are thought to extend processes mainly to the pial surface of the nerve and to blood vessels, but at least some of them also extend processes to nodes.[6] It has proved difficult to study type-2 astrocytes *in situ,* because the antibodies that help to identify them in culture (e.g., antibody A2B5) do not do so in tissue sections. Consequently, it is not known what type-2 astrocytes look like, how they are distributed in the intact nerve, or even if they exist *in vivo.* In this article we describe our studies of the factors, particularly platelet-derived growth factor (PDGF), that influence the proliferation of O-2A progenitors and their differentiation into oligodendrocytes. The alternative differentiation pathway into type-2 astrocytes is described elsewhere in this volume (Fulton *et al.*).

PDGF INFLUENCES O-2A PROGENITOR PROLIFERATION AND THE TIMING OF OLIGODENDROCYTE DIFFERENTIATION *IN VITRO*

In the developing rat optic nerve, O-2A progenitor cells first appear in small numbers around embryonic day 15 (E15)[7] and proliferate rapidly after this, first giving rise to postmitotic oligodendrocytes on the day of birth (E21/P0).[8] O-2A progenitor proliferation and oligodendrocyte differentiation continue in parallel for several weeks after birth.[8] In contrast, when perinatal rat optic nerve cells are cultured *in vitro* in defined medium containing 0.5% fetal calf serum (low-serum medium), the O-2A progenitors stop dividing straightaway and differentiate into

[a]Supported by the Medical Research Council, the Wellcome Trust, and the Multiple Sclerosis Society of Great Britain and Northern Ireland.

oligodendrocytes within a day or so, regardless of the age of the animal from which they were taken. This is true even if a single O-2A progenitor is cultured in low-serum medium in the absence of any other cells[9]; oligodendrocyte differentiation thus appears to be the default response when an O-2A progenitor is deprived of signals from other cells. These observations suggest that there is a mitogen(s) present in the intact optic nerve that keeps O-2A progenitor cells dividing and delays the onset of oligodendrocyte differentiation. Substantial evidence now exists that PDGF is a major player in this process. Purified PDGF is a potent mitogen for cultured O-2A progenitor cells,[10,11] and it recreates the normal *in vivo* timing of oligodendrocyte differentiation *in vitro.*[12] For example, when optic nerve cells from E17 rat embryos are cultured in low-serum medium containing PDGF, the O-2A progenitors divide a number of times and the first oligodendrocytes appear after 4 days, the *in vitro* equivalent of the day of birth. When optic nerve cells taken from an E18 rat embryo are cultured under the same conditions, the first oligodendrocytes appear on schedule after 3 days *in vitro,* and so on. PDGF mRNA is present in the developing rat brain[11] and optic nerve[13] at the time O-2A progenitors are proliferating rapidly *in vivo,* and PDGF protein can be found in postnatal optic nerve extracts,[12] so it seems likely that PDGF is important for stimulating O-2A progenitor cell proliferation during normal development *in vivo.*

PDGF RECEPTORS ON O-2A PROGENITOR CELLS

PDGF is a covalently linked dimer of A and B chains, with the structure AA, AB, or BB, depending on its source. PDGF from human platelets, for example, is a mixture of all three dimeric isoforms, of which PDGF-AB is the major component. PDGF exerts its biologic effects by binding to cell surface receptors with extracellular ligand-binding domains and intracellular tyrosine kinase domains. The unoccupied receptors are monomeric and inactive, but PDGF binding induces receptor dimerization and activation of their intrinsic tyrosine kinase activity. There are two distinct receptors with different ligand specificities; the α receptor binds both A and B chains of PDGF, whereas the β receptor binds only the B chain. Consequently, the response of a given cell to PDGF depends both on the receptor types that it expresses and the PDGF isoforms that it encounters (FIG. 1). (See ref. 14 for a review of PDGF.) O-2A progenitor cells express predominantly, if not exclusively, PDGF α receptors.[15] Consequently, they respond to all three dimeric isoforms of PDGF, but somewhat better to PDGF-AA and PDGF-AB than to PDGF-BB[13] because of the slightly higher affinity of the α receptor for PDGF A chains. Human or rat fibroblasts, in contrast, possess mainly PDGF β receptors and respond poorly to PDGF-AA.[13,16]

SOURCES OF PDGF IN THE CNS: NEURONS *VERSUS* GLIA

Where does the PDGF in the optic nerve originate? The two most obvious potential sources are glial cells in the nerve and the axons of retinal ganglion neurons that traverse the nerve. The idea that axons might be a source of mitogens for oligodendrocyte precursors is an attractive one for teleologic reasons, and because it has long been known that peripheral nervous system (PNS) axons carry surface-bound mitogens for Schwann cells, the PNS equivalent of oligodendrocytes (reviewed in ref. 17). However, the results of optic nerve transection experiments seemed to refute the notion that retinal ganglion axons are important for O-2A progenitor cell proliferation *in vivo.*[18,19] The optic nerves of newborn rats were unilaterally transected just behind the eye; this resulted in the degeneration of the axons in the distal nerve within a day or so. When the surviving nerve stump was examined 1 week later, a large reduction of O-2A lineage cells (about ninefold) was noted relative to the contralateral, untransected nerve, whereas the reduction in the number of type-1 astrocytes was less than twofold.[18] However, the mitotic index of the remaining O-2A progenitors did not differ between transected and untransected nerves, suggesting that axons are important, not for proliferation, but rather for survival of O-2A lineage cells. An alternative interpretation might be that axons are

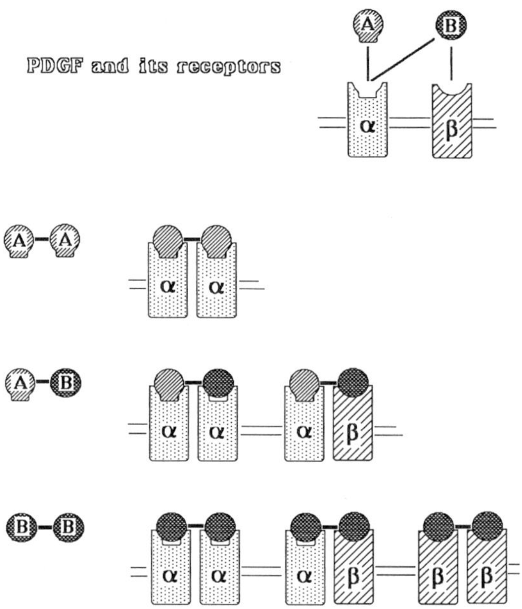

FIGURE 1. Schematic diagram of the different isoforms of PDGF and PDGF receptors and their interactions. Functional PDGF is a dimeric molecule of A and B chains, with the structure AA, BB, or AB. The PDGF β receptor (PDGFβR) binds PDGF B chains but not A chains, whereas the PDGF α receptor (PDGFαR) binds both A and B chains. Thus, the response of a given cell to PDGF depends both on the receptor type(s) that it expresses and the dimeric PDGF isoforms that it encounters. Note that PDGF-AB can bind to but cannot cross-link PDGFβR. PDGF-AB can therefore act either as a PDGF receptor agonist (for cells with PDGFαR) or an antagonist (for cells with only PDGFβR).

required for migration of O-2A progenitors into the optic nerve from the optic chiasm (see below). Regardless of the reason for the reduction in total O-2A lineage cells, the axons of retinal ganglion neurons apparently were not critical for O-2A progenitor proliferation. A likely alternative source of mitogens for O-2A progenitors in the optic nerve was type-1 astrocytes, because these cells constitute more than half the total glial cells in the newborn nerve. This possibility was supported by the demonstration that cultured astrocytes from rat cerebral cortex, which resemble type-1 astrocytes in optic nerve cell cultures, produced a mitogen for O-2A progenitor cells *in vitro.*[20] Moreover, astrocyte-conditioned medium could reconstitute the correct timing of oligodendrocyte development *in vitro.*[21] We subsequently showed that cultured rat cortical astrocytes synthesize and secrete PDGF, probably PDGF-AA, and that antibodies against human PDGF can neutralize up to 90% of the mitogenic activity for O-2A progenitors in astrocyte-conditioned medium.[11] Thus, the available evidence supports the view that O-2A progenitors in the optic nerve are stimulated to divide by PDGF-AA produced by type-1 astrocytes.

Recently it was shown that many neurons in the CNS, as well as some white matter glia (type-1 astrocytes?), contain mRNA encoding the PDGF A chain.[22] Many CNS neurons, but not glia, also contain the PDGF B chain.[23] These observations, together with experiments showing that cultured cerebellar interneurons produce PDGF-like mitogenic activity for O-2A progenitors from cerebellum or optic nerve,[24] suggest that neurons might be a major producer of PDGF in the CNS. Whether neurons secrete PDGF where it would be available to stimulate O-2A progenitor proliferation remains to be seen. Most likely, this would require that PDGF be released from axons, for which no mechanism is known to exist at present. The function(s) of neuron derived PDGF is intriguing and is likely to be a focus of future research.

NEWLY FORMED OLIGODENDROCYTES POSSESS FUNCTIONAL PDGF RECEPTORS BUT DO NOT DIVIDE IN RESPONSE TO PDGF

Knowing one of the factors that drives O-2A progenitors to proliferate during development raises a further question: Why do they eventually stop dividing and differentiate into oligodendrocytes? O-2A progenitors differentiate *in vitro* when they are deprived of PDGF, but for several reasons this is unlikely to be the reason that they drop out of division *in vivo.* First, PDGF mRNA is present in the brain and optic nerve from before birth to adulthood,[11,13] suggesting that PDGF may be continuously available there. Second, O-2A progenitors do not proliferate indefinitely in the presence of PDGF *in vitro;* no matter how much or how often PDGF is added to the culture, most progenitors eventually stop dividing and differentiate. This *in vitro* behavior mirrors what happens in the early postnatal optic nerve, where some O-2A progenitors drop out of division and differentiate into oligodendrocytes at the same time as others continue to divide. This suggests that the decision to differentiate does not depend on environmental signals, but is an intrinsic property of O-2A progenitor cells. This conclusion is supported by experiments in which single O-2A progenitors are cultured on their own in microwells. When grown in low-serum medium conditioned by type-1 astrocytes, a single O-2A progenitor from the perinatal rat optic nerve often divides a (variable) number of times to give rise to a clone of progenitors, most of which then drop out of division and differentiate more or less synchronously into oligodendrocytes. If a single progenitor is allowed to divide once and the daughter cells are segregated into separate wells, the two sibling cells often go through the same number of additional divisions before their progeny stop dividing and differentiate together.[25] These experiments provide evidence for an internal "clock" in each O-2A progenitor cell that sets an upper limit on the number of times that a given progenitor can divide in response to astrocyte-derived growth factors, presumably PDGF.

How might this intrinsic timer operate? One simple hypothesis may be that the number of PDGF receptors on each O-2A progenitor cell is reduced with time or with each cell division until there are no longer enough receptors to mediate a mitogenic response. The progenitor might then drop out of the division cycle and differentiate into an oligodendrocyte by default. This, however, is unlikely, for we have shown by ^{125}I-PDGF binding that at least some newly formed oligodendrocytes continue to express PDGF receptors at similar levels as the dividing O-2A progenitor cells that gave rise to them.[15] The postmitotic oligodendrocytes lose their PDGF receptors after a few days in culture, but this is a consequence of differentiation, not

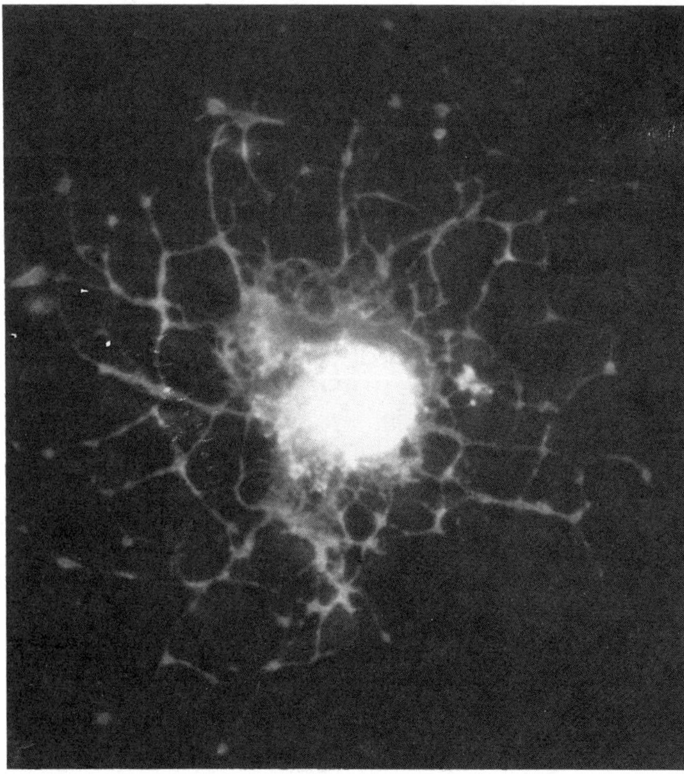

FIGURE 2. Newly formed oligodendrocytes express the nuclear protooncogene *c-jun* in response to stimulation with PDGF. Newborn optic nerves were dissociated and plated on coverslips in defined medium containing 0.5% fetal calf serum. After 24 hours, PDGF (recombinant PDGF-AA, 10 ng/ml) was added and incubation continued for a further 1 hour, after which the cells were fixed and double-labeled with anti-GC serum and an antiserum directed against *c-jun* (kindly provided by G. Evans, ICRF). Newborn optic nerves contain few oligodendrocytes, so the majority of GC-positive cells in these cultures will have differentiated from GC-negative O-2A progenitors during the 24-hour culture period. The figure is an immunofluorescence micrograph of a newly formed oligodendrocyte double-labeled with anti-GC and anti-*c-jun;* anti-GC stains the processes, and anti-*c-jun* the cell nucleus. Similar results were obtained with an antibody against *c-fos* (Hart, Richardson, and Raff, submitted). Thus, at least some newly formed oligodendrocytes possess functionally intact PDGF receptors.

the cause.[15] Furthermore, the PDGF receptors on newly formed oligodendrocytes respond to PDGF binding by mediating an increase in cytosolic Ca^{2+}[26] and elevated levels of *c-fos* and *c-jun* in the cell nucleus (FIG. 2). Thus, the block that develops in the mitotic signal transduction pathway(s) of an O-2A progenitor cell, causing it to become unresponsive to the mitogenic effects of PDGF and differentiate into an oligodendrocyte, presumably lies downstream of *c-fos* and *c-jun* activation or, alternatively, in an unidentified parallel signaling pathway.

Attempts to understand the intracellular events that trigger oligodendrocyte differentiation have been hampered by our ignorance of the biochemical pathways that link growth factor receptor activation to DNA synthesis and cell division, and by our inability to obtain O-2A progenitors in sufficient purity and numbers for biochemical analysis. Recently it was reported that PDGF and basic fibroblast growth factor (bFGF) can cooperate to keep O-2A progenitors proliferating for extended periods *in vitro,* in the apparent absence of differentiation.[27] It is not yet clear how this observation relates to events *in vivo,* because O-2A progenitors do not proliferate indefinitely during normal development. Nevertheless, this finding is important for it allows us to propagate O-2A progenitor cells, purified by immunoselection from suspensions of cells from optic nerve or cerebral cortex, in defined culture conditions. Thus, it may be possible in the future to apply a range of conventional biochemical and molecular biologic procedures to the study of O-2A lineage cells *in vitro.*

IS THE PDGF α RECEPTOR A MARKER FOR O-2A LINEAGE CELLS *IN SITU*?

O-2A progenitors are migratory cells *in vitro* whereas oligodendrocytes are not.[7,10] There is indirect evidence that they are migratory cells *in vivo* also, invading developing white matter tracts from nearby germinal regions.[7,28,29] The location of these germinal zones is not known, although it seems likely that they lie around the ventricles in the brain because some ventricular regions remain mitotically active after birth.[30] It is difficult to follow the development of O-2A progenitors *in situ,* because most of the antibodies that can be used to identify them in neuron-free cultures also react with neurons or their precursors in brain sections. Therefore, a need exists for more specific markers of O-2A lineage cells. We have been investigating the possibility that the PDGF α receptor may be useful as a marker, because we found that O-2A lineage cells seem to be the only glial cells in the optic nerve to express these receptors.[15] To this end we have performed *in situ* hybridizations on sections of developing rat CNS tissue, with a [^{35}S]-RNA probe corresponding to the extracellular domain of the rat PDGF α receptor (PDGFαR). PDGFαR$^+$ cells are widely and fairly uniformly distributed in the early postnatal brain, with a slight accumulation over the developing corpus callosum (FIG. 3a). In the P10 cerebellum, PDGFαR$^+$ cells are mainly found in the developing white matter tracts and the granular layer, and are excluded from the molecular layer (FIG. 3b). PDGFαR$^+$ cells are also found in the postnatal optic nerve (FIG. 3c). The apparent enrichment of PDGFα$^+$ cells in developing white matter and the fact that the silver grains overly cells with small dense nuclei (not shown) suggest to us that PDGFαR may be restricted to glial cells in the CNS (Pringle, Collarini, and Richardson, in preparation). It is possible that these cells represent O-2A lineage cells, but we cannot at this point rule out other glial or neuronal cell lineages. Looking earlier in development, we find that small numbers of PDGFαR$^+$ cells appear in the brain around E16; these cells do not appear to be concentrated in any regions that might correspond to germinal zones; indeed, they appear to be absent from the ventricular and subventricular zones, suggesting that they may migrate away from the germinal zones before or shortly after they start to express PDGFαR. Specific antisera to PDGFαR are now

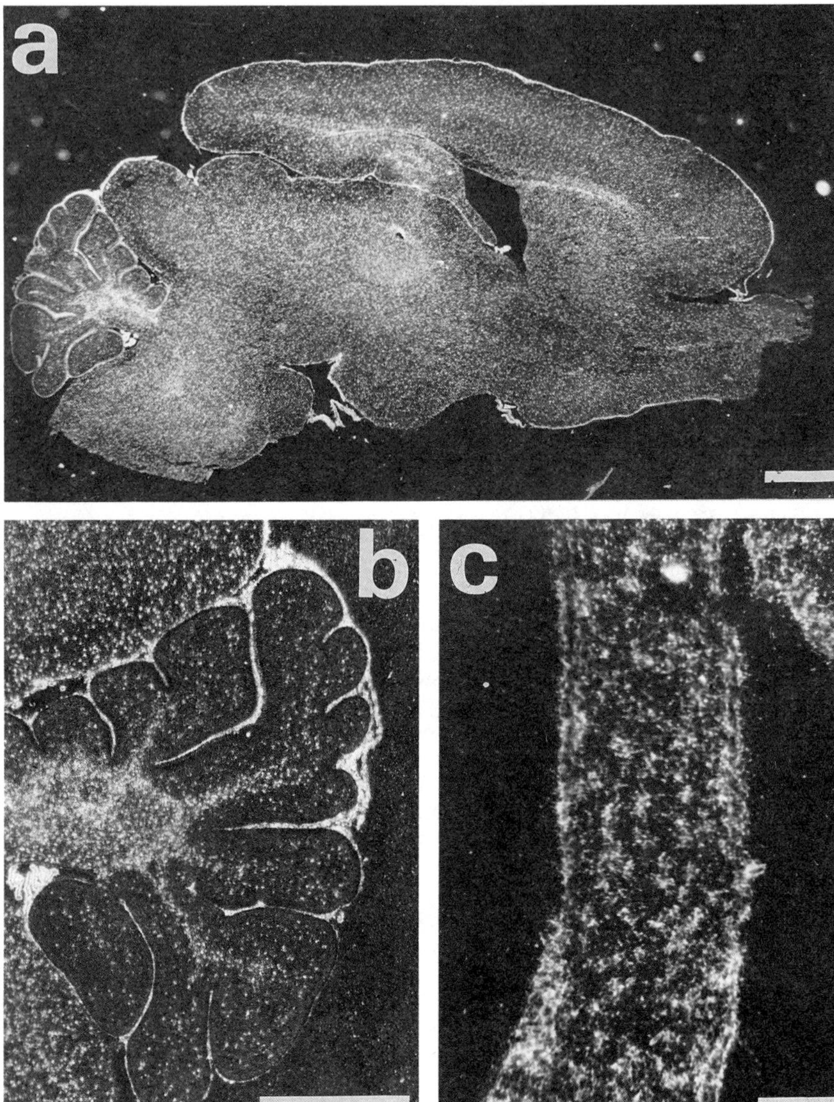

FIGURE 3. PDGF α receptor-positive cells in the developing rat CNS. *In situ* hybridization was performed on sections of P7-P10 rat CNS tissue, using a [^{35}S]-RNA probe corresponding to the extracellular domain of the rat PDGF α receptor (PDGFαR). The sections were subjected to autoradiography for 14 days. (*Panel A*) Sagittal section of a P10 rat brain, showing PDGFαR$^+$ cells distributed widely throughout the brain. (*Panel B*) Higher magnification view of a P10 cerebellum, showing PDGFαR$^+$ cells concentrated in the developing white matter tracts. (*Panel C*) Longitudinal section through a P7 rat optic nerve; PDGFαR$^+$ cells are distributed along the length of the nerve at this age. Scale bars: *panels A* and *B,* 100 μm; *panel C,* 10 μm.

required to enable us to see if PDGFαR colocalizes with other glial or neuronal-specific antigens in the CNS.

REFERENCES

1. Richardson, W. D., M. Raff & M. Noble. 1990. The oligodendrocyte-type-2 astrocyte lineage. Semin. Neurosci. **2:** 445–454.
2. Raff, M. C. 1989. Glial cell diversification in the rat optic nerve. Science **243:** 1450–1455.
3. Raff, M. C., R. H. Miller & M. Noble. 1983. A glial progenitor cell that develops *in vitro* into an astrocyte or an oligodendrocyte depending on the culture medium. Nature **303:** 390–396.
4. ffrench-Constant, C. & M. C. Raff. 1986. The oligodendrocyte-type-2 astrocyte cell lineage is specialized for myelination. Nature **223:** 335–338.
5. ffrench-Constant, C., R. H. Miller, J. Kruse, M. Schachner & M. C. Raff. 1986. Molecular specialization of astrocyte processes at nodes of Ranvier in rat optic nerve. J. Cell Biol. **102:** 844–852.
6. Suarez, I. & M. C. Raff. 1989. Subpial and perivascular astrocytes associated with nodes of Ranvier in the rat optic nerve. J. Neurocytol. **18:** 577–582.
7. Small, R. K., P. Riddle & M. Noble. 1987. Evidence for migration of oligodendrocyte-type-2 astrocyte progenitor cells into the developing rat optic nerve. Nature **328:** 155–157.
8. Miller, R. H., S. David, R. Patel, E. R. Abney & M. Raff. 1985. A quantitative immunohistochemical study of macroglial cell development in the rat optic nerve: *In vivo* evidence for two distinct astrocyte lineages. Dev. Biol. **111:** 35–41.
9. Temple, S. & M. C. Raff. 1985. Differentiation of a bipotential glial progenitor cell in single cell microculture. Nature **313:** 223–225.
10. Noble, M., K. Murray, P. Stroobant, M. D. Waterfield & P. Riddle. 1988. Platelet-derived growth factor promotes division and motility and inhibits premature differentiation of the oligodendrocyte/type-2 astrocyte progenitor cell. Nature **333:** 560–562.
11. Richardson, W. D., N. Pringle, M. J. Mosley, B. Westermark & M. Dubois-Dalcq. 1988. A role for platelet-derived growth factor in normal gliogenesis in the central nervous system. Cell **53:** 309–319.
12. Raff, M. C., L. E. Lillien, W. D. Richardson, J. F. Burne & M. Noble. 1988. Platelet-derived growth factor from astrocytes drives the clock that times oligodendrocyte development in culture. Nature **333:** 562–565.
13. Pringle, N., E. J. Collarini, M. J. Mosley, C.-H. Heldin, B. Westermark & W. D. Richardson. 1989. PDGF A chain homodimers drive proliferation of bipotential (O-2A) glial progenitor cells in the developing rat optic nerve. EMBO J. **8:** 1049–1056.
14. Heldin, C.-H. & B. Westermark. 1989. Platelet-derived growth factor: Three isoforms and two receptor types. Trends Genet. **5:** 108–111.
15. Hart, I. K., W. D. Richardson, C.-H. Heldin, B. Westermark & M. C. Raff. 1989. PDGF receptors on cells of the oligodendrocyte-type-2 astrocyte (O-2A) cell lineage. Development **105:** 595–603.
16. Nistér, M., A. Hammacher, K. Mellström, A. Siegbahn, L. Rönnstrand, B. Westermark & C.-H. Heldin. 1988. A glioma-derived PDGF A-chain homodimer has different functional activities than a PDGF AB heterodimer purified from human platelets. Cell **52:** 791–799.
17. Lemke, G. 1990. Glial growth factors. Semin. Neurosci. **2:** 437–443.
18. David, S., R. H. Miller, R. Patel & M. Raff. 1984. Effects of neonatal transection on glial cell development in the rat optic nerve: Evidence that the oligodendrocyte-type 2 astrocyte cell lineage depends on axons for its survival. J. Neurocytol. **13:** 961–974.
19. Privat, A., J. Valat & J. Fulcrand. 1981. Proliferation of neuroglial cell lines in the degenerating optic nerve of young rats. J. Neuropathol. Exp. Neurol. **40:** 46–60.
20. Noble, M. & K. Murray. 1984. Purified astrocytes promote the *in vitro* division of a bipotential glial progenitor cell. EMBO J. **3:** 2243–2247.

21. RAFF, M. C., E. R. ABNEY & J. FOK-SEANG. 1985. Reconstitution of a developmental clock *in vitro:* A critical role for astrocytes in the timing of oligodendrocyte differentiation. Cell **42:** 61–69.
22. YEH, H.-J., K. G. RUIT, Y.-X. WANG, W. C. PARKS, W. D. SNIDER & T. F. DEUEL. 1991. PDGF A-chain gene is expressed by mammalian neurons during development and in maturity. Cell **64:** 209–216.
23. SASAHARA, M., J. W. U. FRIES, E. W. RAINES, A. M. GOWN, L. E. WESTRUM, M. P. FROSCH, D. T. BONTHRON, R. ROSS & T. COLLINS. 1991. PDGF B-chain in neurons of the central nervous system, posterior pituitary, and in a transgenic model. Cell **64:** 217–227.
24. LEVINE, J. M. 1989. Neuronal influences on glial progenitor cell development. Neuron **3:** 103–113.
25. TEMPLE, S. & M. C. RAFF. 1986. Clonal analysis of oligodendrocyte development in culture: Evidence for a developmental clock that counts cell divisions. Cell **44:** 773–779.
26. HART, I. K., W. D. RICHARDSON, S. R. BOLSOVER & M. C. RAFF. 1989. PDGF and intracellular signaling in the timing of oligodendrocyte differentiation. J. Cell Biol. **109:** 3411–3417.
27. BÖGLER, O., D. WREN, S. C. BARNETT, H. LAND & M. NOBLE. 1990. Cooperation between two growth factors promotes extended self-renewal and inhibits differentiation of O-2A progenitor cells. Proc. Natl. Acad. Sci. USA **87:** 6368–6372.
28. REYNOLDS, R. & G. P. WILKIN. 1988. Development of macroglial cells in rat cerebellum. II. An *in situ* immunohistochemical study of oligodendroglial lineage from precursor to mature myelinating cell. Development **102:** 409–425.
29. LEVINE, S. M. & J. E. GOLDMAN. 1988. Spatial and temporal patterns of oligodendrocyte differentiation in rat cerebrum and cerebellum. J. Comp. Neurol. **277:** 441–455.
30. ALTMAN, J. 1966. Proliferation and migration of undifferentiated precursor cells in the rat during postnatal gliogenesis. Exp. Neurol. **16:** 263–278.

DISCUSSION PAPER

Glial-Neuronal Interaction during Development

JOHN SCHOLES

MRC Muscle and Cell Motility Unit
King's College London
26–29 Drury Lane
London WC2B 5RL, UK

An oddity of evolutionary progress in the nervous system is that while neurons diversify, the repertoire of glial cells appears to narrow. Thus adult vertebrates have only two types of macroglia in the central nervous system (CNS), whereas in small invertebrate animals with named neurons, numerous apparently distinct glial cells can be recognized by their appearance and positions, and take their place among guidepost neurons in the early scaffold for neural development. In warm blooded vertebrates the glia of the adult CNS appears in a kind of metamorphosis at birth, when the glial framework of the fetus is replaced by a new one consisting of astrocytes and oligodendroglia, which is refractory to further neural growth.

During embryonic development in the CNS, the substrate for neural pattern formation is provided by nonneuronal cells in a sequence of developmental states, from early neuroepithelium to extended radial glia. It is thought that they present a regionally differentiated molecular blueprint, guiding neuronal migrations and extending neurites. In the peripheral nervous system (PNS), the earliest pioneer axons lead the glia, but once in place the Schwann cells form nerve tissue sustaining further patterned outgrowth as well as the repair of damage in the adult.

With the use of cell-adhesion and antibody-perturbation assays, much progress has been made in the last decade in the identification of molecules involved in recognition and guidance in neural development. The preceding six chapters have dealt with their mediation between glia and neurons in development and repair. Firstly, three major molecular families subserving adhesive interactions in the nervous system are reviewed by Melitta Schachner and Kevin Tomaselli: (1) adhesion molecules of the IgG superfamily, mediating cell-cell interactions; (2) varied components of the extracellular matrix (ECM); and (3) their specific receptors at the cell surface, the integrin family of matrix adhesion molecules.

PICK AND MIX IN MATRIX RECEPTORS

Tomaselli reviews molecular diversity and regulation in the matrix receptor system. Macromolecules of the ECM, such as collagen, laminin, and fibronectin, potently support neuritic outgrowth *in vitro*[1] and are expressed by glial cells *in situ,* in developing fiber tracts such as the visual pathway. They are recognized by growth cone glycoproteins belonging to the integrin family,[2] heterodimeric receptor complexes, with potential intracellular signaling sites and interactions with the cytoskeleton. The β_1-subunits expressed by neural cells are conserved, and ligand-binding specificity is conferred by different α-subunits. Thus $\alpha_6\beta_1$ is a discriminate receptor for laminin, whereas $\alpha_3\beta_1$ interacts promiscuously with varied matrix components. The possible repertoire for neural specificity escalates indefinitely when it is considered that different neurons express differently composed receptors, that the receptors discriminate ligand domains within matrix molecules (e.g., the cross *vs* the long-arm domains of laminin), that they have multiple affinity states that can be regulated at the cell surface,[3] and that they act in unison with cognate cell-adhesion molecules. Much remains to be found in the developmental patchwork of the matrix and its receptors, but the means exists to test their activities by transfecting tailored genes into neurons and glia, as can be finely targetted *in situ* using lipofection with cDNAs.[4]

GUIDANCE IN PERIPHERAL NERVES

Damaged peripheral nerves are repaired more effectively when their fascicle patterns can be reknitted in detail surgically. This channels the fibers via preexisting Schwann tubes towards targets that are too remote to exert any guiding influence. In addition, specific affinities for correctly matched fascicles may exist, and Melitta Schachner describes important new findings that suggest that the neural antigen, L2/HNK-1, is part of an addressing mechanism of this kind. L2 is a sulfated carbohydrate residue present on two antigenic glycolipids of the PNS (and also the CNS in fish) that have autoimmune activity in some peripheral neuropathies.[5] It is also present on various glycoproteins of the nervous system, most notably adhesion molecules of the IgG superfamily, L1, P_0, and so forth. L2 is localized specifically in ventral roots and ventral root-derived bundles in peripheral nerves. Neurites from explanted motoneurons extended further on cryosections of ventral roots than of dorsal roots, and the L2 epitope is involved, thus constituting a possible signpost for nerve fiber regrowth, because anti-L2 antibodies block this preference.

REGULATING NEURONAL-GLIAL INTERACTIONS

Growth-permissive glial responses, related to those operating during development, are called upon during peripheral nerve regeneration, and the adhesion glycoprotein L1 exemplifies some of the regulatory principles involved. L1 expression is up-regulated in the distal stump after damage by two agents of the Schwann cell response, nerve growth factor and glia-derived nexin. Melitta Schachner also describes how L1-mediated adhesion is enhanced by interactions between this molecule and N-CAM, thus illustrating the subtle possibilities available at the cell surface for regulating adhesive recognition processes. None at present seems subtler than those suggested by recent findings on the adhesion molecule on glia (AMOG), an astrocytic glycoprotein mediating neuronal attachment. AMOG has major tracts of sequence homology with the β-subunit of the Na, K, and ATP-ase and is complexed with an α_2-subunit of the pump. Although AMOG adhesion appears to be independent of the ATP-ase, the pump can be regulated via the adhesive moiety of the AMOG complex.

RING-FENCING DEVELOPMENT

In the rapidly expanding universe of brain development in warm-blooded vertebrates, targets recede as neurites extend. In many cases, connections are achieved by neuronal overgrowth of chaotic proportions. Thus, curtailing exuberance may be as important for precise neural patterning as promoting and guiding outgrowth. For example, no-go factors pattern the visual map in the chick mid-brain,[6] and nonpermissive factors appearing in myelin at birth act to curtail neural outgrowth (Cadelli and Schwab, this volume).

An example of pattern formation by containment is the whisker-barrel field in mouse somatosensory cortex, discussed by Dennis Steindler, in which barrel boundaries in the *muscululus* are transiently delineated by J1/tenascin during development. Tenascin is a six-armed cell-binding glycoprotein of the extracellular matrix, important in the development of diverse tissues such as kidney and cartilage. It is abundant in the CNS, where it is synthesized by astrocytes, and has ambivalent adhesive activity for neurites *in vitro* due to the anti-adhesive activity on one arm of the hexal structure.[7] It may confine afferents to their barrels during development. The astrocytes in the barrel walls show reactive indications, much as in gliosis after wounding, which Steindler felt may recapitulate aspects of normal barrel development. It is also conceivable that these boundary specializations may be related to deployment of the barrel infrastructure, the vascular bed, which sensibly runs through the walls.

GLIAL MONORAILS

Whereas neurulation seems to involve cohesive cell movements, neuroblasts later withdraw from epithelial surface contact and commence the prodigious internal migrations that build the layered cytoarchitecture of the brain. These movements were brought to life some years ago by Pasko Rakic's drawings of the tall palisade of radial glial fibers in the developing brain, guiding the perfect ascent of neuroblasts to the cortical plate and the descent of granule cells to their layer in the cerebellum. Mary Hatten describes how she and her colleagues have reconstituted these interactions for molecular analysis *in vitro,* using granule neurons on cultured Bergmann glial fibers. It usually occurs the other way round, but their beautiful experiments first characterized the motile behavior of the neuroblasts and then identified the molecule, astrotactin, a glycoprotein on the leading edge and soma of the migrating granule cells. It is necessary for attachment and motility, but it is distinct from the major classes of adhesion molecules so far defined, namely, the IgG superfamily and the integrins. Granule neurons express L1 and N-CAM *in situ,* but only on the developing parallel fibers, an important suggestion that quite different molecular mechanisms underlie neuronal migration and neurite outgrowth.

SCULPTING THE *DROSOPHILA* NERVOUS SYSTEM

In vertebrate embryos, the radial glia provide a framework for neuroblast movements and neurite extension, but except in the case of the visual pathway,[8,9] few direct signs of the molecular blueprint have yet emerged. For example, astrotactin supports granule cell movements on neocortical glial fibers as well as on Bergmann glia of the cerebellum. In *Drosophila* embryos, however, Christian Klämbt describes intricate patterns and executive developmental roles for identified glial cells. In invertebrates the glia appear to be more diverse and less polarized functionally than in the vertebrate CNS,[10] and in the insect embryo, varied named glial cells take their place among the identified neurons, guiding early neuritic extension.

At CNS delamination in *Drosophila* embryos, small files of midline glia are set aside to enwrap the commissures of the segmental ganglia and act as guideposts sampled by extending growth cones. *Drosophila* lines were isolated in which P-element insertions containing a *lacZ* reporter were expressed in the midline glia specifically, enabling characterization of the lineage. The behavior of the cells suggests that their developmental role as glia goes much further than the provision of guideposts and enwrapments: they are motile, and leap-frog into new patterns before undertaking an executive role in parting the anterior and posterior commissures, thus sculpting a salient neural pattern in the future ganglion. In agreement, mutations in the genes that identify the midline glia, deleting these cells, lead to failure of the connectives to separate.

GLIAL MITOGENS

In contrast with insects, CNS development in mammals often involves crowd events lacking precision at the level of single cells. This seems to be true in myelin formation, when the subventricular germinal plate of the CNS is reactivated for a final episode of cell division generating oligodendrocyte precursor cells.[11,12] What are the cell signals initiating this step? The rat optic nerve myelin-forming lineage runs autonomously *in vitro,* controlled by factors produced by astrocytes. W. D. Richardson described recent progress in characterizing these important signaling molecules that sustain in isolated cells the mitotic program initiated *in situ.* A step of great interest was the identification of platelet-derived growth factor (PDGF) as a potent endogenous mitogen for oligodendrocyte (O-2A) precursor cells. PDGF itself consists of two protein chains existing in homodimeric (AA and BB) or heterodimeric (AB) forms with widespread activity in wound healing, atherosclerosis, and neoplasia. Astrocytes in developing rat optic nerve express A-chains. The PDGF receptor exists in α and β forms and α-type receptors on O-2A cells mediate the response to A-PDGF.

Specificity of PDGF action in oligodendroglial proliferation presumably depends on the chain composition of the molecule and its receptor, and there are indications that cooperativity with other signaling factors may also be involved. At the conclusion of oligodendroglial proliferation, when mitosis is arrested, the signal pathway in differentiating cells still responds to PDGF; therefore, an outstanding question is the nature of the stop signal. One possibility is TGF-β, which is also produced by astrocytes and down-regulates the mitogenic effect of PDGF on the oligodendroglial lineage. Colonizing presumptive white matter may be a crowd event, but achieving functional stoichiometry between oligodendrocytes and axons must need more precise controls than secreted mitogens. In this connection, it is of interest that PDGF can be presented in mitogenic form at the cell surface,[13] and one wonders if it can operate in tethered form in the CNS as well, on cells, or in the matrix.

REFERENCES

1. MANTHORPE, M., E. ENGVALL, E. RUOSLAHTI, F. M. LONGO, G. E. DAVIS & S. VARON. 1983. Laminin promotes neuritic regeneration from cultured peripheral and central neurons. J. Cell Biol. **97:** 1882–1890.
2. HYNES, R. 1987. Integrins: A family of cell surface receptors. Cell **48:** 549–554.
3. NEUGEBAUER, K. M. & L. F. REICHARDT. 1991. Cell surface regulation of β1-integrin activity on developing retinal neurons. Nature **350:** 68–71.

4. HOLT, C. E., N. GARLICK & E. CORNEL. 1991. Targeted lipofection of cDNAs in the embryonic vertebrate CNS. Neuron **5:** in press.
5. DENISE, K. H., A. A. ILYAS, J. E. EVANS, C. COSTELLO, R. H. QUARLES & F. B. JUNGALWALA. 1986. Structure of sulphated glucuronyl glycolipids in the nervous system reacting with HNK-1 antibody and some IgM paraproteins in neuropathy. J. Biol. Chem. **261:** 11717–11725.
6. WALTER, J., S. HENKE-FAHLE & F. BONHOEFFER. 1987. Avoidance of posterior tectal membranes by temporal retinal axons. Development **101:** 909–913.
7. SPRING, J., H. BECK & R. CHIQUET-EHRISMANN. 1989. Two contrary functions of tenascin: Dissection of the active sites by recombinant tenascin fragments. Cell **59:** 325–334.
8. SILVER, J., M. POSTOM & U. RUTISHAUSER. 1987. Axon pathways in the developing brain. I. Cellular and molecular determinants that separate the optic and olfactory projections. J. Neurosci. **7:** 2264–2272.
9. COHEN, J., J. F. BYRNE, C. MCKINLAY & J. WINTER. 1987. The role of laminin and the laminin/fibronectin receptor complex in the outgrowth of retinal ganglion cell axons. Dev. Biol. **122:** 407–418.
10. HOYLE, G. 1986. Glial cells of an insect ganglion. J. Comp. Neurol. **246:** 85–103.
11. LEVINE, S. M. & J. F. GOLDMAN. 1988. Embryonic divergence of oligodendrocyte and astrocyte lineages in developing rat cerebrum. J. Neurosci. **8:** 3992–4006.
12. HARDY, R. & R. REYNOLDS. 1991. Proliferation and differentiation potential of rat forebrain oligodendroglial progenitors both *in vitro* and *in vivo*. Development **111:** 1061–1080.
13. ROBBINS, K. C., F. LEAL & J. A. PIERCE & S. A. AARONSON. 1985. The *v-sis*/PDGF-2 transforming gene product localises to cell membranes but is not a secretory protein. EMBO J. **4:** 1783–1792.
14. HELDIN, C. H. & B. WESTERMARK. 1989. Platelet-derived growth factor: Three isoforms and two receptor types. Trends Genet. **5:** 108–111.

INTRODUCTION

Myelin Structure and Demyelination in Multiple Sclerosis

A. N. DAVISON[a]

Department of Neurochemistry
Institute of Neurology
The National Hospital for Neurology & Neurosurgery
London WC1N 3BG, England

Myelin is the lipoprotein multimembrane insulation that prevents the flow of ionic currents across the axonal membrane and facilitates conduction velocity between nodes. In the peripheral nervous system one Schwann cell myelinates a single internode, but in the central nervous system (CNS) the oligodendrocyte maintains many internodes, some on different axons. The metabolic demand on glia during development is enormous. In the rat[1] from 15–30 postnatal days, approximately 20 mg of myelin are synthesized (brain weight 1.2 g). During myelination there is a large concentration of ribosomes and Golgi complex within the oligodendrocyte cytoplasm, and the nuclei have prominent nucleoli. These findings correlate temporally with the increased rate of synthesis of myelin proteins and lipid. In the adult CNS the oligodendrocyte cytoplasm has less endoplasmic reticulum than does that in the developing animal, and mature cells have aggregates of chromatin around their rim. These "dark" cells are smaller and have compact electron-dense nuclei and cytoplasm consistent with the reduced metabolic demand on the cell. It has proved difficult to trace the many connections to mature myelin sheaths, but the glial cell is clearly essential for myelin maintenance as damage to the oligodendrocyte affects myelin integrity.

MYELIN COMPOSITION

In the initial stages of myelination, loose whorls of "early" myelin with a similar composition to that of glial plasma membrane are formed. As the membrane undergoes compaction, the composition is modified. The characteristic myelin lipids are galactocerebroside and its sulfate ester (sulfatide) and the phospholipid ethanolamine plasmalogen (phosphatydal ethanolamine). The proportion of lecithin (phosphatidylcholine) is lower than it is in other membranes. About a third of the membrane dry weight is made up of protein, half of which is an important intrinsic membrane constituent, proteolipid protein (PLP). The protein has a molecular weight of about 30 kD with distinct hydrophobic and hydrophilic domains. The amino acid sequence of the 276 residues of bovine PLP has been determined (see review by Lees and Brostoff[2]). The hydrophobic character of the protein may be ascribed to residues 60–90, 151–181, and 239–269 which are postulated to span the membrane bilayer. About 3% (by weight) of the protein is covalently bound fatty acid (60% palmitic, 25% oleic, and 10% stearic acid). These long-chain fatty acids

[a] Present address: Department of Pathology, Royal College of Surgeons, 35–43 Lincoln's Inn Fields, London WC2A 3PN, England.

may contribute to the overall stability of PLP. The next most plentiful protein is the myelin protein which is localized at the major dense line where the paired cytoplasmic faces of the inert membranes meet. The myelin basic proteins for a number of species have been sequenced. The human protein has 170 residues with a high proportion of basic amino acids. Of special importance is the tryptophan residue, because encephalitogenic epitopes contain this amino acid (eg, Phe-Ser-Trp-Gly-Ala-Glu-Gly-Gln-Lys). The remaining 30% of total myelin protein consists of a heterogeneous group of high molecular weight proteins including the Wolfgram proteins, myelin-associated enzymes and glycoproteins such as the myelin-associated glycoprotein and the M2 protein.

METABOLIC STABILITY OF MYELIN COMPONENTS

In early work we found that some CNS myelin components have a surprisingly high degree of metabolic stability (e.g., cerebroside and sulfatide), whereas other lipids such as lecithin and especially the polyphosphoinositides underwent relatively rapid turnover. When [4-^{14}C]-cholesterol was incorporated into the white matter of chickens during development, it was possible to recover the unchanged radioactive molecule years afterwards in the adult brain.[3] Thus the sterol skeleton is structurally very stable within the CNS. Further work showed that the cholesterol molecules are only loosely bound in myelin and can undergo exchange within the membrane and oligodendrocyte cytoplasm.

Studies on myelin turnover (reviewed by Benjamins and Smith[4]) show that some components undergo a short-term period of rapid changes in specific radioactivities within 1–2 weeks after maximal labeling (eg, high molecular weight proteins $t_{0.5} \sim 20$ days, a fraction of the basic protein $t_{0.5} \sim 20$ days). This rapid phase may be mediated by exchange or during development by degradation of newly synthesized myelin before compaction. Other components of myelin undergo slow metabolism which may be explained by exchange after compact myelin formation by lateral movement (e.g., sterols) or exposure of molecules at cytoplasmic loops or even Schmidt-Lantermann incisures. Of the myelin proteins, about half the basic protein and most of the proteolipid protein are extremely stable,[5] indicating that the intrinsic proteolipid protein may provide an inert structural framework for other membrane proteins and lipid. Similar metabolic stability has been reported in long-term studies of peripheral nervous system myelin.[6]

DEMYELINATION IN MULTIPLE SCLEROSIS AND REMYELINATION

In multiple sclerosis (MS) primary damage to myelin occurs, although the exact mechanism is not established. It is possible that the satellite cell is also implicated[7] for there is loss of oligodendrocytes in mature plaques, but until recently most authors found no changes in these glial cells in acute lesions. Prineas and his colleagues[8] reported the loss of oligodendrocytes from recent areas of demyelination where macrophages containing myelin basic protein debris were also present. Some data show hyperactivity of oligodendrocytes at the edge of lesions.[7,9] Oligodendrocyte precursor cells that express cyclic nucleotide phosphohydrolase and the cell adhesion HNK-1 epitope are glial fibrillary acidic protein (GFAP) negative. That these markers are expressed during normal oligodendrocyte differentiation suggests that myelin regeneration may be occurring.[8] The presence of oligodendrocyte 0-2A progenitors in the adult CNS[10] supports the possibility that these cells may serve to remyelinate damaged tissue. In experimental animals in which glial cells have been destroyed by injection of ethidium bromide, demyelinated axons devoid of satellite cells are produced. Injection of cell populations into demyelinated regions shows that oligodendrocytes and their precursors can remyelinate axons in these areas.[11]

MECHANISMS OF DEMYELINATION

Evidence indicates that MS is an autoimmune disease involving cellular immune activity. Activated T cells are found in greater concentration in the spinal fluid than in the peripheral blood, and these cells are also seen in acute lesions. Activated T cells presumably initiate an inflammatory reaction by releasing cytokines, leading to demyelination by such soluble factors and by elicited macrophages. A major question is whether the response occurs initially within the CNS or whether activated cells are sequestered into the brain and spinal cord from the blood. There are indications that autoaggressive T cells are sensitized to myelin basic protein or to encephalitogenic epitopes.[12] There is a genetic association with MS; the T-cell receptor VB gene is associated with an immunodominant region of the basic protein (regions 84–102).[13]

CONCLUSION

Recent findings on the immunopathology of MS indicate that the myelin basic protein is involved in the pathogenesis of the disease. Therefore, research on the molecular biology of this encephalitogenic polypeptide is relevant to our understanding of its possible role in demyelinating disease. Although attention has hitherto focused on the attack on the myelin membrane, there is new interest in the oligodendrocyte as a primary target.[8] The possibility also exists that the oligodendrocyte may be able to remyelinate damaged axons in the adult CNS, emphasizing the relevance of the basic neurobiology of glial cells to human pathology.

REFERENCES

1. Norton, W. T. & S. E. Poduslo. 1973. Myelination in rat brain: Changes in myelin composition during brain maturation. J. Neurochem. **21:** 759–773.
2. Lees, M. B. & S. W. Brostoff. 1984. Proteins of myelin. *In* Myelin. P. Morrell, ed.: 197–224. Plenum Press. New York.
3. Davison, A. N. & M. Wajda. 1959. Persistence of cholesterol (4^{14}-C) in the central nervous system. Nature **183:** 1606–1607.
4. Benjamins, J. A. & M. E. Smith. 1984. Metabolism of myelin. *In* Myelin. P. Morrell, ed.: 225–258. Plenum Press. New York.
5. Sabri, M. I., A. H. Bone & A. N. Davison. 1974. Turnover of myelin and other structural proteins in the developing rat brain. Biochem. J. **142:** 499–507.
6. Davison, A. N. 1961. Metabolically inert proteins of the central and peripheral nervous system, muscle and tendon. Biochem. J. **78:** 272–282.
7. Raine, C. S., L. Scheinberg & J. M. Waltz. 1981. Multiple sclerosis. Oligodendrocyte survival and proliferation in an active established lesion. Lab. Invest. **45:** 534–536.
8. Prineas, J. W., E. E. Kwon, P. Z. Goldenberg, A. A. Ilyas, R. H. Quarles, J. A. Benjamins & T. J. Sprinkle. 1989. Multiple sclerosis. Oligodendrocyte proliferation and differentiation in fresh lesions. Lab. Invest. **65**(5): 489–502.

9. IBRAHIM, M. Z. M. & C. W. M. ADAMS. 1965. The relationship between enzyme activity and neuroglia in early plaques of multiple sclerosis. J. Pathol. Bacteriol. **90:** 239–243.
10. WOLSWIJK, G. & M. NOBLE. 1989. Identification of an adult-specific glial progenitor cell. Development **105:** 387–400.
11. BLAKEMORE, W. F. & A. J. CRANG. 1988. Extensive oligodendrocyte remyelination following injection of cultured central nervous system cells into demyelinating lesions in adult central nervous system. Dev. Neurosci. **10:** 1–11.
12. ALLEGRETTA, M., J. A. NICKLAS, S. SRIRAM & R. J. ALBERTINI. 1990. T cells responsive to myelin basic protein in patients with multiple sclerosis. Science **247:** 718–721.
13. WUCHERPFENIG, K. W., N. ENDO, A. ROSENZWEIG, J. G. SEIDMAN, H. L. WEINER & D. A. HAFLER. 1990. Shared T cell receptor VB gene usage in T cells specific for immunodominant regions of human myelin basic protein. Science **248:** 1016–1018.

Posttranscriptional Regulation of Myelin Protein Gene Expression[a]

ANTHONY T. CAMPAGNONI,[b] JOSEPH M. VERDI, A. NEIL VERITY, SHASHI AMUR-UMARJEE, AND SUJATHA BYRAVAN

Mental Retardation Research Center
U.C.L.A. Center for the Health Sciences
Los Angeles, California 90024

Regulation of gene expression may occur at a number of points between the transcription of a gene and the appearance of a functional protein in its appropriate location within the cell. In recent years we have begun to learn a great deal about the factors that regulate the activity and transcription of genes and the intracellular targeting of proteins after synthesis. However, less is known about the factors that affect the stability and translation of mRNAs, the role that specific mRNAs may play in the intracellular translocation of macromolecules, and the importance of mRNA structure in influencing these processes. Certainly, the importance of the structure of the 5′-untranslated region in the general regulation of translation[1] as well as *cis*-acting elements within this region that regulate the translation of specific mRNAs, such as that encoding ferritin,[2] is being recognized. Evidence also indicates that the 3′-untranslated region of mRNAs may play a role in regulating the translation and stability of mRNAs and, possibly, their translocation within the cell.[3]

The expression of the genes encoding the major myelin proteins — myelin basic proteins (MBP), proteolipid proteins (PLPs), myelin-associated glycoprotein (MAG), and 2′,3′-cyclic nucleotide-3′-phosphodiesterase (CNP) — have been the subject of intense investigation over the last several years. (See ref. 4 for a review.) Expression of all of these genes is developmentally regulated and probably intimately associated with the differentiation of the oligodendrocyte. All of these genes undergo alternative splicing, producing at least two mRNAs, and the proportions of the alternatively spliced mRNAs and/or the protein isoforms produced change significantly with development, in most cases. Recent investigations suggest that the expression of one of these genes, the MBP gene, may be regulated at a number of different levels including (1) promoter utilization, (2) splicing, (3) mRNA translocation, (4) translation, and (5) posttranslational events.

This paper presents further evidence for the posttranscriptional regulation of MBP gene expression, with most examples coming from work done on the expression of the MBP gene.

[a]This work was supported by NIH grants NS 23022 and NS 23322 and a generous donation from the Ciba-Geigy Corporation, Summit, New Jersey.

[b]Address for correspondence: Dr. A. T. Campagnoni, Mental Retardation Research Center, U.C.L.A. Center for the Health Sciences, 760 Westwood Plaza, Los Angeles, CA 90024.

MBP mRNA TRANSLOCATION

It has been known for about 10 years that the MBPs were synthesized on "free" polyribosomes[5,6] and the myelin PLP was synthesized on ribosomes bound to the endoplasmic reticulum.[6] Other experiments, both *in vivo* and in tissue slices, indicated that although MBPs were incorporated very rapidly (within minutes) into myelin, newly synthesized PLP was incorporated into myelin much more slowly (~ 30 minutes).[7] Furthermore, the incorporation of PLP, but not the MBPs, into the membrane could be interrupted by treatment with monensin, suggesting that assembly of PLP into myelin involved transport through the Golgi apparatus.[8] The available biochemical evidence supported the notion that the two classes of proteins were assembled into myelin by separate routes.

Subcellular fractionation experiments demonstrated a 20-fold enrichment of the MBP mRNAs in myelin fractions compared to PLP mRNAs.[6] This observation suggested that ribosomes containing MBP mRNAs might be located within the cytoplasmic processes connecting the oligodendrocyte somas with the myelin sheaths and, perhaps, the cytoplasmic channels infiltrating the sheaths. If MBP synthesis occurred in these processes and/or channels, then it could explain why incorporation of these proteins into the myelin membrane occurred faster than did incorporation of proteins synthesized in the cell bodies which would require posttranslational transport out to the sites of myelin assembly.

Proof of such a model could be obtained by *in situ* hybridization histochemistry (ISH) through the localization of MBP and PLP mRNAs, but it had to await the isolation of appropriate cDNA probes. Recently, several groups have used ISH to localize myelin-specific mRNAs *in vivo*[9–12] and *in vitro*.[12–16] *In vivo,* the pattern of distribution of the MBP mRNAs is quite different from that of the PLP mRNAs. With radioactive probes, the distribution of silver grains generated with the PLP probes tended to be clustered over cell bodies, whereas those generated with the MBP probes were diffusely distributed over myelinated regions in the brain.[9,10] This difference in gross distribution of these two mRNAs was consistent with the MBP mRNA translocation hypothesis. Further evidence was provided by developmental studies. Verity and Campagnoni[10] examined the distribution of MBP and PLP mRNAs in the medulla oblongata of the mouse brain at daily intervals beginning at birth. They observed that before the appearance of histologically stained myelin in this region (0–2 days), silver grains representing MBP mRNAs and PLP mRNAs were clustered over cell bodies. Between 2 and 3 days postpartum the distribution of MBP mRNAs changed dramatically, from a clustered to a diffuse distribution, and this change in distribution coincided with the appearance of histologically recognizable myelin.

Additional proof that MBP mRNAs are translocated from the cell bodies of oligodendrocytes to their processes has come from studies performed in primary brain cell cultures.[12–15] Although localization of silver grains to oligodendrocyte processes has been a bit more apparent *in vitro* than *in vivo,* the most convincing localizations have been obtained *in vitro* with nonradioactive probes labeled with digoxigenin[15] or biotin,[16] which are detected immunochemically. In these studies, MBP mRNAs could be clearly localized within oligodendrocyte processes and the translocation of these mRNAs visualized with time in culture.[15] In contrast, PLP mRNAs, MAG mRNAs, and greater than 99% of CNP mRNAs remained associated with the oligodendrocyte cell bodies.[15] Thus, substantial biochemical and cell biologic evidence has now accumulated indicating that MBP mRNAs are translocated within oligodendrocytes from the cell somas to their processes and, presumably, the cytoplasmic channels infiltrating the myelin sheaths. The protein products of the other major myelin protein genes appear to be targeted to myelin by other mechanisms subsequent to translation.

The intracellular compartmentalization of mRNAs has been described for a number of mRNAs including those that encode mitochondrial,[17] dendritic,[18] and cytoskeletal proteins.[19] The mechanisms by which this intracellular translocation of mRNAs occurs is not yet known, but they could involve the participation of specific carrier proteins binding to cytoskeletal elements in a fashion similar to the transport of messenger ribonucleoproteins from the nucleus to the cytoplasm.[20] It is also unclear whether the MBP mRNAs or the nascent MBP polypeptide chains play a role in the translocation process. In this regard, MBP has been shown to interact with cytoskeletal proteins,[21] and there is evidence that "free" polyribosomes can associate with cytoskeletal elements through their mRNAs.[22]

MBP mRNA AND PLP mRNA TRANSLOCATION IN *JIMPY* AND *QUAKING* MICE

Quaking is an autosomal, recessive mutation that maps to mouse chromosome 17 and does not appear to be a mutation in one of the major myelin protein genes. The mutation is characterized by pronounced hypomyelination of the central nervous system.[23] Work from this laboratory has shown that expression of the MBP gene is altered in the *quaking* brain up to about 18–21 days, after which the levels of MBP mRNA and the synthesis of MBP appear to be close to normal.[24] In marked contrast, the steady-state levels of MBP, as measured by radioimmunologic techniques, are only 5–25% of normal in the central nervous system. Inasmuch as unbound MBP is readily susceptible to protease attack, these results are consistent with an earlier hypothesis that *quaking* mice are defective in their ability to incorporate newly synthesized MBPs into myelin.[25]

In view of evidence indicating that MBP mRNAs are translocated within the oligodendrocyte, it was of interest to determine if this inability of the MBPs to be incorporated into *quaking* myelin was due to a defect in this translocation phenomenon or to some other posttranslational event. ISH examination of the *quaking* brain indicated that MBP mRNAs could be detected at approximately normal levels in the 20-day-old *quaking* brain (FIG. 1) and that the distribution of mRNA was as diffuse as that in the controls. These results suggest that MBP mRNA translocation is unimpaired in the *quaking* brain.

A corollary to these results is that yet another regulatory point must exist in the assembly of the MBPs into the myelin sheath. It has been suggested that as the MBPs are located on the cytoplasmic face of the unit bilayer comprising the myelin sheath, the assembly of MBPs into the membrane occurs by diffusion shortly after their synthesis in the oligodendrocyte processes.[6] Our data suggest that synthesis of MBPs close to the growing myelin sheath is not sufficient to assure proper assembly of the proteins into the membrane. Some additional event, such as a posttranslational modification of the protein, must occur for proper assembly of the MBPs into the membrane. Alternatively, the presence of a receptor for MBP in the membrane might be required before incorporation of the protein.[26]

The *jimpy* mutation appears to be a point mutation in the splice acceptor site of the fifth exon of the PLP gene.[27,28] The phenotype is somewhat complex, affecting the oligodendrocyte and the expression of several of the myelin protein genes, probably through an alteration in the differentiation/maturation/death of the oligodendrocyte.[29–32] The levels of MBP mRNA, even though severely reduced in this mutant, have generally been higher than the steady-state levels of polypeptide in *jimpy* brains.[33,34] This finding suggests that the small amount of protein made in the mutant brain is not incorporated effectively into the membrane. ISH studies[32] have confirmed other biochemical data[34] that MBP mRNA levels are reduced in *jimpy* brains. However, in the mutant MBP mRNA-specific silver grains appear more clustered on oligodendrocyte cell bodies and less diffuse than in either normal or *quaking* brains (FIG. 1). This pattern of labeling was most prominent in the caudate putamen and corpus callosum, but the pattern was similar in all regions of the brain and persisted until the animals died at 3–4 weeks. These data suggest impairment in the translocation of MBP mRNAs in *jimpy* brain. It is unlikely that impairment of MBP mRNA translocation is due to a reduction or truncation of oligodendrocyte cellular processes because two groups have independently reported that the numbers and general appearance of processes with antigalactocerebroside staining appear normal in the mutant.[32,35] It is noteworthy that *jimpy* oligodendrocytes contain numerous lipid inclusions and have a substantial reduction in microtubules.[36] If MBP mRNA translocation involves the cytoskeletal network, both of these factors might contribute to impaired translocation of MBP mRNA in the mutant.

REGULATION OF MYELIN PROTEIN mRNA TRANSLATION BY STEROIDS

Steroids are well known to control gene expression at the transcriptional level; however, a number of studies indicate that steroids also may regulate gene expression posttranscriptionally.[37–39] It has been suggested that steroids can alter the stability of the mRNAs[40] and/or influence the rates of translation of responsive mRNAs.[37,41]

None of the myelin protein genes appears to be transcriptionally regulated by steroids, but two recent studies suggest that steroids might act at the posttranscriptional level. Kumar *et al.*[42] observed that hydrocortisone can stimulate the expression of MBP and PLP in primary oligodendrocyte cultures, without any increase in the transcription rates of these genes. We have found that the translation of MBP and PLP mRNAs is stimulated by hydrocortisone and other steroids in cell-free systems.[43] Our most recent efforts have been aimed toward determining (1) if steroids could exert a direct effect on the translation of mRNAs, (2) if sequences within the MBP mRNA were involved in this phenomenon, and (3) which protein synthetic step is modulated by the steroid.

Toward this end we developed a cell-free translation system that could be programmed with synthetic mRNAs to determine the effects of steroids on the translation of specific synthetic mRNAs. The use of this system permitted us (1) to assess the effect of steroids on mRNA translation independent of effects on transcription, (2) to assess directly the effect of steroids on the rates of translation of specific mRNAs, and (3) to determine possible regulatory sequences in the 5′-untranslated region by programming the system with structurally modified mRNAs.

Using this system we found that several classes of steroids could stimulate the translation of MBP mRNAs in this system, with glucocorticoids being more effective than sex steroids. These same steroids served to inhibit the translation of the CNP mRNA in an equal, but opposite direction to the effects observed on the MBP mRNA. No effect was observed on the translation of either message with the mineralocorticoid, aldosterone, or the sterols cholesterol and desmosterol. Transla-

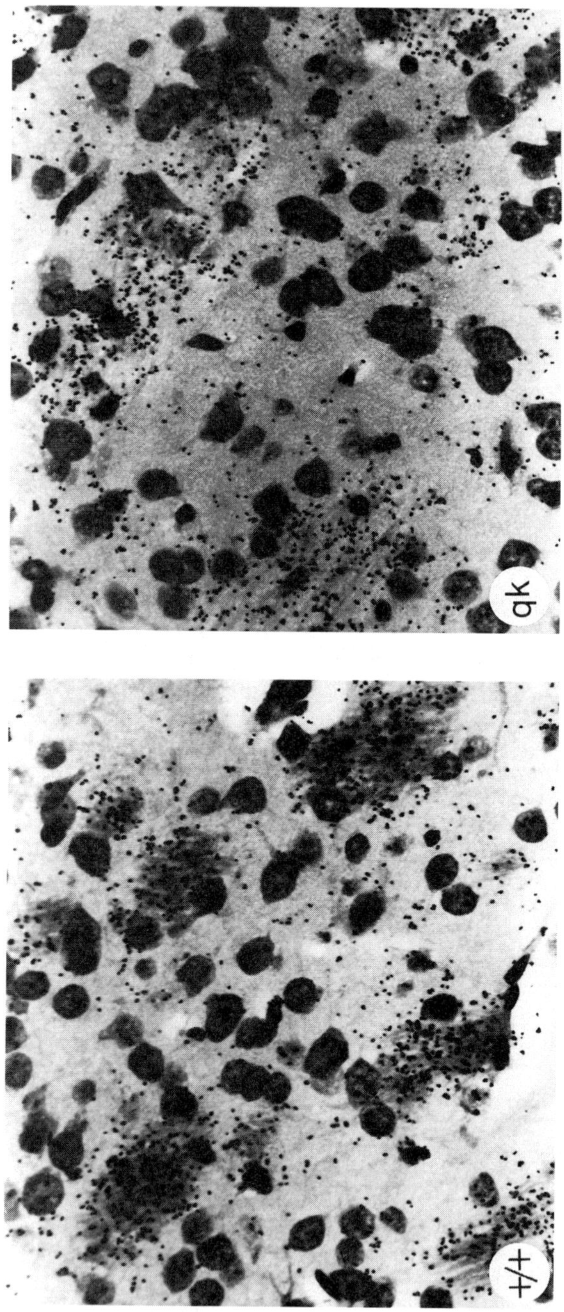

FIGURE 1. See legend on facing page

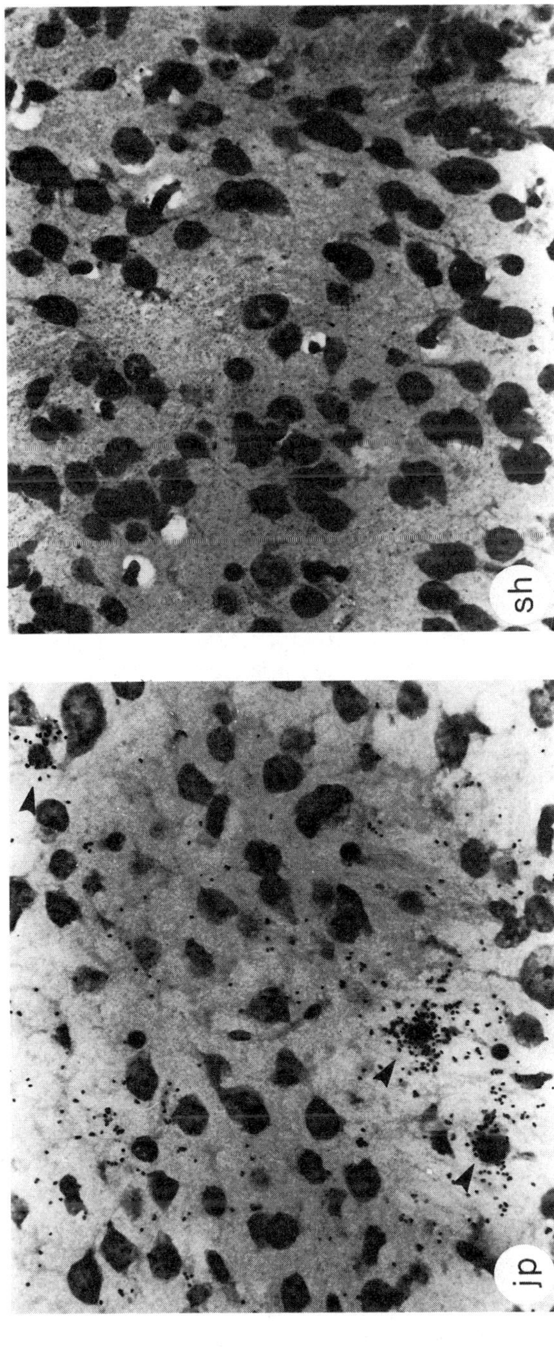

FIGURE 1. *In situ* hybridization histochemistry of MBP mRNAs within the caudate putamen of 15-day-old normal (+/+) mice and the dysmyelinating mutants *quaking* (qk), *jimpy* (jp), and *shiverer* (sh). In normal and *quaking* mice, silver grains representing MBP mRNAs are diffusely distributed over myelinated tracts indicative of normal translocation of MBP mRNAs. In contrast, in the *jimpy* mutant, MBP mRNA translocation is impaired as seen by increased clustering of silver grains over oligodendrocyte cell bodies (*arrows*). As expected, no detectable MBP transcripts are observed in *shiverer* mice.

tion of the PLP mRNA was also stimulated by hydrocortisone, but the translation of MAG mRNA was not affected.

Kinetic analysis of the data indicated that in this system the steroids were affecting the *rates* of translation of the responsive mRNAs. Furthermore, experiments suggested that steroids were acting on translation through a novel mechanism that did not involve the "classic" glucocorticoid receptor: (1) RU 38486, an antagonist of the "classic" glucocorticoid receptor, did not inhibit the stimulation of MBP mRNA translation by glucocorticoids. (2) The order of effectiveness among the glucocorticoids was different from that of the glucocorticoid receptor-mediated effect on transcription. (3) The dose response curves were more complex than observed with the classic glucocorticoid receptor, reaching maxima in the physiologically relevant range of 10^{-8}–10^{-9} M steroid. (4) The nucleotide sequence of the element in the MBP mRNA that is responsive to hydrocortisone is unlike that of the glucocorticoid-responsive element in genes that are transcriptionally regulated by this steroid.

We have found that the translation of all alternatively spliced forms of the MBP mRNAs transcribed from the major promoter of the gene are stimulated by hydrocortisone.[43] Transcription from the major promotor of the MBP gene begins 48 nt upstream of the initiator codon. However, the MBP gene also possesses a second, minor upstream transcription start site. Transcription from this secondary promoter produces MBP mRNAs with a longer 5′-untranslated region that consists of the 48 nt leader plus an additional upstream sequence of at least 350 nt.[44] The translation of these minor MBP mRNAs with a longer 5′-untranslated region is not stimulated by hydrocortisone.[43] These results suggested that the 5′-untranslated region might be important in the steroid modulation of MBP translation. This notion was supported by studies in which the translation of an MBP mRNA, which contained little of the 3′-untranslated region but a normal 5′-untranslated region, was stimulated by hydrocortisone.

This finding prompted us to investigate further the role of the 5′-untranslated region in this phenomenon.[45] A series of cDNA clones encoding the 14 kD and 18.5 kD MBP isoforms, but differing in the lengths of their 5′-untranslated regions were generated using the polymerase chain reaction and subcloned into a transcription vector. Site-directed changes were also introduced into selected nucleotides within the 5′-untranslated region of the MBP mRNAs using this technique. The transcribed RNAs were examined in the cell-free translation system in the presence and absence of hydrocortisone, permitting us to localize a nine-base region within the 5′-untranslated region of MBP mRNA between −29 and −37 that appears necessary for the stimulation of translation by hydrocortisone. Several lines of evidence indicate the importance of this region: (1) Removal of this region of the message eliminated the translational response to steroid. (2) Site-directed mutagenesis of several bases within this region reduced or eliminated the response. (3) A synthetic oligonucleotide of the same sequence could entirely compete out the response at a 10:1 molar ratio of oligonucleotide:message, but no such competition was observed with an oligonucleotide containing only two base changes. (4) Transferral of the sequence to a nonresponsive message rendered the translation of that mRNA responsive to hydrocortisone.

The steroid modulatory element in the 5′-untranslated region of the MBP mRNA is included within the sequence AGAAGACCC. However, this sequence alone probably does not account for the entire translational response to steroid because translation of the minor MBP mRNA with a longer 5′-untranslated region was not influenced by steroids, even though it contains the nine nucleotide element. Examination of the most stable secondary structure of the 5′-untranslated region of this unresponsive message indicates that the element is involved in a base paired stem, whereas in the "normal" MBP message six of the nine nucleotides (AGAAGA) are found in an accessible loop structure. This group of six nucleotides, but not the entire nine nucleotide region, is found twice in the 5′-untranslated region of the PLP mRNA, the translation of which is also stimulated by hydrocortisone.

The position of the glucocorticoid modulatory element in the 5′-untranslated region of the MBP mRNA, the many reports of alterations in the peptide chain initiation step of protein synthesis by steroids, and the fact that the most common form of translational regulation occurs during peptide chain initiation[46] suggested that this is the most likely step that is being affected.

To examine this further we measured directly the formation of 80S initiation complexes by programming reticulocyte lysates with MBP mRNAs containing "normal" and altered 5′-untranslated regions in the presence of sparsomycin (to prevent the peptide elongation step). With mRNAs, such as that encoding globin, which are nonresponsive to translational stimulation by steroids, the formation of the 80S initiation complex could easily be seen on glycerol gradients. When the reticulocyte lysates were programmed with normal MBP mRNAs, the appearance of a complex of 100–110S could be seen in addition to the 80S initiation complex. The formation of the 100–110S complex was dependent upon the presence of the nine nucleotide steroid modulatory element in the MBP mRNA. It was abolished if the sequence was deleted from the message or if site-directed changes were introduced into the sequence. Formation of the 100–110S complex could also be prevented if an oligomer of the nine nucleotide sequence was added to the incubation medium.

The 100–110S complex appears to be composed of an 80S initiation complex and an additional 40S ribosomal subunit bound to the 5′-untranslated region of the message. Because elongation is inhibited by sparsomycin, the complex, which represents the beginning of an additional initiation round, is "frozen" in place with further initiation rounds being prevented.

These data are consistent with a model in which the steroid serves to promote an increase in the initiation rate of protein synthesis in those messages of appropriate structure, such as the MBP mRNA. We propose that the steroids bind to some normal component of the protein synthetic apparatus (a protein factor) and that this steroid-protein complex interacts with the steroid-responsive element in such a way as to facilitate more frequent rounds of chain initiation by enhancing the interaction of the 40S ribosomal subunit with the 5′ end of the message.

It is possible that the effects of steroids on message stability and rates of translation might be related mechanistically. For example, a steroid-induced increase in polypeptide chain initiation would be expected to result in increased ribosomal density per message, thereby leading to increased message stability. For messages whose translation was inhibited by steroids, ribosome density would be expected to decrease in the presence of steroid, thereby leading to decreased message stability. There is evidence that transcripts that are not undergoing active translation or have fewer ribosomes per message are good substrates for degradation.[47]

The magnitude of the effects of steroids on the posttranscriptional modulation of myelin protein gene expression observed by us and others is only two- to threefold. Nonetheless, the importance of this form of regulation should not be underestimated. During most active myelination, the oligodendrocyte produces substantial amounts of myelin estimated at more than three times the weight of the cell body each day.[48] During this period, expression of the myelin protein genes is extremely active, and the myelin proteins produced, when incorporated into myelin, are quite stable, with low turnover rates.[49] Over this same developmental period (10–40 days in the mouse) the myelin membrane undergoes major compositional changes, with the proportion of MBP and PLP increasing significantly relative to CNP. Consequently, a two- to threefold increase in the synthesis of the MBP and PLP polypeptides with an equivalent decrease in the amount of CNP could produce a profound shift in the composition of the membrane during this developmental period. Thus, even though the steroid effects on translation of the myelin protein mRNAs are relatively small, they could result in significant changes in the proportions of proteins produced, and, consequently, the composition of the myelin membrane.

SUMMARY

Regulation of myelin protein gene expression occurs at many different levels including transcription, mRNA translocation, translation, and posttranslational modification of myelin proteins prior to their assembly into the membrane. Translocation of myelin basic protein (MBP) mRNAs into oligodendrocyte processes was observed *in vivo* and in primary cultures, but no such translocation was observed for the mRNAs encoding the proteolipid protein (PLP) or myelin-associated glycoprotein. More than 99% of the mRNAs encoding 2′3′-cyclic nucleotide phosphodiesterase (CNP) remained associated with cell bodies. In the *jimpy* mutant, MBP mRNA translocation appeared to be impaired, but translocation occurred normally in *quaking* brains *in vivo.* We have found that steroids, such as glucocorticoids, stimulate the translation of MBP and PLP mRNAs in cell-free systems and inhibit the translation of CNP mRNA. This pattern of regulation is consistent with compositional changes noted in myelin during development. We have localized a nine nucleotide segment within the 5′-untranslated region of the MBP mRNA that is involved in the action of steroids on translation of this mRNA. We have also determined that the protein synthetic step modulated by the steroids is chain initiation, enhancing the rate at which new ribosomal subunits bind to the MBP mRNAs.

REFERENCES

1. KOZAK, M. 1989. Circumstances and mechanisms of inhibition of translation by secondary structure in eucaryotic mRNAs. Mol. Cell. Biol. **9:** 5134–5142.
2. KLAUSNER, R. D. & J. B. HARFORD. 1989. *Cis-trans* models for post-transcriptional gene regulation. Science **246:** 870–872.
3. JACKSON, R. J. & N. STANDART. 1990. Do the poly(A) tail and 3′ untranslated region control mRNA translation? Cell **62:** 15–24.
4. CAMPAGNONI, A. T. & W. B. MACKLIN. 1988. Cellular and molecular aspects of myelin protein gene expression. Mol. Neurobiol. **2:** 41–89.
5. CAMPAGNONI, A. T., G. D. CAREY & Y.-T. YU. 1980. *In vitro* synthesis of the myelin basic proteins: Subcellular site of synthesis. J. Neurochem. **34:** 677–686.
6. COLMAN, D. R., G. KREIBICH, A. B. FREY & D. SABATINI. 1982. Synthesis and incorporation of myelin polypeptides into CNS myelin. J. Cell. Biol. **95:** 598–608.
7. BENJAMINS, J. A. & P. MORELL. 1978. Proteins of myelin and their metabolism. Neurochem. Res. **3:** 137–174.
8. TOWNSEND, L. E. & J. A. BENJAMINS. 1983. Effects of monensin on posttranslational processing of myelin proteins. J. Neurochem. **40:** 1333–1339.
9. TRAPP, B. D., T. MOENCH, M. PULLEY, E. BARBOSA, G. TENNEKOON & J. GRIFFIN. 1987. Spatial segregation of mRNA encoding myelin-specific proteins. Proc. Natl. Acad. Sci. USA **84:** 7773–7777.

10. VERITY, A. N. & A. T. CAMPAGNONI. 1988. Regional expression of myelin protein genes in the developing mouse brain: *In situ* hybridization studies. J. Neurosci. Res. **21:** 238–248.
11. JORDAN, D., V. FRIEDRICH, JR. & M. DUBOIS-DALCQ. 1989. *In situ* hybridization analysis of myelin gene transcripts in developing mouse spinal cord. J. Neurosci. **9:** 248–257.
12. SHIOTA, C., M. MIURA & K. MIKOSHIBA. 1989. Developmental profile and differential localization of mRNAs of myelin proteins (MBP and PLP) in oligodendrocytes in the brain and in culture. Dev. Brain Res. **45:** 83–94.
13. ZELLER, N. K., T. N. BEHAR, M. E. DUBOIS-DALCQ & R. A. LAZZARINI. 1985. The timely expression of myelin basic protein gene in cultured rat brain oligodendrocytes is independent of continuous neuronal influences. J. Neurosci. **5:** 2955–2962.
14. HOLMES, E., G. HERMANSON, R. COLE & J. DE VELLIS. 1988. Developmental expression of glial-specific mRNAs in primary cultures of rat brain visualized by *in situ* hybridization. J. Neurosci. Res. **19:** 389–396.
15. AMUR-UMARJEE, S. G., L. HALL & A. T. CAMPAGNONI. 1990. Spatial distribution of mRNAs for myelin proteins in primary cultures of mouse brain. Dev. Neurosci. **12:** 262–272.
16. GHANDOUR, M. S. & R. P. SKOFF. 1991. Double labeling *in situ* hybridization analysis of mRNAs for carbonic anhydrase II and myelin basic protein: Expression in developing cultured glial cells. Glia **4:** 1–10.
17. CHANG, D. D. & D. A. CLAYTON. 1987. A mammalian mitochondrial RNA processing activity contains nucleus-encoded RNA. Science **235:** 1178–1184.
18. DAVIS, L., G. A. BANKER & O. STEWARD. 1987. Selective dendritic transport of RNA in hippocampal neurons in culture. Nature **330:** 477–479.
19. LAWRENCE, J. B. & R. H. SINGER. 1986. Intracellular localization of messenger RNAs for cytoskeletal proteins. Cell **45:** 407–415.
20. SCHRODER, H. D., M. BACHMAN, B. DIEHL-SIEFERT & W. E. MULLER. 1987. Transport of mRNA from nucleus to cytoplasm. Prog. Nucleic Acid Res. Mol. Biol. **43:** 89–142.
21. BARYLKO, B. & Z. DOBROWOLSKI. 1984. Ca^{2+}-calmodulin-dependent regulation of F-actin-myelin basic protein interaction. Eur. J. Cell Biol. **35:** 327–335.
22. LENK, R., L. RANSOM, Y. KAUFMAN & S. PENMAN. 1977. A cytoskeletal structure with associated polyribosomes obtained from HeLa cells. Cell **10:** 67–78.
23. HOGAN, E. L. & S. GREENFIELD. 1984. Animal models of genetic disorders of myelin. *In* Myelin. P. Morell, ed.: 489–534. Plenum Press. New York.
24. SORG, B. J. A., M. M. SMITH & A. T. CAMPAGNONI. 1987. Developmental expression of the myelin proteolipid protein and basic protein mRNAs in normal and dysmyelinating mutant mice. J. Neurochem. **49:** 1146–1154.
25. GREENFIELD, S., S. BROSTOFF & E. HOGAN. 1977. Evidence for defective incorporation of proteins in myelin of the *quaking* mutant mouse. Brain Res. **120:** 507–515.
26. BARBARESE, E., C. BARRY, C. H. CHOU, D. J. GOLDTEIN, G. A. NAKOS, R. HYDE-DERUYSCHER, K. SCHELD & J. H. CARSON. 1988. Expression and localization of myelin basic protein in oligodendrocytes and transfected fibroblasts. J. Neurochem. **51:** 1737–1745.
27. MACKLIN, W. B., M. V. GARDINIER, K. D. KING & K. KAMPF. 1987. An AG → GG transition at a splice site in the myelin proteolipid protein gene in *jimpy* mice results in the removal of an exon. FEBS Lett. **223:** 417–421.
28. NAVE, K.-A., F. E. BLOOM & R. J. MILNER. 1987. A single nucleotide difference in the gene for myelin proteolipid protein defines the jimpy mutation in mouse. J. Neurochem. **49:** 1873–1877.
29. KNAPP, P. E., R. P. SKOFF & D. W. REDSTONE. 1986. Oligodendroglial cell death in *jimpy* mice: An explanation for the myelin deficit. J. Neurosci. **6:** 2813–2822.
30. SKOFF, R. P. 1982. Increased proliferation of oligodendrocytes in the hypomyelinated mouse mutant-*jimpy*. Brain Res. **248:** 19–31.
31. KNAPP, P. E. & R. P. SKOFF. 1987. A defect in the cell cycle of neuroglia in the myelin deficient *jimpy* mouse. Dev. Brain Res. **35:** 301–306.
32. VERITY, A. N., M. S. LEVINE & A. T. CAMPAGNONI. 1990. Gene expression in the *jimpy* mutant: Evidence for fewer oligodendrocytes expressing myelin protein genes and impaired translocation of myelin basic protein mRNA. Dev. Neurosci. **12:** 359–372.

33. CARSON, J. H., N. N. HERSCHKOWITZ & P. E. BRAUN. 1975. Synthesis and degradation of myelin basic protein in normal and *jimpy* mouse brain. Trans. Am. Soc. Neurochem. **6:** 207.
34. SORG, B. A., M. M. SMITH & A. T. CAMPAGNONI. 1987. Developmental expression of the myelin proteolipid protein and basic protein mRNAs in normal and dysmyelinating mutant mice. J. Neurochem. **49:** 1146–1154.
35. GHANDOUR, M. S. & R. P. SKOFF. 1988. Expression of galactocerebroside in developing normal and *jimpy* oligodendrocytes *in situ.* J. Neurocytol. **17:** 485–498.
36. MEIER, C., N. HERSCHKOWITZ & A. BISCHOFF. 1974. Morphological and biochemical observations in the *jimpy* spinal cord. Acta Neuropathol. **27:** 349–362.
37. PENNEQUIN, P., D. M. ROBINS & R. T. SCHIMKE. 1978. Regulation of translation of ovalbumin messenger RNA by estrogens and progesterone in oviduct of withdrawn chicks. Eur. J. Biochem. **90:** 51–58.
38. MEYUHAS, O., A. E. THOMPSON, JR. & R. P. PERRY. 1987. Glucocorticoids selectively inhibit translation of ribosomal protein mRNAs in P1798 lymphosarcoma cells. Mol. Cell. Biol. **7:** 2691–2699.
39. SACEDA, M., M. E. LIPPMAN, P. CHAMBON, R. L. LINDSEY, M. PONGLIKITMONGKOL, M. PUENTE & M. B. MARTIN. 1988. Regulation of the estrogen receptor in MCF-7 cells by estradiol. Mol. Endocrinol. **2:** 1157–1162.
40. SHAPIRO, D. J., J. E. BLUME & D. A. NIELSEN. 1985. Regulation of messenger RNA stability in eukaryotic cells. Bioessays **6:** 221–226.
41. LABATE, M. E., S. M. WHELLY & K. L. BARKER. 1986. Ribosome-associated estradiol binding components in the uterus and their relationship to the translational capacity of uterine ribosomes. Endocrinology **119:** 140–151.
42. KUMAR, S., R. COLE, F. CHIAPPELLI & J. DE VELLIS. 1989. Differential regulation of oligodendrocyte markers by glucocorticoids: Post-transcriptional regulation of both proteolipid protein and myelin basic protein and transcriptional regulation of glycerol phosphate dehydrogenase. Proc. Natl. Acad. Sci. USA **86:** 6807–6811.
43. VERDI, J. M., K. KAMPF & A. T. CAMPAGNONI. 1989. Translational regulation of myelin protein synthesis by steroids. J. Neurochem. **52:** 321–324.
44. KITAMURA, K., S. L. NEWMAN, C. W. CAMPAGNONI, J. M. VERDI, T. MOHANDAS, V. W. HANDLEY & A. T. CAMPAGNONI. 1990. Expression of a novel transcript of the myelin basic protein gene. J. Neurochem. **54:** 2032–2041.
45. VERDI, J. M. & A. T. CAMPAGNONI. 1990. Translational regulation by steroids: Identification of a steroid modulatory element in the 5′ untranslated region of the myelin basic protein messenger RNA. J. Biol. Chem. **265:** 20314–20320.
46. PAIN, V. M. 1986. Initiation of protein synthesis in mammalian cells. Biochem. J. **235:** 625–637.
47. KELLY, R., D. R. SHAW & H. L. ENNIS. 1987. Role of protein synthesis in decay and accumulation of mRNA during spore germination in the cellular slime mold *Dictyostelium discoideum.* Mol. Cell. Biol. **7:** 799–805.
48. MORELL, P. & A. D. TOEWS. 1984. *In vivo* metabolism of oligodendroglial lipids. *In* Oligodendroglia. W. T. Norton, ed.: 47–86. Plenum Press. New York.
49. BENJAMINS, J. A. 1984. Protein metabolism of oligodendroglial cells *in vivo. In* Oligodendroglia. W. T. Norton, ed.: 87–124. Plenum Press. New York.

Expression and Activity of the Transcription Factor SCIP during Glial Differentiation and Myelination

GREG LEMKE,[a] RAINER KUHN, EDWIN S. MONUKI, AND GERRY WEINMASTER

Molecular Neurobiology Laboratory
The Salk Institute for Biological Studies
San Diego, California 92138

The differentiation of Schwann cells, the myelin-forming glia of the vertebrate peripheral nervous system, provides a striking example of glial-neuronal interaction. Derived from the neural crest, Schwann cell precursors migrate into developing peripheral nerves and enter a differentiation pathway whose signature features of postnatal proliferation and myelination are entirely specified by axons. Our laboratory has chosen to investigate the axonal control of Schwann cell differentiation by examining the regulated expression of the major and minor myelin genes. These genes encode (among others) the myelin-specific structural proteins myelin basic protein (MBP), P_0, proteolipid protein (PLP), myelin-associated glycoprotein (MAG), and P_2.[1-5] Our efforts over the last 2 years have centered on the identification of *trans*-acting factors that mediate transcriptional control of these genes. This report summarizes recent experiments on one such protein, a homeobox-containing transcription factor whose developmental expression and regulatory activity place it near the midpoint of Schwann cell development.

AXONAL CONTROL OF SCHWANN CELL GENE EXPRESSION

Developing Schwann cells require contact with axons for the initial induction of myelin-specific genes.[6] In addition, mature Schwann cells require sustained contact with axons to maintain expression of these genes.[7] We have demonstrated this latter requirement *in vivo,* by depriving Schwann cells of axonal contact through transection of myelinated peripheral nerves,[8] and *in vitro,* by monitoring myelin gene expression in purified Schwann cells cultured in the absence of neurons.[9] *In vivo* transections of sciatic nerves were performed in rats at postnatal day 35, a time at which peripheral myelination is at a relatively advanced stage. We assayed for myelin gene expression in Schwann cells distal to the transection site (where axons degenerate) on successive days following surgery, using *in situ* hybridization and RNA blot hybridization with MBP and P_0 cDNA probes. Hybridization signals for the MBP and P_0 mRNAs fall dramatically in these axon-deprived cells; significant decreases are evident at 2 days following transection, and by 5 days after surgery, hybridization signals have fallen over 40-fold, reaching a low basal level that is maintained until reinnervation of the distal segment occurs.[8]

[a] Address for correspondence: Greg Lemke, MNL-East, The Salk Institute, P.O. Box 85800, San Diego, CA 92186.

A similarly dramatic decrease in major myelin gene expression is observed if the P_0 and MBP mRNAs levels present in 2–3-day sciatic nerve are compared to the levels present in Schwann cells dissociated from this nerve and cultured for several days in the absence of axons. Cultured Schwann cells deprived of axons express ~40-fold less P_0 and MBP mRNA than do Schwann cells in the early stages of myelination in the 2–3-day nerve. This low basal level of expression is reached within 5 days after Schwann cells are placed in culture and is maintained indefinitely thereafter.[9] Several lines of study, some of which are discussed below, indicate that these examples of axonal dependence reflect changes in the instantaneous rate of Schwann cell transcription of the major myelin genes rather than changes in the turnover rate of the mRNAs transcribed from these genes.

CYCLIC AMP REGULATION OF MYELIN GENES

In cultured Schwann cells, the requirement for continuous axonal contact can be overridden in part by treatment with any agent that stimulates the synthesis or mimics the action of intracellular cAMP. Thus, both the major and minor myelin genes are induced by cAMP analogs, cholera toxin, and the reversible adenyl cyclase activator forskolin,[9,10] and the relative sensitivity of these genes to induction by cAMP parallels their relative level of expression in myelinating Schwann cells. The P_0 gene, for example, is approximately 10-fold more sensitive to forskolin induction than is the MBP gene, and P_0 is expressed at ~10-fold higher levels than is MBP in actively myelinating Schwann cells.[9]

Several lines of evidence suggest that cAMP regulation of myelin gene expression is exercised at the level of transcription. The cloned rat P_0 regulatory region, for example, functions as a positive transcriptional regulator of heterologous reporter genes only when hybrid reporter plasmids are transfected into Schwann cells cultured in the presence of cAMP-elevating agents. When introduced into non-forskolin-treated Schwann cells, such constructs are inactive.[11] This cloned regulatory region (approximately 1 kb of DNA) also exhibits the cell-type specificity of the endogenous P_0 gene. In cultured cells, P_0 promoter-reporter constructs are only expressed in Schwann cells and Schwann cells lines (not in fibroblast, melanoma, neuronal, or muscle lines),[11] and in transgenic mice, expression is confined to myelinating nerves among all tissues examined.[12]

Although cAMP induction of the major myelin genes is robust, it is not of the form most commonly studied. Unlike cAMP regulation of immediate-early response genes such as *c-fos* and NGFI-A,[13] induction of myelin gene expression is neither rapid nor transient. Instead, up-regulated levels of the P_0, MBP, and P_2 mRNAs are not evident until 18–24 hours after cAMP elevation; once induced, these mRNAs remain high as long as cAMP levels remain elevated.[10] These kinetics suggest that cAMP induction of myelin gene expression occurs indirectly, at the end of a cascade of intermediate, cAMP-modulated regulatory molecules.

cAMP REGULATION OF A NEW TRANSCRIPTION FACTOR

We recently identified and cloned one such molecule, a cAMP-induced protein whose structure places it within the "POU" family of transcription factors.[10] This family includes the DNA binding proteins Pit-1/GHF-1, Oct-1, Oct-2, Oct-3, and unc-86, among several others.[14–18] This new protein, which we have named SCIP (pronounced "skip," for suppressed cAMP inducible POU), is expressed in the developing central and peripheral nervous systems and at particularly high levels in developing optic and sciatic nerves. It is not detectably expressed in the liver, kidney, heart, lung, or spleen of 6-day-old rats. Among cultured cells and under appropriate culture conditions (to be described), SCIP mRNA is present in Schwann cells and Schwann cells lines, but not in cultured cortical astrocytes.[10]

Like the major and minor myelin genes, expression of the SCIP gene by cultured Schwann cells is acutely sensitive to induction by cAMP. Up-regulation in the steady-state level of SCIP mRNA is seen with low to moderate concentrations of forskolin (1–10 μM) that result in physiologically significant elevation of intracellular cAMP. To assess its potential regulatory role, we examined the kinetics of SCIP induction relative to cAMP induction of the major and minor myelin genes. Using steady-state (Northern blot) analyses, we first detected induction of SCIP mRNA in cultured Schwann cells at 1 hour after the addition of forskolin. SCIP mRNA levels rise dramatically between 1 and 3 hours, and increase more gradually up to 24 hours after the addition. As for myelin-specific mRNAs, expression of SCIP mRNA is stable as long as cAMP levels remain elevated, but it rapidly returns to baseline (within 3 hours) upon withdrawal of forskolin from the Schwann cell culture medium. Thus, cAMP induction of the SCIP gene precedes cAMP induction of the major and minor myelin genes by $\sim$12 hours.[10]

cAMP does not induce expression of the SCIP gene through prior induction of a transcriptional activator. Co application of the protein synthesis inhibitor cycloheximide does not block induction by forskolin (as would be predicted if *de novo* synthesis of an activator were required); instead, this co-treatment significantly *potentiates* cAMP induction. In addition, treatment of Schwann cells with cycloheximide alone results in weak induction of SCIP mRNA, even in the absence of cAMP elevation.[10] Together, these results indicate that the SCIP gene is suppressed in the absence of forskolin treatment and that this treatment serves to relieve suppression. Recent nuclear run-off experiments, in which the instantaneous rate of transcript elongation was measured in forskolin-treated and untreated Schwann cells, indicate that suppression is probably exercised at the level of transcriptional repression of the SCIP gene (E. S. M. and G. L., unpublished observations). The relevant transcriptional repressor is as yet unidentified.

SCIP EXPRESSION *IN VIVO*

If SCIP participates in the regulation of myelin gene expression during normal development, then its appearance must precede activation of these genes. We addressed this question by isolating RNA from sciatic nerves dissected at various postnatal days and then assaying these RNA preparations for expression of SCIP, P_0, and MBP mRNA by Northern blot. These analyses indicate that as for cAMP regulation in cultured Schwann cells, SCIP expression precedes expression of the major and minor myelin genes. A critical difference, however, is that in peripheral nerve development *in vivo* SCIP expression is transient rather than sustained. In the sciatic nerve, peak expression of SCIP mRNA is observed at 1 day after birth and falls to $<$50% of this level by 7 days after birth, the time at which expression of myelin mRNAs is maximal. By 21 days after birth, SCIP mRNA levels are $\sim$10% of the level present on day 1, whereas major myelin mRNAs remain high.[10] These kinetics of SCIP expression are not consistent with the protein functioning as a direct transcriptional activator of myelin gene expression in the way that Pit-1 and Oct-2 appear to function as activators of the prolactin and immunoglobulin genes, respectively, because these latter regulators are stably expressed in terminally differentiated cells. The phase of Schwann cell development with which SCIP expression *is* well correlated is the highly proliferative blast period that separates pre-myelinating and myelinating phases and immediately precedes full Schwann cell differentiation as evidenced by myelination. The relative time course of SCIP mRNA expression and Schwann cell proliferation is illustrated in FIGURE 1.

In response to nerve injury, Schwann cells exhibit the ability to reversible transit between myelinating and pre-myelinating phenotypes. We investigated the potential role of SCIP in this developmental plasticity with the use of the peripheral nerve transection model just mentioned. We examined Schwann cell gene expression in the region distal to the transection site at 2, 5, 10, and 20 days after transection and observed that although expression of SCIP mRNA was extremely low in the intact nerve, this expression was transiently up-regulated after transection.[19] This pattern differed markedly from those of genes specifically expressed by either myelinating or pre-myelinating cells. As noted already, the major myelin genes encoding P_0 and MBP are stably repressed following transection. In contrast, the reciprocally regulated nerve growth factor (NGF) receptor gene, whose expression marks pre-myelinating Schwann cells, is stably induced.[19] Whereas transient activation of the SCIP gene was clearly distinct from these stable responses, this activation was well correlated with the onset, duration, and relative magnitude of a burst of Schwann cell proliferation previously demonstrated to closely follow sciatic nerve transection (FIG. 2). The time course of SCIP expression, both during normal development and as reversed in response to nerve transection, in thus consistent with the hypothesis that high level SCIP expression is restricted to a rapidly proliferating population of Schwann cell progenitors interposed between myelinating and pre-myelinating cells.

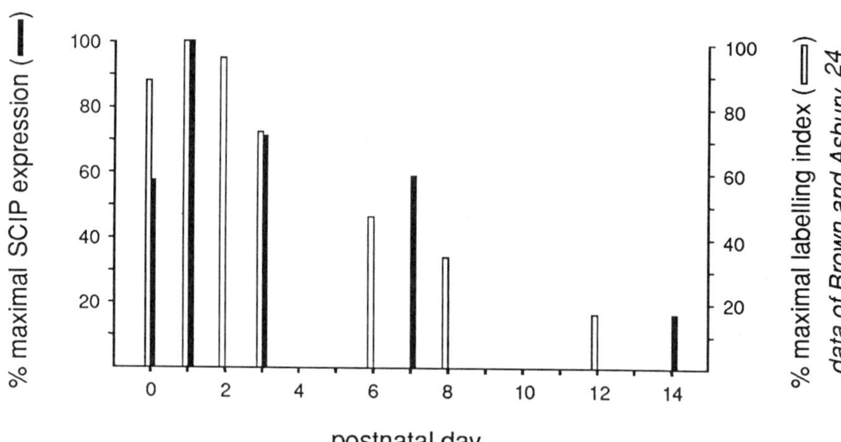

FIGURE 1. Comparison of SCIP mRNA expression and Schwann cell division in early postnatal sciatic nerves. SCIP mRNA levels in developing rat sciatic nerve (*dark bars*) are taken from Monuki *et al.*[10] and have been normalized to the maximal expression observed on postnatal day 1. Schwann cell labeling indexes are taken from the tritiated thymidine (^{3}H-TdR) studies of developing mouse sciatic nerve by Brown and Asbury[24] and have been normalized to the maximal Schwann cell nuclear labeling observed on postnatal day 1. (Approximately 24% of nuclei were labeled with a ^{3}H-TdR pulse administered on this day.)

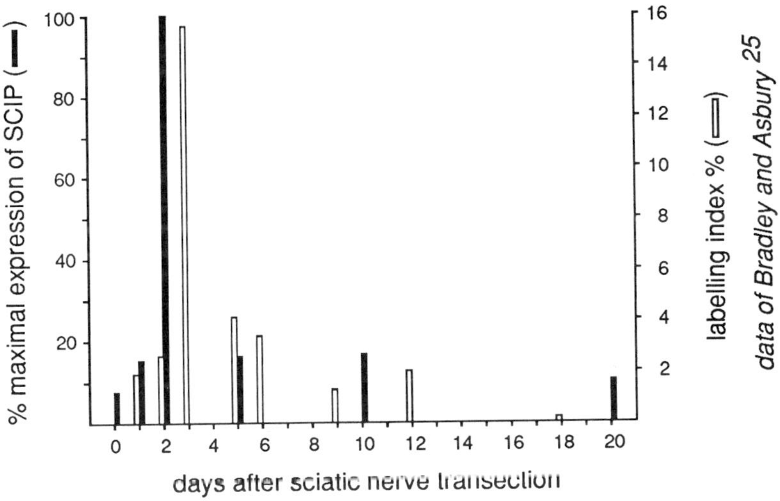

FIGURE 2. Comparison of SCIP mRNA expression and Schwann cell division in transected sciatic nerves. SCIP mRNA levels in transected rat sciatic nerves are taken from Monuki *et al.*[19] and have been normalized to the maximal mRNA expression observed 2 days after transection. Schwann cell division rates in transected mouse sciatic nerve are given as a labeling index (percent of Schwann cell nuclei labeled with a pulse of [^{3}H]-TdR) and are taken from Bradley and Asbury.[25]

SCIP AS A TRANSCRIPTIONAL REGULATOR

To investigate the regulatory properties of SCIP, we transfected cultured Schwann cells with an SV40-based SCIP expression plasmid together with plasmids in which transcription of the choramphenicol acetyltransferase (CAT) reporter gene was driven by different eukaryotic promoters. In accordance with its expression during normal Schwann cell development, SCIP functioned as a transcriptional repressor of the major myelin promoters in these experiments.[19] Cotransfection of the SCIP expression construct (at a mass ratio of 6:1, expression plasmid:reporter plasmid) resulted in moderate repression of the mouse MBP promoter and complete repression of both the rat P_0 and human NGF receptor promoters. These effects appeared to be specific, inasmuch as less than twofold differences in promoter activity (in the presence and absence of SCIP) were seen for CAT reporters driven by the mouse *c-jun* and SV40 early promoters or the Rous sarcoma virus LTR. SCIP repression of the P_0 promoter required an intact DNA binding domain; introduction of a frameshift mutation within the SCIP homeobox resulted in complete loss of repressor activity.[19]

The results of recent experiments are consistent with the hypothesis that SCIP repression of the P_0 gene results from direct binding of SCIP to the P_0 promoter. Recombinant, bacterially expressed SCIP binds with moderate to high affinity to several AT-rich sequences within the first 1 kb of the P_0 regulatory region (R. K., E. S. M., and G. L., unpublished observations). We are currently mutating these binding sites, both singly and in combination, to assess their relative importance with respect to the ability of SCIP to repress P_0 transcription. SCIP is the first identified POU domain transcription factor demonstrated to function as a transcriptional repressor. An important outstanding question is whether this repression is achieved passively, by binding alone (i.e., by competition for the binding sites of transcriptional activators), or actively, by both binding and subsequent negative interaction with regulatory proteins of the core promoter.

SCIP AND OTHER POU PROTEINS IN DEVELOPMENT

The developmentally restricted expression and repressor activity that we have observed for SCIP may reflect the general role played by POU transcription factors in early vertebrate development. A recently described POU protein, closely related to SCIP in its DNA binding domain and variously designated Oct-3, Oct-4, or NF-A3,[20–22] is expressed by embryonic stem cell lines only when these lines are cultured under conditions that permit cell proliferation but prevent differentiation. When induced to differentiate by retinoic acid, embryonic stem cells extinguish expression of Oct-3. In addition, nearly all of the POU proteins analyzed to date are expressed at relatively high levels in regions of the early embryo, such as the ventricular zone of the neural tube, that are populated by rapidly proliferating progenitors.[23] As development proceeds, this early expression is extinguished and replaced in the adult by a pattern in which individual POU domain genes are stably expressed in restricted sets of fully differentiated cells. Together with our studies of SCIP expression during postnatal glial development, these findings suggest that POU proteins may first function as transiently expressed transcriptional repressors in proliferative progenitors and subsequently as stably expressed activators in terminally differentiated cells. We are currently extending our studies of SCIP expression to glial differentiation in the central nervous system, to determine if this expression marks the O-2A progenitors that give rise to both oligodendrocytes and type-2 astrocytes.

REFERENCES

1. Lemke, G. 1988. Unwrapping the genes of myelin. Neuron **1:** 535–543.
2. Roach, A., K. Boylan, S. Horvath, S. B. Prusiner & L. E. Hood. 1983. Characterization of a cloned cDNA representing rat myelin basic protein: Absence of expression in *shiverer* mutant mice. Cell **34:** 799–806.
3. Lemke, G. & R. Axel. 1985. Isolation and sequence of a cDNA encoding the major structural protein of peripheral myelin. Cell **40:** 501–508.
4. Milner, R. J., C. Lai, K.-A. Nave, D. Lenoir, J. Ogata & J. G. Sutcliffe. 1985. Nucleotide sequences of two mRNAs for rat brain myelin proteolipid protein. Cell **42:** 931–939.
5. Lai, C., M. A. Brow, K.-A. Nave, A. B. Noronha, R. H. Quarles, F. E. Bloom, R. J. Milner & J. G. Sutcliffe. 1987. Two forms of 1B236/myelin-associated glycoprotein, a cell adhesion molecule for postnatal neural development, are produced by alternative splicing. Proc. Natl Acad. Sci. USA **84:** 4337–4341.
6. Jessen, K. R., L. Morgan, H. J. S. Stewart & R. Mirsky. 1990. Three markers of adult non-myelin-forming Schwann cells, 217c(Ran-1), A5E3 and GFAP: Development and regulation by neuron-Schwann cell interactions. Development **109:** 91–103.
7. Bray, G. M., M. Rasminsky & A. J. Aguayo. 1981. Interactions between axons and their sheath cells. A. Rev. Neurosci. **4:** 127–162.
8. Trapp, B. D., P. Hauer & G. Lemke. 1988. Axonal regulation of myelin protein mRNA levels in actively myelinating Schwann cells. J. Neurosci. **8:** 3515–3521.
9. Lemke, G. & M. Chao. 1988. Axons regulate Schwann cell expression of the major myelin and NGF receptor genes. Development **102:** 499–504.

10. MONUKI, E. S., G. WEINMASTER, R. KUHN & G. LEMKE. 1989. SCIP: A glial POU domain gene regulated by cAMP. Neuron **3:** 783–793.
11. LEMKE, G., E. LAMAR & J. PATTERSON. 1988. Isolation and analysis of the gene encoding peripheral myelin protein zero. Neuron **1:** 73–83.
12. MESSING, A., R. BEHRINGER, R. PALMITER, R. BRINSTER & G. LEMKE. In preparation.
13. MILBRANDT, J. 1987. Nerve growth factor rapidly induces a gene which encodes a possible transcriptional regulatory factor. Science **238:** 797–799.
14. INGRAHAM, H. A., R. CHEN, H. J. MANGALAM, H. P., ELSHOLTZ, S. E. FLYNN, C. R. LIN, D. M. SIMMONS, L. SWANSON & M. G. ROSENFELD. 1988. A tissue-specific transcription factor containing a homeodomain specifies a pituitary phenotype. Cell **55:** 519–529.
15. BODNER, M., J.-L. CASTRILLO, L.-E. THEIL, T. DEERINCK, M. ELLISMAN & M. KARIN. 1988. The pituitary-specific transcription factor GHF-1 is a homeobox-containing protein. Cell **55:** 505–518.
16. STURM, R. A., G. DAS & W. HERR. 1988. The ubiquitous octamer-binding protein Oct-1 contains a POU domain with a homeobox subdomain. Genes Dev. **2:** 1582–1599.
17. MÜLLER, M. M., S. RUPPERT, W. SCHAFFNER & P. MATTHIAS. 1988. A cloned octamer transcription factor stimulates transcription from lymphoid-specific promoters in non-B cells. Nature **336:** 544–551.
18. FINNEY, M., G. RUVKUN & H. R. HORVITZ. 1988. The *C. elegans* cell lineage and differentiation gene unc-86 encodes a protein with a homeodomain and extended similarity to transcription factors. Cell **55:** 757–769.
19. MONUKI, E. S., R. KUHN, G. WEINMASTER, B. D. TRAPP & G. LEMKE. 1990. Expression and activity of the POU transcription factor SCIP. Science **249:** 1300–1303.
20. OKAMOTO, K., H. OKAZAWA, A. OKUDA, M. SAKAI, M. MURAMATSU & H. HAMADA. 1990. A novel octamer binding transcription factor is differentially expressed in mouse embryonic cells. Cell **60:** 461–472.
21. ROSNER, M. H., M. A. VIGANO, K. OZATO, P. M. TIMMONS, F. POIRIER, P. W. J. RIGBY & L. M. STAUDT. 1990. A POU-domain transcription factor in early stem cells and germ cells of the mammalian embryo. Nature **345:** 686–692.
22. SCHÖLER, H. R., S. RUPPERT, N. SUZUKI, K. CHOWDHURY & P. GRUSS. 1990. New type of POU domain in germ line-specific protein Oct-4. Nature **344:** 435–439.
23. HE, X., M. N. TREACY, D. M. SIMMONS, H. A. INGRAHAM, L. W. SWANSON & M. G. ROSENFELD. 1989. Expression of a large family of POU-domain regulatory genes in mammalian brain development. Nature **340:** 35–42.
24. BROWN, M. J. & A. K. ASBURY. 1981. Schwann cell proliferation in the postnatal mouse: Timing and topography. Exp. Neurol. **74:** 170–186.
25. BRADLEY, W. G. & A. K. ASBURY. 1970. Duration of synthesis phase in neurilemma cells in mouse sciatic nerve during degeneration. Exp. Neurol. **26:** 275–282.

Immune-Mediated Oligodendrocyte Injury

D. A. S. COMPSTON AND N. J. SCOLDING

University of Cambridge Clinical School
Department of Neurology
Addenbrooke's Hospital
Cambridge CB2 2QQ, England

DEMYELINATING DISEASES OF THE HUMAN CENTRAL NERVOUS SYSTEM

Several diseases of the human central nervous system (CNS) depend on injury of myelinated axons. Clinically, the symptoms and signs of demyelination are an expression of impaired saltatory conduction through the affected parts, the clinical features depending on the anatomical location of individual lesions. The commonest and most socioeconomically significant of these diseases is multiple sclerosis which differs from the acute disseminated encephalomyelopathies and syndromes of isolated demyelination not so much in the anatomic distribution of the lesions but in their temporal dissemination.

Neuropathological findings and immunological observations suggest the following sequence of events in the evolution of individual demyelinating lesions. First, activated T lymphocytes and macrophages express the cell surface adhesion molecules that allow them to breach the endothelial lining of cerebral blood vessels[1] which together with the basement membrane and foot processes of type-1 astrocytes make up the blood brain barrier.[2] The increased permeability permits a range of cellular and soluble immune mediators to become concentrated on the abluminal surface of cerebral capillaries, and extensive local tissue edema is noted[3]; oligodendrocyte numbers are reduced, surviving cells appearing morphologically abnormal,[4] and myelin is removed by macrophages from otherwise intact axons[5]; phagocytic cells infiltrating developing plaques are seen to contain myelin debris associated with antibody and complement[6]; the subacute lesion is now reconstituted by oligodendrocytes or their progenitors and there is extensive remyelination[4,7] but this seems not to last, and oligodendrocyte depletion with demyelination is seen in a large proportion of chronic lesions.[8] Functionally, this sequence initially is associated with stable symptomatic recovery, but events tend to recur at the same anatomic site, and with time many individuals switch from the characteristic relapsing and remitting course of the disease to a slowly progressive phase. The extent to which resolution of edema, remyelination, or, more interestingly, repair of acute lesions that never reach the stage of demyelination account for clinical remission is uncertain.[9]

These pathological observations, based on reconstructing the putative sequence of events from autopsy or biopsy findings, have now been supplemented by information on the evolution of lesions in life using gadolinium-enhanced magnetic resonance imaging; these spectacular studies confirm that the blood-brain barrier breaks down well in advance of any parenchymal abnormality and show clear dissociation between these two phases in the development of plaques; cycles of blood-brain barrier damage may recur repeatedly at one site without parenchymal damage, and when damage does occur, there is a significant interval between the vascular and white matter lesions.[10] Taken together, the evidence suggests that damage to the blood-brain barrier is necessary but not sufficient for the development of demyelination.

Although much about the etiology and pathogenesis of multiple sclerosis has been learned in recent years, important information is lacking, and no treatment has yet been identified that is effective, safe, and appropriate in a situation in which prolonged periods of clinical inactivity are common and the prognosis is usually uncertain. Thus, while research aimed at defining the nature of susceptibility and identifying events that induce or precipitate patches of inflammatory demyelination may eventually lead to improved treatment for individuals with demyelinating disease, it can be argued that the critical problem, and the one that lends itself most obviously to therapeutic intervention, is understanding the molecular basis for injury of oligodendrocytes and their myelin membranes; the available evidence suggests that this is macrophage mediated and critically dependent on the action of complement and antibody.

MACROPHAGE ACTIVATION AND FUNCTION; MICROGLIA

Macrophages constitutively express receptors for the Fc portion of immunoglobulin and for the C3b component of complement.[11] In the presence of specific antibody, Fc binding allows macrophages to adhere to target cells; this is one mechanism of antibody-dependent cell-mediated cytotoxicity. C3b is one of the short-lived products of both classical and alternative pathway complement activation, but unlike C3a and C5a it remains membrane bound; binding is transient even if macrophages adhere to the surface of cells that have undergone complement activation, unless the macrophage is in a state of activation after exposure to various cytokines, especially gamma interferon. Macrophage activation is accompanied by several changes in the morphological appearance of the cell and in its capacity to secrete soluble inflammatory and immune mediators, including free oxygen radicals and tumor necrosis factor. Although activated macrophages are able to cross the blood-brain barrier and enter the CNS, the brain has its own resident population of phagocytic cells or microglia, cells that may be derived from macrophages infiltrating during neonatal life.[12] The morphological changes associated with microglial activation closely resemble those seen in macrophages and were first observed several decades ago; these changes are now known to be accompanied by alterations in the expression of major histocompatibility antigens that are a feature of immunological activation.

ROLE OF CELL BIOLOGY IN HUMAN DEMYELINATING DISEASE

Although an improved understanding of the pathogenesis of human demyelination at the cell biological and molecular level may result from magnetic resonance imaging and spectroscopy of individuals during life, an alternative approach is to exploit the opportunity for growing purified populations of CNS glial cells and to study their immunological and biological behavior *in vitro.* Applying the principles of glial neuroscience using techniques pioneered in rodents, and extrapolating from biological properties of the neonatal rat optic nerve to human demyelinating disease may not be appropriate, but this approach has gone some way towards explaining the basis for immunological injury and repair of oligodendrocytes, observations that appear relevant to multiple sclerosis and other human demyelinating diseases.

In vitro observations suggest that neuroepithelial cells differentiate in the developing optic nerve at embryonic day 16 into type-1 astrocytes; these cells contribute to formation of the blood-brain barrier and secrete growth factors that influence the differentiation of the other main lineage, the oligodendrocyte-type-2 astrocyte (O-2A) progenitors.[13,14] O-2A progenitors cannot be isolated from the optic nerve until embryonic day 17 and are followed by the appearance of oligodendrocytes at the time of birth; platelet-derived growth factor produced by type-1 astrocytes may be important in their differentiation. Furthermore, it has been suggested that O-2A progenitors migrating into the neonatal optic nerve, and perhaps also those that persist into adult life, encounter an altered environment and differentiate into type-2 astrocytes, cells that contact the bare parts of axons between the territories of neighboring oligodendrocytes (the nodes of Ranvier), under the influence of ciliary neurotrophic factor which is also produced by type-1 astrocytes. Leaving aside the validity of these distinctions for glial organization in the intact human and animal nervous system, manipulating their developmental profiles and growth requirements has allowed pure populations of each cell type to be cultured and investigated *in vitro.*

COMPLEMENT ACTIVATION

Living intact rat optic nerve oligodendrocytes and adult O-2A progenitors are unique among glia in that they express a cell surface component, not yet fully characterized, that fixes and activates complement through the classical pathway but in the absence of antibody.[15,16] The lytic property of autologous serum on oligodendrocytes grown in tissue culture has been encountered by many investigators, and others have demonstrated that purified myelin activates complement, also through the classical pathway.[17,18] This effect is heat labile, is affected by freezing, and decays with time; it is not present in serum obtained from animals depleted of complement using cobra venom factor or in serum specifically depleted of C1 or C9, reconstitution with each restoring cytotoxicity. Blocking classical pathway complement activation with EGTA abolishes the cytotoxic effect, but antibody is not involved because serum preabsorbed with rat myelin retains its cytotoxicity. C9 but not immunoglobulin can be identified on the surface of oligodendrocytes following exposure to serum as a source of complement. Thus, the two cell types most closely involved in myelin synthesis each potentially initiate their own destruction when exposed to normal serum, a property seen only to a limited extent in neonatal progenitors and not shared by type-1 and type-2 astrocytes or Schwann cells.[16,19] Immunocytochemical studies indicate that cytotoxicity involves the formation of the membrane attack complexes of complement formed by the sequential addition of the late complement components, especially C9, following C3 cleavage, the step at which the classical and alternate pathways converge. Membrane attack complexes form 100 angstrom pores in the cell membrane which act initially as calcium ionophores, later lysing the cell by osmotic disruption.

Complement activation is necessarily associated with the production of surface-bound C3b and should provide an opportunity for investigating adherence of macrophages through constitutively expressed C3b receptors. This is made possible by an additional property of oligodendrocytes, which is their capacity to recover from complement attack despite the formation of membrane attack complexes; this not only makes possible the study of macrophage adherence using the C3b receptor but also suggests new ideas about the nature of reversible oligodendrocyte injury, remyelination, and repair of demyelinated lesions *in vivo.*

OLIGODENDROCYTE RECOVERY FROM COMPLEMENT ATTACK

The cytotoxic effect of autologous serum on neonatal rat optic nerve oligodendrocytes is dose dependent and is not seen using dilutions > 1:30 even though complement fixation and activation occur at these concentrations. Cell damage is lacking because oligodendrocytes have an inherent capacity to recover using a process in which the pores formed by membrane attack complexes are concentrated within the semifluid surface membrane and coalesce into vesicles that are then extruded from the cell surface, subsequently appearing in the culture supernatant.[20] This mechanism begins within 1–2 minutes of exposure to sublytic concentrations of serum and is complete within 12 minutes. Thereafter, the oligodendrocyte is viable and, on the present evidence, can resume many specialized functions, but it is vulnerable to rapid repeated attack and cannot withstand more than three exposures separated by up to 6 hours.

Cell lysis mediated by the membrane attack complexes of complement is associated with an uncontrolled rise in intracellular calcium, whereas vesicular repair is associated with a transient calcium flux.[20,21] The change in intracellular calcium depends mainly on the passage of calcium across the cell membrane, but there is some release from intracellular stores; the calcium transient is followed by a brief reduction in intracellular levels of ATP. Two lines of evidence indicate that the vesicular repair process is the direct result of these changes in intracellular calcium. First, an identical vesicular response and sequence of changes in intracellular calcium are seen after exposure to T-cell–derived perforins. These soluble mediators are released by cytotoxic T cells and operate over a short range, producing pores in the surface membrane of target cells that have morphological similarities and some structural homology with the membrane attack complexes of complement.[22] The ability of the calcium ionophore A23187 closely to mimic these effects further suggests that changes in calcium are central in orchestrating the oligodendrocyte response to injury, indicating that vesicular repair can be mediated immunologically and by nonimmune calcium ionophores.

Secondly, cytotoxicity after complement activation is enhanced in the presence of substances that block metabolic pathways responsible in eukaryotic cells for mediating the consequences of an intracellular calcium surge. Calcium entering the cell or released from intracellular stores interacts with and is bound by calmodulin, which thereafter activates various calcium-dependent enzymes. The calcium-binding properties of calmodulin and other intracellular messengers interacting with calcium are inhibited by W7. W7 is not directly toxic to oligodendrocytes in tissue culture but, by inhibiting calmodulin and interrupting the sequence of calcium-triggered changes, it increases the cytotoxic properties of normal serum by inhibiting vesiculation and by replacing reversible sublytic injury by cell lysis (unpublished observations).

These two lines of evidence suggest that the fate of oligodendrocytes exposed to pore-forming agents lies in the ability to withstand the rise in intracellular calcium, calcium transients initiating vesicular repair with survival of the cell, but higher concentrations exceeding the threshold for recovery and resulting in cell lysis.

COMPLEMENT ACTIVATION IN HUMAN DEMYELINATING DISEASE

Cerebrospinal fluid analysis has confirmed that intrathecal complement activation occurs in patients with isolated demyelinating lesions and multiple sclerosis. The complement activation products C3a and C4a can be detected in spinal fluid from patients but not controls,[23] whereas the concentration of C9, which is consumed as part of the activation process and not resynthesized within the nervous system, is reduced[24]; the reduction in C9 seen after even a single episode of demyelination[25] correlates inversely with the rate of IgG synthesis and increases following the use of high dose intravenous methylprednisolone. The neoantigen that forms with insertion of poly C9 into the C5b-9 complex that makes up the membrane attack complex can be detected in patients with demyelinating disease and those with some other inflammatory diseases.[26,27] Finally, particulate material, identical to that appearing in the supernatant of oligodendrocytes grown in tissue culture, is also present in cerebrospinal fluid from patients with multiple sclerosis but not controls; these vesicles bind antibodies to C8, C9, and the neoantigen of membrane attack complexes but not C3, and transmission electron microscopy shows them to have a high concentration of membrane pores; galactocerebroside but not myelin basic protein is present on their surface.[20]

These findings, indicating that complement activation in human demyelinating disease involves the formation of membrane attack complexes, are supported by more direct immunocytochemical observations demonstrating that complement is activated in areas of myelin breakdown[6,28]; complement is associated with the myelin debris incorporated within macrophages, and material rich in membrane attack complexes accumulates in the adventitia surrounding cerebral endothelial cells on the abluminal surface of the blood-brain barrier. The possibility arises that this vesicular material also is of oligodendrocyte origin and identical to that in cerebrospinal fluid, awaiting export from the CNS.

It is important to emphasize that complement activation is not unique to multiple sclerosis and is seen in a variety of conditions associated with intrathecal inflammation. The same is true of intrathecal IgG synthesis, the other characteristic immunological feature of multiple sclerosis. The oligoclonal antibodies can be detected electrophoretically and by quantitative methods, but it has not proved possible to identify their antigenic targets; absorption studies have demonstrated heterogeneity in the specificity within and between individuals, even though the synthesis and banding characteristics seem to remain stable within an individual over time whether or not fluctuations in disease activity occur. Although these antibodies are known to be synthesized by B-lymphocytes and plasma cells associated with areas of myelin damage in patients with demyelinating disease, their significance for the pathogenesis of multiple sclerosis has remained enigmatic for several decades; this and the role played by T lymphocytes in the development of experimental allergic encephalomyelitis have increasingly focused the attention of immunologists on cell-mediated mechanisms of demyelination.[29] However, although experimental allergic encephalomyelitis can be passively transferred in intact animals using T lymphocytes, synergistic use of cells activated against CNS antigens and anti-myelin antibodies is required for precise modeling of the clinical and histological features of acute human demyelination[30]; and demyelination cannot be induced in animals that have been depleted of complement or macrophage function. The study of interactions between oligodendrocytes and macrophages *in vitro* has provided further evidence that even though cellular mechanisms, involving activated lymphocytes, may trigger damage to the blood-brain barrier, thus altering the immunological environment on the abluminal surface of the cerebral vasculature, myelin injury represents a secondary phase in the evolution of individual plaques and seems to depend critically on antibody-dependent macrophage-mediated cytotoxicity.

MACROPHAGES AND OLIGODENDROCYTES

Resting autologous macrophages, obtained by peritoneal lavage, do not interact *in vitro* with neonatal rat optic nerve oligodendrocytes or increase their membrane

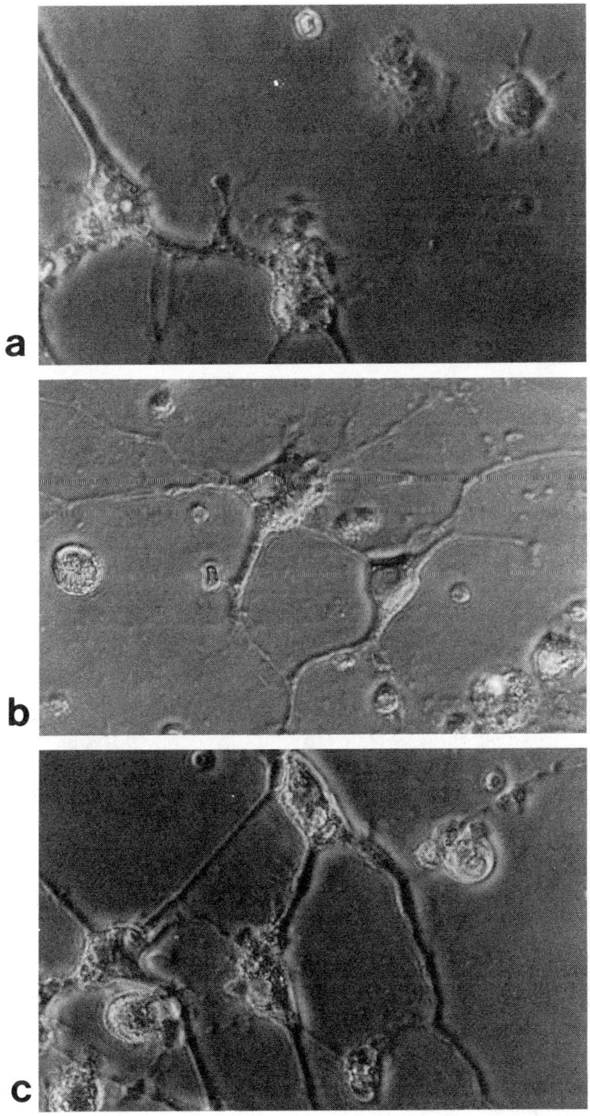

FIGURE 1. Cocultures of oligodendrocytes and macrophages. Oligodendrocyte-macrophage cocultures were prepared as previously described.[31] In untreated cultures (**a**), there are no interactions between the two cell types; the same findings obtain in the presence of 1:10 diluted homologous C9-depleted serum (**b**). However, in the presence of antigalactocerebroside antibody (**c**), macrophages become adherent to oligodendrocytes. Similar adhesion occurs in the presence of anti-MOG antibody (not shown) (original magnification × 400).

permeability. To assess the effects of complement activation and the potential in these cocultures for macrophage binding to oligodendrocytes through the C3b receptor, serum has to either be used at a sublytic concentration or be depleted of C9 to prevent cell lysis. However, under these circumstances, and despite the possibility of adherence through the constitutively expressed C3b receptor, no increase in macrophage/oligodendrocyte interaction occurs. Thus, complement activation by oligodendrocytes is not of itself an adequate stimulus for recruiting resting macrophages or microglia. However, incubating rat oligodendrocytes with antibody directed against structural components of the cell surface affects the subsequent fate of these cells. After approximately 24 hours in culture, macrophages adhere to the surface of oligodendrocytes coated with antibody directed against myelin oligodendrocyte glycoprotein or galactocerebroside. No difference in the appearance of adherent macrophages and oligodendrocytes was discernible with these two antibodies, but attachment does not occur in the presence of anti-progesterone antibody used as a control; in addition, the macrophage-mediated phagocytosis of myelin basic protein is much enhanced in the presence of cells coated with specific antibody.[31] These results are consistent with the view that oligodendrocytes and macrophages do not ordinarily interact *in vivo,* but antibodies to oligodendrocyte surface antigens opsonize the oligodendrocyte myelin unit, stimulating attachment through the Fc receptor. It is perhaps relevant that any one of several antibodies will stimulate macrophage-mediated oligodendrocyte injury in tissue culture. This may explain why the search for a single antigenic determinant within the oligoclonal bands in cerebrospinal fluid from patients with multiple sclerosis has been unsuccessful; no one antigen may be critical because several antigens have the identical effect of attracting macrophages onto the oligodendrocyte cell surface.

IMPLICATIONS FOR HUMAN DEMYELINATING DISEASE

The hypothesis to emerge from the results of these human and experimental studies[32] is that activated T cells and macrophages, probably responding to environmental triggers, increase their normal surveillance of the nervous system and disrupt the blood-brain barrier, causing immune mediators to accumulate within the CNS. Oligodendrocytes are unduly sensitive to contact with complement and other pore-forming molecules, especially when antibody directed against surface components of the oligodendrocyte or its myelin processes is also present, perhaps as a result of previous oligodendrocyte injury and vesicular repair. The interaction of these soluble mediators increases membrane permeability and leads to a rise in intracellular calcium, the degree and duration of which determine whether or not reversible cell injury ensues. As complement activation proceeds, and when antibody also is present, macrophages are recruited and contribute to cell injury by releasing cytokines[33] and phagocytosing the cell body or its myelin lamellae. Multiple sclerosis can be regarded as a low-grade vasculitis mediated by activated lymphocytes, with secondary demyelination resulting from the presence of macrophages and local antibody synthesis within the CNS and initiated by the tendency for oligodendrocytes to activate complement.

REFERENCES

1. Wekerle, H., C. Linington, H. Lassmann & R. Meyerman. 1986. Cellular immune reactivity within the CNS. Trends in Neurosci. **9:** 271–277.
2. Bradbury, M. W. 1983. *In* The Concept of a Blood-Brain Barrier. John Wiley & Sons. London.
3. Adams, C. W. M. 1977. Pathology of multiple sclerosis: Progression of the lesion. Brit. Med. Bull. **33:** 15–20.

4. PRINEAS, J. W., E. KWON, P. Z. GOLDENBERG, *et al.* 1989. Multiple sclerosis: Oligodendrocyte proliferation and differentiation in fresh lesions. Lab. Invest. **61:** 489–503.
5. ESIRI, M. M. & M. C. READING. 1987. Macrophage populations associated with multiple sclerosis plaques. Neuropathol. & Appl. Neurobiol. **13:** 451–465.
6. GAY, D. & M. M. ESIRI. 1991. Blood brain barrier damage in acute multiple sclerosis plaques: An immunocytological study. Brain **114:**
7. PRINEAS, J. W. & F. CONNELL. 1979. Demyelination in multiple sclerosis. Ann. Neurol. **5:** 22–31.
8. PRINEAS, J. W. 1985. The neuropathology of multiple sclerosis. *In* Handbook of Clinical Neurology—Demyelinating Diseases. J. Koetsier, ed. **3**(47): 213–257. Elsevier. Amsterdam.
9. MCDONALD, W. I. 1986. The pathophysiology of multiple sclerosis. *In* Multiple Sclerosis. W. I. McDonald & D. H. Silberberg, eds.: 112–133. Butterworth, London.
10. KERMODE, A. G., P. S. TOFTS, A. J. THOMPSON, *et al.* 1990. Heterogeneity of blood brain barrier changes in multiple sclerosis: An MRI study with gadolinium-DPTA enhancement. Neurology **40:** 229–235.
11. GRIFFIN, F. M. 1984. Activation of macrophage complement receptors for phagocytosis. *In* Contemporary Topics in Immunobiology. D. O. Adams & M. G. Hanna, eds. **13:** 50–70. Plenum Press. New York.
12. PERRY, V. H. & S. GORDON. 1988. Macrophages and microglia in the nervous system. Trends in Neurosci. **11:** 273–277.
13. RAFF, M. C., R. H. MILLER & M. NOBLE. 1983. A glial cell precursor that develops *in vitro* into an astrocyte or an oligodendrocyte depending on culture medium. Nature **303:** 390–396.
14. MILLER, R. H., C. FFRENCH CONSTANT & M. C. RAFF. 1989. The macroglial cells of the rat optic nerve. Ann. Rev. Neurosci. **12:** 517–534.
15. SCOLDING, N. J., B. P. MORGAN, A. HOUSTON, *et al.* 1989. Normal rat serum cytotoxicity against syngeneic oligodendrocytes. Complement activation and attack in the absence of antimyelin antibodies. J. Neurol. Sci. **89:** 289–300.
16. WREN, D. R. & M. NOBLE. 1989. Oligodendrocytes and adult specific O-2A progenitor cells are uniquely susceptible to the lytic effects of complement in the absence of antibody. Proc. Natl. Acad. Sci. USA **86:** 9025–9029.
17. CYONG, J.-C., S. S. WITKIN, B. REIGER, *et al.* 1982. Antibody independent complement activation by myelin via the classical complement pathway. J. Exp. Med. **155:** 587–597.
18. VANGURI, P., C. L. KOSKI, B. SILVERMAN & M. L. SHIN. 1982. Complement activation by isolated myelin: Activation of the classical pathway in the absence of myelin specific antibodies. Proc. Natl. Acad. Sci. USA **79:** 3290–3294.
19. SCOLDING, N. J., B. P. MORGAN, A. K. CAMPBELL & D. A. S. COMPSTON. 1990. Complement mediated serum cytotoxicity against oligodendrocytes; a comparison with other cells of the oligodendrocyte—type 2 astrocyte lineage. J. Neurol. Sci. **97:** 155–162.
20. SCOLDING, N. J., B. P. MORGAN, W. A. J. HOUSTON, *et al.* 1989. Vesicular removal by oligodendrocytes of membrane attack complexes formed by activated complement. Nature **339:** 620–622.
21. SCOLDING, N. J., W. A. J. HOUSTON, B. P. MORGAN, *et al.* 1989. Reversible injury of cultured rat oligodendrocytes by complement. Immunology **67:** 441–446.
22. SCOLDING, N. J., J. JONES, D. A. S. COMPSTON & B. P. MORGAN. 1990. Oligodendrocyte susceptibility to perforin mediated cell injury. Immunology **70:** 6–10.
23. HARTUNG, H.-P., B. SCHAFER, K. HENINGER, *et al.* 1989. Intrathecal complement activation in inflammatory and demyelinating diseases of the peripheral and central nervous systems. *In* Recent Advances in Multiple Sclerosis Therapy. R. E. Gonsette & P. Delmotte, eds.: 243–247. Elsevier, Amsterdam.
24. MORGAN, B. P., A. K. CAMPBELL & D. A. S. COMPSTON. 1984. Terminal component of complement (C9) in cerebrospinal fluid of patients with multiple sclerosis. Lancet **ii:** 251–254.
25. COMPSTON, D. A. S., B. P. MORGAN, D. OLEESKY, *et al.* 1986. Cerebrospinal fluid C9 in demyelinating disease. Neurology **36:** 1503–1506.
26. SANDERS, M. E., C. L. KOSKI, D. ROBBINS, *et al.* 1986. Activated terminal complement in

cerebrospinal fluid in Guillain-Barré syndrome and multiple sclerosis. J. Immunol. **136:** 4456–4459.
27. Hartung, H.-P. & K. Heininger. 1989. Non-specific mechanisms of inflammation and tissue damage in MS. Ann. Inst. Pasteur **140:** 226–233.
28. Compston, D. A. S., B. P. Morgan, A. K. Campbell, *et al.* 1989. Immunocytochemical localisation of the terminal complement complex in multiple sclerosis. Neuropathol. Appl. Neurobiol. **15:** 307–316.
29. Waksman, B. H. & W. E. Reynolds. 1984. Multiple sclerosis: A disease of immune regulation. Proc. Soc. Exp. Biol. Med. **175:** 282–294.
30. Linington, C., M. Bradl, H. Lassmann, *et al.* 1988. Augmentation of demyelination in rat acute allergic encephalomyelitis by circulating mouse monoclonal antibodies directed against a myelin oligodendrocyte glycoprotein. Am. J. Pathol. **130:** 443–454.
31. Scolding, N. J. & D. A. S. Compston. 1991. Oligodendrocyte-macrophage interactions *in vitro* triggered by specific antibodies. Immunology **77:** 127–132.
32. Compston, D. A. S., N. J. Scolding, D. Wren & M. Noble. 1991. The pathogenesis of demyelinating disease: Insights from cell biology. Trends Neurosci. **14:** 175–182.
33. Selmaj, K. W. & C. S. Raine. 1988. Tumour necrosis factor mediates myelin and oligodendrocyte damage *in vitro.* Ann. Neurol. **23:** 339–346.

DISCUSSION PAPER

Myelin and Demyelination

PETER J. BROPHY

Department of Biological and Molecular Sciences
University of Stirling
Stirling FK9 4LA, Scotland, UK

The program by which Schwann cells and oligodendrocyte precursors differentiate into myelin-forming cells is the result of an interplay between intrinsic gene expression and environmental cues. The development of well-characterized tissue culture systems in the laboratories of Bunge *et al.*[1] and Raff[2] has been particularly important in defining the effects of potential modulators in glial differentiation, but so far little is known about the specific genes or gene products that are their intracellular targets. A more profound understanding of these interactions would inform attempts to mobilize the endogenous glial progenitor populations of patients with multiple sclerosis in order to repair their demyelinated nerve fibers.

The maturation of glial cells is likely to be determined by the sequential transcriptional activation and repression of specific genes by means of DNA binding proteins. *Drosophila* has served as the paradigm for molecular biological studies on development in other animals, because so much is known about the proteins that regulate gene transcription during morphogenesis. Therefore, it was extremely exciting when proteins that are closely related to the homeodomain proteins of *Drosophila* were discovered in vertebrates.

Homeodomain proteins share a conserved 60 amino acid homeodomain and regulate transcription during *Drosophila's* development by interacting with specific DNA sequences. A growing number of DNA-binding proteins that possess the homeodomain have now been identified in embryonic and adult cells from the tissues of rodents and humans.[3-5] These proteins include Pit-1, Oct-1, Oct-2, Oct-3, Oct-4, Oct-5, Oct-6, Brn-1, Brn-2, Brn-3, and Tst-1/SCIP; together with the *unc*-86 gene of *Caenorhabditis elegans,* they constitute the POU family of transcriptional regulatory factors. In addition to the homeodomain the POU proteins share a 76–78 amino acid POU-specific domain. The POU homeodomain, the POU-specific domain, and the region that links them are together known as the POU domain. Sequence similarities between POU domains have allowed POU proteins to be classified further into four groups.[3] The POU-III group of proteins presently comprises Brn-1, Brn-2, Tst-1/SCIP, and Oct-6, each of which is expressed in the nervous system of rodents during development.[3,5] Tst-1/SCIP and Oct-6 are also expressed in testicular tissue.

A common feature of the type III group is that although they are widely distributed in the cells of the early neural tube, their regional expression becomes more restricted at later stages of nervous system development. For example, Oct-6 is found in the developing cerebral cortex but not in the cerebellum.[5] Considerable differences can also be demonstrated in the expression of these transcriptional regulators among various cell types in a particular region of the nervous system. In the cerebellum, Brn-1 and Brn-2 are expressed in Purkinje cells, whereas Tst-1/SCIP is found in granule neurons.[3] Tst-1/SCIP is also found in myelinating glia of the central and peripheral nervous systems, where the transient nature of its expression

[a] Much of the research in my laboratory has been supported by the Multiple Sclerosis Society of Great Britain.

has led to the suggestion that it has a function in the early events that precede the induction of myelin-specific genes.[6]

The Oct group of proteins are so named because they bind to an octameric DNA motif,[7] and regulation by both activation[8] and repression[9] of transcription has been observed at this binding site. Therefore, an intriguing question is, do the POU proteins act to stimulate or to inhibit transcription of differentiation-specific genes? At the time of this writing, only Tst-1/SCIP has been shown to act by repression.[6] Identification of the signals that induce the expression or influence the binding of POU proteins to their regulatory elements will also prove to be an extremely interesting field of study. The discovery that Tst-1/SCIP expression in Schwann cells is induced by raising intracellular levels of cAMP, a procedure that instigates some early stages of differentiation in tissue culture, suggests that new and valuable insights will spring from a molecular biological analysis of differentiation of glial cells *in vitro*.[6,10] The recent discovery by Mirsky's group that Schwann cell differentiation can be stimulated most effectively *in vitro* by a combination of cAMP induction and growth inhibition provides an even more effective means by which early events in Schwann cell differentiation can be mimicked in tissue culture.[11]

Tst-1/SCIP is also expressed by oligodendrocytes but not by astrocytes.[6] Therefore, the careful evaluation by Noble's group (in this volume) of the roles of fibroblast growth factor (FGF) and platelet-derived growth factor (PDGF) in effecting the switch from oligodendrocyte precursor to myelin-forming cell in tissue culture should lay the groundwork for a concerted search for the target genes that trigger the differentiation of these cells in the central nervous system. Inasmuch as elevation of cAMP levels in the early differentiation of Schwann cells is believed to mimic the effects of axonal contact,[11] it is particularly significant that recent studies show that expression of myelin-specific genes in oligodendrocytes is also influenced by axons *in vivo*.[12,13] Therefore, the mechanisms by which transcription is regulated during the differentiation of oligodendrocytes will probably prove to bear many similarities to those of their counterparts in the peripheral nervous system.

Myelinogenesis requires the coordinated biosynthesis and assembly of large quantities of proteins into membrane over a relatively short period. Regulation of this process at translational and posttranslational levels focused attention on mRNA translocation in the oligodendrocyte. As Campagnoni *et al.*[14] point out in this volume, this phenomenon was proposed to occur in oligodendrocytes to account for the sequestration of mRNAs for the myelin basic proteins (MBPs) in myelinating processes.[15] The MBPs are synthesized on free polysomes.[14] The machinery responsible for the movement of mRNA or polysomes or both has not yet been identified, but it will likely involve the cytoskeleton. Several proteins of the myelin membrane including the MBPs show a strong propensity to interact with microfilaments and microtubules in both purified myelin and cultured oligodendrocytes.[16–18] The association of polysomes with the cytoskeleton via nascent polypeptides is plausible, but it must be very specific because two other polypeptides that are also synthesized on free polysomes in oligodendrocytes have mRNAs that do not segregate into myelin processes.[19,20]

The involvement of specific regions of MBP polypeptides in mediating interactions with the cytoskeleton is a distinct possibility and amenable to experimental verification by molecular biological means. At this point it is worth noting that transfection and expression of four variants of rodent MBP have revealed that the two isoforms that include exon II, that is, the 21.5- and 17.0-kD polypeptides, distribute in HeLa cells in a way quite different from that of the other two isoforms.[21] Interestingly, these same two exon II-containing isoforms show a marked preference for cytoskeletal interaction over the 18.5- and 14.0-kD polypeptides.[16] A particularly significant role for MBP-cytoskeletal interactions may therefore be indicated at early stages of myelination because the MBP transcripts that include exon II predominate during this period.[22]

The impetus behind much of the neurobiological research concerned with myelination comes from a desire to understand demyelination, as it occurs in multiple sclerosis. In this volume Compston brings us up to date on the toxic effects of sera and purified complement on rat oligodendrocytes and their precursors in culture, together with the abundant evidence for the existence of activated complement in the central nervous system of patients with multiple sclerosis.[23]

Bornstein pioneered the use of rat organotypic cultures for investigating the demyelinating effects of human sera.[24] This demyelinating activity is not confined to sera from patients with multiple sclerosis, leading to the view that oligodendrocytes are peculiarly susceptible to the lytic effects of complement. Nevertheless, the necessity for caution in interpreting the results of toxicity testing of sera on heterologous cells (e.g., human sera and rat oligodendrocytes) is underscored by the recent observations of Ruijs *et al.*[25] They found that heat-inactivated sera from patients and normal controls are toxic to rat but not human oligodendrocytes in culture.

REFERENCES

1. Wood, P., F. Moya, C. Eldridge, G. Owens, B. Ranscht, M. Schachner, M. Bunge & R. Bunge. 1990. Studies of the initiation of myelination by Schwann cells. Ann. N.Y. Acad. Sci. **605:** 1–14.
2. Raff, M. C. 1989. Glial cell diversification in the optic nerve. Science **243:** 1450–1455.
3. He, X., M. N. Treacy, D. M. Simmons, H. A. Ingraham, L. W. Swanson & M. G. Rosenfeld. 1989. Expression of a large family of POU-domain regulatory genes in mammalian brain development. Nature **340:** 35–42.
4. Rosner, M. H., M. A. Vigano, K. Ozato, P. M. Timmons, F. Poirier, P. W. J. Rigby & L. M. Staudt. 1990. A POU-domain transcription factor in early stem cells and germ cells of the mammalian embryo. Nature **345:** 686–692.
5. Suzuki, N., H. Rohdewohld, T. Neuman, P. Gruss & H. R. Schöler. 1990. Oct-6: A POU transcription factor expressed in brain stem cells and in the developing brain. EMBO J. **9:** 3723–3732.
6. Lemke, G., R. Kuhn, E. S. Monuki & G. Weinmaster. 1991. Expression and activity of the transcription factor SCIP during glial differentiation and myelination. N. Y. Acad. Sci. This volume.
7. Falkner, F. G., R. Mocikat & H. G. Zachau. 1986. Sequences closely related to an immunoglobulin gene promoter enhancer element occur also upstream of other eukaryotic and prokaryotic genes. Nucl. Acid Res. **14:** 8819–8827.
8. Schöler, H. R., S. Ruppert, N. Suzuki, K. Chowdhury & P. Gruss. 1990. New type of POU domain in germ line-specific protein Oct-4. Nature **344:** 435–439.
9. Lenardo, M. J., L. Staudt, P. Robbins, A. Kuang, R. C. Mulligan & D. Baltimore. 1989. Repression of the IgH enhancer in teratocarcinoma cells associated with a novel octamer factor. Science **243:** 544–546.
10. Sobue, G. & D. Pleasure. 1984. Schwann cell galactocerebroside induced by derivatives of adenosine 3′,5′-monophosphate. Science **224:** 72–74.
11. Morgan, L., K. R. Jessen & R. Mirsky. 1991. The effects of cyclic AMP on differentiation of cultured Schwann cells: Progression from an early phenotype (04^+) to a myelin phenotype (P_0^+, $GFAP^-$, NGF-receptor$^-$) depends on growth inhibition. J. Cell Biol. **112:** 457–468.
12. Goto, K., T. Kurihara, Y. Takahashi & H. Kondo. 1990. Expression of genes for the myelin-specific proteins in oligodendrocytes *in vivo* demands the presence of axons. Neurosci. Letts. **117:** 269–274.

13. KIDD, G. J., P. E. HAUER & B. D. TRAPP. 1990. Axons modulate myelin protein messenger RNA levels during central nervous system myelination *in vivo.* J. Neurosci. Res. **26:** 409–418.
14. CAMPAGNONI, A. T., J. M. VERDI, A. N. VERITY, S. AMUR-UMARJEE & S. BYRAVAN. 1991. Post-transcriptional regulation of myelin protein gene expression. This volume.
15. COLMAN, D. R., G. KREIBICH, A. B. FREY & D. SABATINI. 1982. Synthesis and incorporation of myelin polypeptides into CNS myelin. J. Cell Biol. **95:** 598–608.
16. GILLESPIE, C. S., R. WILSON, A. DAVIDSON & P. J. BROPHY. 1989. Characterization of a cytoskeletal matrix associated with myelin from rat brain. Biochem. J. **260:** 689–696.
17. PEREYRA, P. M., E. HORVATH & P. E. BRAUN. 1988. Triton X-100 extractions of central nervous system myelin indicate a possible role for the minor myelin proteins in the stability of lamellae. Neurochem. Res. **13:** 583–595.
18. WILSON, R. & P. J. BROPHY. 1989. Role for the oligodendrocyte cytoskeleton in myelination. J. Neurosci. Res. **22:** 439–448.
19. GILLESPIE, C. S., L. BERNIER, P. J. BROPHY & D. R. COLMAN. 1990. Biosynthesis of the myelin 2′,3′-cyclic nucleotide 3′-phosphodiesterases. J. Neurochem. **54:** 656–661.
20. GILLESPIE, C. S., B. D. TRAPP, D. R. COLMAN & P. J. BROPHY. 1990. Distribution of myelin basic protein and P_2 mRNAs in rabbit spinal cord oligodendrocytes. J. Neurochem. **54:** 1556–1561.
21. STAUGAITIS, S. M., P. R. SMITH & D. R. COLMAN. 1990. Expression of myelin basic protein isoforms in nonglial cells. J. Cell Biol. **110:** 1719–1727.
22. JORDAN, C., V. FRIEDRICH, JR. & M. DUBOIS-DALCQ. 1989. *In situ* hybridization analysis of myelin gene transcripts in developing mouse spinal cord. J. Neurosci. **9:** 248–257.
23. COMPSTON, D. A. S. & N. J. SCOLDING. 1991. Immune mediated oligodendrocyte injury. This volume.
24. BORNSTEIN, M. B. & C. RAINE. 1977. Multiple sclerosis and experimental allergic encephalomyelitis: Specific demyelination of CNS in culture. Neuropathol. Appl. Neurobiol. **3:** 359–367.
25. RUIJS, T. C. G., A. OLIVIER & J. P. ANTEL. 1990. Serum cytotoxicity to human and rat oligodendrocytes in culture. Brain Res. **517:** 99–104.

PART IV. GLIAL-NEURONAL INTERACTION IN REGENERATION

INTRODUCTION

Glia, Neurons, and Plasticity

G. RAISMAN

Norman and Sadie Lee Research Centre
Laboratory of Neurobiology
National Institute for Medical Research
The Ridgeway
Mill Hill
London NW7 1AA, UK

The central nervous system (CNS) is made up of neurons and glia, the glia being by far the more numerous in all regions and in all species. Although in development the glia are the first to be formed, the vast majority do not arise until very late, after the majority of neurons have developed, around the time of birth or in the early postnatal period.[1] In an ascending phylogenetic scale, the numeric preponderance of glia over neurons, already marked in rodents, greatly increases in primates and reaches its peak in the human; the human brain has both relatively and absolutely the greatest number of glia.[2]

An innocent faced with these facts might be excused for thinking that the highest brain functions depend on glia. In this respect, the history of our concept of glia has been a great deal less than innocent. The ability of nineteenth century scientists to perform accurate measurement of minute changes in electrical signals ensured that the nerve cell became top dog. It was the nerve cell that received and integrated information, made decisions, and dispatched instructions through its axon. There was no place for glia: the vital functions of the brain would be totally understood via nerve cells.

How much different are our opinions today? Neuroglia means nerve glue (Virchow, 1846, in ref. 3). It was imagined that glia formed a matrix in which the nerve cells were supported during the performance of their noble neuronal functions. So little was thought of this gluey glial function that glial cells have never received the honor of an individual name. To this day, there is no singular form of the word neuroglia; we have the word "neuron" but there is no "glion."

Even as late as the 1960s the best that could be done to explain the function of neuroglia was that they acted as a sort of blotting paper to mop up excess potassium and other ions that nerve cells might release into their environment during periods of especially frantic activity or to release their stores of glial glycogen to reenergize tired neurons whose needs temporarily exceeded what the blood could supply.[4]

Anatomists, however, could hardly ignore this vast population of glial cells. Ramon y Cajal[5] illustrates the immense variety and regional specificity of glia. Take a look at his plates of the cerebral cortex (Figs. 345 and 373), the dentate gyrus (Fig. 478), or the olfactory bulb (Fig. 417). The neurons dance through them in fascinating, intricate forms. Each plate is matched by a similar plate, in identical layout, drawn to the same scale, showing the glia (Figs. 380, 483, and 422). They clearly exceed the neurons in the variety of their different types, their complexity, and their intricate regional specialization.

The first clue about glial function may have come from their behavior in disease or injury. They were exquisitely sensitive to disturbance, modifying their structure and disposition in response to the most minor changes. The discovery of glial fibrillary acidic protein[6] showed that in their response to injury,[7] the astroglial cells acted as a vast, *quasi*-syncytial network, originally described by Penfield[8] as a sort of fishing net. Disturbances at any one point were rapidly reflected throughout the entire system. The astrocytes formed a communication system, like a vast aerial scaffolding, arching over the neurons below, recruiting and transmitting their own responses over immense distances through the network.

Later, in a study of synaptic plasticity after partial denervation, we observed[9] that the process of replacement of one presynaptic terminal by another went through the essential intermediary stage of engulfment by an astrocytic process. Neither is this astrocytic response passive, because after phagocytosis of a degenerating terminal, the astrocytic process had to be prepared to recognize and give way to the ingrowth of a newly forming nerve terminal (FIG. 1). The glia seemed to be playing active and intelligent roles in synapse formation after injury to the adult brain. More recently,

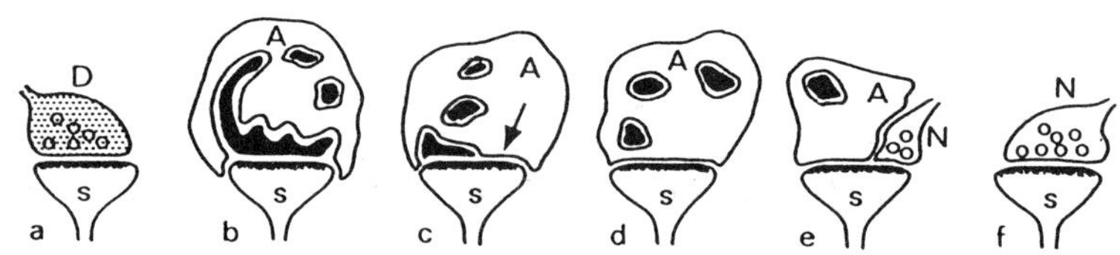

FIGURE 1. A schematic representation of the sequence of structures found in apposition to the dendritic spines in the adult rat lateral septal nucleus after a fimbrial transection destroying the ipsilateral hippocampal input. These configurations have been arranged in sequence from **a** to **f** to illustrate a series of changes that would lead to reinnervation of the denervated dendritic spine. In **a**, the axon terminal (D), degenerating as a result of axotomy, lies in contact with the dendritic spine (s). In **b**, the degenerating terminal, now much darker and more shrunken, has become surrounded by a swollen astrocytic process (A) that is indenting its surface and has partially engulfed two detached fragments of degenerating terminal cytoplasm. In **c**, the reactive astrocytic process has now partially displaced (*arrow*) the degenerating terminal from the region of the synaptic thickening, and in **d**, this displacement is complete. However, this stage is only transient: in **e**, the reactive astrocytic process itself now starts to be displaced from the synaptic thickening by a nondegenerating axon terminal (N), which in **f**, completes the reinnervation cycle by completely occupying the formerly denervated terminal site. The entire population of degenerating synapses passes through this cycle over the first 4–6 weeks after axotomy. Reproduced (slightly modified), with permission, from Raisman and Field.[9]

our work[10,11] has demonstrated that this glial response is able to discriminate among different types of neuronal input. In addition, the recent demonstration of neurotransmitter receptors on glial cells[12–15] now offers a clue as to how such specificity could be obtained.[16]

At the same time, studies of development were showing the crucial roles of the early embryonic radial glia in guiding the migration of neurons from the germinal epithelium to their adult locations[17–19] and also in providing glial end-feet along which axons progress during development.[20] Consider the optic nerve. At the time of outgrowth of the optic vesicle from the forebrain, only nonneuronal elements connect the future retina with the diencephalon. Therefore, in the course of their long growth back to the brain, the retinal ganglion cell axons can only grow along glial pathways; there is nothing else. This seems to be the pattern with early axonal development in a number of systems that have been examined.

Moreover, in the adult, glia remain the scaffolding along which axons run. It is the failure of axonal regeneration that underlies the current irreparability of brain and spinal cord injuries. Any strategy to devise methods of repair will have to take as much account of glia as of neurons.

If we accept that glia have a role in development and also in the reconstruction of the damaged adult brain, it is not a large step to entertaining the possibility that glia may play a role in guiding neuronal plasticity throughout adult life, that is, in the continuous process of learning, adaptation, and memory.

The CNS is the most complex system in nature, reaching its peak in humans, the most complex of all creations. Fundamental to its makeup is the division into neurons and glia. For some, still unknown reason, brain cells must become one or the other; this commitment is absolute. Although the electrical events in neurons are reflected in glia, no glia are integrated into the synaptic network and no neurons into the glial arrays.[3,4] The reason for this universal exclusiveness, this absence of half-and-half cells, is a mystery, but together the neurons and glia do form an integrated community. Ramon y Cajal's drawings[5] show that the regional specializations of glia and neurons are complementary, as if during development each was influencing the form of the other, imprinting on each other the regional diversity and specialization.[16,21]

Glia fall into three large families, astrocytes (and their earlier form, radial glia[22]), oligodendrocytes, and microglia (probably of mesodermal origin[23]). But rather than trying to find ever new and distinct species of rare glia and introducing ever more complex taxonomies of subfamilies, we may have to accept that glial cells, like neurons, defy rigid classification, because within these three broad families, they are infinitely variable and infinitely responsive to changes in their environment.

Current experimental work[24] makes it unlikely that the genome conveys to each individual cell a complete master plan of its entire future life, and that the cell, thus briefed, grimly sets off through the maelstrom of developing brain tissue, intransigent, determined to carry out its unavoidable destiny. Rather the cell may set out with a small number of simple messages, each of a provisional nature. At each crossroad, the cell reads the signals and is empowered to make contingent decisions. In the life history of each cell, nature (the genome) and nurture (the environment) interact inextricably in a continuous dialogue. As the cells develop receptors, they become responsive; to the right cell, poised at the right moment of development, nerve growth factor[25] or laminin[20] may mean "Yes," and Schwab's oligodendrocyte-associated inhibitory factor[26] or other molecules[27] may mean "No."

The earlier, electrophysiologically dominated view of brain function not only did little justice to glia, but also presented a rather dull picture of the neurons. Although the excellent work on critical periods of development[28] has shown very precise examples of how environmental factors influence the development of neuronal circuitry, the very term "critical period" implies an end. After that early postnatal sowing of its wild oats, does the brain settle down to its long adult life as a mere telephone exchange or does it remain, in different ways, plastic and adaptive to its environment?

Common sense suggests the second. If so, it is more than likely that the glia play an essential role in these plastic properties. Maybe it is no coincidence that the first demonstration[29] that plasticity can be restored to the adult cat visual system at the end of the critical period occurred by injection of neonatal *glia.*

This looks, however, beyond the yet available evidence, towards possible wider roles that glia may be shown to play in the future. Maybe, it is giving glia rather too noble a métier, too high a profile. But they have had such a poor deal in the past, that a little indulgence to the historically underprivileged can surely be but little sin.

REFERENCES

1. BRIZZEE, K. R., J. VOGT & X. KHARETCHKO. 1964. Postnatal changes in glial/neuron index with a comparison of methods of cell enumeration in the white rat. Prog. Brain Res. **4:** 136–149.
2. REICHENBACH, A. 1989. Glia:neuron index: Review and hypothesis to account for different values in various mammals. Glia **2:** 71–77.
3. KUFFLER, S. W., J. G. NICHOLLS & A. R. MARTIN. 1984. From Neuron to Brain. A Cellular Approach to the Function of the Nervous System. Sinauer. Massachusetts.
4. KUFFLER, S. W. & D. D. POTTER. 1964. Glia in the leech central nervous system: Physiological properties and neuron-glia relationship. J. Neurophysiol. **27:** 290–320.
5. RAMON Y CAJAL, S. 1911. Histologie du Système Nerveux de l'Homme et des Vertébrés. Maloine. Paris.
6. BIGNAMI, A. & D. DAHL. 1974. Astrocyte specific protein and neuroglial differentiation. An immunofluorescence study with antibodies to glial fibrillary acidic protein. J. Comp. Neurol. **153:** 27–38.
7. BIGNAMI, A. & D. DAHL. 1976. The astroglial response to stabbing. Immunofluorescence studies with antibodies to astrocyte-specific protein (GFA) in mammalian and submammalian vertebrates. Neuropathol. Appl. Neurobiol. **2:** 99–110.
8. PENFIELD, W. 1927. The mechanism of cicatricial contraction in the brain. Brain **50:** 499–517.
9. RAISMAN, G. & P. M. FIELD. 1973. A quantitative investigation of the development of collateral reinnervation after partial deafferentation of the septal nuclei. Brain Res. **50:** 241–264.
10. FIELD, P. M. & G. RAISMAN. 1983. Relative slowness of heterotypic synaptogenesis in the septal nuclei. Brain Res. **272:** 83–99.
11. ZHOU, C.-F., Y. LI & G. RAISMAN. 1989. Embryonic entorhinal transplants project selectively to the deafferented entorhinal zone of adult mouse hippocampi, as demonstrated by the use of Thy-1 allelic immunohistochemistry. Effect of timing of transplantation in relation to deafferentation. Neuroscience **32:** 349–362.
12. CULL-CANDY, S. G. & D. J. A. WYLLIE. 1991. Glutamate-receptor channels in mammalian glial cells. This volume.
13. ERNSBERGER, P., L. IACOVITTI & D. J. REIS. 1990. Astrocytes cultured from specific brain regions differ in their expression of adrenergic binding sites. Brain Res. **517:** 202–208.
14. SOMOGYI, P., N. ESHHAR, V. I. TEICHBERG & J. D. B. ROBERTS. 1990. Subcellular localization of a putative kainate receptor in Bergmann glial cells using a monoclonal antibody in the chick and fish cerebellar cortex. Neuroscience **35:** 9–30.
15. WILKIN, G. P., D. R. MARRIOTT, J. N. WOOD, G. W. TAYLOR, G. J. STEPHENS & M. B. A. DJAMGOZ. 1991. Glial receptors and intracellular second messengers. This volume.
16. SHAIN, W., D. S. FORMAN, V. MADELIAN & J. N. TURNER. 1987. Morphology of astroglial cells is controlled by beta-adrenergic receptors. J. Cell Biol. **105:** 2307–2314.
17. RAKIC, P. 1971. Neuron-glia relationship during granule cell migration in developing cerebellar cortex. A Golgi and electron microscopic study in *Macacus rhesus.* J. Comp. Neurol. **141:** 283–312.
18. GASSER, U. E. & M. E. HATTEN. 1990. Neuron-glia interactions of rat hippocampal cells *in vitro:* Glial-guided neuronal migration and neuronal regulation of glial differentiation. J. Neurosci. **10:** 1276–1285.
19. STITT, T. N., U. E. GASSER & M. E. HATTEN. 1991. Molecular mechanisms of glial-guided neuronal migration. This volume.
20. COHEN, J., J. F. BURKE, C. MCKINLAY & J. WINTER. 1987. The role of laminin and the laminin/fibronectin receptor complex in the outgrowth of retinal ganglion cell axons. Dev. Biol. **122:** 407–418.
21. CHAMAK, B., A. FELLOUS, J. GLOWINSKI & A. PROCHIANTZ. 1987. MAP2 expression and neuritic outgrowth and branching are coregulated through region-specific neuroastroglial interactions. J. Neurosci. **7:** 3163–3170.
22. VOIGT, T. 1989. Development of glial cells in the cerebral wall of ferrets: Direct tracing of their transformation from radial glia into astrocytes. J. Comp. Neurol. **289:** 74–88.

23. PERRY, V. H. & S. GORDON. 1988. Macrophages and microglia in the nervous system. Trends. Neurosci. **11:** 273–277.
24. LILLIEN, L. E. & M. C. RAFF. 1990. Differentiation signals in the CNS: Type-2 astrocyte development *in vitro* as a model system. Neuron **5:** 111–119.
25. HEUMANN, R., B. HENGERER, M. BROWN & H. PERRY. 1991. Molecular mechanisms leading to lesion-induced increases in nerve growth factor (NGF) synthesis. This volume.
26. CADELLI, D. S. & M. E. SCHWAB. 1991. CNS myelin associated inhibitors of neurite outgrowth: Their role in regeneration. This volume.
27. PATTERSON, P. H. 1988. On the importance of being inhibited, or saying no to growth cones. Neuron **1:** 263–268.
28. PURVES, D. & J. W. LICHTMAN. 1985. Principles of Neural Development. Sinauer. Massachusetts.
29. MÜLLER, C. M. & J. BEST. 1989. Ocular dominance plasticity in adult cat visual cortex after transplantation of cultured astrocytes. Nature **342:** 427–430.

Neuronal and Nonneuronal Influences on Retinal Ganglion Cell Survival, Axonal Regrowth, and Connectivity after Axotomy[a]

GARTH M. BRAY, MARIA PAZ VILLEGAS-PÉREZ, MANUEL VIDAL-SANZ, DAVID A. CARTER, AND ALBERT J. AGUAYO

Centre for Research in Neuroscience
The Montreal General Hospital and McGill University
Montréal, Québec, Canada, H3G 1A4

Trauma or disease that interrupts axons in the central nervous system (CNS) of adult mammals usually leads to persistent functional deficits. Although numerous factors contribute to these deficits, attempts to understand or ameliorate this situation must address two main injury-related effects that contribute to the impairment of neural connectivity in the CNS of adult mammals: (1) the retrograde loss of neurons that follows axotomy and (2) the failure of axons to regrow the distances necessary to reconnect widely separated parts of the nervous system.

In the embryo, complex cell-cell and axon-substrate interactions are thought to play key roles in the formation of neuronal pathways.[1] The connectivities that ultimately result from these interactions involve matrix substrata, cell surface components, and diffusible factors that act on receptive neurons and their growth cones to influence their survival, the overall course and direction followed by their axons, the confinement of their terminal arbors to specific regions such as clusters, laminae, or columns of cells, and the establishment of synaptic contacts with appropriate domains of target neurons.

With maturation, the nonneuronal milieu of the mammalian CNS no longer facilitates, and may actively inhibit, the growth of axons, a capacity that is retained by the mammalian peripheral nervous system as well as the CNS of anamniotes. Peripheral nerve grafts, which change the nonneuronal milieu of neurons at the site of CNS injury, permit the regrowth of the axons of CNS neurons in adult mammals. (For a review, see ref. 2.) Thus, with this experimental model, it is possible to guide regenerating CNS axons to regions of the nervous system that contain appropriate or inappropriate groups of target neurons, to determine the extent to which synaptic connectivity can be reestablished (for a review, see refs. 3 and 4), and to investigate the possibility that this novel axonal regeneration can influence the survival of injured neurons.[5]

[a] M.P. V.-P. was supported by a Multiple Sclerosis Society of Canada fellowship. M. V.-S. and D.C. were supported by Medical Research Council fellowships. The Medical Research Council of Canada, the Multiple Sclerosis Society of Canada, the Spinal Cord Research Foundation, the Daniel Heumann Fund for Spinal Cord Research, and the M. Wiesberg Vision Regeneration Fund provided financial support. The laboratories in which many of the studies reviewed here were carried out are part of the Canadian Network for the Study of Neural Regeneration and Functional Recovery.

The regrowth of interrupted CNS axons along peripheral nerve grafts and back into CNS targets involves interactions of neurons with different nonneuronal environments and with other CNS neurons. In this report, we summarize experiments that emphasize these neuronal and nonneuronal influences as they relate to each of the main responses to injury in the CNS of adult mammals.

METHODS

When interrupted CNS axons are placed in contact with a segment of peripheral nerve, many classes of axotomized CNS neurons can regrow lengthy axons into the

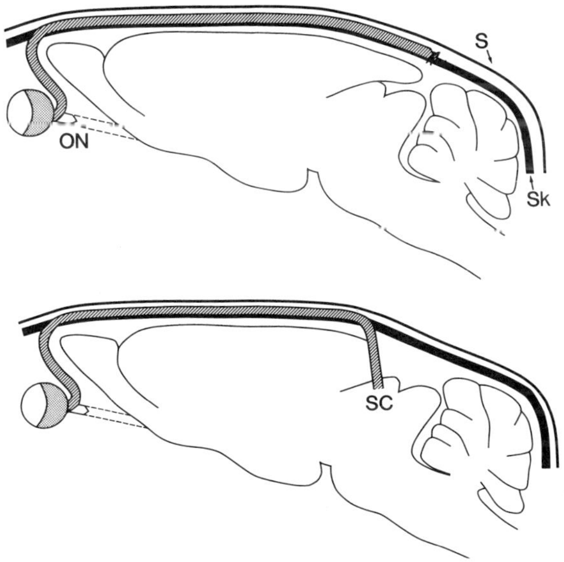

FIGURE 1. Diagram of a peripheral nerve (PN) graft (*cross-hatched*) connected to the ocular stump of an optic nerve (ON) transected in adult rats or hamsters. **(Top)** At the time of graft attachment close to the eye, the distal end of the autologous PN graft was placed blind-ended between the scalp (S) and the skull (Sk). After intervals of time to allow the retinal ganglion cell (RGC) axons to grow into the graft, these preparations were used to investigate the survival of the injured RGCs or the growth of axons along the grafts. **(Bottom)** After 8–10 weeks in other animals, the unconnected end of the PN graft was inserted into the superior colliculus (SC). Two to 18 months later, orthogradely transported tracers were injected intravitreally to label regenerating RGC axons and their terminals. (Reproduced with permission from Vidal-Sanz *et al.*[7])

peripheral nerve grafts. In such grafts, the glial substrate of the adult CNS is replaced by a milieu that contains degenerating axons and myelin sheaths, multiplying Schwann cells, other nonneuronal cells, and an extensive extracellular matrix. (For a review, see ref. 2.) For our current studies of the survival, regrowth, and reconnection of neurons after axotomy, we have used the retinocollicular pathway in adult rats or hamsters. (For a review, see ref. 4.) For these experiments (FIG. 1), an autologous segment of peripheral nerve was attached to the ocular stump of the optic nerve transected near the eye and the remainder of the peripheral nerve graft placed beneath the scalp. These animals were used to study the survival of retinal ganglion cells (RGCs)[5] or the growth of axons within blind-ended grafts in the absence of terminal connectivity.[6,7] In other animals, the distal end of the graft was inserted into the superior colliculus,[7,8] the target of most RGC axons in rodents, or into an inappropriate target: the cerebellum.[9]

To document axonal growth in the peripheral nerve graft or terminal connectivity in the superior colliculus, the RGCs were labeled by intravitreal injection of horseradish peroxidase, fluorescent tracers such as rhodamine (RITC), or tritiated amino acids. The peripheral nerve graft and the superior colliculus were then surveyed by light and electron microscopy to identify axons or axon terminals containing these orthogradely transported markers. In other animals, the peripheral nerve grafts or the portions of the superior colliculus near the termination of the grafts were investigated electrophysiologically for responses to flashes of light onto the retina.[10,11]

SURVIVAL OF AXOTOMIZED NEURONS

Axotomy near the perikaryon is a prerequisite for CNS axons to grow into peripheral nerve grafts.[12–16] Paradoxically, such lesions are likely to be lethal to many neurons.[17] For example, optic nerve transection near the eye causes the loss of many RGCs.[18–21] In our study of adult rats with RGC axotomy close to the optic disc, the densities of surviving RGCs decreased to 18% at 1 month and 5% at 3 months.[21] However, in animals with optic nerve lesions approximately 10 mm from the eye, which are not conducive to RGC regrowth into peripheral nerve grafts, 54% of RGCs survived at 3 months.[21]

In the RGCs of adult rats, transection of the optic nerve near the eye, but not at distances of 8–10 mm, also leads to enhanced immunoreactivity for the growth-associated protein GAP-43[22] and increased transport of this protein along the proximal axonal stump.[23] The relationship of axonal growth and GAP-43 expression to the location of the axonal injury suggests that the expression of this molecule may be involved in the sequence of changes that modulate the responsiveness of regenerating axons to molecules in the nonneuronal environment.

Several experimental strategies have been used to augment the number of neurons that survive axotomy. RGC survival *in vitro* can be enhanced by coculture with fetal explants of RGC target regions of brain.[24–26] *In vivo,* grafts of target fetal brain have been shown to increase the survival of neurons axotomized in the red nucleus[27] and the lateral geniculate nucleus[28] of neonatal rats. In adult rats, the placement of embryonic thalamic and tectal tissue at the site of axotomy 2–3 mm from the eye led to a threefold enhancement in the survival of RGCs at 1 month[29] and the preservation of retinal responses to stimuli at 5 months.[30] Similar effects were observed after the placement of fetal retina adjacent to retinal lesions in adult rats.[31] The extent of RGC loss after optic nerve transection close to the eye can also be reduced in adult rats, at least temporarily, by the apposition of a segment of peripheral nerve to the ocular stump of the transected optic nerve. One month after attaching such grafts, Berry *et al.*[20] observed a 10% increase in RGC survival. Using a different method of RGC identification, Villegas-Pérez *et al.*[5] observed that the survival of these neurons was enhanced nearly fourfold, although in these grafts in which terminal connectivity was not permitted, the effect declined to less than threefold by 3 months after graft placement (FIG. 2). The apposition of peripheral nerve segments to small retinal lesions that interrupted RGC axons also delayed the onset of RGC death by 2 weeks.[32] It has been reported that the introduction of Schwann cells into the eye can also enhance the survival of axotomized RGCs.[33]

In the examples cited in the preceding paragraph, it is postulated that cells and tissues that have some capacity to enhance the survival of injured neurons, such as axotomized RGCs, express their effects through the production of trophic factors that are taken up by the injured axons and retrogradely transported to the perikarya (for review, see ref. 34), where they regulate cell functions that apparently prevent

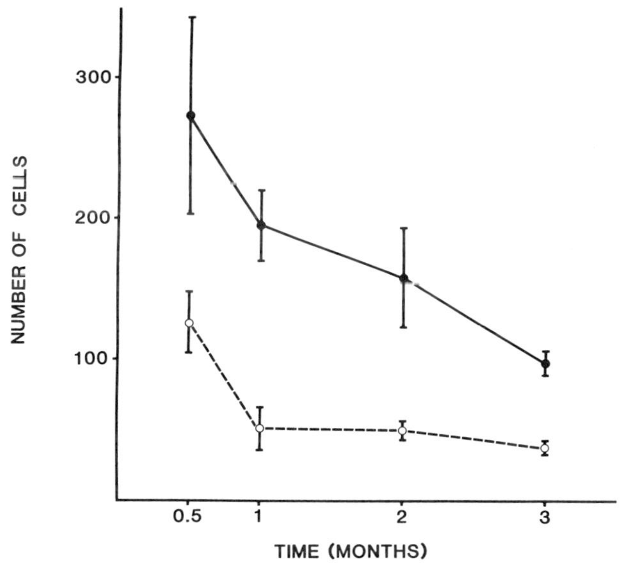

FIGURE 2. Survival of ganglion cells (RGCs) in the retinas of adult rats following optic nerve transection with or without attachment of peripheral nerve grafts. At the time of optic nerve transection, RGCs were labeled with the persistent tracer diI. Surviving (diI-labeled) RGCs were counted in standard areas of each retina and were expressed as numbers of cells per square millimeter. By 1 month after axotomy, the densities of RGCs in the retinas without PN grafts (*open circles*) were decreased to less than 10% of control densities. In the retinas with attached PN grafts (*closed circles*), there was a fourfold enhancement of RGC densities at 1 month after axotomy, but it decreased to less than threefold by 3 months. (Reproduced with permission from Villegas-Pérez *et al.*[5])

cell death.[35] Support for the involvement of target-derived trophic substances in the survival of axotomized neurons in adult animals is suggested by analogy with the situation in the developing nervous system. During development, contacts between axon terminals and sources of trophic molecules in their targets appear to be necessary for neuronal survival. (For a review, see ref. 34.) Thus, for example, the massive loss of RGCs that follows optic nerve section in newborn rodents[19,36] may be related to the interruption of the supply of brain-derived neurotrophic factor (BDNF) or related molecules at a time when neuronal survival is critically dependent on such factors.

Peripheral nerve segments, such as those used as peripheral nerve grafts, contain several factors that apparently have the capacity to promote the survival of injured neurons. Such peripheral nerve segments synthesize nerve growth factor (NGF)[37] and NGF receptors.[38] The administration of NGF increases the survival of axotomized RGCs,[39] an effect that may be related to the possible transient expression of NGF receptors by injured RGCs[33] or to sequence homologies between NGF and BDNF.[40,41] For many RGCs, BDNF is the most likely trophic factor.[42] BDNF mRNA, which is widely distributed in brain,[43] has also been detected in Schwann cells and fibroblasts *in vitro.*[44] The survival of adult RGCs *in vitro* is enhanced by BDNF,[45] although it remains to be shown that such an effect occurs *in vivo* and is sustained. Another important neurotrophic molecule synthesized in peripheral nerve is ciliary neurotrophic factor, which promotes the survival of certain peripheral neurons *in vitro*[46] and plays a role in the differentiation of type-2 astrocytes in the optic nerve.[47] Although the application of ciliary neurotrophic factor to the proximal stump of the rat facial nerve resulted in the survival of most of the axotomized facial neurons,[48] its possible effects on the survival of RGCs or other CNS neurons *in vivo* remain a matter of speculation. Fibroblast growth factor, which increases the survival of adult RGCs *in vitro,*[49] also limits the retrograde loss of axotomized RGCs.[50] Because fibroblast growth factor is also a mitogen for nonneuronal cells,[51] it has been postulated that fibroblast growth factor may enhance neuron viability through indirect effects on other cells.[34] Moreover, RGCs, which exist in an environment rich in basic fibroblast growth factor,[51] actively internalize and transport this molecule to their targets.[52] The presence of several trophic molecules in peripheral nerves suggests that the capacity of peripheral nerve grafts to promote the survival of injured neurons is the net effect of multiple trophic mechanisms.

The effectiveness of peripheral nerve grafts in limiting the loss of axotomized RGCs is partial and temporary[5] (FIG. 2). Thus, as occurs during development,[34] putative trophic substance(s) may be limited in quantity or duration of expression in such substitute sources. Indeed, as regenerating axons become associated with Schwann cells in the distal stump of an injured peripheral nerve, the production of NGF decreases[53] and retinal levels of NGF mRNA are only increased for a few days after optic nerve transection (I. Black, personal communication). Alternatively, the limited effects of peripheral nerve grafts or the administration of certain trophic molecules on neuronal survival could be an indication that injuries such as axotomy near the neuronal somata trigger other alterations that can lead to neuronal death.

The long-term survival of developing neurons appears to depend on the formation of stable synaptic contacts. Such an effect could involve trophic molecules or some aspect of synaptic activity. In adult rats and hamsters in which peripheral nerve grafts connect the eye and the superior colliculus, we have found no apparent reduction in the extent of superior colliculus reinnervation or synapse formation between 2 and 18 months after introducing the regenerated axons into the superior colliculus.[54,55] These results provide indirect support for the hypothesis that synaptic contacts promote neuronal survival during regeneration as well as development.

AXONAL GROWTH IN PERIPHERAL NERVE GRAFTS

Within peripheral nerve grafts, the regenerating axons of CNS neurons can elongate without branching for distances of 3–4 cm.[7,12,15,56] The growing tips (growth cones) of these axons extend in close apposition to Schwann cells or their basal lamina (FIG. 3), suggesting that cell membrane contacts are important components of the interactions between the axon tips and the nonneuronal environment. The regenerating axons of CNS neurons become unsheathed by Schwann cells as myelinated or unmyelinated fibers. Thus, after the nonneuronal milieu of the peripheral nerve grafts is entered, further growth of the axons of CNS neurons resembles that of regenerating peripheral nerve fibers. For example, RGC axons grow along peripheral nerve grafts at rates of 1–2 mm per day[6,57] and can extend for distances that are approximately twice the length of the normal retinocollicular projection in adult rats.[7] These rates of axonal growth are paralleled by similar rates in the transport of cytoskeletal proteins.[58]

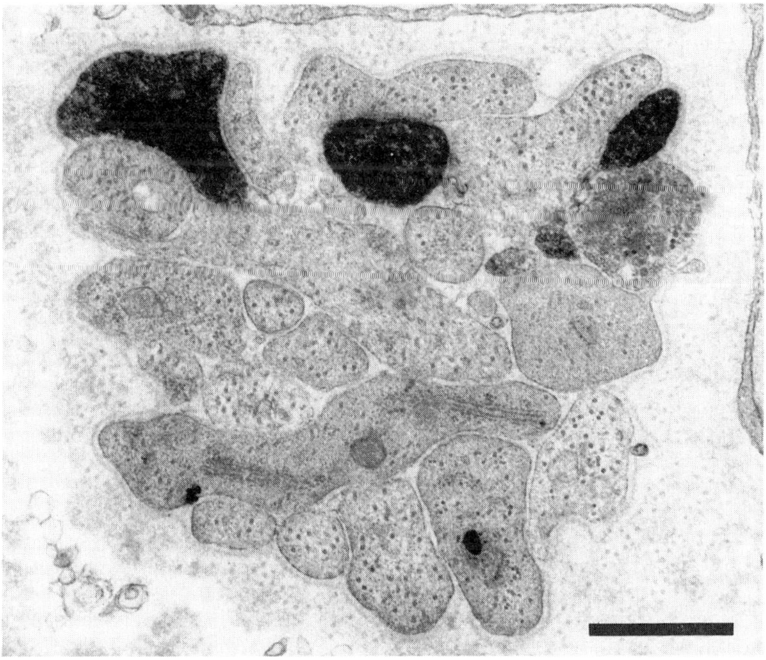

FIGURE 3. Regenerating retinal ganglion cell (RGC) axons in a peripheral nerve graft attached to the retina of an adult rat are densely labeled with horseradish peroxidase (HRP) orthogradely transported from the eye. The RGC axons contact Schwann cell processes and basal lamina. (Electron micrograph. Bar = 1 μm)

The growth of injured CNS axons into and along peripheral nerve grafts involves exchanging the glial substrate of the CNS for that of the distal stump of a transected peripheral nerve whose nonneuronal components permit or facilitate the growth of interrupted axons. The segments of peripheral nerve used as grafts contain: rapidly degenerating axons and myelin (that may stimulate Schwann cell multiplication[59]); macrophages (that influence Schwann cell synthesis of NGF through the expression of interleukin-1[60]); Schwann cells that multiply vigorously, dedifferentiate, become aligned within tubes of basal lamina, synthesize NGF,[37] NGF receptors,[38] and BDNF[44]; and fibroblasts and extracellular matrix components, including basal lamina and collagen.

Several of the cellular or extracellular components of peripheral nerve may help promote axonal regrowth; in addition, the peripheral nerve milieu appears to lack the putative growth-inhibiting molecules that are found in the CNS of adult mammals.[61–63] As proposed by Ramón y Cajal,[64] living Schwann cells appear to be necessary for RGC axons to elongate in peripheral nerve segments such as those used *in vivo* as grafts.[65,66] Axonal outgrowth *in vitro* is also promoted by: (1) diffusible molecules such as NGF[67] and BDNF,[45,67] that are present in peripheral nerve[44,53]; (2) various components of the extracellular matrix[68] (for a review, see ref. 69) that interact with transmembrane receptors, including the integrins[70]; and (3) cell adhesion molecules that are either Ca^{2+}-dependent (cadherins) or Ca^{2+}-independent (CAMs). (For a review, see ref. 70.)

The molecular interactions that involve neurons, Schwann cells, and the extracellular matrix have been reviewed by Reichardt *et al.*[70] *In vitro,* the extension of axons on Schwann cells is rapid and involves interactions between specific integrins of the neuronal growth cone and components of the extracellular matrix, particularly laminin. In addition, there is homotypic binding of *N*-cadherin and L1/Ng-CAM molecules on the surfaces of neurons and Schwann cells. Because these cell surface glycoproteins are expressed at higher levels in developing than in adult animals, it is postulated that the integrins, cadherins, and CAMs regulate axonal growth *in vivo* and may be reexpressed during peripheral nerve regeneration.[70] Experimental evidence supporting a role for the extracellular matrix and cell adhesion molecules in peripheral nerve regeneration includes the demonstration that laminin is involved in peripheral nerve regeneration *in vivo*[71,72] and that this regeneration is inhibited by a monoclonal antibody to an integrin receptor.[73] During peripheral nerve regeneration, immunoreactivity for L1/Ng-CAM and N-CAM appears to recapitulate that of development.[74]

Changing the adult mammalian CNS nonneuronal environment of injured axons to that of the peripheral nerve system, which is more permissive to axonal growth, may also influence transport of cytoskeletal proteins that are essential for the survival and regrowth of neurons. In the ocular stumps of intracranially transected optic nerves, there is a 10-fold decrease in the transport of tubulin and neurofilaments.[75] This decrease in rates of slow transport was not accompanied by comparable changes in fast axonal transport. In contrast, when the optic nerve was transected near the eye and replaced by a peripheral nerve graft, the rates of tubulin and neurofilament transport doubled and the rate of actin transport decreased,[62] a pattern that occurs in the developing optic nerve.[76] These changes in the rates of slow axonal transport during regeneration presumably reflect rather than determine growth-associated changes in the axonal cytoskeleton.

Thus, the results of experiments in which peripheral nerve grafts were substituted for CNS pathways, considered in light of *in vitro* and developmental studies of axonal extension, suggest that adult mammalian CNS neurons, which have been primed for axonal regrowth by proximal axotomy, interact with critical molecules associated with Schwann cells and other elements of the peripheral nerve milieu. As a result, the axons of CNS neurons can elongate in experimental animals for distances several-fold greater than those achieved during development.

INNERVATION OF CNS TARGETS

To investigate the possibility of reconnecting regenerated RGC axons with CNS neurons, autologous peripheral nerve grafts attached to the ocular stump of transected optic nerves in adult rats or hamsters (FIG. 1) were inserted into selected regions of the brain. For studies of terminal connectivity in a normal target, peripheral nerve grafts containing RGC axons were inserted into the superior colliculus.[7,8] After different time intervals, RGCs were labeled intravitreally with tracer substances (horseradish peroxidase [HRP], tritiated amino acids, or fluorescent tracers), and the CNS near the site of peripheral nerve graft insertion was surveyed by light or electron microscopy to identify axon terminals containing the orthogradely transported markers. In other animals, the CNS target regions were investigated electrophysiologically for responses to flashes of light onto the retina.[11]

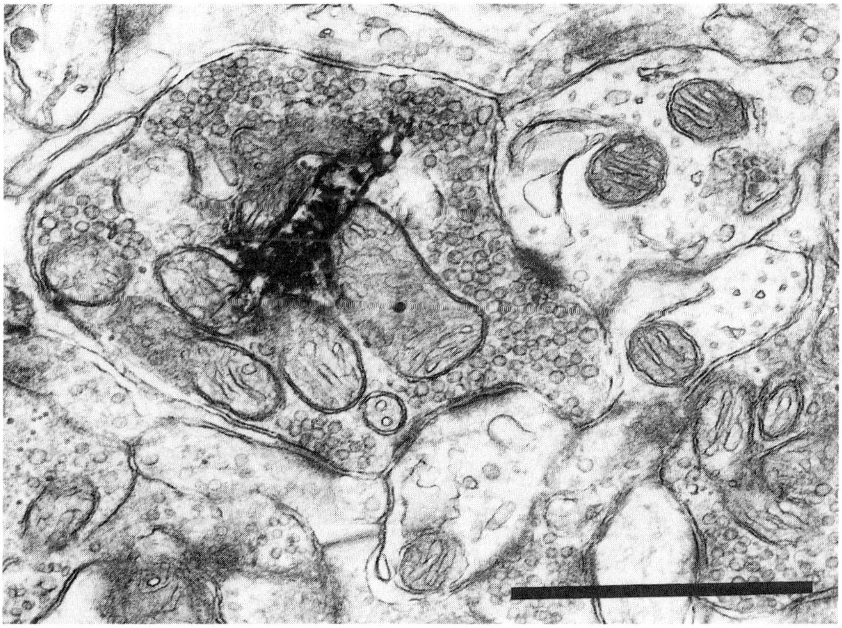

FIGURE 4. Terminal of a regenerated retinal ganglion cell axon in the superficial superior colliculus 4 months after connecting the eye and the midbrain of an adult hamster with a peripheral nerve graft. The terminal, labeled with clumped reaction product of HRP transported from the eye, forms asymmetric synapses with dendrites of superior colliculus neurons. (Bar = 1 μm) (Reproduced with permission from Aguayo *et al.*[3])

Regenerating RGC axons left the growth-enhancing or growth-permissive milieu of the peripheral nerve grafts, penetrated the narrow zone of glial fibrillary acidic protein (GFAP)-positive astrocyte processes at the end of the graft, and reentered the CNS environment where once again their capacity for growth was subject to the effects of CNS glia. In the hamsters with bridging grafts connecting the retina and midbrain, the extension of RGC axons in the superficial superior colliculus was limited to 500 μm[8] by 2 months and 1,000 μm by 8–10 months.[55] This limited, but relatively normal appearing growth of regenerated axons into the CNS target could have been influenced, at least temporarily, by short-range effects arising from graft components or by changes in the growth-promoting properties of glial cells such as astrocytes, which can support axonal growth during development and *in vitro.* (For a review, see ref. 70). Within the narrow zone of growth beyond the end of the peripheral nerve graft, RGC axons became ensheathed by myelin of central type (Vidal-Sanz *et al.,* unpublished observations); it is not known if such CNS myelin sheaths originated from previously differentiated or newly formed oligodendrocytes. The regenerating RGC axons extended toward the superficial layers of the superior colliculus that normally receive most retinal projections and formed normal appearing terminals and synapses (FIGS. 4 and 5) with appropriate postsynaptic domains (dendritic shafts and spines) on superior colliculus neurons[8] (Vidal-Sanz *et al.,* unpublished observations). These terminals persisted for up to 10 months in hamsters and up to 18 months in rats (FIG. 6). From 15–18 weeks after graft placement, extracellular recordings in the superior colliculus revealed excitatory and inhibitory postsynaptic responses to visual stimuli,[11] indicating that some of the regenerated axons had established functional synapses with neurons in the superior colliculi of these adult mammals.

Although it is not yet known if these regenerated retinocollicular connections in adult rodents are retinotopically deployed in the superior colliculus, it may now be possible to explore if the mechanisms thought to regulate terminal connectivity of axons *in vitro* or during development (for reviews, see refs. 77–79) are also operational during regeneration in adult mammals. Although incompletely understood, these mechanisms include: (1) pre-terminal axon guidance, probably involving contact with the surfaces of glial cells and neighboring axons, to the target region[77]; (2) diffusible chemoattractants,[80] including NGF,[81] that may guide axons over short distances; (3) molecules on the surfaces of target neurons that attract or repulse growth cones[82,83]; and (4) a later stage of activity-dependent synaptic refinement and

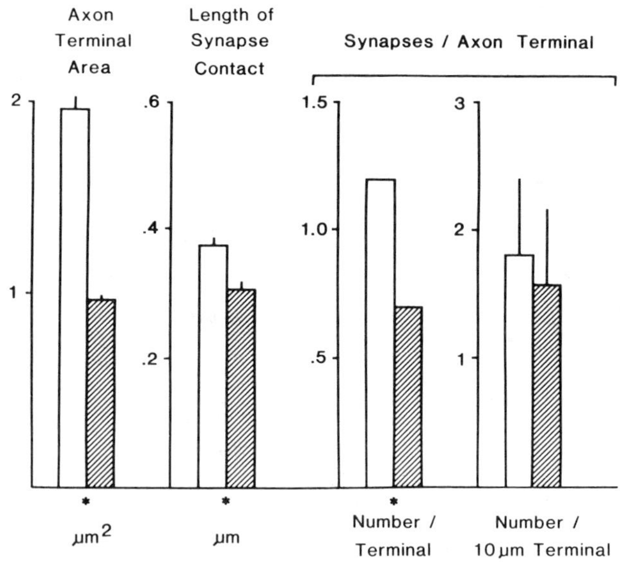

FIGURE 5. Morphometric comparisons of regenerated (*open bars*) and control (*cross-hatched bars*) retinocollicular axon terminals in adult hamsters. The regenerated terminals were observed in the superficial layers of the superior colliculi 2 months after peripheral nerve graft placement between the eye and the midbrain in four animals. *indicates significant difference from control. (Reproduced with permission from Aguayo *et al.*[3])

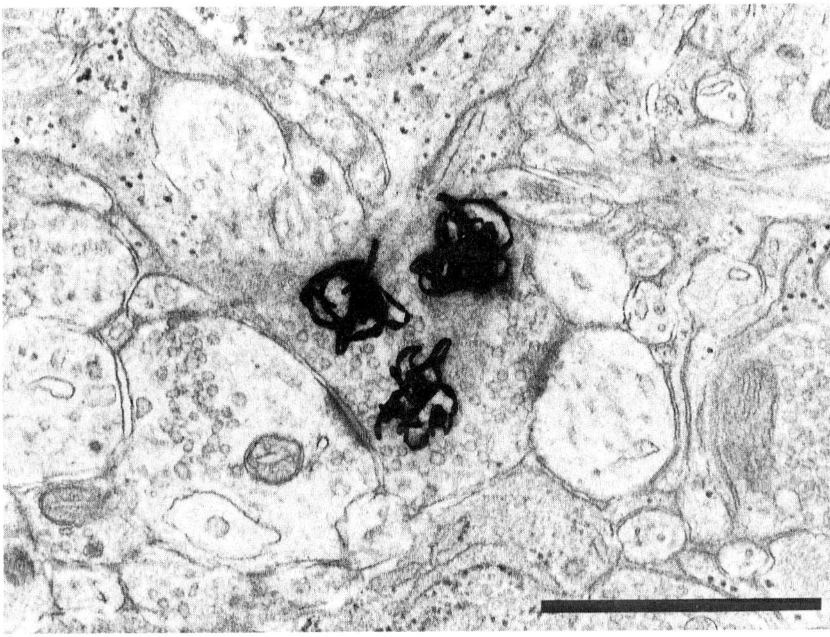

FIGURE 6. Terminal of a regenerated retinal ganglion cell axon in the superficial superior colliculus 16 months after the eye and midbrain of an adult rat were connected with a peripheral nerve graft. The terminal, which is labeled with silver grains 2 days after intravitreal injection of [^{3}H]-leucine/proline, forms asymmetric synapses with dendrites of superior colliculus neurons. (Bar = 1 μm) (Reproduced with permission from Aguayo *et al.*[4])

stabilization that appears to be mediated by receptors for neurotransmitters. (For a review, see ref. 79.)

SUMMARY

In contrast to the abortive regrowth that occurs when axons are interrupted in the adult mammalian CNS, exposure of injured CNS axons to the nonneuronal milieu of a peripheral nerve can lead to extensive axonal elongation. With the application of this experimental approach to the retinocollicular pathway in adult rodents, it has been possible to investigate the influences of neuron-glia and other interactions on the capacity of axotomized CNS neurons to survive injury, to elongate the distances necessary to reach specific targets, and to form connections in the CNS in adult rodents. The results of these investigations indicate that the changed glial environment provided by peripheral nerve grafts permits the guided regeneration of RGC axons to their CNS targets. Back in the CNS glial environment, regenerated axons penetrate their targets for short distances and re-form normal appearing synapses that can excite or inhibit postsynaptic neurons. Further studies will require a better understanding of intrinsic neuronal properties and of the interactions of these neurons with other neurons and with the cellular and noncellular components of the extraneural milieu.

ACKNOWLEDGMENTS

The technical assistance of M. David, J. Laganière, S. Shinn, J. Trecarten, and W. Wilcox is gratefully acknowledged.

REFERENCES

1. Dodd, J. & T. M. Jessell. 1988. Axonal guidance and the patterning of neuronal projections in vertebrates. Science **242:** 692–699.
2. Aguayo, A. J. 1985. Axonal regeneration from injured neurons in the adult mammalian central nervous system. *In* Synaptic Plasticity. C. W. Cotman, ed.: 457–484. Guilford Press. New York, NY.
3. Aguayo, A. J., D. A. Carter, T. Zwimpfer, M. Vidal-Sanz & G. M. Bray. 1990. Axonal regeneration and synapse formation in the injured CNS of adult mammals. *In* Brain Repair. A. Björklund, A. J. Aguayo & D. Ottoson, eds. Wenner-Gren International Symposium Series. Vol. 56: 252–272. Stockton Press. New York, NY.
4. Aguayo, A. J., G. M. Bray, M. Rasminsky, T. Zwimpfer, D. A. Carter & M. Vidal-Sanz. 1990. Synaptic connections made by axons regenerating in the CNS of adult mammals. J. Exp. Biol. **153:** 199–224.
5. Villegas-Perez, M. P., M. Vidal-Sanz, G. M. Bray & A. J. Aguayo. 1988. Influences of peripheral nerve grafts on the survival and regrowth of axotomized retinal ganglion cells in adult rats. J. Neurosci. **8:** 265–280.
6. Trecarten, M. J., M. P. Villegas-Perez, M. Vidal-Sanz, S. Thanos & A. J. Aguayo. 1986. Growth of retinal axons along peripheral nerve grafts inserted into the retina of adult rats. Soc. Neurosci. Abstr. **12:** 701.
7. Vidal-Sanz, M., G. M. Bray, S. Thanos, M. P. Villegas-Perez & A. J. Aguayo. 1987. Axonal regeneration and synapse formation in the superior colliculus by retinal ganglion cells in the adult rat. J. Neurosci. **7:** 2894–2907.
8. Carter, D., G. M. Bray & A. J. Aguayo. 1989. Regenerated retinal ganglion cell axons can form well-differentiated synapses in the superior colliculus of adult hamsters. J. Neurosci. **9:** 4042–4050.
9. Zwimpfer, T. J., A. J. Aguayo, G. M. Bray & D. G. Lawrence. 1989. Synapse formation by regenerating retinal ganglion cell axons directed into an inappropriate target (the cerebellar cortex) in adult hamsters. Soc. Neurosci. Abstr. **15:** 458.
10. Keirstead, S. A., M. Vidal-Sanz, M. Rasminsky, A. J. Aguayo, M. Levesque & K.-F. So. 1985. Responses to light of retinal neurons regenerating axons into peripheral nerve grafts in the rat. Brain Res. **359:** 402–406.
11. Keirstead, S. A., M. Rasminsky, Y. Fukuda, D. A. Carter, A. J. Aguayo & M. Vidal-Sanz. 1989. Electrophysiologic responses in hamster superior colliculus evoked by regenerating retinal axons. Science **246:** 255–258.
12. David, S. & A. J. Aguayo. 1981. Axonal elongation into peripheral nervous system "bridges" after central nervous system injury in adult rats. Science **214:** 931–933.
13. Richardson, P. M., V. M. K. Issa & S. Shemie. 1982. Regeneration and retrograde degeneration of axons in the rat optic nerve. J. Neurocytol. **11:** 949–966.
14. Richardson, P. M., V. M. K. Issa & A. J. Aguayo. 1984. Regeneration of long spinal axons in the rat. J. Neurocytol. **13:** 165–182.
15. So, K.-F. & A. J. Aguayo. 1985. Lengthy regrowth of cut axons from ganglion cells after peripheral nerve transplantation into the retina of adult rats. Brain Res. **328:** 349–354.
16. Friedman, B. & A. J. Aguayo. 1985. Injured neurons in the olfactory bulb of adult rat grow new axons along peripheral nerve grafts. J. Neurosci. **5:** 1616–1625.
17. Lieberman, A. R. 1974. Some factors affecting retrograde neuronal responses to axonal lesions. *In* Essays of the Nervous System. R. Bellairs & E. G. Gray, eds.: 71–105. Clarendon. Oxford.
18. Perry, V. H. 1981. Evidence for an amacrine cell system in the ganglion cell layer of the rat retina. Neuroscience **6:** 931–944.
19. Allcutt, D., M. Berry & J. Sievers. 1984. A quantitative comparison of the reactions of

retinal ganglion cells to optic nerve crush in neonatal and adult mice. Dev. Brain Res. **16:** 219–230.

20. BERRY, M., L. REES & J. SIEVERS. 1986. Regeneration of axons in the mammalian visual system. Exp. Brain Res. (Suppl.) **13:** 18–33.
21. VILLEGAS-PEREZ, M. P., M. VIDAL-SANZ, G. M. BRAY & A. J. AGUAYO. 1989. The distance of axotomy from the neuronal cell body influences the rate of retrograde degeneration but not long-term survival of retinal ganglion cells (RGCs). Soc. Neurosci. Abstr. **15:** 457.
22. LOZANO, A. M., K. S. DOSTER, A. J. AGUAYO & M. B. WILLARD. 1987. Immunoreactivity to GAP-43 in axotomized and regenerating retinal ganglion cells of adult rats. Soc. Neurosci. Abstr. **13:** 1389.
23. DOSTER, K. S., A. M. LOZANO, A. J. AGUAYO & M. B. WILLARD. 1990. Expression of the growth-associated protein, GAP-43, in adult rat retinal ganglion cells following axon injury. Neuron **6:** 635–647.
24. NURCOMBE V. & M. R. BENNETT. 1981. Embryonic chick retinal ganglion cells identified *in vitro.* Their survival is dependent on a factor from the optic tectum. Exp. Brain Res. **44:** 249–258.
25. MCCAFFERY, C. A., M. R. BENNETT & B. DREHER. 1982. The survival of neonatal rat ganglion cells *in vitro* is enhanced in the presence of appropriate parts of the brain. Exp. Brain Res. **48:** 377–386.
26. ARMSON, P. F. & M. R. BENNETT. 1983. Neonatal retinal ganglion cell cultures of high purity. Effect of superior colliculus on their survival. Neurosci. Lett. **38:** 181–186.
27. BREGMAN, B. S. & P. REIER. 1986. Neural tissue transplants rescue axotomized rubrospinal cells from retrograde death. J. Comp. Neurol. **244:** 86–95.
28. HAUN, F. & T. J. CUNNINGHAM. 1987. Specific neurotrophic interactions between cortical and subcortical visual structures in developing rats: *In vivo* studies. J. Comp. Neurol. **256:** 561–569.
29. SIEVERS, J., B. HAUSMANN & M. BERRY. 1989. Fetal brain grafts rescue adult retinal ganglion cells from axotomy-induced cell death. J. Comp. Neurol. **281:** 467–478.
30. GRAVINA, A., L. DOMENICI, N. BERARDI, L. GALLI & L. MAFFEI. 1990. Transplant of embryonal nervous tissue preserves the responses of rat retinal ganglion cells after section of the optic nerve. Exp. Brain Res. **80:** 631–634.
31. LAEDTKE, T. W. & J. E. TURNER. 1989. Embryonic grafts have a beneficial effect on the damaged host retina. Brain Res. **500:** 61–66.
32. TURNER, J. E., J. R. BLAIR & E. T. CHAPPEL. 1987. Peripheral nerve implant effects on survival of retinal ganglion layer cells after axotomy initiated by a penetrating lesion. Brain Res. **419:** 46–54.
33. MAFFEI, L., G. CARMIGNOTO, V. H. PERRY, P. CANDEO & G. FERRARI. 1990. Schwann cells promote the survival of rat retinal ganglion cells after optic nerve section. Proc. Natl. Acad. Sci. USA **87:** 1855–1859.
34. BARDE, Y.-A. 1989. Trophic factors and neuronal survival. Neuron **2:** 1525–1534.
35. MARTIN, D. P., T. L. WALLACE & E. M. JOHNSON, JR. 1990. Cytosine arabinoside kills postmitotic neurons in a fashion resembling trophic factor deprivation: Evidence that a deoxycytidine-dependent process may be required for nerve growth factor signal transduction. J. Neurosci. **10:** 184–193.
36. MUCHNICK-MILLER, N. & M. OBERDORFER. 1981. Neuronal and neuroglial responses following retinal lesions in neonatal rats. J. Comp. Neurol. **202:** 493–504.
37. HEUMANN, R., S. KORSCHING, C. BANDTLOW & H. THOENEN. 1987. Changes of nerve growth factor synthesis in non-neuronal cells in response to sciatic nerve transection. J. Cell Biol. **104:** 1623–1631.
38. TANIUCHI M., H. B. CLARK & E. M. JOHNSON, JR. 1986. Induction of nerve growth factor receptor in Schwann cells after axotomy. Proc. Natl. Acad. Sci. USA **83:** 4094–4098.
39. CARMIGNOTO, G., L. MAFFEI, P. CANDEO, R. CANELLA & C. COMELLI. 1989. Effect of NGF on the survival of rat retinal ganglion cells following optic nerve section. J. Neurosci. **9:** 1263–1272.
40. LEIBROCK J., F. LOTTSPEICH, A. HOHN, M. HOFER, B. HENGERER, P. MASIAKOWSKI, H. THOENEN & Y.-A. BARDE. 1989. Molecular cloning and expression of brain-derived neurotrophic factor. Nature **341:** 149–152.

41. RODRIGUEZ-TEBAR, A., G. DECHART & Y.-A. BARDE. 1990. Binding of brain-derived neurotrophic factor to the nerve growth factor receptor. Neuron **4:** 487–492.
42. JOHNSON, J. E., Y.-A. BARDE, M. SCHWAB & H. THOENEN. 1986. Brain-derived neurotrophic factor supports the survival of cultured rat retinal ganglion cells. J. Neurosci. **6:** 3031–3038.
43. HOFER, M., S. R. PAGLIUSI, A. HOHN, J. LEIBROCK & Y.-A. BARDE. 1990. Regional distribution of brain-derived neurotrophic factor mRNA in the adult mouse brain. EMBO J. **9:** 2459–2464.
44. BARKER, P., A. ACHESON, S. PAREEK, F. D. MILLER & R. A. MURPHY. 1990. Detection of brain-derived neurotrophic factor (BDNF)-like biological activity and mRNA in sciatic nerve fibroblasts and Schwann cells. Soc. Neurosci. Abstr. **16:** 1136.
45. THANOS, S., M. BAHR, Y.-A. BARDE & J. VANSELOW. 1989. Survival and axonal elongation of adult rat retinal ganglion cells. Eur. J. Neurosci. **1:** 19–26.
46. MANTHORPE, M., S. D. SKAPER, L. R. WILLIAMS & S. VARON. 1986. Purification of adult rat sciatic nerve ciliary neuronotrophic factor. Brain Res. **367:** 282–286.
47. HUGHES, S. M., L. E. LILLIEN, M. C. RAFF, H. ROHRER & M. SENDTNER. 1988. Ciliary neurotrophic factor induces type-2 astrocyte differentiation in culture. Nature **335:** 70–73.
48. SENDTNER, M., G. W. KREUTZBERG & H. THOENEN. 1990. Ciliary neurotrophic factor prevents the degeneration of motor neurons after axotomy. Nature **345:** 440–441.
49. BAHR, M., J. VANSELOW & S. THANOS. 1989. Ability of adult rat ganglion cells to regrow axons *in vitro* can be influenced by fibroblast growth factor and gangliosides. Neurosci. Lett. **96:** 197–201.
50. SIEVERS, J., B. HAUSMANN, K. UNSICKER & M. BERRY. 1987. Fibroblast growth factors promote the survival of adult rat retinal ganglion cells after a transection of the optic nerves. Neurosci. Lett. **76:** 157–162.
51. GOSPODAROWICZ, D., G. NEURFELD & L. SCHWEIGERER. 1986. Molecular and biological characterization of fibroblast growth factor, an angiogenic factor which also controls the proliferation and differentiation of mesoderm and neuroectoderm derived cells. Cell Differentiation **19:** 1–17.
52. FERGUSON, I. A., J. B. SCHWEITZER & E. M. JOHNSON, JR. 1990. Basic fibroblast growth factor: Receptor mediated internalization, metabolism and anterograde axonal transport in retinal ganglion cells. J. Neurosci. **10:** 2176–2189.
53. HEUMANN, R., D. LINDHOLM, C. BANDTLOW, M. MEYER, M. J. RADEKE, T. P. MISKO, E. SHOOTER & H. THOENEN. 1987. Differential regulation of nerve growth factor (NGF) and NGF-receptor mRNA in the rat sciatic nerve during development, degeneration, and regeneration; role played by macrophages. Proc. Natl. Acad. Sci. USA. **84:** 8735–8739.
54. BRAY, G. M., M. VIDAL-SANZ & A. J. AGUAYO. 1988. Regenerated retino-collicular synapses eighteen months after substitution of the optic nerve by a peripheral nerve graft in adult rats. Soc. Neurosci. Abstr. **14:** 654.
55. CARTER, D., G. M. BRAY & A. J. AGUAYO. 1989. Extension and persistence of regenerated retinal ganglion cell axons in the superior colliculus of adult hamsters. Soc. Neurosci. Abstr. **15:** 872.
56. BENFEY, M. & A. J. AGUAYO. 1982. Extensive elongation of axons from rat brain into peripheral nerve grafts. Nature **296:** 150–152.
57. CHO, E. Y. P. & K.-F. SO. 1987. Rate of regrowth of damaged retinal ganglion cell axons regenerating in a peripheral nerve graft in adult hamsters. Brain Res. **419:** 369–374.
58. MCKERRACHER, L., M. VIDAL-SANZ & A. J. AGUAYO. 1990. Slow transport rates of cytoskeletal proteins change during regeneration of axotomized retinal neurons in adult rats. J. Neurosci. **10:** 641–648.
59. SALZER, J. L. & R. P. BUNGE. 1980. Studies of Schwann cell proliferation. I. An analysis in tissue culture of proliferation during development, Wallerian degeneration, and direct injury. J. Cell Biol. **84:** 739–752.
60. LINDHOLM, D., R. HEUMANN, M. MEYER & H. THOENEN. 1987. Interleukin-1 regulates synthesis of nerve growth factor in non-neuronal cells of rat sciatic nerve. Nature **330:** 658–659.

61. CARONI, P. & M. E. SCHWAB. 1988. Two membrane protein fractions from rat central myelin with inhibitory properties for neurite growth and fibroblast spreading. J. Cell Biol. **106:** 1281–1288.
62. SCHWAB, M. E. & P. CARONI. 1988. Oligodendrocytes and CNS myelin are nonpermissive substrates for neurite growth and fibroblast spreading *in vitro.* J. Neurosci. **8:** 2381–2393.
63. PESHEVA, P., E. SPIESS & M. SCHACHNER. 1989. J1-160 and J1-180 are oligodendrocyte-secreted nonpermissive substrates for cell adhesion. J. Cell Biol. **109:** 1765–1778.
64. RAMÓN Y CAJAL, S. 1914. Estudios Sobre la Degeneración y Regeneración del Sistema Nervioso, Vol. 2. Imprenta de Hijos de Nicolás Moya. Madrid.
65. BERRY, M., L. REES, S. HALL, P. YIU & J. SIEVERS. 1988. Optic axons regenerate into sciatic nerve isografts only in the presence of Schwann cells. Brain Res. Bull. **20:** 223–231.
66. SMITH, G. V. & J. A. STEVENSON. 1988. Peripheral nerve grafts lacking viable Schwann cells fail to support central nervous system axonal regeneration. Exp. Brain Res. **69:** 299–306.
67. LINDSAY, R. M. 1988. Nerve growth factors (NGF, BDNF) enhance axonal regeneration but are not required for survival of adult sensory neurons. J. Neurosci. **8:** 2394–2405.
68. MANTHORPE, M., E. ENGVALL, E. RUOSLAHTI, F. M. LONGO, G. E. DAVIS & S. VARON. 1983. Laminin promotes neuritic regeneration from cultured peripheral and central neurons. J. Cell Biol. **97:** 1882–1890
69. DOUVILLE, P. & S. CARBONETTO. 1989. Extracellular matrix adhesive glycoproteins and their receptors in the nervous system. *In* Neurobiology of Glycoconjugates. R. U. Margolis & R. K. Margolis, eds.: 383–409. Plenum Press. New York, NY
70. REICHARDT, L. F., J. L. BIXBY, D. E. HALL, M. J. IGNATIUS, K. M. NEUGEBAUER & K. J. TOMASELLI. 1989. Integrins and cell adhesion molecules: Neuronal receptors that regulate axon growth on extracellular matrices and cell surfaces. Dev. Neurosci. **11:** 332–347.
71. SANDROCK, A. W. & W. D. MATTHEW. 1987. Identification of a peripheral nerve neurite growth-promoting activity by development and use of an *in vitro* bioassay. Proc. Natl. Acad. Sci. USA **84:** 6934–6938.
72. MADISON, R. C., C. DA SILVA, P. DIKKES, R. L. SIDMAN & T.-H. CHIU. 1987. Peripheral nerve regeneration with entubulation repair: Comparison of biodegradable nerve guide versus polyethylene tubes and the effects of a laminin-containing gel. Exp. Neurol. **95:** 378–390.
73. TOYOTA, B., S. CARBONETTO & S. DAVID. 1990. A dual laminin/collagen receptor acts in peripheral nerve regeneration. Proc. Natl. Acad. Sci. USA **87:** 1319–1322.
74. MARTINI, R. & M. SCHACHNER. 1988. Immunoelectron microscopic localization of neural cell adhesion molecules (L_1/Ng-CAM, N-CAM, and myelin-associated glycoprotein) in regenerating adult mouse sciatic nerve. J. Cell Biol. **106:** 1735–1746.
75. MCKERRACHER, L., M. VIDAL-SANZ, C. ESSAGIAN & A. J. AGUAYO. 1990. Selective impairment of slow axonal transport after optic nerve injury in adult rats. J. Neurosci. **10:** 2834–2841.
76. WILLARD, M. B. & C. SIMON. 1983. Modulations of neurofilament axonal transport during the development of rabbit retinal ganglion cells. Cell **35:** 551–559.
77. UDIN, S. B. & J. W. FAWCETT. 1988. Formation of topographic maps. Ann. Rev. Neurosci. **11:** 289–327.
78. FRASER, S. E. & N. A. O'ROURKE. 1990. The role of neuronal dynamics and positional cues in the patterning of nerve connections. *In* Brain Repair. A. Björklund, A. J. Aguayo & D. Ottoson, eds. Wenner-Gren International Symposium series. Vol. **56:** 225–235. Stockton Press. New York, NY.
79. CONSTANTINE-PATON, M., H. T. CLINE & E. DEBSKI. 1990. Patterned activity, synaptic convergence, and the NMDA receptor in developing visual pathways. Ann. Rev. Neurosci. **31:** 129–154.
80. BOLZ, J., N. NOVAK, M. GOTZ & T. BONHOEFFER. 1990. Formation of target-specific neuronal projections in organotypic slice cultures from rat visual cortex. Nature **346:** 359–362.

81. LEVI-MONTALCINI, R. 1976. The nerve growth factor: Its role in growth, differentiation and function of the sympathetic adrenergic neuron. Progr. Brain Res. **45:** 235–258.
82. BONHOEFFER, F. & J. HUF. 1980. Recognition of cell types by axonal growth cones *in vitro.* Nature **288:** 162–164.
83. COX, E. C., B. MULLER & F. BONHOEFFER. 1990. Axonal guidance in the chick visual system: Posterior tectal membranes induce collapse of growth cones from the temporal retina. Neuron **2:** 31–37.

Schwann Cells in Central Regeneration[a]

RICHARD P. BUNGE

The Miami Project to Cure Paralysis
University of Miami School of Medicine
Miami, Florida 33136

The demonstration of the ability of segments of peripheral nerve to promote regeneration after damage to central nervous system (CNS) axons[1] has revolutionized thinking regarding the capacity of the CNS to regenerate. Transplants of peripheral nerve used to induce CNS regeneration in these pioneering experiments contained a variety of extracellular matrix components and cellular constituents.

Several research groups have undertaken experiments to determine which of these constituents are essential for the nerve to promote regeneration and neuronal survival. These studies have centered on retinal ganglion cell regeneration after optic nerve injury. Berry and his collaborators[2] have examined the capability of segments of peripheral nerve, which have been treated to retain matrix material but to remove the viable cellular content, to promote regeneration of optic nerve axons in adult rats. These experiments clearly showed that the peripheral nerve segments devoid of viable Schwann cells were unable to promote the regenerative activity observed when these peripheral nerve segments were used intact.

Our approach to the study of the relative contribution of the various constituents utilized new tissue culture techniques that allow the various components of the peripheral nerve to be established either separately or together in tissue culture dishes; this permits direct observation of the ability of constituents to influence the regeneration of central neurons.[3] Initially, the central neuronal population was derived from the 15-day embryonic rat retina (cultured as fragments).[4] Substrata for growth were provided which contained either Schwann cells or Schwann cells plus the extracellular matrix that they deposit over their external surfaces, or this extracellular matrix alone, free of Schwann cell content. Inasmuch as these experiments were undertaken during a period in which the marked ability of extracellular matrix components, particularly laminin, to promote neurite growth was first being observed,[5] the result that we obtained was unexpected. Embryonic rat retinal ganglion cells provide a vigorous outgrowth on suitable substratum. The fragments, however, produced no neuritic outgrowth on extracellular matrix components that are deposited by Schwann cells, but they showed a remarkable ability to grow on a substratum containing viable Schwann cells. They also grew vigorously when provided with a substratum containing both Schwann cells and extracellular matrix components.

We subsequently undertook a series of experiments to determine if it would be possible to identify the surface molecule on the Schwann cell surface that was responsible for the promotion of embryonic rat retinal ganglion cell axonal growth.[6] These experiments also provided unexpected results. It has generally been observed that axonal growth on Schwann cell surfaces involves the interaction of a number of surface moieties including calcium-dependent or calcium-independent cell adhesion

[a]Work in the author's laboratory is supported by grants NS09923 and NS19923 from the National Institutes of Health of the U.S. Public Health Service and by The Miami Project to Cure Paralysis.

molecules and a number of integrins.[7] In this culture system, however, in which embryonic rat retinal ganglion cell axons were provided with Schwann cell surfaces to promote regeneration, we found that provision of antibodies against the surface of the cell adhesion molecule L1 was effective in blocking essentially all retinal ganglion cell growth.[6]

During the period in which these experiments were being undertaken, methods were devised that allowed the culture of adult rat retina; thus, we were able to make direct comparisons between the ability of Schwann cells to elicit outgrowth from embryonic vis-à-vis adult rat retinal ganglion cells.[8] Culture of the adult retina was facilitated by crushing the optic nerve approximately 5 days before harvesting the retina. This conditioned lesion promoted an early outgrowth of retinal ganglion cell neurites from the retinal fragments placed in culture. In these experiments too, we observed that when the substratum contained Schwann cells alone or Schwann cells plus extracellular matrix, rat retinal ganglion cell regenerative growth was promoted. Again, we observed that the provision of extracellular matrix generated by Schwann cells in culture dishes was ineffective in promoting this neuritic growth. Inasmuch as data was available from the culture of retinal explants from a number of species, including fish and chick, we were able to make comparisons with the growth requirements for retinal ganglion cells across several species and at different ages. These comparisons clearly show that the agents that promote retinal ganglion cell growth in embryonic periods differ from species to species, and furthermore that the requirements for adult retinal ganglion cell axonal growth show differences from the requirements for the growth of embryonic neurites. (For a review, see ref. 9.)

More recently we were able to make similar observations on the growth of human retinal ganglion cell neurons that were obtained from the eyes of human donors to eye banks.[10] This adult retinal material showed relatively little viability or propensity to grow in culture for some days, but if the cultures are maintained for more than a week, neuritic growth from the explants is observed when a substratum of rat Schwann cells is provided. Once again, we observed that the Schwann cell surface, rather than the extracellular matrix generated by Schwann cells, was effective in promoting regeneration. The initial delay in growth corresponds to the period between a "conditioning" lesion and the onset of a regenerative effort. It should be noted that related experiments have been conducted *in vivo* in a second CNS neuronal system, the cholinergic neurons of the medial forebrain.[11] Here, too, the cellular constituents, rather than extracellular matrix components, were shown to be critical.

Because of the remarkable ability of the Schwann cell to influence regeneration in the CNS, and because of the multiple role it is known to play in peripheral nerve degeneration, we questioned whether or not transplantation of Schwann cells would be an effective way to improve recovery after spinal cord injury in experimental animals and in humans. Our long-term goal was to learn how to obtain cultures of purified populations of Schwann cells from segments of adult peripheral nerve. This approach would allow autografting of cellular materials into regions of spinal cord injury in a situation in which the cellular material was derived from the individual with a spinal cord injury. Our efforts to devise effective ways to obtain large numbers of purified populations of Schwann cells from peripheral nerve segments from the adult were derived from the earlier work of Askanas *et al.*[12]

If implants of Schwann cells, or prosthetic devices containing Schwann cells, prove to be a practical approach to treating either central or peripheral nervous system injury, it would be advantageous to be able to harvest and grow a patient's own Schwann cells for autologous grafting. We recently began studies of methods of purifying and amplifying Schwann cells obtained from adult peripheral nerve, comparing methods of primary dissociation with explant culture, followed by dissociation.[13] Adult rat sciatic nerves were removed and enzymatically dissociated either immediately or after a period as 1 × 1 mm explants in tissue culture. Immediate dissociation yielded populations of cells containing less than 20% Schwann cells. Contaminating cells were mostly endoneurial fibroblasts. Conversely, serial explantation of explants over 4–5 weeks decreased fibroblast contamination; these fibroblasts grow out from the explants and were left behind when explants were replanted. After these explantations, dissociation yielded up to 98% pure Schwann cells. Schwann cells from this source retained functional capacity in that they associated with sensory neurons in culture and formed myelin under appropriate conditions. These cells proliferated in response to axonal mitogens as well as in the presence of glial growth factor and forskolin. Studies are underway to determine the most efficient way to acquire large numbers of these cells.

We are concurrently applying these techniques to adult human peripheral nerve. Preliminary results indicate that serial explanation yields a purer population of Schwann cells than does primary dissociation. It thus appears possible to obtain pure populations of Schwann cells by simple culture methods; it remains to be determined, however, how extensively this population can be expanded while retaining growth control and full functional capacity.[14]

Whereas these techniques have begun to provide purified populations of Schwann cells in rats, monkeys, and humans, the essential step of expanding the Schwann cell populations by using effective mitogens has so far been accomplished only in the rat. Here, provision of glial growth factor in the presence of forskolin is effective, and it is possible to obtain several million Schwann cells from culture of fragments of adult rat sciatic nerve. Although the purification procedures work well in monkeys and humans, Schwann cells derived from adult nerves from these species do not respond in the same way when provided with a number of potential mitogens that we have tried up to this point. Thus, the challenge of learning to expand primate Schwann cell populations after purification from adult peripheral nerve is still before us.

We are presently exploring the ability of transplanted Schwann cell populations to influence regeneration by creating lesions in the thoracic region of the adult rat spinal cord. We had previously[15,16] found no axonal ingrowth into implants containing sensory neurons as well as Schwann cells. In the current studies lesions are made by a photochemical method[17] in which a central cavity of tissue loss is created with retention of the meningeal covering and some of the ventral cord.

We then tested whether orphaned Schwann cells raised in culture elicit axonal ingrowth from lesioned cord tissue.[18] Schwann cells were allowed to ensheathe and myelinate parallel arrays of axons from dorsal root ganglion neurons in culture. Bands of longitudinally oriented Schwann cells within basal lamina tubes were created by severing the fascicles. After undergoing Wallerian degeneration for 4 days, the collagen substratum supporting these bands of Büngner was rolled to form an implant. Rolls of collagen seeded with dissociated Schwann cells or with no cells were also tested. Adult female rats (170–200 g) were lesioned through the T9 lamina with a laser beam of 562 nm after the intravenous application of Rose Bengal. This procedure reliably causes a cystic lesion which affects the dorsal two thirds of the spinal cord at the epicenter and extends 4–6 mm rostrocaudally. Five or 28 days after lesioning, implants were placed into cyst cavities where they remained for 14, 28, or 90 days. Light microscopy of spinal cords revealed good graft-host integration with a highly vascularized interface, minimal astrocytic scarring, and numerous Schwann cells. Electron microscopy confirmed the survival of Schwann cells and the presence of both myelinated and unmyelinated axons within the implant after 14 days. We conclude from these observations that Schwann cell implants devoid of neurons provide an effective bridge for axonal ingrowth.

These experiments indicate that there are neural populations whose axons course in the spinal cord and which respond to the presence of Schwann cells in the region of the lesion with both regenerative axonal growth as well as myelination by the Schwann cells provided. It remains to be determined which of the various spinal cord axonal pathways that are interrupted by this lesion may respond to the presence of Schwann cells by mounting a regenerative effort and it also remains to be determined whether this regenerative activity will have useful functional consequences.

REFERENCES

1. VIDAL-SANZ, M., G. M. BRAY, M. P. VILLEGAS-PEREZ, S. THANOS & A. J. AGUAYO. 1987. Axonal regeneration and synapse formation in the superior colliculus by retinal ganglion cells in the adult rat. J. Neurosci. **7:** 2894–2909.
2. BERRY, M., S. HALL, R. FOLLOWS, L. REES, N. GREGSON & J. SIEVERS. 1988. Response of axons and glia at the site of anastomosis between the optic nerve and cellular or acellular sciatic nerve grafts. J. Neurocytol. **17:** 727–744.
3. ARD, M. D., R. P. BUNGE & M. B. BUNGE. 1987. Comparison of the Schwann cell surface and Schwann cell extracellular matrix as promoters of neurite growth. J. Neurocytol. **16:** 539–555.
4. KLEITMAN, N., P. WOOD, M. I. JOHNSON & R. P. BUNGE. 1988. Schwann cell surfaces but not extracellular matrix organized by Schwann cells support neurite outgrowth from embryonic rat retina. J. Neurosci. **8:** 653–663.
5. ROGERS, S. L., P. C. LETOURNEAU, S. L. PALM, J. MCCARTHY & L. T. FURCHT. 1983. Neurote extension by peripheral and central nervous system neurons in response to substratum-bound fibronectin and laminin. Dev. Biol. **98:** 212–220.
6. KLEITMAN, N., D. K. SIMON, M. SCHACHNER & R. P. BUNGE. 1988. Growth of embryonic retinal neurites elicited by contact with Schwann cell surfaces is blocked by antibodies to L1. Exp. Neurol. **102:** 298–306.
7. BIXBY, J. L., J. LILIEN & L. F. REICHARDT. 1988. Identification of the major proteins that promote neuronal process outgrowth on Schwann cells *in vitro.* J. Cell Biol. **107:** 353–361.
8. HOPKINS, J. M. & R. P. BUNGE. 1991. Regeneration of axons from adult rat retinal ganglion cells on cultured Schwann cells is not dependent on basal lamina. Glia **4:** 46–55.
9. BUNGE, R. & J. HOPKINS. 1990. The role of peripheral and central neuroglia in regeneration in vertebrates. *In* Seminars in the Neurosciences. Neurobiology of Glia. K. R. Jessen & R. Mirsky, eds. **2:** 509–518. W.B. Saunders Co. Philadelphia, PA.
10. HOPKINS, J. M. & R. P. BUNGE. 1991. Regeneration of axons from adult human retina *in vitro.* Exp. Neurol. **112:** 243–251.
11. KROMER, L. F. & C. J. CORNBROOKS. 1985. Transplants of Schwann cell cultures promote axonal regeneration in the adult mammalian brain. Proc. Natl. Acad. Sci. USA **82:** 6330–6334.
12. ASKANAS, V., W. K. ENGEL, M. C. DALAKAS, J. V. LAWRENCE & L. S. CARTER. 1980. Human Schwann cells in tissue culture. Arch. Neurol. **37:** 329–337.
13. MORRISSEY, T. K., N. KLEITMAN & R. P. BUNGE. 1991. Isolation and functional characterization of Schwann cells derived from adult peripheral nerve. J. Neurosci. In press.
14. PORTER, S., M. B. CLARK, L. GLASER & R. P. BUNGE. 1986. Schwann cells stimulated to proliferate in the absence of neurons retain full functional capability. J. Neurosci. **6**(10): 3070–3078.
15. KUHLENGEL, K. R., M. B. BUNGE & R. P. BUNGE. 1990. Implantation of cultured sensory

neurons and Schwann cells into lesioned neonatal rat spinal cord. I. Methods for preparing implants from dissociated cells. J. Comp. Neurol. **293:** 63–73.
16. Kuhlengel, K. R., M. B. Bunge & R. P. Bunge. 1990. Implantation of cultured sensory neurons and Schwann cells into lesioned neonatal rat spinal cord. II. Implant characteristics and examination of corticospinal tract growth. J. Comp. Neurol. **293:** 74–91.
17. Cameron, T., R. Prado, B. D. Watson, M. Gonzalez-Carvajal & V. R. Holets. 1990. Photochemically induced cystic lesion in the rat spinal cord. Exp. Neurol. **109:** 214–223.
18. Paino, C. L. & M. B. Bunge. 1991. Induction of axon growth into Schwann cell implants grafted into lesioned adult rat spinal cord. Exp. Neurol. In press.

Myelin-Associated Inhibitors of Neurite Outgrowth and Their Role in CNS Regeneration

D. S. CADELLI AND M. E. SCHWAB[a]

Brain Research Institute
University of Zürich
CH-8029 Zurich, Switzerland

The differentiated central nervous system (CNS) of higher vertebrates bears very limited capacity to regenerate interrupted fiber tracts after a lesion, as was already stated in 1928 by Ramon y Cajal in his book, "Degeneration and Regeneration of the Nervous System."[1] Extensive regenerative processes take place, however, in the peripheral nervous system (PNS) where they can eventually lead to functional restoration.

Grafting experiments by Aguayo's group were decisive in showing that if placed in a peripheral nerve environment, CNS neurons of rats can regenerate lesioned neurites over long distances. (For examples, see refs. 2 and 3.) If the peripheral nerve bridge is placed in a target area, these regenerated fibers are even able to restore functional connections, as was recently shown for the optic system.[4] Conversely, PNS neurons failed to extend processes into CNS tissue. Thus, CNS neurons seem to have the intrinsic capability to regenerate, but the adult CNS seems to express unfavorable environmental conditions for that process.

One possible reason for the absence of CNS neurite regeneration, as already suggested in 1911 by Tello[5] on the basis of his transplantation experiments, could be a lack of neurotrophic factors in the CNS. *In vitro* experiments were therefore devised with a three-compartment chamber culture and bridges of peripheral and optic nerves between these chambers.[6] Neurons were grown in the presence of optimal levels of neurotrophic factors, nerve growth factor (NGF) for sensory and sympathetic neurons and brain-derived neurotrophic factor (BDNF) for retinal cells. These neurons were plated in the central chamber and had access to optic and sciatic nerve explants to grow in. A dramatic difference in neurite ingrowth was observed between peripheral and central explants for both living and frozen tissues; sciatic nerves contained up to several hundred axons, but optic nerves were totally devoid of neurites.[6] These findings showed a nonpermissive substrate effect of the CNS tissue, which could not be overcome by the strong stimulatory effect of NGF or BDNF.

Frozen sections of adult peripheral nerves or various CNS regions were then used as tissue culture substrates for rat sensory and sympathetic neurons or neuroblastoma cells. Neurons attached well and grew processes on the PNS tissue, and although less well, also on CNS gray matter. On the other side cell adhesion was poor and the growing neurites clearly avoided the white matter.[7,8]

[a]Address for correspondence: M. E. Schwab, Brain Research Institute, University of Zurich, August-Forel Strasse 1, CH-8029 Zurich, Switzerland.

OLIGODENDROCYTES AND MYELIN ARE INHIBITORY SUBSTRATES

The results just described prompted experiments focused on the neuron-glia interaction. Dissociated peripheral sympathetic and sensory or central retinal neurons were cocultured with optic nerve glial cells. Glial fibrillary acidic protein (GFAP)-positive astrocytes were a good substrate for cell adhesion and neurite outgrowth, as already observed by several investigators. (For example, see refs. 9, 10, or 11.) In contrast, differentiated oligodendrocytes (positive for the antigens 01/GalC and 04; negative for the precursor marker A2B5), with their typical highly branched morphology, clearly exerted inhibitory substrate properties.[12] Growing neurites avoided the oligodendrocytes, leaving "windows" around them. This was true not only for neurons; astrocytes and fibroblasts were also prevented from spreading and migrating over differentiated oligodendrocytes. Immature oligodendrocytes (01-negative; 04-, GD3-, and A2B5-positive), on the contrary, were covered by the neuronal network.

Video time lapse cinematography allowed direct observation of growth cone behavior in those cultures.[13] Growth cones grew over astrocytes with unchanged velocity. On the other hand, as soon as they touched an oligodendrocyte process with the tip of their exploring filopodia, dorsal root ganglion (DRG) growth cones became immobilized. Filopodial activity ceased within minutes after contact, neurite extension ceased, and the growth cones either remained immobile for hours (time of observation), collapsed, or eventually retracted towards the cell body. This growth inhibition required direct contact with the oligodendrocytes and was localized to the neurites interacting with the cell processes; the other neurites of the affected neuron continued to grow. No "habituation" did occur within the limit of the observation period. Oligodendrocytes thus specifically inhibit neuronal growth, and a signal arising from the filopodia-oligodendrocyte interaction seems to locally paralyze the growth cone movement machinery.

The substrate properties of CNS and PNS myelin were tested by adsorbing purified adult rat spinal cord myelin on culture dishes and observing neurite growth or fibroblast spreading. CNS myelin strongly impaired neurite outgrowth, whereas PNS myelin extracts from sciatic nerve and extracts from a number of nonneuronal tissues were good substrates.[12,14] The inhibitory property of the CNS is thus associated with differentiated oligodendrocytes and their product, CNS myelin.

INHIBITORY MYELIN COMPONENTS NI-35 AND NI-250: BIOCHEMICAL CHARACTERIZATION

As a rich source of inhibitory components, myelin provided a good material for biochemical studies.[14] The inhibitory effect was at first recognized to be linked to proteins as it was abolished by proteases but not by extraction of the lipids. Proteins could be solubilized from the myelin membranes with detergents under denaturing conditions, and inhibitory activity was recovered after reconstitution of myelin proteins into liposomes and adsorption onto culture dishes. Myelin proteins were fractionated by polyacrylamide gel electrophoresis in SDS, and the protein fractions were tested for their substrate quality by reconstitution into liposomes. Two minor protein fractions with a molecular weight of 35 and 250 kD and having strong inhibitory effects on fibroblast spreading and neurite outgrowth were isolated.[14] They were called neurite growth inhibitors NI-35 and NI-250, the second being possibly a high molecular weight complex of the first (C. Bandtlow, personal communication). They were recognized as the crucial inhibitory components of CNS myelin, because selective removal of the 35- and 250-kD fractions resulted in CNS myelin protein with good substrate properties. On the other hand, favorable substrates like liver or sciatic nerve proteins supplemented with NI-250 or NI-35 became inhibitory. In the cell biology experiments, the addition of NI-35 liposomes to growing DRG neurons in culture rapidly arrested growth cone movements. NI-35 and NI-250 were not found in PNS myelin or in fish and frog CNS myelin, where sets of CNS axons are known to regenerate.

These membrane-bound proteins do not bind the lectin peanut agglutinin (C. Bandtlow, unpublished observations); these properties as well as the time course and the expressing cell type distinguish them from the repulsive component described in the chicken embryonic posterior tectum opticum[15,16] and from the nonpermissive factor present in the posterior half somites in early chick embryos.[17,18]

Cloning of the NI-35 molecule is presently under way in our laboratory.

ANTIBODIES NEUTRALIZE THE INHIBITORY ACTIVITY OF NI-35 AND NI-250

Two monoclonal antibodies (IN-1 and IN-2) were raised against gel-purified 250-kD and 35-kD proteins. Both effectively neutralize the inhibitory activity of both inhibitors and of the entire CNS myelin.[19] In neuron-oligodendrocyte coculture experiments, antibodies (IN-1) successfully prevented the inhibition of the growth cone by the oligodendrocyte, and growth cones were able to further extend processes over the oligodendrocytes.[13] A large reduction of the myelin inhibitory effect was seen in the fibroblast spreading assay, for neurite outgrowth on CNS myelin and for neurons cultured on CNS frozen sections.[7,19] Finally, in a decisive experiment, antibodies were injected into optic nerve explants, and a profuse neurite ingrowth from cocultured sensory neurons could be followed for several millimeters, whereas in control optic nerves injected with a control antibody (antigalactocerebroside) the majority of the neurites stopped in less than 1 mm.[19]

These results confirmed that identified inhibitory components NI-35/250 are the major inhibitory factors of oligodendrocytes and myelin and a crucial determinant of the inhibitory substrate property of CNS tissue.

INHIBITORS AND REGENERATION

The foregoing results raised the crucial question of the *in vivo* implication of the neurite growth inhibitors for the lack of regeneration in the CNS. The rat corticospinal tract (CST) was chosen for regeneration experiments as an anatomically well characterized, easily labeled, and readily accessible tract, and two experimental models were designed.

A series of experiments were performed with young rats irradiated on the thoracic and lumbar spinal cord.[20] Irradiation at birth eliminates the developing oligodendrocytes but preserves the neurons and astrocytes that are postmitotic at this age. Spinal cords of these rats were thus unmyelinated and were lesioned with mechanical transection (see below). In another series of experiments, IN-1 antibody-producing hybridoma cells were implanted into the rat brain.[21] On immunosuppression, small tumors developed next to the ventricles and constantly supplied the cerebrospinal fluid with neutralizing antibodies. For both series, the dorsal two thirds of the spinal cord were transected at 2–7 weeks of age. At 10 days postnatally, corticospinal neurons have finished growing, and regeneration is known to be totally absent after postnatal day 4–6.[22] Two to four weeks after the injury, lesioned corticospinal neurons were stained with wheat germ agglutinin horseradish peroxidase (WGA-HRP) injection into the frontal and parietal cortex. To be able to reconstruct the regrowth of the neurons, spinal cords were cut in complete longitudinal series of sections. In most of the animals, sprouting was seen at the lesion site, and some corticospinal tract axons could circumvent the lesion. All regrowth measurements were taken from the distal end of the lesion. In myelin-free spinal cords or in IN-1–treated rats, these fibers were seen to successfully regenerate over long distances: most axons elongated over 5 mm, up to 18 mm. In control rats (no treatment, or implantation of anti-HRP–producing hybridoma cells) only abortive regeneration not exceeding the sprouting distance of about 1 mm was observed.[20,21]

When IN-1–treated rats were lesioned at older ages (5–7-weeks), the same results of successful regeneration were obtained. The age thus does not seem to be critical, and adult corticospinal tract neurons bear the capacity to regrow.

These observations clearly pointed to a determinant role of neurite growth inhibitors in the known lack of fiber regeneration in the CNS *in vivo*.

In all these experiments the percentage of successfully regenerating axons was small. Several reasons could be responsible, either alone or in combination:

1. Problems at the lesion site: Large lesions often form cavities, and neurons have to find their way through tortuous routes along the caverns. Moreover, astrocytes wall off the lesion area and may form a glial scar that can act as a barrier. (For examples, see refs. 23 and 24.)

2. Although nerve cell bodies in the adult CNS often survive lesions for very long times, the axons often retract from the lesion site.[1] Trophic factors may increase the survival and stimulate sprouting. A good trophic and sprouting effect has been obtained with embryonic spinal cord grafts,[25] and specific trophic factors are presently being tested.

3. Turning on the regeneration/growth program in neural cells, including increased synthesis of cytoskeletal proteins and GAP 43 may necessitate particular signals or metabolic and environmental conditions. Neurons or neuronal subsets may differ in their intrinsic capacity to regrow and in their response to or requirement for particular environmental stimuli, such as neurite growth factors.

A further unsolved question concerns the connections formed by the regenerating cerebrospinal tract fibers and their functional consequences. The first question will be addressed through detailed light and electron microscopic analysis of the terminal branch pattern and the synapses formed; for the second question, behavioral tests for fine motor control are planned.

Other CNS tracts are also being investigated. In the septohippocampal system, large fimbria-fornix lesions have been made by aspiration in 3-week-old rats, and the lesions were bridged with laminin-rich placenta basement membranes on a nitrocellulose filter support placed between the septum and the rostral part of the hippocampus. Rats were implanted with either IN-1 antibody or anti-HRP–producing cells. After a 2–5-week survival time, acetylcholinesterase (ACHE) staining was used to detect cholinergic fibers. Regenerating fibers and fascicles were observed to cross the bridge and enter the hippocampus. Once within the hippocampus, fibers developed profuse arborizations in both control and experimental animals. The axonal network was arranged in layers and partly reconstituted its original anatomical distribution, as reported in the literature for the regrowth of lesioned septal neurons through an embryonic hippocampal bridge[26] or for embryonic grafts implanted next to a lesioned hippocampus. (See, for example, refs. 27 and 28.) In the controls, ACHE-positive fibers extended within 1 mm caudally and laterally from the bridge. In IN-1-treated animals, AchE-positive fibers elongated over 2–3.5 mm and in some cases even more, in both directions.[29]

These results indicate, first, that NI-35/250 also play a role in the hippocampus and, second, that the presence of a largely denervated target area may induce terminal arborization and synapse formation, thus considerably slowing down fiber growth despite the presence of inhibitor-neutralizing antibodies.

PHYSIOLOGICAL HINTS: INHIBITORS IN DEVELOPMENT

Growing neurons do not encounter mature oligodendrocytes during the early stages of CNS development. It has been found that inhibitor proteins appear just before the onset of myelination (for instance, in the newborn rat optic nerve), where myelin formation starts at P5-7.[30] The onset of myelination is slightly before birth in the rat CNS and varies in the various fiber tracts of a given CNS region. At that time of development most neuronal tracts have already reached their terminal area, while some late growing tracts are still elongating. The late forming growth cones may contact oligodendrocytes or myelin sheath surrounding previously formed tracts. Inhibitor-expressing oligodendrocytes could form a "protective girdle" or barrier around a given pathway or region, preventing the ingrowth of late arriving fibers. Guidance functions may thus result. Sprouting and branching from the tracts may also be influenced or restricted.

To test this hypothesis, the cerebrospinal tract was particularly interesting. The cerebrospinal tract is a late growing tract: its fibers enter the rat cervical spinal cord around birth and grow down the cord during the first 10 days postnatally.[31,32] In its position in the dorsal funiculus, the ingrowing cerebrospinal tract is surrounded by the myelinating sensory fascicles cuneatus and gracilis.[33] A sharp boundary separates the ascending and descending tracts in these rostral halves of the spinal cord. The growth pattern of the cerebrospinal tract was then studied under experimental conditions, in which the NI-35/250 were either suppressed through oligodendrocyte elimination or neutralized with IN-1 antibodies.[33] Rats that had been rendered myelin free by X-irradiation or rats injected with IN-1-producing cells at birth were examined at 4–12 days of age. A clear aberration of cerebrospinal tract fibers was found; the cerebrospinal tract axons occupied a larger area in experimental animals, and corticospinal fibers and fascicles grew into ectopic locations intermingling for long distances with the afferent tracts, a feature never seen in the control rats.

OUTLOOK

The question of the possible homology and relatedness of neurite growth inhibitors NI-35/250 to other membrane proteins still awaits their cloning and molecular characterization. Molecular data will also show if families of inhibitors/repulsors exist. It will then be possible to focus on the putative neuronal receptor mediating growth arrest and to further dissect the intracellular signals and the mechanism of inhibition of the growth cone movement machinery. The *in vivo* meaning of such inhibitors of neurite outgrowth in the CNS of vertebrates is still a matter of speculation. In addition to guidance of late growing tracts, stabilization of newly formed tracts in a still developing system could guarantee the preservation of the highly ordered architecture of the vertebrate CNS. This process could even bear an economical advantage because the same local recognition cues and guidance mechanisms can then be used in parallel by different tracts in different CNS areas without danger of confusion between neighboring paths. The complexity thus achieved does not require a huge amount of information for its development. In the adult, the inhibitors could act as stabilizers of the system; plasticity is known to be limited to a critical period after birth, and this is notably also the period of myelin formation. In the fully adult CNS, plastic changes in dendritic and axonal trees are limited to very short (below 1 mm!) distances.

REFERENCES

1. RAMON Y CAJAL, S. 1928. Degeneration and Regeneration of the Nervous System. Translation, 1959. Hafner. New York.
2. DAVID, S. & A. J. AGUAYO. 1981. Axonal elongation into peripheral nervous system 'bridges' after central nervous system injury in adult rats. Science **214:** 931–933.
3. BENFEY, M. & A. J. AGUAYO. 1982. Extensive elongation of axons from rat brain into peripheral nerve grafts. Nature **296:** 150–152.
4. KEIRSTEAD, S. A., M. RASMINSKY, Y. FUKUDA, D. A. CARTER, A. J. AGUAYO & M. VIDAL-SANZ. 1989. Electrophysiologic responses in hamster superior colliculus evoked by regenerating retinal axons. Science **246:** 255–257.
5. TELLO, F. 1911. La influencia del neurotropismo en la regeneracion de los centros nerviosos. Trab. Lab. Invest. Biol. **9:** 123–159.
6. SCHWAB, M. E. & H. THOENEN. 1985. Dissociated neurons regenerate into sciatic but not optic nerve explants in culture irrespective of neurotrophic factors. J. Neurosci. **5:** 2415–2423.
7. SAVIO, T. & M. E. SCHWAB. 1989. Rat CNS white matter, but not gray matter, is nonpermissive for neuronal cell adhesion and fiber outgrowth. J. Neurosci. **9:** 1126–1133.
8. CRUTCHER, K. A. 1989. Tissue sections from the mature rat brain and spinal cord as substrates for neurite outgrowth *in vitro:* extensive ingrowth on gray matter but little growth on white matter. Exp. Neurol. **104:** 39–54.
9. LINDSAY, R. M. 1979. Adult rat brain astrocytes support survival of both NGF-sensitive and -insensitive neurons. Nature **282:** 80–82.
10. BANKER, G. A. 1980. Trophic interactions between astroglial cells and hippocampal neurons in culture. Science **209:** 209–210.
11. NEUGEBAUER, K. M., K. J. TOMASELLI, J. LILIEN & L. F. REICHARDT. 1988. N-cadherin, NCAM, and integrins promote retinal neurite outgrowth on astrocytes *in vitro.* J. Cell Biol. **107:** 1177–1187.
12. SCHWAB, M. E. & P. CARONI. 1988. Oligodendrocytes and CNS myelin are nonpermissive substrates for neurite growth and fibroblast spreading *in vitro.* J. Neurosci. **8:** 2381–2393.
13. BANDTLOW, C. E., T. ZACHLEDER & M. E. SCHWAB. 1990. Oligodendrocytes arrest neurite growth by contact inhibition. J. Neurosci. **10:** 3937–3948.
14. CARONI, P. & M. E. SCHWAB. 1988. Two membrane protein fraction from rat central myelin with inhibitory properties for neurite growth and fibroblast spreading. J. Cell Biol. **106:** 1281–1288.
15. WALTER, J., R. KERN-VEITS, J. HUF, B. STOLZE & F. BONHOEFFER. 1987. Recognition of position-specific properties of tectal cell membranes by retinal axons *in vitro.* Development **101:** 685–696.
16. COX, E. C., B. MUELLER & F. BONHOEFFER. 1990. Axonal guidance in the chick visual system: Posterior tectal membranes induce collapse of growth cones from the temporal retina. Neuron **2:** 31–37.
17. KEYNES, R. J. & C. D. STERN. 1984. Segmentation in the vertebrate nervous system. Nature **310:** 786–787.

18. DAVIES, J. A., G. M. W. COOK, C. D. STERN & R. J. KEYNES. 1990. Isolation from chick somites of a glycoprotein fraction that causes collapse of dorsal root ganglion growth cones. Neuron **2:** 11–20.
19. CARONI, P. & M. E. SCHWAB. 1988. Antibody against myelin-associated inhibitor of neurite growth neutralizes nonpermissive substrate properties of CNS white matter. Neuron **1:** 85–96.
20. SAVIO, T. & M. E. SCHWAB. 1990. Lesioned corticospinal tract axons regenerate in myelin-free rat spinal cord. Proc. Natl. Acad. Sci. USA **87:** 4130–4133.
21. SCHNELL, L. & M. E. SCHWAB. 1990. Axonal regeneration in the rat spinal cord produced by an antibody against myelin-associated neurite growth inhibitors. Nature **343:** 269–272.
22. BERNSTEIN, D. R. & D. J. STELZNER. 1983. Plasticity of the corticospinal tract following midthoracic injury in the postnatal rat. J. Comp. Neurol. **221:** 382–400.
23. LINDSAY, R. M. 1986. Reactive gliosis. *In* Astrocytes. S. Fedoroff & A. Verdanakis, eds. Vol. 3: 231–262. Academic Press, Inc. New York.
24. REIER, P. J. 1986. Gliosis following CNS injury: The anatomy of astrocytic scars and their influence on axonal elongation. *In* Astrocytes. S. Fedoroff & A. Verdanakis, eds. Vol. 3: 263–324. Academic Press, Inc. New York.
25. BREGMAN, B. S., E. KUNKEL-BAGDEN, M. MCATEE & A. O'NEILL. Extension of the critical period for developmental plasticity of the corticospinal pathway. J. Comp. Neurol. **15:** 355–370.
26. SEGAL, M., U. STENEVI & A. BJÖRKLUND. 1981. Reformation in adult rats of functional septo-hippocampal connections by septal neurons regenerating across an embryonic hippocampal tissue bridge. Neurosci. Lett. **27:** 7–12.
27. BJÖRKLUND A. & U. STENEVI. 1984. Intracerebral neural implants: Neuronal replacement and reconstruction of damaged circuitries. Ann. Rev. Neurosci. **7:** 279–308.
28. NILSSON, O. G., D. J. CLARKE, P. BRUNDIN & A. BJÖRKLUND. 1988. Comparison of growth and reinnervation properties of cholinergic neurons from different brain regions grafted to the hippocampus. J. Comp. Neurol. **268:** 204–222.
29. CADELLI, D. & M. E. SCHWAB. 1991. Regeneration of lesioned septohippocampal acetylcholinesterase-positive axons is improved by antibodies against the myelin-associated neurite growth inhibitors NI-35/250. Eur. J. Neurosci., in press.
30. CARONI, P. & M. E. SCHWAB. 1989. Codistribution of neurite growth inhibitors and oligodendrocytes in rat CNS: Appearance follows nerve fiber growth and precedes myelination. Dev. Biol. **136:** 287–295.
31. SCHREYER, D. J. & E. G. JONES. 1982. Growth and target finding by axons of the corticospinal tract in prenatal and postnatal rats. Neuroscience **7:** 1837–1853.
32. GRIBNAU, A. A., E. J. DE KORT, P. J. DEDEREN & R. NIEUWENHUYS. 1986. On the development of the pyramidal tract in the rat. II. An anterograde tracer study of the outgrowth of the corticospinal fibers. Anat. Embryol. **175:** 101–110.
33. SCHWAB, M. E. & L. SCHNELL. 1991. Channelling of developing rat corticospinal tract axons by myelin-associated neurite growth inhibitors. J. Neurosci. **11:** 709–722.

DISCUSSION PAPER

Is Functional Repair in the CNS a Realistic Possibility?

NIGEL HOLDER

Developmental Biology Research Centre
Division of Biomedical Sciences
King's College
26–29 Drury Lane
London WC2B 5RL, UK

During development differentiating neurons from the central and peripheral nervous systems send out axons that must grow and make contact with specific target cells. In some cases, growing axons must cover enormous distances; some grow into the periphery, whereas others stay confined within the CNS. In amniotes (lizards to man) this stage of neurogenesis is completed during the embryonic phase of life, but in anamniotes (fish and amphibia) this is often not the case, as these animals are continuously growing and adding neurons to the organized "adult" nervous system.[1,2] It is not a coincidence that these continuously growing animals can show remarkable degrees of functional recovery following damage. A complete spinal cord section in a salamander is functionally repaired in a few weeks,[3] whereas such a lesion in any amniote would result in permanent paralysis.

The topic of the papers in this section is not to ask why are lower vertebrates able to show such remarkable regenerative powers, but to establish why higher vertebrates do not. It is an applied science, focused on the crucial clinical issue for which the prize is high, curing people who have damaged their central or peripheral nervous systems. The need is to turn back evolutionary and developmental clocks and discover how to make adult nerves grow axons that make contact with their target cells.

One important observation, made in the last century, has contributed greatly to experimental approaches made to this central question. When a peripheral nerve of a mammal is cut or crushed, axons grow readily into the distal nerve stump. The problem is that they grow randomly, making contact with inappropriate targets and causing abnormal function. If a motor nerve is damaged, for example, regrowing axons will contact the wrong muscles, resulting in paralysis. In contrast, following damage to the CNS, axons do not grow, highlighting the major difference between responses to damage in the two regions. In recent years, progress has been made towards the clinical goal by exploiting the fact that axons can grow in a peripheral environment but not in a central environment. The basic experiment, then, is to place a piece of peripheral nerve into the damaged CNS to see if CNS axons can grow into the graft that provides a permissive environment. The answer is that they can, as shown elegantly in the paper by Bray and his colleagues. This experiment has two very important consequences, one basic, one applied. Crucially, it shows that CNS axons are fully capable of regrowing axons and, furthermore, that the block to axon growth in the CNS is due to an interaction between the axon and its environment. The hope, clinically, is that a piece of peripheral nerve can be used to bridge a CNS lesion and allow axons to make it back to their normal target area, where they will make contacts in a manner that will allow at least some functional recovery. This experiment also focuses our attention on the detailed mechanism of the interaction of axons with their environment. What is it about the environment of a peripheral axon that allows it to regrow? The other side of the coin is, what is in the CNS environment of a central axon that prevents its regrowth? These questions lie at the heart of the remaining three papers in this section. Two of them, by Bunge and by Heumann,[a] examine the interactions between axons and Schwann cells that result in axon regrowth; the third, by Cadelli and Schwab, describes the presence of proteins that actively inhibit axon growth in the CNS.

So what do we know of the interaction between the axon and its environment? Raisman, in his introductory comments, draws attention to the glial cell as the unsung hero of the nervous system. There is good evidence that in the periphery the Schwann cell plays a crucial role in stimulating and maintaining axonal outgrowth following damage, but in the CNS glia seem to be the villains of the piece because the proteins that prevent axonal regrowth are associated with the myelin produced by the glial subtype, the oligodendrocyte. Let us first examine the positive role played by Schwann cells.

The paper presented by Bunge describes a series of experiments designed to characterize the interaction between peripheral (dorsal root ganglia) and central (retinal) derived neurons and Schwann cells in culture. Working *in vitro* presents the opportunity to break down the components of a peripheral nerve, the Schwann cells, fibroblasts, and extracellular matrix proteins, and recombine them to establish the important interactions necessary for axonal growth. A lot of work remains to characterize the best conditions for axonal growth in such cultures because neurite outgrowth is variable depending upon the exact source and age of the neuron population being examined. Despite this, a great deal of progress has been made towards achieving the clinically directed goal of obtaining a purified population of Schwann cells that can be grafted to the site of a central lesion in an attempt to stimulate and maintain regrowth of damaged central neuronal processes. Obtaining a pure population of Schwann cells in culture is no trivial matter, because contaminating fibroblasts must be removed. This can be achieved after prolonged culture periods with remarkable success for both rodent and human cells. The next challenge for Bunge and his colleagues is to make the purified human Schwann cells divide and produce a continuous source of cells, sufficient for clinical applications. At that time we shall find out if human Schwann cells will be capable of stimulating regrowth of central axons in the damaged human CNS.

There is no doubt that Schwann cells are directly involved in maintaining axonal growth in peripheral nerves, and it has become clear in recent years that a key player in the interaction between Schwann cells and the growing axon is nerve growth factor (NGF). One of the first isolated growth factors,[4] NGF has been studied in great detail. Much is known about the protein itself, its cellular receptors, and the NGF gene. An approach taken by Heumann and his colleagues (this volume) is to understand in detail the control of NGF expression in damaged peripheral nerve tissue. The basic observation is that Schwann cells express NGF mRNA immediately after nerve damage. If the rodent sciatic nerve is damaged, NGF mRNA up-regulation occurs very quickly, but declines within 24 hours, only to up-regulate again by 72 hours. This biphasic NGF response is now being dissected by molecular methods: experiments with transfected fibroblasts indicate that the early response of NGF expression involves the rapid response of protooncogenes c-*fos* and c-*jun* and that an AP-I consensus sequence in the NGF gene promoter is involved in the regulation of NGF transcription. These elegant experiments dissect the molecular control of NGF expression, but it has all been carried out in transfected fibroblasts

[a] Paper presented at conference but not submitted for publication.

and it remains to be shown that this mechanism of regulation occurs in the Schwann cells of damaged peripheral nerves. What of the late response of NGF mRNA up-regulation? This response presumably produces the NGF required by axons to regrow, so an understanding of the mechanism of its control may be of great importance in the longer term with respect to clinical advances. The response is clearly different from the early up-regulation of NGF mRNA and appears to involve a trigger, interleukin-1, which is released by macrophages that themselves invade the damage site as part of the injury response. Here again lies an important mechanism for control of NGF expression. The link between the macrophage and the Schwann cell has been strengthened in recent experiments in which NGF mRNA up-regulation in damaged peripheral nerve has been examined in a mutant rat that does not produce invading macrophages. In these mutants, the long-term NGF response in Schwann cells is dramatically reduced. There is still a long way to go before a complete understanding of the control of NGF up-regulation is achieved. Also much is to be learned about the neuronal response to NGF in terms of regrowth and the mechanism of action of NGF receptors, itself an area of intense research activity recently because of the long sought after identification of the high-affinity NGF receptor.[5]

The work with Schwann cells and the mechanism of action of NGF in nerve repair are both relevant to the eventual controls of neuronal repair in humans if ways can be found to employ the factors that permit axonal regrowth in the periphery to force central axons to regrow. An alternative approach, mentioned earlier, is to identify the factors that block regrowth of central axons and inactivate them. This is the line that has been taken by Schwab and his colleagues. Using culture assays, it was made clear that axons from CNS neurons would extend over various substrates, including CNS gray matter or purified astrocytes, but not others including CNS white matter or oligodendrocytes. Schwab concluded that the common denominator for inhibition was CNS myelin produced by oligodendrocytes and present as a major constituent in the white matter. Furthermore, it was established that CNS neurons would grow axons over purified myelin from peripheral nerve but not over purified myelin from the CNS. Using this culture system as a bioassay, the components of CNS myelin that have inhibitory effects on axonal growth have been purified as two bands of 35 and 250 kD on a gel. The inhibitory function of the proteins has been taken a step further by adding them to liposomes and using these to analyze their effects on axonal growth in culture. When such liposomes are added to actively growing neurites, they cause growth cone collapse, an effect that can be blocked by antibodies raised against the original proteins. These advances are perhaps the most exciting in recent years for those interested in solving the clinical problems of permanent damage to the human CNS, but there is a long way to go. The proteins must be sequenced and the genes cloned and analyzed. It may then be possible to engineer a cure, allowing growth of axons centrally. Schwab and his colleagues have taken a first stab at overcoming the effects of the inhibitory proteins *in vivo* by implanting the hybridoma cells producing antibodies to the proteins into the brain ventricles of a rat. Following a lesion to the corticospinal tract, some limited axonal regrowth was obtained.[6] The result, which is promising but not definitive, may be considered the best one yet, as the search for a clinical cure for paralysis following damage to the human CNS heats up.

CONCLUSIONS

Is functional repair in the CNS a possibility? Will it be possible to roll back the evolutionary and developmental clocks and have humans responding to a damaged

CNS in the same way as does a salamander? I think there is much cause for optimism, as the molecular understanding of NGF makes our knowledge of peripheral nerve regrowth more complete, methods for human Schwann cell culture are improved, and inhibitory proteins in the CNS are isolated and their normal function eventually overcome.

REFERENCES

1. HOLDER, N. & J. D. W. CLARKE. 1988. Is there a correlation between continuous neurogenesis and directed axon regeneration in the vertebrate nervous system? TINS **11:** 94–99.
2. HOLDER, N., J. D. W. CLARKE, N. STEPHENS, S. WILSON, C. ORSI, T. BLOOMER & D. A. TONGE. 1991. Continuous growth of the motor system in the axolotl. J. Comp. Neurol. **303:** 534–550.
3. CLARKE, J. D. W., R. ALEXANDER & N. HOLDER. 1988. Regeneration of descending axons in the spinal cord of the axolotl. Neurosci. Letts. **89:** 1–6.
4. LEVI-MONTALCINI, R. & P. ANGELETTI. 1968. Nerve growth factor. Physiol. Rev. **48:** 534–568.
5. HEMPSTEAD, B., D. MARTIN-ZANCA, D. R. KAPLAN, L. PARADA & M. CHAO. 1991. High affinity NGF binding requires co-expression of the *trk* proto-oncogene and the low affinity NGF receptor. Nature **350:** 678–683.
6. SCHNELL, L. & M. SCHWAB. 1990. Axonal regeneration in the rat spinal cord produced by an antibody against myelin-associated neurite growth inhibitors. Nature **343:** 269–272.

PART V. GLIAL ELECTROPHYSIOLOGY AND TRANSPORT

INTRODUCTION

Glial Electrophysiology and Transport

RICHARD K. ORKAND

Institute of Neurobiology and Department of Physiology
University of Puerto Rico Medical Sciences Campus
San Juan, Puerto Rico 00901

Study of the membrane properties of glial cells is motivated by a need both to define and to understand glial function. Observations with the electron microscope were sufficient to establish unequivocally the role of oligodendrocytes in forming the myelin sheath which serves to increase the conduction velocity of nerve impulses.[1] Similarly, the role of glial cells in forming the blood brain barrier of higher invertebrates and elasmobranch fishes[2] was delineated primarily through anatomical studies. In contrast, specifying the functions of astrocytes has required less direct methods. It is difficult to observe astrocytes in action.

The general approach used by electrophysiologists to study glial function, which has its origins in the work of Kuffler and Potter[3] in the early 1960s, has been to define the properties of glial cell membranes and to use this knowledge to suggest hypotheses of function. Thus, the search for astrocyte functions has involved the kinds of fishing expeditions often frowned upon by funding agencies. Nevertheless, the progress evident from the work presented in this session testifies to the success of this strategy. In recent years, the application of more sophisticated technology, especially the patch-clamp technique and the use of ion-selective electrodes, has further contributed to our understanding of the membrane properties of astrocytes. The results of these studies have raised the possibility of a whole new range of physiological roles and neuron-glia interactions. The immediate challenge is to show how these properties play a role in the function of astrocytes.

This problem is well illustrated by the results presented in the papers by Barres[4] and Waxman *et al.*[5] Elegant descriptions of the properties of voltage-sensitive channels in glial membranes under a variety of conditions are given. The existence of such channels was first demonstrated by Ritchie[6] and his colleagues over 5 years ago. Their presence in glial cells is puzzling; the discoverers suggested that the channels are manufactured by glia and inserted into neuronal membranes. Indeed, macromolecular transfer occurs between glia and neurons and is discussed in the paper by Grossfeld.[7] The reason it is difficult to propose a physiological role for most voltage-dependent channels in glial cells is that the glial membrane potential is dominated by a high resting potassium permeability that tends to hold the potential near the potassium equilibrium potential of -80 to -90 mV.[3] Intracellular recordings from glial cells in the intact nervous system indicate that the membrane potential does not vary by more than a few millivolts during normal activity; much greater depolarizations are required to change significantly the opening of most of these channels.[8] Moreover, the syncytial nature of glial cells and the apparently low channel density preclude the appearance of regenerative electrical responses in glial cells under all but the most unusual circumstances.[9] Glial cells are inexcitable.

The high permeability of glial cells to potassium suggested a role for these cells in potassium homeostasis and led directly to the hypothesis that glial cells help redistribute focally accumulated potassium ions by a passive mechanism termed spatial buffering.[10] The quantitative significance of this process depends critically on the anatomic arrangement of the cells as well as the distribution[11] and properties of the potassium channels.[8] These questions are addressed in the paper by Reichenbach.[12] Similarly, as discussed by Kempski and colleagues,[13] because glial cells are known to take up KCl[14] and to have an electrogenic Na/K ATPase[15] as well as a Na/H antiporter,[16] they must undergo variations in cell volume that have important implications in the cause and prevention of brain edema.[17] The demonstration of these properties leads to the question of their physiological significance.[18] This question is not yet fully resolved.

Studies of pH regulation in glial cells using ion-selective electrodes by Schlue[19] and the patch-clamp analysis of electrogenic glutamate uptake by Attwell *et al.*[20] go a long way toward establishing additional dynamic functions for neuroglia in the control of the neuronal microenvironment.[21] The regulation of intracellular pH in glial cells involves a variety of specific membrane transporters and exchangers, including Na/H exchange[16] and electrogenic $NaHCO_3$ cotransport.[22] These processes require ion movements across the glial membrane and, therefore, they also serve as pH regulators for the narrow system of clefts that constitute the extracellular space of neurons as well as glia. The glutamate studies provide an electrophysiological method in living cells to study the apparently vital role of glial cells in controlling the extracellular levels of neurotransmitter amino acids. They provide an important addition to the original autoradiographic anatomical techniques that first suggested this function.[23,24]

A striking feature of the presentations is that each group successfully exploits a particular experimental preparation that best demonstrates the phenomena being investigated. These include preparations from invertebrates like the crayfish and leech, a lower vertebrate, like the salamander, as well as primary cultures of mammalian brain and a glioma cell culture line. The value of this comparative approach to understanding glial function was emphasized in a highly successful meeting, held 10 years ago, organized and led by John Treherne, and was a central theme in his important contributions to this field.[25]

REFERENCES

1. Bunge, R. P. 1968. Glial cells and central myelin sheath. Physiol. Rev. **48:** 197–251.
2. Abbott, N. J., N. J. Lane & M. Bundgaard. 1986. The blood-brain interface in invertebrates. Ann. N.Y. Acad. Sci. **481:** 20–42.
3. Kuffler, S. W. & J. G. Nicholls. 1966. The physiology of neuroglial cells. Ergeb. Physiol. **57:** 1–90.
4. Barres, B. A. 1991. Five electrophysiologic properties of glial cells. Ann. N.Y. Acad. Sci. This volume.
5. Minturn, J. E., H. Sontheimer, J. A. Black, B. R. Ransom, J. M. Ritchie & S. G. Waxman. 1991. Membrane-associated sodium channels and cytoplasmic precursors in glial cells: Immunocytochemical, electrophysiologic and pharmacologic studies. Ann. N.Y. Acad. Sci. This volume.
6. Gray, P. T. A. & J. M. Ritchie. 1985. Ion channels in Schwann and glial cells. TINS **8:** 411–415.
7. Grossfeld, R. M. 1991. Axon-glia exchange of macromolecules. Ann. N.Y. Acad. Sci. This volume.
8. Barres, B. A., L. L. Y. Chun & D. P. Corey. 1990. Ion channels in vertebrate glia. Ann. Rev. Neurosci. **13:** 441–474.
9. MacVicar, B. A. 1984. Voltage-dependent calcium channels in glial cells. Science **226:** 1345–1347.

10. ORKAND, R. K., J. G. NICHOLLS & S. W. KUFFLER. 1966. The effect of nerve impulses on the membrane potential of the glial cells in the central nervous system of amphibia. J. Neurophysiol. **29:** 788–806.
11. NEWMAN, E. A. 1986. High potassium conductance in astrocyte endfeet. Science **233:** 453–454.
12. REICHENBACH, A. 1991. Glial K^+ permeability and CNS K^+ clearance by diffusion and spatial buffering. Ann. N.Y. Acad. Sci. This volume.
13. KEMPSKI, O., S. VON ROSEN, H. WEIGT, F. STAUB & A. BAETHMANN. 1991. Glial ion transport and volume control. Ann. N.Y. Acad. Sci. This volume.
14. COLES, J. A., R. K. ORKAND & C. L. YAMATE. 1989. Chloride enters glial cells and photoreceptors in response to light stimulation in the retina of the honey bee drone. GLIA **2:** 287–297.
15. TANG, C. M., M. W. COHEN & R. K. ORKAND. 1980. Electrogenic pumps in axons and neuroglia and extracellular potassium homeostasis. Brain Res. **194:** 283–286.
16. ASTION, M. L., A. CHVATAL & R. K. ORKAND. 1989. Na^+/H^+ exchange in glial cells of Necturus optic nerve. Neurosci. Lett. **107:** 167–172.
17. KIMELBERG, H. K. & B. R. RANSOM. 1986. Physiological and pathological aspects of astrocytic swelling. *In* Astrocytes, Vol. 3: Cell Biology and Pathology. S. Federoff & A. Vernadakis, eds.: 129–166. Academic Press. New York.
18. KARWOSKI, C. J., L. HONG-KAI & E. A. NEWMAN. 1989. Spatial buffering of light-evoked potassium increases by retinal Muller (glial) cells. Science **244:** 578–580.
19. SCHLUE, W.-R. 1991. Glial H^+ transport and control of pH. Ann. N.Y. Acad. Sci. This volume.
20. ATTWELL, D., B. BARBOUR, H. BREW, M. SARANTIS & M. SZATKOWSKI. 1991. Electrogenic glutamate uptake: Basic properties, modulation and failure in anoxia. Ann. N.Y. Acad. Sci. This volume.
21. CSERR, H. F. (Ed.) 1986. The neuronal microenvironment. Ann. N.Y. Acad. Sci. **481:** 1–391.
22. ASTION, M. L. & R. K. ORKAND. 1988. Evidence for electrogenic Na/HCO_3 cotransport in glial cells in Necturus optic nerve. Glia **1:** 355–357.
23. ORKAND, P. M. & E. KRAVITZ. 1971. Localization of the sites of γ-aminobutyric acid (GABA) uptake in lobster nerve-muscle preparations. J. Cell Biol. **49:** 75–89.
24. CURRIE, D. N. & J. N. KELLY. 1981. Glial versus neuronal uptake of glutamate. J. Exp. Biol. **95:** 181–193.
25. TREHERNE, J. E. (Ed.) 1981. Glial-neurone interactions. J. Exp. Biol. **95:** 1–240.

Five Electrophysiological Properties of Glial Cells

B. A. BARRES[a]

University College London
Department of Biology
London WC1E 6BT, England

Why has neurobiology failed to figure out what glia do? Guesses about glial function have been difficult to test. Central problems have included the relative inaccessibility and complexity of the nervous system, the small size of mammalian glial cells, and the difficulty of identifying the cell type studied. Early glial electrophysiologists managed to circumvent these difficulties in a variety of clever ways: large neuroglial cells in invertebrates and relatively accessible regions of brain containing only glial cells, such as optic nerve, were studied. These studies, which utilized microelectrode-recording techniques, established that glial cells are generally electrically coupled, have more negative resting potentials than do neurons, have predominantly a potassium membrane permeability, and are not excitable.[1]

More recently, advances in cell culture, cell dissociation techniques, and the identification of glial cell types, as well as the development of patch-clamp recording, have allowed further electrophysiological characterization of identified types of mammalian glial cells from many regions of the nervous system. (Results from a large variety of preparations have recently been reviewed.[2]) These studies have continued to extend our understanding of glial membrane properties and have once again begun to reveal electrophysiological properties common to many or all glial cell types. Five of these newly uncovered properties will be described.

GLIAL CELLS EXPRESS VOLTAGE-DEPENDENT ION CHANNELS *IN VITRO*

Voltage-dependent ion channels in glia were first observed in Schwann cells in culture[3] and in cortical astrocytes in culture.[4] Since then, they have been found in every glial cell type in culture that has been examined with patch-clamp recording.[2] A large amount of data from a variety of glial cells has now accumulated, and several themes are recurrent. First, many types of ion channels are found in glia. All of the main types of voltage-dependent ion channels found in neurons have been described in glia, although all channel types are not always found in one glial cell type. Thus, sodium, potassium, and chloride channels are all found in glia, including several different types of each. Recently, neurotransmitter-gated channels, including glutamate and GABA-gated channels, have also been found in glia.[5,6] As occurs in neurons, specific ion channel types are localized to glial processes.[7–9]

A second theme is that each glial cell type (where a homogeneous population of glial cells can be identified) has been found to express its own unique complement of

[a] Address for correspondence: University College London, Department of Biology, Medawar Building B3, Gower Street, London WC1E 6BT, England.

ion channel types.[2] It is not clear yet whether these phenotypes will differ across species because most studies have been performed with rodent cells. However, it is interesting that oligodendrocytes, the one glial cell type so far extensively studied in several species (rodents, sheep, cows, and humans), have membrane properties that appear to be remarkably conserved.[10–13]

Heterogeneity of ion channel phenotype is well illustrated by studies of rat optic nerve glia. Here, three types of glial cells can be identified by morphology in culture and antigenic phenotype.[14] Type-1 astrocytes in culture express mainly a delayed rectifying potassium channel, K_D.[9] In contrast, type-2 astrocytes express at least seven different voltage-dependent ion channels including two forms of sodium current, Na_N and Na_G, two types of calcium current, Ca_L and Ca_T, and three types of potassium current, K_{IR}, K_D, and K_A.[10] Oligodendrocytes express three types of potassium current, K_{IR}, K_D, and K_A.[10,16] A fourth type of cell in optic nerve, the O-2A glial progenitor cell,[17] also has its own distinct ion channel phenotype.[16] The electrophysiological properties of each of these cells are sufficiently characteristic that by themselves they allow accurate identification of each optic nerve glial cell type.

As previously reported for some neurons, variations in composition of the culture medium may alter the relative densities or types of glial ion channels detected,[16,17] raising the issue of how *in vitro* and *in vivo* phenotypes compare.

GLIAL CELLS EXPRESS VOLTAGE-DEPENDENT ION CHANNELS *IN VIVO*

Are voltage-dependent and neurotransmitter-gated ion channels present in glia *in vivo?* Glial cells *in situ* have been successfully studied using microelectrodes, but the extensive coupling of glial cells limits the amount of information that can be obtained this way. Thus, to address this question, recordings have generally been obtained from acutely isolated cell preparations: channels present immediately after dissociation and prior to culture are assumed to have been present *in vivo.* (Fortunately the functionally important parts of voltage-gated channels are buried in the cell membrane and thus protected from enzymatic digestion.) In addition, in a few instances slice preparations have been successfully utilized[18,19] (and these have not yielded results significantly different from the acutely isolated cell preparations).

In all cases where acutely isolated glial cells have been studied with the patch-clamp technique, a large variety of voltage-dependent ion channels (and in some cases neurotransmitter-gated channels[16,20]) have been observed.[2] For instance, voltage-dependent ion channels are found in each glial cell type in the rat optic nerve including type-1 astrocytes, oligodendrocytes, and O-2A glial progenitor cells.[9,16] Because of the specific localization of some ion channel types to glial cell processes, these studies used a new type of enzymatic cell dissociation procedure called tissue printing that yields cells still bearing processes.[9] Each acutely isolated optic nerve glial cell type continues to express a characteristic complement of ion channel types.

It is not yet clear in Schwann cells, Müller cells, or cortical astrocytes how this ion channel expression in acutely isolated cells compares with expression by the same cell type *in vitro.* This question has been addressed for optic nerve glial cells in the rat optic nerve: Cells of the O-2A lineage, O-2A progenitors and oligodendrocytes, express the same ion channel types *in vitro* and *in vivo.*[9,16] However, type-1 astrocytes express a simple ion channel phenotype *in vitro,* but they were found to express a more complex phenotype *in vivo* (see below).[9]

THE GLIAL RESTING POTENTIAL IS DETERMINED BY INWARDLY RECTIFYING POTASSIUM CHANNELS AND NOT "LEAKAGE" CHANNELS

Many of the early microelectrode studies of glial cells observed large, apparently voltage-independent potassium currents and suggested that they could be accounted for by a leakage potassium conductance uniquely present in glia. However, convincing evidence of a leakage potassium current for any glial cell type has never been reported when patch recording has been used and where only voltage-dependent currents have been found. (Several early reports of voltage-independent potassium currents in oligodendrocytes and Müller cells were followed in each instance by a report that the corresponding channels were actually voltage dependent.) Studies involving microelectrode recordings from glia in a coupled glial syncytium may have simply been unable to achieve the large depolarizations or hyperpolarizations necessary to reveal the underlying rectifying potassium currents.[2]

Despite this, the current/voltage relationships of many glial cell types studied with patch-clamp recording do appear roughly linear in the physiological range. One explanation may be that the inwardly rectifying and outwardly rectifying potassium channels sum to give a linear response. Evidence for this possibility has been obtained in recent studies of rat optic nerve glia: specific blockers of the inwardly rectifying or outwardly rectifying potassium current components reveal the underlying asymmetric voltage-dependent currents. (For oligodendrocytes and type-2 astrocytes, see Fig. 4 in ref. 16; for type-1 astrocytes, see Fig. 7 in ref. 9.)

The conductances underlying the resting potential in neural cells have been poorly understood. An early suggestion, that in glia the "leakage" voltage-independent potassium channels formed the resting conductance, is no longer tenable. The resting potential of glial cells has generally been measured to be about -70 to -90 mV. To date, the only type of voltage-dependent ion channel found in glia that is active at these potentials is the inwardly rectifying potassium channel K_{IR}. K_{IR} has been observed in all glial cell types studied so far[2] and therefore seems virtually certain to underlie the glial resting potential. The main evidence is that cesium, a specific blocker of the inwardly rectifying potassium channel, virtually eliminates glial resting conductance *in vitro* and in acutely isolated glia.[9,16] Definitive proof that this is true *in vivo,* however, has not been presented.

Could inwardly rectifying channels also underlie the resting potential in neurons? Many neurons have been reported to express inwardly rectifying potassium channels, and in some cases these appear similar to those in glia, having similar voltage dependence and potassium selectivity. However, another type of inwardly rectifying channel is found in at least some neurons that is permeable to both sodium and potassium.[21,22] Inasmuch as this channel is also active at negative membrane potentials, it likely contributes to the resting potential of neurons: this makes sense, because their resting potential is generally positive to that of glia, a phenomenon caused by a higher relative resting sodium permeability.

In whole-cell patch studies of both glia and neurons, there has been an unfortunate tendency to (inadvertently) subtract away the inwardly rectifying currents as leakage current, because they often appear linear when extracellular potassium is 3–5 mM. Hopefully, as it becomes increasingly clear that these currents are often present in glia and in neurons, we shall rapidly learn more about their cell type distribution.

SOME GLIAL ION CHANNELS DIFFER FROM THEIR NEURONAL COUNTERPARTS AND APPEAR TO BE FUNCTIONALLY SPECIALIZED

Not surprisingly, all of the glial channels discovered so far have properties that are qualitatively similar to those of the channel types found in neurons. In most cases, the properties of glial and neuronal channels have not yet been subjected to a rigorous comparison. There are, however, three instances in which they have been compared. In each case the same general results have been obtained: glial and neuronal channels are qualitatively similar but quantitatively distinct.[5,6,23]

Two of these instances involve neurotransmitter-gated channels. Both glutamate- and GABA-gated ion channels are found on glial cells: glutamate-gated channels of the non-NMDA type are found on type-2 astrocytes and O-2A glial progenitor cells.[5] Non-NMDA agonist activate at least five substates of this channel, but which substates are preferentially activated depends on whether the cell is neuronal or glial.[5] A different kind of variation has been reported in glial $GABA_A$-gated channels. Inverse agonists such as beta-carboline modulate the gating kinetics of the chloride channels activated by GABA: in neurons net currents are decreased, whereas in cortical astrocytes currents are increased.[6]

Differences between voltage-gated sodium channels in neurons and glia have been described.[23] Whole-cell and single-channel currents of acutely isolated retinal ganglion cells and optic nerve type-1 astrocytes were compared under identical recording conditions. Both channels have voltage-dependent activation and inactivation, identical tetrodotoxin sensitivity, and single-channel conductance. But the voltage dependence of the glial channel is shifted in a hyperpolarizing direction, by 10 mV for activation and by 25 mV for steady-state inactivation. The glial channels also have slower kinetics; they activate four times more slowly and inactivate two times more slowly. In type-2 astrocytes, both the glial and the neuronal forms of the sodium channel are present.[23] Interestingly, in both cases the voltage at which half of the glial and neuronal sodium channels is available for activation is near the resting potential of each cell type, −80 and −55 mV, respectively. This finding suggests that the glial sodium channel may be functionally specialized.

It is possible that other glial voltage-dependent channels also differ from those in neurons. Already mentioned herein, the type of inwardly rectifying potassium channel found in glia and in neurons tends to differ. Another interesting difference is that when inactivating outward potassium channels, K_A, have been described in glia, they invariably have steady-state inactivation curves that center at −70 mV; that is, the channels are half available for activation at −70 mV. In neurons, the half-inactivation potential has generally been reported to be more negative, at about −90 mV. Thus, at the resting potential of glia, most K_A channels will be available for activation, whereas in neurons only a fraction of them will be. It is unclear what functional significance this has.

Although some differences between neuronal and glial channels have been described, these studies are just beginning. We still do not understand their molecular basis or functional significance, nor do we know if these different types of channels are actually confined to glial and neuronal cells or if some neuron types will turn out to express what are currently thought of as glial forms of the channel and vice versa.

NEURONS UP-REGULATE EXPRESSION OR ACTIVITY OF CERTAIN ION CHANNEL TYPES IN GLIAL CELLS

Some glial cells in culture have been shown to express the same ionic currents *in vitro* as after acute isolation, suggesting that their basic ion channel phenotype is intrinsically programmed and not sensitive to extracellular influences.[9,16] In the case of several glial cell types, however, differences have been observed in channel expression (or activity) depending on whether they are *in vitro* and *in vivo,* and recent evidence in several different preparations has traced this phenomenon to an up-regulatory influence of neurons on glial ion channels.[9,24–26]

The first indication that neurons might up-regulate glial ion channels came from studies of glia in the peripheral nervous system. Transection of sciatic nerve results in the appearance of sodium channels in myelinating Schwann cells and in the loss of sodium channels from nonmyelinating Schwann cells.[24] Recently, a similar effect was demonstrated *in vitro:* Embryonic neurons cocultured with cortical astrocytes induced the appearance of two types of voltage-dependent channels, Ca_T and Ca_L, that are not otherwise detectable in purified astrocytes.[25]

Similar findings were recently observed in type-1 astrocytes from rat optic nerve. In cultured and acutely isolated type-1 astrocytes from early postnatal nerves, the main current present is the delayed rectifying potassium current, K_D.[9] In contrast, type-1 astrocytes isolated after postnatal day 10 (P10) have a more complex phenotype: K_D continues to be present, and in addition the glial form of the sodium current, Na_G, and an inwardly rectifying potassium current, K_{IR}, is present in all cells.[9]

To determine whether the change was purely intrinsic to astrocyte aging, P2 type-1 astrocytes were cultured for 10 days and then studied.[9] They continued to express the simpler phenotype, suggesting that extrinsic factors are involved. To test this, optic nerves were transected at P3, and type-1 astrocytes were studied at P10. K_{IR} and Na_G were not present, suggesting that an extrinsic factor, most likely emanating from or dependent on neurons, is required for the developmental change in astrocyte phenotype. This factor must be required to maintain the change once it has taken place: when transections were performed at P12 and the type-1 astrocytes studied 8 days later, Na_G and K_{IR} were once again absent.

To test directly whether neurons can up-regulate expression of glial ion channel types, P10 retinal ganglion cells purified by panning to near homogeneity, were cocultured with P10 type-1 astrocytes.[9] After 2 days of culture, sodium currents were found in 95% of the type-1 astrocytes contacting neurons, but in only 25% of cells in the neuron-free cultures. The density of K_{IR} increased sixfold in type-1 astrocytes contacting neurons in cocultures. Preliminary experiments suggest that these effects may be contact mediated: in 10 cells, type-1 astrocytes that did not contact neurons lacked both Na_G and K_{IR}.

Remarkably similar results were recently obtained using immunohistochemical rather than electrophysiological techniques to follow the expression of sodium channels.[26] Sodium channel immunoreactivity progressively diminished with time in culture in optic nerve type-1 astrocytes, but not in type-2 astrocytes. These results suggest that it is the expression of the channel rather than its activity that is modulated by neurons.

Why is ion channel expression by type-1 astrocytes so plastic? A possibility is that the channels arise, temporally and spatially, in response to neuronal requirements. For instance, the inwardly rectifying potassium channel K_{IR}, implicated in glial potassium homeostasis,[2] develops in type-1 astrocytes at about the same time as the nerve develops the ability to buffer neuronally released potassium.[27]

Is it possible that expression of ion channels by neurons is regulated by glia?

CONCLUSION

How do the results of the microelectrode studies and the patch-clamp studies of glia compare? Both sets of studies indicate that the glial membrane is mainly permeable to potassium ions, that glia have a resting potential generally negative to that of neurons, and that glia in general are inexcitable. Patch-clamp studies not only have confirmed the microelectrode work, but also have continued to extend our knowledge of glial membrane properties. They have revealed that although the glial membrane is mainly permeable to potassium, this conductance can be attributed to voltage-dependent potassium channels rather than to "leakage" voltage-independent potassium channels. In addition, they have shown that all glial cells are not electrophysiologically homogeneous; rather, each glial cell type has its own distinct voltage-dependent ion channel phenotype. Thus glia do not appear quite as electrophysiologically distinct from neurons as initially thought.

The microelectrode studies, on the other hand, have allowed study of glial cells *in situ* in relatively intact preparations, in the absence or presence of neuronal activity, a feat not yet achieved with patch electrodes (but see ref. 28). A problem with *in situ* recordings has been that the extensive coupling of glial cells prevents adequate voltage control of the cell studied. Thus events occurring in glial processes will be more difficult to detect. It has often been suggested that in some neuron types information processing occurs at distal processes where the activity is electrically remote from the soma and thus not easy to detect. The specific localization of many voltage-dependent and neurotransmitter-gated ion channel types to glial processes suggests that this possibility must also be considered for glial cells. Thus an important goal is to develop new preparations and techniques that allow activity occurring at distal processes to be measured.

The presence of voltage-dependent and neurotransmitter-gated channels in glial cells has not yet taught us more about glial function, but it does suggest that previously unsuspected neuronal-glial signaling processes may be occurring. Thus, electrophysiological studies of glial cells and their interactions with neurons should continue to be an informative approach. Deducing the function of glia is likely to remain one of the most formidable—and inviting—neurobiological questions for a long time to come.

REFERENCES

1. KUFFLER, S. W. & D. D. POTTER. 1964. Glia in the leech central nervous system: Physiologic properties and neuron-glia relationship. J. Neurophysiol. **27:** 290–320.
2. BARRES, B. A., L. L. Y. CHUN & D. P. COREY. 1990. Ion channels in vertebrate glia. Ann. Rev. Neurosci. **13:** 441–474.
3. CHIU, S. Y., P. SCHRAGER & J. M. RITCHIE. 1984. Neuronal-type sodium and potassium channels in rabbit cultured Schwann cells. Nature **311:** 156–157.
4. BEVAN, S., S. Y. CHIU, P. T. A. GRAY & J. M. RITCHIE. 1985. The presence of voltage-gated sodium, potassium and chloride channels in rat cultured astrocytes. Proc. R. Soc. Lond. B **225:** 299–313.
5. USOWICZ, M. M., V. GALLO & S. G. CULL-CANDY. 1989. Multiple conductance channels in type-2 cerebellar astrocytes activated by excitatory amino acids. Nature **339:** 380–383.
6. BORMANN, J. & H. KETTENMANN. 1988. Patch-clamp study of GABA receptor Cl channels in cultured astrocytes. Proc. Natl. Acad. Sci. USA **85:** 9336–9340.
7. NEWMAN, E. A. 1984. Regional specialization of retinal glial cell membrane. Nature **309:** 155–157.
8. BREW, H., P. T. A. GRAY, P. MOBBS & D. ATTWELL. 1986. Endfeet of retinal glial cells have higher densities of ion channels that mediate K buffering. Nature **324:** 466–468.
9. BARRES, B. A., W. J. KOROSHETZ, L. L. Y. CHUN & D. P. COREY. 1990. Ion channel expression by white matter glia: The type-1 astrocyte. Neuron, in press.
10. BARRES, B. A., L. L. Y. CHUN & D. P. COREY. 1988. Ion channel expression by white matter glia: I. Type 2 astrocytes and oligodendrocytes. Glia **1:** 10–30.
11. SOLIVEN, B., S. SZUCHET, B. G. ARNASON & D. J. NELSON. 1988. Voltage-gated potassium currents in cultured ovine oligodendrocytes. J. Neurosci. **8:** 2131–2141.

12. McLarnon, J. G. & S. U. Kim. 1989. Existence of inward potassium currents in adult human oligodendrocytes. Neurosci. Lett. **101:** 107–112.
13. McLarnon, J. G. & S. U. Kim. 1989. Single channel potassium currents in cultured adult bovine oligodendrocytes. Glia **2:** 298–307.
14. Miller, R. H., C. ffrench-Constant & M. C. Raff. 1989. The macroglial cells of the rat optic nerve. Ann. Rev. Neurosci. **12:** 517–534.
15. Raff, M. C., R. H. Miller & M. Noble. 1983. A glial progenitor cell that develops *in vitro* into an astrocyte or an oligodendrocyte depending on culture medium. Nature **303:** 390–396.
16. Barres, B. A., W. J. Koroshetz, K. J. Swartz, L. L. Y. Chun & D. P. Corey. 1990. Ion channel expression by white matter glia: The O2A glial progenitor cell. Neuron **4:** 507–524.
17. Barres, B. A., L. L. Y. Chun & D. P. Corey. 1989. A calcium current in cortical astrocytes: Induction by cAMP and neurotransmitters, and permissive effect of serum factors. J. Neurosci. **9:** 3169–3175.
18. Newman, E. A. 1985. Voltage-dependent calcium and potassium channels in retinal glial cells. Nature **317:** 809–811.
19. Kettenmann, H., G. V. Blankenfeld & J. Trotter. 1990. Ion channel expression in the oligodendrocyte lineage. Ann. N.Y. Acad. Sci. This volume.
20. Malchow, R. P., H. Qian & H. Ripps. 1989. GABA-induced currents of skate Muller (glial) cells are mediated by neuronal-like GABA-A receptors. Proc. Natl. Acad. Sci. USA **86:** 4326–4330.
21. Baker, M., H. Bostock, P. Grafe & P. Martius. 1987. Function and distribution of three types of rectifying channel in rat spinal root myelinated axons. J. Physiol. **383:** 45–67.
22. Eng, D. L., T. R. Gordon, J. D. Kocsis & S. G. Waxman. 1990. Current-clamp analysis of a time-dependent rectification in rat optic nerve. J. Physiol. **421:** 185–202.
23. Barres, B. A., L. L. Y. Chun & D. P. Corey. 1989. Glial and neuronal forms of the voltage-dependent sodium channel: Characteristics and cell-type distribution. Neuron **2:** 1375–1388.
24. Chiu, S. Y. 1988. Changes in excitable membrane properties in Schwann cells of adult rabbit sciatic nerves following nerve transection. J. Physiol. **396:** 173–188.
25. Corvalan, V., R. Cole, J. de Vellis & S. Hagiwara. 1990. Neuronal modulation of calcium channel activity in cultured rat astrocytes. Proc. Natl. Acad. Sci. USA **87:** 4345–4348.
26. Minturn, J. E., J. A. Black, K. J. Angelides & S. G. Waxman. 1990. Sodium channel expression detected with antibody 7493 in A2B5-positive and A2B5-negative astrocytes from rat optic nerve *in vitro.* Glia **3:** 358–367.
27. Ransom, B. R., C. L. Yamate & B. W. Connors. 1985. Activity-dependent shrinkage of the extracellular space in rat optic nerve. J. Neurosci. **5:** 532–535.
28. Marrero, H., M. L. Astion, J. A. Coles & R. K. Orkand. 1989. Facilitation of voltage-gated ion channels in frog neuroglia by nerve impulses. Nature **339:** 378–380.

Membrane-Associated Sodium Channels and Cytoplasmic Precursors in Glial Cells

Immunocytochemical, Electrophysiological, and Pharmacological Studies[a]

J. E. MINTURN,[b] H. SONTHEIMER,[b] J. A. BLACK,[b]
K. J. ANGELIDES,[d] B. R. RANSOM,[b] J. M. RITCHIE,[c]
AND S. G. WAXMAN[b,c,e]

Departments of Neurology[b] and Pharmacology[c]
Yale University School of Medicine
New Haven, Connecticut 06510

[b]PVA/EPVA Neuroscience Research Center
VA Medical Center
West Haven, Connecticut 06516

[d]Department of Molecular Physiology and Biophysics
Baylor College of Medicine
Houston, Texas 77030

Voltage-sensitive sodium channels have been demonstrated in astrocytes from a number of sites in the central nervous system (CNS) studied *in vitro.*[1-4] These astrocyte sodium channels display kinetic and gating properties and pharmacological characteristics that are qualitatively similar to those of sodium channels in neurons. However, despite the ubiquitous distribution of astrocytes in both the gray matter and white matter of the CNS and their contribution to the response of the CNS to injury, little is known about the spatial distribution and time course of expression of astrocyte sodium channels and the mechanisms that regulate ion channel deployment in astrocytes.

As part of an investigation of ion channels in astrocytes, we have used new immunocytochemical methods[5-7] to study sodium channel expression in cultured astrocytes derived from the rat optic nerve, a white matter tract.[8] In parallel studies using the patch-clamp method to measure ionic currents, we studied electrophysiologic aspects of sodium channel expression in these cells. The present report summarizes these morphophysiological studies, which demonstrate that different types of astrocytes display different patterns of sodium channel expression, and it examines the relation between the immunocytochemical and physiological expression of sodium channels in optic nerve astrocytes. We then summarize data that

[a]This work was supported in part by grants from the National Institutes of Health, the National Multiple Sclerosis Society, and the Medical Research Service, Department of Veterans Affairs. J.E.M. was a Medical Student Research Fellow of the American Heart Association. H.S. was a Spinal Cord Research Fellow of the EPVA.

[e]Address for correspondence: Stephen G. Waxman, M.D., Ph.D., Department of Neurology, LCI 707, Yale University School of Medicine, 333 Cedar Street, New Haven, CT 06510.

indicate that sodium channels are expressed in astrocytes *in situ* in rat optic nerve.[5,6] These immunocytochemical studies suggest that a subpopulation of astrocyte sodium channels, or precursor molecules, are located not only in the plasma membrane, but also in a cytoplasmic pool. We therefore also review recent results[9] indicating that at least two populations of sodium channels or channel precursors are present in glial cells, one associated with the plasma membrane and the other localized within the cytoplasm.

IMMUNOCYTOCHEMICAL DEMONSTRATION OF SODIUM CHANNELS IN OPTIC NERVE ASTROCYTES *IN VITRO*

Two populations of astrocytes have been described in tissue culture preparations derived from developing rat optic nerve on the basis of morphology, growth characteristics, and the ability to bind the monoclonal antibody A2B5.[10] Moreover, it has been suggested that these cultured astrocytes originate from distinct populations of precursor cells. Raff *et al.*[10,11] suggest that A2B5$^-$ astrocytes (which they term type 1) are derived from neuroepithelial cells of the optic stalk, whereas A2B5$^+$ astrocytes (termed type 2) are derived from O-2A cells, which also yield oligodendrocytes. In the immunocytochemical studies[8] described here, we examined sodium channel expression in A2B5$^+$ and A2B5$^-$ astrocytes from developing optic nerve, maintained in culture for up to 28 days.

For these studies, optic nerves (excluding the chiasm) were dissected from Wistar rats on postnatal day 0 (P0) or postnatal day 7 (P7) and were incubated for 45 minutes at 37°C in an enzyme solution containing Earle's salts, 30 units/ml papain, 1.5 mM calcium chloride, 0.5 mM EDTA, and 1.65 mM L-cysteine. Following trituration, the cell suspension was plated onto collagen/poly-D-lysine-coated coverslips (100,000 cells/ml). The cells were maintained in Earle's MEM containing 5% fetal bovine serum, serum extender (Collaborative Research), penicillin/streptomycin (500 units/ml), and glucose (20 mM) at 37°C in a 5% CO_2/95% air atmosphere.

Sodium channel expression in cultured cells was examined using immunocytochemical staining with polyclonal antibody 7493,[5,6] which was generated against purified rat brain sodium channel protein.[12] Previous studies demonstrated that this antibody immunoblots and immunoprecipitates, from labeled membrane proteins, a 260-kD polypeptide corresponding to the sodium channel α subunit, and immunoprecipitates sodium channels saturated with [^{3}H]-saxitoxin in a concentration-dependent manner; this polyclonal antibody, however, does not recognize sodium channels in cardiac or skeletal muscle.[7] FIGURE 1 is a Western blot demonstrating that antibody 7493 selectively immunostains a protein with an M_r of 260 kD, corresponding in migration and mobility to the α subunit of the rat brain sodium channel, after separation of crude glycoproteins from adult rat optic nerve by SDS-PAGE. Preimmune sera did not react with the separated glycoproteins of optic nerve. Our earlier immunoultrastructural studies[5] demonstrated that antibody 7493 specifically stains the axon membrane at the node of Ranvier, where sodium channels are known to be clustered in high density.[13] Antibody 7493 thus provides a specific immunocytochemical marker that permits localization of sodium channels at both the light and electron microscopic level.

In the present investigation of cultured optic nerve astrocytes, cells were fixed on ice for 5 minutes with 4% paraformaldehyde-0.01% glutaraldehyde and then prepared for indirect immunofluorescence microscopy after exposure to polyclonal antibody 7493, mouse monoclonal anti-GFAP (an IgG, Boehringer-Mannheim), and, in some experiments, mouse monoclonal A2B5 (an IgM, American Type

Culture Collection). Appropriate secondary antibodies were applied simultaneously and included fluorescein-conjugated goat anti-rabbit IgG, rhodamine-conjugated goat anti-mouse IgG or 7-amino-4 methylcoumarin-3-acetic acid-conjugated goat anti-mouse IgG, and rhodamine-conjugated goat anti-mouse IgM.

POPULATION OF CELLS DISSOCIATED FROM P7 OPTIC NERVE *IN VITRO* CHANGES OVER TIME

P7-derived optic nerve cultures contain a dynamic population of astrocytes and astrocyte precursor cells that changes over time. When cells obtained from P7 rats were examined at 2 days *in vitro* (DIV), the population, as determined by indirect immunofluorescence, consisted of ~30–50% GFAP$^+$ cells (astrocytes), with the remainder of cells being round, bi-, and tripolar GFAP$^-$/A2B5$^+$ cells (O-2A progenitor cells according to the terminology of Raff *et al.*[10]) At this early stage in culture, 50–60% of the cells expressing GFAP were A2B5$^-$ and had a flat morphology, whereas 40–50% were A2B5$^+$ and were process bearing. At 4 DIV, ~80% of cells expressed GFAP, and the majority of astrocytes were A2B5$^-$/GFAP$^+$, with a fibroblast-like morphology, the remainder being process-bearing or stellate A2B5$^+$/GFAP$^+$ cells.

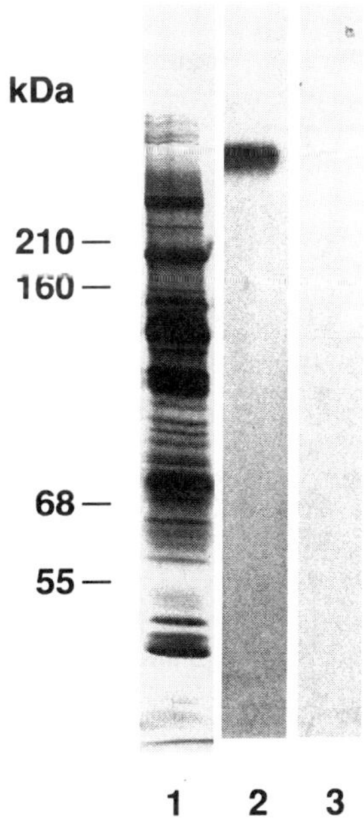

FIGURE 1. Immunoblot of adult rat optic nerve. *Lane 1:* Optic nerve glycoproteins were separated by SDS polyacrylamide gel electrophoresis and silver stained. Note the complex mixture of glycoproteins in this preparation. *Lane 2:* Glycoproteins of *lane 1* were electrophoretically transferred to nitrocellulose paper and reacted with 7493 antiserum. The antiserum selectively immunostains a protein with apparent M_r 260 kD. *Lane 3:* Glycoproteins of *lane 1* were electrophoretically transferred to nitrocellulose paper and reacted with preimmune serum. No immunostaining is present. Modified from Black *et al.*[5]

By 6 DIV, flat A2B5$^-$/GFAP$^+$ astrocytes became the dominant cell type in the cultures, with process-bearing A2B5$^+$/GFAP$^+$ astrocytes comprising 5–10% of the cell population. At 8 DIV and thereafter, the flat A2B5$^-$ astrocytes constituted >95% of the cells in culture, and these cells became confluent unless cytosine arabinoside (10 μM) was added to the medium.

A2B5⁺ AND A2B5⁻ ASTROCYTES *IN VITRO* EXHIBIT DIFFERENT PATTERNS OF SODIUM CHANNEL IMMUNOREACTIVITY

Both $A2B5^+$ and $A2B5^-$ astrocytes in P7-derived cultures display 7493 immunoreactivity at early stages *in vitro,* but with increasing time in culture, they show different patterns of sodium channel immunoreactivity. Thus, when astrocyte cultures obtained from P7 optic nerves were examined at early stages (2 and 4 DIV) using indirect immunofluorescence with antibody 7493, all cells ($A2B5^+$ and $A2B5^-$ astrocytes and $GFAP^-$ progenitor cells) exhibited 7493 immunoreactivity (FIG. 2).

By 6 DIV, however, the pattern of 7493 staining changed dramatically. At this time, $A2B5^+/GFAP^+$ astrocytes continued to exhibit 7493 immunoreactivity which was distributed throughout the cytoplasm and cell processes. However, in these same cultures, sodium channel immunoreactivity was essentially lost from cells that were $A2B5^-/GFAP^+$. This differential pattern of 7493 staining in $A2B5^+$ and $A2B5^-$ astrocytes persisted for up to 28 DIV (the oldest cultures examined). Thus, sodium channel staining of the $A2B5^+/GFAP^+$ astrocytes was present in cultures examined up to 28 DIV (FIG. 3). $A2B5^-/GFAP^+$ cells in the same cultures, however, did not reexpress sodium channel immunoreactivity with antibody 7493 with extended time in culture.

LOSS OF SODIUM CHANNEL IMMUNOSTAINING IS RELATED TO TIME *IN VITRO* AND NOT TO BIOLOGICAL AGE

The results just described indicate that 7493 immunoreactivity in $A2B5^-$ astrocytes, in P7-derived cultures, is attenuated after 6 DIV. Loss of sodium channels in astrocytes could be related to some aspect of culture conditions or could reflect a biologically timed event related to postnatal age. To distinguish between these possibilities, we studied sodium channel expression in cultured astrocytes dissociated from P0 rat optic nerves,[8] which permitted us to compare astrocytes of different postconception ages maintained for the same time *in vitro.* In these P0-derived cultures, at 2 and 4 DIV, all astrocytes exhibited sodium channel immunoreactivity. By 6 days *in vitro,* however, sodium channel staining with antibody 7493 was virtually absent from $A2B5^-$ astrocytes (FIG. 4). Moreover, immunoreactivity to antibody 7493 did not reappear in these cells at any of the later time points examined (up to 15 DIV). In contrast, $A2B5^+$ astrocytes ($<1\%$) that were present in P0-derived cultures displayed immunoreactivity to antibody 7493 at all observed times (up to 15 DIV). At 6 DIV, the P0-derived astrocytes corresponded roughly, in terms of postconception age, to newly cultured astrocytes in P7-derived cultures, which exhibit intense immunoreactivity to antibody 7493 (see above). Thus, time *in vitro,* and not biological age, appears to determine loss of 7493 immunoreactive astrocyte sodium channels in $A2B5^-$ astrocytes.

Our observation of attenuation of 7493 immunoreactivity in $A2B5^-$ astrocytes in both P0- and P7-derived cultures, after a similar latency period (6 DIV), suggests that the loss of immunostaining reflects some aspect of the culture conditions, such as the absence of a factor(s) present *in vivo,* possibly neuronal in origin, that is necessary for the expression of sodium channels that are recognized by antibody 7493.[8] In this regard, Yarowsky and Krueger[14] observed an increase in the number of high-affinity saxitoxin-binding sites associated with changing cortical astrocyte cultures from medium containing serum to a chemically defined medium. Rudy *et al.*[15] demonstrated an increase in the number of functional sodium channels induced by

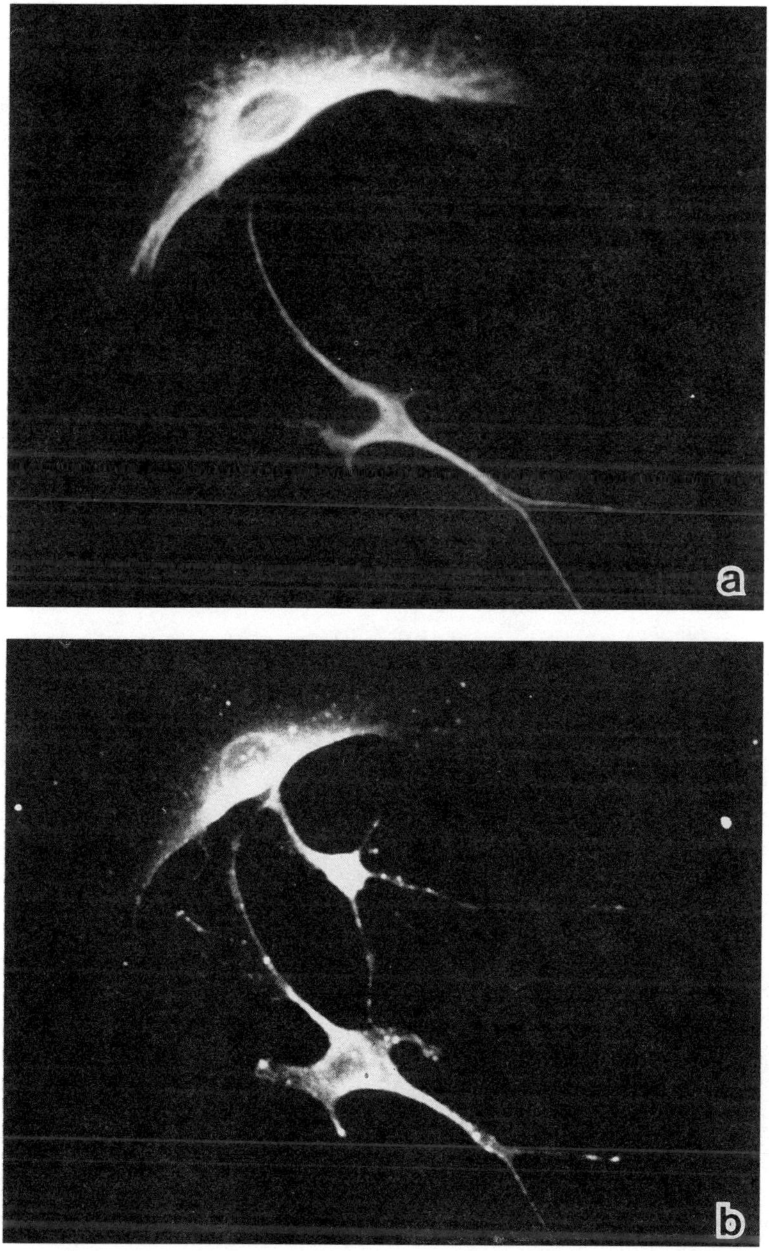

FIGURE 2. Astrocytes from P7 rat optic nerve at 2 days *in vitro* (DIV) immunostained with antibody 7493 and anti-GFAP antibody. GFAP was visualized with rhodamine and 7493 with fluorescein-conjugated secondary antibodies. At 2 DIV, three cell types are present: fibroblast-like and stellate $GFAP^+$ cells (**a**) and tripolar $GFAP^-$ cells (**b**). All cells exhibit positive immunostaining with antibody 7493 (**b**). Magnification × 900.

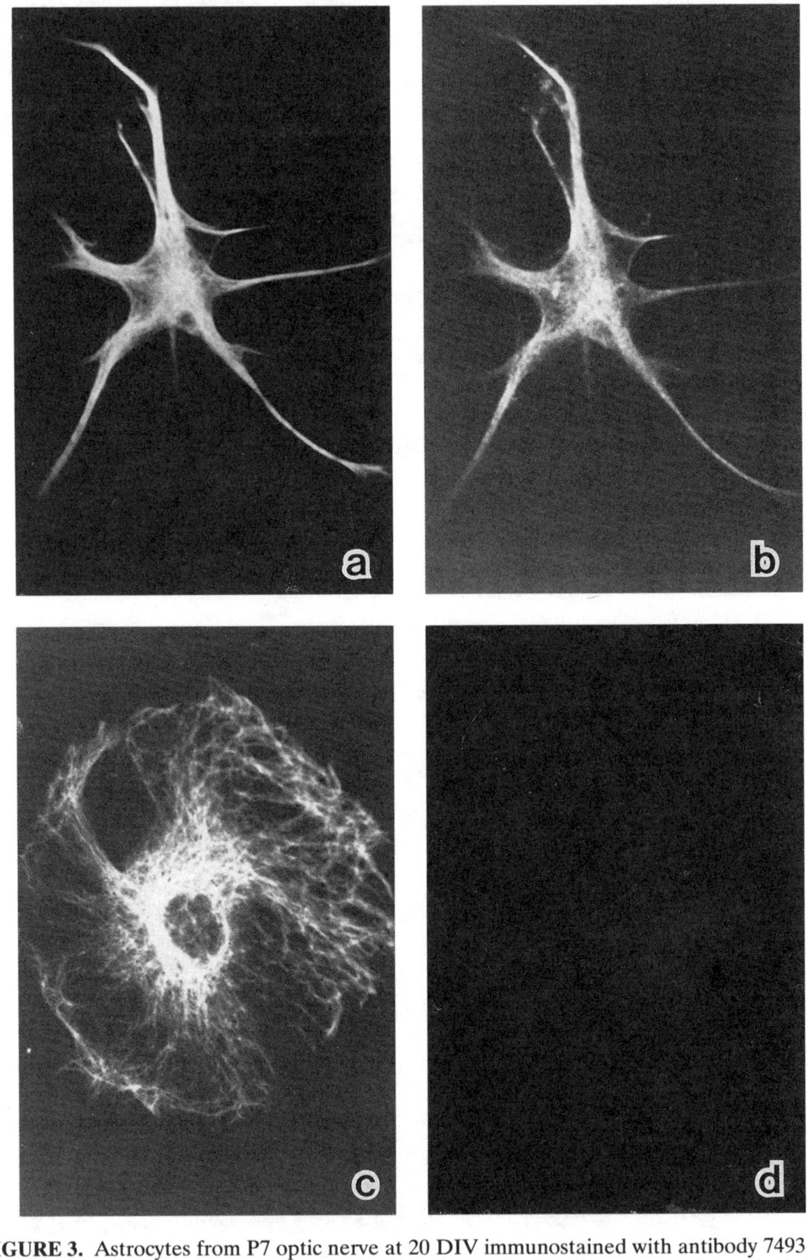

FIGURE 3. Astrocytes from P7 optic nerve at 20 DIV immunostained with antibody 7493 and anti-GFAP antibody. Astrocytes positively identified with GFAP immunostaining (**a,c**) show differential expression of sodium channels immunoreactive to antibody 7493. Stellate astrocytes continued to exhibit immunoreactivity to antibody 7493 (**b**), whereas flat-shaped astrocytes failed to react to the antibody 7493 at 20 DIV (**d**). Magnification × 700.

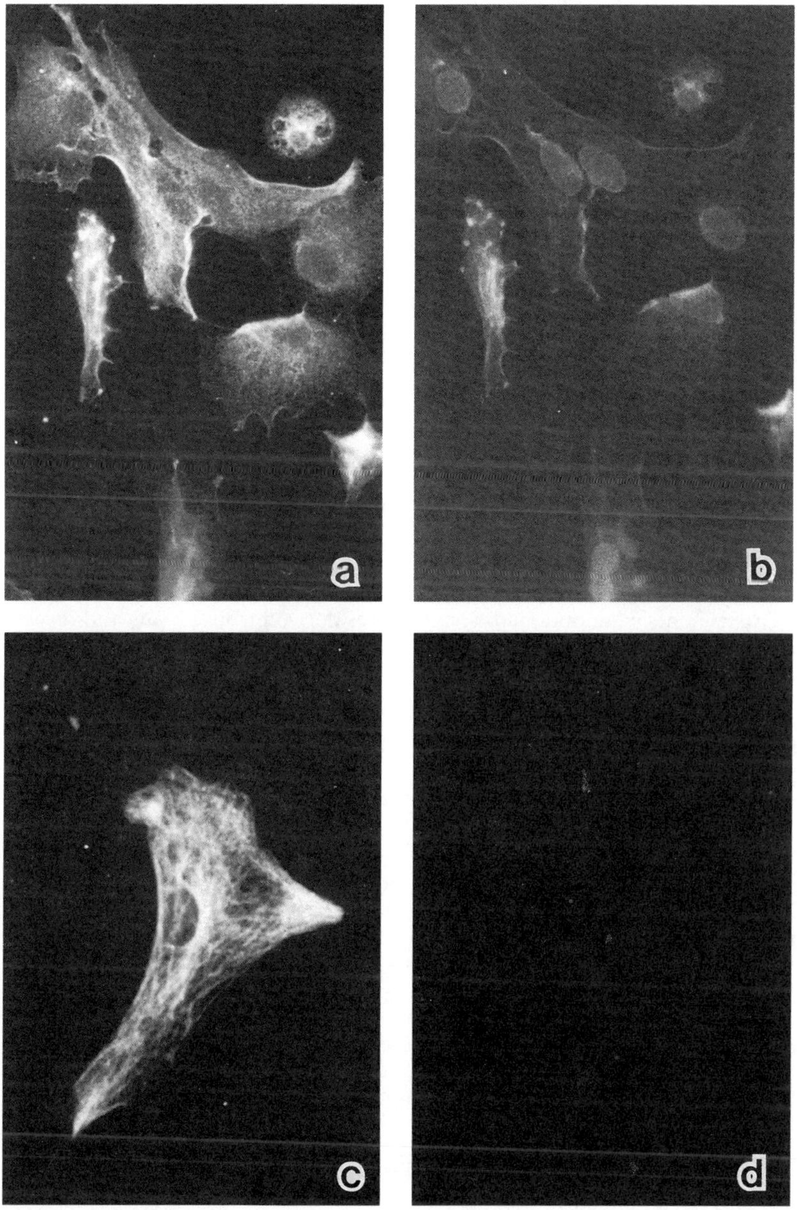

FIGURE 4. Astrocytes from P0 optic nerve at 4 DIV. Cells are labeled with anti-GFAP antibody and visualized with rhodamine (**a**). All cells at this number of days *in vitro* were immunoreactive to the 7493 antibody (**b**). Astrocyte from P0 optic nerve at 11 DIV. Fibroblast-like astrocyte stained with anti-GFAP antibody (**c**) similar to $A2B5^-/GFAP^+$ astrocytes in P7 optic nerves. Diminished immunoreactivity to antibody 7493 (**d**) was apparent in all $A2B5^-/GFAP^+$ cells examined at this time *in vitro*. Magnification × 900.

nerve growth factor (NGF) in PC12 pheochromocytoma cells in culture, and Kalman *et al.*[16] showed that NGF acts through cAMP-dependent protein kinase to increase the number of sodium channels in these cells. Moreover, recent experiments demonstrate that calcium channels are only expressed in cortical astrocytes *in vitro* when these cells are cocultured with neurons, suggesting neuronal modulation of calcium channel activity in these cells.[17] As noted below, *in situ,* all astrocytes within the rat optic nerve maintain some sodium channel immunoreactivity.[6] Taken together with the present findings and observations in other *in vitro* models, the *in situ* observations on optic nerve suggest the existence of a factor(s) in the optic nerve *in vivo* that is necessary for expression of sodium channels in $A2B5^-$ astrocytes. Interestingly, our electrophysiological results confirm the loss of voltage-sensitive sodium currents in $A2B5^-$ optic nerve astrocytes in culture, although the loss of sodium currents (which presumably reflects the loss of functional sodium channels inserted in the plasma membrane) lags behind the loss of sodium channel immunoreactivity by 4–6 days.

ELECTROPHYSIOLOGICALLY MEASURED SODIUM CURRENTS CHANGE OVER TIME IN $A2B5^-$ ASTROCYTES IN CULTURE, BUT LAG BEHIND IMMUNOCYTOCHEMICAL CHANGES

Using the whole cell mode of the patch-clamp technique, we studied sodium current expression in cultured rat optic nerve astrocytes during their development *in vitro.* Experiments were obtained from cultures derived from P7 rats and recordings were obtained throughout the first 20 DIV. Both $A2B5^+$ and $A2B5^-$ astrocytes were typically encountered and studied. To confirm morphological identification of cell types from which recordings were obtained, cells were filled with Lucifer yellow, which was included in the recording pipette, and then fixed and immunostained with antibody A2B5. Pipette and bath solutions used for current recording were designed to isolate inward currents carried by Na^+ ions. The pipette solution contained (in mM): *N*-methyl-D-glucamine (NmDg) 105, NaCl 10, TEA-Cl 20, $MgCl_2$ 1, $CaCl_2$ 0.2, EGTA 10, Hepes (sodium salt) 20, and Lucifer yellow 2 (dilithium salt, Sigma). The pH was adjusted to 7.4 using hydrochloric acid. The bath solution contained (in mM): NaCl 130.0, $MgSO_4$ 1.2, $CaCl_2$ 1.0, Na_2HPO_4 1.6, NaH_2PO_4 0.4, glucose 10.5, and Hepes (acid) 32.5. The pH was adjusted to 7.4 with NaOH. For all measurements, series resistance errors were minimized to reduce residual voltage errors to less than 5%. The recording configuration used eliminated liquid junction potentials.

To activate inward currents, astrocytes were voltage clamped at −70 mV and voltage steps were applied to more positive membrane voltages between −60 and 60 mV. Voltage steps were preceded by a 200-ms step to −110 mV to remove sodium current inactivation. Depolarizing voltage steps activated inward currents as the test potential became more positive than −50 mV, and peak currents were recorded in response to voltage steps between −20 and −10 mV. For more positive potentials, current amplitude decreased and currents reversed at 40 mV, close to the sodium equilibrium potential for the imposed ion gradients (E_{Na^+} = 42 mV), clearly identifying currents as sodium currents.

To characterize these currents further, we studied steady-state current inactivation. Cells were held at −70 mV, and 30-ms voltage steps were applied to −20 mV, the potential of peak current activation. These (activating) voltage steps were preceded by a 200-ms step to various prepulse potentials ranging from −110 to −30 mV. Peak currents were typically recorded in response to very negative prepulse potentials, such as −110 mV. As the prepulse potential became more positive, current amplitude decreased. To compare the steady-state inactivation properties, we analyzed the h_∞-curves of both types of astrocytes. To facilitate comparisons between cells, the current amplitudes elicited by a series of voltage steps with varying prepulse potentials were normalized to the largest recorded amplitude (I/I_{max}, typically recorded at −110 mV) and plotted as a function of prepulse potential. The data were then fitted to a modified Boltzmann equation of the form:

$$h(V) = (1 + \exp[V - V^{1/2*} a^{-1}])^{-1}$$

using a least-squares fit algorithm, and midpoints derived from the fitted curves were used for comparison.

A2B5$^+$ and A2B5$^-$ astrocytes consistently showed differences with respect to their h_∞-curves. In A2B5$^-$ astrocytes, the midpoint was always close to −85 mV (mean −84.2, SD 6.8, n = 47), and currents were almost completely inactivated at prepulse potentials more positive than −60 mV. In contrast, in A2B5$^+$ astrocytes the entire curve appeared to be shifted almost 20 mV along the voltage axis, with a midpoint close to −70 mV (mean −68.3, SD 6.8, n = 41).

Further differences were observed between the types of astrocytes with respect to changes in sodium current densities with time in culture. Current density was measured as the quotient of peak current and membrane capacity (which gives an estimate of membrane area) and was monitored for both cell types. Initially (1 DIV), current densities were close to 10 pA/pF in both A2B5$^+$ and A2B5$^-$ astrocytes (FIG. 5a and b), a value that would correspond (if we assume a single channel conductance of 20 ps and an open probability of about 0.25, as estimated for optic nerve type-1 astrocytes[18] and for Schwann cells[19]) to a surface density of about 0.3 sodium channels per square micrometer. This value only changed a little during development of A2B5$^+$ astrocytes in culture for the time period studied (15 DIV, FIG. 5 d and f). In contrast, A2B5$^-$ astrocytes showed a marked decrease in current densities after 10 DIV, and current densities fell by 12 DIV so that they were close to the level of detection (1 pA/pF) (FIG. 5c and e). This decrease was due to both a substantial reduction in the percentage of cells expressing any sodium current as well as a decrease in the current density in current-expressing cells, which was always below 5 pA/pF after 12 DIV.

Qualitatively the time course of current expression in the two types of astrocytes was similar to the staining pattern observed with 7493 antibody staining, with an attenuation of sodium current expression in A2B5$^-$ astrocytes and maintained expression of sodium currents in A2B5$^+$ astrocytes. However, when loss of sodium channel immunoreactivity (at 6 DIV) in A2B5$^-$ astrocytes was compared with loss of sodium currents in these cells (at 10–12 DIV), a lag period of 4–6 days was noted during which no sodium channel immunoreactivity was apparent, but measurable low sodium currents were still detected.

The difference in time course, for loss of sodium channel immunoreactivity versus loss of electrophysiologically measured sodium current, raises the interesting possibility that staining of astrocytes with antibody 7493 is predominantly due to the presence of sodium channel precursors located in the cytoplasm of these cells. We previously demonstrated that antibody 7493 does not stain premyelinated axons in rat optic nerve[20]; [^{3}H]-STX binding experiments indicate that sodium density is $\approx 2/\mu m^2$ in these premyelinated fibers.[21] As just described, the electrophysiological recordings in optic nerve astrocytes suggest sodium channel densities lower than $1/\mu m^2$, a density of membrane-associated sodium channels that is unlikely to be detected by immunocytochemical methods. Why then do optic nerve astrocytes in

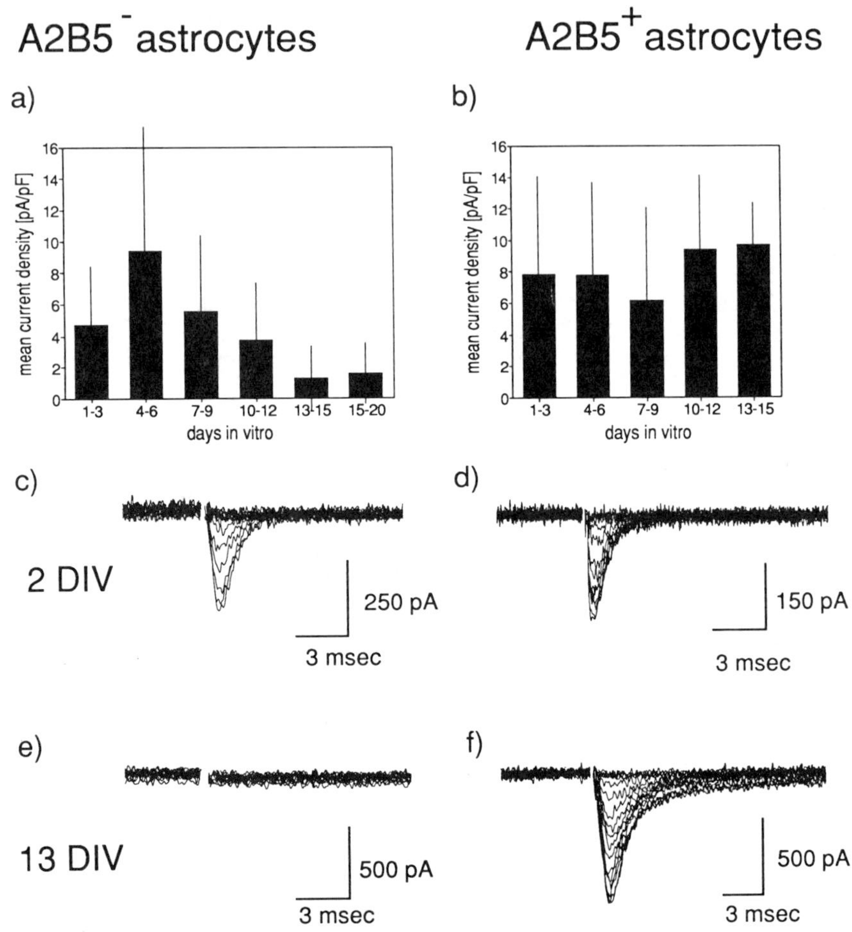

FIGURE 5. Time course of Na^+-current expression in astrocytes from rat optic nerve. Whole-cell currents were recorded in A2B5$^+$ and A2B5$^-$ astrocytes over the first 20 DIV. Na^+-current densities were pooled for astrocytes of six age groups: 1–3, 4–6, 6–9, 10–12, 13–15, and 15–20 DIV. Mean values were plotted as a function of time in culture (**a,b**) for both current-expressing astrocytes and cells that lacked current expression. Error bars indicate standard deviation. (**a**) In A2B5$^-$ astrocytes, mean current densities were highest between 3 and 6 DIV (close to 10 pA/pF) and showed a decline thereafter because of both a reduction in the number of cells expressing detectable Na^+ currents and much smaller current densities in cells still showing measurable currents. Currents were barely detectable after 13 DIV. (Note that only current densities in excess of 1 pA/pF could have been detected in our experiments.) The number of cells represented in the six time groups were (left to right): 21, 30, 22, 30, 27, and 24. (**b**) Mean current densities for A2B5$^+$ astrocytes as a function of time in culture. Current densities showed little variation within 15 DIV and remained close to 10 pA/pF. A2B5$^+$ astrocytes were rarely encountered after 16 DIV. Therefore, no corresponding value could be given for this age group. The number of cells from which data were obtained were (left to right): 22, 8, 14, 9, and 8. (**c–f**) Representative examples of current recordings obtained from A2B5$^-$ (**C,E**) and A2B5$^+$ astrocytes (**d,f**) after 2 DIV (**c,d**) and 13 DIV (**e,f**). Current traces were obtained by stepping the membrane to −20 mV from various prepulse potentials ranging

culture reveal sodium channel immunoreactivity? Moreover, why is this immunoreactivity lost 4–6 days prior to attenuation of sodium currents in A2B5$^-$ astrocytes?

As shown in FIGURE 2 and 3, sodium channel immunoreactivity in cultured astrocytes appears to be distributed throughout the cytoplasm and cell processes. This observation suggests that antibody 7493 may be detecting sodium channels, or channel precursors, that are not yet inserted into the plasma membrane (and thus are not yet functional). Assuming that this cytoplasmic pool contains precursors that are subsequently inserted into the plasma membrane, depletion of the cytoplasmic pool would be expected to lead, after an appropriate lag time, to loss of membrane-associated sodium channels. The estimated half-life of sodium channels incorporated into the plasma membrane of cultured rabbit Schwann cells is 2.2 days.[22] The 4–6-day lag between loss of sodium channel immunoreactivity and loss of electrophysiological evidence for the presence of membrane-associated sodium channels in cultured A2B5$^-$ astrocytes is thus consistent with the following sequence of events: following plating of A2B5$^-$ optic nerve astrocytes, the cytoplasmic pool of sodium channel precursors is depleted at 6 DIV. Subsequently, inasmuch as cytoplasmic channel precursors are not available for insertion into the membrane, a loss of membrane-associated sodium channels occurs as these channels turn over with a half-life of 2–3 days.

SODIUM CHANNEL EXPRESSION IN OPTIC NERVE ASTROCYTES *IN SITU*

An important question concerns the degree to which phenotypic differentiation of cells in culture is similar to, or different from, that of similar cells *in vivo.* To study sodium channel expression in astrocytes *in situ,* we carried out immunoultrastructural studies on astrocytes in the rat optic nerve.[5,6] In these studies we examined astrocytes in the rat optic nerve *in situ* after perfusion fixation with 4% paraformaldehyde/0.01% glutaraldehyde. Immunoelectron microscopic localization of sodium channels was demonstrated as an electron-dense reaction product after incubation of the tissue with polyclonal antibody 7493, biotinylated secondary antibody, avidin-horseradish peroxidase complex, and diaminobenzidine-hydrogen peroxide.

The immunoelectron microscopic studies demonstrate clearly that astrocytes in rat optic nerve *in situ* express sodium channels. As shown in FIGURE 6a, dense sodium channel immunoreactivity is present at nodes of Ranvier, with nodal axon membrane displaying heavy staining as expected from the high density of sodium channels in this part of the axon membrane.[23] Adjacent paranodal and internodal axon membranes, which are known to contain a much lower density of sodium channels than is the nodal membrane,[13] do not display immunoreactivity with antibody 7493. Similarly, sodium channel immunoreactivity was not present in the

---

[FIGURE 5 *continued*] from −110 mV to −30 mV (in 5-mV increments), a pulse protocol routinely used to record h_∞-curves. Sodium currents were detectable in almost all A2B5$^-$ (**c**) and A2B5$^+$ (**d**) astrocytes at 2 DIV. In contrast, such currents were rarely observed in A2B5$^-$ astrocytes after 13 DIV (**e**), whereas almost all A2B5$^+$ astrocytes continued to express Na^+ currents (**F**) in unchanged density (**b**) up to the latest time points examined (15 DIV). Note that currents were already inactivated at a prepulse potential of −85 mV in A2B5$^-$ astroctyes (**C**, sixth trace), whereas in A2B5$^+$ astrocytes, currents were only inactivated at potentials more positive than −50 mV (**d,f;** 13th trace), indicating a substantial difference between h_∞-curves of the two astrocyte types.

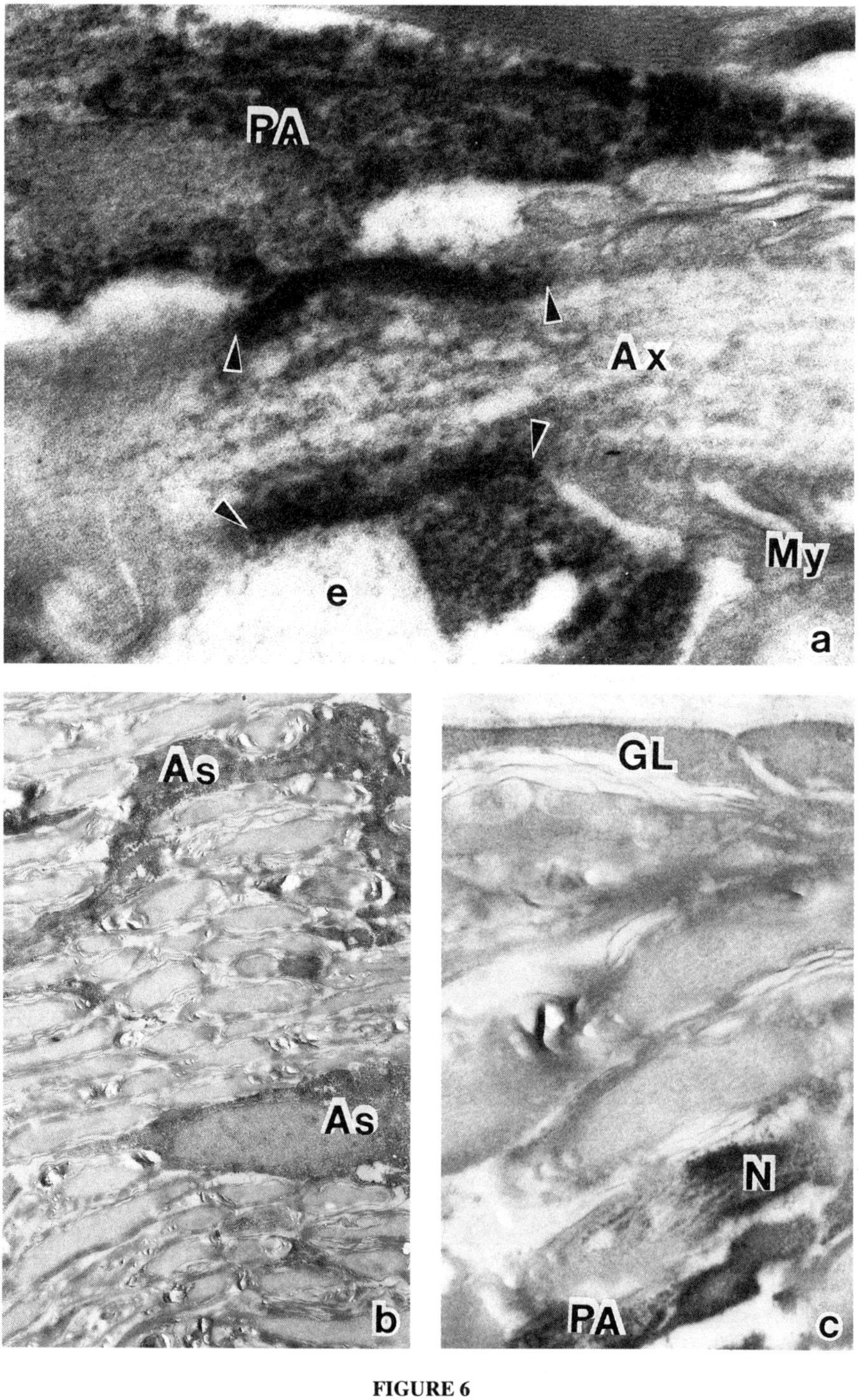

FIGURE 6

plasma membrane of terminal oligodendroglial paranodal loops bounding the node of Ranvier or elsewhere along oligodendrocyte plasma membrane. The axoplasm at the node of Ranvier also characteristically displayed some degree of sodium channel immunoreactivity, which appeared often to be associated with microtubules.

A number of studies have demonstrated the presence of perinodal astrocyte processes, which establish a specific relationship with the node of Ranvier.[24–26] When incubated with antibody 7493, perinodal astrocyte processes *in situ* within the optic nerve display distinct sodium channel immunoreactivity (FIG. 6a). In most cases, intense sodium channel immunostaining in perinodal astrocyte processes was distributed through the cytoplasm, and cytoplasmic immunostaining was usually more prominent than was staining of the plasma membrane.

Astrocyte somata and major processes also display immunostaining with antibody 7493. The extent of sodium channel immunoreactivity can be seen in FIGURE 6b, which shows a longitudinally sectioned field from the optic nerve at low magnification. Sodium channel immunoreactivity can be followed from the perikaryal region into the major processes that run longitudinally, parallel to the myelinated fibers within the nerve. Immunostaining is present in the astrocyte cytoplasm surrounding the nucleus, but not within the nucleus. Within the astrocyte cytoplasm, multiple focal accumulations of sodium channel immunostaining are present; these zones of immunostaining are not present within bundles of glial filaments, but it is not possible to determine whether they are associated with intracellular organelles (e.g., Golgi apparatus or lysosomes).

Sodium channel immunostaining appears to be heterogeneously distributed in optic nerve astrocytes. In contrast to perinodal astrocyte processes, astrocyte cell bodies, and major processes that show dense sodium channel immunostaining, astrocyte processes comprising the glia limitans characteristically exhibit substantially reduced sodium channel immunoreactivity.[6] This reduction was most apparent when, within a single field observed with the electron microscope, dense sodium channel immunostaining of a node of Ranvier and perinodal astrocyte could be observed within several microns of less densely stained astrocyte processes within the glia limitans (FIG. 6c). Despite this, occasional regions of the glia limitans display robust 7493 immunoreactivity, associated with both the astrocyte plasma membrane and the cytoplasm. Perivascular astrocyte processes also exhibited a moderate degree of sodium channel immunoreactivity (not shown), similar to that displayed by subpial astrocyte processes.

A CYTOPLASMIC POOL OF SODIUM CHANNELS OR PRECURSORS IN GLIAL CELLS

Inasmuch as the immunocytochemical results suggest that sodium channels are present within glial cytoplasm as well as in the plasma membrane, we recently carried

---

FIGURE 6. (**a**) Antibody 7493 immunoreactivity at node of Ranvier. The axon membrane at the node of Ranvier (*between arrowheads*) and subjacent axoplasm of the axon (Ax) display intense sodium channel immunoreactivity. A perinodal astrocyte process (PA) also exhibits 7493 immunostaining. The axon membrane beneath the myelin sheath (My) does not display sodium channel staining. e = extracellular space. Magnification × 60,000. (**b**) 7493 immunostaining of rat optic nerve. Longitudinally oriented processes and cell bodies of astrocytes (As) display sodium channel immunoreactivity. Magnification × 5000. (**c**) 7493 immunostaining of glia limitans. The intensity of the sodium channel staining of the glia limitans (GL) is much reduced compared to that displayed by an adjacent node of Ranvier (N) and associated perinodal process (PA). Magnification × 18,000.

out saxitoxin (STX) binding studies[9] which demonstrate that a significant subpopulation of sodium channels is located within the cytoplasm in some glial cells. For these studies, we used Schwann cell cultures derived from neonatal rabbit sciatic nerves, which provide the large number of cells required for STX binding experiments. Schwann cells were prepared for culture 6 days previously, as described by Shrager *et al.*[27] Immediately before the experiment, one half the cells were homogenized in a Duall glass homogenizer. Aliquots containing about 70 μg of protein (determined by the Lowry method[28]) were equilibrated with various concentrations of [^{3}H]-STX (apparent specific activity 89 dpm/mol, purity 65%) for 20 minutes at 4°C. Intact-cell pellets were prepared by centrifugation at 1,000 *g* for 15 minutes; homogenates were centrifuged at 200,000 *g* for 1 hour. Uptake of [^{3}H]-STX was determined at each of a variety of concentrations[29] by scintillation spectroscopy; double labeling of the incubation solution with [^{14}C]-mannitol permitted correction for extracellular [^{3}H]-STX. The uptakes were fitted to the sum of a rectangular hyperbolic component, representing specific uptake by sodium channels, and a linear component, representing nonspecific uptake.

STX binding in intact Schwann cells and after homogenization is shown in FIGURE 7. Uptake of [^{3}H]-STX was measured in the absence of (●) and presence of (◐), a very high concentration (8 μM) of unlabeled STX. The former curve gives the total uptake of STX as a function of concentration and the latter curve the nonspecific uptake. Saturable uptake (FIG. 7B) is represented by the difference between these two curves.

In the first set of experiments, illustrated in FIGURE 7 (continuous lines), intact Schwann cells were exposed to the [^{3}H]-STX prior to scintillation spectroscopy. In the second set of experiments (interrupted lines) aliquots of homogenized cells containing the debris of the same number of cells as in the first set were similarly equilibrated with [^{3}H]-STX and then pelleted. Notably, total binding, linear binding, and saturable binding are all significantly increased by exposure of the intracellular contents by homogenization. Thus, intact Schwann cell experiments gave an estimated value for M (maximal saturable binding capacity) of 214 ± 21 fmol/mg of protein, whereas homogenized Schwann cell experiments produce a significantly higher value of 410 ± 37 fmol/mg of protein ($p < 0.001$). Equilibrium dissociation constant K was significantly greater ($p < 0.002$) for homogenized Schwann cells (1.77 ± 0.37 nM) than for intact Schwann cells (1.06 ± 0.29 nM).

These results demonstrate that binding of [^{3}H]-STX to homogenized Schwann cells is almost twice as great as binding to intact cells. This finding provides evidence for a substantial population of intracellular sodium channels or channel precursors. An intracellular pool of sodium channel α subunits was previously observed in cultured neuroblastoma cells[30] and in neurons cultured from neonatal rat brain[31]; however, in these cases the intracellular sodium channel α subunits were reportedly inactive and did not bind STX with high affinity.

The increase in nonspecific binding in the homogenized preparation presumably reflects the additional binding of the labeled STX to newly exposed intracellular membranes; nonspecific binding of labeled STX apparently reflects primarily an accumulation of the divalent toxin in the vicinity of cell membranes because of the fixed negative charges on them. The increase in the equilibrium dissociation constant in the homogenized preparation may reflect a precursor-product relationship and the possible maturation of the sodium channel as it enters the plasma membrane of the Schwann cell; a similar increase in the equilibrium dissociation constant has been reported for the pool of intracellular sodium channels found in developing growth

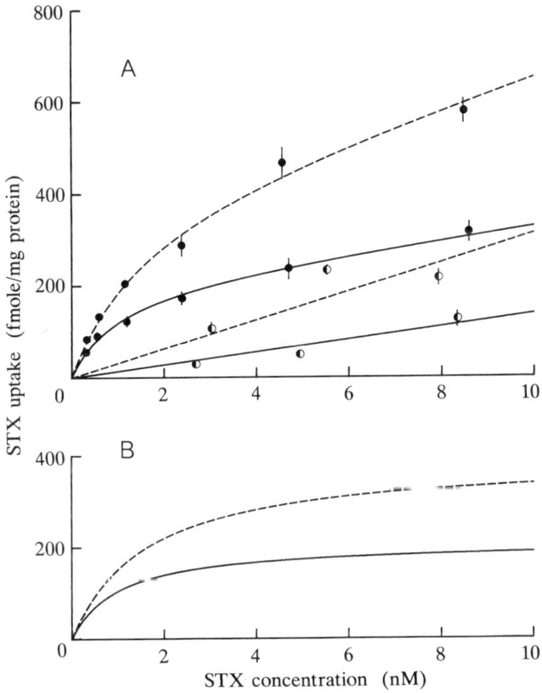

FIGURE 7. [^{3}H]-saxitoxin (STX) uptake as a function of saxitoxin concentration from binding experiments on primary cultures of Schwann cells derived from neonatal rabbit sciatic nerve. Both intact and homogenized Schwann cells were studied to examine binding of [^{3}H]-STX to cytoplasmic channels or precursors. Filled circles (●) are total uptakes. Half-filled circles (◐) are uptakes in the presence of 8 μM unlabeled saxitoxin. The unbroken curve through the *solid circle* depicts the best least-squares fit to the results for intact cells according to the rectangular hyperbola:

$$\text{Uptake} = M\,\{[STX]/[STX] + K\} + B[STX]$$

where M is the maximal saturable uptake, K is the equilibrium dissociation constant, B is the constant defining the linear nonspecific uptake, and [STX] is the external [^{3}H]-saxitoxin concentration. Each point is the mean of 6–8 experiments. The *bars* (sometimes obscured by the symbols) represent ± 1 SEM. The interrupted lines are from a companion set of experiments on homogenized Schwann cells from the same batch. (**B**) The saturable components of binding in intact and homogenized Schwann cells determined by subtraction of the linear from the total components of binding shown in **A.** Modified from Ritchie *et al.*[9]

cones in which both an increase in channel density and the exposure of additional lower affinity saxitoxin binding sites occur with exocytosis or permeabilization.[32]

The presence of an intracellular pool of sodium channels or channel precursors within glial cells is consistent with a steady intracellular synthesis and buildup of sodium channels that, once inserted in the plasma membrane, have a relatively short half-life.[30–22] Ritchie and his colleagues[1,27,34] have suggested that Schwann and glial cells may function as extraneuronal sources of sodium channels which are subsequently transferred to the axon membrane to supplement the synthesis of channels in the neuronal cell body, which are axonally transported.

IMPLICATIONS FOR ION CHANNEL EXPRESSION IN GLIA

The results summarized in this paper have a number of implications for expression of ion channels in glial cells. First, on the basis of our observations on astrocytes from optic nerve, we believe that cultured astrocytes are a heterogeneous group of cells in terms of ion channel expression. $A2B5^+$ and $A2B5^-$ astrocytes derived from rat optic nerve clearly exhibit different patterns of ion channel expression. In other studies[4] we showed that hippocampal astrocytes display a pattern of sodium channel expression that is different from that of $A2B5^+$ or $A2B5^-$ astrocytes from optic nerve. Although the degree to which astrocytes differ in terms of ion channel expression remains to be further explored, we believe that comparative studies in which physiological properties of astrocytes from differing regions in the CNS are examined will provide novel and useful information.

Second, our data suggest the possibility that astrocytic properties are modulated by factors present *in vivo* that are not present in culture. One attractive speculation is that these factors are neuron derived and possibly mediated by contact with neurons. As focal modulation of the axon surface, mediated by contact with glial cells, has been demonstrated,[35] it will be interesting to determine whether similar focal interactions can occur in the opposite direction. In this regard, we expect that coculture experiments, in which glial cells and neurons are permitted to differentiate within the same cultures, will be instructive. In addition, it will be important to examine the roles of various trophic factors in modulating glial expression of ion channels.

Third, our results provide evidence for the presence of a substantial pool of intracellular sodium channels or channel precursors in glial cells. Although we believe that the cytoplasmic channels are destined for insertion in the plasma membrane, a number of important questions remain. How are ion channels transported from sites where they are synthesized and undergo posttranslational modification, to the plasma membrane? What targets ion channels to particular domains in the axon membrane? What factor(s) regulate the density of ion channels in the plasma membrane? We expect that these and related questions will be answered as new subtype-specific probes are developed, so that various types of sodium channels can be tracked from their origin to their destination in the cell membrane.

REFERENCES

1. Bevan, S., S. Y. Chiu, P. T. A. Gray & J. M. Ritchie. 1985. Proc. R. Soc. Lond. (Biol.) **225:** 299–313.
2. Nowak, L., P. Ascher & Y. Berwald-Netter. 1987. J. Neurosci. **7:** 101–109.
3. Barres, B. A., L. L. Y. Chun & D. P. Corey. 1988. Glia **1:** 10–30.
4. Sontheimer, H., B. R. Ransom, A. H. Cornell-Bell, J. A. Black & S. G. Waxman. 1991. J. Neurophysiol. **65:** 3–19.
5. Black, J. A., B. Friedman, S. G. Waxman, L. W. Elmer & K. J. Angelides. 1989a. Proc. R. Soc. Lond. (Biol.) **238:** 39–51.
6. Black, J. A., S. G. Waxman, B. Friedman, L. W. Elmer & K. J. Angelides. 1989b. Glia **2:** 353–369.
7. Elmer, L. W., J. A. Black, S. G. Waxman & K. J. Angelides. 1990. Brain Res. **532:** 222–232.
8. Minturn, J. E., J. A. Black, K. J. Angelides & S. G. Waxman. 1990. Glia **3:** 358–368.
9. Ritchie, J. M., J. A. Black, S. G. Waxman & K. J. Angelides. 1990. Proc. Natl. Acad. Sci. USA **87:** 9290–9295.

10. RAFF, M. C., E. R. ABNEY, J. COHEN, R. LINDSAY & M. NOBLE. 1983. J. Neurosci. **3:** 1289–1300.
11. RAFF, M. C., E. R. ABNEY & R. H. MILLER. 1984. Dev. Biol. **106:** 53–60.
12. ELMER, L. W., B. O'BRIEN, J. J. NUTTER & K. J. ANGELIDES. 1985. Biochemistry **24:** 8128–8137.
13. WAXMAN, S. G. & J. M. RITCHIE. 1985. Science **228:** 1502–1507.
14. YAROWSKY, P. J. & B. K. KRUEGER. 1989. J. Neurosci. **9:** 1055–1061.
15. RUDY, B., B. KIRSCHENBAUM, A. RUKENSTEIN & L. A. GREENE. 1987. J. Neurosci. **7:** 1613–1625.
16. KALMAN, E., B. WONG, A. E. HORVAI, M. J. CLINE & P. H. O'LAGUE. 1990. Neuron **2:** 355–366.
17. CORVALAN, V., R. COLE, J. DE VELLIS & S. HAGIWARA. 1990. Proc. Natl. Acad. Sci. USA **87:** 4345–4348.
18. BARRES, B. A., L. L. Y. CHUN & D. P. COREY. 1989. Neuron **2:** 1375–1388.
19. HOWE, J. R. & J. M. RITCHIE. 1990. J. Physiol. **425:** 169–210.
20. BLACK, J. A., B. FRIEDMAN, A. CORNELL-BELL, K. J. ANGELIDES, J. M. RITCHIE & S. G. WAXMAN. 1990. *In* Cellular and Molecular Biology of Myelination. G. Jeserich *et al.*, eds. Vol. H43: 81–97. Springer-Verlag. Berlin, Heidelberg.
21. WAXMAN, S. G., J. A. BLACK, J. D. KOCSIS & J. M. RITCHIE. 1989. Proc. Natl. Acad. Sci. USA **86:** 1406–1410.
22. RITCHIE, J. M. 1988. Proc. R. Soc. Lond. B **233:** 423–430.
23. BLACK, J. A., J. D. KOCSIS & S. G. WAXMAN. 1990. TINS **13:** 48–54.
24. HILDEBRAND, C. 1971. Acta Physiol. Scand. **364:** 43–71.
25. WAXMAN, S. G. & J. A. BLACK. 1984. Brain Res. **308:** 77–87.
26. RAINE, C. S. 1984. J. Neurocytol. **13:** 21–27.
27. SHRAGER, P., S. Y. CHIU & J. M. RITCHIE. 1985. Proc. Natl. Acad. Sci. USA **82:** 948–952.
28. LOWRY, O. H., N. J. ROSEBROUGH, A. L. FARR & R. J. RANDALL. 1951. J. Biol. Chem. **193:** 265–275.
29. RITCHIE, J. M., R. B. ROGART & G. STRICHARTZ. 1976. J. Physiol. Lond. **261:** 477–494.
30. WAECHTER, C. J., J. W. SCHMIDT & W. A. CATTERALL. 1985. J. Biol. Chem. **258:** 5117–5123.
31. SCHMIDT, J., S. ROSSIE & W. A. CATTERALL. 1985. Proc. Natl. Acad. Sci. USA **82:** 4847–4851.
32. WOOD, M. R., G. STRICHARTZ & K. H. PFENNINGER. 1989. J. Cell Biol. **109:** 335a.
33. SCHMIDT, J. W. & W. A. CATTERALL. 1986. Cell **46:** 437–445.
34. GRAY, P. T. A. & J. M. RITCHIE. 1985. TINS **8:** 411–415.
35. WAXMAN, S. G. 1987. *In* Glial-Neuronal Communication in Development and Regeneration. H. Althaus & W. Seifert, eds.: 711–736. Springer-Verlag. Germany.

Glial K^+ Permeability and CNS K^+ Clearance by Diffusion and Spatial Buffering

ANDREAS REICHENBACH

Carl Ludwig Institute of Physiology
Leipzig University
Liebigstrasse 27
7010 Leipzig, Germany

Any electrical activity of neurons leads to the release of K^+ ions into the extracellular space (ECS) where, as a consequence, the K^+ concentration ($[K^+]$) increases.[1] This increase may cause severe alterations in neuronal excitability.[1] Gardner-Medwin[2] has proposed a fundamental division of the possible processes through which glial cells may control the extracellular accumulation of K^+ into two groups, a space-independent (with net uptake of K^+) and a space-dependent group.

Space-dependent processes require a concentration gradient of $[K^+]$ within tissue compartments not larger than 300–500 μm in size.[2] These processes include (1) extracellular diffusion down the concentration gradient and (2) the transport of ions by flow of electric current through the glial cell cytoplasm, called spatial buffering. Both mechanisms will be discussed in more detail. Inasmuch as some excellent recent reviews exist,[2-4] the following considerations will focus on the particular aspect that central nervous system (CNS) tissue is not homogeneous but consists of disparate subdivisions, often organized as layers, with peculiar geometry. Geometric features, however, should play an important role in space-dependent processes.

TYPES OF CNS TISSUE WITH DISTINCT GEOMETRY

In classical terms, CNS tissue is subdivided into white and gray matter. Interestingly, it has long been known that these compartments are occupied by two different types of astroglial cells, white matter by fibrous astrocytes and gray matter by protoplasmic astrocytes.[5] More recently, a third—"velate"—type of astrocyte has been distinguished[6]; it was found in a particular structure, the granular layer of the cerebellum. Similar granular layers exist in hippocampal structures and in the retina and are characterized by densely packed, mostly small, neuronal somata without many neuronal processes and synapses, surrounded by thin, velate glial cell sheaths. Thus, three different basic types of CNS tissue correspond to three types of astroglial cells: (1) "Neuropile" tissue consists of a more or less irregular network of interwoven fine neuronal cell processes and similar glial cell processes and sheaths. This type of tissue forms "molecular" or "plexiform" layers and is dominated by protoplasmic astrocytes with short feathery processes. (2) "White matter" consists of nerve fibers, mostly parallel aligned, and glial elements. Whereas thicker axons may be enwrapped by extensive oligodendrocytic sheaths, typical fibrous astrocytes (the dominating astroglia of this tissue) have rather smooth, long processes. Such tissue may form fiber tracts or layers, or even nerves. (3) "Nuclear layers," which have already been characterized here, occur only in specialized regions. Classical gray matter contains both neuronal somata and cell processes, but because somata are rather sparsely distributed, the general features are determined by the neuropil.

The retina, an example of thin CNS tissue, is characterized by two particular properties. Thin layers with very different properties are closely apposed, and the entire thickness of the tissue is spanned by a particular type of (astro-) glia, namely, the Müller cell. The processes of these cells were shown to express features of protoplasmic, fibrous, and velate astrocytes along their course through the retinal layers, adapting to the properties of the surrounding tissue[7] (FIG. 1). How both tissue

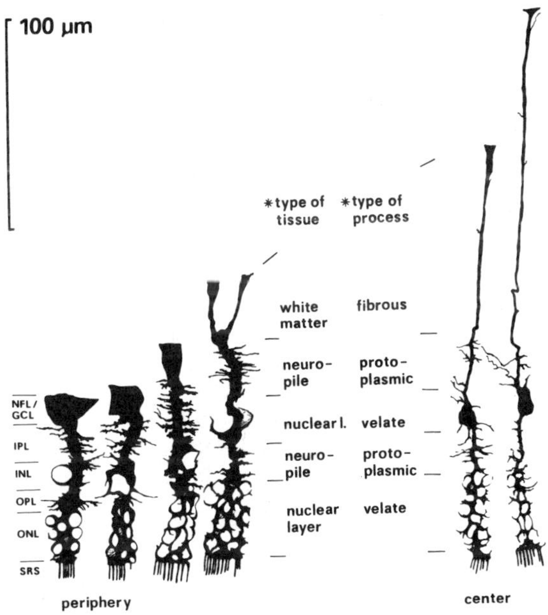

FIGURE 1. Camera-lucida drawings of Golgi-stained Müller cells from various regions of the rabbit retina. (**Left**) Far periphery; (**right**) medullary rays. NFL = nerve fiber layer; GCL = ganglion cell layer; IPL = inner plexiform layer; INL = inner nuclear layer; OPL = outer plexiform layer; ONL = outer nuclear layer; and SRS = subretinal space.

geometry and adapted glial morphology may influence space-dependent K^+ clearance will be discussed in the following sections.

EXTRACELLULAR K^+ DIFFUSION

Diffusion of ions in aqueous solution is a well-defined physical process. Unfortunately, however, the ECS of the CNS is a region with complex geometry, and the direction(s) of the concentration gradient(s) can usually not be determined. Most models of the brain are based on a point source of elevated $[K^+]_e$. On the other hand, physiological neuronal activity—and thus K^+ release—in brain cortex seems to be organized in columns rather than in points.[8] In another well-studied CNS tissue, the retina, physiological uniform (ganzfeld) stimulation was shown to cause planar K^+ sources within the two plexiform layers.[9] Here, diffusion should occur mainly in one direction, that is, toward the vitreous body which is believed to act as a K^+ sink.[4] In any case, the geometry of the ECS affects ion movements in two ways: (1) through the tortuosity of the ECS "channels" and (2) through the restricted space available for ion movement.

Extracellular Volume Fraction

Extracellular volume fraction α represents that volume of tissue that is available for ion diffusion. Evidence is increasing that α is not uniform throughout the CNS compartments.

1. Within the neuropile, the diameter of many cellular elements is at most one or two orders of magnitude above the width of the extracellular clefts, and relatively large α values should occur. This tissue geometry seems to determine the ECS of typical gray matter where α values of about 0.2 (i.e., 20%) have been measured by different methods.[10,11]

2. White matter and similar structures, such as the retinal nerve fiber layer, have another basic geometry (FIG. 2). The parallel alignment of cylindrical fibers leaves open triangular spaces in addition to the ECS "film" surrounding each fiber, particularly where most fibers have uniform diameters. On the other hand, the diameter of nerve fibers (d) is much greater than the width of the extracellular films (w). Thus, α values of white matter should be lower than those of gray matter; indeed, many authors report values of about 0.1 for white matter even if their method produces larger values for gray matter.[12,13] Assuming a uniform diameter of fibers, for geometric reasons α depends on the relation between d and w as follows[14]:

$$\alpha = 1 - \frac{\pi}{\sqrt{12}} \frac{d^2}{(d + w)^2}. \qquad (1)$$

This leads to a calculated α of about 0.1 where the fibers are thicker than about 1 μm, but it produces considerably higher values for thinner fibers. The latter prediction might be of interest for tracts of thin axons, such as the olfactory nerve.

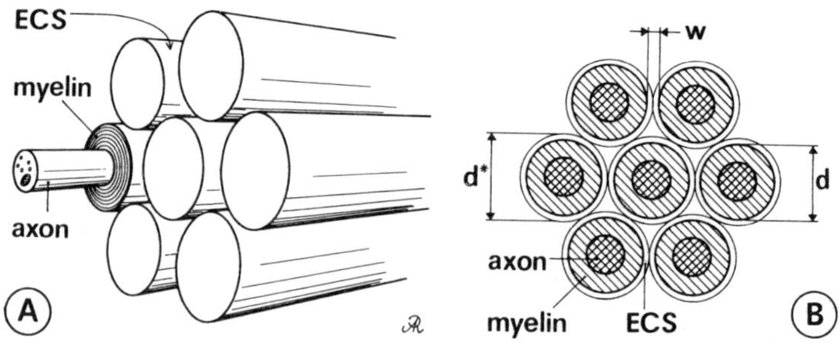

FIGURE 2. Schematic geometry of white matter. w = mean width of ECS clefts; d = mean diameter of fibers. (From Reichenbach *et al.*[7] with permission from *Z. Mikrosk.-Anat. Forsch. Leipz.*)

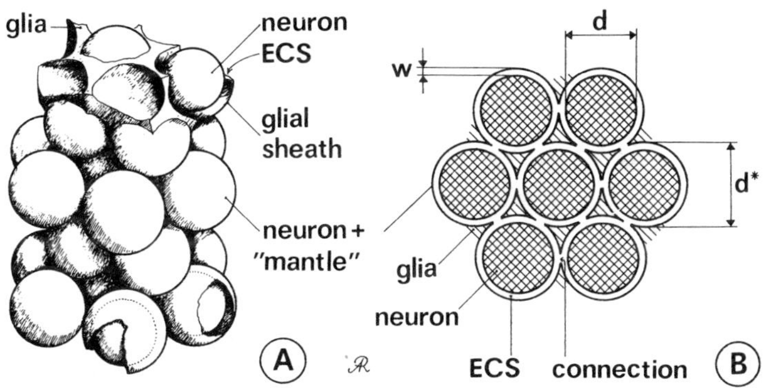

FIGURE 3. Schematic geometry of the outer nuclear layer of the retina. (From Reichenbach *et al.*[7] with permission from *Z. Mikrosk.-Anat. Forsch. Leipz.*)

3. In a simplified model (FIG. 3), nuclear layers consist of densely packed globes, each formed by a neuronal soma and its ECS "mantle" and by glial processes filling the spaces between these spherules. As neuronal somata have a diameter (d) that is far greater than the ECS width (w), such tissue compartments should have very small ECS volume fractions. Indeed, values of 0.02–0.06 have been evaluated for retinal nuclear layers by different methods.[14,15] In the geometric model, α is strongly dependent on the relation between d and w[14]:

$$\alpha = \frac{\pi}{\sqrt{18}}\left(1 - \frac{d^3}{(d + 2w)^3}\right). \qquad (2)$$

FIGURE 4 compares calculated and measured α values for different nuclear layers in retinas of various species. (For details, see Reichenbach *et al.*[14]) Both theoretical and experimental data show that tissues with densely packed, relatively large somata may express extremely low ECS volume fractions. This, as well as the fact that it must also cause severe restriction of extracellular ion movements by diffusion, should be taken into consideration in studies of current source density analysis.

SPATIAL BUFFERING

Electric currents will occur when a gradient of membrane potential exists along the processes of a cell or between different cells that are electrically coupled. The current flow may serve to redistribute ions from one region of the ECS to another, a process called the "spatial buffer mechanism" by Orkand and coworkers.[16] Currents flowing through the membranes and fluid spaces of cells must flow in complete loops; the redistribution of ions occurs because the current in different segments of the loop may be carried by different ions. Thus, most of the current across the cell membranes and within the cells is carried by K^+, whereas that in the extracellular loop is predominantly carried by Na^+ and Cl^-.

The driving forces for those currents are local differences between the K^+ equilibrium potential (E_K) and the membrane potential (E_m). In regions where elevated $[K^+]_e$ causes reduction of E_K whereas E_m remains "clamped" by other

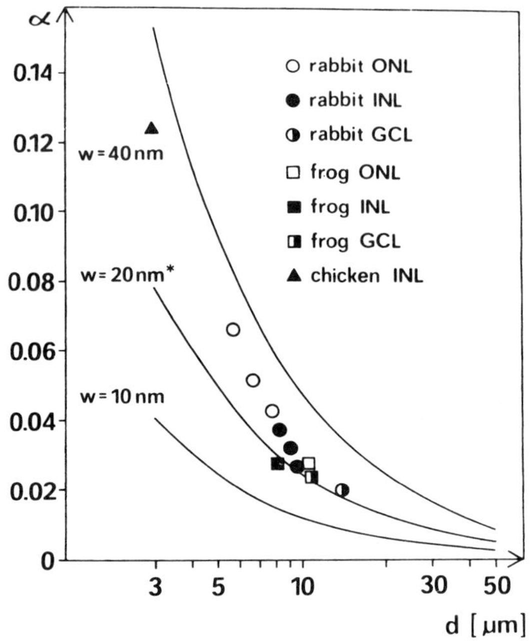

FIGURE 4. Dependence of ECS volume fraction α on the mean diameter of neuronal somata in nuclear layers of the retina. *Symbols:* Values estimated on the basis of experimental data; *lines:* values calculated according to equation 2, based on different ECS width w. (From Reichenbach *et al.*[7] with permission from *Z. Mikrosk.-Anat. Forsch. Leipz.*)

membrane regions not exposed to high $[K^+]_e$, K^+ ions enter the cell membrane. The intracellular K^+ current leaves the cell where E_K is still above E_m (see also Orkand[3]). Conversely, a local decrease in $[K^+]_e$ may enhance local K^+ outflow; this must enforce K^+ influx at other membrane regions exposed to normal $[K^+]_e$ but reached by the electrotonically spreading hyperpolarization of E_m.

The resistances against those driving forces are: (1) membrane resistances at both influx and outflow regions of K^+ ions, (2) the intracellular resistance, and (3) the resistance of the ECS within the extracellular loop. The latter is determined by the same geometric rules as those for extracellular diffusion described in the previous section; the former will be discussed subsequently below.

THE CABLE MODEL

Spatial buffering currents may be described by the cable model in which a core conductance connects sources and sinks of currents (or, in other terms, of K^+ ions). FIGURE 5 shows a simplified "glial cable" in which the main K^+ currents are given as thick arrows. This figure also demonstrates that the cable is imperfect; along the cable, shunting currents as well as secondary sources may occur. The efficiency of such a cable can be estimated by calculating the length constant λ as follows:

$$\lambda = \sqrt{\frac{r_m}{r_i}\frac{V}{S}} \tag{3}$$

with r_m being the membrane resistivity against shunting currents, r_i the cytoplasmic core resistivity, and V/S the volume-to-surface ratio of the cell process(es). Cables that are longer than λ are thought to carry insufficient current (compare also Rall[17]) to act as spatial buffers. Taken together, the following features should optimize the function of a spatial buffering cable: (1) low resistances against K^+ influx and outflow, that is, R_{in} and R_{out}, (2) low longitudinal resistances, R_{long}, which may consist of cytoplasmic and junctional resistances, (3) high shunt resistances, R_{shunt}, and (4) low resistances against recurrent extracellular currents, R_{rec}. It should be noted that the same membrane region may serve different functions in different situations; if, for example, there is a secondary source along the cable length, the same membrane provides both R_{shunt} for the primary current and R_{in} for the secondary one. Thus, aforementioned "optimum" conditions 1 and 3 may become contradictory (see also below).

SOURCES AND SINKS

Obviously, sources of spatial buffering currents are regions of high $[K^+]_e$ and sinks are regions of low $[K^+]_e$. It should be kept in mind, however, that concentrations are quotients of amounts (e.g., K^+ ions released by active neurons) and volumes, that is, the available ECS. Thus, ECS volume fraction α is an important variable, determining the efficiency of sources and sinks. Although it may also play a role in sources in different types of CNS tissue (see section on extracellular volume fraction), it is of particular importance for sinks. Different types of sinks have been considered. (1) Unspecialized ECS regions in the neighborhood of the source, without increased $[K^+]_e$, were first proposed to act as sinks.[16] Since, after active periods, neuronal Na^+ K^+ pumps reestablish intracellular K^+ stores, causing decreased $[K^+]_e$, the efficiency of such sinks may even be enhanced. Pump processes, however, are rather slow, and the available ECS within the neuronal tissue is very restricted, even in typical neuropile. Thus, alternative sinks have been proposed that might express particularly suitable properties such as large volume. (2) For the retina, the vitreous body, which is actually a big "ocean" compared to the small retinal ECS, is available as a sink.[4] (3) For the brain, Paulson and Newman[18] suggested a mechanism by which the glial syncytium may release K^+ into the perivascular space as a primary sink. The Na^+ K^+ pumps of the endothelial cells could transport excess K^+ ions from there into the lumen of blood vessels which would then act as secondary sinks. (4) Big reservoirs of

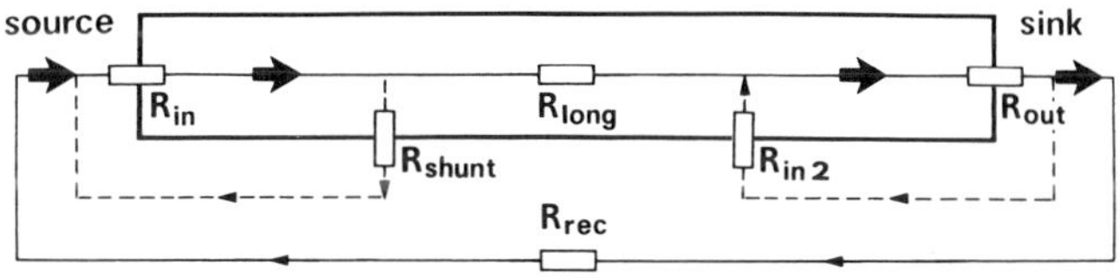

FIGURE 5. Schematic account of the cable model. *Large arrows* indicate those parts of the current loop that are carried by K^+.

fluid could be available in the form of cerebrospinal fluid (CSF) either in the subarachnoidal space or within the ventricular system.[19]

Distribution of K^+ Channels

If such specialized sinks exist, it would be advantageous to direct the currents within the cable (network) toward those preferential ends. This could be achieved by decreasing R_{out} at the adjacent end compared to the membrane resistance elsewhere. In addition, "specialized sinks" offer their services by maintaining low $[K^+]_e$. Indeed, membranes of Müller cell endfeet abutting the vitreous body[20–22] and of astrocytic endfeet abutting blood vessels and/or the pial border[23] have been shown to express much greater K^+ conductivity than do the remaining cell membranes. Whereas in amphibian Müller cells this high endfoot conductance originates from a greatly enhanced density of uniform, rather small K^+ channels with inward rectification,[21] there is evidence that rabbit Müller cells express symmetric K^+ "maxi channels" within their endfoot membrane[24] (FIG. 6). This may indicate some phylogenetic progress and may be reflected in the occurrence of orthogonal arrays of intramembranous particles in endfoot membranes of mammalian Müller cells.[25]

In contrast to the specialized endfoot membrane, the remaining glial cell membrane may have to serve varying functions; the "mantle" of the core conducting cable has to provide both R_{shunt} and R_{in}. As just mentioned, this leads to a discrepancy

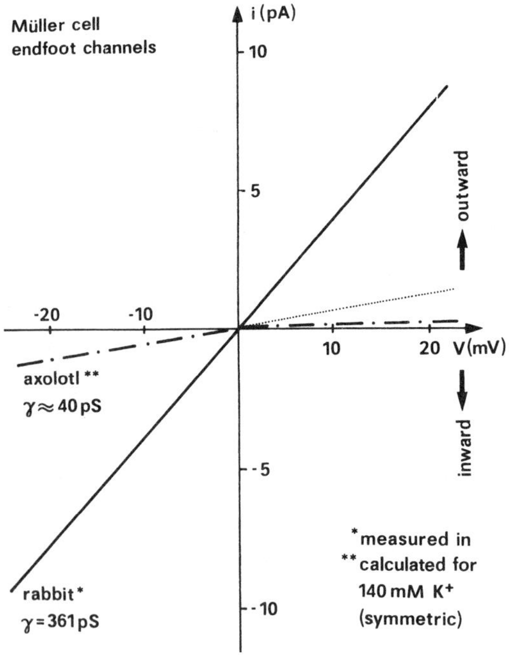

FIGURE 6. Comparison of current-voltage relationships of K^+ channels in endfoot membranes of axolotl[21] and rabbit[24] Müller cells.

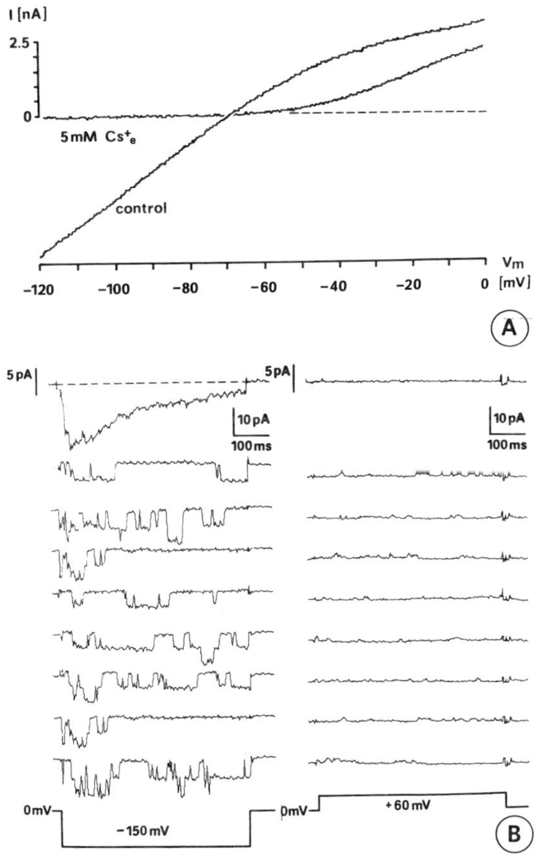

FIGURE 7. Inward rectification in the membrane of enzymatically isolated rabbit Müller cells. (**A**) Current voltage curve obtained by applying a voltage ramp changing the membrane potential (V_m) from −120 to 0 mV while recording current (I) in the whole-cell recording mode of the patch-clamp technique. The *I/V* curve before application of 5 mM extracellular caesium (control) and 2 minutes after Cs^+ application is shown. (**B**) Single-channel currents and averaged currents (*top of each column*) through a high conductance rectifier (HCR)[24] channel of the vitread process membrane in response to hyperpolarizing (*left column*) and depolarizing (*right column*) voltage steps; cell attached patch.

because R_{shunt} (for outward currents) should be high and R_{in} (for inward currents) should be low. Very probably, the glial membrane overcomes that problem by expressing inwardly rectifying K^+ channels. This kind of channel has been found in almost every glial cell type studied so far.[26] Even the membrane conductance of rabbit retinal Müller cells, despite being dominated by the endfoot conductance with symmetrical channels,[22,24] shows slight inward rectification in whole-cell voltage clamp (FIG. 7A). In the membrane of the vitread process where a large secondary sink is passed (FIG. 9), rabbit Müller cells express specialized K^+ channels that are comparable to perfect technical, inward rectifiers[24] (FIG. 7B).

Of course, such inward rectification should complicate outward fluxes of K^+ ions which are expected to occur after periods of K^+ release at sites of former sources then becoming sinks. On the other hand, neuronal K^+ reaccumulation is much slower than K^+ release, and the cooperation of delayed, backward spatial buffering and diffusion may be sufficient.

Geometry of Glial Cell Processes

It becomes obvious from the cable model that the geometry of glial cell processes determines much of the current efficiency. Glial cell processes within different types of CNS tissue have disparate morphologies (FIG. 1). (1) Fibrous astrocytes of white matter tracts have smooth processes, that is, low surface-to-volume ratios. As a result, λ should be rather high. Indeed, fibrous astrocytes have long processes radiating over considerable distances (Langstrahler); likewise, tanycytes in mammalian white matter can reach lengths of almost 1 mm. On the other hand, low surface-to-volume ratio means not only high R_{shunt} but also, inevitably, high R_{in}. Perhaps in white matter a need for much K^+ influx across R_{in} does not exist because neuronal activity causes much less of an increase of $[K^+]_e$ in white matter than in gray matter.[1] (2) The feathery processes of protoplasmic astrocytes in the neuropile, as well as the "blistered" processes of velate astrocytes in nuclear layers, are characterized by a high surface-to-volume ratio. Consequently, such cells have processes radiating only over short distances (Kurzstrahler); the same is true for tanycytes within gray matter.[27]

Despite the same general morphology of a given cell type such as Bergmann glial cells, great quantitative differences remain between different species of animals. To span the entire molecular layer, processes of human Bergmann glial cells have to be as long as 300 μm, whereas in the small native African mouse (*Leggada minutoides*), that length is hardly more than 100 μm. Of course, the long path should greatly impair spatial buffering currents through human cells. Fortunately (for human beings), two compensatory effects can be observed (FIG. 8): (1) Human radial stem processes bear much fewer lateral protrusions than do these processes in mice, indicating that the surface-to-volume ratio of human cell processes is reduced. (2) The diameter of the radial stem processes in man is about 1.0 μm, whereas that in mice is only 0.7 μm. Inasmuch as R_{long} is determined by the cross-sectional area of the processes (depending on the square of the diameter), both effects together may be sufficient to compensate for the greater length of human cell processes. Thus, the geometry of (astro-)glial cell processes probably is adapted to achieve optimum spatial buffering capacity.

MATHEMATICAL MODELING OF K^+ CLEARANCE

Much work has been done to describe both diffusion and spatial buffering in brain and retina by mathematical models; thus, just a few remarks will be given on the properties of our own model of the rabbit retina. This model incorporates the results of electron optic morphometry[14] as well as electrophysiological data at both whole-cell[22] and single-channel[24] levels and is illustrated in FIGURE 9. The model was applied to various regions of the rabbit retina because the thickness of the tissue varies greatly and because Müller cells have different morphologies; in particular, cells of the myelinated retinal center have long fibrous processes which are missing in cells of the retinal periphery (cf. FIG. 1).

FIGURE 10 shows how the half-time ($t_{1/2}$) of K^+ clearance after a step increase in $[K^+]_e$ depends on retinal geometry (defined by retinal thickness, i.e., Müller cell length l in microns). In the range between 60 and 140 μm, where all Müller cells are dominated by protoplasmic and velate surfaces, increasing the thickness of the tissue causes increased delay of both clearance processes (although spatial buffering leads to considerable acceleration compared to diffusion). This is exactly the result expected when the pathway for diffusion is elongated and when spatial buffering cables become longer and thinner. In contrast, the very thick central retina (l = 200 and 350 μm) shows rather fast K^+ clearance. This is attributed to the white matter

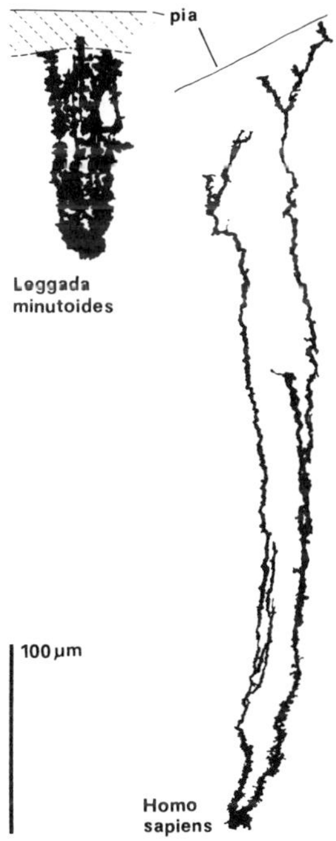

FIGURE 8. Camera-lucida drawings of Golgi-stained Bergmann glial cells of the cerebellum of a small African native mouse (*left cell*) and a human being (*right cell*); the same calibration is valid for both cells.

properties of the thick myelinated nerve fiber layer crossed by smooth fibrous processes of Müller cells with a large λ. Combining results, suitable mathematical models seem to describe K^+ clearance processes in a reasonable manner.

AN OUTLOOK: DYNAMIC REGULATION OF SPATIAL BUFFERING

In contrast to the aforementioned models, *in vivo* conditions of spatial buffering are thought to be dynamic rather than static. During the flow of spatial buffering currents there seem to be dramatic and fast changes of many of the parameters such as the ECS volume fraction α.[28] Furthermore, dynamic long-term changes in tissue structure and function are coupled with learning, aging, and other events. Such changes may require regulatory responses of glial cells.

Short-Time Regulatory Events

The most obvious shortcoming of the simplified cable model presented in FIGURE 5 is its linearity; in the brain *in vivo* exists a very complex three-dimensional network of electrotonically coupled astrocytic processes that may carry currents in almost every direction. Moreover, most tissue compartments may act as sinks or sources in turn, depending on the state of neuronal activity.[16] Thus, another, still simplified model is depicted in FIGURE 11. Field A may be a dominant source such as

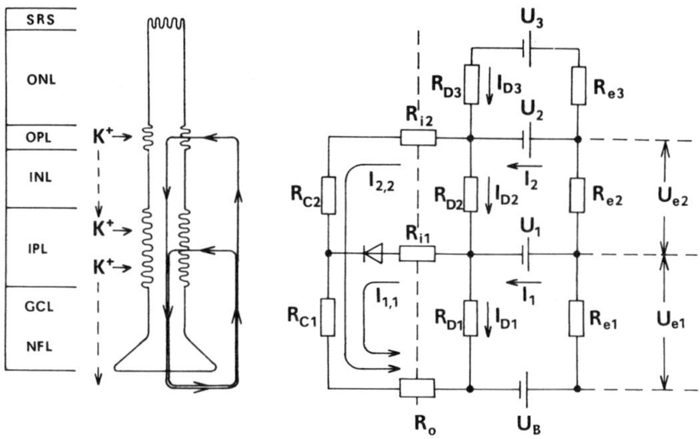

FIGURE 9. K^+ clearance in the rabbit retina. (*Left side*) Retinal layers (compare FIG. 1) and sites of $[K^+]_e$ rise during illumination (K^+).[9] Excess K^+ ions are removed by both extracellular diffusion (*dotted lines*) and spatial buffering currents through Müller cells (*continuous lines*). (*Right side*) Modeling circuit. Spatial buffering currents flow through cytoplasmic resistances (R_{c1}, R_{c2}) and, backward, through extracellular resistances (R_{e1}, R_{e2}); R_i corresponds to R_{in} and R_{in2} in FIGURE 5; R_o means R_{out} of FIGURE 5, that is, the resistance of the endfoot membrane. Extracellular diffusion is modeled as a current flowing through resistances (R_{D1} etc). Driving forces are represented by the batteries, whereas the extracellular voltage drops (which lead to field potentials) are given by U_{e1} and U_{e2}.

a region with spontaneous neuronal activity, field B represents a specialized sink (see above), and fields C and D are regions that may act as either sources or sinks.

When only field A is a source, the current will flow either mainly to field B, if there is low R_{out}, that is, high endfoot conductance, and if R_{long} is low, or to fields B, C, and D if, for example, the path between x and B has a high longitudinal resistance. When there is additional activity in field C, only fields B and D remain available as potential sinks because high $[K^+]_e$ blocks K^+ outflux at C.

There are other ways in which a field may be protected from being a sink or may be separated from the spatial buffering network. (1) The cable between the crossroads (x) and that field may be disconnected by decoupling the gap junctions between the cell processes involved. It has been shown that weak acids, such as CO_2, block ionic coupling between glial cells[29]; changes of extracellular Ca^{2+} and some

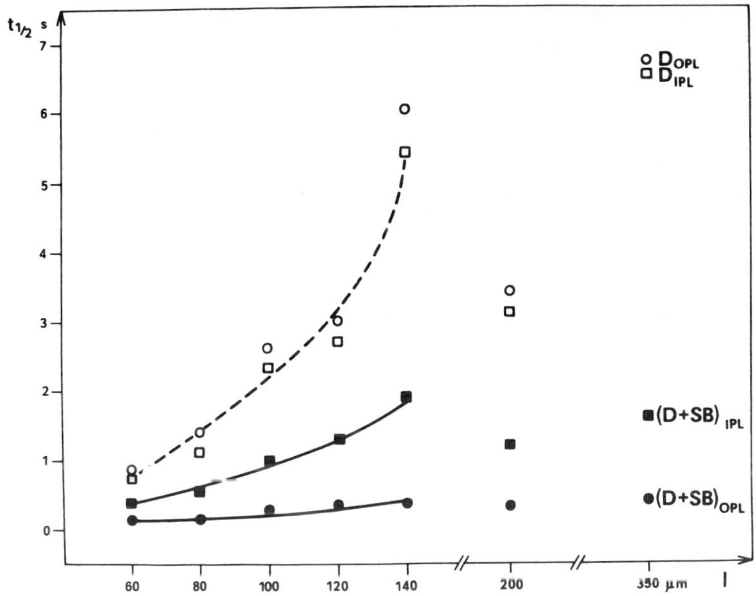

FIGURE 10. Calculated half-times ($t_{1/2}$) of K^+ clearance in inner (IPL) and outer (OPL) plexiform layer of the rabbit retina, depending on the local thickness of the tissue (given as Müller cell length in µm). D = diffusion alone; (D+SB) = combined effect of diffusion and spatial buffering.

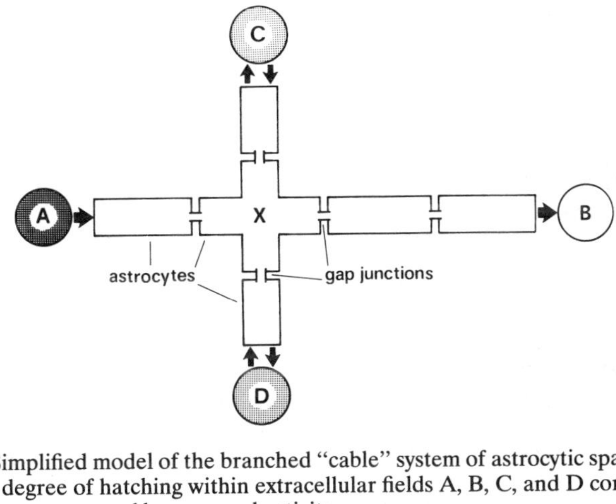

FIGURE 11. Simplified model of the branched "cable" system of astrocytic spatial buffering in the brain. The degree of hatching within extracellular fields A, B, C, and D corresponds to the level of $[K^+]_e$ increase caused by neuronal activity.

neurotransmitters may have effects on glial cell coupling as well. (2) The resistance of the membrane, R_{out} as well as R_{in}, may be increased. It has been proposed that the neuropeptide receptors on astrocytes could guide spatial buffering currents by closing local K^+ channels.[30] Other neurotransmitters may exert similar effects.[31] Some evidence also exists that glial cells express ATP-sensitive K^+ channels (own unpublished results) which could depend on the metabolic state of the cells. (3) Some glial K^+ channels show time- and/or voltage-dependent inactivation[24,26] and their conductance may be down-regulated during the process of spatial buffering. (4) Finally, other (non-K^+) receptor-activated conductances may reduce the actual conductance of K^+ channels by acting on voltage-dependent gates.[26] On the other hand, (5) a given region of the spatial buffering network may be isolated when the K^+ conductance of the remaining membranes is markedly enhanced, thus shunting any currents directed into the "dead end."

Most of the aforementioned events are thought to occur *in vivo* and may be regulated by the actual activity of the tissue. Such processes may even be influenced

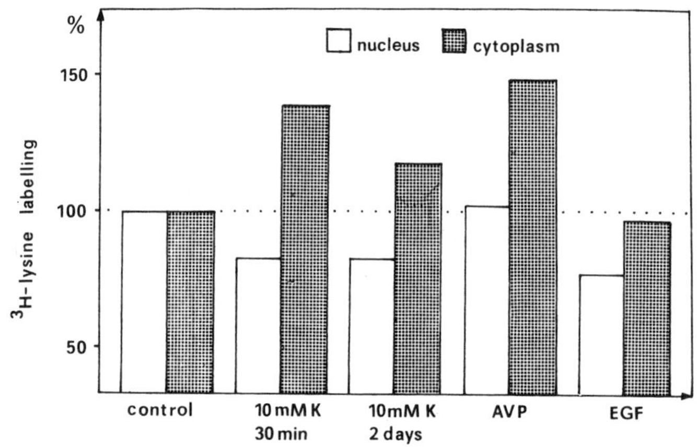

FIGURE 12. Autoradiographic [^{3}H]-lysine label overlying the nucleus (*empty columns*) and the cytoplasm (*hatched columns*) of cultured rabbit Müller cells exposed to various media. The control medium contained 5.8 mM K^+; AVP = arginine-vasopressin; EGF = epithelial growth factor.

by pharmacological drugs; these may become powerful tools in the treatment of states with enhanced $[K^+]_e$ such as epilepsy and migraine.

Long-Term Adaptation

During development and because of long-lasting enhanced neuronal activity caused by either physiological (such as learning) or pathological (such as epilepsy) events, some regions of CNS tissue may express enhanced demands in terms of K^+ clearance. Glial cells could respond with adaptive changes in their spatial buffering capacity. A simple way to do this would be to thicken and/or to multiply their processes, that is, the "cables" for K^+ currents; it should be noted that reactive astrocytes are indeed characterized by hypertrophy. Another way would be to express a greater density and/or more suitable types of K^+ channels or more intensive cell-to-cell coupling by gap junctions. All these processes would require enhanced protein synthesis; indeed, we could show that cultured rabbit Müller cells respond to high $[K^+]_e$ media with increased incorporation of radioactively labeled amino acids[32] (FIG. 12). On the other hand, active neuronal elements may exert some attractive effect on growing glial cell processes[33] that may lead to the formation of glial sheaths around such neuronal elements. This attraction may be mediated by neuromodulatory substances (cf. also FIG. 12) or by elevated $[K^+]_e$. As soon as newly grown astroglial processes reach the ECS around those active neurons, they will act as cables buffering the very same high $[K^+]_e$ that had enforced their growth. This kind of self-regulating mechanism should play an important role in the organization of adequate glial structure and function. For example, it could account for the fact that glial processes within the white matter, where no large increases in $[K^+]_e$ are observed,[1] extend almost no side branches, whereas protoplasmic astrocytes in gray matter have many finely elaborated protrusions that ensheath the sites of K^+ release.

ACKNOWLEDGMENTS

I wish to thank my colleagues Wolfgang Eberhardt, Winfried Reichelt, André Henke, and Albrecht Siegel who contributed much to the results presented and A. Diener for reliable technical assistance.

REFERENCES

1. SYKOVA, E. 1983. Prog. Biophys. Mol. Biol. **42:** 135–189.
2. GARDNER-MEDWIN, A. R. 1983. J. Physiol. (Lond.) **335:** 393–426.
3. ORKAND, R. K. 1986. Ann. N. Y. Acad. Sci. **481:** 269–272.
4. NEWMAN, E. A. 1985. Trends Neurosci. **8:** 156–159.
5. ANDRIEZEN, W. L. 1893. Br. Med. J. **2:** 227–230.
6. CHAN-PALAY, V. & S. L. PALAY. 1972. Z. Anat. Entw. Gesch. **138:** 1–19.
7. REICHENBACH, A., H. SCHNEIDER, L. LEIBNITZ, W. REICHELT, P. SCHAAF & R. SCHÜMANN. 1989. Anat. Embryol. **180:** 71–79.
8. MOUNTCASTLE, V. B. 1979. An organizing principle for cerebral function: The unit module and the distributed system. *In* The Neurosciences: Fourth Study Program. F. O. Schmitt & F. G. Worden, eds.: 21–42. MIT Press. Cambridge, MA.
9. KARWOSKI, C. J., E. A. NEWMAN, H. SHIMAZAKI & L. M. PROENZA. 1985. J. Gen. Physiol. **86:** 189–213.
10. NICHOLSON, C., J. M. PHILLIPS & A. R. GARDNER-MEDWIN. 1979. Brain Res. **169:** 580–584.
11. LEVIN, V. A., J. D. FENSTERMACHER & C. S. PATLAK. 1970. Am. J. Physiol. **219:** 1528–1533.
12. COLE, K. S., C.-L. LI & A. F. BAK. 1969. Exp. Neurol. **24:** 459–473.
13. FENSTERMACHER, J. D. & C. S. PATLAK. 1975. The exchange of material between cerebrospinal fluid and brain. *In* Fluid Environment of the Brain. H. F. Cserr, J. D. Fenstermacher & V. Pencl, eds.: 201–214. Academic Press. New York, NY.
14. REICHENBACH, A., E. HAGEN, K. SCHIPPEL & W. EBERHARDT. 1988. Z. mikrosk.-anat. Forsch. Leipz. **102:** 721–755.
15. KARWOSKI, C. J., D. A. FRAMBACH & L. M. PROENZA. 1985. J. Neurophysiol. **54:** 1607–1619.
16. ORKAND, R. K., J. G. NICHOLLS & S. W. KUFFLER. 1966. J. Neurophysiol. **29:** 788–806.
17. RALL, W. 1977. Core conductor theory and cable properties of neurons. *In* Handbook of Physiology, Section 1. J. M. Brookhart & V. B. Mountcastle, eds. Vol. I (Part 1): 39–97. American Physiology Society. Bethesda, MD.
18. PAULSON, O. B. & E. A. NEWMAN. 1987. Science **237:** 896–898.

19. CONNORS, B. W. & B. R. RANSOM. 1987. J. Physiol. (Lond.) **385:** 287–306.
20. NEWMAN, E. A. 1984. Nature **309:** 155–157.
21. BREW, H., P. T. A. GRAY, P. MOBBS & D. ATTWELL. 1986. Nature **324:** 466–468.
22. REICHENBACH, A. & W. EBERHARDT. 1988. Glia **1:** 191–197.
23. NEWMAN, E. A. 1986. Science **233:** 453–454.
24. NILIUS, B. & A. REICHENBACH. 1988. Pflügers Arch. **411:** 654–660.
25. WOLBURG, H. & K. BERG. 1987. Neurosci. Lett. **82:** 273–277.
26. BARRES, B. A., L. L. Y. CHUN & D. P. COREY. 1990. Ann. Rev. Neurosci. **13:** 441–474.
27. REICHENBACH, A., M. NEUMANN & G. BRÜCKNER. 1987. Neurosci. Lett. **73:** 95–100.
28. DIETZEL, I., U. HEINEMANN, G. HOFMEIER & H. D. LUX. 1980. Exp. Brain Res. **40:** 432–439.
29. TANG, C.-M., P. M. ORKAND & R. K. ORKAND. 1985. Neurosci. Lett. **54:** 237–242.
30. BACKUS, K. H., T. BERGER & H. KETTENMANN. 1991. Activation of neurokinin receptors modulates K^+ and Cl^- channel activity in cultured astrocytes from rat cortex. Brain Res. **541:** 103–109.
31. HERTZ, L. 1990. Regulation of potassium homeostasis by glial cells. *In* Differentiation and Functions of Glial Cells. G. Levi, ed.: 225–234. Alan R. Liss, Inc. New York, NY.
32. REICHELT, W., D. DETTMER, G. BRÜCKNER, P. BRUST, W. EBERHARDT & A. REICHENBACH. 1989. Cell. Signalling **1:** 187–194.
33. REICHENBACH, A. & W. REICHELT. 1986. Neurosci. Lett. **71:** 125–130.

Glial H^+ Transport and Control of pH

W.-R. SCHLUE, R. DÖRNER, L. REMPE, AND B. RIEHL

Institut für Zoologie / Neurobiologie
Heinrich-Heine-Universität
D-4000 Düsseldorf, Germany

The appropriate regulation of intracellular pH (pH_i) is crucial because of the large number of pH-sensitive processes.[1–5] In excitable cells, changes in pH_i modify various ionic conductances and the electrical coupling between adjacent cells is blocked by decreased pH_i.[6] In addition to its effects on other ionic conductances, H^+ might itself act as a charge carrier through H^+-specific channels.[7–9]

In the nervous system the pH of the interstitial space (pH_o) has been studied extensively. The pH_o decreases during neuronal activity, often after an alkaline shift.[10,11] In rat spinal cord, prolonged interstitial acid shifts arise as a result of physiological stimuli.[12] Specifically, noxious stimulation of the hind paw causes acidification as great as 0.05 pH units. In contrast, a decrease in pH_o of up to 0.2 units occurs during intense stimulation of afferent pathways and during direct stimulation of nervous tissue.[11] More dramatic changes in pH_o have been observed in pathological situations. The magnitude of the pH_o change in complete ischemia is usually 0.5 units, but it can be significantly greater in animals rendered hyperglycemic before the ischemic challenge.[13]

Artificial changes in pH_o by changing the pH of extracellular fluid or bathing medium alter pH_i in neurons,[14,15] and glial cells (W.-R. Schlue, unpublished observation). In crayfish-stretch receptor neurons a decrease in pH_o from 7.35 to 6.8 in HEPES-buffered solution caused internal acidification by 0.1–0.2 pH units within 10 minutes.[14] Furthermore, reduction of pH_o by about 1.4 units slowed down (or even reversed) acid extrusion. Thus, pH_o is an important determinant of pH_i and of its regulation in these receptor neurons, as in barnacle muscle[16] and salamander renal proximal tubule.[17]

With electrical stimulation of the cortical surface, rat astrocytes depolarized and became more alkaline by 0.05–0.40 pH units, and the glial pH_i shifted more alkaline by 0.11–0.78 pH units during spreading depression.[18,19] It was shown that the alkalinization was triggered by membrane depolarization, suggesting that glial pH_i is normally modulated by the level of local neuronal activity.

Neurons may acidify after prolonged activity. For example, Ahmed and Connor[20] found that in *Archidoris* neurons trains of action potentials produced a pH_i decrease, and recovery of the pH_i change required periods >10 minutes. Illumination of *Balanus* photoreceptors acidified the pH_i by as much as 0.2 pH units.[21]

A full understanding of the role of pH_i and pH_o changes in affecting the function of the nervous system requires a detailed knowledge of the mechanisms by which pH_i is regulated. In recent years, my laboratory has intensively studied these mechanisms in the central nervous system (CNS). The present report summarizes the research on acid-base physiology of glial cells and presents new data with the goal of characterizing ion transporters involved in pH shifts.

Leech Preparation

The leech nervous system is particularly suitable for this kind of study, because all connectives and ganglia show an identical organization. It consists of three types of glial cells that can be identified and distinguished according to their location.[22,23] Six packet glial cells in each ganglion surround the neuron cell bodies, and two neuropile glial cells enclose the neuron processes and synaptic regions. In the connectives are two connective glial cells that are located between two ganglia. They are characterized by many radial cytoplasmic processes that surround the axons of nerve cells.

METHODS

Localization of Carbonic Anhydrase

Hansson's technique[24] was used to localize carbonic anhydrase (CA) in tissue sections. This is based on the precipitation of cobalt salts that occurs as the result of alterations in pH value. In addition, the enzyme was detected with 1-dimethylamino-naphthalene-5-sulfonamide (DNSA). Pochhammer *et al.*[25] first used DNSA, and Dermietzel *et al.*[26] improved the use of this substance, which has an affinity for the catalytic center of CA.[27] Together they form a fluorescent enzyme-substrate complex that is microscopically visible. The distribution of CA isoenzymes I and II in leech glial cells was investigated by indirect immunofluorescence.[28]

pH Measurements

The involvement of different pH_i-regulating mechanisms has been investigated in neuropile, connective, and packet glial cells using ion-selective microelectrodes for intracellular recording of ion activities. These glial cells can be penetrated successfully by several types of ion-selective microelectrodes for measuring rapid ion activity changes.

The preparation and dissection procedures used to isolate single ganglia of the leech, *Hirudo medicinalis,* and the selection of cells have been described before.[29]

Solutions: Normal leech saline solution (nominally HCO_3^--free) was composed of: NaCl, 115 mM; KCl, 4 mM; $CaCl_2$, 1.8 mM; HEPES (adjusted to pH 7.4 with NaOH), 10 mM; and glucose, 11 mM. HCO_3^- solutions were equilibrated with 2% CO_2 and 98% oxygen and contained 11 mM $NaHCO_3$ instead of HEPES (pH 7.4). Na^+-free solutions were made by replacing Na^+ with *N*-methyl-D-glucamine neutralized with HCl and/or CO_2 as appropriate. The Cl^--free solutions were made by equimolar replacements of the Cl^- salts with gluconate salts.

Active pH_i regulation has been investigated by inducing intracellular acidification by the addition and subsequent removal of NH_4Cl. Recovery from intracellular alkalosis was studied after removal of 60 mM acetate from the superfusing saline solution. Solutions containing ammonium or acetate were made by equimolar replacement of NaCl with either NH_4Cl or Na acetate, respectively.

Amiloride (3,5-diamino-6-chloropyrazinoyl guanidine; Sigma), SITS (4-acetamido-4′-isothiocyanatostilbene-2,2′-disulfonic acid; Sigma), DIDS (4,4′-diisothiocyanato-stilbene-2,2′-disulfonic acid; Sigma), acetazolamide (5-acetamido-1,3,4-thiadiazole-2-sulfonamide; Cyanamid), and ethoxzolamide (6-ethoxy-2-benzothiazole sulfonamide; a gift of Dr. Siffert) were used within a few minutes of dissolving in solutions.

Microelectrodes: The pH-selective microelectrodes were made as described by Schlue and Thomas.[30] The silanized barrel was filled at the tip with a proton cocktail (tri-*n*-dodecylamine as H^+ ionophore[31]) and behind that with a Cl^--containing citrate buffer, pH 6, saturated with CO_2. Finally the reference barrel was filled with 3 M KCl, and the microelectrode was tested. The Na^+-selective microelectrodes were made as described by Deitmer and Schlue.[32] The tip of the silanized barrel was filled with a column of Na^+ cocktail (ETH 227 as Na^+ ionophore[33]). The Na^+-selective microelectrode barrel was filled with 100 mM NaCl + 10 mM HEPES, pH 7.0, and the reference barrel with 3 M KCl. These electrodes were calibrated in solutions with different Na^+ concentrations and a constant ionic background containing: KCl, 100 mM; $MgCl_2$, 2 mM; $CaCl_2$, 0.01 mM; HEPES, 2 mM (pH 7.0).

RESULTS

Localization of Carbonic Anhydrase by Application of Histochemical and Immunocytochemical Methods

The enzyme carbonic anhydrase (CA) catalyzes the reversible conversion of carbon dioxide (CO_2) and water to bicarbonate and hydrogen ions in a large number of tissues. The reactants influence intra- and extracellular pH and ion concentrations. Therefore, CA is an essential factor in acid-base balance and a transport mechanism for CO_2 and ions.[34,35]

CA was localized in frozen sections of the leech CNS.[36] Hansson's cobalt precipitation method and the use of DNSA to build a fluorescent enzyme-substrate complex showed that glial cells are the sites of CA activity in the leech. Neuropile and connective glial cells surrounding the axons had strong CA activity, whereas packet glial cells, which surround neuron cell bodies, and neurons themselves remained unstained. Furthermore, glial cells reacted markedly with FITC-coupled antibodies against CA isoenzyme II, but experiments with antibodies against CA isoenzyme I showed no reaction.

Giacobini[37] and Korhonen *et al.*,[38] using biochemical and histochemical methods, respectively, were the first to show that glial cells are the sites of CA activity in the brain. Some doubt still exists as to whether CA activity in mammalian brain is limited to oligodendrocytes and glial progenitor cell[39–42] or whether astrocytes have to be included.[43,44]

Extracellular pH Shifts

FIGURE 1 shows a recording with the double-barreled pH-sensitive microelectrode in the extracellular space of a leech ganglion under different experimental conditions. High-frequency firing, typical for neurites, recorded upon further advancement or withdrawal of the electrode tip, indicated that the electrode had entered the neuropile in HEPES-buffered solution (pH 7.4). With the electrode tip located in the extracellular space of the neuropile a pH_o of 7.23 was measured. In nominally HCO_3^--free, HEPES-buffered solution the pH_o was 7.19 ± 0.07 (n = 35) in the extracellular space. Perfusion with CO_2/HCO_3^--buffered saline solution produced an alkalinization by 0.22 pH units to 7.45. After removal of the CO_2/HCO_3^- the pH returned to about 7.2. In saline solution buffered with 2% CO_2/11 mM HCO_3^- (pH

7.4) the pH_o in the extracellular space was 7.40 ± 0.05 ($n = 25$). Hence, CO_2/HCO_3^- buffer caused an alkaline shift of 0.2 pH units in the extracellular space.

Subsequently shown are the pH transients of the extracellular space during and after perfusion of the ganglion with 20 mM NH_4Cl saline solution. On introduction of NH_4Cl the pH_o decreased by 0.3 pH units, which was partially reversed. Removal of NH_4Cl produced a large alkalinization in the extracellular space followed by transient acidosis. We will demonstrate later that the pH_o measured in the microenvironment of glial cells (and neurons) changes in the opposite direction to the glial pH_i during and after NH_4Cl superfusion.[30,45]

Steady-State pH_i

Potential and pH profiles of two impalements into a leech ganglion recorded by a double-barreled neutral-carrier pH-selective microelectrode are shown in FIGURE 2. When the microelectrode tip passed through the ganglion capsule, a stepwise change

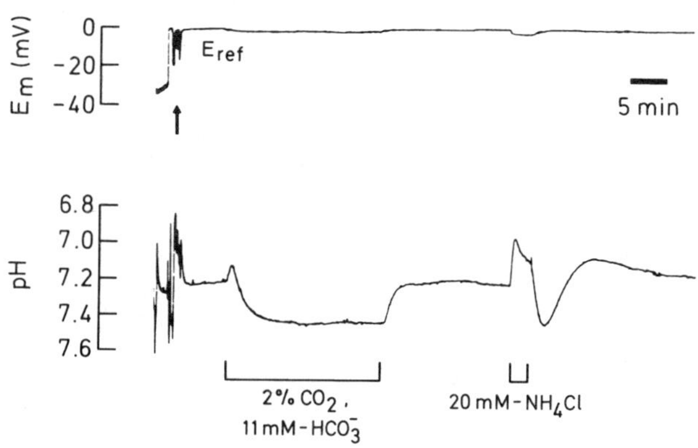

FIGURE 1. Measurement of pH in the extracellular space of the neuropile, and pH changes produced by 2% CO_2, 11 mM HCO_3^-, and 20 NH_4Cl.

in the membrane potential, −45 mV, of a penetrated Retzius (RZ) cell occurred, the soma of which was visible in the microscope; it discharged spontaneous action potentials (FIG. 2A). The pH_i in this HEPES-buffered, nominally CO_2/HCO_3^--free saline solution was 7.3. As the electrode advanced through the ganglion, the microelectrode tip emerged on the other side of the Retzius cell in the extracellular space, where a pH of 7.17 and a reference potential of −3 mV were measured. An acid pH in extracellular spaces and in those near the cell surface has been reported also for other preparations.[11,46,47] Penetration by the microelectrode tip of the anterior neuropile glial (NG,ant) cell elicited a stepwise change in the membrane potential of −64 mV; the pH_i was 6.93. The posterior neuropile glial cell (NG,post) could be penetrated without puncturing other identified cells (FIG. 2B). The pH_i of the posterior neuropile glial cell of the same leech ganglion was 7.0 and the membrane potential was −56 mV.

The intracellular pH, pH_i, membrane potential E_m, and the equilibrium potential for H^+ ions, E_{H^+}, in glial cells of the leech nervous system are summarized in TABLE 1. In leech glial cells the distribution of H^+ ions across the membrane is such that the pH_i is much higher than that predicted if H^+ were passively distributed. Therefore,

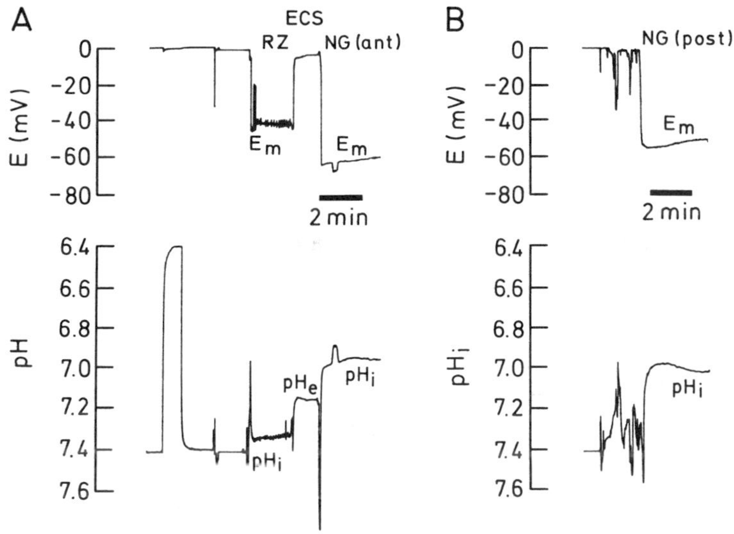

FIGURE 2. Potential profiles of double-barreled pH-sensitive microelectrode during impalement of a ganglion. (**A**) After calibration of the electrode in saline solution buffered to pH 6.4, the tip impaled a Retzius neuron (RZ), the extracellular space (ECS), and finally the anterior neuropile glial cell (NG [ant]). (**B**) Impalement of the posterior neuropile glial cell (NG [post]) of the same ganglion.

the glial cells have mechanisms for H^+ extrusion that maintain pH_i at a value that is well above equilibrium.

Mechanisms of pH_i Regulation

pH_i regulation from intracellular acidosis appears to be mediated by mechanisms driven by the large inwardly directed Na^+ gradient. For recovery from acidification, three separate mechanisms occur: Na^+/H^+ exchange, Na^+-dependent Cl^-/HCO_3^-

TABLE 1. Intracellular pH (pH_i), Membrane Potential (E_m), and H^+ Equilibrium Potential (E_{H^+}) in Glial Cells of Leech Nervous System

Cells	pH_i	E_m (mV)	E_{H^+} (mV)
Neuropile glial cells			
Buffer			
HEPES	6.85 ± 0.06 (n = 23)	−68.4 ± 7.5 (n = 23)	−32.2
2% CO_2/11 mM HCO_3^-	7.18 ± 0.13 (n = 25)	−73.5 ± 5.6 (n = 17)	−12.9
Connective glial cells			
Buffer			
HEPES	7.09 ± 0.03 (n = 35)	−61.2 ± 1.5 (n = 35)	−18.1
2% CO_2/11 mM HCO_3^-	7.37 ± 0.03 (n = 22)	−66.0 ± 1.8 (n = 22)	−1.8

exchange, and Na^+-HCO_3^- cotransport. The Na^+-independent Cl^-/HCO_3^- exchange may mediate some of the pH_i recovery from intracellular alkalinization.

Recovery of pH_i from Intracellular Acidosis in the Absence of CO_2/HCO_3^-

Evidence for Na^+/H^+ Exchange

The Na^+/H^+ exchange process depends on the presence of Na^+ in the external solution and involves the entry of Na^+ into the cell.[15] It is independent of the concentration of HCO_3^- and Cl^- in the external solution and is not blocked by SITS. Furthermore, the K^+-sparing diuretic amiloride has been shown to block the Na^+/H^+ exchanger in most preparations, but not in all.

To test for the presence of a Na^+/H^+ exchanger, we monitored the recovery of pH_i from an acute acid load in HEPES-buffered, nominally HCO_3^--free solutions. The recovery of pH_i from acidification is inhibited completely by removal of external Na^+ (exchanged by *N*-methyl-D-glucamine) or by application of 2 mM amiloride (FIG. 3). Thus, pH_i recovery is mediated by a Na^+/H^+ exchanger in the glial cell membrane, as seen in many vertebrate preparations[48] and also in invertebrate preparations, such as crayfish[15] and leech neurons.[30]

The recovery of pH_i after acidification in neuropile glial cells often occurred more rapidly at similar pH_i values when external Na^+ was re-added after a period of superfusion with Na^+-free saline solution. Sometimes the pH_i recovery was more than twice as fast under these conditions. It can be speculated that intracellular Na^+ affects the Na^+/H^+ exchanger, as was reported for snail neurons[49] and for lymphocytes.[50] After several minutes in Na^+-free solution, the intracellular Na^+ activity of neuropile glial cells decreases rapidly (W. Wuttke and W.-R. Schlue, unpublished observations). A low intracellular Na^+ level may help to accelerate Na^+/H^+ exchange if intracellular Na^+ interferes with this exchanger, as by competing with H^+ for the intracellular binding sites at the cell membrane or by reducing the electrochemical gradient for Na^+.

In leech connective glial cells the ability of the Na^+/H^+ exchanger to regulate pH_i is very poor with recovery from acidosis, in the nominal absence of CO_2/HCO_3^-, being at least 50% slower than in the neuropile glial cell[45] and orders of magnitude slower than that in cultured glial cells.[51,52] Its ineffectiveness in moving protons in either direction (removal of extracellular Na^+ also results in no change in steady-state pH_i in the nominal absence of CO_2/HCO_3^-), at first sight, seems to indicate poor proton mobility in the extracellular space, resulting in either rapid proton accumulation or depletion, depending on the direction of the exchange. Coggeshall and Fawcett[22] have shown that the true extracellular space is very narrow and not immediately accessible to the outside of the axonal bundle. Vanheel *et al.*[53] have shown that if cells are surrounded by areas where movement of protons is limited to simple diffusion, then the rate of recovery from intracellular acidosis can be enhanced by increasing the extracellular buffering power. It was shown that by raising the extracellular HEPES concentration from 10–30 mM, the rate of pH_i recovery from intracellular acidosis increased 2.5 times. An increase in extracellular buffering power can not only increase steady-state pH at the membrane surface but will also reduce extracellular acid accumulation during pH_i regulation.[53] Such maintenance of high pH_o could promote the activity of the Na^+/H^+ exchanger.[54]

In tracer flux studies primary astrocytes in culture have also been shown to possess an external Na^+-dependent, amiloride-sensitive process that results in H^+

ion secretion, probably Na^+/H^+ exchange.[55,56] These findings were subsequently confirmed for cultured C6 glioma cells[57,51] and NN hamster astrocytes.[51] Benos and Sapirstein[57] demonstrated that the transport system of glioma cells did not saturate with external Na^+, which is similar to what Villereal[58] reported for human fibroblasts. In additional experiments, they found that these cells increase the rate at which they acidify the external medium in response to growth factors. Furthermore, this increased acidification could be prevented by amiloride.

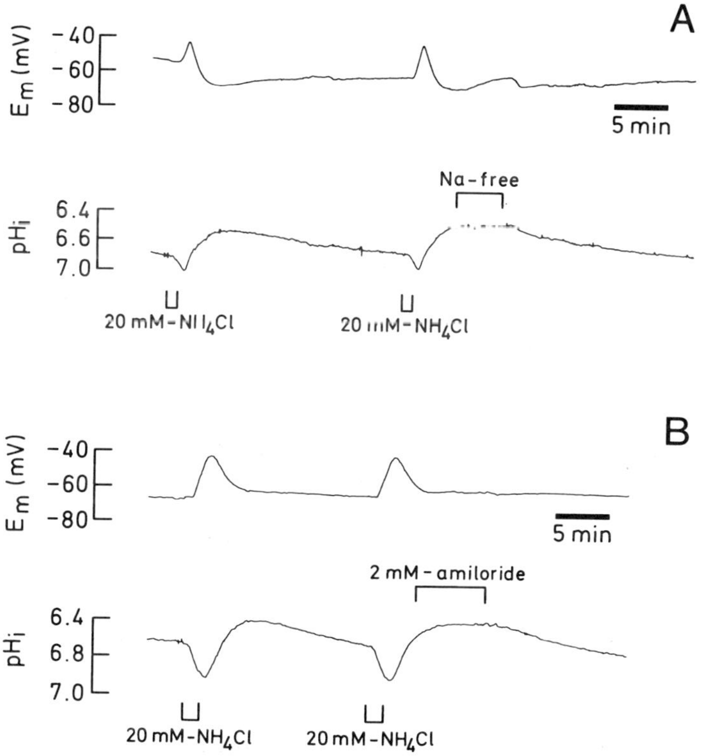

FIGURE 3. Inhibition of pH_i recovery from NH_4Cl-induced acidification in neuropile glial cells by removal of external Na^+ or application of 2 mM amiloride. All solutions were HCO_3^- free.

Measurement of pH_i in cultured NN hamster astrocytes and C6 glioma cells was also based on the fluorescence of the intracellularly trapped fluorescent pH indicator BCECF.[51,59] In both cell types, the Na^+/H^+ exchanger was demonstrated to be the major mechanism by which these cells recover their pH_i after intracellular acidification. Recovery from acidification was dependent on external Na^+ and inhibited by amiloride or its derivatives (ethylisopropylamiloride or benzamil). In cultured mouse oligodendrocytes, based on pH-sensitive microelectrode measurements, recovery of pH_i from acid loads in HCO_3^--free solutions was also found to be mediated by amiloride-sensitive Na^+/H^+ exchange.[52]

Recovery of pH_i from Intracellular Acidosis in the Presence of CO_2/HCO_3^-

The presence of CO_2/HCO_3^- increases the rate of pH_i recovery from intracellular acidosis in leech glial cells. This CO_2/HCO_3^--stimulated recovery appears to be mediated by two further mechanisms: a Na^+-dependent exchange of Cl^- and HCO_3^- and an electrogenic cotransport of Na^+ and HCO_3^- into the glial cells, which is Cl^- independent. The regulation of pH_i in CO_2/HCO_3^- buffer was not significantly affected by the application of amiloride, suggesting that in physiological conditions any pH_i regulatory action of Na^+/H^+ exchange is totally overwhelmed by these HCO_3^--dependent mechanisms.

Evidence for Na^+-Dependent Cl^-/HCO_3^- Exchange

Although the Na^+-dependent Cl^-/HCO_3^- exchanger involves the efflux of Cl^- ions, is dependent on HCO_3^-, and is SITS or DIDS sensitive, it has an absolute requirement for Na^+ and has a steep pH_i dependence.

In leech glial cells, removal of Cl^- from CO_2/HCO_3^--buffered solution results in intracellular alkalosis and its subsequent readdition in intracellular acidosis, suggesting the presence of a Cl^-/HCO_3^- exchanger. This exchange may be in part Na^+ dependent. First, changes in pH_i seen on Cl^- readdition to previously Cl^--depleted cells were found to be reduced in the absence of Na^+. Second, in all cells studied, removal of Cl^- reduced the maximum rate of pH_i recovery from intracellular acidosis on switching from HEPES- to CO_2/HCO_3^--buffered solutions. As all recovery of pH_i from acidosis was Na^+ dependent and predominantly HCO_3^- dependent, the data suggest the presence of a Na^+-dependent Cl^-/HCO_3^- exchanger.

The tight coupling of HCO_3^- transport to Na^+ has been reported for a variety of different animal cells.[2,60] A Na^+-dependent Cl^-/HCO_3 exchanger was first proposed to explain acid extrusion by snail neurons[61] and has subsequently been found in several other invertebrate preparations (e.g., barnacle muscle fibers,[16] squid axon,[62] crayfish central neurons,[15] and crayfish stretch receptor sensory neuron[14]). This acid extrusion system is thought to be electroneutral because pH_i recovery occurs with no change in E_m, and changes in the latter do not affect the rate of pH_i recovery.[63] In both squid axon[64] and snail neurons,[49] oppositely directed acid fluxes were obtained by reversing the ionic electrochemical gradients.

Evidence for Na^+-HCO_3^- Cotransport

The Na^+-HCO_3^- cotransporter depends on the presence of Na^+ and HCO_3^- ions in the external solution, should be blocked by the stilbene derivatives SITS or DIDS, but does not involve the efflux of Cl^- ions from the cell. This cotransporter is electrogenic, and thought to tightly couple the movement of one Na^+ ion to the transport of two or three HCO_3^- ions, thus leading to a transfer of net negative charge.

Evidence for the activation of Na^+-HCO_3^- cotransport across the membrane of leech glial cells is based on analysis of two experimental observations: (1) increase of pH_i recovery from acidification in CO_2/HCO_3^- buffer and (2) alkalinization after the change from HEPES- to CO_2/HCO_3^--buffered saline solution.

Increase of pH_i recovery from acidification in CO_2/HCO_3^- buffer. In the presence of CO_2/HCO_3^- buffer, pH_i recovery from intracellular acidosis of leech glial cells seems to be mediated not only by Na^+-dependent Cl^-/HCO_3^- exchange but also by

Na^+-HCO_3^- cotransport. Evidence that this cotransport mechanism also exists in leech glial cells as a pH_i regulator is: (1) The addition of CO_2/HCO_3^- to the superfusing saline solution always stimulates the rate of recovery from intracellular acidosis and results in an overall increase in steady-state pH_i and E_m (FIG. 4); (2) alkalosis and the associated increase in E_m are inhibited in the absence of Na^+ and in the presence of 0.5 mM DIDS; and (3) recovery from intracellular acidosis still occurs in the absence of Cl^-, although the rate of pH_i recovery is slowed. Furthermore, this remaining recovery is unaffected by the presence of amiloride.

Alkalinization in CO_2/HCO_3^--buffered saline solution. It had been shown previously that alkalinization occurred in neuropile glial cells when HEPES-buffered saline solution was exchanged for saline solution buffered with 2% CO_2/11 mM HCO_3^-.[45,65] Alkalinization occurs in the opposite direction as would be expected from CO_2 movement across the cell membrane and its hydration to form carbonic acid which dissociates into H^+ and HCO_3^- ions.

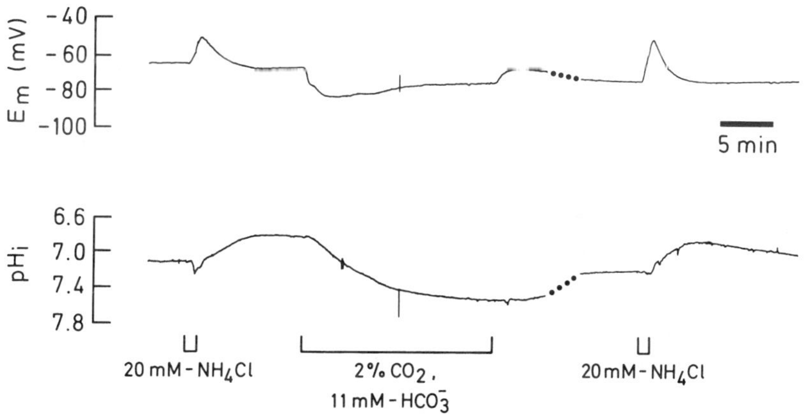

FIGURE 4. pH_i recovery from NH_4Cl-induced acidification in a neuropile glial cell in the absence and presence of 2% CO_2, 11 mM HCO_3^-.

With the pH of the bathing medium kept constant at 7.4, the CO_2/HCO_3^--induced alkalinization was not restricted to the neuropile or connective glial cells, but was also observed in packet glial cells surrounding the nerve cell bodies located at the ganglion periphery. One of the six packet glial cells can be penetrated just between the two Retzius cells. In the sample recording shown (FIG. 5) the pH_i of the packet glial cells was 6.85, and the membrane potential was near −60 mV. The cell could regulate its pH_i, as shown by recovery from the large acidification produced by the addition and subsequent removal of NH_4Cl. Thereafter, HEPES-buffered saline solution was twice replaced by CO_2/HCO_3^--buffered saline solution, which produced alkalinizations from 6.85–7.1 and membrane hyperpolarizations of 8–12 mV. The more alkaline pH_i was maintained during the entire presence of CO_2/HCO_3^-, and only reversed to its initial more acid value when CO_2/HCO_3^--buffered saline solution was replaced by HEPES-buffered saline solution. Thus, the CO_2/HCO_3^--induced alkalinization is a true shift of steady-state pH_i and not a transient phenomenon. Such observations have already been made in smooth muscle[66] and bovine corneal endothelial cells.[67]

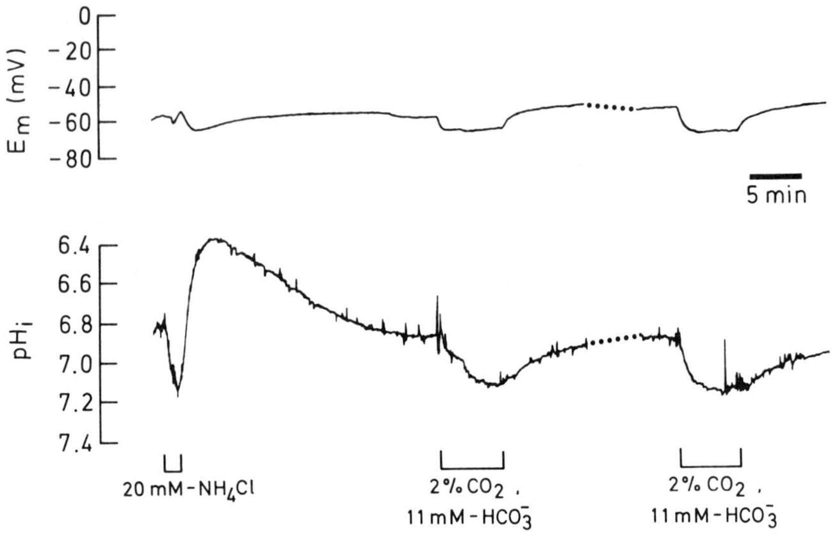

FIGURE 5. Recording from packet glial cell.

It should be emphasized that the alkalinization was typical for leech glial cells and was not observed in neurons, where pH_i decreased and then increased following the change from HEPES- to CO_2/HCO_3^--buffered saline solution.[30] Hence, the alkalinization seems to be unique to glial cells in the nervous system.

As the carbonic anhydrase has been localized in leech glial cells,[36] the possible involvement of this enzyme during glial and extracellular space alkalinization was investigated. The alkalinization after the change from HEPES- to CO_2/HCO_3^--buffered saline solution was not affected by 2 μM ethoxzolamide (FIG. 6). Acetazolamide (0.2 mM) was also without effect (not shown). The membrane hyperpolarization produced by CO_2/HCO_3^- was reduced by 20–30% in some cases, indicating

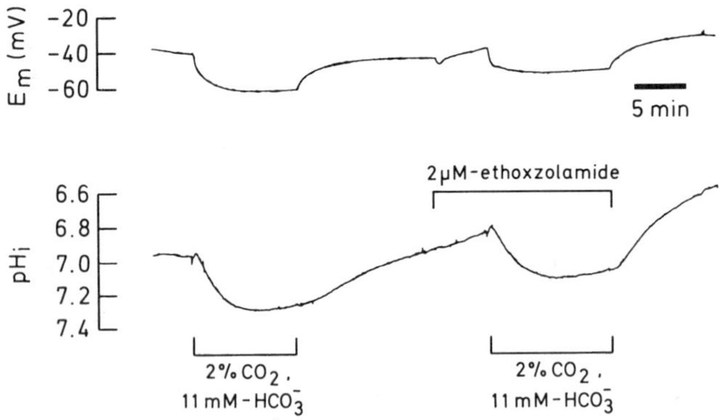

FIGURE 6. Effect of carbonic anhydrase inhibitor ethoxzolamide on the alkaline shift produced by 2% CO_2, 11 mM HCO_3^- in a neuropile glial cell.

that the activity of carbonic anhydrase may be involved in the maintenance of an alkaline pH_i of glial cells.

The alkaline shift produced by CO_2/HCO_3^- in the extracellular space was greatly reduced by carbonic anhydrase inhibitors. The large pH transient of 0.4 pH units produced by CO_2/HCO_3^- was fully inhibited by 0.2 mM acetazolamide (FIG. 7), which itself appeared to cause slow alkalinization of the pH in the extracellular space.

Alkalinization could be due to the activity of an inwardly directed Na^+-HCO_3^- cotransporter,[45,65,68] which determines the pH_i and thereby affects the intracellular buffering power of the glial cells. A coupled transport of Na^+ and HCO_3^- has been suggested to operate in a variety of cells involved in transepithelial transport of HCO_3^- ions.[69–74.]

Evidence for the activation of Na^+-HCO_3^- cotransport similar but reversed compared with that originally suggested by Boron and Boulpaep[69] in the salamander proximal tubule is based on the following findings: (1) the addition of CO_2/HCO_3^- as

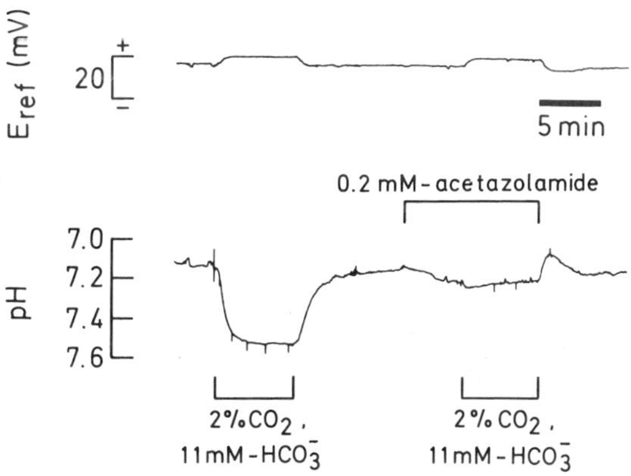

FIGURE 7. Effect of carbonic anhydrase inhibitor acetazolamide on the alkalinization produced by 2% CO_2, 11 mM HCO_3^- in the extracellular space.

an external buffer produced intracellular alkalinization (indicative of an increase in intracellular HCO_3^-; FIG. 5), a rise in intracellular Na^+ (FIG. 8), and membrane hyperpolarization; (2) removal of either HCO_3^- or Na^+ from the saline solution or the addition of the stilbene DIDS abolished or reversed increases in pH_i or intracellular Na^+ activity, and membrane hyperpolarization; (3) CO_2/HCO_3^--induced alkalinization was unaffected by amiloride, the inhibitor of Na^+/H^+ exchange, and by depletion of external and intracellular Cl^-.

It was also shown that the membrane hyperpolarization that accompanied CO_2/HCO_3^--induced changes in pH_i and intracellular Na^+ activity was not due to an increase in membrane conductance, as, for example, an increase in K^+ conductance. The CO_2/HCO_3^- induced membrane hyperpolarization was studied in the presence of Ba^{2+}, which reduces the K^+ permeability of the cell membrane. The addition of 5 mM Ba^{2+} depolarized the glial membrane (FIG. 9). Hyperpolarization on exposure to CO_2/HCO_3^--buffered saline solution, however, increased in the presence of Ba^{2+}. The greater membrane potential change is probably due to increased membrane

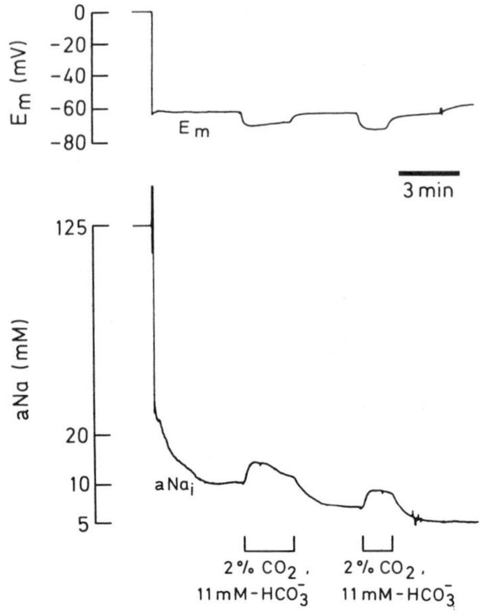

FIGURE 8. Penetration of a neuropile glial cell with a Na^+-selective double-barreled microelectrode and measurement of intracellular Na^+ activity. Two exposures to 2% CO_2, 11 mM HCO_3^--buffered saline solution each induced a rise in intracellular Na^+.

resistance produced by Ba^{2+}. Intracellular Na^+ increased on introduction of CO_2/HCO_3^--buffered saline solution in both the absence and the presence of Ba^{2+}. This confirms that CO_2/HCO_3^--induced hyperpolarization was Na^+ dependent and unlikely to be the result of an increase in membrane K^+ conductance.

Membrane hyperpolarization was believed to be due to an electrogenic process.

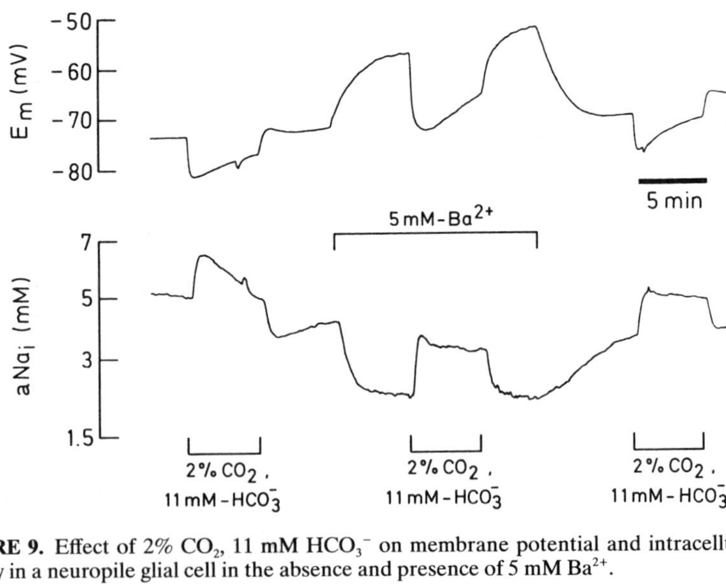

FIGURE 9. Effect of 2% CO_2, 11 mM HCO_3^- on membrane potential and intracellular Na^+ activity in a neuropile glial cell in the absence and presence of 5 mM Ba^{2+}.

Inasmuch as the inhibitor of the Na^+-K^+ pump ouabain did not abolish hyperpolarization,[68] it was concluded that it is the Na^+-HCO_3^- cotransport itself that is electrogenic. Previous studies using radioactive ion fluxes have suggested only electroneutral anion transporters in glial cells.[75] In vertebrate epithelial cells, Na^+/HCO_3^- cotransport was reported to be electrogenic.[69,73,71,76]

If electrogenic Na^+-HCO_3^- cotransport is indeed responsible for alkalinization and membrane hyperpolarization in the presence of CO_2/HCO_3^- buffer, it might be expected that imposed membrane potential changes would result in steady-state pH_i changes. The experiment illustrated in FIGURE 10 shows that if the membrane was depolarized by increasing the external K^+ concentration, and the electrical gradient for HCO_3^- entry on the cotransporter was enhanced, then pH_i increased. FIGURE 10 also shows that the effect of membrane depolarization on pH_i was inhibited by DIDS.

The stoichiometry of the inwardly directed cotransport in leech glial cells was calculated to be 2 HCO_3^-:1 Na^+ under steady-state conditions with an equilibrium potential of −90 mV using the measured values of intracellular Na^+ and pH of the glial cells.[68] A stoichiometry of 2 HCO_3^-:1 Na^+ was suggested for Na^+-HCO_3^- cotransport in the renal proximal tubule of the salamander,[69] whereas a stoichiometry of 3:1 was favored in mammalian renal epithelia.[73,77] In corneal endothelial cells a coupling ratio of 2:1 or 3:1 was compatible with the results.[70]

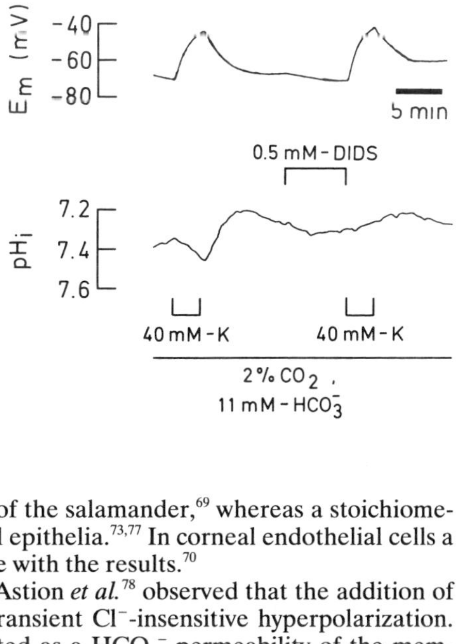

FIGURE 10. Effect of membrane depolarization on pH_i before and after the application of DIDS in a connective glial cell in the continuous presence of 2% CO_2, 11 mM HCO_3^-.

In glial cells of *Necturus* optic nerve, Astion *et al.*[78] observed that the addition of HCO_3^- to the bathing medium caused transient Cl^--insensitive hyperpolarization. The underlying mechanism was interpreted as a HCO_3^- permeability of the membrane of the same order as that for K^+. In subsequent studies the possibility that hyperpolarization could be due to Na^+-HCO_3^- cotransport was examined.[79,80] It was shown that in the presence of Ba^{2+}, the addition of HCO_3^- to the bathing medium caused Na^+-dependent and SITS-sensitive membrane hyperpolarization. Removal of Na^+ resulted in HCO_3^- -dependent, SITS-sensitive depolarization.

These results, based entirely on recording of membrane potential, were indeed in agreement with electrogenic Na^+-HCO_3^- cotransport in which the HCO_3^-: Na^+ stoichiometry was greater than 1. Furthermore, in the presence of Ba^{2+} and HCO_3^-, the addition of SITS caused membrane depolarization, suggesting that the cotransporter is inwardly directed.

In subsequent studies with pH-sensitive microelectrodes it was demonstrated that pH_i recovery from acidification was mediated by Na^+/H^+ exchange and an electrogenic process that was SITS sensitive, Na^+ and HCO_3^- dependent, and Cl^- independent, probably Na^+-HCO_3^- cotransport.[81]

The Na^+- and HCO_3^--mediated pH_i regulation by cultured oligodendrocytes from mouse spinal cord is also not linked to Cl^-. Most likely, a stilbene-insensitive Na^+-HCO_3^- cotransport is responsible for inward transport of HCO_3^- in this preparation, but the stoichiometry of this mechanism is unknown.

Recovery of pH_i from Intracellular Alkalosis in the Presence of CO_2/HCO_3^-

Evidence for Na^+-Independent Cl^-/HCO_3^- Exchange

In addition to the leech glial cell acid extrusion mechanisms just described, evidence for a Na^+-independent Cl^-/HCO_3^- exchanger was also found. At alkaline pH_i, this exchanger most likely operates in the direction of net acid loading (i.e., HCO_3^- efflux). The simple Cl^-/HCO_3^- exchanger is inhibited by SITS and DIDS, by removing external Cl^-, or by depleting internal HCO_3^-, but it does not involve the influx of Na^+ ions into the cells.

FIGURE 11A shows an experiment in which a connective glial cell was twice made alkaline by the addition and subsequent removal of 60 mM acetate during prolonged exposure to 2% CO_2/11 mM HCO_3^-. Recovery in CO_2/HCO_3^--buffered saline solution from the intracellular alkalosis was rapid and averaged a maximum rate of 7.72 ± 0.48 pH units/hour at a pH_i of 7.62 ± 0.03 (n = 10). In Cl^--free saline solution, however, pH_i recovery was slowed, averaging 4.75 ± 0.30 pH units/hour (n = 4) at a mean pH_i of 7.80 ± 0.02.

The experiment in FIGURE 11B demonstrates the effect of SITS on the rate of pH_i recovery from intracellular alkalosis. In the presence of CO_2/HCO_3^--buffered saline solution the rate of pH_i recovery was rapid, but application of 0.5 mM SITS reduced the rate of pH_i recovery from the induced alkalosis considerably. On an average the maximum rate of recovery from intracellular alkalosis in the presence of the drug was reduced from 7.62 ± 0.32 (pH_i 7.61 ± 0.04) to 5.45 ± 0.74 pH units/hour (pH_i 7.53 ± 0.03) (n = 4).

There is strong evidence for the existence of a SITS-sensitive, Cl^-- and HCO_3^--dependent, but Na^+-independent mechanism that may have a role in acid transport across the connective glial cell membrane. It can be predicted that if Cl^- is passively distributed, as in neuropile glial cells,[82] then indeed the inwardly directed Cl^- gradient would be sufficient to drive HCO_3^- out of the cell, resulting in net intracellular acidification. These findings are consistent with observations in other cell types that Cl^-/HCO_3^- exchange may mediate some of the pH_i recovery from intracellular alkalosis.[83–86] The nature of the remaining mechanism(s) that promote pH_i recovery from intracellular alkalosis in leech glial cells are unknown.

Indirect evidence by tracer flux studies for the existence of Na^+-dependent Cl^-/HCO_3^- exchange has been obtained for cultured primary astrocytes[55,87] and LRM55 glioma cells originally isolated from rat spinal tumor[88]. In both preparations the uptake and efflux of $^{36}Cl^-$ were stimulated in the presence of HCO_3^- and inhibited by the stilbene derivative SITS. There is no evidence that Na^+-independent Cl^-/HCO_3^- exchange occurs in cultured mouse oligodendrocytes.[52]

Effect of Neurotransmitter on pH_i

At present it is unknown how neuronal activity may alter the activation of the different pH_i regulatory systems in leech glial cells. Dörner *et al.*[89] demonstrated that bath applications of the glutaminergic agonists L-glutamate, kainate, and quisqualate induced basically similar membrane depolarizations and alterations in intracellular Na^+ and K^+ activities in leech neuropile glial cells. In contrast, *N*-methyl-D-aspartic acid (NMDA) did not affect glial membrane potential and ion activities. It

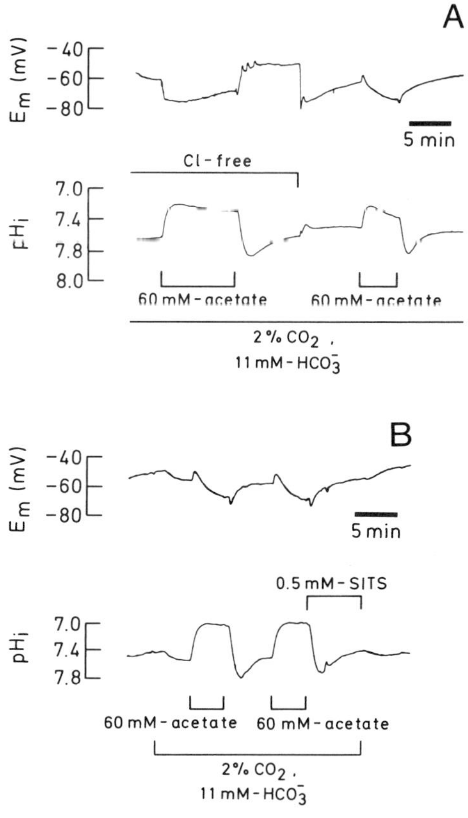

FIGURE 11. Effects of Cl-free solution (**A**) and 0.5 mM SITS (**B**) on pH_i recovery from intracellular alkalosis in connective glial cells after removal of 60 mM acetate in the presence of 2% CO_2, 11 mM HCO_3^-.

was concluded that leech glial cells have a non-NMDA, kainate/quisqualate-preferring glutamate receptor.

The kainate depolarization was accompanied by a transient acidification of the glial cells in HEPES-buffered solution, very similar to that during activation of acetylcholine receptors in the same preparation.[90] The mean pH_i in standard HEPES-buffered saline (nominally CO_2/HCO_3^--free) was 6.85 ± 0.1 (mean ± SD, $n = 12$) and in 2% CO_2/11 mM HCO_3^- buffered saline solution 7.22 ± 0.13 ($n = 10$). During application of 10^{-4} M kainate (1 minute), pH_i transiently fell by 0.26 ± 0.07 units in HEPES and by 0.22 ± 0.07 in CO_2/HCO_3^- buffer.

FIGURE 12 shows examples of the effect of removing external Na^+ and application of amiloride on the pH_i recovery. It is evident that the pH_i returned to normal in both cases, and that its recovery was somewhat faster under both experimental conditions. This result suggests that the activation of Na^+/H^+ exchange is probably not involved in the pH_i recovery from the kainate-induced acidification, and since the experiments were performed in the absence of HCO_3^-, the involvement of HCO_3^--dependent mechanisms can also be excluded. Therefore, it may be that the neurotransmitter agonists affect an unidentified acid extrusion mechanism.

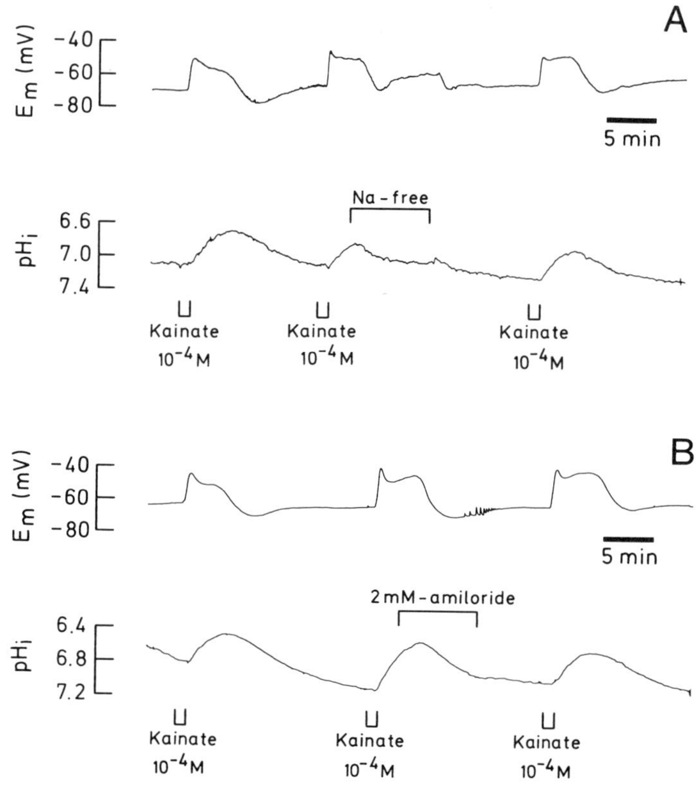

FIGURE 12. Effects of Na-free solution (**A**) and amiloride (**B**) on pH_i recovery in neuropile glial cells from kainate-induced acidification.

CONCLUSIONS

The data demonstrate that leech glial cells have three acid extruders that contribute to pH_i regulation. In the presence of CO_2/HCO_3^- buffer, pH_i recovery from intracellular acidosis is mediated by Na^+-HCO_3^- cotransport. In addition, there is evidence for Na^+-dependent Cl^-/HCO_3^- exchange and Na^+/H^+ exchange across the glial cell membrane. Simultaneous action of Na^+-HCO_3^- cotransport and Na^+/H^+ exchange was found in cultured mouse spinal cord oligodendrocytes[52] and in mudpuppy optic nerve glial cells.[79–81] Na^+/H^+ exchange was found in cultured rat primary astrocytes,[55,56] NN hamster astrocytes,[51] and C6 glioma cells.[51,57,59]

The recovery of pH_i from intracellular alkalosis of leech glial cells in the presence of CO_2/HCO_3^- buffer apparently is mediated by Na^+-independent Cl^-/HCO_3^- exchange, which is an acid loader. This exchanger was also found in cultured rat astrocytes[55,87] and in LRM55 glioma cells.[88]

Activation of Na^+-HCO_3^- cotransport and pH_i increase in glial cells are associated with HCO_3^- uptake, which increases the buffering power of glial cytoplasm. If the concentration of CO_2 in the nervous tissue increases, as may be caused by increased neuronal activity, the pH in glial cells would go alkaline because of stimulation of Na^+-HCO_3^- cotransport. Furthermore, glial membrane depolarization during neuronal activity could facilitate electrogenic Na^+-HCO_3^- cotransport and induce an increase in glial pH_i. Indeed, alkalinization was recently measured in rat astrocytes after evoked electrical activity of nerve cells.[18,19] Hence, stimulation of Na^+-HCO_3^- cotransport into the glial cells could be responsible for the depolarization-induced alkaline shift in mammalian glial cells.

REFERENCES

1. Roos, A. & W. F. Boron. 1981. Physiol. Rev. **61:** 296–434.
2. Boron, W. F. 1983. J. Membrane Biol. **72:** 1–16.
3. Busa, W. B. & R. Nuccitelli. 1984. Am. J. Physiol. **246:** R409–R438.
4. Frelin, C., P. Vigne, A. Ladoux & M. Lazdunski. 1988. Eur. J. Biochem. **174:** 3–14.
5. Madshus, I. H. 1988. Biochem. J. **250:** 1–8.
6. Moody, W. 1984. Ann. Rev. Neurosci. **7:** 257–278.
7. Thomas, R. C. & R. W. Meech. 1982. Nature **299:** 826–828.
8. Meech, R. W. & R. C. Thomas. 1987. J. Physiol. **390:** 433–452.
9. Thomas, R. C. 1988. J. Physiol. **398:** 313–327.
10. Urbanics, R., E. Leniger-Follert & D. W. Lübbers. 1978. Pflügers Arch. **378:** 47–53.
11. Kraig, R. P., C. R. Ferreira-Filho & C. Nicholson. 1983. J. Neurophysiol. **49:** 831–850.
12. Sykova, E. & J. Svoboda. 1990. Brain Res. **512:** 181–189.
13. Hansen, A. J. 1985. Physiol. Rev. **65:** 101–148.
14. Moser, H. 1985. J. Physiol. **362:** 23–38.
15. Moody, W. 1981. J. Physiol. **316:** 293–308.
16. Boron, W. F., W. C. McCormick & A. Roos. 1979. Am. J. Physiol. **237:** C185–C193.
17. Boron, W. F. & E. L. Boulpaep. 1983. J. Gen. Physiol. **81:** 29–52.
18. Chesler, M. & R. P. Kraig. 1987. Am. J. Physiol. **253:** R666–670.
19. Chesler, M. & R. P. Kraig. 1989. J. Neurosci. **9:** 2011–2019.
20. Ahmed, Z. & J. A. Connor. 1980. J. Gen. Physiol. **75:** 403–426.
21. Brown, H. M. & R. W. Meech. 1979. J. Physiol. **297:** 73–93.
22. Coggeshall, R. E. & D. W. Fawcett. 1964. J. Neurophysiol. **27:** 229–289.
23. Gray, E. G. & R. W. Guillery. 1963. Z. Zellforsch. **60:** 826–849.
24. Hansson, H. P. J. 1967. Histochemie **11:** 112–128.
25. Pochhammer, C., P. Dietsch & P. R. Siegmund. 1979. J. Histochem. Cytochem. **27:** 1103–1107.
26. Dermietzel, R., A. Leibstein, W. Siffert, N. Zamboglou & G. Gros. 1985. J. Histochem. Cytochem. **33:** 93–98.
27. Chen, R. F. & J. C. Kernohan. 1967. J. Biol. Chem. **242:** 5813–5823.
28. Sternberger, L. A. 1979. Immunocytochemistry. 2nd. Ed.: 24–58. Wiley. New York.
29. Schlue, W.-R. & J. W. Deitmer. 1980. J. Exp. Biol. **87:** 23–43.
30. Schlue, W.-R. & R. C. Thomas. 1985. J. Physiol. **364:** 327–338.
31. Ammann, D., F. Lanter, R. A. Stiner, P. Schulthess, Y. Shijo & W. Simon. 1981. Anal. Chem. **53:** 2267–2269.
32. Deitmer, J. W. & Schlue, W.-R. 1983. Pflügers Arch. **397:** 195–201.
33. Steiner, R. A., M. Oehme, D. Ammann & W. Simon. 1979. Anal. Chem. **51:** 351–353.

34. TASHIAN, R. E. & D. HEWETT-EMMET. 1984. Ann. N. Y. Acad. Sci. **429**.
35. MAREN, T. H. 1967. Physiol. Rev. **47:** 595–781.
36. RIEHL, B. & W.-R. SCHLUE. 1990. J. Histochem. Cytochem. **38:** 1173–1178.
37. GIACOBINI, E. 1962. J. Neurochem. **9:** 169–177.
38. KORHONEN, K. L., E. NÄÄTÄNEN & M. HYYPPÄ. 1964. Acta Histochem. **18:** 336–347.
39. BENJELLOUN, S., J. P. DELAUNOY, D. GOMES, F. DE VITRY, D. LANGUI & P. DUPOUNEY. 1986. Dev. Neurosci. **8:** 150–159.
40. GHANDOUR, M. S., O. K. LANLEY, G. VINCENDON & G. GOMBOS. 1979. J. Histochem. Cytochem. **27:** 1634–1637.
41. GHANDOUR, M. S., G. VINCENDON & G. GOMBOS. 1980. Dev. Biol. **77:** 73–83.
42. YANDRASITZ, J. R., S. A. ERNST & L. SALGANICOFF. 1976. J. Neurochem. **27:** 707–715.
43. CAMMER, W. & F. A. TANSEY. 1988. J. Neurochem. **50:** 319–322.
44. KIMELBERG, H. K., P. E. STIEG & J. E. MAZURKIEWICZ. 1982. J. Neurochem. **39:** 734–742.
45. DEITMER, J. W. & W.-R. SCHLUE. 1987. J. Physiol. **388:** 261–283.
46. DE HEMPTINNE, A. & F. HUGUENIN, 1984. J. Physiol. **347:** 581–592.
47. CHESLER, M. 1986. J. Physiol. **381:** 241–261.
48. GRINSTEIN, S. & A. ROTHSTEIN. 1986. J. Membrane Biol. **90:** 1–12.
49. EVANS, M. G. & R. C. THOMAS. 1984. J. Physiol. **346:** 143–154.
50. GRINSTEIN, S., J. D. GOETZ & A. ROTHSTEIN. 1984. J. Gen. Physiol. **84:** 585–600.
51. JEAN, T., C. FRELIN, P. VIGNE & M. LAZDUNSKI. 1986. Eur. J. Biochem. **160:** 211–219.
52. KETTENMANN, H. & W.-R. SCHLUE. 1988. J. Physiol. **406:** 147–162.
53. VANHEEL, B., A. DE HEMPTINNE & I. LEUSEN. 1986. Am. J. Physiol. **250:** C748–C760.
54. ARONSON, P. S., M. A. SUHM & J. NEE. 1983. J. Biol. Chem. **258:** 6767–6771.
55. KIMELBERG, H. K., S. BIDDLECOME & R. S. BOURKE. 1979. Brain Res. **173:** 111–124.
56. KIMELBERG, H. K. & C. RICARD. 1981. Am. Soc. Neurochem. **13:** 112.
57. BENOS, D. J. & V. S. SAPIRSTEIN. 1983. J. Cell. Physiol. **116:** 213–220.
58. VILLEREAL, M. 1981. J. Cell. Physiol. **108:** 251–259.
59. JAKUBOVICZ, D. E., S. GRINSTEIN & A. KLIP. 1987. Brain Res. **435:** 138–146.
60. THOMAS, R. C. 1984. J. Physiol. **354:** 3P-22P.
61. THOMAS, R. C. 1977. J. Physiol. **273:** 317–338.
62. BORON, W. F. & J. M. RUSSELL. 1983. J. Gen. Physiol. **81:** 373–399.
63. THOMAS, R. C. 1978. Respir. Physiol. **33:** 63–73.
64. RUSSELL, J. M., W. F. BORON & M. S. BRODWICK. 1983. J. Gen. Physiol. **82:** 47–78.
65. SCHLUE, W.-R. & J. W. DEITMER. 1988. *In* Proton Passage Across Cell Membranes. Ciba Found. Symp. **139:** 47–69. Wiley. Chichester.
66. AICKIN, C. C. 1984. J. Physiol. **349:** 571–585.
67. JENTSCH, T. J., C. KORBMACHER, I. JANICKE, D. G. FISCHER, F. STAHL, H. HELBIG, E. J. CRAGOE, S. K. KELLER & M. WIEDERHOLT. 1988. J. Membr. Biol. **103:** 29–40.
68. DEITMER, J. W. & W.-R. SCHLUE. 1989. J. Physiol. **411:** 179–194.
69. BORON, W. F. & E. L. BOULPAEP. 1983. J. Gen. Physiol. **81:** 53–94.
70. JENTSCH, T. J., S. K. KELLER, M. KOCH & M. WIEDERHOLT. 1984. J. Membr. Biol. **81:** 189–204.
71. JENTSCH, T. J., B. S. SCHILL, P. SCHWATZ, H. MATTHES, S. K. KELLER & M. WIEDERHOLT. 1985. J. Biol. Chem. **260:** 15554–15560.
72. ALPERN, R. J. 1985. J. Gen. Physiol. **86:** 613–636.
73. YOSHITOMI, K., B. C. BURCKHARDT & E. FRÖMTER. 1985. Pflügers Arch. Eur. J. Physiol. **405:** 360–366.
74. BIAGI, B. A. & M. SOHTELL. 1986. Am. J. Physiol. **250:** F267–272.
75. KIMELBERG, H. K. & M. V. FRANGAKIS. 1986. Adv. Biosci. **61:** 177–186.
76. JENTSCH, T. J., H. MATTHES, S. K. KELLER & M. WIEDERHOLT. 1986. Am. J. Physiol. **251:** F954–F968.
77. SOLEIMANI, M., S. M. GRASSL & P. S. ARONSON. 1987. J. Clin. Invest. **79:** 1276–1280.
78. ASTION, M. L., J. A. COLES & R. K. ORKAND. 1987. Neurosci. Lett. **76:** 47–52.
79. ASTION, M. L. & R. K. ORKAND. 1988. Glia **1:** 355–357.
80. ASTION, M. L., A. L. OBAID & R. K. ORKAND. 1989a. J. Gen. Physiol. **93:** 731–744.
81. ASTION, M. L., A. CHVATAL, J. A. COLES & R. K. ORKAND. 1989b. Acta Physiol. Scand. **136**(Suppl. 582): 64.

82. BALLANYI, K. & W.-R. SCHLUE. 1990. J. Physiol. **420:** 325–336.
83. VAUGHAN-JONES, R. D. 1979. J. Physiol. **295:** 111–137.
84. JENTSCH, T. J., I. JANICKE, D. SORGENFREI, S. K. KELLER & M. WIEDERHOLT. 1986. J. Biol. Chem. **261:** 12120–12127.
85. KORBMACHER, C., H. HELBIG, F. STAHL & M. WIEDERHOLT. 1988. Pflügers Arch. **412:** 29–36.
86. BOYARSKY, G., M. B. GANZ, R. B. STERZEL & W. F. BORON. 1988. Am. J. Physiol. **255:** C857–C869.
87. KIMELBERG, H. K. 1981. Biochim. Biophys. Acta **464:** 179–184.
88. WOLPAW, E. & D. L. MARTIN. 1984. Brain Res. **297:** 317–327.
89. DÖRNER, R., K. BALLANYI & W.-R. SCHLUE. Brain Res. **523:** 111–116.
90. BALLANYI, K. & W.-R. SCHLUE. 1989. Glia **2:** 330–345.

Glial Ion Transport and Volume Control

O. KEMPSKI,[a,b] S. VON ROSEN,[c] H. WEIGT,[c] F. STAUB,[c]
J. PETERS,[c] AND A. BAETHMANN[c]

[a] *Institute for Neurosurgical Pathophysiology*
Johannes Gutenberg-University Mainz
Mainz, Germany

[c] *Institute for Surgical Research*
Ludwig-Maximilians University
Munich, Germany

Ion homeostasis in the central nervous system (CNS) to a great extent is achieved by glial cells.[1,2] Extracellular K^+ concentrations are closely controlled at 3 mM. Because of glial uptake and buffering mechanisms, extracellular K^+ levels seldom surpass 12–15 mM. Only during pathological events such as anoxia, hypoxia, ischemia, or hypoglycemia and during spreading depression, K^+ increases slowly first to a ceiling of 12–15 mM and then it suddenly reaches 60–80 mM. (For a review see refs. 2 and 3.) During this phase, extensive glial swelling is regularly observed. If the pathological situation does not last long, the increases of extracellular K^+ are completely reversible as is glial swelling. The molecular mechanisms involved in the dramatic increase of extracellular potassium during pathophysiological conditions as well as the factors leading to glial swelling are not well understood.[2] The same holds true for the normalization of cell volume and extracellular K^+ after short-lasting insults.

Considerable scientific effort has been devoted to the role of glial cells in potassium uptake from the extracellular space. Only a little information, however, is available on whether and how elevated extracellular concentrations of K^+ cause glial swelling, which may be explained by two reasons: (1) accurate measurements of relatively discrete changes of cellular volume are not possible using conventional electrophysiological equipment, and (2) under *in vivo* conditions, pathophysiological events such as cerebral ischemia or elevated extracellular K^+ are accompanied by a multitude of changes, such as release of transmitters, altered cerebral metabolism, or changes of cerebral blood flow, which by themselves may cause glial swelling.

For the current study, therefore, an *in vitro* system[4–8] was used that permits close control of extracellular conditions and detection of cell volume changes of less than 1% by the use of an advanced flow cytometer. The cell volume response to 30 mM K^+ was evaluated in C6 glioma cells used as model cells with glial characteristics and in astrocytes from primary culture.

MATERIAL AND METHODS

Cell Culture

C6 glioma cells were grown at 37°C in Dulbecco's minimal essential medium supplemented with 10% fetal calf serum (Boehringer, Mannheim, Germany) in a

[b] Address for correspondence: Univ.-Prof. Dr. med. O. Kempski, Institute for Neurosurgical Pathophysiology, Johannes Gutenberg-University Mainz, Langenbeckstraße 1, 6500 Mainz, Germany.

humidified atmosphere of 5% CO_2 and 95% room air in plastic petri dishes. The cells were subcultured 2–3 times per week. Confluent cultures obtained 2 days after subcultivation were used for the experiments. Glial cells from primary culture were prepared from 3-day-old DB9-rats according to a method modified from a procedure described by Frangakis and Kimelberg.[9] In brief, cerebral hemispheres were isolated and freed from meninges under a dissecting microscope. The tissue was minced with scissors and then repeatedly exposed to dispase (Boehringer, Mannheim, Germany) for 10 minutes. Single cells were harvested from the supernatant by centrifugation and seeded into petri dishes. Culture and subcultivation conditions were identical to those described for C6 glioma cells. For the experiments, only confluent cultures obtained 2 days after subcultivation were used. The cells were harvested and suspended in serum-free medium as previously described.[4]

Experimental Setup

Suspended cells were transferred to a Plexiglass incubation chamber[4] that allowed for continuous control of pH, PO_2, and temperature, using respective electrodes. Permeable silicon rubber tubing inside the chamber served as a membrane oxygenator supplying the cell suspension with CO_2, O_2, and N_2 in any desired mixture. Cell sedimentation was prevented by a magnetic stirrer. Drugs were injected and cell samples collected via cannulas inserted into the chamber.

Experimental Groups

The experiments were performed after a 45-minute control period, necessary for measurements of normal cell volume, medium osmolality, and cell viability. Preparations with an unstable cell volume or a cell viability below 90% were discarded. Viability was determined by trypan blue exclusion. Three cell volume measurements, obtained during the last 15 minutes of the control phase, were taken as reference values in order to express experimental data on a percentage basis.

To study the effect of high extracellular potassium concentrations, the cell suspension in the chamber ($[K^+_e]$ = 5 mM) was diluted with isotonic medium of 55 mM K^+, keeping the sum of Na^+ and K^+ constant (sodium ions replaced by potassium ions). The final potassium concentration in the suspension was 30 mM. Drugs (furosemide, amiloride, iodoacetate, and ouabain) were added 15 minutes before the increase of extracellular potassium and were also present in the high potassium solution. Quercetin was added 30 minutes before the potassium concentration was increased.

In a subset of experiments, anoxia was induced before the elevation of extracellular K^+. Anoxia was initiated by disconnecting the oxygen supply to the membrane oxygenator, leading to a decrease in PO_2 to 0 mm Hg within approximately 5 minutes. The incubation chamber, in addition, was placed in an air-tight plastic tent filled with N_2 to prevent leakage of room air into the chamber. Details of the methodology have been published previously.[5]

Cell Volume Measurements

Cell volume was determined by flow cytometry (Metricell®) according to the Coulter principle with additional hydrodynamic focusing.[10] The accuracy of the method allows detection of cell size alterations of less than 1%. The system is calibrated electrically as well as mechanically by the use of latex beads of known diameter.

Statistics

The results were evaluated statistically by the Kruskal-Wallis test for nonparametric analysis of variance and multiple comparisons on ranks of several independent samples.[11]

RESULTS

Exposure of C6 glioma cells to 30 mM K^+ was followed by very stereotypic changes of cell volume: a phase of gradual swelling to a maximum of 108.3 ± 0.91% of the control volume after 10 minutes was followed by slow volume normalization (100.84 ± 1.26% after 45 minutes), although extracellular K^+ remained elevated during the entire experimental period (FIG. 1A). Comparable swelling kinetics were found in astrocytes from primary culture, with a slightly higher swelling maximum of 109.97 ± 2.09% after 12 minutes and a less rapid and complete normalization of cell volume (FIG. 1B). Preincubation of the cell suspension with 1.0 mM furosemide or 0.5 mM Ba^{2+} ions (TABLE 1) reduced the swelling maximum without reaching statistical significance. With 0.1 mM amiloride the swelling maximum was decreased to 103.45 ± 0.69% of control ($p < 0.05$, TABLE 1). Pretreatment with 1.0 mM ouabain or 2.5 mM iodoacetate completely prevented the swelling response (FIG. 2). Elevation of extracellular K^+ during anoxia was followed by earlier swelling of 108.6 ± 2.68 after 2 minutes ($p < 0.05$ *vs.* normoxia) and 110.2 ± 1.85% after 8 minutes (TABLE 1). As in normoxia, iodoacetate prevented the swelling response to K^+ in anoxia (TABLE 1, $p < 0.05$).

Quercetin (25 μM), an inhibitor of lactate export from the cell, by itself caused considerable swelling of 114.1%, reaching a plateau after 30 minutes (FIG. 3A). To evaluate K^+-dependent swelling after quercetin treatment, the volume measured during the plateau phase was chosen as 100% control value. Increasing extracellular K^+ during this plateau phase slightly augmented the swelling maximum and, most importantly, nearly abolished the recovery of cell volume (FIG. 3B). Upon elevation of extracellular K^+, cell volume increased to 109.27 ± 1.69 within 15 minutes. After 30 minutes cell volume was still increased to 108.76 ± 1.57% of the volume during the quercetin-induced plateau phase; after 45 minutes, 106.60 ± 1.27% was found (FIG. 3B). Cell viability was not affected by any treatment.

DISCUSSION

The role of glial cells in CNS K^+ homeostasis has been studied by a number of investigators. (For a review see ref. 2.) Several mechanisms have been proposed by which elevated extracellular potassium concentrations are reduced by glia: active uptake via glial Na^+/K^+-ATPase and furosemide-sensitive NaCl, KCl, or (K, Na, 2Cl)-cotransport, extracellular diffusion, or transcellular potassium migration involving current flow and spatial buffering.[2,12,13] Despite intense research, however, the molecular interactions involved are not clear yet. A phenomenon regularly observed

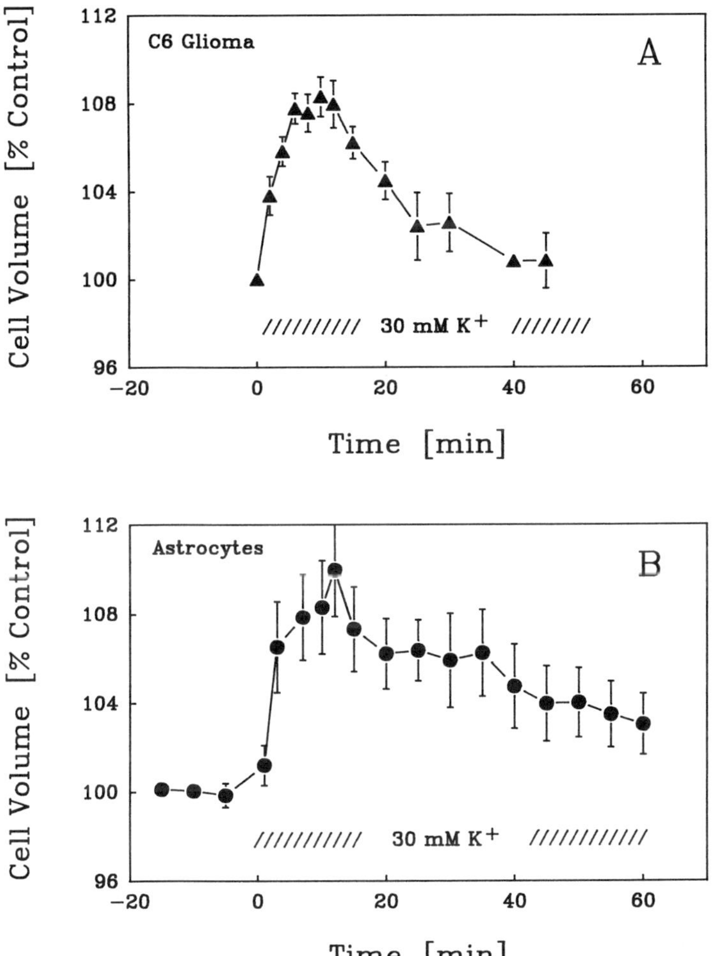

FIGURE 1. Cell volume response of C6 glioma cells (**A**) and astrocytes from primary culture (**B**) during exposure to isotonic medium containing 30 mM K^+. Potassium concentration was elevated suddenly at time zero. Cell volume is expressed as a percentage of the mean of three consecutive cell volume measurements from the preceding control phase at 5 mM K^+. Results are expressed as the means ± SEM from 10 (**A**) and 6 (**B**) experiments. Cell viability was not affected during the course of the experiments.

after pronounced elevation of extracellular K^+ is glial swelling.[13–18] During pathophysiological events such as anoxia, ischemia, or spreading depression, any increase in glial cell volume may present a serious clinical problem because it contributes to brain swelling and therefore to the rise in intracranial pressure that often determines the fate of a patient. The current study was designed to evaluate the sequelae of glial volume changes at high K^+ and the mechanisms of K^+ homeostasis.

The *in vitro* model used has the clear advantage that even subtle alterations in cell

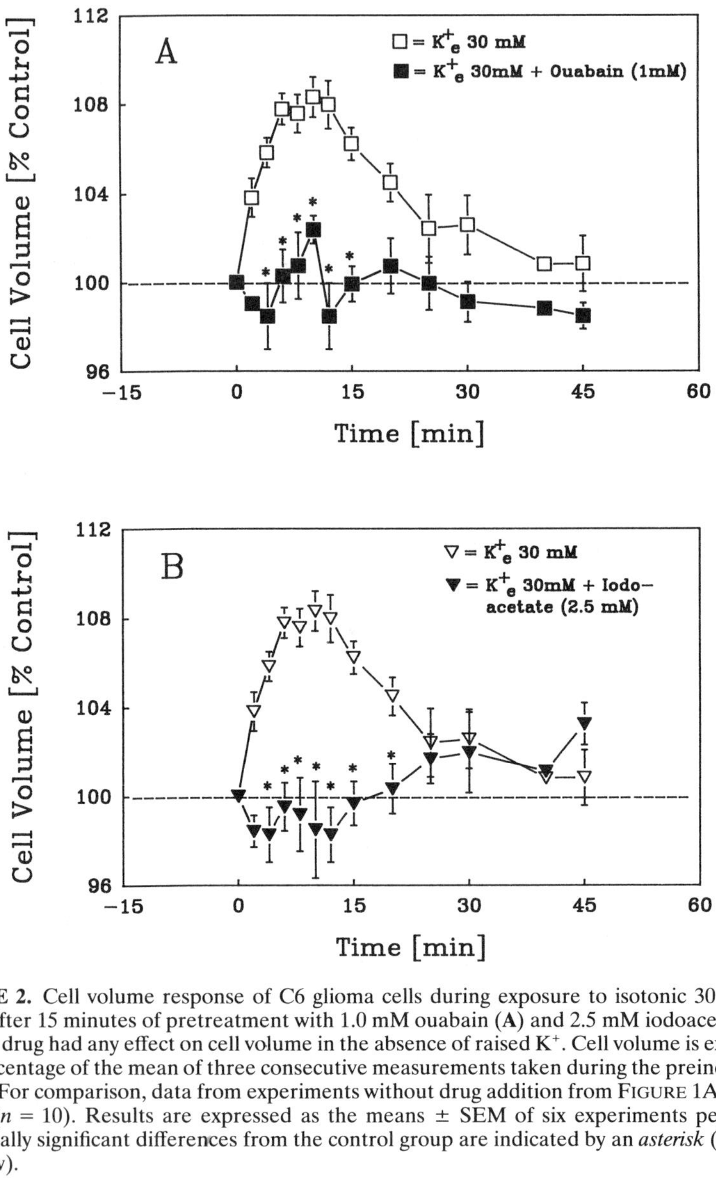

FIGURE 2. Cell volume response of C6 glioma cells during exposure to isotonic 30 mM K^+ media after 15 minutes of pretreatment with 1.0 mM ouabain (**A**) and 2.5 mM iodoacetate (**B**). Neither drug had any effect on cell volume in the absence of raised K^+. Cell volume is expressed as a percentage of the mean of three consecutive measurements taken during the preincubation period. For comparison, data from experiments without drug addition from FIGURE 1A are also shown ($n = 10$). Results are expressed as the means ± SEM of six experiments per group. Statistically significant differences from the control group are indicated by an *asterisk* ($p < 0.05$ or below).

volume of less than 1% may be detected, a sensitivity never reached by radioactive tracers to determine intracellular water space.[18] The model also has been used for related studies on the effects of hypotonic media,[4] anoxia,[5] ouabain,[7] or acidosis[6] in C6 glioma cells or astrocytes from primary culture. The current study is based on those previous investigations that demonstrated that glial cell volume is unaffected by either anoxia, blockage of anaerobic glycolysis by iodoacetate, or inhibition of Na^+/K^+-ATPase by ouabain, whereas acidosis below pH 7.0 caused glial swelling. In conclusion, glial swelling observed *in vivo* during cerebral ischemia cannot be a direct consequence of interrupted oxygen supply, but rather results from secondary events such as acidosis[6] or increased extracellular potassium concentrations, the object of the present investigation.

The *in vitro* model has provided very precise information on glial volume regulatory phenomena. C6 glioma cells were again used as model cells with proper-

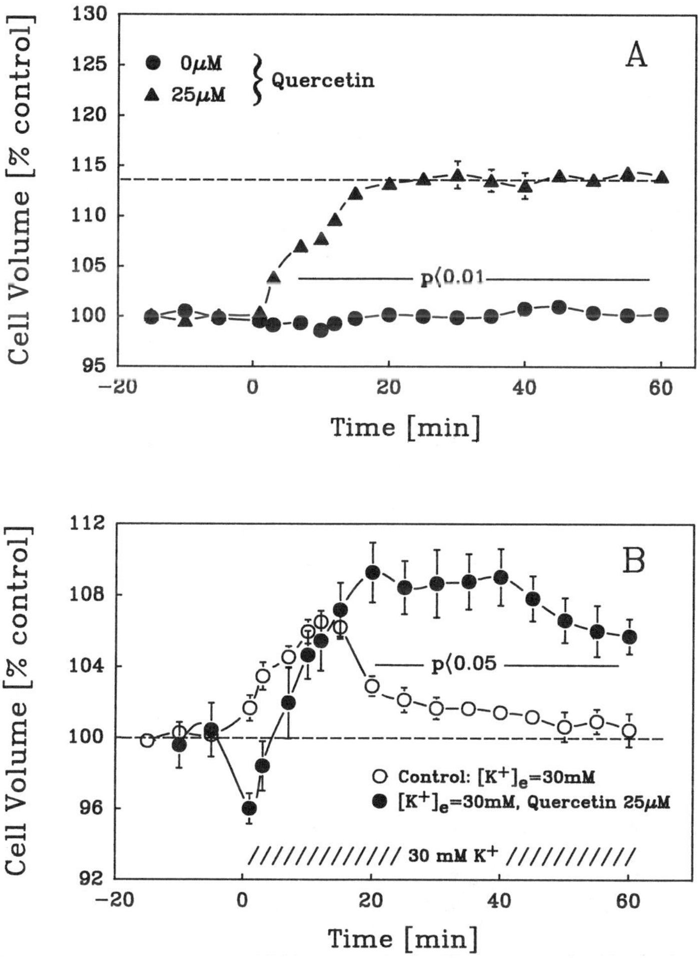

FIGURE 3. (**A**) Effect of quercetin (25 μM) on C6 glial volume. Quercetin was added at time zero; *symbols* represent the means ± SEM of six experiments per group. Absence of error bars indicates standard errors smaller than symbol size. *Filled circles* represent experiments with an exposure to ethyl alcohol that was used to dissolve quercetin in the other experiments. The *dotted line* marks the cell volume plateau reached after a 30-minute exposure to quercetin. (**B**) Cell volume response of C6 glioma cells during exposure to isotonic 30 mM K^+ media after a 30-minute pretreatment with 25 μM quercetin ($n = 5$). Cell volume is expressed as a percentage of the mean of three consecutive measurements taken during the plateau phase marked in **A.** For comparison, data from experiments without drug addition from FIGURE 1A are also shown ($n = 10$). Results are expressed as the means ± SEM.

ties specific for glia.[6,19] Because of the nearly unlimited supply of cells, the use of C6 glioma allows the experiments to be repeated many times for statistical evaluation and a variety of pharmacological inhibitors to be studied. Up to now, results of studies obtained with C6 glioma have been confirmed with glial cells from primary culture (FIG. 1).

One drawback of the model is that the use of suspended cells prohibits the study of spatial buffering effects. The current data, therefore, are pertinent only for effects of active K^+ accumulation or changed membrane permeability to K^+ ions.

K^+ Uptake As a Cause of Glial Swelling?

Studies of cerebral Na^+/K^+ ATPase have consistently shown that activity is greater in glial cells than in neurons (for a review see ref. 2) and that the enzyme is stimulated by elevated extracellular concentrations of potassium.[20] Uptake via Na^+/K^+ ATPase alone still seems insufficient to explain the total amount of K^+ that can be accumulated by glia.[21] Therefore, other pathways of K^+ entry, that is, through K^+-channels and ouabain-resistant uptake have been postulated.[21–23] The contribution of these alternative pathways to net uptake of potassium apparently is species dependent[24] and varies with the extracellular concentration of potassium.[2]

The current data prove that activation of Na^+/K^+ ATPase is a key factor in K^+-induced swelling, because in the presence of ouabain, cell volume was not affected by 30 mM K^+ (FIG. 2). Activation of the enzyme, however, is the beginning of a chain of events rather than the final cause of swelling: Na^+/K^+ exchange alone cannot explain the volume changes observed. The driving force for the observed swelling of 7–9% in the absence of ouabain cannot be solely a buildup of osmotic pressure from actively accumulated K^+ ions accompanied by Cl^-. Assuming an intracellular osmotic load of 300 mOsm/l under control conditions, 8% swelling is equal to an intracellular gain of 24 mOsm/l, that is, 12 mM K^+ plus the necessary anions. With a cellular Na^+ content of 16.5 mmol/l cell mass in C6 glioma cells,[5] not enough Na^+ ions are available to fuel the necessary 3:2 exchange of Na^+ versus K^+. In fact, Walz and Hertz[25] demonstrated that far more K^+ ions were accumulated than Na^+ was lost by astrocytes exposed to 54 mM extracellular K^+. Moreover, a simple exchange of K^+ versus Na^+ does not necessarily lead to an increase in net extracellular osmotic load and therefore alone cannot explain swelling. The reverse situation, inhibition of Na^+/K^+ ATPase by ouabain at normal extracellular K^+, does not cause glial swelling[7] because the passive gain of intracellular Na^+ is counteracted by a more than equal loss of K^+. So, additional mechanisms are necessary to explain the changes in cell volume observed during exposure to 30 mM K^+.

A more sophisticated model could postulate that high extracellular K^+ causes an increase in Na^+ permeability that would supply Na^+ for Na^+/K^+ exchange. Alternatively, glycolysis due to the increased metabolic demand at high extracellular K^+[26,27] may acidify the cell and stimulate Na^+/H^+ exchange.[25] Na^+ ions entering the cell through Na^+/H^+ exchange would then permit Na^+/K^+ countertransport. With this mechanism, a net gain of cations would result. The latter explanation is supported by the current observation that amiloride, an inhibitor of Na^+/H^+ exchange, significantly reduced the swelling response (TABLE 1).

A minor contribution to the swelling process may be attributed to KCl or (Na, K, 2Cl) cotransport[28] and to increased K^+ permeability as seen from the small reduction in swelling in the presence of furosemide or Ba^{2+} ions (TABLE 2). (Na, K, 2Cl) cotransport is fueled by the Na^+ gradient that is diminished at high extracellular K^+,[13] which has led Walz and Hertz[28] to conclude that passive KCl influx by DONNAN forces is more important under these conditions.

Anions Contributing to Glial Swelling

Walz and Hertz,[25] from their data on astrocytic K^+ uptake, calculated that the increased Cl^- content during K^+ uptake is not sufficient to ensure electroneutrality. Cl^- permeability is comparatively low in glia under these circumstances.[29] The investigators postulate that the production of endogenous substances could be associated with the metabolic response of glial cells and could contribute to the increase in intracellular osmolytes. Ritchie *et al.*[17] also observed an "anion gap" and hypothesized that lactate or bicarbonate ions accumulate in glia during K^+-induced swelling. In another publication Walz and Mukerji[30] actually demonstrated an intracellular accumulation of lactate at 60 mM K^+. The current data are in line with this observation.

Under resting conditions, lactate is produced by glial cells *in vitro.*[30,31] *In vivo* there is a continuous release into the extracellular space of the brain.[32] Neuronal activity stimulates the release *in vivo,*[32] as does exposure to high extracellular K^+ *in vitro.*[30] Moreover, glial glycogen stores as a source of glucose for glycolysis are reduced by exposure of brain slices or astrocytes to high K^+ media.[33,34]

The present findings with inhibitors of lactate production or export suggest that the production of lactate increases suddenly in glial cells after exposure to high extracellular K^+ concentrations. Lactate accumulates intracellularly, thereby contributing to glial swelling.

Iodoacetate, an inhibitor of sulfhydryl enzymes, blocks the activity of glyceraldehyde-6-phosphate dehydrogenase and prevents K^+-induced swelling. The substance reduced K^+ uptake during *in vivo* superfusion of the rat brain cortex.[35] In our studies iodoacetate at 2.5 mM completely interrupted lactate production in C6 cells.[5]

In another investigation iodoacetate effectively inhibited glial swelling caused by quercetin.[8] Quercetin, a flavonoid, is an inhibitor of lactate outward transport which, up to now, has been studied mostly in tumor cells from sources other than brain.[36] When lactate export is inhibited, the anion accumulates intracellularly and augments the osmotic load of the cell. Whenever the kinetics of lactate outward transport are slower than its production, the cell should increase in volume. Assuming that this is the mechanism underlying K^+-induced glial swelling, complete block of lactate transport should also prevent the *normalization of glial volume* after prolonged exposure to K^+, which is exactly what was found in the current experiments (FIG. 3B). This hypothesis, however, includes two assumptions. First, the kinetics of quercetin-inhibitable lactate transport in glia are not well studied. Hence, it is unknown if lactate export is really the limiting step responsible for intracellular lactate accumulation. Second, the observed normalization of cell volume after a 15–30 minute exposure to 30 mM K^+ implies that stimulation of glycolysis is a short-lasting event. A slowdown of lactate production after prolonged exposure to potassium has not been shown. "Normalization" of lactate production after initial stimulation might be deduced from observations of Belt *et al.*[36] Studying Ehrlich ascites tumor cells, the authors found that the intracellular accumulation of lactate (after quercetin treatment) caused a decrease in intracellular pH and concomitantly inhibited glycolysis. Walz and Mukerji,[30] on the other hand, using 60 mM K^+ in astrocytes, found that the increased lactate production was a delayed phenomenon, taking place when swelling should have terminated. Therefore, additional

TABLE 1. Effect of Anoxia and Various Inhibitors on Swelling of C_6 Glioma Cells during Exposure to 30 mM K^+ (% of control volume, means ± SEM)

Treatment	n	Time (min) of Exposure to 30 mM K^+: 0	2	6	8	10	12	15	30	45
None	10	100	103.8 ± 0.9	107.8 ± 0.7	107.6 ± 0.9	108.3 ± 0.9	108.0 ± 1.1	106.2 ± 0.7	102.6 ± 1.3	100.8 ± 1.3
Anoxia	5	100	108.6 ± 2.7[a]	108.9 ± 1.3[a]	110.2 ± 1.9	107.7 ± 1.6	106.4 ± 3.1	104.2 ± 2.2	103.7 ± 1.4	—
Anoxia + Iodoacetate (2.5 mM)	5	100	101.3 ± 2.5	102.6 ± 1.5	—	101.5 ± 1.7[a]	102.7 ± 0.8[a]	102.7 ± 1.0	105.3 ± 1.5	105.7 ± 1.9
Amiloride (0.1 mM)	6	100	102.7 ± 0.8	—	103.1 ± 0.7[a]	103.4 ± 0.8[a]	103.4 ± 0.7[a]	103.0 ± 0.8	101.5 ± 1.0	99.3 ± 0.8
Furosemide (1.0 mM)	6	100	100.6 ± 1.5	105.0 ± 2.0	105.1 ± 2.7	104.6 ± 1.6	106.9 ± 1.7	105.4 ± 1.4	105.0 ± 1.5	102.9 ± 1.8
Ba^{2+}-ions (0.5 mM)	6	100	99.4 ± 1.1	103.9 ± 1.1	104.5 ± 1.4	104.7 ± 1.2	104.4 ± 1.2	103.5 ± 1.0	102.8 ± 1.7	102.5 ± 1.1

[a] $p < 0.05$ vs. no treatment.

experiments are necessary to further elucidate the exact time course of lactate production and the kinetics of its transporter.

The significance of intracellular lactate accumulation is further supported by experiments performed in anoxia. With a functioning oxygen supply, the metabolic demand will be met mostly by aerobic glycolysis, whereas in anoxia, anaerobic breakdown of glucose is the only energy source. Therefore, exposure to high extracellular K^+ in anoxia will cause a more rapid accumulation of intracellular lactate, which explains the even faster swelling kinetics observed under these conditions (TABLE 1). That iodoacetate also blocked K^+-induced swelling in anoxia is again confirming evidence for the significance of lactate as an intracellular osmolyte.

SUMMARY AND CONCLUSIONS

K^+-induced glial swelling results from an intricate interaction of transport and diffusion processes and metabolic stimulation, with many open questions remaining. Our concept of the major mechanisms involved can be summarized as follows: high extracellular K^+ causes a burst-like stimulation of Na^+/K^+ ATPase and, hence, increases the metabolic demands. Lactate is produced; the cell is slightly acidified. To maintain a normal intracellular pH, the Na^+/K^+ antiporter extrudes protons and supplies Na^+ for further Na^+/K^+ exchange. In addition, K^+ ions enter the cell via membrane channels or furosemide-inhibitable transport. K^+, Cl^-, and lactate$^-$ ions accumulate as the osmotic basis for cell swelling. Later, cell volume normalizes slowly, a process involving lactate export and other, so far unidentified mechanisms.

Taken together, the temporary swelling of glia at high K^+ concentrations is the result of a homeostatic function, for the maintenance of a constant extracellular potassium concentration. Ion control ranges over volume control. In pathophysiologic states the loss of cell volume regulation may become a clinical problem, if cerebral swelling leads to an increase in intracranial pressure. It should be kept in mind, however, that elevation of the extracellular K^+ concentration is not the only cause of glial swelling. Tissue acidosis, the release of neurotransmitters, especially glutamate, or free fatty acids are other mediator mechanisms initiating the swelling of glial elements.[37] Only under controlled *in vitro* conditions can the individual significance of these factors be evaluated on a quantitative basis. Therapeutic approaches should be selected very carefully in order to maintain homeostatic mechanisms that are of utmost importance, especially after an insult to the brain.

ACKNOWLEDGMENTS

The excellent technical and secretarial assistance of Barbara Kempski and Monika Westenhuber is gratefully acknowledged.

REFERENCES

1. HERTZ, L. 1978. An intense potassium uptake into astrocytes, its further enhancement by high concentrations of potassium, and its possible involvement in potassium homeostasis at the cellular level. Brain Res. **145:** 202–208.
2. WALZ, W. & L. HERTZ. 1983. Functional interactions between neurons and astrocytes. II. Potassium homeostasis at the cellular level. Progr. Neurobiol. **20:** 133–183.

3. SYKOVA, E. 1983. Extracellular K^+ accumulation in the central nervous system. Prog. Biophys. Molec. Biol. **42:** 135–189.
4. KEMPSKI, O., L. CHAUSSY, U. GROSS, M. ZIMMER & A. BAETHMANN. 1983. Volume and metabolism of suspended C6 glioma cells: An *in vitro* model to study cytotoxic brain edema. Brain Res. **279:** 217–228.
5. KEMPSKI, O., M. ZIMMER, A. NEU, F. VON ROSEN, M. JANSEN & A. BAETHMANN. 1987. Control of glial volume in anoxia: *In vitro* studies on ischemic cell swelling. Stroke **18:** 623–628.
6. KEMPSKI, O., F. STAUB, M. JANSEN, F. SCHÖDEL & A. BAETHMANN. 1988. Glial swelling during extracellular acidosis *in vitro.* Stroke **19:** 385–392.
7. KEMPSKI, O., F. STAUB, F. V. ROSEN, M. ZIMMER, A. NEU & A. BAETHMANN. 1988. Molecular mechanisms of glial swelling *in vitro.* Neurochem. Pathol. **9:** 109–125.
8. WEIGT, H., F. STAUB, A. BAETHMANN & O. KEMPSKI. 1991. Inhibition of lactate transport causes glial swelling. *In* Funktionsanalyse Biolog. Systeme. J. Grote & E. Witzleb, eds. G. Fischer. Stuttgart, New York. In press.
9. FRANGAKIS, M. V. & H. K. KIMELBERG. 1984. Dissociation of neonatal rat brain by dispase for preparation of primary astrocyte cultures. Neurochem. Res. **9:** 1689–1698.
10. KACHEL, V. 1976. Basic principles of electrical sizing of cells and particles and their realization in the new instrument "metricell." J. Histochem. Cytochem. **24:** 211–230.
11. THEODORSSON-NORHEIM, E. 1986. Kruskal-Wallis Test: BASIC computer program to perform one-way analysis of variance and multiple comparisons on ranks of several independent samples. Comput. Methods Program Biomed. **23:** 57–62.
12. GRAFE, P. & K. BALLANYI. 1987. Cellular mechanisms of potassium homeostasis in the mammalian nervous system. Can. J. Physiol. Pharmacol. **65:** 1038–1042.
13. WALZ, W. & S. MUKERJI. 1988. KCI movements during potassium-induced cytotoxic swelling of cultured astrocytes. Exp. Neurol. **99:** 17–29.
14. BOURKE, R. S., K. M. NELSON, R. A. NAUMANN & O. M. YOUNG. 1970. Studies of the production and subsequent reduction of swelling in primate cerebral cortex under isosmotic conditions *in vivo.* Exp. Brain Res. **10:** 427–446.
15. LODIN, Z., J. HARTMAN, M. P. KAGE, P. KORINKOVA & J. BOOHER. 1971. Potassium-induced hydration in cultured neural tissue. Neurobiology **1:** 69–85.
16. TRUBATCH, J. & A. V. LOUD. 1977. Quantitative stereological evaluation of KCI-induced ultrastructural changes in frog brain. Neuroscience **2:** 963–974.
17. RITCHIE, T., D. J. PACKEY, M. C. TRACHTENBERG & R. HABER. 1981. K^+-induced ion and water movements in the frog spinal cord and filum terminale. Exp. Neurol. **71:** 356–369.
18. WALZ, W. 1986. Swelling and potassium uptake in cultured astrocytes. Can. J. Physiol. Pharmacol. **65:** 1051–1057.
19. PFEIFFER, S. E., B. BETSCHART, J. COOK, P. MANCINI & R. MORRIS. 1977. Glial cell lines. *In* Cell, Tissue, and Organ Cultures in Neurobiology. S. Fedoroff & L. Hertz, eds.: 287–346. Academic Press. New York.
20. HENN, F. A., H. HALJAMÄE & A. HAMBERGER. 1972. Glial cell function: Active control of extracellular K^+ concentration. Brain Res. **43:** 437–443.
21. WALZ, W. & L. HERTZ. 1982. Ouabain-sensitive and ouabain-resistant net uptake of potassium into astrocytes and neurons in primary cultures. J. Neurochem. **39:** 70–77.
22. WALZ, W., M. SHARGOOL & L. HERTZ. 1984. Barium-induced inhibition of K^+ transport mechanisms in cortical astrocytes: Its possible contribution to the large Ba^{2+}-evoked extracellular K^+ signal in brain. Neuroscience **13:** 945–949.
23. WALZ, W. & E. HINKS. 1985. Carrier-mediated KCl accumulation accompanied by water movements is involved in the control of physiological K^+ levels by astrocytes. Brain Res. **343:** 44–51.
24. WALZ, W. & H. K. KIMELBERG. 1985. Differences in cation transport properties of primary astrocyte cultures from mouse and rat brain. Brain Res. **340:** 333–340.
25. WALZ, W. & L. HERTZ. 1983. Intracellular ion changes of astrocytes in response to extracellular potassium. J. Neurosci. Res. **10:** 411–423.
26. ORKAND, P., H. BRACHO & R. K. ORKAND. 1973. Glial metabolism: Alteration by potassium levels comparable to those during neural activity. Brain Res. **55:** 467–471.

27. HERTZ, L., L. DITTMANN & P. MANDEL. 1973. K^+ induced stimulation of oxygen uptake in cultured cerebral glial cells. Brain Res. **60:** 517–520.
28. WALZ, W. & L. HERTZ. 1984. Intense furosemide-sensitive potassium accumulation in astrocytes in the presence of pathologically high extracellular potassium levels. J. Cereb. Blood Flow Metab. **4:** 301–304.
29. WALZ, W. & L. HERTZ. 1983. Comparison between fluxes of potassium and of chloride in astrocytes in primary cultures. Brain Res. **277:** 321–328.
30. WALZ, W. & S. MUKERJI. 1988. Lactate production and release in cultured astrocytes. Neurosc. Lett. **86:** 296–300.
31. WALZ, W. & S. MUKERJI. 1988. Lactate release from cultured astrocytes and neurons: A comparison. Glia **1:** 366–370.
32. KUHR, W. G. & J. KORF. 1988. Extracellular lactic acid as an indicator of brain metabolism: Continuous on-line measurement in conscious, freely moving rats with intrastriatal dialysis. J. Cerebr. Blood Flow Metab. **8:** 130–137.
33. CAMBRAY-DEAKIN, M., B. PEARCE, C. MORROW & S. MURPHY. 1988. Effects of extracellular potassium on glycogen stores of astrocytes *in vitro.* J. Neurochem. **51:** 1846–1851.
34. HOF, P. R., E. PASCALE & P. J. MAGISTRETTI. 1988. K^+ at concentrations reached in the extracellular space during neuronal activity promotes a Ca^{2+}-dependent glycogen hydrolysis in mouse cerebral cortex. J. Neurosci. **8:** 1922–1928.
35. RAFFIN, C. N., T. J. SICK & M. ROSENTHAL. 1988. Inhibition of glycolysis alters potassium ion transport and mitochondrial redox activity in rat brain. J. Cereb. Blood Flow Metab. **8:** 857–865.
36. BELT, J. A., J. A. THOMAS, R. N. BUCHSBAUM & E. RACKER. 1979. Inhibition of lactate transport and glycolysis in Ehrlich ascites tumor cells by bioflavonoids. Biochemistry **18:** 3506–3511.
37. BAETHMANN, A., K. MAIER-HAUFF, O. KEMPSKI, A. UNTERBERG, M. WAHL & L. SCHÜRER. 1988. Mediators of brain edema and secondary brain damage. Crit. Care Med. **16:** 972–978.

Axon-Glia Exchange of Macromolecules[a]

ROBERT M. GROSSFELD

Zoology Department and Physiology Program
North Carolina State University
Raleigh, North Carolina 27695–7617

Glia occupy a strategic position in regulating interactions among neurons and between neurons and the circulation (FIG. 1).[1–4] Their close proximity to axons creates a barrier that can influence transport of substances from and to the axonal surface. In addition, this close association provides unique opportunities for intercellular interactions that allow neurons to develop and function in harmony with neighboring and distant cells[5] and to exchange molecules that compensate for metabolic or physiological deficiencies.[6–8] In these respects, neurons and glia represent an interdependent partnership that insures proper neuronal homeostasis, growth, and repair.

Many of the axon-glia interactions are mediated by direct surface contact between complementary macromolecules or by activation of plasma membrane receptors through secretion of the appropriate activators, such as growth factors or transmitters.[9–11] Vital interactions occur also through direct exchanges of molecules, the most well known being transfers of potassium that maintain efficient electrical signaling.[8,12] What has not been clear is whether larger molecules also are exchanged and, if so, the role of such transfers in neural function. This manuscript reviews evidence that axons and glia share molecules larger than simple ions and discusses the functional implications of such exchanges.

Direct axon-glia exchanges of macromolecules might be expected to be particularly important for large diameter, long axons, such as the giant axons that are commonly found in the nervous system of invertebrates and lower vertebrates. The volume of axoplasm in these processes is relatively large compared to the volume of the cell body, and the distance between the axon terminus and the cell body is great. These geometric features tax the ability of the perikaryon to maintain cell structure and function via transport of protein to the ribosome-deficient axon.[13] To compensate for this difficulty, some large axons may use additional means to insure their integrity under normal conditions and especially when the supply of substances from the perikaryon is compromised. These adaptations may include conservation of axonal macromolecules to reduce turnover or supplementation of the perikaryal supply of axonal protein by an alternate cellular source such as the periaxonal glia, which actively synthesize protein.[14,15] The long-term survival (LTS) of some giant axons following disconnection from their perikarya attests to their ability to preserve or replace their contents independent of the cell body.[14,15] Glial participation in LTS has been proposed, from indirect evidence, for two large axons of crayfish, the medial giant axon (MGA) in the central nervous system (CNS)[14,16] and the inhibitory motor axons in the peripheral nervous system (PNS).[17] The more rapid deterioration of most lesioned axons of higher, endothermal vertebrates implies that they cannot effectively mount a comparable response, at least not at their normal body temperatures.[15]

[a]This work was supported by NSF Grant BNS-8517184, NIMH Grant R03MH32075, N. C. Agricultural Research Foundation Grants 5582 and 5702, and US Sea Grant R/MG8902.

Chemical exchanges between axons and glia are most likely to occur where the axoplasm and gliaplasm are readily accessible to each other. The multilamellar organization of the glial sheath encapsulating some invertebrate giant axons may represent a specialization that juxtaposes the axon to a relatively large amount of glial cytoplasm. This could explain why a giant axon such as the crayfish MGA should be invested with a myelin-like sheath although the axonal specializations of large diameter and insulating sheathing are redundant in accelerating action potential propagation. The primary function of ensheathment for these axons may be to provide trophic support that insures the axon's metabolic stability and/or to shield the axon from potentially deleterious changes in the ambient chemical environment. This notion is consistent with the morphological observation that these sheaths are loosely packed in a way that retains multiple layers of glial cytoplasm near the axon,[4,18] whereas the insulating type of myelin sheath that supports rapid electrical

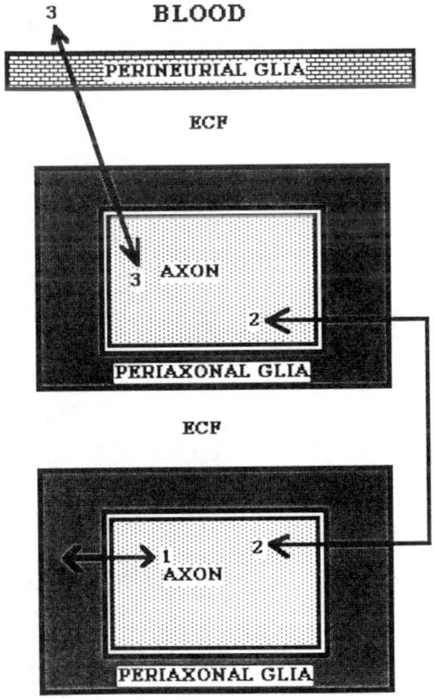

FIGURE 1. Strategic anatomic location of crayfish glia for regulating interactions among axons and between axons and the circulation. The periaxonal glia ensheath CNS axons with alternating layers of glial cytoplasm and extracellular space. This close association permits efficient exchanges of molecules between these cells (*arrow 1*). The permeability of the periaxonal sheath to many molecules provides opportunities for chemical interactions among axons (*arrow 2*) and between axons and glia of the perineurial sheath (*arrow 3*), which regulates exchanges of molecules between the CNS and the circulation.

signaling in vertebrates is tightly packed and excludes most of the electrically conducting glial cytoplasm, except that at the nodes of Ranvier and Schmidt-Lantermann clefts.[13,19] Because the invertebrate giant axons perform a crucial role in activating escape from predators,[20] the organism's survival is likely to be as dependent on adaptations that sustain the axons metabolically as on adaptations that promote their capacity for rapid signaling.

Transfers of molecules between axons and glia have been reported for several invertebrate[6,7,21] and vertebrate[22–24] species but have been demonstrated most conclusively for giant axons of the crayfish and squid.[16,25–33] These latter preparations can be cannulated for intracellular injection of molecules or collection of axonal molecules derived from the glia. This special technical advantage has facilitated experiments that have indicated transfers of small and large molecules from giant axons to their ensheathing glial cells and in the reverse direction as well.

AXON-TO-GLIA TRANSFERS OF MOLECULES

Within minutes to hours after their injection into a crayfish medial giant axon (MGA), tracer molecules are present in the associated periaxonal sheath but not in the adjacent lateral giant axon or the contralateral MGA. This translocation is observed for small dye molecules such as Lucifer yellow (LY), acridine orange, ethidium bromide, and fast green, and for large molecules such as fluorescein-conjugated dextrans of 10 and 18 kD (f-D), horseradish peroxidase (HRP, 40 kD), or fluorescent bovine serum albumin (f-BSA, 67 kD) (FIG. 2B).[27,31–33] Some of the tracer enters glial cells of the periaxonal sheath, where it can be identified by its accumulation in the glial nuclei.[27,32] In a whole mount of a CNS in which one MGA has been injected with LY, the MGA exhibits a characteristic speckled appearance as the nuclei of all of the glial cells of the innermost (adaxonal) layer of the sheath acquire tracer.[32] After injection of f-BSA, the MGA has a similar appearance in a whole-mount preparation (FIG. 2A), whereas after injection of f-D, a greater number of fluorescent spots can be observed in a whole mount preparation, corresponding to tracer in adaxonal and other glial cells in tissue sections. These results indicate that the adaxonal glia have rapid access to a variety of molecules introduced into the axon. With LY or HRP, tracer is not observed in the outer layers of the periaxonal

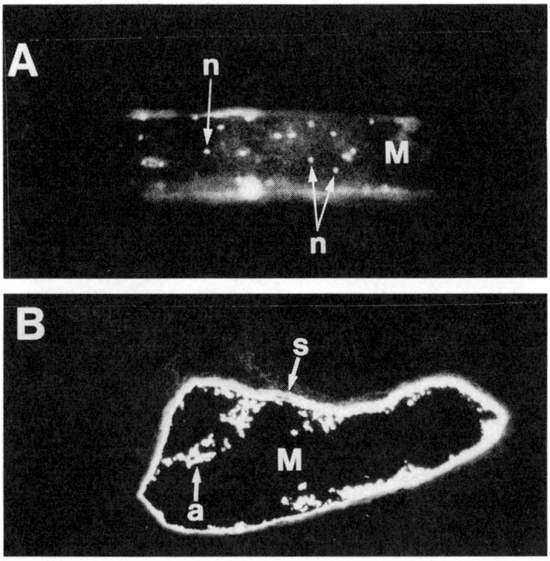

FIGURE 2. Axon-to-glia transfer of a protein. One of the two medial giant axons (MGA) in the crayfish CNS was injected with fluorescein-conjugated bovine serum albumin, after which the tissue was fixed with paraformaldehyde and cleared with methyl salicylate for examination of the cells in a whole mount preparation (**A**) or was fixed in paraformaldehyde and embedded in plastic for examination of thick sections (**B**). (**A**) A portion of an injected MGA (M) is viewed longitudinally from the dorsal surface. The small bright fluorescent circles (n) correspond in size and appearance to the nuclei of periaxonal glial cells that accumulate Lucifer yellow after its injection into the MGA. (**B**) A transverse section of an injected MGA exhibits fluorescent tracer associated with axoplasm (a) of the MGA (M) and with the periaxonal sheath (s). In both **A** and **B,** the adjacent, uninjected lateral giant axon was not fluorescent.

TABLE 1. Axon to Glia Transfers: Controls[a]

Procedure	Result
(1) Dialyze tracer	Tracer in sheath
(2) Axon, glial membrane potentials	No change
(3) Gel electrophoresis axoplasm, sheath	Tracer intact
(4) Aldehyde, nonaldehyde fixatives	Tracer in sheath
(5) Cross-linkable dextran derivative	Tracer in sheath

[a]Control experiments to validate axon-to-glia transfers in crayfish, using fluorescent derivatives of dextran and bovine serum albumin as exogenous tracers.[27] Procedures: (1) controls for transfer of free fluorophore; (2) controls for transfer across damaged cell membranes; (3) controls for transfer of degraded tracer; (4) controls for transfer artifactually produced by a particular type of fixative; (5) controls for movement of dextran during histological processing.

sheath or in the sheaths of nearby axons,[32,33] but with f-D and f-BSA, it is.[27] The discrepancy between the transport distance of these tracers is unexplained but may reflect subtle differences in the experimental procedures or animals used in different laboratories. The greater penetration by f-D and f-BSA could result from artifactual leakage of these tracers beyond the adaxonal layer, but the rapid passage of similar tracers across the entire sheath in the opposite direction (see subsequent section on Mechanisms) suggests that it may be a valid result. If the tracers do normally penetrate beyond the MGA sheath, then the results indicate that many types of molecules may move among neighboring neurons and/or between neurons and the circulation.

The validity of these observations of axon-to-glia transfer is supported by several control experiments that indicate that such exchange occurs under physiological conditions. These experiments virtually exclude several possible technical artifacts that could influence the results (TABLE 1). For example, these experiments demonstrate that: (1) the tracer transferred is the intact molecule, not a degraded component or the free fluorophore; (2) the membrane potentials of the axons and glia are unaltered during the transfers; and (3) consistent results are produced with a variety of fixatives that insure tracer immobilization while the tissue is being processed. Elimination of fixation artifacts is the most difficult criterion to satisfy, because tissue fixation can create holes through which tracers can diffuse rapidly.[34] This condition has been met for LY,[32] as equivalent transfers are recognizable even in whole mounts of unfixed tissues. Comparable experiments with larger molecules in unfixed tissues will be required to confirm axon-to-glia transfer of macromolecules. In my laboratory, we are pursuing this issue at the present time.

GLIA-TO-AXON TRANSFERS OF MOLECULES

Although transfers of molecules from glia to axons in several invertebrate and vertebrate species have been reported they have been demonstrated most conclusively for the squid giant axon,[25,26,29] and the phenomenon has been confirmed by comparable experiments on the crayfish MGA.[16,31] Because of the narrow width of the glial processes that envelop these axons, it has not been possible to introduce diverse exogenous tracers directly into the gliaplasm to study their movement into the axon. However, the active synthesis of protein by the glia has permitted examination of glia-to-axon transfers of endogenous macromolecules. Within minutes of adding radioactive amino acid to the saline solution bathing the isolated nerve preparation, labeled protein is present in a perfusate of the axon.[16,26,31] The newly synthesized axonal protein appears to be derived from the periaxonal glia rather than from axonal synthesis or posttranslational modification of axoplasmic proteins because (a) most of the axoplasm is eliminated by the perfusion procedure, (b) the axon lacks adequate ribosomal-like RNA for *de novo* synthesis of protein, (c) incubation of axoplasm with the radioactive amino acid either *in situ* or *in vitro* does not result in incorporation of label into the same proteins, (d) injection of RNase into the squid axon does not alter the result, and (e) bath addition of puromycin, which inhibits ribosomal-derived protein synthesis (in the glia), blocks the appearance of radioactive protein in a perfusate of the squid axon.

FUNCTIONAL ROLES OF AXON-GLIA EXCHANGES

Studies of axon-glia exchanges have revealed that some molecules leaving the axon enter surrounding glial cells and that a large fraction of the molecules leaving the periaxonal glia are destined to enter the axon. The importance of these

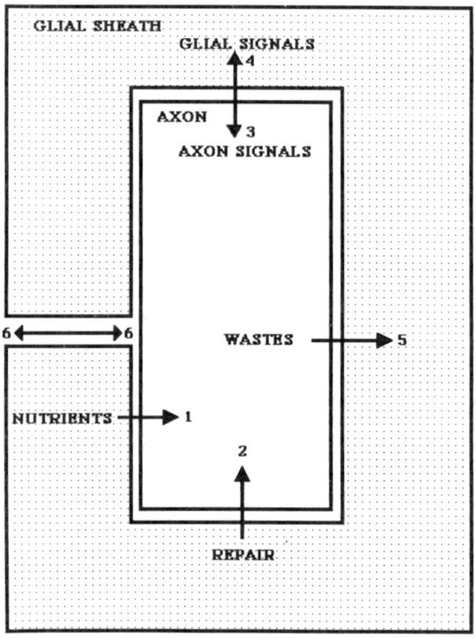

FIGURE 3. Putative functions of axon-glia exchanges of molecules. Periaxonal glia may provide their companion axons with substances essential for sustenance (*arrow 1*), repair (*arrow 2*), or signaling of alterations in the extracellular environment (*arrow 3*). The axons, in turn, may supply the glia with substances that signal alterations in the internal environment of the neuron (*arrow 4*) or that are byproducts of metabolism (*arrow 5*). Some of these substances, as well as other chemicals received from outside the axon-glia unit, may traverse the periaxonal sheath (*arrow 6*).

intercellular transfers is unknown, but it is useful to consider some potential roles (FIG. 3).

Molecules that pass from axon to glia may include chemical signals to inform the glia of changes in the neuronal intracellular environment and waste products that the glia are better prepared to metabolize. With the exception of phospholipid transfer in avian autonomic nerves,[23] axon-to-glia molecular transfers have been studied only with exogenous tracers that have been useful for examining transport in nonneural and neural cells or tissues.[35-39] Ideally, similar molecular cytochemical investigations should be performed on living, intact nervous tissue with probes that are biologically active derivatives of axoplasmic macromolecules. As this has not been done for axon-to-glia transport, the relevance of the model studies to normal function is unclear at present.

Molecules synthesized by the glia for the axon are likely to provide nutrients, restorative or reparative substances, and chemical signals to inform the neurons of changes in the ambient extracellular environment. Better understanding of their value awaits further studies to determine their identities and functions in the axon. The ability to label native glial constituents *in situ* and to recover them in axoplasm has permitted identification of several of the 50–100 proteins (transferred glial polypeptides or TGPs) that are donated by the glia to the squid giant axon.[40–42] Among the TGPs are actin, a fodrin-like protein, and several heat shock proteins (stress proteins). The characteristic appearance of these radioactive proteins in the axoplasm in a fixed ratio suggests that they are transferred intercellularly as a conglomerate, perhaps associated with a particle such as a vesicle or with a piece of gliaplasm ingested by the axon (see subsequent section on Mechanisms). An analogous set of proteins is specifically released by cultured mammalian fibroblasts.[43] This may indicate that the secretion of such proteins is common to many cell types and may represent an adaptation that compensates for specialization of function by providing to deficient cells the essential constituents that they lack.

STRESS PROTEINS: GLIA-AXON EXCHANGES AND ADAPTATION OF NERVOUS TISSUE TO STRESS

The presence of stress proteins (SPs) among the TGPs provides a clue as to one practical benefit of intercellular molecular exchange between glia and axons, namely, an enhanced neuronal tolerance for stress. This transfer might be an efficient means by which axons resist acute stress that would not be quickly compensated by the relatively slow transport of molecules, including SPs, from the perikaryon.[41,44–46] Some subset of these SPs is synthesized in all organisms studied to date, including microbes, plants, and animals.[47–50] SPs are made by neural and nonneural cells in response to a variety of stresses, including heat shock, exposure to chemical intoxicants, oxygen or glucose deprivation, and traumatic injury. Several SPs, notably the 8-kD (ubiquitin), 70-kD, and 90-kD families, are among the most highly conserved proteins known, attesting to their importance for stress adaptations of all organisms.[47,48,51,52] The 70-kD SPs are particularly important in promoting stress tolerance of many cell types, although their specific functions in nervous tissue are uncertain. In many nonneural cells, deletions of SP genes or specific inhibition of SP expression or function results in loss of an organism's ability to adapt to a potentially lethal heat shock.[53–57] Conversely, in some cells, an increased amount of SP confers increased thermotolerance.[58–61] The contribution of SPs to recovery from stress is probably to stabilize vulnerable proteins during stress, to promote the correct folding, assembly, and intracellular translocation of newly synthesized proteins after stress, and to target defective proteins for degradation.[62–65]

If SPs are important participants in cellular resistance to stress, then might their transfer between glia and axons represent an adaptation that is especially beneficial to neurons during times of stress? The crayfish MGA offers an accessible biological preparation in which this issue can be addressed. The MGA and peripheral motor axons of the crayfish are notable for their longevity following a traumatic injury that separates them from their cell bodies.[14,66] By contrast, crayfish peripheral sensory axons degenerate rapidly in response to a comparable lesion.[67] One possibility is that the remarkable stress resistance of the crayfish MGA derives largely from glial supply of essential molecules to the isolated axon segments and that the differential survival among cells reflects differences in axon-glia interactions. To test this hypothesis, we have initiated a series of collaborative studies designed to define the roles of SPs in crayfish nervous tissue and the contribution of glia-to-axon and axon-to-glia interactions.

We have found that the stress response of crayfish CNS tissue is like that of neural and nonneural tissues from many species, including mammals.[68–71] The main SPs are similar not only in size and immunochemical structure, but also in their induction by heat shock (FIG. 4A), sodium arsenite, or crush injury (FIG. 4B).[72,73] Although the primary SPs expressed during heat or surgical stresses have slightly different apparent molecular weights (FIG. 4), they probably include related members of multigene families. For example, the 66-kD heat shock band and the 73-kD trauma-induced component (FIG. 4) each include several proteins recognized by an antibody raised against the mammalian 70-kD family of SPs (Xue and Grossfeld, preliminary results; not shown). Increased expression of these proteins after stress may promote the tissue's recovery. With acute hyperthermia, SP synthesis remains elevated for hours after the normal acclimation temperature is restored,[72] whereas after a localized crush injury, SP levels remain elevated in the treated region and beyond for days or weeks.[73] This prolonged biochemical change accompanying a surgical crush may reflect the morphological changes underlying wound healing of the tissue. During the first few postoperative days, most axons are lost from the damaged region, but the tissue regains near normal volume. This recovery is characterized by the acquisition of tissue elements and large numbers of cells of unknown origin.[74] This response resembles the proliferation of glia or macrophages during wound healing in mammalian CNS tissue,[75] and we surmise that similar reparative changes in CNS glia or infiltrating cells, involving SPs, underlie the crayfish injury response.

Although definitive immunocytochemical evidence is lacking for the cellular distribution of SPs in normal or stressed crayfish nervous tissues, biochemical experiments suggest that SPs are made by both neurons and glia, including glia of both the periaxonal and perineurial sheaths (Martin and Grossfeld, unpublished results). These determinations are made by gel electrophoresis and gel autoradiography of heat-stressed CNS tissue dissected after the stress to separate its component parts, that is, the connectives, ganglia, and perineurial sheath, for comparative analysis. Each of these three cell fractions synthesizes the major SPs (FIG. 5). In the connectives, protein synthesis occurs primarily in the periaxonal glia, as the axons lack ribosomes.[76,77] In the ganglia, however, the neuronal perikarya are likely to account for the greatest fraction of the total protein synthesis (Grossfeld and Bittner, unpublished results). Similarity between the SPs synthesized by the connectives and ganglia suggests that the periaxonal glia represent a potential reservoir of SPs for associated neurons that might be deficient in these important molecules during times of stress, as when chemical transport from the neuronal cell body becomes inadequate to sustain the axon. Indirect support for the notion of glia-to-axon transfer of SP after injury comes from studies of crushed mammalian spinal cord, where glia possess SPs, and SPs are found in axons sooner than would be predicted from their slow rate of axonal transport.[78]

Synthesis of SPs by the perineurial sheath suggests that the perineurial glia also participate in the crayfish CNS stress response. These glia are the primary cell type[4] that can account for the sheath's active synthesis of protein.[79] SPs, by insuring glial cell viability, may enable the perineurial and periaxonal sheaths to indirectly protect neurons from harsh ambient conditions (FIG. 1). Studies with exogenous tracers indicate that the periaxonal glia normally are relatively indiscriminate in regulating molecular transport toward or away from the axon (see subsequent Mechanisms

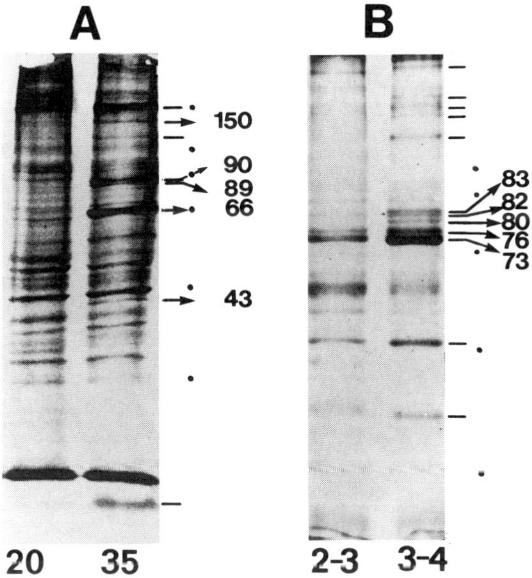

FIGURE 4. Stress protein induction by hyperthermia and traumatic injury of crayfish CNS tissue. (**A**) The abdominal portion of the CNS was isolated and divided into two approximately equivalent pieces. One piece was heat shocked *in vitro* for 5 hours at 35°C, whereas the other piece was incubated similarly for 5 hours at the acclimation temperature of 20°C, with radioactive methionine present. Equal amounts of tissue protein radioactivity were fractionated by electrophoresis in a 6–15% gradient polyacrylamide gel, with sodium dodecyl sulfate (SDS) present, to produce the gel autoradiograph illustrated. This panel shows the changes in protein synthesis induced by hyperthermia. The major heat shock proteins are denoted by *arrows,* and visual estimates of their molecular weights are indicated, along with *circles* that mark the locations of reference standards of 205, 116, 97.4, 66, 45, and 29 kD (*from top to bottom*) separated in the same gel. (**B**) The connective that links the third and fourth abdominal ganglia was crushed *in vivo* and the exoskeletal wound was closed. Two days later, the abdominal portion of the CNS was removed and incubated with radioactive methionine for 5 hours at the acclimation temperature of 20°C. Equal amounts of tissue protein were fractionated by electrophoresis in 10% polyacrylamide gels containing AcrylAide, with SDS present. This panel illustrates the changes in the relative amounts of Coomassie blue-stained tissue proteins induced by a traumatic injury. In comparison with the adjacent, nonlesioned connective (C2-3) between the second and third abdominal ganglia of the same animal, the crushed connective (C3-4) exhibited a pronounced increase in several proteins. The major crush-induced stress proteins are denoted by *arrows,* and calculated estimates of their molecular weights are indicated, along with *circles* that mark the locations of reference standards of 116, 97.4, 66, 45, and 29 kD (*from top to bottom*) separated in the same gel.

section). Stress-induced changes in these glia, perhaps involving SPs, could enhance the capacity of the periaxonal sheath to shield axons from harmful chemicals that penetrate the CNS or that are generated within it. Similar stress adaptations in the perineurial glia may enhance the restrictive properties of the perineurial sheath,[1,2,80–83] and thereby improve its ability to protect CNS axons and periaxonal glia from blood-borne intoxicants. Because the PNS perineurial sheath is leakier than the CNS perineurial sheath,[2,82,83] peripheral axons normally may be more vulnerable to toxic

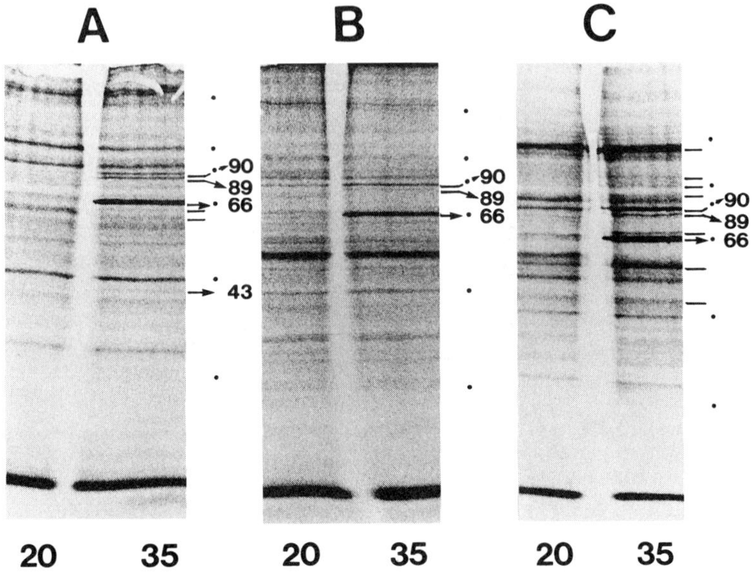

FIGURE 5. Stress protein induction by periaxonal and perineurial glia of crayfish CNS. The abdominal portion of the CNS was isolated and divided into two approximately equivalent pieces. One piece was heat shocked *in vitro* for 5 hours at 35°C, whereas the other piece was incubated similarly for 5 hours at the acclimation temperature of 20°C, with radioactive methionine present. After the tissues were rinsed briefly in saline solution, the perineurial sheath was dissected free of the underlying connectives and ganglia. Equal amounts of tissue protein radioactivity were fractionated by electrophoresis in 6–15% polyacrylamide gels, with sodium dodecyl sulfate present, to produce the gel autoradiograph illustrated. The panels illustrate the responses of connectives (**A**), ganglia (**B**), and perineurial sheath (**C**). The major heat shock proteins are denoted by *arrows,* and visual estimates of their molecular weights are indicated, along with *circles* that mark the locations of reference standards of 205, 116, 97.4, 66, 45, and 29 kD (*from top to bottom*) separated in the same gel.

substances or may require additional or alternative compensatory mechanisms to insure their survival. Evidence to support these proposals awaits further studies of the relationship of SPs to the transport properties of the glia and of the axon-glia units.

MECHANISMS OF AXON-GLIA EXCHANGES

Microscopic studies of giant axons in crayfish and squid have suggested several possible mechanisms of axon-glia exchanges of molecules. The model described here (FIG. 6) is a modification of the one proposed more than a decade ago for glia-to-axon transfers in invertebrates[25] and includes not only the more recent evidence of comparable processes in the vertebrates but also evidence that the process is bidirectional.

In crayfish, a tubular lattice system spans the width of the adaxonal layer of the periaxonal sheath and relatively large channels directly link the extracellular layers of the rest of the multilamellar sheath.[18,84] This system permits molecules to move rapidly through the sheath, probably in both directions.[18,27,85] Molecules may be exchanged by the axons and glia through finger-like projections of the adaxonal glia that invaginate the axolemma, and comparable projections of the axon that invaginate the adaxonal glia.[33,85] Model tracers are found in vesicles associated with these projections, suggesting that the tracers are transferred either by coupled exocytosis-endocytosis or by mutual microphagocytosis. Several pieces of evidence argue against the former mechanism, namely, (1) the TGPs arrive in squid axoplasm with kinetics different from the entry of protein added to the bathing solution[26] and (2) the newly synthesized proteins that are secreted into the bath by the tissue are different from the TGPs.[29,31] The alternative mechanism, that is, sampling of cortical cytoplasm of one cell by its neighbor, could account for the coordinated intercellular transfer of several cytoplasmic TGPs in squid[42] and crayfish nervous tissues[31] and could facilitate intercellular exchanges in some vertebrate nervous tissues. The adaxonal layer of the oligodendroglia in rodent and monkey brain has a similar morphological relationship with the axon that it envelops, and serial reconstructions from tissue sections indicate a true link between the invaginations and formation of vesicles.[86] It is interesting that these cellular specializations in mammalian CNS tissue frequently are associated with the Schmidt-Lantermann clefts,[86] where the otherwise compact myelin sheath retains layers of glial cytoplasm,[13] much like the periaxonal sheath of the crayfish MGA. The interspecies similarities in morphological interactions of axons and glia suggest a common mode of glia-to-axon transfer in invertebrates and vertebrates.[86,87] The same may be true for exchanges in the opposite direction, from axon to glia, as axonal invaginations of glia occur (although at lesser frequency) in both invertebrates and vertebrates. Because axon-glia interactions involving microphagocytosis are contact mediated, some of the vertebrate neuron-glia interactions that require cell-cell contact or that employ a soluble neuronal factor[88-91] may involve direct exchange of molecules.

SUMMARY

The periaxonal and perineurial glia of crayfish and squid are strategically situated to regulate the neuronal microenvironment. Diverse molecules rapidly traverse the periaxonal sheath and a fraction of them enters the axons from glia or the glia from axons. The significance of these intercellular exchanges has not been tested directly.

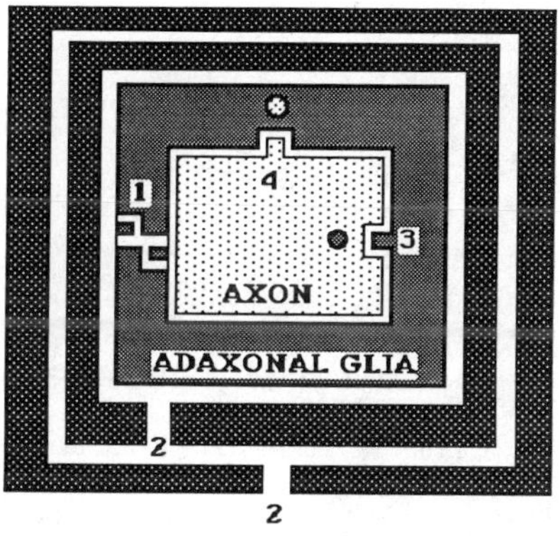

FIGURE 6. Morphological specializations underlying axon-glia exchanges of molecules. The periaxonal sheath possesses a tubular lattice system (1) in the adaxonal layer and transglial channels (2) in adjacent layers through which substances traverse the sheath. The adaxonal glia invaginate the axolemma (3) and the axon invaginates the adaxonal glia (4) at locations where axon-glia exchanges of molecules may occur (depicted as vesicles containing cytoplasm from the adjacent cell).

However, recent reports suggest that stress proteins, which probably are synthesized by both types of glia and transferred to axons, may be essential components by which the glia directly and indirectly assist neurons in tolerating ambient stress.

ACKNOWLEDGMENTS

Important contributions to this research were made by Maura Klinge, Julie Martin Rochelle, Zheng-yu Xue, Louise Stewart, and Douglas Hansen in my laboratory and by collaborators Edward Lieberman, Michael Tytell, and Barney Dwyer. I also acknowledge the major contributions to this field by George Bittner and his colleagues at the University of Texas, including Rebecca Sheller and Terry Viancour, and their encouragement of our efforts. I am grateful to George Barthalmus, George Bittner, Rebecca Sheller, Michael Tytell, and Herbert Underwood for their help in reviewing this manuscript.

REFERENCES

1. Abbott, N. J., N. J. Lane & M. Bundgaard. 1986. Ann. N. Y. Acad. Sci. **481:** 20–41.
2. Abbott, N. J. & Y. Pichon. 1987. J. Physiol., Paris **82:** 304–313.
3. Goldstein, G. W. 1988. Ann. N. Y. Acad. Sci. **529:** 31–39.
4. Lane, N. J. & N. J. Abbott. 1975. Cell Tiss. Res. **156:** 173–187.
5. Bunge, R. P. 1987. J. Exp. Biol. **132:** 21–34.
6. Coles, J. A. & M. Tsacopoulos. 1981. J. Exp. Biol. **95:** 75–92.
7. Griffiths, G. W. 1979. J. Cell Sci. **36:** 361–389.
8. Kuffler, S. W. & J. G. Nicholls. 1966. Ergebn. Physiol. **57:** 1–90.
9. Lauder, J. & K. McCarthy. 1986. *In* Biochemistry, Physiology, and Pharmacology of Astrocytes. S. Fedoroff & A. Vernadakis, eds. Vol **2:** 295–314. Academic Press. New York.
10. Lieberman, E. M., N. J. Abbott & S. Hassan. 1989. Glia **2:** 94–102.
11. Villegas, J. 1981. J. Exp. Biol. **95:** 135–151.
12. Orkand, R. K. 1986. Ann. N. Y. Acad. Sci. **481:** 269–272.
13. Peters, A., S. L. Palay & H. F. Webster. 1970. The Fine Structure of the Nervous System. Harper & Row. Hagerstown, MD.
14. Bittner, G. D. 1988. Am. Zool. **28:** 1165–1179.
15. Blundon, J. A., R. A. Sheller, J. W. Moehlenbruck & G. D. Bittner. 1990. J. Comp. Neurol. **297:** 1–14.
16. Sheller, R. A. & G. D. Bittner. 1991. J. Cell. Biol., in press.
17. Sarne, Y., B. K. Schrier & H. Gainer. 1976. Brain Res. **110:** 91–97.
18. Shivers, R. R. & M. W. Brightman. 1976. J. Comp. Neurol. **167:** 1–26.
19. Bunge, R. P. 1968. Physiol. Revs. **48:** 197–251.
20. Krasne, F. B. & J. J. Wine. 1977. *In* Identified Neurons and Behavior of Arthropods. G. Hoyle, ed.: 275–291. Plenum Press. New York.
21. Goldstein, R. S., K. R. Weiss & J. H. Schwartz. 1982. J. Neurosci. **2:** 1567–1577.
22. Berkley, K. J. & N. Contos. 1987. Brain Res. **414:** 49–67.
23. Droz, B. 1979. TINS **2:** 146–148.
24. Singer, M. & M. N. Salpeter. 1966. J. Morphol. **120:** 281–316.
25. Gainer, H. 1978. TINS **1:** 93–96.
26. Gainer, H., I. Tasaki & R. J. Lasek. 1977. J. Cell Biol. **74:** 524–530.
27. Grossfeld, R. M., M. W. Klinge, E. M. Lieberman & L. C. Stewart. 1988. Glia **1:** 292–300.
28. Lasek, R. J., H. Gainer & R. J. Przybylski. 1974. Proc. Natl. Acad. Sci. USA **71:** 1188–1192.
29. Lasek, R. J., H. Gainer & J. L. Barker. 1977. J. Cell Biol. **74:** 501–523.

30. LASEK, R. J. & M. A. TYTELL. 1981. J. Exp. Biol. **95:** 153–165.
31. SHELLER, R. A. 1989. Ph.D. Thesis. The University of Texas at Austin.
32. VIANCOUR, T. A., G. D. BITTNER & M. L. BALLINGER. 1981. Nature **293:** 65–67.
33. VIANCOUR, T. A., K. R. SESHAN, G. D. BITTNER & R. A. SHELLER. 1987. J. Neurocytol. **16:** 557–566.
34. BENNETT, M. V. L., N. FEDER, T. S. REESE & W. STEWART. 1973. J. Gen. Physiol. **61:** 254–255.
35. HULSTROM, D., L. MALMGREN, D. GILSTRING & Y. OLSSON. 1983. Acta Neuropathol. (Berl.) **59:** 53–62.
36. LANG, I., M. SCHOLZ & R. PETERS. 1986. J. Cell Biol. **102:** 1183–1190.
37. PARR, M. B. & E. L. PARR. 1986. Biol. Reprod. **34:** 393–403.
38. THORBALL, N. 1981. Histochemistry **71:** 209–233.
39. WILLIAMS, W. M., S.-T. LU, M. DEL CERRO & S. M. MICHAELSON. 1984. Brain Res. Rev. **7:** 191–212.
40. TYTELL, M. & R. J. LASEK. 1984. Brain Res. **324:** 223–232.
41. TYTELL, M., S. G. GREENBERG & R. J. LASEK. 1986. Brain Res. **363:** 161–164.
42. TYTELL, M. 1987. *In* Glial-Neuronal Communication in Development and Regeneration. NATO ASI Series H, Vol. **2:** 248–261. Springer-Verlag. Berlin; Heidelberg.
43. HIGHTOWER, L. E. & P. T. GUIDON, JR. 1989. J. Cell. Physiol. **138:** 257–266.
44. CLARK, B. D. & I. R. BROWN. 1985. Proc. Natl. Acad. Sci. USA **82:** 1281–1285.
45. TYTELL, M. & M. F. BARBE. 1987. Neurol. Neurobiol.-Axonal Transport. **25:** 473–492.
46. WAEGH, S. & S. T. BRADY. 1989. J. Neurosci. Res. **23:** 433–440.
47. BURDON, R. H. 1986. Biochem. J. **240:** 313–324.
48. LINDOUIST, S. & E. A. CRAIG. 1988. Annu. Rev. Genet. **22:** 631–677.
49. SUBJECK, J. R. & T. T. SHYY. 1986. Am. J. Physiol. **250:** C1–C17.
50. WELCH, W. J., L. A. MIZZEN & A.-P. ARRIGO. 1989. *In* Stress-Induced Proteins. M. L. Pardue, J. R. Feramisco & S. Lindquist, eds. UCLA Symp. Molec. Cell Biol. Vol. **96:** 187–202. Alan R. Liss. New York.
51. FINLEY, D. & A. VARSHAVSKY. 1985. TIBS **10:** 343–347.
52. HARTL, F.-U. & W. NEUPERT. 1990. Science **247:** 930–938.
53. FINLEY, D., E. OZLAYNAK & A. VARSHAVSKY. 1987. Cell **48:** 1035–1046.
54. JOHNSTON, R. N. & B. L. KUCEY. 1988. Science **242:** 1551–1554.
55. LOOMIS, W. F. & S. A. WHEELER. 1982. Dev. Biol. **90:** 412–418.
56. RIABOWOL, K. T., L. A. MIZZEN & W. J. WELCH. 1988. Science **242:** 433–436.
57. SANCHEZ, Y. & S. L. LINDQUIST. 1990. Science **248:** 1112–1115.
58. BARBE, M. F., M. TYTELL, D. J. GOWER & W. J. WELCH. 1988. Science **241:** 1817–1820.
59. CHOPP, M., H. CHEN, K.-L. HO, M. O. DERESKI, E. BROWN, F. W. HETZEL & K. M. A. WELCH. 1989. Neurol. **39:** 1396–1398.
60. JOHNSON, A. D., P. A. BERBERIAN & M. G. BOND. 1990. Atherosclerosis **84:** 111–119.
61. PELHAM, H. R. B. 1984. EMBO J. **3:** 3095–3100.
62. BECKMANN, R. P., L. A. MIZZEN & W. J. WELCH. 1990. Science **248:** 850–853.
63. CHIANG, H.-L., S. R. TERLECKY, C. P. PLANT & J. F. DICE. 1989. Science **246:** 382–385.
64. FLYNN, G. C., T. G. CHAPPEL & J. E. ROTHMAN. 1989. Science **245:** 385–390.
65. JINN, T. L., Y. C. YEH, Y. M. CHEN & C. Y. LIN. 1989. Plant & Cell Physiol. **30:** 463–469.
66. HOY, R. R. 1973. *In* Developmental Neurobiology of Arthropods. D. Young, ed.: 203–232. Cambridge University Press. Cambridge, England.
67. NORDLANDER, R. & M. SINGER. 1973. J. Comp. Neurol. **152:** 175–192.
68. BROWN, I. R., D. G. LOWE & L. A. MORAN. 1985. Neurochem. Res. **10:** 1277–1284.
69. BROWN, I. R., S. RUSH & G. O. IVY. 1989. Neuron **2:** 1559–1564.
70. NOWAK, T. S., JR. 1985. J. Neurochem. **45:** 1635–1641.
71. NOWAK, T. S., JR., U. BOND & M. J. SCHLESINGER. 1990. J. Neurochem. **54:** 451–458.
72. MARTIN, J., R. M. GROSSFELD, D. L. BUNTING, M. TYTELL, B. E. DWYER & Z.-Y. XUE. 1991. Neurochem. Res. **16:** 533–542.
73. XUE, Z.-Y. & R. M. GROSSFELD. 1990. Neurochem. Res. Submitted.
74. GROSSFELD, R. M. & D. B. HANSEN. 1987. Neurochem. Res. **12:** 977–983.
75. RUDGE, J. S., G. M. SMITH & J. SILVER. 1989. Exp. Neurol. **103:** 1–16.
76. SARNE, Y., E. A. NEALE & H. GAINER. 1976. Brain Res. **110:** 73–89.

77. Meyer, M. R. & G. D. Bittner. 1978. Brain Res. **143:** 195–211.
78. Gower, D. J., C. Hollman, K. S. Lee & M. Tytell. 1989. J. Neurosurg. **70:** 605–611.
79. Grossfeld, R. M., G. D. Bittner & M. A. Raymond. 1982. J. Neurobiol. **13:** 191–197.
80. Butt, A. M., P. T. Hargittai & E. M. Lieberman. 1990. Neuroscience. In press.
81. Hargittai, P. T., A. M. Butt & E. M. Lieberman. 1990. Neuroscience. In press.
82. Kristensson, K., E. Stromberg, R. Elofsson & Y. Olsson. 1972. J. Neurocytol. **1:** 35–48.
83. Lane, N. J., L. S. Swales & N. J. Abbott. 1977. J. Cell Sci. **23:** 315–324.
84. Shivers, R. R. 1976. Brain Res. **108:** 47–58.
85. Nordlander, R. H., J. A. Masnyi & M. Singer. 1975. J. Comp. Neurol. **161:** 499–514.
86. Novotny, G. E. K. 1984. Acta Anat. (Basel) **119:** 106–112.
87. Holtzman, E. 1971. Phil. Trans. R. Soc. Lond. B **261:** 407–421.
88. Bologna, L., Y. Aizenman, F. Chiapelli & J. De Vellis. 1986. J. Neurosci. Res. **15:** 521–528.
89. Fisher, M. & R. J. Mullen. 1988. Neuron **1:** 151–157.
90. Giulian, D., K. Vaca & B. Johnson. 1988. J. Neurosci. Res. **21:** 487–500.
91. Mokuno, K., J. Kamholz, T. Behrman, C. Black, M. Sessa, D. Feinstein, V. Lee & D. Pleasure. 1989. J. Neurosci. Res. **23**: 396–405.

DISCUSSION PAPER

Current Perspectives in Glial Electrophysiology[a]

J. MURDOCH RITCHIE

Department of Pharmacology
Yale University School of Medicine
New Haven, Connecticut 06510

The session on glial electrophysiology was a testament to the remarkably rapid development of this field since voltage-gated sodium and potassium channels were first described in rabbit Schwann cells in 1984 by Chiu *et al.*[1] Quickly following this in 1985 came the first full paper by Bevan *et al.*[2] describing similar channels in rat cortical astrocytes as well as a hitherto unsuspected voltage-gated chloride channel. Abstracts and other papers published then and since made it clear that similar work was being carried out independently elsewhere (Harvard by Barres *et al.*[3]; College de France by Berwald-Netter *et al.*[4]). Since then several things have become clear. First, several different types of glial cell can be distinguished embryologically and immunologically.[5–8] Second, any given satellite cell (Schwann or glial) may express more than one kind of sodium or potassium current. Third, the pattern of expression of different kinds of current in a given class of glial cells may differ consistently from the pattern in another class of glia. Finally, even within a given class of glial cell the pattern of expression at any given time may well be determined by its developmental history. This then was the framework for the exciting discussion of the burgeoning findings presented. Differences in the electrophysiological expression of the currents, because of phenotypic and developmental factors and as a result of technological differences in the methods used by different groups, inevitably led to apparent differences in experimental findings. Nevertheless, the heartening aspect of this symposium was the general agreement on the properties of these intriguing voltage-dependent sodium and potassium channels.

SODIUM CHANNELS

The first full description of voltage-gated glial sodium channels was made in rat *cortical* astrocytes by Bevan *et al.*[2] These cultures were derived from many different areas of the brain whose astrocytes might well express different electrophysiological properties. A major recent advance, therefore, came from the study of specific phenotypes of glial cells from one specific area of the central nervous system, (CNS), namely, glial cells of the rat optic nerve studied *in vitro.* (See refs. 9 and 10.) This particular preparation has the advantage that the different phenotypic channel types have been characterized extensively on the basis of both their morphology and growth characteristics and their ability to bind antibodies, particularly monoclonal antibody A2B5.[5] Thus, A2B5-negative astrocytes *in vitro,* derived from neural epithelial cells of the optic nerve, are termed type-1 astrocytes and react with antibody RAN2 but not with another antibody A2B5; cortical astrocytes (protoplasmic astrocytes found generally in gray matter), which are also A2B5 negative, seem

[a]This work was supported in part by grants NS08304 and NS12327 from the U.S.P.H.S.

largely to share the electrophysiological and immunological properties of type-1 cells. On the other hand, A2B5-positive astrocytes, which have been termed type 2, are derived from O-2A progenitor cells, which also can develop into oligodendrocytes. The different groups consistently use their own special terminologies; for example, Barres and her colleagues[6–9] use "type-1" (but sometimes A2B5-negative), whereas Waxman and his colleagues[10–12] always use the term A2B5-negative. As the latter term is used by all groups, it will usually be used here. The first two papers of the symposium reported the remarkable profusion of data obtained from this preparation by the two main groups (Barres and colleagues at Harvard/University College London; Waxman and colleagues at Yale). Considering the large separation in both distance and experimental approach, we cannot be but pleased by the remarkable degree of agreement in their findings.

Plasticity of Sodium Channel Expression

The clearest finding is that the expression of voltage-gated channels by a given population of cells is not fixed. Early experiments on cortical astrocytes showed no clear dependence of sodium channel expression on the age of the culture,[2,4,13] but a large variation existed in the amplitude of the currents in individual astrocytes. The same is true for Schwann cells from rabbit sciatic nerve: these express a highly variable sodium current[1,14–16] whose pattern, despite reports to the contrary,[17,18] does not substantially change with development or with time in culture, at least over a 2-week period.[15] Furthermore, immunological studies clearly indicate the presence of voltage-gated sodium channels in the Schwann cells of freshly dissected rabbit nerve *in situ.*[11,18] In contrast to this relative uniformity in cortical astrocytes and Schwann cells, the various phenotypic rat optic nerve astrocytes exhibit a rich diversity. Thus, A2B5-negative astrocytes express sodium currents after only 1–3 days in culture but gradually lose this ability over the next 14 days; in contrast, the considerably less abundant A2B5-positive astrocytes continue (perhaps even increase) their expression of sodium currents.[11] These electrophysiological changes are nicely paralleled by sodium channel antibody studies on the two types.[10] What is not clear is whether similar temporal changes occur *in vivo.* Sodium channel reactivity is clearly seen in optic nerve astrocytes *in vivo* from adult rats.[19] Definite sodium channel staining is present in astrocytes of newborn to 10-day-old astrocytes (Waxman, unpublished), but whether the staining is in A2B5-negative or A2B5-positive cells is unclear. Furthermore, it will be intriguing to examine the *in vivo* expression of sodium channels in view of the absence of sodium channels in optic nerve astrocytes that are axon deprived by enucleation[12] and in the light of the finding that contact with neurons greatly increases the expression of sodium currents in isolated A2B5-negative cells.[9,20]

What Do Different Kinds of Sodium Current Reflect?

The sodium currents expressed by A2B5-negative and A2B5-positive cells do seem to differ. Although no pharmacological differences between the two currents have been noted, electrophysiological differences clearly exist. The kinetics of inactivation of sodium currents in A2B5-negative cells are slower than those of the A2B5-positive cells (by a factor of about two[7]). Furthermore, the midpoint of the h-infinity curve for A2B5-negative optic nerve astrocytes is set at about -80 to -90 mV,[6,7,9,11] whereas that of the A2B5-positive cells is set at about the same voltage as that of their associated neurons, namely, −50 to −60 mV. Barres *et al.*[7] take this pattern to indicate two distinct morphological types of channel: one in which the midpoints of the *h*-infinity and *I*/*V* curves are at about −80 and −35 mV, respectively; the other in which the corresponding points are at about −55 and −25 mV, respectively. The former is characteristic of type-1 rat optic nerve astrocytes, whereas the latter is characteristic not only of type-2 astrocytes, but also of O-2A progenitor cells and neurons. A special point is made of the facts that the *I*/*V* curve of the type-1 rat optic nerve astrocyte is shifted by about 10 mV along the voltage axis in the *hyperpolarizing* direction, relative to that of the type-2 astrocyte, and, furthermore, that the shift in the same direction is even greater for the *h*-infinity curve. However, this distinction may be special for rat optic nerve astrocytes and not for astrocytes in general. For example, the position of the *I*/*V* curve for cortical (presumably A2B5-negative) astrocytes studied by Bevan *et al.*[2] is almost identical with, and is not shifted relative to, that of the *I*/*V* curve of rat optic nerve neurons and so might be classified as "neuronal." But the midpoint of the *h*-infinity curve in these cortical astrocytes was at −72 mV (i.e., closer to type-1, A2B5-negative, than to type-2, A2B5-positive rat optic nerve astrocytes) and so might be classified as "glial." This point will be continued further below.

Some of the acrimony that clearly flowed beneath the surface in the discussion of these findings stems from taking the *I*/*V* curve of a cell as a structural criterion. This has led Barres and her group[6–9] to regard a difference in an *I*/*V* curve to reflect a difference in molecular structure. But how standard is this "standard"? The midpoint of the type-2 rat optic nerve astrocyte and retinal ganglion cell *I*/*V* curves (i.e., the potential at which half-maximal current is elicited) are at about −25 mV, and it is also there for AB2B5-positive astrocytes, O-2A progenitor cells. Barres and colleagues[6–9] emphasize this similarity and contrast it with the *I*/*V* curve of the A2B5-negative cell which is shifted by 10 mV in the *hyperpolarizing* direction. But cortical astrocytes have their midpoint at about −25 mV (i.e., where the "neuronal" midpoint is), and in other systems studied, quite a different pattern emerges. Thus, in satellite cells of the peripheral nervous system, (PNS) where the midpoint of the neuronal *I*/*V* curve is about −55 mV,[21,22] the *I*/*V* curve for the associated Schwann cell is shifted by about 25–30 mV, but in the *depolarizing* direction. (See ref. 15 for detailed comparison.) Furthermore, in the hippocampus, where the midpoint of the neuronal *I*/*V* curve is at about −40 mV,[23] the corresponding *I*/*V* curve of the associated astrocytes is again shifted in the depolarizing direction to about −30 mV. Why would all neuronal *I*/*V* curves be expected to be identical? Why would all satellite cell *I*/*V* curves be expected to be identical? Is every class of neuron and glial cell expected to have its own special molecular form of sodium channel? These questions immediately lead into the second point.

It has been assumed (e.g., by Barres *et al.*[6–9]) that the different kinds of current expressed by the various cells reflect different molecular forms of the sodium channel. This assumption is at present unwarranted. To be categorical, except in the few cases in which site-directed mutagenesis of the sodium channel has been accomplished and the product (*de facto* an altered structure) then expressed in receptive cells, there is no evidence to date that different molecular structures necessarily follow from the finding of sodium currents with different electrophysiological characteristics. Pharmacologists are noted for their propensity to alter such characteristics quite extensively with drugs, much too quickly for any structural (as opposed to conformational) change to occur. For example, scorpion toxin completely changes the kinetics of sodium channel inactivation virtually as soon as it is applied. Does this reflect an immediate change in the molecular structure? Furthermore, it is generally believed that the sensors governing voltage-dependent inactivation and activation are probably in separate parts of the molecule, and there is at the moment no *a priori* reason from biophysical experiments on sodium channels in axons and in planar lipid bilayers for excluding shifts in voltage dependence with different membrane composition and no reason for expecting any such shift in the inactivation curve to parallel exactly a shift in the activation curve; rather the contrary. Substantial shifts in the voltage dependence of activation and inactivation are known to occur merely as a result of patch excision (see, for example, ref. 24), a process that also occurs in rabbit Schwann cells.[15] In the opinion of this reviewer, the quite different positions of the I/V curve in retinal ganglion cells, myelinated peripheral axons, and hippocampal neurons, together with the fact that the I/V curves of the satellite cells bear no standard relation to those of their associated neurons, argue less in favor of three different molecular species of neuronal sodium channel and more in favor of different membrane composition of the different kinds of cell. But it is wrong and misleading (in my opinion) to assume that measurements made exclusively of different *current* characteristics are necessarily expressions of different *forms* of the sodium channel unless molecular biological evidence is presented. Furthermore, the classification into "glial" and "neuronal" forms is of questionable utility: not only do some glia sometimes express the "glial" and sometimes the "neuronal" forms, but also if the hyperpolarizing shift of the I/V curve is taken as a criterion of the "glial" form, PNS axons and neurons from parts of the CNS other than the optic nerve would have to be considered as expressing "glial" rather than "neuronal" sodium channels. This is not to say that differences in sodium channel structure in glia do not exist: they just have not been demonstrated. The molecular biological techniques needed to do this are indeed, in principle, now available, and this issue is a challenge to the molecular biologist.

Functional Aspects

The function of sodium channels in astrocytes remains up in the air. It is well established that astrocytic processes (from perinodal astrocytes) contact the axon at the node of Ranvier.[25] But what does this mean as far as a role for astrocytic sodium channels is concerned? The speculation,[14,26] that they may be transferred to, and be subsequently used by, their associated neuron, remains just an intriguing possibility. Some support for this idea does come from the presentation of Grossfeld[27] demonstrating that transfer of molecules between axons and glia does occur in several invertebrate and vertebrate species and does so particularly clearly in the giant axons of crayfish and squid. Both axon-to-glia and glia-to-axon transfer of molecules (even as large as proteins) seems reasonably well established. However, cytoplasm-cytoplasm exchange is a different process from the plasmalemmal-neurolemmal exchange envisaged by the Schwann cell-axonal transfer of the sodium channel transfer hypothesis; and so the hypothesis, although not discarded, remains unproven.

In their seminal study, Orkand *et al.*[28] did consider the possible significance of glial potentials as indicators of overall nervous activity. Glial potentials would achieve this by registering (indiscriminately) impulse traffic in inhibitory and excitatory pathways as well as synaptic events. But how such potentials and their associated currents might, in turn, influence neurons was, and still remains, unknown. However, it is now clear that membrane depolarization of glial cells by potassium can release neuroactive substances, such as acetylcholine and possibly VIP-like peptide[29]; it can also inhibit glial cell uptake of glutamate, leading possibly to neurotoxic extracellular concentrations of the amino acid.[30] These and other similar phenomena could, in principle, provide the necessary signals influencing neuronal activity. One possible function of the glial sodium channels might be that their local responses would give a phasic extra depolarization to an otherwise steady, potassium-induced, glial depolarization (and so provide a phasic component to the putative signal). But, again, this is mere speculation.

Technical Points

Some of the discussion centered around technical points. For example, any biophysical analysis of ionic currents is generally taken to require a knowledge of defined equilibrium potentials. The chairman of the session noted that some sodium currents were being measured in the presence of zero intracellular sodium, that is, with an infinite sodium equilibrium potential, and, similarly, the outward potassium currents were measured in the presence of zero external potassium. Although this in no way detracts from the descriptive value of the observations, the criticism could easily have been avoided. It is also worth remarking in this connection that we know little about the important factors that determine the equilibrium potentials of the various ions, namely, their intracellular concentrations *in vivo* in the various experimental preparations. An estimate can be made for Schwann cells. In the rabbit vagus nerve, where about half the cell volume is attributable to Schwann cell cytoplasm and half to the axoplasm of the nonmyelinated fibers,[31] direct flame photometry gave average concentrations of intracellular potassium, sodium, and chloride ions of 145, 63, and 39 mM, respectively (at corresponding external concentrations of 5.6, 154, and 165 mM, respectively). These yield values for E_K, E_{Na}, and E_{Cl} of −82, +22, and −36 mV, respectively. The corresponding intracellular concentrations for guinea pig glia, calculated from the activities measured by Ballanyi *et al.*,[32] are 66, 25, and 6 mM, respectively, with extracellular concentrations of 3, 144, and 128 mM (i.e., E_K, E_{Na}, and E_{Cl} for these mammalian glia are −75, +43, and −74 mV, respectively). These values, if they apply to astrocytes *in vivo,* are not necessarily matched by the values that obtain in practice in the patch-clamp experiments; therefore, some differences in this regard between *in vitro* and *in vivo* conditions might well be expected. In general, these concentrations are consistent with the finding of a lower resting potential for Schwann cells (−40 mV[15]) than for astrocytes (−75 mV[33]).

It is a pity too that the intriguing A2B5-positive astrocytes and O-2A progenitor cells are so relatively rare that statistical variation in the cited mean values is relatively large. Even with the abundant A2B5-negative cells, it is not clear at times just how well some descriptions and statements of differences between cells would stand up to rigorous statistical analysis of significance. For example, the finding that 10 cells expressed an average sodium current of −150 ± 50 pA (mean ± SD)[7] loses some weight when one finds that sodium currents are only present in about 20–50% of cells. The fact that they are now found in 100% of the cells isolated by a new technique called "tissue printing" does not allay this reviewer's confusion, particularly until it is made clear that the new technique is not by chance selecting a particular population of cells. The application of this technique to glial studies is new and different from other procedures used to date. But it still requires enzymatic treatment with papain. Why such cells that survive being pressed (albeit gently) on a nitrocellulose film can necessarily be taken as being more likely than cells otherwise obtained to resemble cells *in vivo* escapes the present discussant. All the different methods used so far are important because they exemplify the multifaceted approaches necessary for the solution of any scientific problem (specific sodium-channel ligand binding; cultured cells; transformed cells; freshly dissociated cells;

tissue printing; antibody staining *in vitro* and *in situ*). To single out one as clearly better than others seems (to this reviewer) an example of hubris.

Another technological difference deserves comment. We are clearly in an area where more and more different groups are amassing bodies of experimental information that, because of differences in the individual techniques involved, may appear to be disparate. It is important, therefore, that the various techniques used be specified fairly exactly. For example, because the basis for understanding electrode potentials has been recognized (at least since the time of Max Planck) to be somewhat axiomatic, and because most electrophysiologists are fairly successful in practice in allowing for, or eliminating (hopefully), contact potentials, this seems to be a minor problem and should not be made a point at issue. A more important problem with the patch-clamp is to specify the electrophysiological conditions clearly such as: the actual series resistance of the pipette in the cell and not just in saline solution; whether or not series resistance compensation was used and to what extent; the total cell capacity, that is, fast plus slow. The latter two measurements are important because they give a time constant for the system that may be quite different from that of the recording amplifier, and knowing them allows some assessment of the faithfulness of the recording.

POTASSIUM CHANNELS

The problem of sorting out the various kinds of sodium currents pales in relation to the very much larger number of voltage-gated potassium channels that have been described for Schwann cells and glial cells.[6,9,14,32,34] (See also refs. 35 and 36.) These different channels, which are described fully elsewhere, include an inward rectifier (K_{ir}), a delayed rectifier (K_D), an A-current (K_A), a calcium-activated cation channel that is permeable to both potassium and sodium, and other potassium channels activated by intracellular components that surely are yet to be described. These are well reviewed in the literature (only a few examples being cited above) and will not be touched on here except in the consideration that follows in the role of potassium uptake by glia, which was a major theme of the meeting.

A main topic of the presentations, discussion, and subsequent discussion around the posters concerned the role of potassium channels in the regulation of extracellular potassium ions. The issue concerns how the extracellular potassium concentration that is raised as a result of neuronal activity can be prevented from so depolarizing the neuron that its activity is prematurely terminated. The major idea dominating this issue was put forward two and a half decades ago in the insightful work of Orkand *et al.*[28] (together with other colleagues) who suggested that the high potassium conductance of glial cells (that they had just discovered) might play an important role in allowing potassium ions to be driven into or taken up by the glial cells. That neuronal activity and high extracellular potassium each does indeed increase glial potassium uptake is now well-established experimentally. A second, more minor addition came with the discovery of chloride channels in glia by Bevan *et al.*[2] (and the later demonstration in Schwann cells by Howe and Ritchie[35]) who suggested that a concurrent uptake of chloride ions by astrocytes would aid in this process; they also pointed out that water would necessarily accompany this process. That chloride does enter glial cells during neuronal activity was soon experimentally demonstrated in both vertebrate and invertebrate glia[32,37]; and volume changes with neuronal activity have long been known. (See refs. 38 and 39.) So the central idea, as well as its experimental background, is well established; it seems unnecessary, therefore, to rediscover it (as sometimes seemed to be being claimed).

An unresolved question, however, concerns the details of the process by which the demonstrable rise in glial potassium concentration is brought about in response to a rise in extracellular potassium. Some must be taken up to satisfy Donnan requirements (see ref. 40), but this is relatively small. Some is taken up by active uptake, and some by being driven inwards by a difference between the local membrane and potassium equilibrium potentials.

Once the potassium is taken up by the cell, the question arises of how it is dispersed throughout the glial syncytium. Is it stored locally in the astrocyte with chloride and water (a process sometimes dubbed "potassium accumulation") except insofar as diffusion along the syncytium occurs, or is it driven to a more remote part of the syncytium by a standing intraglial current (by a process often called "spatial buffering")? The theoretical aspects of this issue have been dealt with in detail in the past by Gardner-Medwin,[41] for example, and beautifully summarized in the rigorous analysis presented in the symposium by Reichenbach.[42] But the *kind* of channel involved is irrelevant in this connection. All mechanisms (except that involving a pump) require a high potassium conductance through which potassium ions can move in accordance with clear thermodynamic principles, namely, there must be a difference between the membrane potential, E_m, and the potassium equilibrium potential, E_K. The presence of any particular potassium conductance cannot discriminate between the various possibilities.

In the light, therefore, of the well-established hypothesis of glial potassium uptake, the rigorous analytical models that have been proposed, and the wealth of available experimental data, it is difficult for the present reviewer to understand some of the jostling at the meeting concerning whether the newly discovered astrocytic potassium channels favor one or other of the two possibilities (even to the point of seemingly claiming priority). The real question is not whether astrocytic potassium uptake occurs; it is known to. Neither is it whether a potassium conductance is involved; it must be. So it is not sufficient just to show that astrocytes express an inwardly rectifying potassium channel (as indeed was already known for glia[34]). Rather, the important question is to determine which of the presently described channels[6–9] have the biophysical properties that fit the bill, a question that is still up in the air.

The voltage-dependent potassium and chloride channels described by Bevan *et al.*[2] cannot contribute much to potassium and chloride conductances at rest (although, as was pointed out by Bevan *et al.*,[2] they must contribute at a very high external potassium concentration), because they activate only at −40 to −50 mV; at a resting potential of −75 mV (see, for example, ref. 33) very few channels would be open. Orkand *et al.*,[28] however, did note that intense electrical stimulation raised the extracellular potassium concentration around glia from its normal value of 3 mM to 20 mM, which corresponds to a depolarization of over 45 mV. And Ballanyi *et al.*[32] have recorded glial depolarizations as large as 34 mV on stimulation of slices of guinea-pig olfactory cortex. Finally, to play devil's advocate, if it is generally accepted that the electrophysiological properties of a cell in culture may differ considerably from those *in vivo,* might not the activation of the potassium and chloride currents be shifted *in vivo* to more negative potentials?

The inward-rectifier potassium channel, which has been described both for Schwann cells[43] and glial cells,[34] must clearly be a candidate for astrocytic potassium uptake. An inward-rectifier channel in astrocytes has indeed been found experimentally[6–9]; and it is open at a patch-clamp potential near the putative resting potential, which must be somewhat more positive than the potassium equilibrium potential of about −85 mV and is taken, for the sake of argument, to be about −75 mV.[33] However, some of the properties of inward rectifiers must be considered. A characteristic of an inward rectifier is that the slope of its I/V curve, that is, the conductance, diminishes by up to several orders of magnitude at potentials more positive than a critical voltage (or small range of voltages), E_{crit}, which is usually close to E_K. However, as Hille[44] has pointed out, the voltage dependence of the classical inward rectifier depends on $[K]_o$ and shifts along the voltage axis in the depolarizing direction according to the quantity $RTln[K]_o$. But all the presently available measurements of the inward-rectifier currents on astrocytes by Barres *et al.*[6–9] have been made with an external potassium concentration of 20 mM. This means that at any given patch potential the I/V curve of the typical inward rectifier in astrocytes must have already been shifted about 35 mV in the depolarizing direction along the voltage axis (relative to its position with $[K]_o$ = 5 mM, and nearly 50 mV relative to a more appropriate value for $[K]_o$ of 3 mM[32]). The inward rectifier may thus have been opened adventitiously by the experimental conditions. The question remains, therefore, whether the astrocytic inward rectifier really is open at the *normal* resting potential under *normal* values of $[K]_o$; and if it is, is it sufficiently open at rest to carry the required inward current either to mediate potassium uptake or to help maintain the normal resting potential? It may well be; but there is an important loose end. The existence of an as yet unidentified "potassium-preferring" leakage channel, similar to that adumbrated by Hille[45] for axons, remains tenable for astrocytes. It is surely premature to propose a mechanism for potassium uptake by astrocytes[9] based on an as yet incompletely characterized inwardly rectifying potassium channel and an as yet unidentified chloride channel.

Barres[9] suggests that the inward rectifier may also underlie the glial resting potential, giving as the main evidence the fact that cesium virtually completely eliminates the resting glial conductance *in vitro*. Again, it has to be shown that this channel is open at the resting potential under normal conditions. Moreover, cesium blocks too many other potassium channels, including some esoteric nonneuronal potassium channels described by Hille,[44] for it to be taken as a specific blocker of inward-rectifier channels.

These considerations are not meant to belittle the role of inward rectifiers in glial potassium uptake, for they clearly operate in some situations. Thus, Newman[46] (whose presence at this meeting was missed) clearly showed experimentally in Müller cells (the glial cells of the retina) that potassium ions surrounding a local region of high potassium conductance can indeed be driven into the main body of the cell (and driven out at the endfeet). This occurs when the membrane potential in the local region of high potassium conductance is maintained electrotonically more negative than is the local potassium equilibrium potential by the higher membrane potential of the large area of surrounding membrane that is relatively more impermeable to potassium, and that is thus not so much depolarized by any raised external potassium (being supported possibly by electrogenic pumping as well as a chloride potential). The process depends on an uneven distribution of potassium conductance along the surface of the single cell, which has been demonstrated by the recording of single-channel currents in retinal ganglion cells.[34] In uniform distribution along a glial membrane, simply raising the external potassium concentration along a given stretch of it will not cause any substantial change in the intracellular potassium concentration (other than the small amount required to satisfy Donnan considerations and to transfer charge appropriate to the membrane depolarization). Bulk transfer of potassium ions into the glial cell requires a driving force: the local membrane potential in the region of high potassium conductance must be maintained more negative than E_K and *mutata mutandis* in sink regions.

CHLORIDE CHANNELS

It should be emphasized that when the external potassium concentration is increased, the positive shift in the potassium equilibrium potential, E_K, will sooner or later be followed by a depolarization, so that the difference between E_K and the membrane potential, E_m, will no longer drive potassium into the glia. The decrease in resting potential, however, can be delayed or even prevented by any of several mechanisms, thus allowing substantially greater potassium accumulation. For example, as Newman[46] has shown in Müller cells, surrounding nondepolarized membrane may electrotonically maintain the membrane potential of a high-conductance region. Alternatively, the electrogenic sodium pump may sustain a hyperpolarization, and the pump activity itself may in turn be modulated by a stimulation by high extracellular potassium of glial sodium/proton and sodium-bicarbonate antiporter and cotransport systems, respectively (see refs. 39 and 47). An important additional (perhaps major) mechanism is the simultaneous uptake of chloride ions, whose equilibrium potential, E_{Cl}, is near that of E_m.[32] Any decrease in resting potential due to potassium entry will immediately lead to a difference between E_m and E_{Cl} that will drive chloride ions into the cell, thus limiting depolarization and permitting further potassium uptake. The chloride channel involved, however, has not yet been characterized electrophysiologically.

THE IMPENDING IMPORTANT ROLE OF MOLECULAR BIOLOGY

A most interesting line of approach to glial sodium channels and their function was brought out more in the poster sessions than in the main session itself. This emphasized the immense contribution that molecular biology in association with electrophysiology is clearly about to make. For example, Berwald-Netter and her colleagues[48] suggested that astrocytes are capable of expressing forms of voltage-gated sodium channels distinct from those of neurons. Using RNA extracts from cultured neuronal or glial rat brain cells with cDNA probes specific for each of the three known channel types in rat brain, they showed that the three Na channel types capable of being expressed in neurons are not expressed in cortical astrocytes. The important issue that needs to be settled is whether these channels (presently expressed only in oocytes) are indeed the ones expressed in neurons and glia, and this does require proof. In other words, do the cloned cDNA sequences actually code for the functional membrane channels in the way that it has been shown for the acetylcholine receptor (see ref. 49)?

It is perhaps germane to the foregoing considerations that all the neuronal sodium channels so far sequenced have been derived from extracts of whole brain. Inasmuch as there are 10 times as many glial cells as neurons in brain, it would seem important to isolate and sequence a channel obtained from a purely neuronal source and a channel from a purely glial source if the comparison between glial and neuronal channels is to be made secure. In either case, the actual sodium channel protein in the neuronal and glial cell membranes must be sequenced. There is clearly a rich experimental field for molecular biologists. Already, site-directed changes in specific molecules have indicated at least part of the sodium channel that is involved in inactivation.[50,51]

CONCLUSIONS

The criticisms offered herein, which arise mainly from the overenthusiasm of the experimenters in interpreting the wealth of valuable data being collected, should not detract from the excitement over the vast amount of new information that is being generated, from which several specific conclusions can be drawn: (1) Astrocytes do express voltage-gated channels. (2) These channels express a spectrum of currents. (3) Some of the different currents flow in distinctly separate kinds of channel (sodium, potassium, chloride, etc.), but a given cell may express currents with different electrophysiological characteristics under different *in vivo* and *in vitro* conditions. (4) In particular, the characteristics of the current expressed by a given type of astrocyte may well change with time in culture. (5) There is a strong hint that the pattern of current expression by an astrocyte depends on neuronal contact. (6) There is increasing support for the idea, originally put forward by Orkand, Nicholls, and Kuffler 25 years ago,[28] that astrocytes play a major role in the control of the potassium concentration in the extracellular space, and there are hints of a variety of other functions.

It is hoped that the information being collected will ultimately be incorporated into some functional corpus. This will require an interdisiplinary approach. The developmental expression of the various channels, their control by neurons perhaps through a medium of cytokines, and the exact molecular structure of the sodium channels in glial cell plasmalemma and axolemma are all fruitful lines of inquiry for electrophysiologists, morphologists, and molecular biologists alike. Clearly, the "facts" must be unequivocally established. Wilkie[52] once began a review with, "Facts and theories are natural enemies." One might add: "Without facts no theory is worth fighting for." At the moment the need for facts outweighs the need for speculation. Let us emulate Isaac Newton[53]: *"Hypotheses non fingo"* (in context usually taken to mean, "I do not indulge in (idle) speculation)."

REFERENCES

1. Chiu, S. Y., P. Shrager & J. M. Ritchie. 1984. Neuronal-type Na^+ and K^+ channels in rabbit cultured Schwann cells. Nature **311:** 156–157.
2. Bevan, S., S. Y. Chiu, P. T. A. Gray & J. M. Ritchie. 1985. The presence of voltage-gated sodium, potassium and chloride channels in rat cultured astrocytes. Proc. R. Soc. Lond. B **225:** 299–313.
3. Barres, B. A., L. L. Y. Chun & D. P. Corey. 1985. Voltage-dependent ion channels in glial cells. Soc. Neurosci. (Abstr.) **11:** 147.
4. Berwald-Netter, Y. A., A. Koulakoff, F. Couraud, L. Nowak & P. Ascher. 1986. Ionic Channels in Glial Cells. S. Federoff & A. Vernadakis, eds. Academic Press. New York.
5. Raff, M. C., E. R. Abney, J. Cohen, R. Lindsay & N. Noble. 1983. Two types of astrocytes in cultures of developing rat white matter; differences in morphology, surface gangliosides, and growth characteristics. J. Neurosci. **3:** 1289–1300.
6. Barres, B. A., L. L. Y. Chun & D. P. Corey. 1988a. Ion channel expression by white matter glia. I. Type 2 astrocytes and oligodendrocytes. Glia **1:** 10–30.
7. Barres, B. A., L. L. Y. Chun & D. P. Corey. 1989. Glial and neuronal forms of the voltage-dependent sodium channel: Characteristics and cell-type distribution. Neuron **2:** 1375–1388.
8. Barres, B. A., L. Y. Chun & D. P. Corey. 1990. Ion channels in vertebrate glia. Annu. Rev. Neurosci. **13:** 441–474.
9. Barres, B. A. 1991. Five electrophysiological properties of glial cells. Ann. N.Y. Acad. Sci. This volume.

10. MINTURN, J. E., J. A. BLACK, S. G. WAXMAN & K. J. ANGELIDES. 1990. Sodium channel expression detected with antibody 7453 in A2B5$^+$ and A2B5$^-$ astrocytes from rat optic nerve *in vitro.* Glia **3:** 358–368.
11. MINTURN, J. E., H. SONTHEIMER, J. A. BLACK, K. J. ANGELIDES, B. R. RANSOM, J. M. RITCHIE & S. G. WAXMAN. 1991. Membrane-associated sodium channels and cytoplasmic precursors in glial cells: immunological, electrophysiological and pharmacological studies. Ann. N.Y. Acad. Sci. This volume.
12. MINTURN, J. E., J. A. BLACK, H. SONTHEIMER, J. R. EMANNUEL, B. R. RANSOM & S. G. WAXMAN. 1991. Voltage-sensitive channels of astrocytes deprived of axons: Peptides, mRNA, and sodium current expression. J. Cell Biol. **111:** 489a.
13. NOWAK, L., P. ASCHER & Y. BERWALD-NETTER. 1987. Ionic channels in mouse astrocytes in culture. J. Neurosci. **7:** 101–109.
14. SHRAGER, P., S. Y. CHIU & J. M. RITCHIE. 1985. Voltage-dependent sodium and potassium channels in mammalian cultured Schwann cells. Proc. Natl. Acad. Sci. USA **82:** 948–952.
15. HOWE, J. R. & J. M. RITCHIE. 1990. Sodium currents in Schwann cells from myelinated and non-myelinated nerves of neonatal and adult rabbits. J. Physiol. **425:** 169–210.
16. CHIU, S. Y. 1987. Sodium currents in axon-associated Schwann cells from adult rabbits. J. Physiol. **386:** 181–203.
17. CHIU, S. Y. 1988. Changes in excitable membrane properties in Schwann cells of adult rabbit sciatic nerves following nerve transection. J. Physiol. **396:** 173–188.
18. RITCHIE, J. M., J. A. BLACK, S. G. WAXMAN & K. J. ANGELIDES. 1990. Sodium channels in the cytoplasm of Schwann cells. Proc. Natl. Acad. Sci. USA **87:** 9290–9294.
19. BLACK, J. A., B. FRIEDMAN, S. G. WAXMAN, L. W. ELMER & K. J. ANGELIDES. 1989. Immuno-ultrastructural localization of sodium channels in nodes of Ranvier and perinodal astrocyte processes in rat optic nerve. Proc. R. Soc. Lond. B **238:** 39–51.
20. BARRES, B. A., W. J. KOROSHETZ, L. L. Y. CHUN & D. P. COREY. 1990. Ion channel expression by white matter glia: The type-1 astrocyte. Neuron **5:** 527–544.
21. CHIU, S. Y., J. M. RITCHIE, R. B. ROGART & D. STAGG. 1979. A quantitative description of membrane currents in rabbit myelinated nerve. J. Physiol. **292:** 149–166.
22. NEUMCKE, B. & R. STÄMPFLI. 1982. Sodium currents and sodium-current fluctuations in rat myelinated nerve fibres. J. Physiol. **329:** 163–184.
23. SONTHEIMER, H., B. R. RANSOM, A. H. CORNELL-BELL, J. A. BLACK & S. G. WAXMAN. 1991. Na$^+$ current expression in rat hippocampus astrocytes *in vitro:* Alterations during development. J. Neurophysiol. **65:** 3–19.
24. HORN, R. & C. A. VANDENBERG. 1986. Inactivation of single sodium channels. *In* Ion Channels in Neural Membranes. J. M. Ritchie, R. D. Keynes & L. Bolis. :71–83. Alan R. Liss. New York.
25. WAXMAN, S. G. 1985. Perinodal astrocytic processes at nodes of Ranvier in developing and glial deficient rat spinal cord. Brain Res. **337:** 321–333.
26. GRAY, P. T. A. & J. M. RITCHIE. 1986. A voltage-gated chloride conductance in rat cultured astrocytes. Proc. R. Soc. Lond. B **228:** 267–288.
27. GROSSFELD, R. M. 1991. Axon-glia exchange of macromolecules. Ann. N.Y. Acad. Sci. This volume.
28. ORKAND, R. K., J. G. NICHOLLS & S. W. KUFFLER. 1966. Effect of nerve impulses on the membrane potential of glial cells in the central nervous system of amphibia. J. Neurophysiol. **29:** 788–806.
29. VILLEGAS, J. 1991. Mechanisms of axon Schwann cell signalling in the squid nerve fiber. Ann. N.Y. Acad. Sci. This volume.
30. ATTWELL, D., B. BARBOUR, H. BREW, M. SARANTIS & M. ZATKOWSKI. 1991. Electrogenic glutamate uptake: Basic properties, modulation and failure in anoxia. Ann. N.Y. Acad. Sci. This volume.
31. KEYNES, R. D. & J. M. RITCHIE. 1965. The movements of labelled ions in mammalian non-myelinated nerve fibres. J. Physiol. **179:** 745–793.
32. BALLANYI, K., P. GRAFE & G. TEN BRUGGENCATE. 1987. Ion activities and potassium uptake mechanisms of glial cells in guinea-pig olfactory cortex slices. J. Physiol. **382:** 159–174.

33. RANSOM, B. R. & W. G. CARLINI. 1986. Electrophysiological properties of astrocytes. *In* Astrocytes. S. Federoff & A. Vernadakis, eds. : 1–47. Academic Press. London.
34. BREW, H., P. T. A. GRAY, P. MOBBS & D. ATTWELL. 1986. Endfeet of retinal glial cells have higher densities of ion channels that mediate K buffering. Nature **324:** 466–468.
35. HOWE, J. R. & J. M. RITCHIE. 1988. Two types of potassium current in rabbit cultured Schwann cells. Proc. Roy. Soc. B. **235:** 119–127.
36. HOWE, J. R. & J. M. RITCHIE. 1991. On the active form of 4-aminopyridine in rabbit Schwann cells. J. Physiol. **433:** 183–205.
37. COLES, J. A., R. K. ORKAND & C. L. YAMATE. 1989. Chloride enters glial cells and photoreceptors in response to light stimulation in the retina of the honey bee drone. Glia **2:** 287–297.
38. RANSOM, B. R., C. L. YAMATE & B. W. CONNORS. 1985. Activity-dependent shrinkage of extracellular space in rat optic nerve: A developmental study. J. Neurosci. **5:** 532–535.
39. KEMPSKI, O., S. VON ROSEN, H. WEIGT, F. STAUB, J. PETERS & A. BAETHMANN. 1991. Glial ion transport and volume control. Ann. N.Y. Acad. Sci. This volume.
40. HODGKIN, A. L. & P. HOROWICZ. 1959. The influence of potassium and chloride ions on the membrane potential of single muscle fibers. J. Physiol. **148:** 127–160.
41. GARDNER-MEDWIN, A. R. 1986. A new framework for assessment of potassium-buffering mechanisms. Ann. NY Acad. Sci. **481:** 287–302.
42. REICHENBACH, A. 1991. Glial K^+ permeability and CNS K^+ clearance by diffusion and spatial buffering. Ann. N.Y. Acad. Sci. This volume.
43. WILSON, G. F. & S. Y. CHIU. 1990. Potassium channel regulation in Schwann cells during early developmental myelinogenesis. J. Neurosci. **10:** 1615–1625.
44. HILLE, B. 1984. Ionic Channels in Excitable Membranes. Sinauer. Sunderland, MA.
45. HILLE, B. 1973. Potassium channels in myelinated nerve. J. Gen. Physiol. **61:** 559–686.
46. NEWMAN, E. A. 1984. Regional specialization of retinal glial cell membrane. Nature **309:** 155–157.
47. SHLUE, W.-R., R. DORNER, L. REMPE & B. RIEHL. 1991. Glial H^+ transport and control of pH. Ann. N.Y. Acad. Sci. This volume.
48. BERWALD-NETTER, Y., G. DOS SANTOS, S. GAUTRON, A. KOULAKOFF & D. PINTO-HENRIQUE. Astrocytes and neurons express distinct forms of voltage-gated Na channels. This volume.
49. KRUEGER, B. K. 1988. Toward an understanding of structure and function of ion channels. F.A.S.E.B.J. **3:** 1906–1914.
50. VASSILEV, P. M., T. SCHEUR & W. CATTERALL. 1988. Identification of an intracellular peptide segment involved in sodium channel inactivation. Science **241:** 1658–1661.
51. STÜHMER, W., C. METHFESSEL, B. SAKMANN, M. NODA & S. NUMA. 1987. Patch-clamp characterization of sodium channels expressed from rat brain cDNA. Eur. Biophys. J. **14:** 131–138.
52. WILKIE, D. R. 1954. Facts and theories about muscle. Prog. Biophys. Mol. Biol. **4:** 288.
53. NEWTON, I. 1687. Principia Mathematica. S. Peyps. London.

INTRODUCTION

Implication of Astroglia in the Blood-Brain Barrier

MILTON BRIGHTMAN

Department of Health & Human Services
National Institutes of Health
Bethesda, Maryland 20892

The recognition of both a direct and an indirect role of the perivascular astrocyte in the maintenance of the passive blood-brain barrier to hydrophilic solutes will be briefly traced in this commentary. In certain vertebrates and invertebrates, it is the perivascular astroglial sheath that forms the endothelium as the barrier to the extracellular movement of substances between blood and the interstitial fluid of the brain. The same technique that revealed the endothelial site of the blood-brain barrier to circulating horseradish peroxidase (HRP) in mammals[1,2] and urodeles[3] also demonstrated, several years later, an equivalent glial barrier in some elasmobranchs. The infusion of HRP into the peripheral blood of the nurse shark, *Ginglymostoma cirratum,* and the guitar fish, *Rhinobatus lentiginosus,* consistently led to rapid passage of the glycoprotein through open junctions between adjacent endothelial cells and into a collagen-containing perivascular space. The narrowest portion of the open cleft between adjacent endothelial cells in the nurse shark was about 3 nm wide. The gradient from the perivascular to the interstitial compartments of the brain was extremely sharp. The reaction product abruptly ended at the perivascular, astroglial border, and no detectable peroxidase penetrated beyond this astroglial sheath. It was inferred that the observed tight junctions between perivascular glial cells, rather than the open junctions between endothelial cells, were the site of the barrier to the extracellular diffusion of peroxidase. The astroglial tight junctions were arranged in tandem with gap junctions between contiguous glial cells, but it was thought that the fixation was inadequate to convincingly support the inference that the junctions were zonular.[4] In the brain tissue of three species of another elasmobranch, the skate, *Raja,* examined by Bundgaard and Cserr,[5] HRP was also depicted as stopping at junctions between perivascular glial cells. The only evidence of junctional zonularity was, as reported earlier,[4] the sharp gradient in HRP between perivascular space and interstitial clefts. Although the clefts between endothelial cells in the skate brain were usually interrupted by punctate junctions, a few of the clefts were open throughout their length, the clefts being about 15–20 nm wide. A striking characteristic, restricted in other vertebrate brains to the vessels of the circumventricular organs and the dura mater, was the fenestration of parenchymal vessels within the skate brain.[5] Fenestrated vessels are passively permeable to large solutes, and exogenous peroxidase was able to cross the skates' endothelium within 1 minute of circulation. The perivascular glial cells of the skate not only impeded the extracellular movement of a macromolecule, but also were able to perform another function of a barrier type of epithelium: electrolyte homeostasis.[5] No transcellular channels could be detected across the shark or skate brain endothelium. However, the brain endothelium in several species of shark does have numerous pits and vesicular profiles that have been interpreted as forming transendothelial channels, purported to transfer circulating peroxidase from blood to the interstitial fluid of the brain.[6] An endothelium that has open clefts, however, would permit the glycoprotein to pass into the periendothelial space via the clefts and thence the retrograde labeling of those pits on the abluminal side of the endothelium. A more suitable cell to test the notion that transcellular channels provide a route for macromolecular transport is an endothelium with numerous pits, vesicular profiles, and tubules. Capillaries with such an endothelium supply the brain of the primitive vertebrate, the cyclostome, hagfish, *Myxine glutinosa.* Blood-borne peroxidase, being prevented from reaching the abluminal face of the endothelium by zonular tight junctions, did not penetrate many abluminal pits or tubules within the endothelial cytoplasm, and no detectable amount reached the perivascular basal lamina.[7] The brain capillary of the hagfish, like that of higher forms, stood as a barrier to blood-interstitial fluid exchange. Thus, although endothelium may be riddled with pits, vesicles, and tubules, these structures are generally not be involved in the transfer of large, hydrophilic molecules across the cells.

At about the time that the perivascular astroglia rather than the vascular endothelium was first recognized as the site of the blood-brain interstitial fluid barrier in certain elasmobranchs, a comparable glial barrier was defined morphologically and physiologically in insects.[8–10] Insects, the nervous systems of which have some of the most evolutionarily advanced physiological behavior of all invertebrates, do not have blood vessels within their ganglia. They do have a sheath of "perineurial" glial cells that envelopes their nervous system. The relatively small tracer, ionic lanthanum, is impeded by junctional complexes in its passage through the interstitial clefts of the sheath, and it is the occluding tight junctions of these complexes that comprise the barrier.[8,9] Thus, in certain arthropods, some adjacent glial cells of the perineurium are fastened together by apparently zonular tight junctions that are responsible for establishing permeability gradients between hemolymph and the interstitial fluid of ganglia and connectives.[10–12] The neuroglial cells of other invertebrates lack such junctions and do not form a barrier. However, certain molluscs, like the insect members of the arthropods, also have a glial-interstitial fluid barrier without apparently the intervention of the usual tight junction. In cephalopod molluscs, such as cuttlefish, squid, and octopus, the cerebral ganglia are invaded by microvessels, an arrangement more akin to higher forms. Their endothelium can be crossed by HRP, but further progress of the glycoprotein is halted by perivascular glial cells. In many places, the glial cells of the sheath apparently form a continuous, "seamless cuff" around the blood space.[11] However, where clefts do occur, the barrier appears to be formed by a reticulum of extracellular fibrils rather than tight junctions. The barrier, then, may be in the form of zonular intercellular junctions, as in some arthropods, or extracellular matrix, as in some molluscs. The perceptive conclusion has been drawn that these barriers function not only to protect neurons from the vagaries of solute concentration in blood, but also to bring about the subtle, local control or "fine-tuning" of the microenvironment around specific neuronal groups. Such discrete isolation would be required by the more complex nervous system of these higher invertebrates.[12]

The mammalian perivascular astrocyte appears to enhance the development of tight junctions only between cerebral endothelial cells. By the same token, can the perivascular astrocyte of certain elasmobranchs influence the formation of a CNS endothelium with *open* junctions, and does this endothelium, in turn, enhance the formation of zonular, tight junctions between astroglial cells? This question should be amenable to *in vitro* tests.

Can cerebral endothelium also affect the structure and function of the nonjunctional portion of the astrocytic cell membrane? That such a structural influence may be exerted by brain endothelium has been suggested by *in vitro* experiments.

Orthogonal arrays or assemblies of particles, visualized within the freeze-fractured plasmalemma of astrocytes, may be modified by brain endothelial cells *in vitro.*[13] The coculture of bovine brain endothelial cells with secondary cultures of rat astrocytes leads to an increase in the number of assemblies, their aggregation, and a loss in randomness, that is, an alignment of the assemblies into train-like arrays that resemble those of perivascular astrocytes *in situ.* The function of the assemblies is unknown, but they are believed to be associated with K^+ channels.[14] Although the mammalian perivascular astrocyte endfoot is not "zippered" tightly to its neighbor and, consequently, does not block intercellular diffusion between interstitial fluid and pericapillary space, it very likely influences the composition of the brain's interstitial fluid by, for example, acting as a spatial buffer of extracellular K^+ bathing the neuropil.[15] The effect on the conductance and redistribution of K^+ ions by the rearrangement of assemblies under the influence of brain endothelium is a question that would be interesting, but difficult to explore.

There are other indications that perivascular astroglia may directly affect the barrier properties of endothelium. In most vertebrates, the passive, extracellular blood-brain barrier resides in the endothelium of cerebral vessels rather than their astroglial sheath where the intercellular links are circumventable gap junctions rather than tight junctions.[2,3] Although the suggestion was made about 25 years ago[16] that the perivascular processes belonging to astrocytes might influence the permeability of adjacent capillary endothelium, the demonstration of such an influence has only been recent. The elegant experiments of Stewart and Wiley[17] led them to the significant generalization that the milieu presented by brain cells could influence the permeability of capillaries that invaded it. Thus, when a piece of avian embryonic avascular brain was placed on the chorioallantois, the permeable vessels from this membrane became impermeable to Evans blue-albumin once they had entered the brain graft. Conversely, the impermeable vessels of brain became permeable when they entered the domain of tissues that lay outside the CNS.

The brain cells that were responsible for the development of endothelial impermeability were identified as astroglial by *in vivo* and *in vitro* approaches. When rat astrocytes in tissue culture were harvested and injected into the anterior chamber of the chick eye or on the chorioallantois,[18] the permeable vessels from either the iris or the chorioallantois became impermeable to Evans blue-albumin when they had entered the transplanted bolus of astrocytes. At the same time, there appeared another line of evidence for the astrocyte as a modulator of endothelial permeability. When endothelial cells were cocultured alone, their tight junctions, examined in freeze-fracture replicas, appeared as fragmentary, discontinuous strands intermingled with gap junctions. The latter junctions do not prevent extracellular diffusion of solutes and are not part of a fully developed, zonular tight junction. When, however, the brain endothelial cells were cocultured with astrocytes, the length, breadth, and complexity of tight junctions between the endothelial cells were enhanced, whereas the gap junctions were drastically reduced in frequency and size. The tight junctions more closely resembled those *in situ.*[19] When cells other than astrocytes were cocultured with brain endothelial cells, their tight junctions were not enhanced. These studies gave rise to the idea that astrocytes could induce tight junctions in endothelium. However, the results of some control experiments were unexpected. Thus, tight junctions did not develop in the endothelium from other organs, such as umbilical vein, aorta, and pulmonary artery, when their endothelial cells were cocultured with astrocytes.

Astrocytes also modulate functional attributes of brain endothelium. Gamma glutamyl transpeptidase activity is no longer expressed by brain endothelial cells that have been maintained in solo culture, but it is reexpressed when astrocyte-like cells, members of a C6 glioma cell line, are added to the culture.[20] The addition of astrocytes to brain endothelial cells in culture also augments the incorporation of neutral amino acids by the endothelial cells.[21]

The recent successes in growing monolayers of brain endothelial cells *in vitro* with a relatively high electrical resistance lend further support to the inference that astroglia augment the impermeability of brain endothelium. The considerable resistance, which is about a third that *in situ,* has been attained by two different methods, both of which require the presence of astrocytes or their conditioned medium.[22]

One technique of obtaining endothelial layers of this high resistance is based on the cloning of endothelium from capillaries, in distinction to larger microvessels, and plating them on a filter coated with an appropriate extracellular matrix and with an astrocyte layer on the opposite side of the filter (Cecchelli *et al.,* this volume). The second method entails the continued presence of medium conditioned by astrocytes, the stimulation of adenylyl cyclase, and the immunological removal of pericytes (Rubin *et al.,* this volume). With both methods, the resistance has an average value of about 600 $\Omega \cdot cm^2$. Both systems provide a ready means of screening substances for their ability to perturb the endothelial barrier and to define the conditions by which monocytes may cross it. The enhancement of the resistance attained by both methods by either astrocytes themselves or their conditioned medium lends additional support to the possible role of astrocytes in the development of an endothelial barrier.

Despite reinforcement of the notion that astrocytes may form and maintain the endothelial barrier, there are reasons for doubt. First, as discussed here, the cerebral endothelium of some elasmobranchs is ensheathed by astrocytes, but their endothelial cells are joined not by tight junctions but by open ones permeable to horseradish peroxidase.[4,5] Either these particular astrocytes ensure that the endothelial junctions remain open or the junctional configuration is an inherent one that is not determined by the astrocytic investment. Abbott (this volume) raises the idea that a glial barrier may be the primitive condition. Secondly, astrocyte-like cells, the pituicytes of the neural lobe, are neighbors of an endothelium that does not act as a blood-brain barrier; instead, this endothelium is fenestrated and permeable to peroxidase. Thirdly, dissociated, cloned brain endothelial cells *in vitro* can establish a continuous, tight layer with a resistance ranging from 157–783 $\Omega \cdot cm^2$ without benefit of astroglial cells or their conditioned medium.[23] Thus, the precise role of perivascular astrocytes in the induction and maintenance of brain endothelium as a structural and functional barrier has yet to be fully elaborated.

REFERENCES

1. REESE, T. S. & M. KARNOVSKY. 1967. Fine structural localization of a blood-brain barrier to exogenous peroxidase. J. Cell Biol. **34:** 207–217.
2. BRIGHTMAN, M. W. & T. S. REESE. 1969. Junctions between intimately apposed cell membranes in the vertebrate brain. J. Cell Biol. **40:** 648–677.
3. BODENHEIMER, T. S. & M. W. BRIGHTMAN. 1968. A blood-brain barrier to peroxidase in capillaries surrounded by perivascular spaces. Am. J. Anat. **122:** 249–267.
4. BRIGHTMAN, M. W., T. S. REESE, Y. OLSSON & I. KLATZO. 1971. Morphological aspects of the blood-brain barrier to peroxidase in elasmobranchs. Progr. Neuropathol. **1:** 146–161.
5. BUNDGAARD M. & H. F. CSERR. 1981. A glial blood-brain barrier in elasmobranchs. Brain Res. **226:** 61–73.

6. HASHIMOTO, P. H. 1972. Intracellular channels as a route for protein passage in the capillary endothelium of the shark brain. Am. J. Anat. **134:** 41–58.
7. BUNDGAARD, M., H. CSERR & M. MURRAY. 1979. Impermeability of hagfish cerebral capillaries to horseradish peroxidase. Cell Tissue Res. **198:** 65–77.
8. TREHERNE, J. E. & Y. PICHON. 1972. The insect blood-brain barrier. Adv. Insect Physiol. **9:** 257–313.
9. LANE, N. J. & J. E. TREHERNE. 1972. Studies on perineurial junctional complexes and the sites of uptake of microperoxidase and lanthanum in the cockroach central nervous system. Tissue & Cell **4:** 427–436.
10. LANE, J. C. 1981. Invertebrate neuroglia-junctional structure and development. J. Exp. Biol. **95:** 7–33.
11. ABBOTT, N. J., N. J. LANE & M. BUNDGAARD. 1986. The blood-brain interface in invertebrates. Ann. N. Y. Acad. Sci. **481:** 20–42.
12. ABBOTT, N. J., Y. PICHON & N. J. LANE. 1977. Primitive forms of potassium homeostasis: Observations on crustacean central nervous system with implications for bertebrage brain. Exp. Eye Res. (Suppl). **25:** 259–271.
13. TAO-CHENG, J.-H., Z. NAGY & M. BRIGHTMAN. 1990. Astrocytic orthogonal arrays of intramembranous particle assemblies are modulated by brain endothelial cells *in vitro.* J. Neurocytol. **19:** 143–153.
14. WOLBURG, H., J. NEUHAUS, B. PETTMANN, G. LABOURDETTE & M. SENSENBRENNER. 1986. Decrease in density of orthogonal arrays of particles in membranes of cultured rat astroglial cells by the brain fibroblastic growth factor. Neurosci. Lett. **72:** 25–30.
15. NEWMAN, E. A. 1985. J. Neurosci. **5:** 2225–2239.
16. OLDENDORF, W. & H. DAVSON 1967. Transport in the central nervous system. Proc. R. Soc. **60:** 326–328.
17. STEWART, P. A. & M. J. WILEY. 1981. Developing nervous tissue induces formation of blood-brain characteristics in invading endothelial cells: A study using quail-chick transplantation chimeras. Dev. Biol. **84:** 183–192.
18. JANZER, R. C. & M. C. RAFF. 1987. Astrocytes induce blood-brain properties in endothelial cells. Nature (Lond.) **325:** 253–257.
19. TAO-CHENG, J. H., Z. NAGY & M. W. BRIGHTMAN. 1987. Tight junctions of brain endothelium *in vitro* are enhanced by astrocytes. J. Neurosci. **7:** 3293–3299.
20. DEBAULT, L. E. & P. A. CANCILLA. 1980. Gamma-glutamyl transpeptidase in isolated brain endothelial cells: Induction by glial cells *in vitro.* Science **207:** 653–655.
21. CANCILLA, P. A. & L. E. DEBAULT. 1983. Neutral amino acid transport properties of cerebral endothelial cells *in vitro.* J. Neuropathol. Exp. Neurol. **42:** 191–199.
22. MERESSE, S., M-P. DEHOUCK, P. M. DELORME, J.-P. BENSAID, C. TAUBER, C. DELBART, J.-C. FRUCHART & R. CECCHELLI. 1989. Bovine brain endothelial cells express tight junctions and monoamine oxidase activity in longterm culture. J. Neurochem. **53:** 1363–1371.
23. RUTTEN, M. J., R. L. HOOVER & M. J. KARNOVSKY. 1987. Electrical resistance and macromolecular permeability of brain endothelial monolayer cultures. Brain Res. **425:** 301–310.

Morphology of Glial Blood-Brain Barriers

NANCY J. LANE[a]

AFRC Laboratory of Molecular Signalling
Department of Zoology
Cambridge University
Downing Street
Cambridge CB2 3EJ, UK

Blood-brain barriers are found in all vertebrates, but only those invertebrates with highly developed central neural functions seem to possess comparable barrier systems; it would appear that the brain microenvironment must be isolated for the effective integration of information. Within the invertebrate nervous system, which is often avascular, the function of forming the barrier system has been taken on by the glial cells rather than the endothelial cells which provide the sites of occlusion in the highly vascularized vertebrate central nervous system (CNS).

There have been a number of general reviews on the structure and function of glial cells in invertebrates[1–3] as well as several that deal specifically with the neuroglia of arthropods[4,5] or insects.[6–8] Invertebrate glial cells possess many functions and exhibit a range of structural modifications, but we will consider here their capacity, in these "lower" or less highly evolved organisms, to become modified to form the morphologic basis of the blood-brain barrier. This contrasts with the situation in most vertebrates in which the endothelial cells lining the blood vessels of the capillary network are the site of this barrier. One exception is the elasmobranchs in which, although an endothelial layer is present, the barrier resides in the perivascular glial layer, formed from astrocytic foot processes.[9,10]

MORPHOLOGIC BASIS OF GLIAL BLOOD-BRAIN BARRIERS

In invertebrates, often little or no vascularization of the CNS occurs, although intracerebral blood channels are found in some large ganglia, such as those of crustacea and arachnids. The neurons and glial cells that comprise the CNS frequently are simply ensheathed by a variety of protective layers, acellular and cellular. Most commonly an outermost connective tissue layer, the neural lamella, overlies a cellular perineurial sheath, which surrounds the avascular nervous tissues; the perineurial layer is composed of modified glial cells that are sometimes in the form of overlapping "bracelet" cell processes. Exogenous substances attempting to attain the underlying neuronal and glial cell surfaces can only do so, therefore, by transport across this perineurium either by selective passage across the membranes and through the cells or via the paracellular pathway of intercellular clefts. This latter path is the site of restricted entry of molecules. If such paracellular movement

[a] This chapter is dedicated to the memory of the late John E. Treherne, my colleague for over 20 years, with whom I collaborated in much of the early Cambridge work mentioned in this report.

is blocked, some intercellular structure must either be occluding the cleft or straddling it in such a way as to restrict, either partially, selectively, or completely, the passage of substances through it.

Many earlier structural studies were made on the existence of glial cells modified into perineurial layers in the invertebrates,[11–13] but some of the original observations leading to the postulation of a blood-brain barrier came from physiologic observations. Desheathing intact insect CNS preparations led to a disappearance of the barrier,[14] whereas the intact CNS could still function well in Na^+-free solutions.[15] Fine structural studies using tracers on insect CNS showed that the morphologic basis for this observed barrier was in the complex interdigitating clefts between the adjacent perineurial cells.[13,16] It then emerged that the clefts contained septate and gap junctions[17] as well as tight junctions.[16,18] These last are infrequent in some cases, so that in certain species they have not even been seen in thin sections.[8] Freeze-fracture studies are preferable to thin sections for revealing tight junctions; in thin sections they appear as occasional punctate appositions, whereas in freeze-fracture replicas they are seen as distinct intramembranous ridges and grooves. It has been

TABLE 1. Perineurial Glial Junctions

Junctional Type	Associated Peripheral Cytoskeletal Proteins	Integral Membrane Proteins
Tight junctions	Actin ZO-1? Cingulin?	Asymmetric fibrils of lipoprotein?
Linker junctions	Actin?	?
Gap junctions	?	16–18-kD protein
Septate junctions	Actin? to Microtubules?	23–24 kD protein
Adhering junctions A. *Zonulae adhaerentes*-like	Actin to microtubules	?
B. *Maculae adhaerentes*-like	Intermediate filaments?	?
Scalariform junctions	Actin? To ER cisternae or mitochondria	?

suggested that the complex interdigitations of the perineurial bracelet cells where occlusion to tracer entry has occurred[19] produce such narrow and tortuous clefts that they are in themselves capable of restricting the entry of ions and molecules.[20] In most cases, however, as with the insect blood-testis barrier,[21] thin sections give little information, whereas freeze-fracture studies reveal the existence of parallel intramembranous ridges that are the basis of simple tight junctions; these junctions are even more apparent in the CNS of arachnids where they form the blood-brain barrier by means of more complex intramembranous networks.[22,23]

In arthropods, peripheral nerves, in contrast to the CNS, often lack a continuous layer of perineurial glial cells, and no restriction to the entrance of tracer molecules is observed.[24]

In elasmobranch fish, the vascular endothelium in the CNS forms an incomplete layer, and the underlying astrocytic glial cells restrict tracer entry[9,25] by these tight junctions.[10] Such tight junctions also occur between astrocytic processes forming the glia limitans at the brain surface in elasmobranchs.[10]

GROUPS WITH GLIAL BARRIERS

The groups whose blood-brain barrier is made up of glial cells thus include the elasmobranchs as well as some arthropods with more sophisticated nervous systems, such as insects and arachnids, all of which possess zonular tight junctions. Some crustacea exhibit fasciar tight junctions, whereas myriapods and *Limulus* display

TABLE 2. Junctional Types Found between Membranes of Adjacent Glial Cells in Invertebrate Tissues[a]

Junction	Type
Zonula occludens (Tight Junctions)	**zonular**
Linker Junctions	**zonular**
Macula communicans (Gap Junctions)	**macular**
Septate Junctions	**zonular**
Zonula adhaerens (Intermediate Junctions)	**zonular**
Macula adhaerens (Desmosomes)	**macular**
Scalariform Junctions	**fasciar**

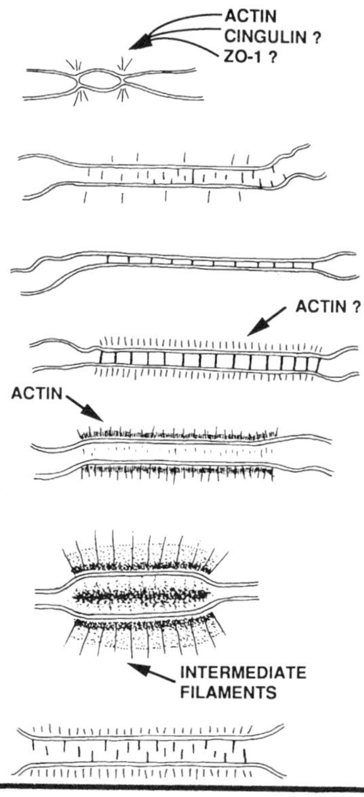

[a]Although the membranes at the point of intercellular contact (shown here in isolation from the rest of the cells' plasma membranes) are structurally very similar to those found between vertebrate cells, they may differ biochemically. In certain cases, as with the intermediate filaments, it is not clear if such components are present in invertebrates, while the chemical nature of the transmembrane and intercellular structures are, for the most part, unknown. Scalariform junctions appear very like linker junctions in thin section, with faint cross-striations in the cleft; however, by freeze-fracture criteria (FIG. 3d), they differ substantially, lacking the ridges of IMPs that are present in some, but not all, linker junctions (FIG. 3b).

TABLE 3. Restrictive or Semirestrictive Intercellular Junctions between Perineurial Glial Cells in Different Invertebrate Groups

Invertebrates	Junctions
Arthropoda	
Arachnids (spiders, scorpions)	Tight junctions (complex)
Insects (cockroaches, moths, flies, locusts,	Tight junctions (simple)
bugs, etc.)	Septate junctions
Crustacea (crabs, crayfish, lobsters)	Tight junctions (fasciar)
Myriapods (centipedes, millipedes)	Linker junctions
Xiphosaurans (*Limulus*)	Linker junctions
Acarina (mites, ticks)	—
Onychophora	
Peripatus	—
Mollusca	
Lamellibranchs (clams, mussels)	—
Gastropods (snails, slugs)	—
Cephalopods (squid, octopus)	Linker junctions

linker junctions between their adjacent glial cells.[26] Other arthropod groups such as the Acarina (ticks) seem to lack junctional types between perineurial cells[27] and have no barrier. Within the phylum Mollusca, the cephalopods also have linker junctions between glial cells which restrict tracer entry and form a blood-brain barrier.[28] However, this is not true of the lamellibranchs and gastropods which seem to lack such junctions and whose glial cells around the periphery of the CNS are not modified into a tight junction-bearing perineurial layer[29,30]; a blood-brain barrier is also lacking in the primitive onychophora,[31] a possible ancestor of the arthropods and annelids.

DIFFERENT TYPES OF GLIAL JUNCTIONS AT SITES OF BLOOD-BRAIN BARRIERS

The different kinds of junctions that may be found between perineurial glial cells are listed in TABLE 1 together with the details of their cytoskeletal associations and their major junctional proteins, when known. Diagrams of their basic structural features are found in TABLE 2. The different types of restrictive junctions in the various invertebrate groups are summarized in TABLE 3. Their thin-section appearances are shown in FIGURE 1, and their freeze-fracture structure in FIGURE 2. Models of major glial junctions are presented in FIGURE 3 in order to establish and clarify their similarities and dissimilarities.

Tight Junctions

As the name suggests, tight junctions are characterized by occlusion of the extracellular space owing to the apparent fusion of membranes associated with the two apposing cell membranes (FIG. 1a). In replicas, they appear in one membrane half leaflet as rows of closely packed intramembranous particles (IMPs) arranged as ridges or networks in a belt-like or zonular configuration, with complementary grooves in the apposing membrane half (FIG. 2a). A model derived from thin-section and freeze-fracture data shows that the fibrils have their basis in an offset two-fibril

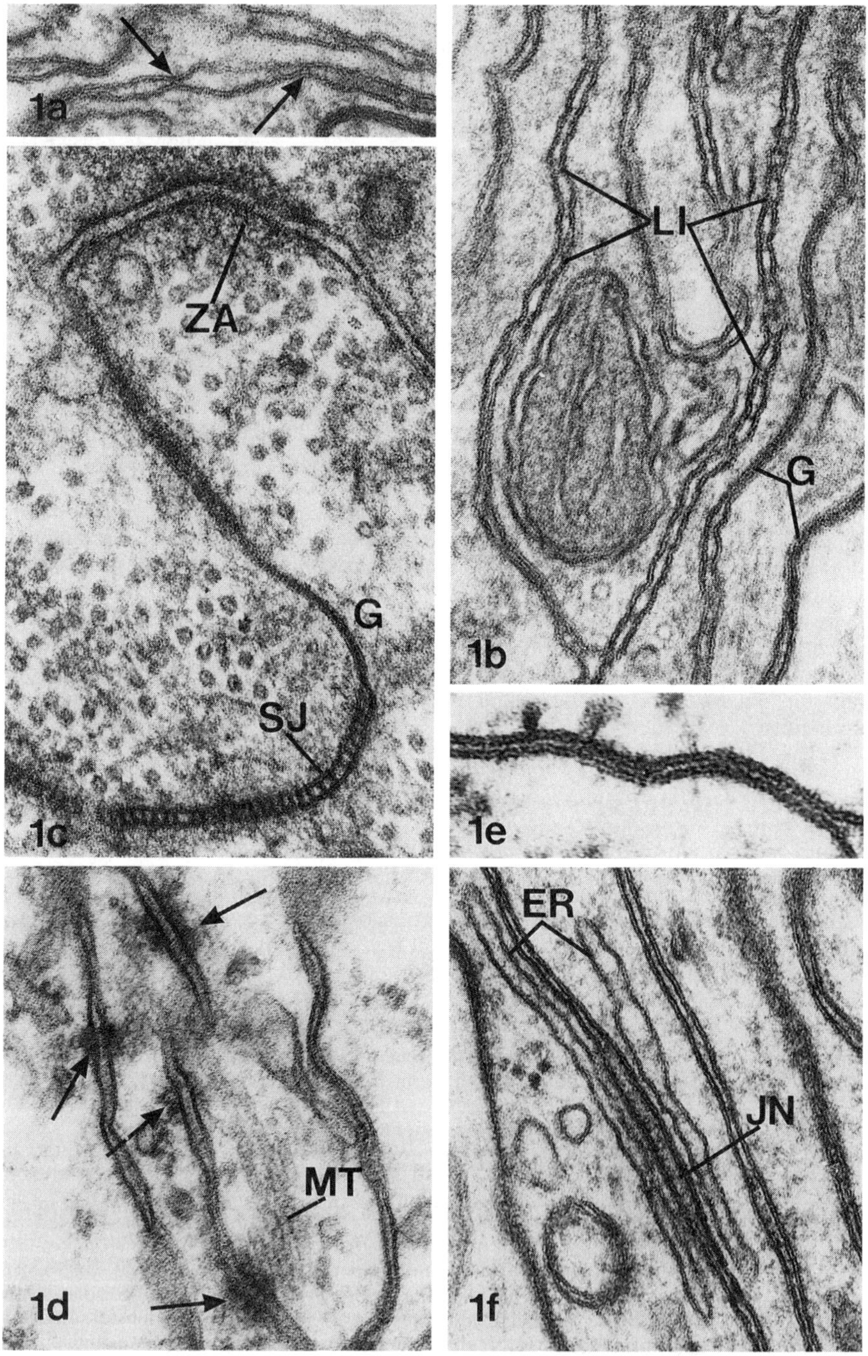

FIGURE 1.

configuration (FIG. 3a). Associated with the peripheral cytoplasm in these junctions are fibrils of actin,[32] ZO-1,[33] and cingulin.[34] The latter two proteins have been found in some vertebrate tissues in which permeability barriers are present, but it is less clear if they are present beside invertebrate tight junctions. Rapid freezing suggests an asymmetric structure with a lipid component associated with a peripheral, cytoplasmic-facing protein moiety[35] (FIG. 4a), but there is as yet no biochemical data to support this model.

Linker Junctions

These junctions are cell-cell associations in which extracellular columns or fibrils straddle the intercellular space between adjacent glial cells. In thin sections there is a fairly regular or slightly reduced cleft across which these components are found (FIG. 1b). In replicas these junctions may exhibit no recognizable profile, as with cephalopod glia,[36] or they may appear as rows of IMPs (FIG. 2b), as with myriapods.[26] A model can be constructed (FIG. 3b) that suggests one possible configuration in myriapod systems. There is a suggestion of microfilaments in association with the cytoplasmic surface of linker junctions, but as yet there is no biochemical data on the nature of their major junctional proteins. When tissues with these structures are incubated in exogenous tracers, the junctions prove to be restrictive in that they slow down tracer entry[26,28] perhaps by charged moieties associated with intercellular columns or extracellular matrix.

Septate Junctions

Septate junctions that occur between modified glial cells in the outer perineurial sheath of invertebrate ganglia have so far always been of the pleated variety. They are characterized by a regular intercellular cleft, circa 18–20 nm in width, straddled by septal ribbons, at irregular intervals (FIG. 1c). In *en face* sections, particularly after tracer impregnation, the undulating patterns of the intercellular septal ribbons are seen. Freeze-fracture reveals rows of individual intramembranous particles (IMPs) aligned with one another (FIG. 2c), and these are likely the sites of insertion of the septal ribbons (see model in FIG. 3c). These junctional membranes have cytoskeletal associations in that actin microfilaments insert into them,[37] either directly or indirectly via other proteins; in some cases, the microfilaments, in turn, are anchored to

←

FIGURE 1. These are all thin sections of intercellular junctions between glial cells in arthropod tissues, where modified glial cells form a perineurial sheath around the CNS. **(a)** Tight junctional appositions (at *arrows*) with associated cytoskeletal fibrils, probably actin as indicated by labeling with the S_1 subunit of heavy meromyosin (× 112,000). From the spider, *Aranaeus.* **(b)** Linker junctions (LI) displaying cross-striations between membranes. Intermingled between these are gap junctions (G) (× 130,000). From the millipede, *Julus.* **(c)** Adhering junctions (*zonula adhaerens,* ZA) with associated fibrils and microtubules, running into a gap junction (G) followed by a pleated septate junction (SJ), which is also intimately associated with cytoskeletal elements (× 93,000). From the locust, *Schistocerca gregaria.* **(d)** Adhering contacts between cells (*arrows*) associated with dense fibrillar mats and microtubules (MT) (× 76,800). From the millipede, *Julus.* **(e)** Gap junctional associations between glial cells displaying the typical 2–3-nm gap between plasma membranes (× 154,000). From the lobster, *Nephrops norvegicus.* **(f)** Scalariform junction (JN) characterized by cross-striations between glial membranes, lying parallel to cisternae of endoplasmic reticulum (ER) (× 120,000). From the locust, *Schistocerca gregaria.*

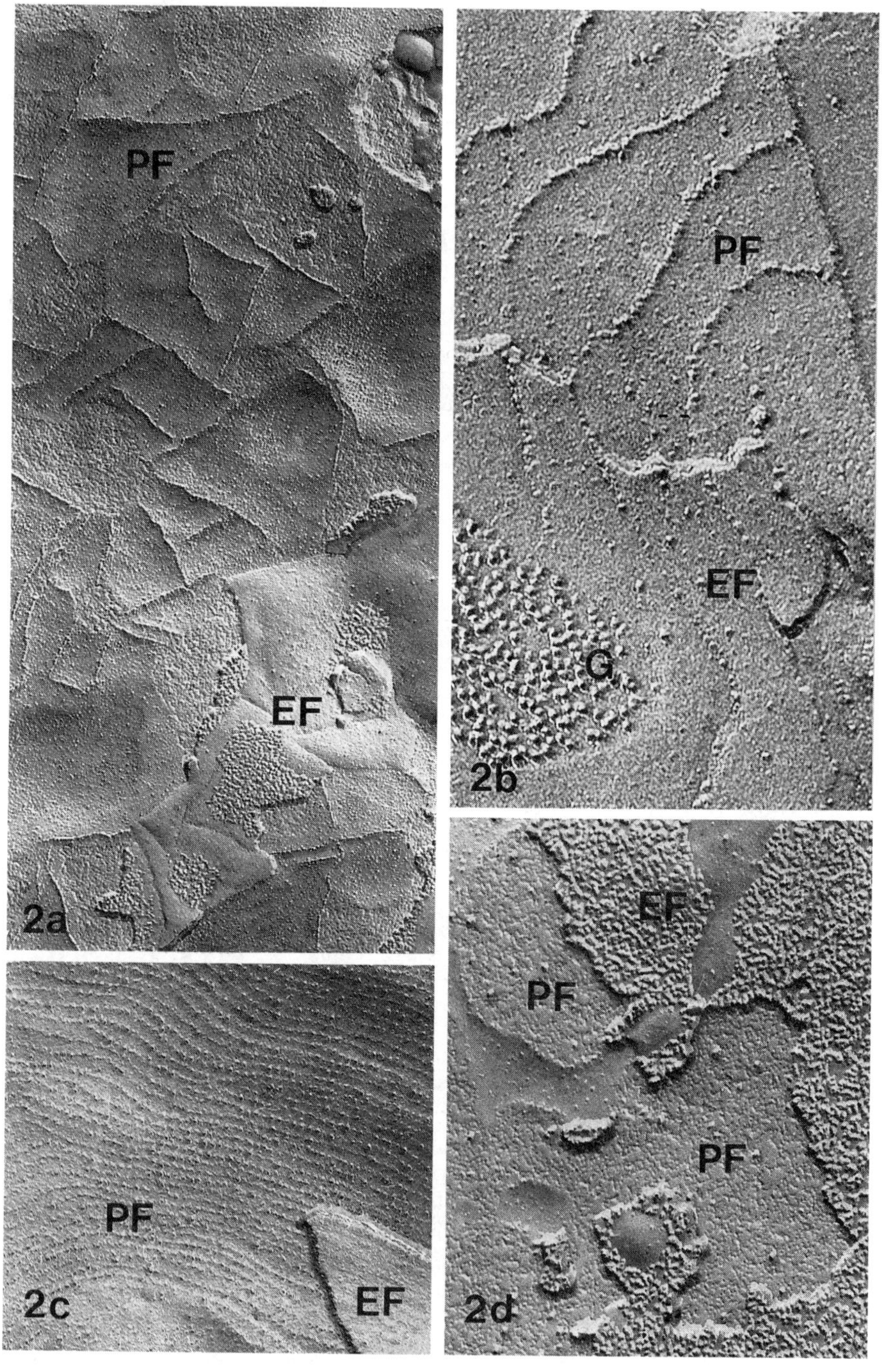

FIGURE 2.

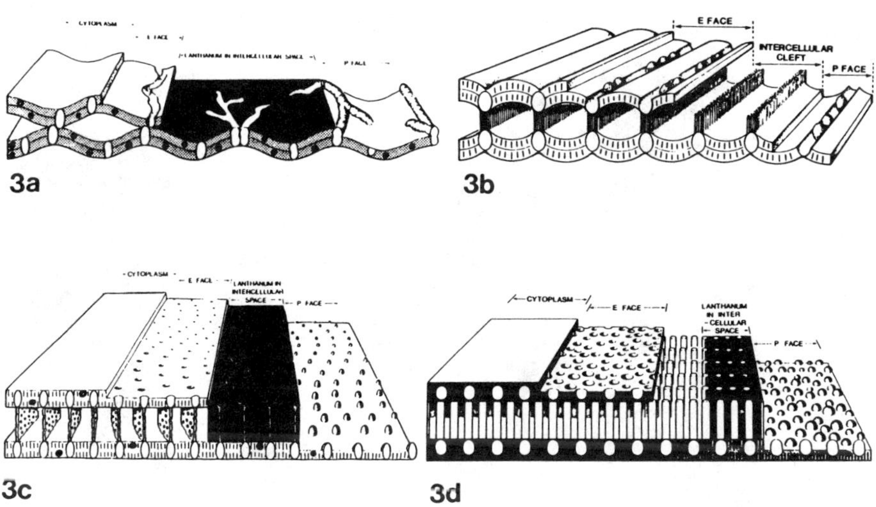

FIGURE 3. Models of some junctions that occur between perineurial glial cells. These indicate the adjacent membranes of two apposing cells, as seen in thin section, after tracer infiltration, and with freeze fracturing, revealing either the inner or outer membrane half leaflet. **(a)** Tight junction wherein the intercellular cleft can be occluded by the fusion of fibrils in adjacent cell membranes. **(b)** Linker junctions from myriapods whereby the intercellular columns may insert into the membranes at the particle rows. **(c)** Pleated septate junctions wherein the septal ribbons between membranes are thought to insert into the IMP rows; the ribbons may be fenestrated and hence leaky. **(d)** Scalariform junctions which show that the intercellular cleft has many intercellular columns, with no obvious insertion points, other than IMPs, in the adjoining membranes. These are very like the linker junctions found between cephalopod perivascular glial cells.[28] In **b, c,** and **d** the intercellular components may be charged or may be surrounded by a charged matrix, which would have the potential to impede the entry of exogenous molecules.

microtubules that lie parallel to junctional membranes. Preliminary studies on the proteins making up these junctions reveal one that has a molecular weight of 22–23 kD associated with isolated preparations (FIG. 4b and c) of pleated septate junctions,[38] which may constitute the intramembranous component. The extracellular septal moities, which may be glycoproteins, are either discontinuous or fenestrated; therefore, although they may slow down the inward passage of exogenous ions and molecules, they do not prohibit their entry entirely[39]; the extent of this partial

←

FIGURE 2. Freeze-fracture replicas from the perineurial sheath of invertebrates in which modified glial cells are found with different kinds of junctional modifications. **(a)** Tight junctional network composed of ridges made up of Pface (PF) particles with complementary Eface (EF) grooves (× 37,000). From the spider, *Tegenaria.* **(b)** Linker junctions do not always exhibit a profile in replicas, but in the horseshoe crab, *Limulus,* as shown here, rows of particles are found on both Pface (PF) and Eface (EF), often intermingled with gap junctional (G) plaques of Eface particles (× 90,000). **(c)** Pleated septate junctions display parallel rows of aligned particles on their Pface (PF), with complementary rows of pits on the Eface (EF) (× 47,000). From the centipede, *Scolopendra.* **(d)** Gap junctions occur often between perineurial cells as large or small plaques of Eface (EF) particles, with complementary Pface (PF) pits (× 45,500). From the locust, *Schistocerca.*

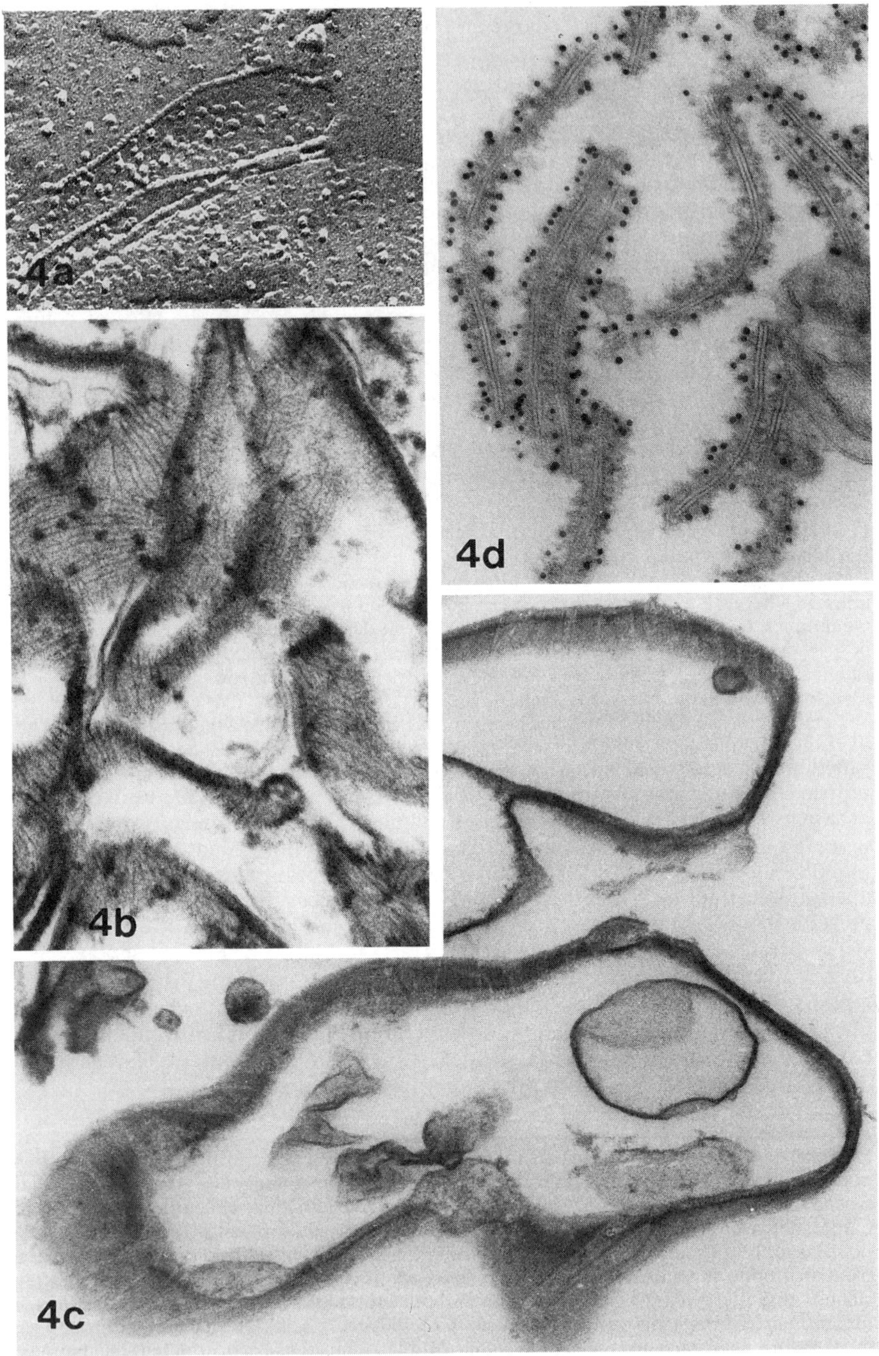

FIGURE 4.

occlusion may depend on the charge and size of the molecules attempting paracellular passage.

Scalariform Junctions

These junctions are only rarely found between glial cells, and they were first observed in the perineurium of the cockroach,[4] although subsequently they have been seen in locust CNS.[40] They feature cross-striations like those in perineurial linker junctions, which appear as columns, at least in the transporting tissues in which they were originally seen[41] (FIG. 1f). In freeze fracture, they exhibit numerous IMPs, with no particular orientation, some of which may be the sites of column insertion (FIG. 3d). Cytoplasmic filaments often insert into their component membranes, which may be actin, but no extensive study has yet been made. They also appear to be anchored at the other end into membranes of mitochondria or endoplasmic reticulum. Unlike linker junctions, which appear capable of restricting the intercellular passage of molecules, the scalariform junctions maintain an uncollapsed intercellular cleft, so that solute and water transport at optimal rates can occur between cells thus associated.[42]

Adhesive Junctions

Variations on the theme of adhesive junctions between glia are numerous. Some of those found between perineurial glial cells may be of the zonular adhering type (FIG. 1c), referred to as *zonulae adhaerentes* or intermediate junctions; these are distinct from the *maculae* or *fascia adhaerentes,* known as desmosomes, which may also occur between the glial cells beneath the perineurium.[13] These adhering junctions are associated with actin fibrils based on specific binding with the S_1 subunit of heavy meromyosin,[43] which in turn may attach onto microtubules (FIG. 1d). Details of their biochemistry in the CNS are unknown, although there are data from nonnervous, mammalian sources[44]; it is assumed that they do not contribute to the blood-brain barrier in the arthropods, because exclusion of tracers has not been associated with them.

Gap Junctions

These communicating junctions serve to couple cells and are frequently found between glial cells. In perineurial glia, they are often numerous and extensive and

←

FIGURE 4. **(a)** Rapid freezing of arthropod tight junctions with no fixation or cryoprotection; note the smooth cylindrical appearance of the structures in the Pface. The Eface reveals particles plastically deformed. This suggests that a lipid component may be present on the outer half leaflet, with a protein component in the inner half leaflet, possibly attached to skeletal proteins in the adjacent cytoplasm[35] (× 127,000). From the spider, *Aranaeus.* (**b** and **c**) Isolated pleated septate junctions showing, where they are sectioned, obliquely, the *en face* septal ribbons that are evident as stained fibrils in both crustacean (*Nephrops norvegicus*) (**b**) and insect (*Periplaneta americana*) (**c**) tissues. **b,** × 65,700; **c,** × 58,500. **(d)** Isolated arthropod gap junctions from crustacean tissues, from which a 16–18 kD protein can be isolated. Incubation of the junctions with antibodies raised against this protein shows specific gap junctional labeling with protein A gold; note the mitochondrion which is unlabeled by the gold (× 115,000).

are identifiable in thin sections by close membrane appositions (FIG. 1e) and in replicas by connexons (IMPs) packed loosely or tightly in macular arrays (FIG. 2d). In models,[45] they are arranged so as to permit the aligning of apposing channels from the connections of one cell to those of the adjacent cell. They are generally considered to have no cytoskeletal associations, and they are unable to form barriers in epithelia. Their biochemical nature is not fully understood; connexons from vertebrate brain have a molecular weight of 28–36 kD[46] with an additional 16-kD component, but it is not clear if these junctional proteins are glial or neuronal in origin. In invertebrates, the CNS, being small, has not yet been analyzed, but gap junction-like structures isolated from other arthropod tissues (FIG. 4d), for both crustacea[47] and insects,[48,49] have a major junctional protein of 16–18 kD molecular weight. This has sequence homology with the 16-kD vacuolar ATPase proton channel protein.[49] Whether this also applies for gap junctional connexons from invertebrate CNS is not yet known, but the 16-kD gap junctional protein from the gut tissues of the moth *Manduca sexta* also shows homology.[49] The significance of this gap junctional homology with the 16-kD proton channel protein is not fully understood, but clearly some very exciting possibilities exist in terms of the evolution of a 16-kD protein capable of forming channels in both cytoplasmic vacuoles and plasma membrane-communicating structures.

DISCUSSION

Development and Evolution of Barrier Layers

In early studies, glial cells could only be categorized into different types on the basis of location and fine structure. However, recently, immunologic probes have been developed that reveal spatial and developmental diversity in insect neuroglia[50]; in the cricket, antibodies bind to the perineurial neuroglia and not to other glial categories. However, this specific perineurial antibody does not bind detectably to any embryonic tissues within the nervous system, although it has been shown[51] that perineurial elements are present in late embryos in the locust. Recent experiments with mutant flies show that the bracelet cells of the perineurium are not mesodermal in origin[52] and are retained in mutants of the gene *twist* that lack the extracellular neural lamella and the outer glial cells that secrete it. Other studies[53] reveal a glial cell-associated glycoprotein in the insect nervous system that is developmentally modulated; it is thought to be a small protoglycan and is found in the embryonic basal lamina in differentiating tissues. As development proceeds, its distribution becomes more restricted, so that in the adult it is found predominantly in the CNS, associated with certain glia as part of the extracellular matrix.

Most of the developmental studies on the junctions comprising glial-based barriers have been carried out on arthropods, in which tight junctional development and resistance to entry of exogenous molecules can be followed in the perineurium of both larval CNS[54] and pupal CNS development[55] of dipteran flies as well as in lepidoptera[19] and arachnids.[56] Septate junctional development has also been followed in the CNS of insects.[57] In all these cases, lateral migration of junctional IMPs occurs across the glial membrane faces to take up their positions in the mature state. The specific stages in assembly vary, but IMPs generally become aligned in rows or ridges, which in turn become aligned in parallel arrays. With the isolation of septate junctional proteins,[38] it is hoped that specific antibodies raised against these will enable us to follow these assembly stages more precisely, both in thin sections and ultimately by immunolabeled replicas. In addition, the extent to which evolutionarily conserved proteins exist in these different restrictive junctions may be elucidated, inasmuch as common components will be revealed by immunocytochemical staining. Such analyses may clarify the manner in which these crucial junctions developed to form blood-brain barriers during evolution, and ultimately to elucidate what relationships exist between glial cells of invertebrate and vertebrate CNS.

SUMMARY

Glial cells, in certain situations in the CNS, may become modified to form the structural basis of the blood-brain barrier. This occurs in more primitive vertebrates, such as the elasmobranch fish, and in some higher invertebrates. In the latter, the outermost glial sheath, often called the perineurium in avascular ganglia, substitutes functionally for the vascular endothelium of higher organisms. The intercellular junctions between the lateral borders of these modified glial or perineurial cells may be of several types. In nearly all cases, adhesive and communicating (gap) junctions are found together with an occluding junctional structure. The latter is assumed to be the morphologic basis of the observed blood brain barrier. It varies in nature and may be one in which the adjacent cell membranes fuse, partially or completely, to form a classical tight junction, or it may be one in which the cell membranes remain separated by a distinct intercellular cleft. If the latter, the cleft may be straddled by columns or septal ribbons, between which a charged matrix substance may be found. Restrictive linker junctions, recently found to be the basis of the interglial barrier in cephalopod CNS, as well as that of myriapods, are characterized by cross-striations or columns which, in combination with charged residues, inherent either in them or in the associated extracellular matrix, slow down the entry of exogenous molecules. Septate junctions, which occur between glial cells in certain other invertebrates, exhibit intercellular septal ribbons, which do not prohibit paracellular transport of all substances but may slow down the passage of some by virtue of charged moieties. There is an association of cytoskeletal components with these septate, linker, and tight junctions; the role of the cytoskeleton in tight junctions, which can be seen by freeze fracture to be based on simple ridges in insects or a more complex network of them in arachnids, may also be important in the regulation of paracellular permeability. The structural details of the junctions in different groups are summarized and their physiologic implications discussed.

ACKNOWLEDGMENTS

I am deeply indebted to my colleague, William M. Lee, who prepared the photographic montages, and Margaret Clements, who prepared the typescript.

REFERENCES

1. RADOJCIC, T. & V. W. PENTREATH. 1979. Invertebrate glia. Prog. Neurobiol. **12:** 115–179.
2. LANE, N. J. 1981. Invertebrate neuroglia: Junctional structure and development. *In* Symposium on Glial-Neuronal Interactions. J. Exp. Biol. **95:** 7–33.
3. PENTREATH, V. W. 1989. Invertebrate glial cells. Comp. Biochem. Physiol. C **93A:** 77–83.
4. LANE, N. J. & J. E. TREHERNE. 1980. Functional organization of arthropod neuroglia. *In* Insect Biology in the Future–V. B. W. 80. M. Locke & D. S. Smith, eds.: 765–795. Academic Press. London.

5. ABBOTT, N. J., M. BUNDGAARD, N. J. LANE & K. MØLLGÅRD. 1988. Parallels between junctions in invertebrate brain and embryonic mammalian brain. Proc. Phys. Soc. **400:** 72P.
6. SAINT MARIE, R. L., S. D. CARLSON & C. CHI. 1982. The glial cells of insects. *In* Insect Ultrastructure. R. C. King & H. Akai, eds. Vol. 2: 435–445. Plenum Publ. New York.
7. HOYLE, G. 1986. Glial cells of an insect ganglion. J. Comp. Neurol. **246:** 83–103.
8. CARLSON, S. D. & R. L. SAINT MARIE. 1990. Structure and function of insect glia. Ann. Rev. Entomol. **35:** 597–621.
9. BUNDGAARD, M. & H. CSERR. 1981. A glial blood-brain barrier in elasmobranchs. Brain Res. **226:** 61–73.
10. GOTOW, T. & P. H. HASHIMOTO. 1984. Plasma membrane organization of astrocytes in elasmobranchs with special reference to the brain barrier system. J. Neurocytol. **13:** 727–742.
11. MADDRELL, S. H. P. & J. E. TREHERNE. 1967. The ultrastructure of the perineurium in two insect species, *Carausius morosus* and *Periplaneta americana.* J. Cell Sci. **2:** 119–128.
12. LANE, N. J. 1968. The thoracic ganglia of the grasshopper, *Malanoplus differentialis:* Fine structure of the perineurium and neuroglia with special reference to the intracellular distribution of phosphatases. Zeit. Zellforsch. **86:** 293–312.
13. LANE, N. J. 1972. Fine structure of a lepidopteran nervous system and its accessibility to peroxidase and lanthanum. Z. Zellforsch. **131:** 205–222.
14. LANE, N. J. & J. E. TREHERNE. 1969. Peroxidase uptake by glial cells in desheathed ganglia of the cockroach. Nature **223:** 861–862.
15. TREHERNE, J. E. & Y. PICHON. 1972. The insect blood-brain barrier. Adv. Insect Physiol. **9:** 257–313.
16. LANE, N. J. & J. E. TREHERNE. 1972. Studies on perineurial junctional complexes and the sites of uptake of microperoxidase and lanthanum in the cockroach central nervous system. Tissue Cell **4:** 427–436.
17. SKAER, H. LEB. & N. J. LANE. 1974. Junctional complexes, perineurial and glia-axonal relationships and the ensheathing structures in the insect nervous system; a comparative study using conventional and freeze-cleaving techniques. Tissue Cell **6(4):** 695–718.
18. LANE, N. J., H. LEB. SKAER & L. S. SWALES. 1977. Intercellular junctions in the central nervous system of insects. J. Cell Sci. **26:** 175–199.
19. LANE, N. J. & L. S. SWALES. 1979. Intercellular junctions and the development of the blood brain barrier in *Manduca sexta.* Brain Res. **169:** 227–245.
20. SHAW, S. R. 1983. Is the blood-brain barrier of insects just a single seal of tight junctions? Soc. Neurosci. Abstr. **9:** 885.
21. TOSHIMORI, K., T. IWASHITA & C. OURA. 1979. Cell junctions in the cyst envelope in the silkworm testis, *Bombyx mori* Linne. Cell Tiss. Res. **202:** 63–73.
22. LANE, N. J. & H. J. CHANDLER. 1980. Definitive evidence for the existence of tight junctions in invertebrates. J. Cell Biol. **86:** 765–774.
23. LANE, N. J., J. B. HARRISON & R. F. BOWERMAN. 1981. A vertebrate-like blood-brain barrier, with intraganglionic blood channels and occluding junctions, in the scorpion. Tissue Cell **13:** 557–576.
24. LANE, N. J. & J. E. TREHERNE. 1973. The ultrastructural organisation of peripheral nerves in two insect species *Periplaneta americana* and *Schistocerca gregaria.* Tissue Cell **5:** 703–710.
25. BRIGHTMAN, M. W., T. S. REESE, Y. OLSSON & I. KLATZO. 1971. Morphologic aspects of the blood-brain barrier to peroxidase in elasmobranchs. Prog. Neuropathol. **1:** 146–161.
26. LANE, N. J. 1989. Novel arthropod cell junctions with restricting intercellular 'linkers'. J. Neurocytol. **18:** 661–669.
27. BINNINGTON, K. C. & N. J. LANE. 1980. Perineurial and glial cells in the tick *Boophilus microplus.* Freeze-fracture and tracer studies. J. Neurocytol. **9:** 343–362.
28. ABBOTT, N. J., N. J. LANE & M. BUNDEGAARD. 1990. A fibre matrix model for the restricting junction of the blood-brain barrier in a cephalopod mollusc: Implications for capillary and epithelial permeability. J. Neurocytol., in press.
29. LANE, N. J. & J. E. TREHERNE. 1972. Accessibility of the central nervous connectives of *Anodonta cygnea* to a large molecular weight compound. J. Exp. Biol. **56:** 492–499.

30. SATTELLE, D. B. & N. J. LANE. 1972. Architecture of gastropod central nervous tissues to ionic movements: A study of *Viviparus connectus* and *Lymnaea stagnalis* (Gastropoda; pulmonata). Tissue Cell **4:** 253–270.
31. LANE, N. J. & S. CAMPIGLIA. 1987. The lack of a structured blood-brain barrier in the onychophoran *Peripatus acacioi.* J. Neurocytol. **16:** 93–104.
32. MEZA, I., G. IBARRA, M. SABANERO, A. MARTINEZ-PALOMA & M. CEREIJIDO. 1980. Occluding junctions and cytoskeletal components in a cultured transporting epithelium. J. Cell Biol. **87:** 746–754.
33. STEVENSON, B. R., J. D. SILICIANO, M. S. MOOSEKER & D. A. GOODENOUGH. 1986. Identification of Z0-1: A high molecular weight polypeptide associated with the tight junction zonula occludens in a variety of epithelia. J. Cell Biol. **103:** 755–766.
34. CITI, S., H. SABANAY, R. JAKES, B. GEIGER & J. KENDRICK-JONES. 1988. Identification and isolation of cingulin, a new cytoplasmic component of tight junctions. Nature, Lond. **333:** 272–275.
35. LANE, N. J., B. KACHAR & T. J. REESE. 1990. Asymmetry of the tight junctional intramembranous fibrils. Anat. Rec., submitted.
36. LANE, N. J. & N. J. ABBOTT. 1990. Freeze-fracture evidence for a novel restrictive junction at the blood-brain barrier of the cuttlefish *Sepia officinalis.* J. Neurocytol., in press.
37. LANE, N. J. & V. FLORES. 1988. Actin filaments are associated with the septate junctions of invertebrates. Tissue Cell **20(2):** 211–217.
38. LANE, N. J. & S. DILWORTH. 1989. Isolation and biochemical characterization of septate junctions. Differences between the proteins in smooth and pleated varieties. J. Cell Sci. **93:** 123–131.
39. SKAER, H. LE B., S. H. P. MADDRELL & J. B. HARRISON. 1987. The permeability properties of septate junctions in Malpighian tubules of *Rhodnius.* J. Cell Sci. **88:** 251–265.
40. POW, D. V. & D. W. GOLDING. 1989. Intercellular junctions in the corpora cardiaca of locusts. Cell Tiss. Res. **258:** 585–591.
41. FAIN-MAUREL, M. A. & P. CASSIER. 1972. Une nouveau type de junction; les junctions scalarieform. Étude ultrastructurale et cytochimique. J. Ultrastruct. Res. **39:** 222–238.
42. LANE, N. J. 1979. Freeze-fracture and tracer studies in the intercellular junctions of rectal tissues in insects. Tissue Cell **11:** 481–506.
43. LANE, N. J. & V. FLORES. 1990. The role of cytoskeletal components in the maintenance of insect junctions. Cell Tiss. Res. **262:** 373–385.
44. GEIGER, B., T. VOLK & R. BENDORI. 1987. Molecular interactions in adherens-type contacts. J. Cell Sci. **8:** 251–272.
45. LANE, N. J. 1984. A comparison of the construction of intercellular junctions in the CNS of vertebrate and invertebrates. Trends Neurosci. **7:** 95–99.
46. DERMIETZEL, R., M. VOLKER, T.-K. HWANG, R. J. BERZBORN & H. E. MEYER. 1989. A 16 kDa protein co-isolating with gap junctions from brain tissue belonging to the class of proteolipids of the vacuolar H^+-ATPases. FEBS. Lett. **252:** 1–5.
47. FINBOW, M. E., J. SHUTTLEWORTH, N. J. LANE & J. D. PITTS. 1983. Gap junctions isolated from arthropod tissue. J. Cell Biol. **97:** 80A, 307.
48. RYERSE, J. S. 1986. Isolation and characterisation of gap junctions from *Drosophila.* J. Cell Biol. **103:** 74A.
49. FINBOW, M. E., N. J. LANE, L. MEAGHER & J. FINDLAY. 1990. Isolation of the 16 kD vacuolar proton channel from insects in a gap junction-like state. J. Cell Biol. **111:** 277.
50. MEYER, M. R., G. R. REDDY & J. S. EDWARDS. 1987. Immunological probes reveal spatial and developmental diversity in insect neuroglia. J. Neurosci. **7:** 512–521.
51. SWALES, L. S. & N. J. LANE. 1985. Embryonic development of glial cells in the locust central nervous system. J. Neurosci. **5:** 117–127.
52. EDWARDS, J. S., L. S. SWALES & C. M. BATE. 1990. Development of non-neural elements in the central nervous system of *Drosophila. In* Glial-Neuronal Interactions. Proc. N.Y. Acad. Sci., This volume.
53. MEYER, M. R., P. BRUNNER & J. S. EDWARDS. 1988. Developmental modulation of a glial cell-associated glycoprotein, 5B 12, in an insect, *Acheta domesticus.* Dev. Biol. **130:** 374–391.
54. LANE, N. J. & L. S. SWALES. 1978. Changes in the blood-brain barrier of the central

nervous system in the blowfly during development, with special reference to the formation and disaggregation of gap and tight junctions. I. Larval development. Dev. Biol. **62:** 389–414.

55. LANE, N. J. & L. S. SWALES. 1978. Changes in the blood-brain barrier of the central nervous system in the blowfly during development, with special reference to the formation and disaggregation of gap and tight junctions. II. Pupal development and adult flies. Dev. Biol. **62:** 415–431.
56. LANE, N. J. 1981. Evidence for two separate categories of junctional particle during the concurrent formation of tight and gap junctions. J. Ult. Res. **77:** 54–65.
57. LANE, N. J. & L. S. SWALES. 1982. Stages in the assembly of pleated and smooth septate junctions in developing insect embryos. J. Cell Sci. **56:** 245–262.

Modulation of a Glial Blood-Brain Barrier

ARTHUR M. BUTT

Sherrington School of Physiology
United Medical and Dental Schools
St. Thomas's Hospital
London SE1 7AH, UK

The primary role of the blood-brain barrier and of glial cells in the central nervous system (CNS) is to provide an extracellular milieu that is compatible with normal neuronal and synaptic activity. In mammals and most vertebrates, the blood-brain barrier is formed by tight intercellular junctions between endothelial cells of the cerebral blood vessels and serves to isolate the brain from changes in the composition of the circulating body fluids.[1] The blood-brain barrier provides a steady baseline against which the brain extracellular microenvironment may be controlled precisely by glial cells.[2] However, the extracellular environment of the brain is not constant,[3] and during neuronal and synaptic activity there are changes in $[K^+]$, $[Ca^{2+}]$, shifts in pH, and the release of neuroactive agents, factors that are known to affect neuronal excitability, transport processes in glial cells, and blood-brain barrier permeability. It is likely that these changes act as extracellular signals helping to integrate the activity of neurons, glial cells, and the blood-brain barrier. It is difficult to study these interactions in the intact mammalian CNS, and most investigations have therefore concentrated on the simpler systems found in invertebrates.

The ventral nerve cord and perineurial blood-brain barrier of the cockroach, *Periplaneta americana,* and the crayfish, *Procambarus clarkii,* have been particularly well studied.[4–13] The blood-brain barrier in insects and crustacea is formed not by endothelial cells but by glial cells and as such provides a useful preparation in which glia may be investigated in their normal relation to neurons. The CNS in the cockroach and crayfish consists of the ventral nerve cord that contains giant axons and many smaller axons and is completely enclosed within a glial perineurial sheath consisting of an outer thick neural lamella of proteoglycans embedded within a matrix of collagen fibers overlaying several layers of glial cells.[5,6,11,13] Tight junctions between the outer layers of perineurial glia restrict the movement of peroxidase and ionic lanthanum and are the site of the permeability barrier.[5,11,13] The combination of low paracellular permeability with highly efficient glial uptake mechanisms provides a level of CNS ion regulation comparable to that of the mammalian endothelial blood-brain barrier.[12,13] In this paper, the sensitivity of perineurial ion regulation to $[K^+]_o$, $[Ca^{2+}]_o$, pH, and neuroactive agents is investigated, and evidence is provided that these effects are determined by glial uptake mechanisms, paracellular permeability, and the cation binding properties of the extracellular matrix. It is likely that changes occurring naturally in the extracellular environment modulate ion regulation by the glial blood-brain barrier so that the brain interstitial fluid is fine tuned to the changing needs of the CNS.

METHODS

Ion transport across the glial perineurium in the isolated intact ventral nerve cord of the crayfish was studied using electrophysiological techniques (FIG. 1). The poten-

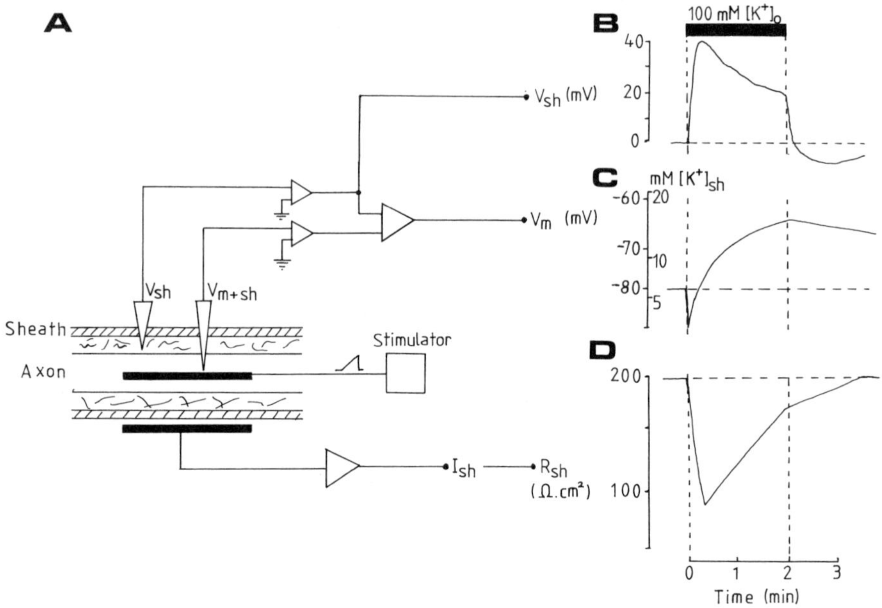

FIGURE 1. (**A**) Diagram of the electrophysiological recording system used to measure (**B**) the transperineurial voltage, V_{sh}, (**C**) the medial giant axon membrane potential, V_m, and (**D**) the transperineurial resistance, R_{sh}. Two microelectrodes were used to measure the voltage across the perineurial sheath, V_{sh}, and the sum of the voltages across the sheath plus the axon membrane potential, V_{m+sh}, so that the axon membrane potential, V_m, could be monitored simultaneously and continuously by differential recording.[12] For current-voltage relations, the medial giant axon was cannulated with an axial wire stimulating electrode, and the transperineurial current, I_{sh}, and voltage, V_{sh}, were measured during application of a ramp current applied across the perineurial sheath.[12] Transperineurial resistance, R_{sh}, was calculated from the slope of the I/V relation. The effects of changing the solution bathing the nerve cord to one containing 100 mM [K^+] on V_{sh}, V_m, and R_{sh} are illustrated. (**B**) The transperineurial V_{sh} was characterized by a rapid transient potential that peaked at around 40 mV, followed by a slowly decaying plateau potential, and on return to normal saline solution, a rapid decrease that undershot the zero baseline; (**C**) there was an initial rapid depolarization of the axon V_m at the time of the peak in V_{sh}, which was followed by a slower depolarization during the plateau phase of the V_{sh}; (**D**) at the time of the peak in V_{sh}, transperineurial R_{sh} fell transiently to 40% of the control level and recovered to 75% during the slowly decaying plateau V_{sh}. The peak in V_{sh} was used to measure the K^+ selective permeability of the perineurial glial barrier, and the decrease in V_{sh} during the plateau phase is an indication of K^+ transport to the inner face of the barrier. The giant axon V_m was used as a biological K^+ electrode to monitor intrasheath potassium ($[K^+]_{sh}$) and was used to determine rate constants for transperineurial K^+ transport. Changes in R_{sh} correspond to changes in perineurial K^+ permeability and most likely represent altered K^+ conductance of the barrier glial cell membranes.

tial across the perineurial sheath (extracellular V_{sh}) and the giant axon membrane potential (intracellular V_m) were measured with microelectrodes, and the current-voltage relations of the perineurium were measured using a space- and current-clamping technique to obtain perineurial resistance (R_{sh}) (FIG. 1A).[12] When the solution bathing the nerve cord is changed for one containing high [K^+], Nernst-like voltage is generated across the perineurium (FIG. 1B). In the crayfish, the transperineurial V_{sh} in 100 mM $[K^+]$ was characterized by a rapid transient potential that peaked at around 40 mV (intracord positive) and fell over 60–90 s to a slowly decaying plateau potential. On return to normal saline solution, the V_{sh} fell rapidly and undershot the zero baseline by 2–5 mV (intracord negative). Transperineurial resistance, R_{sh}, was 260 Ω cm^2 in normal saline solution, and at the time of the early transient V_{sh} in 100 mM $[K^+]_o$ it fell to 113 Ω cm^2 and recovered during the slowly decaying plateau V_{sh} (FIG. 1D). The giant axon, V_m, was used as a biological K^+ electrode to monitor intracord potassium ($[K^+]_{sh}$); during exposure to 100 mM $[K^+]_o$, the giant axon depolarized by 10–15 mV as potassium built up inside the sheath in the vicinity of the axons (FIG. 1C). The initial rapid depolarization of the axon membrane was an indication of the permeability of the perineurial barrier ($[K^+]_{sh}$ increased to 10–15 mM in 100 mM $[K^+]_o$), and the slower depolarization was used to determine the rate constant for bulk K^+ transport through the perineurium[4,12] at the time of the slowly decaying V_{sh} and recovering R_{sh} (FIG. 1). In ion substitution experiments, the perineurium had a 10 times greater permeability to K^+ than to Cl^-, it was effectively impermeable to Na^+, and the high K^+ perm-selectivity was independent of either ion.[12] The K^+-induced change in V_{sh} was highly selective, and of a range of cations tested, only Rb^+, which is transported by most K^+-selective processes, was effective in generating a similar V_{sh} (the selectivity sequence was $K^+ \geq Rb^+ \gg NH_4^+ > Cs^+ > Li^+ \geq Na^+$).[12,13] The glial cells of the outermost layers of the perineurial sheath are the site of the permeability barrier and of the high selectivity,[7,8,9,12,13] and the early transient V_{sh} generated across these cells is therefore a measure of their K^+-selective permeability. The fall in V_{sh} is an indication of K^+ transport across this peripheral barrier, the voltage being offset as K^+ crosses the barrier and builds up at its inner face. The V_{sh} is unaffected by ions other than K^+ (or Rb^+), and the corresponding changes in R_{sh} most likely represent altered K^+ conductance of the barrier glial cell membranes.

Accordingly, V_{sh}, V_m, and R_{sh} were used to measure K^+ selectivity, permeability, conductance, and bulk transport in the crayfish perineurium to determine the effects of extracellular $[K^+]$, Ca^{2+}, pH, and disruption of the paracellular shunt with osmotic shock or enzymatic digestion (Butt, Hargittai, and Lieberman, refs. 12, 13, 22, and 23 and unpublished observations).

RESULTS

Potassium Spatial Buffering

To investigate the regulation of interstitial $[K^+]$ by the crayfish glial perineurium, the effect of changing the $[K^+]$ of the bathing solution on the transperineurial voltage, resistance, and K^+ transport was measured.[12] The perineurial barrier responded to an inwardly directed concentration gradient (3.7–18.5:1, outside $[K^+]$ relative to intracord $[K^+]$) in $[K^+]_o$ above 10 mM with a rapidly developing Nernst-like potential change (FIG. 2A). During a 2-minute pulse of high $[K^+]_o$, intracord $[K^+]$ increased to 10–15 mM (FIG. 1C), indicating low barrier permeability. On return of the bathing solution to normal saline solution (5.4 mM $[K^+]$), an outwardly directed concentration gradient (1:1.9–2.8, $[K^+]_o$:$[K^+]_{sh}$) was set up that generated an undershoot potential of −2 to −3 mV and had the same Nernstian relationship as did the inward gradient (FIG. 2A), indicating equally high K^+ permeability at the inner and outer barrier faces. The sheath potential, however, was relatively insensitive to lower than normal extracellular $[K^+]$, and a decrease in $[K^+]_o$ to 0.1 mM (a 50:1 outwardly directed concentration gradient) generated a potential of only −2 mV, with little

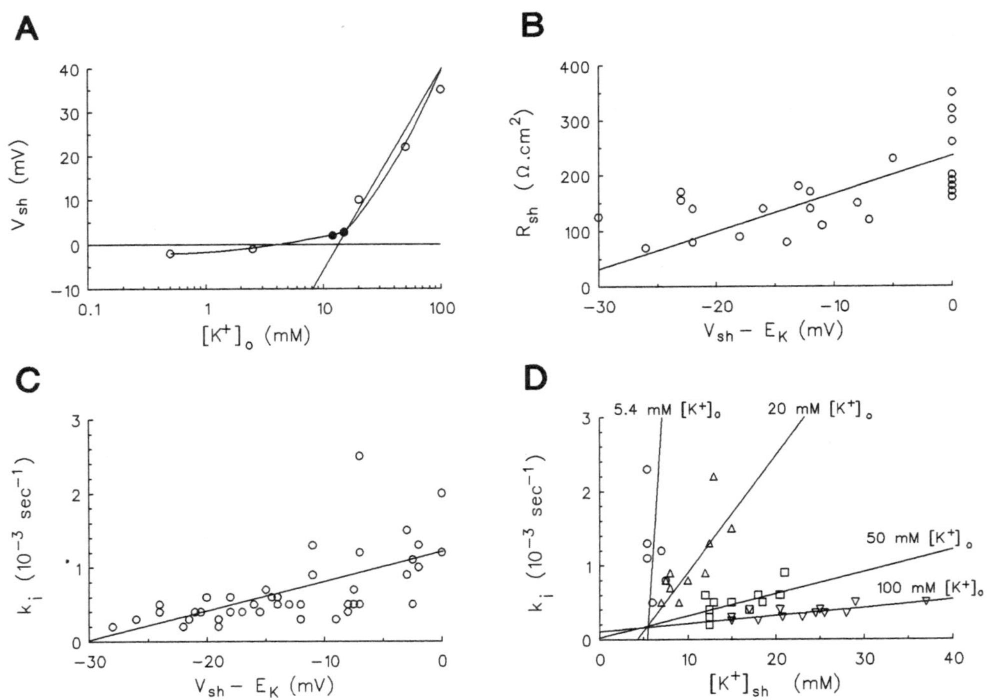

FIGURE 2. Effects of $[K^+]_o$ and the electrochemical gradient on perineurial K^+-selective permeability. (**A**) Semilog plot of the relation between external $[K^+]$ and perineurial peak potential. The peak potentials respond to $[K^+]_o$ above 10 mM in a Nernst-like manner (*straight line*), but deviate sharply below 5.4 mM K^+_o; the undershoot potentials generated on return from 20–100 mM $[K^+]_o$ to 5.4 mM K^+_o (*filled symbols*) fall on the same line as the peak potentials. (**B**) The effect of the K^+ electrochemical gradient on transperineurial resistance. Resistance measurements were made in 20, 50, and 100 mM $[K^+]_o$ at peak, plateau, and recovery of the transperineurial potential, and under all these conditions R_{sh} was inversely related to the electrochemical gradient for K^+. (**C**) The relationship between the electrochemical driving force on K^+ and the rate constant for transperineurial K^+ transport. The rate constants for K^+ influx or efflux in 20, 50, and 100 mM $[K^+]_o$ fell on a single regression line. (**D**) The relationship between intrasheath $[K^+]$ and the rate constant for transperineurial K^+ transport in different external $[K^+]$. With increasing $[K^+]_{sh}$ the rate constant for K^+ transport increased, but with increasing $[K^+]_o$ the sensitivity of the transport to K^+_{sh} decreased. (From Hargittai, Butt, and Lieberman[12] with permission.)

change in intracord [K^+]. These results suggested that the barrier has an equally high K^+ selective permeability at both inner and outer membranes of the glial barrier and that K^+ may be transported rapidly in either direction in response to elevated extracellular [K^+], but it is restricted in low [K^+].

The current-voltage relations of the crayfish perineurium were linear over the range ± 40 mV, indicating that K^+ conductance across inner and outer glial membranes was equal.[12] The changes in R_{sh} were K^+ dependent; the rapid fall in R_{sh} in high external [K^+] was greater with increasing $[K^+]_o$, and the speed of recovery decreased with increasing $[K^+]_o$.[12] The results indicated that the electrochemical gradient was the determining factor in the changes in R_{sh} (FIG. 2B); when the $[K^+]_o$ was increased, R_{sh} decreased in response to the large electrochemical gradient generated across the perineurial barrier, and the increased permeability led to a buildup of intrasheath [K^+], followed by a decrease in the electrochemical gradient and a recovery in R_{sh}. The coincident changes in R_{sh}, V_{sh}, and V_m (FIG. 1), together with their comparable K^+ dependence, suggest that the changes in R_{sh} are determined by changes in the K^+ conductance of the glial cell membranes, and that transglial K^+ transport via these conductance channels is primarily responsible for the initial rapid depolarization of V_m.[12] An increase in glial K^+ conductance appears to be dependent on [K^+] at both the inner and outer aspects of the perineurial glial layer as well as on the electrochemical driving force on K^+.

The bulk transport of K^+ through the perineurium was inversely and linearly related to the electrochemical gradient across the barrier (FIG. 2C). Rate constants for the inward and outward movement of K^+ were similarly dependent on the electrochemical gradient, suggesting that the same transport processes are acting in both directions. The rate constants increased with decreasing external [K^+] or with increasing intracord [K^+] (FIG. 2D), illustrating the importance of both intracord and external [K^+] in regulating interstitial K^+. The results indicated that the rate of transport through the perineurium is most rapid within the physiological range of $[K^+]_o$.

Therefore, the driving force for K^+ transport across the peripheral barrier and through the perineurium in the presence of high extracellular [K^+] was the electrochemical gradient, indicating a passive process that displayed high K^+-selective permeability. These are typical properties of glial cell membranes and of glial K^+ channels.[14–16] The results in the crayfish indicate an effective system for glial uptake and transperineurial spatial buffering of excess K^+ via K^+ channels with concentration- and voltage-sensitive properties,[12] similar to inwardly rectifying K^+ channels described in glial cells that were considered especially appropriate for extracellular K^+ regulation.[16]

Calcium-Dependent Potassium Permeability

A high degree of K^+-selective permeability is a characteristic of glial membrane K^+ channels, and so a range of K^+ channel blockers was tested on the high K_o^+-induced perineurial V_{sh} in the crayfish.[13] A variety of agents was used that between them are known to affect most K^+ channels,[16,17] namely, Cs^+, Ba^{2+}, TEA^+, and 3,4-DAP, and when applied extracellularly, only Ba^{2+} was observed to have an effect. Barium replaced equal amounts of Ca^{2+} in these experiments, and as Ca^{2+} is naturally high (13.5 mM) in crayfish extracellular fluids, its depletion is likely to have significant physiological effects and may have affected perineurial Ca^{2+} conductance and fluxes. To determine if Ca^{2+} is important for perineurial K^+-selective permeability, the effects of Ca^{2+} depletion and Ca^{2+} channel blockers were tested.[13]

Perineurial K^+-selective permeability and transport were reversibly blocked by Ca^{2+} removal, Ca^{2+} channel blocking agents, and Ba^{2+}. Exposure of the ventral nerve cord to 0.1–0.01 times the normal 13.5 mM $[Ca^{2+}]_o$ caused an increase in the baseline R_{sh} (FIG. 7) and blocked the initial rapid changes in conductance, V_{sh}, and V_m in 100 mM $[K^+]_o$, but did not affect the plateau V_{sh} or the bulk transport of K^+ through the perineurium. This effect was only apparent after 30 minutes or more of Ca^{2+} depletion; there was no difference in the K^+-induced V_{sh} when low Ca_o^{2+} + 100 mM K^+_o was given without pretreatment with low $[Ca^{2+}]_o$ and, similarly, after 30–60 minutes of Ca^{2+} depletion, 100 mM K^+_o containing normal $[Ca^{2+}]$ resulted in a V_{sh} not significantly different from that of the controls. These results indicate that external Ca^{2+} must be present for the increase in perineurial K^+ conductance and permeability, and that a compartment with a high affinity for Ca^{2+} is present at the peripheral perineurial barrier. The effect of Ba^{2+} was to reversibly reduce both the initial Nernst-like and steady-state plateau V_{sh} in a concentration-dependent manner, but it did not appear to affect the rate constants for K^+ transport.[13] Substituting Ba^{2+} for Ca^{2+} had a similar effect to that of Ca^{2+} depletion alone, but the effect was more marked than an equivalent decrease in Ca^{2+} in the absence of Ba^{2+}.

Exposure to the Ca^{2+} channel blockers verapamil or La^{3+} had the same effect as did Ca^{2+}_o depletion or Ba^{2+}, and the effects of La^{3+} were investigated in some detail (FIG. 3). There was a rapid and marked increase in baseline R_{sh} by 50% in La^{3+}, and in 100 mM K^+_o both the increase in conductance (FIG. 3A) and the corresponding Nernst-like increase in V_{sh} (FIG. 3B) were blocked. Also, in the presence of La^{3+}, both the initial rapid influx of K^+ across the barrier and the slower bulk transport through the perineurium were considerably reduced (FIG. 3C), and the intracord $[K^+]$ increased by only 3 mM during 2 minutes of exposure to 100 mM $[K^+]_o$ compared to 10–15 mM in controls. These results suggest that the bulk of K^+ influx into the interstitial space is primarily dependent on perineurial glial K^+ permeability and transcellular transport. After La^{3+} washout there was a recovery of both the conductance and the voltage changes at the perineurium, and of the rapid influx of K^+ across the barrier; however, baseline resistance remained high and did not fall to control levels (FIG. 3).

The reversible block of the rapid transient changes in potential and conductance in the absence of external Ca^{2+}, or in the presence of Ba^{2+}, La^{3+}, or verapamil suggest that the high K^+ selective permeability of the perineurium is due to a Ca^{2+}-dependent increase in K^+-uptake by the perineurial glia. Barium may directly block perineurial glial K^+ permeability,[16,17] but it may also act by blocking Ca^{2+} influx through Ca^{2+} conductance channels,[17,18] and the site of action of the agents used in these studies was presumed to be glial voltage-dependent Ca^{2+} channels.[19] Calcium channels, Ca^{2+} influx, and Ca^{2+}-activated K^+ channels have been demonstrated in other glia.[16,18,20,21]

Paracellular Potassium Permeability

Part of the transperineurial potential was not sensitive to Ba^{2+}, verapamil, La^{3+}, or Ca^{2+} depletion, indicating a component of perineurial K^+ permeability by other than the highly selective Ca^{2+}-dependent, $[K^+]_o$- and voltage-sensitive transglial mechanism. This was a nonselective, weakly charged pathway most likely corresponding to the paracellular pathway of the intercellular junctions.[12,13] In addition, the long-term increase in transperineurial resistance following La^{3+} exposure is associated with irreversible La^{3+} binding to the extracellular matrix in the neural lamella and glial basement membranes, suggesting that cation binding sites in the extracellular matrix may also play an important role in interstitial ion homeostasis as well as barrier permeability.[13] To investigate the effects of paracellular permeability on interstitial K^+ regulation in the crayfish, osmotic shock and enzymatic digestion were used to disrupt the intercellular junctions and extracellular matrix, respectively.[22,23]

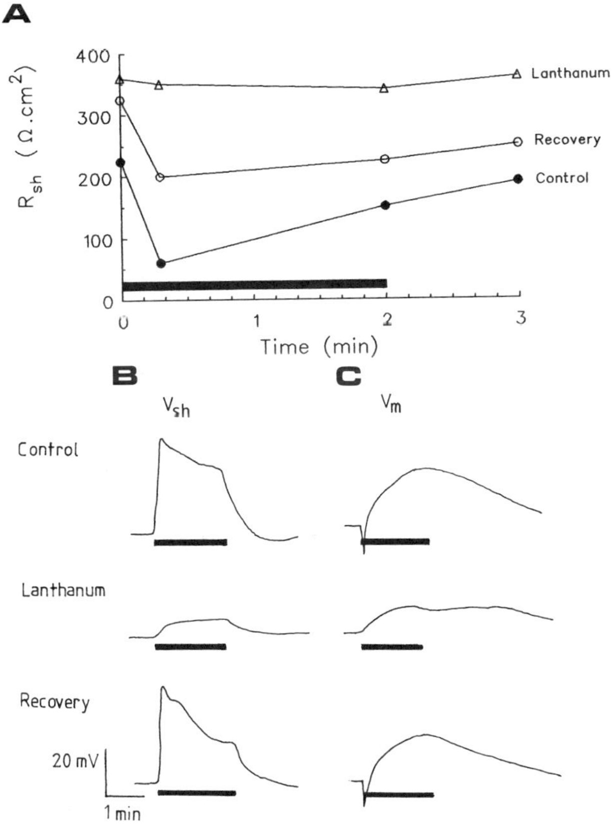

FIGURE 3. The effect of 5 mM La^{3+} on the changes in (**A**) transperineurial resistance, (**B**) transperineurial potential, and (**C**) axon depolarization, induced by a 2-minute pulse of 100 mM $[K^+]_o$ (*bar*). Lanthanum increased basal perineurial resistance and blocked the K^+-dependent increase in conductance and the corresponding increase in V_{sh}. Following La^{3+} washout, the conductance and voltage changes recovered, but the resistance did not return to control levels. Perineurial K^+ permeability and transperineurial transport were considerably reduced in La^{3+} and did not return to control levels following La^{3+} washout. (From Hargittai, Butt and Lieberman[12] with permission.)

The addition of mannitol to the external bathing solution increased perineurial permeability in a concentration-dependent manner, as measured by transperineurial resistance (FIG. 4A). Baseline R_{sh} in normal K^+_o was unaffected by the addition of 100 mM mannitol (total osmolarity 520 mOsm), but it decreased by 25% in 175 mM mannitol and by 50% in 250 mM mannitol. The K^+-dependent increase in V_{sh} was reduced in the higher osmolarities, but the level to which sheath resistance fell was the same as that in controls, and although there was a greater initial rapid influx of K^+, the rate constants for bulk K^+ transport appeared to be largely unaffected. This is illustrated by the results in 100 mM of mannitol, in which the K^+-dependent increase in V_{sh} and glial conductance was the same as that in controls (FIG. 4B), whereas the

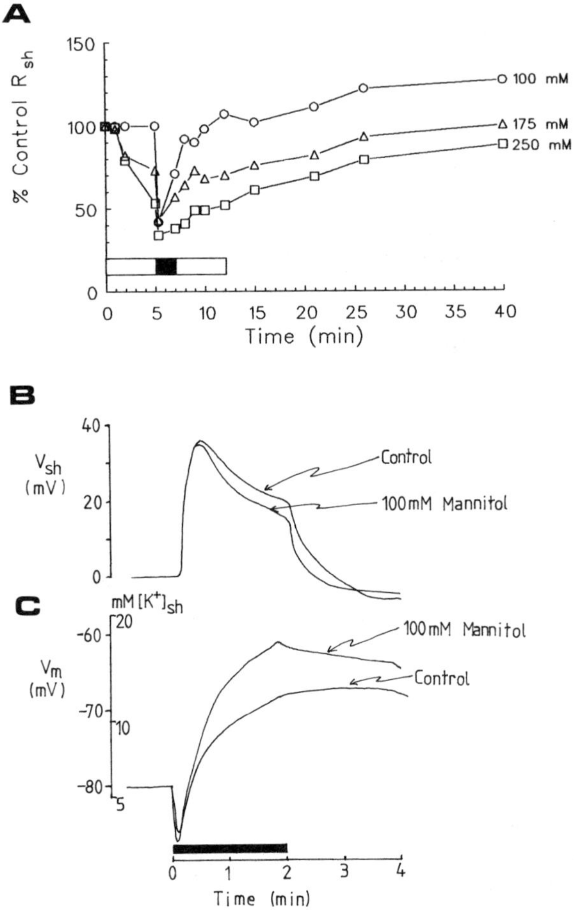

FIGURE 4. Effect of hyperosmolar mannitol solutions (100, 175, and 250 mM mannitol added to normal 420 mOsm crayfish saline solution) on perineurial K^+ permeability. (**A**) The *open bar* represents the period of mannitol exposure and the *filled bar* exposure to 100 mM K^+_o plus mannitol. Basal transperineurial resistance was unaffected by 100 mM mannitol but fell in 175 and 250 mM mannitol to 70% and 50% of control levels, respectively. In mannitol-treated sheaths, there was a reversible decrease in resistance in 100 mM K^+_o to the same level as that observed in controls. On return to normal crayfish saline solution, recovery in transperineurial resistance was slow. (**B** and **C**) illustrate the effect of a 5-minute exposure to 100 mM mannitol on the 100 mM K^+_o -induced change in transperineurial potential and axonal depolarization. The initial rapid rise in V_{sh} was unaffected by mannitol, whereas the decrease in V_{sh} was more rapid and axonal depolarization was greater, indicating an increase in transperineurial K^+ permeability but not in glial K^+ selective permeability.

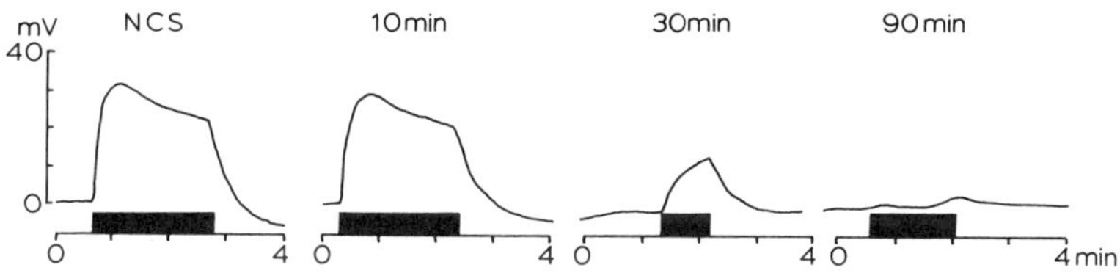

FIGURE 5. Effect of 1% hyaluronidase on transperineurial potential. Hyaluronidase completely abolished the 100 mM K^+_o -induce change in V_{sh} in a time-dependent manner.

depolarization of V_m was greater than that in controls (FIG. 4C), indicating that there was an increase in barrier permeability but the initial rapid transglial transport of K^+ was unaffected. The increased permeability and more rapid buildup of K^+ at the inner face of the barrier would decrease the electrochemical driving force on K^+, resulting in the accelerated recovery in R_{sh} (compare 100 mM of mannitol in FIG. 4A with controls in FIG. 7) and V_{sh} (FIG. 4B). In higher osmolarities the greater leak would shunt off the voltage normally generated across the perineurium so that V_{sh} was reduced. These results suggest that osmotic shock did not directly affect perineurial glial K^+ selectivity but increased the barrier permeability, presumably by disrupting intercellular junctions, and to be most effective the glial K^+ uptake mechanisms require low paracellular permeability.

Treatment of the ventral nerve cord with the enzyme hyaluronidase increased transperineurial resistance by 35% and irreversibly abolished the K^+-dependent increase in perineurial K^+ conductance (see FIG. 7), selectivity, and permeability (FIG. 5),but it had little effect on the rate constant for the bulk transport of K^+ through the perineurial sheath. Hyaluronic acid is the backbone of the mucopolysaccharide intercellular matrix of the insect perineurium,[24] and the results suggest that hyaluronic acid is important in perineurial glial K^+ transport in the crayfish. The striking similarity between these results and those in La^{3+} or low Ca^{2+} (FIG. 7) suggests that the charged binding sites of hyaluronic acid may provide the high affinity store of extracellular Ca^{2+} that is critical for the rapid K^+ uptake mechanism in perineurial glia and for effective spatial buffering.

Effect of pH on Potassium Permeability

Concomitant changes in CNS extracellular pH and K^+ may be important in interstitial ion regulation,[2,3] and in the crayfish perineurium some evidence indicates that K^+ permeability was affected by extracellular pH (FIGS. 6 and 7). The K^+-dependent increase in conductance and selectivity appeared to be largely insensitive to pH, but the subsequent decay in conductance and V_{sh} was modified. In alkaline pH, the initial influx of K^+ across the barrier, and the decay in V_{sh} were slower than those in controls (FIG. 6), whereas the rate constant for bulk transport of K^+ through the perineurium was increased. The opposite was the case in acidic pH, and there was a greater initial influx of K^+ (FIG. 6B) and a more rapid decrease in perineurial conductance (FIG. 7) and in the V_{sh} (FIG. 6A) than in controls. It is not certain that pH does not act directly on glial K^+ uptake, but the lack of effect on the amplitude of the rapid increase in V_{sh} and decrease in R_{sh} suggests that glial K^+-selective permeability was not altered. The results may be explained by changes in permeability of the low selectivity paracellular pathway. A greater paracellular leak in low pH, similar to that observed following osmotic shock (compare FIGS. 5 and 6), would increase the initial rate of K^+ influx to the inner barrier face, which would then decrease the

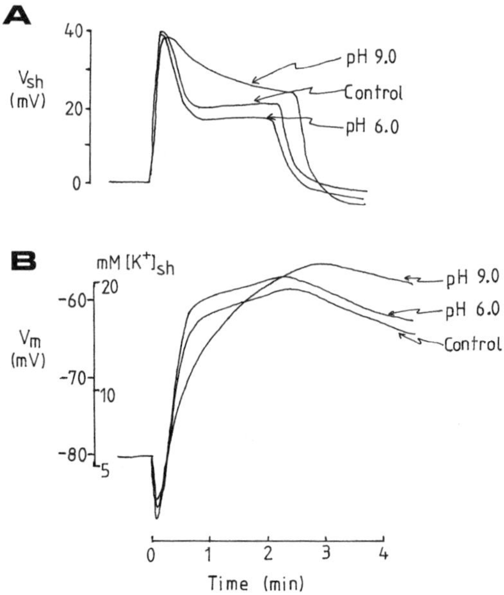

FIGURE 6. Effect of pH on (**A**) the transperineurial potential and (**B**) axonal depolarization induced by 100 mM K^+_o. The initial rapid rise in V_{sh} was largely insensitive to pH, but the decrease in V_{sh} was accelerated in low pH and slowed in high pH. These corresponded to an increase in permeability in low pH and decreased permeability in high pH, as indicated by the axonal V_m.

electrochemical driving force on K^+, resulting in the more rapid decrease in V_{sh} and recovery in R_{sh}; lower paracellular permeability in alkaline pH would have a similar but opposite effect on perineurial K^+ transport. In extreme acidic conditions, perineurial K^+ selectivity was lost, as observed in La^{3+} and hyaluronidase; transperineurial R_{sh} was higher and bulk K^+ transport lower than those in controls, and there was a decrease in barrier permeability (FIG. 7). The results are consistent with a primary effect of pH on the cation-binding properties of the extracellular matrix.

DISCUSSION

The perineurium in insects and crustacea has been shown to be capable of maintaining low extracellular $[K^+]$ and high extracellular $[Na^+]$ in the region of the axons in the presence of extreme concentration gradients, by a combination of low barrier permeability and glial uptake mechanisms.[4–13] The site of the permeability barrier is the tight intercellular junctions between the outer layers of glial cells, and brain interstitial ion regulation is provided by uptake mechanisms on the basolateral membranes of these cells. The glial barrier has a highly efficient K^+ spatial buffering mechanism that is K^+- and voltage-sensitive, Ca^{2+}-dependent, and modulated in altered extracellular Ca^{2+}, pH, or neurohumoral agents. These may act directly on glial uptake or indirectly by changing paracellular permeability, and evidence suggests that the extracellular matrix is important in these effects.

Potassium transport across the crayfish glial blood-brain barrier is mainly transcellular and is both voltage- and concentration-dependent, and the driving force for K^+ transport (influx or efflux) across the barrier is the electrochemical gradient (FIG. 2). These are typical properties of glial membranes[14–16] and have characteristics associated with inwardly rectifying K^+ channels as described in other glia.[16] The effects of Ca^{2+} and Ca^{2+} channel blockers indicated that the rapid K^+ uptake mechanism was also Ca^{2+} dependent and required the influx of Ca^{2+} through voltage-sensitive Ca^{2+} channels[17–19] (FIG. 3). Calcium-activated K^+ channels are also known to have a complex sensitivity to voltage and to the concentrations of both Ca^{2+} and K^+ with characteristics of inward rectifiers,[17] and they have been described in a number of glial cell types.[16,21] Other experiments on the crayfish perineurium in which K^+ was substituted with Rb^+ suggested at least two K^+ channel types involved in maintaining high K^+ selectivity of the perineurial glia, a rapid Ca^{2+}-activated K^+ channel and a voltage-activated K^+ channel.[13] Calcium influx through voltage-dependent Ca^{2+} channels may be the trigger for an increase in glial K^+ permeability. These channels open in response to rapid depolarization of the glial membranes above a certain threshold which is only attained in external $[K^+]$ above 10 mM. Calcium-dependent K^+ channels activated by the influx of Ca^{2+} appear to be important in the increase in perineurial permeability. Potassium permeability changed in response to a combination of glial membrane potential and extracellular $[K^+]$, and the high degree of perineurial K^+ selectivity may be due to inwardly rectifying K^+ channels.

The normally high $[Ca^{2+}]$ in crayfish hemolymph may explain the predominance of Ca-dependent transport at the perineurium, and its voltage and ion dependence at both inner and outer faces of the barrier would be well suited to play a role in K^+ spatial buffering of the CNS. In the presence of high extracellular $[K^+]$, basal K^+ permeability increased, ensuring rapid cellular uptake that would greatly increase the rate of K^+ movement through the perineurium, and the potential generated across the glial perineurium would favor efflux of K^+ at membranes facing a lower $[K^+]$.[14,15] In the case of a large concentration gradient directed inwardly across the perineurium, as could conceivably occur, for example, during strenuous exercise or after feeding (the ventral nerve cord is enclosed within the skeletal muscle of the limbs and tail and is surrounded by the body fluids), a rapid initial increase in $[K^+]$

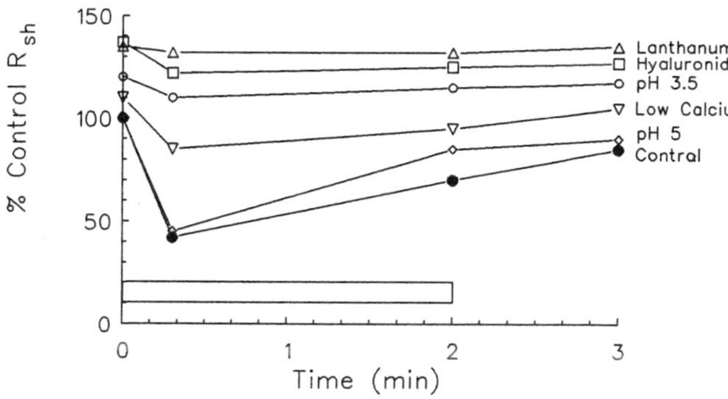

FIGURE 7. Effects of various agents on the 100-mM K^+_o-induced changes in transperineurial resistance. Hyaluronidase, La^{3+}, Ca^{2+}_o depletion, and pH 3.5 had a similar effect by increasing the basal R_{sh} and blocking the increase in K^+ conductance. The increase in K^+ conductance was not affected by pH 5.0, but the recovery was accelerated.

would occur at the inner membranes of the glial barrier, causing a decrease in the electrochemical driving force on K^+, thus decreasing K^+ permeability and helping to maintain a low $[K^+]$ in the vicinity of the axons. With an outwardly directed concentration gradient, which might occur following repetitive action potential generation, K^+ would be transported across the perineurium away from the axons. As interstitial K^+ fell towards normal, a decrease in the electrochemical driving force would reduce the K^+ permeability of the perineurium. The sink action of the hemolymph would ensure continued removal of excess K^+ in the presence of extended periods of high CNS interstitial $[K^+]$. In the event of a small increase in intracord $[K^+]$ due to action potential generation, bulk K^+ transport through the perineurium is high and would serve to remove the excess intracord K^+ by glial uptake and spatial buffering mechanisms. Conversely, the decreased K^+ permeability of the perineurium in the presence of lower than normal external $[K^+]$ would act to conserve intracord $[K^+]$.

It is apparent that the rapid and highly selective transport of K^+ by the perineurial glial barrier is only effective with a low paracellular permeability, as indicated by the effects of hyperosmolar mannitol (FIG. 4). Disruption of the intercellular junctions as a result of cell shrinkage is the likely cause of the increase in permeability observed during osmotic shock. Changes in cell volume have been shown to play a significant part in the regulation of periaxonal K^+ in the crayfish,[25,26] and it is reasonable that Cl^- and H_2O movement following the rapid and selective uptake of K^+ by the perineurial glia would cause cell swelling which may contribute to the low paracellular leak in high external $[K^+]$.[12] Cell volume is sensitive to extracellular K^+ in mammalian glia (e.g., Kempski *et al.,* this volume), and the corresponding changes in extracellular volume associated with K^+ release during neuronal activity may alter the relative contributions to CNS K^+ clearance made by cellular uptake and extracellular diffusion (e.g., Reichenbach, this volume). Mechanisms of glial H^+ transport and pH regulation may also result in cell volume changes (e.g., Kempski *et al.,* and Schlue, this volume). In this study, the results in low pH and 100 mM mannitol indicate a comparable effect on paracellular permeability, which may be interpreted to reflect changes in cell volume due to the action of H^+ regulatory processes in the perineurial glia. However, application of ouabain to inhibit the Na-K pump, bumetanide to inhibit Na-K-Cl cotransport, or cyanide plus iodoacetic acid to inhibit all energy production, or the removal of Cl^-_o had no obvious effect on perineurial K^+ selectivity or permeability,[12,22,23] and it seems unlikely that the effects of pH in this study are due to changes in cell volume. These transport mechanisms are important in periaxonal K^+ regulation and interstitial K^+ conservation in the crayfish,[25,26] and it is possible that at lower concentrations of extracellular K^+, perineurial K^+ transport is primarily by passive diffusion through intercellular clefts and is modulated by active Na-K exchange and Na-K-Cl cotransport.[12]

There is good evidence from radioisotopic, electrophysiological, and ultrastructural studies in the cockroach perineurium that the anionic glycosaminoglycans that fill the intercellular spaces bind cations and provide a reservoir that helps maintain a high extracellular Na^+ and Ca^{2+} in the region of the axons.[6] In the crayfish perineurium, the very low Na^+ permeability,[12] together with the effects of La^{3+} binding and hyaluronidase, indicate a significant role for the extracellular matrix in paracellular ion permeability (FIGS. 3 and 5). It is likely that the extracellular matrix is the site of the high affinity Ca^{2+} store indicated in Ca^{2+} depletion experiments, maintaining a high intrasheath $[Ca^{2+}]$ which is important not only for glial K^+ uptake but also for neuronal excitability and synaptic transmission. The effect of pH on perineurial K^+ regulation was compatible with an effect on the cation-binding properties of the extracellular matrix. In low pH, the increase in the initial rapid influx of K^+ to the inner face of the barrier may be due to H^+ displacing K^+ from binding sites, whereas in extreme acidic conditions perineurial K^+ selectivity and high permeability were lost in a similar manner to Ca^{2+} depletion, treatment with La^{3+} or hyaluronidase,

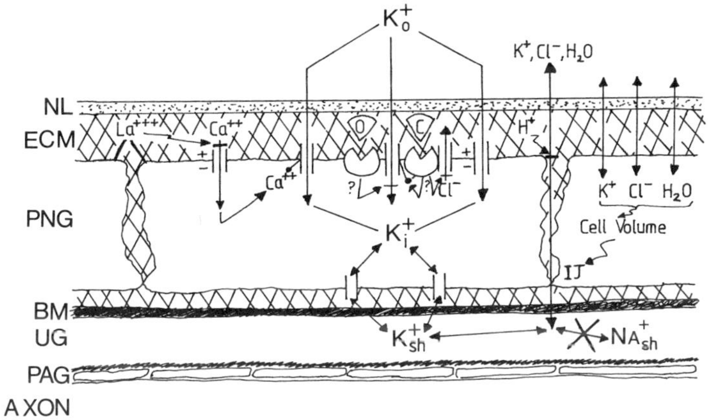

FIGURE 8. A diagrammatic representation of the perineurial sheath and the pathways for ion movement between the interstitial and outside fluid space; for simplicity, the transport processes are illustrated on the outer membrane of the perineurial glia, but in reality appear to be situated on the basolateral membranes of a peripheral layer of glial cells that are the site of the perm-selectivity barrier in the crayfish perineurium. BM = basement membrane; ECM = extracellular matrix; IJ = intercellular junction; NL = neural lamella; PAG = periaxonal glia; PNG = perineurial glia; UG = underlying glia. The major route for K^+ transport is transglial, by a Ca^{2+}-dependent mechanism.[12,13] The effects of K^+ and Ca^{2+} channel blockers indicated that Ca^{2+} influx through voltage-sensitive Ca^{2+} channels results in K^+ uptake via Ca^{2+}-activated and voltage-sensitive K^+ channels. The dependence of K^+ permeability on the electrochemical gradient suggested that inwardly rectifying K^+ channels may also be important. It is reasonable that octopaminergic (O) and cholinergic (C) receptor-mediated changes in K^+ selectivity observed in the cockroach perineurium[10] are also operative in the crayfish perineurial glia and may act directly on K^+ permeability or, based on observations on periaxonal glia in the crayfish,[25,26] may affect Cl^- permeability of the glial membranes. The paracellular movement of ions is restricted by the low permeability of the intercellular junctions, and it is likely that changes in cell volume that result from Cl^- and water following the rapid transport of K^+ across perineurial glia may modify intercellular permeability. The cation-binding properties of the extracellular matrix are important in restricting paracellular ion permeability and in maintaining high intrasheath Ca^{2+} and Na^+ and may be modified by changes in extracellular pH. The voltage and concentration dependence of perineurial K^+ transport is well suited for providing efficient K^+ spatial buffering of the crayfish CNS, and the sensitivity of the glial perineurium to K^+, Ca^{2+}, H^+, and neuroactive agents makes it highly suitable for controlling the brain interstitial fluids during changes in the extracellular environment that may occur under normal conditions of feeding or of neuronal and muscular activity.

suggesting high $[H^+]$ may more completely displace Ca^{2+} from the negatively charged, strongly basic sites in the extracellular matrix. Lanthanum, hyaluronidase, and pH may also act directly on the Ca^{2+} or K^+ channels of the perineurial glial cell membranes by altering cell surface charges that influence the conductance of ion channels.[17,19]

During neuronal or muscular activity, an increase in extracellular [K^+] in the body fluids or brain interstitial fluid, respectively, may be associated with a decrease in [Ca^{2+}] and pH and the release of neuroactive substances.[2,3] In the crayfish perineurium, both Ca^{2+} and pH would affect interstitial K^+ regulation in the CNS but in an opposing manner. A decrease in $Ca^{2+}{}_o$ reduced glial K^+ uptake, whereas a decrease in pH appeared to increase the permeability of the barrier, and both effects in turn modified the response of the glial cells to extracellular K^+. In addition, perineurial glial K^+ selectivity and permeability in the cockroach were reversibly decreased by physiological levels of octopamine, whereas noradrenaline, adrenaline, histamine, and carbamylcholine resulted in an increase in glial K^+ selectivity,[10] and it is reasonable that receptor-mediated changes in glial permeability are also important in the crayfish perineurium. Cholinergic and glutaminergic receptor-mediated changes in permeability have been demonstrated in periaxonal glial cells surrounding the giant axon in the crayfish.[25,26] Clearly, the sensitivity of the glial perineurium to K^+, Ca^{2+}, H^+, and neuroactive agents makes it highly suitable for controlling the brain interstitial fluids during changes in the extracellular environment, which may occur under normal conditions of feeding or neuronal and muscular activity.

SUMMARY

CNS interstitial fluid homeostasis by the glial perineurial blood-brain barrier in the crayfish and cockroach is dependent on glial uptake mechanisms, low paracellular permeability, and the cation-binding properties of the extracellular matrix. Potassium selective permeability of the crayfish perineurium is modulated by a Ca^{2+}-dependent mechanism at the basolateral membranes of the glial barrier and is ion and voltage dependent. In addition, extracellular charged sites are significant in perineurial K^+ and Ca^{2+} homeostasis and may be modified by changes in pH. In the cockroach, and probably the crayfish, perineurial K^+ transport may also be modulated by receptor-mediated changes in glial membrane permeability. The factors acting at the crayfish and cockroach blood-brain barrier are summarized in FIGURE 8 and would be well suited for providing efficient K^+ spatial buffering of the CNS. Analogous processes have been described in vertebrate glial cells and in the endothelial blood-brain barrier, which implies a common primary function. The CNS is protected from large fluctuations in the body fluids by the blood-brain barrier, whereas glial uptake mechanisms control the composition of the brain interstitial fluid, and modulation of both barrier permeability and glial transport by the altered chemical environment following neuronal activity allows precise adjustment of the brain extracellular fluids to the changing needs of the CNS. The insect and crustacean ventral nerve cord and perineurial blood-brain barrier provide an excellent preparation in which the interactions between these factors can be investigated in intact CNS tissue.

REFERENCES

1. BRADBURY, M. W. B. 1979. The Concept of a Blood-Brain Barrier. Wiley. Chichester.
2. WALZ, W. 1976. Role of glial cells in the regulation of the brain ion microenvironment. Progr. Neurobiol. **33:** 309–333.
3. NICHOLSON, C. 1980. Dynamics of the brain cell microenvironment. Neurosci. Res. Progr. Bull. **18:** 177–322.

4. Pichon, Y., R. B. Moreton & J. E. Treherne. 1971. A quantitative study of the ionic basis of extraneuronal potential changes in the central nervous system of the cockroach (*Periplaneta americana* L.). J. Exp. Biol. **54:** 757–777.
5. Lane, N. J. & J. E. Treherne. 1972. Studies on perineurial junctional complexes and the sites of uptake of microperoxidase and lanthanum in the cockroach central nervous system. Tissue & Cell **4:** 427–436.
6. Treherne, J. E., P. K. Schofield & N. J. Lane. 1982. Physiological and ultrastructural evidence for an extracellular anion matrix in the central nervous system of an insect (*Periplaneta americana*). Brain Res. **247:** 255–267.
7. Schofield, P. K. & J. E. Treherne. 1984. Localization of the blood-brain barrier of an insect: Electrical model and analysis. J. Exp. Biol. **109:** 319–331.
8. Schofield, P. K., L. S. Swales & J. E. Treherne. 1984a. Potentials associated with the blood-brain barrier of an insect: Recordings from identified neuroglia. J. Exp. Biol. **109:** 307–318.
9. Schofield, P. K., L. S. Swales & J. E. Treherne. 1984b. Quantitative analysis of cellular and paracellular effects involved in disruption of the blood-brain barrier of an insect by hypertonic urea. J. Exp. Biol. **109:** 333–340.
10. Schofield, P. K. & J. E. Treherne. 1986. Octopamine sensitivity of the blood-brain barrier of an insect. J. Exp. Biol. **123:** 423–439.
11. Lane, N. J., L. S. Swales & N. J. Abbott. 1977. Lanthanum penetration in crayfish nervous system: Observations on intact and 'desheathed' preparations. J. Cell Sci. **23:** 315–324.
12. Hargittai, P. T., A. M. Butt & E. M. Lieberman. 1990. High potassium selective permeability and extracellular ion regulation in the glial perineurium (blood-brain barrier) of the crayfish. Neuroscience **38:** 163–173.
13. Butt, A. M., P. T. Hargittai & E. M. Lieberman. 1990. Calcium-dependent regulation of potassium permeability in the glial perineurium (blood-brain barrier) of the crayfish. Neuroscience **38:** 175–185.
14. Orkand, R. K., J. G. Nicholls & S. W. Kuffler. 1966. Physiological properties of glial cells in the central nervous system of amphibia. J. Neurophysiol. **29:** 788–806.
15. Coles, J. A. & R. K. Orkand. 1983. Modification of potassium movement through the retina of the drone (*Apis mellifera*) by glial uptake. J. Physiol. **340:** 157–174.
16. Newman, E. A. 1985. Voltage-dependent calcium and potassium channels in retinal glial cells. Nature **317:** 809–811.
17. Hille, B. 1984. Ionic Channels of Excitable Membranes. Sinauer Association Inc. Massachusetts.
18. Walz, W. & D. C. Wilson. 1986. Calcium entry into cultured mouse astrocytes. Neurosci. Lett. **67:** 301–306.
19. Hagiwara, S. & L. Byerly. 1981. Calcium channels. Ann. Rev. Neurosci. **4:** 69–125.
20. MacVicar, B. A. 1984. Voltage-dependent calcium channels in glial cells. Science **226:** 1345–1347.
21. Quandt, F. N. & B. A. MacVicar. 1986. Calcium activated potassium channels in glial cells. Neuroscience **19:** 29–41.
22. Butt, A. M., P. T. Hargittai & E. M. Lieberman. 1986. Ion selectivity of the crayfish blood brain barrier: An electrophysiological study. Neurosci. Abstr. **12:** 1259.
23. Butt, A. M., P. T. Hargittai & E. M. Lieberman. 1987. Role of mucopolysaccharides in the ion perm-selectivity of the crayfish nerve perineurium. Fed. Proc. **46:** 1275.
24. Ashhurst, D. E. & N. M. Costin. 1971. Insect mucosubstances. II. The mucosubstances of the central nervous system. Histochem. J. **3:** 297–310.
25. Hassan, S. & E. M. Lieberman. 1988. Studies of axon-glial interactions and periaxonal K^+ homeostasis: II. The effect of axonal stimulation, cholinergic agents and transport inhibitors on the resistance in series with the axon membrane. Neuroscience **25:** 961–969.
26. Lieberman, E. M. & S. Hassan. 1988. Studies of axon-glial interactions and periaxonal K^+ homeostasis: III. The effect of anisosmotic media and potassium on the relationship between the resistance in series with the axon membrane and glial cell volume. Neuroscience **25:** 971–981.

Permeability and Transport of Glial Blood-Brain Barriers[a]

N. JOAN ABBOTT

Physiology Group
Biomedical Sciences Division
King's College
Strand, London WC2R 2LS, UK

In the mammalian brain, tissue is separated from blood by a barrier layer. This layer is formed by the endothelium of microvessels invading the tissue and by epithelial cells in the choroid plexus and arachnoid membrane (reviewed in refs. 1 and 2). However, this pattern is not universal among animal groups; most invertebrates have no blood-brain barrier, whereas in those groups showing restriction at the blood-nervous system interface, the barrier is formed predominantly by glial cells.[3] A glial barrier is also found in elasmobranch fish.[4,5] These glial barriers provide an excellent opportunity to examine some key questions concerning the brain microenvironment and its regulation. This discussion will center on the physiology of the barrier layers, but will also consider their evolution and embryological development. Study of glial barriers also adds an interesting dimension to the question of the mechanism(s) by which glial and endothelial cells induce specific properties in each other. As the blood-tissue barrier, whether endothelial or glial, serves to regulate the brain microenvironment,[6,7] the properties of the barrier have important implications for glial-neuronal interaction.

QUESTIONS

FIGURE 1 summarizes key questions for consideration. These relate to the reasons why our brains need a barrier, whether endothelial and glial cells are equally good at making barrier layers, how the barriers have developed in an evolutionary and embryological sense, and how they operate.

VALUE OF COMPARATIVE STUDIES

For questions as wide-ranging as these, study of a single mammalian species such as the rat can provide few clear-cut answers. It has proved extremely useful to examine the variety of animal strategies for controlling the brain microenvironment as a first step in developing testable hypotheses. FIGURE 2 shows three species that have been particularly useful, the cuttlefish *Sepia,* the octopus *Eledone,* and the dogfish *Scyliorhinus.* The first two are cephalopod molluscs, the last is an elasmobranch, a cartilaginous fish. All these species possess large, well-vascularized brains,

[a]This work was supported by the Wellcome Trust, Royal Society, Nuffield Foundation, British Council, and SERC.

QUESTIONS :

1. What is the BBB for ?
2. Can glia form an effective BBB ?
3. Do glial and endothelial barriers use identical transport mechanisms ?
4. How did the BBB evolve ?
5. How does the BBB develop in the embryo ?
6. What is the source of barrier-inducing agents ?

FIGURE 1.

have a predominantly glial blood-brain barrier, and are a convenient size for experimental comparisons with mammals.[8]

WHAT IS A BLOOD BARRIER FOR?

Formulating the Hypothesis

This question is hard to answer in a mammal. For example, in the rat, the blood-brain barrier is so well developed and all-enveloping that it is difficult to identify the primary stimulus that led to its formation. What critical feature of brain

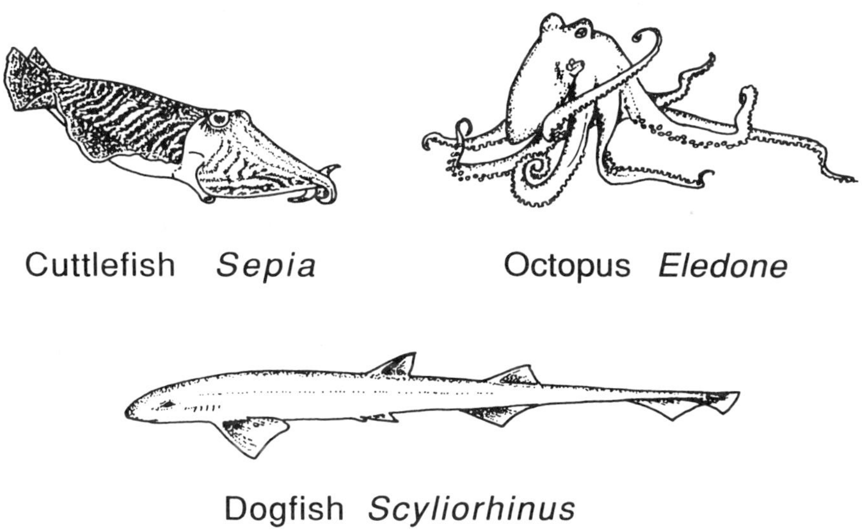

FIGURE 2. Animals that have contributed to comparative studies of the role of glia in the blood-brain barrier: *Sepia officinalis, Eledone cirrhosa,* and *Scyliorhinus canicula.*

function does the barrier promote, so that possession of a barrier gives significant evolutionary advantage?

It is possible to gain some clues by comparing animals within the same Phylum that do and do not have a blood-brain barrier. FIGURE 3 shows the eyes of a marine snail (gastropod) and an octopus (cephalopod), both molluscs. Although the eye of the gastropod is a relatively coarse light-sensing epithelium with a simple lens, the cephalopod eye is equivalent to that of mammals in complexity; it is possible to identify exact analogues of pupil, cornea, iris, and lens, and the thousands of photoreceptors lead to remarkable visual acuity. A blood-brain barrier is absent in gastropods and present in cephalopods.[3] The simple conclusion is that an association exists between the presence of a barrier and the ability to analyze complex sensory inputs. Analogous arguments can be used to show a correlation with complexity of motor performance. These comparisons suggest that the barrier is necessary for the brain to do complicated calculating tasks. We have proposed that a barrier is needed to preserve ionic homeostasis for graded electrical signaling at integrating synapses involved in detailed pattern discrimination (e.g., vision and motor control[6]). It follows that neural elements such as axons conducting "digital" (all-or-none) action potentials should not need such effective control.

This hypothesis is hard to test in the mammalian brain, because as Milton Brightman and colleagues have shown (see comments by M. W. Brightman, this volume), both axonal (white matter tracts) and synaptic (gray matter) parts of the brain have a tight barrier. Perhaps when the brain complexity reaches high levels, it becomes simpler to protect the whole microenvironment. By contrast, the cuttlefish *Sepia* makes a very suitable animal model for testing the hypothesis.

Testing the Hypothesis: The Blood-Brain Barrier Is for Ionic Homeostasis Around Synapses

In studies with Magnus Bundgaard, we used horseradish peroxidase (HRP) as an electron-dense tracer injected intravenously in order to pinpoint the site of the

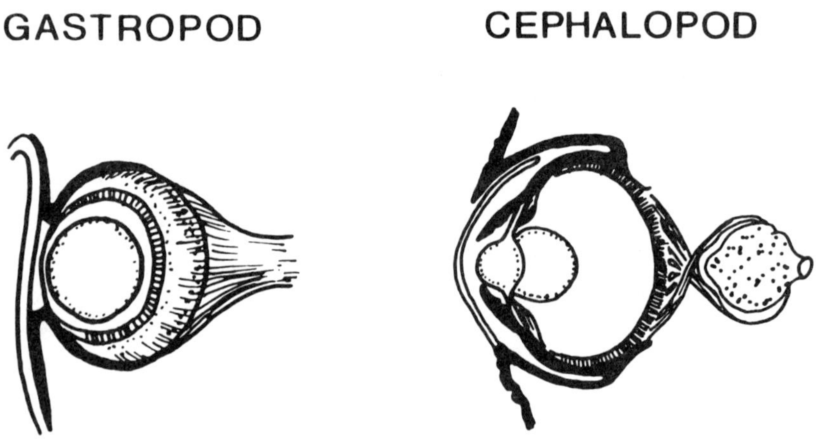

FIGURE 3. Comparison of the eyes of two molluscs: (**a**) a marine snail (prosobranch gastropod) *Murex* and (**b**) the cephalopod *Octopus.* (Based on figures in: Barnes, R. D. 1980. Invertebrate Zoology, 4th Ed. Saunders College. Philadelphia.)

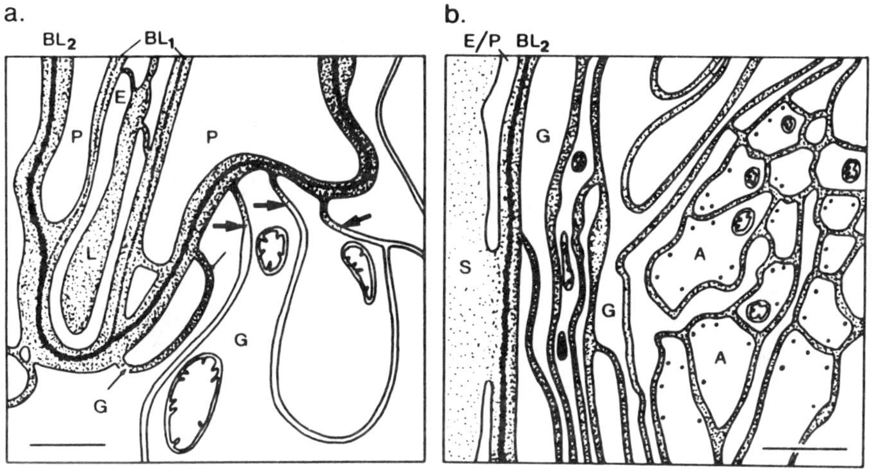

FIGURE 4. *Sepia* nervous tissue. Comparison of the permeability of the blood-nervous tissue interface to the electron-dense tracer, horseradish peroxidase (HRP), in (**a**) brain and (**b**) optic nerve; drawings based on electron micrographs from animals injected with HRP (intravenous) 10 minutes before fixation. Intercellular cleft width has been exaggerated for clarity. (**a**) Blockade of HRP penetration (*stipple*) by perivascular glial cells (G) in an optic lobe microvessel. Basal lamina 1 (BL_1) marks the boundary between endothelium and pericytes, BL_2 that between pericytes and glial end-feet. Although HRP reaction product fills some "loops" of extracellular cleft between glial cells near the vessel lumen (*small arrows*), in clefts that run deeper into the neuropil, tracer is undetectable beyond the characteristic junctional zone (*large arrows*). Scale bar approximately 0.5 μm. (Based on Figure 10 in ref. 10.) (**b**) Permeable blood-nerve interface in the region where optic nerve axons leave the optic lobe. The vascular sinus (S) is to the left, lined by a discontinuous layer of endothelial/pericytes (E/P) and underlying basal lamina (BL_2). HRP reaction product (*stipple*) is seen in the intercellular clefts between glial processes (G) and surrounding optic axons (A). Scale bar approximately 1 μm. (Based on electron micrograph from Bundgaard and Abbott, in preparation.)

barrier layer in *Sepia.* FIGURE 4 summarizes the findings. HRP was blocked from penetrating the interstitium in the neuropil of the optic lobe[9,10] (FIG. 4a), where the major processing of visual information occurs; unlike vertebrates, the cephalopod retina does relatively little peripheral processing. The site of restriction at the capillary level was the perivascular glial layer, not the endothelium. By contrast, the nerves connecting the retina and the optic lobe were permeable to HRP[6] (FIG. 4b). This is good evidence in support of our hypothesis that synapses need a barrier, whereas axons may not, provided the plasma is an adequate medium for neuronal function. These studies add support to the view that the primary function of the blood-brain barrier is to protect synapses by preserving the ionic milieu necessary for the generation of graded (analogue) potentials.[6]

CAN GLIA FORM AN EFFECTIVE BLOOD-BRAIN BARRIER?

There are standard techniques for measuring the permeability of the mammalian blood-brain barrier, and we have simply adapted these for the cuttlefish, octopus, and dogfish. TABLE 1 shows the permeability multiplied by surface area (*PS*)

TABLE 1. Microvessel Permeability to Nonelectrolytes EDTA and Polyethylene Glycol (PEG_{4000}) in Brain Optic Lobe of the Cuttlefish *Sepia*, Compared with the Rat[a,b]

Parameter Estimated	Nonelectrolyte	*Sepia*	Rat
Permeability × surface area product, PS ($ml.g^{-1}.min^{-1} \times 10^4$)	EDTA	2.9	
	PEG_{4000}	1.0	
Microvessel surface area, S ($cm^2.cm^{-3}$)		176	100
Permeability coefficient, P ($cm.s^{-1} \times 10^8$)	EDTA	2.8	0.7–7
	PEG_{4000}	0.9	0.25–2.5

[a]Rat data from similar sized molecules, sucrose (≃EDTA) and inulin (≃PEG_{4000})
[b]Reproduced from ref. 3 with permission.

products measured in *Sepia* optic lobe, using radioactive tracers, work done with Helen Cserr and Magnus Bundgaard.[11,12] The test molecule was introduced into the blood and allowed to circulate, blood samples were taken periodically to establish the time integral of the plasma concentration ($\int C_{pl.} .t$), and finally the concentration of the test substance in brain was measured (C_{br}) to establish how much had escaped from the blood over the experimental period. A correction was made for intravascular marker using known values for regional blood volume in the brain. PS was then calculated from $C_{br}/\int C_{pl} .t$. Independent stereologic analysis of histological sections gave values for the surface area (S) of the blood-brain interface.[13] The table shows that the surface area density of vessels in *Sepia* and rat brain is comparable, and the permeability of the two small nonelectrolyte tracer molecules in *Sepia* is within the range reported for the rat. In analogous experiments in dogfish, with Arthur Butt and Berislav Zloković, we showed that the small ion and nonelectrolyte permeability of the dogfish blood-brain barrier is also similar to that of the rat.[8] Further electron microscopic studies in *Sepia* brain with the smaller electron-dense tracer ionic lanthanum (hydrated radius about 0.4 nm) showed that the glial layer blocks penetration of lanthanum[10] (FIG. 5a), diagnostic of a relatively tight barrier (cf. ref. 14). These studies show that glia can indeed make an effective barrier comparable to endothelial barriers in tightness.

WHAT IS THE NATURE OF THE RESTRICTION?

Glial cells in many animal groups appear able to form tight junctions (*zonulae occludentes*), and the barriers in the brains of insects and elasmobranchs are apparently composed of these structures. However, studies in the cuttlefish *Sepia* show that this is not the universal pattern. In a high proportion of microvessels (about 90%), diffusional restriction is provided by the neat expedient of having a "seamless" layer of perivascular glial end-feet ensheathing the vessel[9]; if there are no intercellular clefts across the boundary layer, it will inevitably be very tight. The mechanism by which this anatomy is produced is unknown. In the remaining 10% of the surface area, where clefts are found, these are long and tortuous, but *zonulae occludentes* are not present; instead, extracellular tracers reveal relatively widely spaced membranes, with an interesting striated junctional zone[10] (FIG. 5b). There is a clear gradient of tracer density, high at the luminal end and undetectable at the tissue end of the cleft. The junction leaves no detectable imprint in freeze-fracture replicas of the vessel wall.[15] Therefore, the junction, although occluding, is clearly not a classical tight junction. We have proposed that this is a novel and possibly primitive junctional type, and the restriction is not due to narrowing of the extracellular cleft, but is conferred by the properties of the extracellular matrix and its associated proteins, forming an occluding fibrillar "matrix filter" in the junction.[16,17] As observations on domestic kitchen sinks illustrate, an outlet blocked by fibers and debris, particularly with a hydrophobic component (fat, grease), can form an effective barrier.

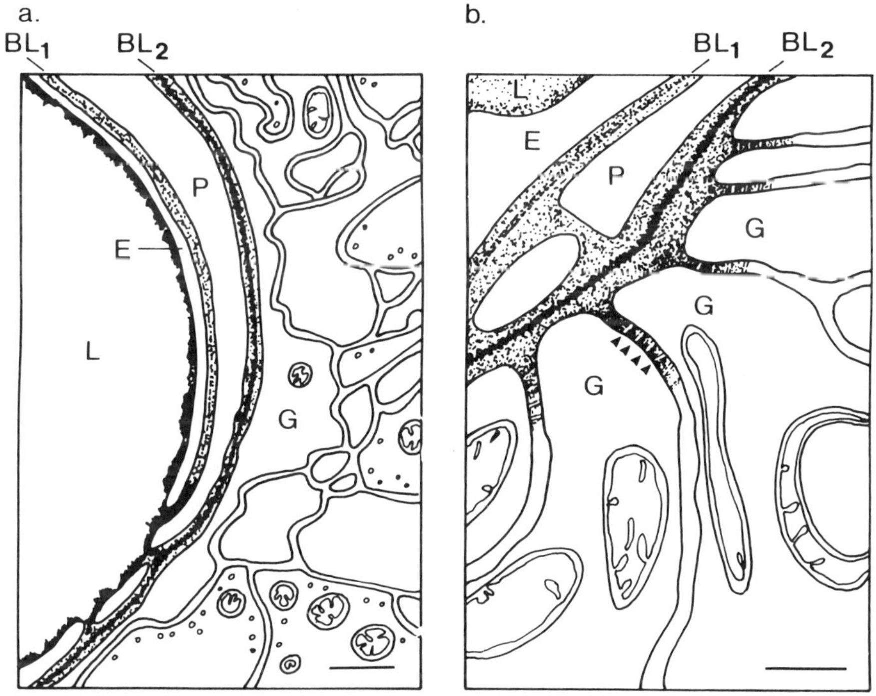

FIGURE 5. *Sepia* brain optic lobe. Drawings based on electron micrographs; width of extracellular clefts exaggerated for clarity. **(a)** Blockade of ionic lanthanum by perivascular glial cells in microvessels of optic lobe. Animals were perfused via the aorta with saline containing $LaCl_3$, then with phosphate-buffered fixative to precipitate lanthanum. Lanthanum deposits (*black patches*) are present on the surface of the endothelial cells (E) facing the vessel lumen (L), in clefts between pericytes (P), and as punctate deposits in both basal laminae (BL_1 and BL_2). Lanthanum (*dots*) is not seen deeper in the parenchyma beyond the level of the perivascular glia (G). Scale bar approximately 0.5 μm. (Based on Figure 44 of ref. 10.) **(b)** Details of glial junctions revealed by HRP in animal injected with the tracer 10 minutes before fixation. Glial end-feet (G) are seen at the surface of the optic lobe where endothelial cells (E) and pericytes (P) with associated basal laminae (BL_1 and BL_2) line the vascular sinus (S). Electron lucent periodicities (*arrowheads*) can be seen spanning the intercellular clefts, in negative contrast against the HRP reaction product (*stippled*). This is the region where extracellular tracers (HRP, lanthanum, and tannic acid) are progressively blocked. Scale bar approximately 0.2 μm. (Based on micrograph used for Figures 39 and 42 in ref. 10.)

DO GLIAL AND ENDOTHELIAL BARRIERS USE IDENTICAL TRANSPORT MECHANISMS?

Glucose Transport

The demonstration that D-glucose but not L-glucose is preferentially removed from blood during a single pass through the mammalian brain was one of the first pieces of evidence that brain microvessels are unlike those elsewhere in the body.[18] It is now known that the Na^+-independent glucose carrier is present in high concentration in the brain endothelium, but not in most other microvascular beds.[19] The carrier appears identical to the glucose carrier of erythrocytes, with significant differences

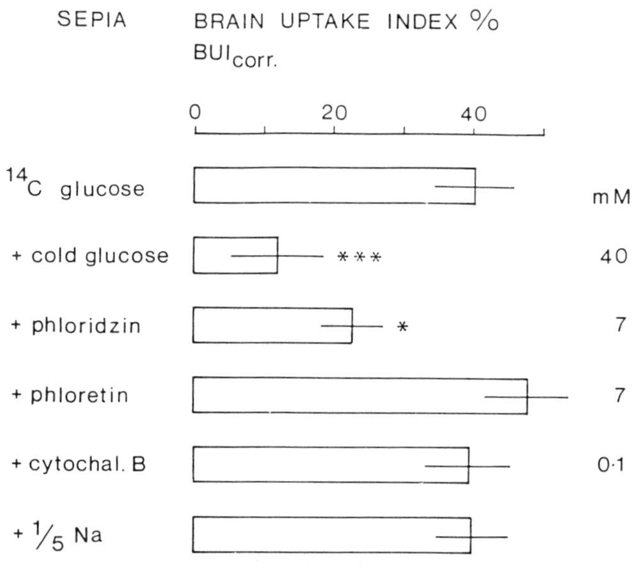

FIGURE 6. *Sepia* optic lobe. Brain uptake index (BUI) for [^{14}C]-glucose, corrected for the background of the method by subtraction of BUI for [^{14}C]-mannitol (giving BUI_{corr}, mean ± SEM, $n = 4–7$), and in the presence of 40 mM cold glucose, 7 mM phloridzin or phloretin, 0.1 mM cytochalasin B, and [Na^+] reduced to one fifth normal (replaced by Tris). $*P < 0.05$, $***P < 0.001$.

from the Na^+-dependent carrier of epithelia such as kidney and intestine. A glucose carrier is needed in brain endothelium because the junctions are too tight to allow diffusional entry, unlike the situation in other capillary beds. If a hexose carrier is therefore a necessary feature of a blood-brain barrier, it would be expected at high density in the glial cells forming the barrier in *Sepia* brain.

FIGURE 6 summarizes studies on the penetration of [^{14}C]-glucose in *Sepia* brain, measured using a modification of the intracarotid bolus injection method of Oldendorf,[20] with tritiated water as a rapidly equilibrating reference. The "brain uptake index" (BUI) is a measure of the penetration of glucose relative to tritiated water and has been corrected (BUI_{corr}) for the BUI for mannitol, an inert, nontransported molecule with similar molecular weight to glucose. The BUI for glucose in the brain optic lobe is high (around 60%; cf. mannitol, 20%) and is reduced by the addition of 40 mM of cold glucose, but not by the reduction of sodium in the bolus. However, the sensitivity of the carrier to inhibitors of the endothelial glucose transporter is different. In *Sepia,* phloridzin is a more effective blocker than phloretin[21] at the dose used in mammals (7 mM); in the rat the sensitivity is greater to phloretin. Cytochalasin B, a potent blocker of the mammalian glucose transporter, has no effect in *Sepia.*

The BUI technique has proved particularly suitable for the dogfish. Because of a quirk of anatomy, the internal carotids meet in the midline in the roof of the mouth, so that it is possible to cannulate one side through a small skin incision, while preserving blood flow through the other.[8] The hexose transporter of the dogfish resembles that of *Sepia,* again with some anomalies compared to the endothelial carrier.[22] These studies identify the Na^+-independent glucose carrier as diagnostic of blood-brain barrier sites, but also point to interesting and subtle differences in the specificity and molecular configuration of the carrier when it is expressed in high concentrations in glial barriers.

Ion Transport and Interstitial Fluid Production

We have used microelectrodes to measure the transendothelial potential of the blood-brain barrier in the frog[8,23] and the rat.[24] In both, the potential is sensitive to alterations in $[K^+]$ on the brain side and includes a component blocked by ouabain, evidence for an electrogenic Na^+-K^+-ATPase on the abluminal membrane. In exactly analogous experiments on the subependymal vessels of the dogfish brain, a potential difference could be measured, generated by the glial end-feet.[25,26] A pump-dependent component was again present and again abluminal. Thus, the Na^+-K^+-ATPase is abluminal in both endothelial and glial barriers, although in the case of the endothelium this membrane would be basal, whereas in the glial barrier it is effectively apical. Thus an Na^+-K^+-ATPase on the brain-facing side of the barrier appears to be a functional necessity. In most epithelia the Na^+-K^+-ATPase has a basolateral location. The presence of the transporter on the apical membrane of the glial layer makes it analogous to the other anomalous epithelium, the choroid plexus[1]; in both cases the brain-facing membrane is the one with the carrier. The fact that the abluminal pump location is fundamental to the secretory function of the choroid plexus suggests that the pump might play an equivalent role at the vessel wall, that is, secretion of interstitial fluid (ISF). Evidence for a flowing ISF compartment in cephalopod molluscs,[12] a group without ventricles or cerebrospinal fluid, suggests that fluid secretion by the capillaries could indeed be the source of ISF. That the machinery for such a secretion exists in both endothelial and glial barriers suggests that a flowing rather than a stagnant ISF is universal. A flowing ISF is expected to confer evolutionary advantage by providing improved intracerebral ionic homeostasis.[8]

Amino Acid Transport

A major question concerns the way in which glial barriers handle amino acids. The predominant amino acid carriers of the vertebrate endothelial barrier are the L system (leucine-preferring, large neutral, luminal membrane) and the Na^+-dependent A system (alanine-preferring, abluminal membrane), with a small component of transport possibly mediated by the ASC system (reviewed in ref. 1). FIGURE 7 shows that the glial barrier of elasmobranchs expresses a similar pattern as far as L

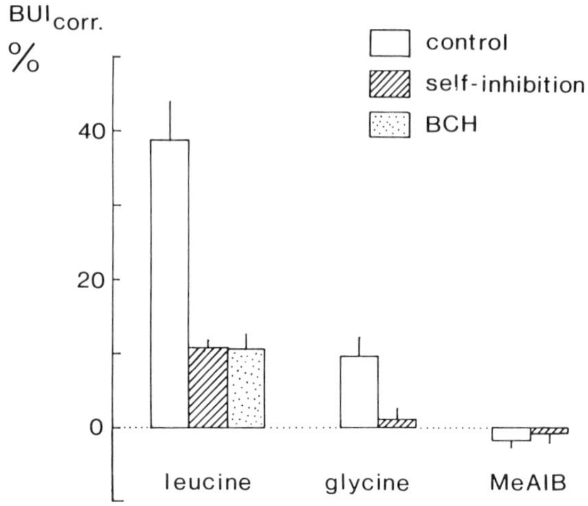

FIGURE 7. Dogfish telencephalon. BUI_{corr} (see FIG. 6) for L-leucine, glycine, and α-methylaminoisobutyric acid (MeAIB), mean ± SEM, $n = 5$–7. Self-inhibition was tested in the presence of 10 mM L-leucine, and 40 mM glycine and MeAIB, respectively. BCH, 2-aminobicycloheptane-2-carboxylic acid, L-system inhibitor, 10 mM.

and A system carriers are concerned. Leucine uptake (BUI_{corr}) is marked, inhibited by cold leucine and by 2-aminobicycloheptane-2-carboxylic acid (BCH),[27] the diagnostic L-system competitor. Methylaminoisobutyric acid (MeAIB) uptake is negligible, but the presence of a small "negative uptake" is suggestive of efflux via an abluminal carrier. Interestingly, the dogfish barrier shows saturable uptake of glycine, negligibly transported by the mammalian endothelial barrier. This could indicate the presence of the Gly transporter in the glial barrier.

The handling of acidic amino acids is of particular interest. Glutamate and GABA show very low levels of uptake by endothelial barriers, with circumstantial evidence that the barrier subserves net efflux, indicating an abluminal carrier. By contrast, where glial cells have been studied in culture, they possess prominent carriers for amino acid neurotransmitters such as glutamate and GABA. A key question was whether barrier glia would conform to the endothelial pattern or retain a glial pattern of carriers. FIGURE 8 shows BUI_{corr} values in dogfish for these amino acids, and it is clear that the glial barrier does express carriers for glutamate and GABA. Glutamate uptake is inhibited by cold glutamate, and by D- and L-aspartate, but not by GABA.[28] GABA uptake is reduced by cold GABA, β-alanine, and taurine, but not by glutamate. The pattern of inhibitory action of agents on GABA uptake shows some interesting similarities with that reported for the glia of the bullfrog sympathetic ganglia,[29] in that δ-amino valeric acid is the most potent inhibitor, although some differences do exist; for example, 2-4 diamino butyric acid (DABA) has little effect in the dogfish.

COMPARISON OF GLIAL AND ENDOTHELIAL BARRIERS

In summary, many similarities are found between endothelial and glial barriers: both express the Na^+-independent glucose carrier and the L-system amino acid carrier on the luminal membrane and the Na^+-K^+-ATPase and A system amino acid carrier on the abluminal membrane. However, in addition to some subtle differences in carrier specificity (glucose transporter), the glial barriers retain certain glial transporters. As one of the functions of the absence of glutamate and GABA carriers in the endothelial barriers is to segregate central and peripheral pools of neurotransmitters, it would be very unlikely that the glial barrier transporters would be designed to deliver glutamate and GABA from the blood to the brain. Most species have high (mM) levels of glutamate in the plasma, which would have neurotoxic actions if allowed free access to the brain. Instead it is likely that the glutamate and GABA taken up by the glial barrier are rapidly metabolized within the glial cells, so that the glial layer exerts an effective enzymic barrier for these substances. Thus the net result of endothelial and glial barriers will be the same, excluding glutamate and GABA. This survey of transport mechanisms shows that endothelial and glial barriers are similar but not identical in their transport properties, but the end result is functionally the same.

HOW DID THE BLOOD-BRAIN BARRIER EVOLVE?

FIGURE 9 summarizes the site of the barrier layer in different animal groups: endothelial in mammals, birds, reptiles, amphibia, and teleost fish[30]; absent in lower molluscs, annelids, and primitive arachnids[3]; glial in the microvessels of cephalopods and elasmobranchs[3,30] and also at the blood-brain interface of insects and crustaceans.[3] However, detailed electron microscopic study of *Sepia* brain vessels reveals a more complex picture. FIGURE 10 is a summary diagram showing the cell layers at the blood-brain interface along the vascular tree in *Sepia* brain, from artery to vein,

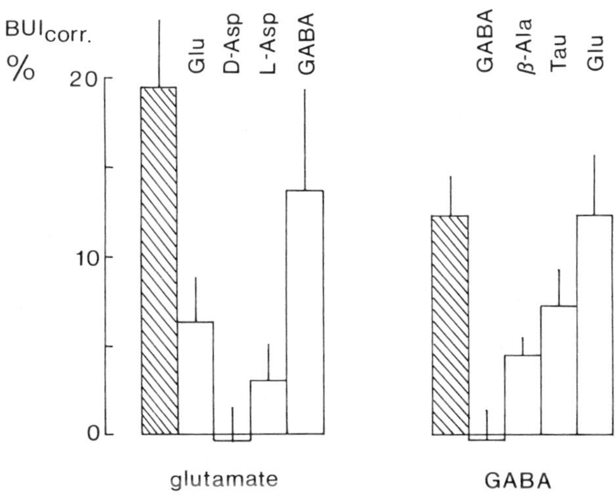

FIGURE 8. Dogfish telencephalon. $BUI_{corr.}$ (see FIG. 6) for [^{14}C]-labeled L-glutamate and GABA, mean ± SEM, n = 3–4. Control uptake (*shaded*) was compared with uptake in the presence of unlabeled amino acids (40 mM) including glutamate (Glu), D- and L-aspartate (D-Asp, L-Asp), β-alanine (β-Ala), and taurine (Tau). Two independent carriers are revealed, characteristic of glial cells, not endothelial cells.

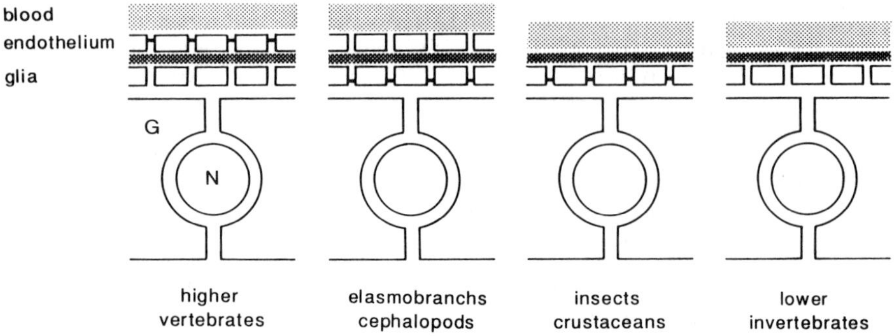

FIGURE 9. Blood-brain interface of different animal groups. Location of barrier layer is shown by bars between adjacent cells. Basal lamina is shown by *dark stipple.* For clarity, pericytes are omitted, but see text. In many invertebrate groups, an endothelial layer is absent. The lower part of the figure shows a schematic relation between glia (G) and neurons (N) of the parenchyma.

illustrating the relative positions of endothelial cells, pericytes, and perivascular glia. In microvessels the distinction between endothelial cells and pericytes disappears, and we have evidence from their relation to each other and to the basal lamina that they are actually subclasses of a single cell type (endothelial/pericyte).[9] In capillaries and veins the endothelial/pericyte layer is leaky, and the barrier layer is formed by

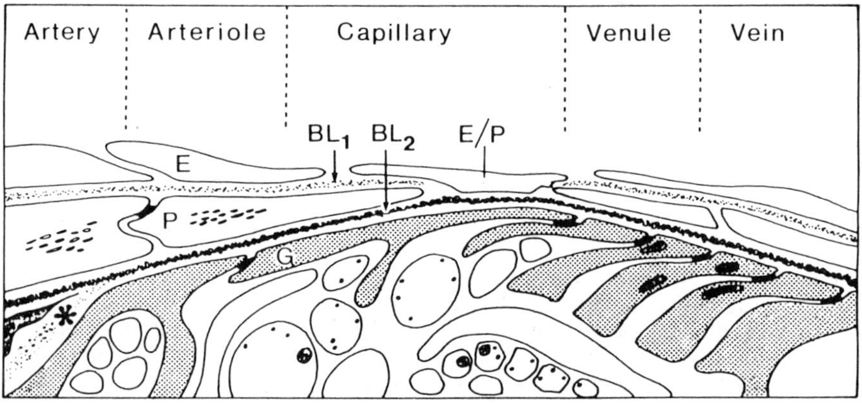

FIGURE 10. *Sepia* blood-brain interface. Schematic diagram summarizing the cell layers present at the blood-brain interface, from the arterial to the venous end of the microvasculature (not to scale). E = endothelial cell; P = pericyte; BL_1 and BL_2 = basal laminae; G (*shaded*) = perivascular glial layer; * collagen-filled extracellular space. In the smallest vessels, only a single basal lamina can be resolved; endothelial cells and pericytes here become indistinguishable[9] and are shown as a merged class of endothelial/pericytes (E/P). Restricting junctions (*black bars*) are formed between pericytes in arterial vessels and between perivascular glial cells in microvessels (capillaries) and venous vessels. A large proportion of the glial sheath covering the microvessel surface is "seamless" (no intercellular clefts),[9] whereas in the venous vessels glial end-feet make a lamellated layer with extensive overlap between adjacent cells. (Reproduced from ref. 10, with permission.)

glial end-feet.[10] However, in arterioles and arteries, it shifts to the pericyte layer.[10] As lower molluscs have no barrier,[3] the barrier seen in *Sepia* must be a later evolutionary addition. In all other invertebrate groups with a barrier (crustaceans, insects, and higher arachnids), the barrier is entirely glial, but cephalopod molluscs are the group with the largest brains, most developed cerebral vasculature, and most complex behavioral repertoire. This pattern of barrier sites suggests that the blood-brain barrier in all these groups was originally glial and that only when the layer of cells lining the vessels became relatively complete did the barrier shift to the endothelial/pericyte layer. The *Sepia* pattern may then be showing us the sequence that occurred in evolution, captured in time.

A similar shift can be hypothesized for vertebrate evolution. According to this scheme, the glial blood-brain barrier of elasmobranchs, a primitive group of cartilaginous fish, preserves the evolutionarily primitive condition. The endothelial barrier of more advanced groups, now the predominant vertebrate pattern, would be a later evolutionary development. The fact that an endothelial barrier is found in the holocephalans,[30] cartilaginous fish thought to be related to the common ancestors of elasmobranchs, together with the evidence from cephalopods, suggests that an endothelial/pericyte barrier evolved independently several times in evolution.

HOW DOES THE BLOOD-BRAIN BARRIER DEVELOP IN THE EMBRYO?

The concept that "ontogeny repeats phylogeny," that is, that developmental stages of the embryo retrace the stages that occurred in evolution,[31] is now regarded as simplistic, but it does help explain why mammalian and chick embryos show features such as gill arches, of no use to either embryo or adult. The concept has been less commonly applied at the cellular level. In view of the evidence presented here that a glial barrier preceded the endothelial barrier in evolution, it is of interest to see if the same sequence can be detected in the embryo. FIGURE 11a, from an electron microscopic study by Kjeld Møllgård, shows the junctional zone between the superficial cells of the neural tube where they face the developing subarachnoid space; similar junctions are found in the ependymal lining of the embryonic ventricles. These junctions are characterized by a wide intercellular cleft spanned by electron-lucent "linkers" or "spacers" and mark the site where a single-stranded tight junction restricting to protein will later develop (FIG. 11b[32]). Thus the primitive glial-ependyma of the developing vertebrate embryo shows junctions remarkably similar to the matrix junctions of adult *Sepia* brain.[33] Taken together, these observations suggest that the linker junction is an early form of junctional structure that can be specialized in one of two different ways, either by insertion of intramembranous proteins to provide the tight seal of the *zonula occludens* or by association with extracellular fibers and proteins to form the matrix junction of modern cephalopods.[16] Furthermore, the junctional specialization, in ontogeny as in phylogeny, is established initially in the glial-ependymal layer and only shifts to the endothelial layer following vascularization of the nervous tissue. In the embryo, as in evolution, a glial barrier precedes the endothelial barrier.

Traces of the barrier-forming potential of glial cells can still be detected in the adult vertebrate brain. The ependyma and choroid plexus of the adult are related to the undifferentiated glial ependyma of the early embryo and retain some glial characteristics. In certain "circumventricular organs" of the adult brain, where tight junctions have failed to develop between endothelial cells of the vessels for reasons associated with the secretory and/or chemosensitive functions of the local regions, tight junctions are instead found in the ependyma, such as in the choroid plexus (secretory) and the median eminence (neurosecretory).[1] The tanycytes, also ependymal derivatives, seal off neurosecretory areas from adjacent brain parenchyma, using an additional belt of tight junctions. The pigment epithelium of the eye is an outpouching of the ependyma and expresses tight junctions that are sufficiently tight to give a transepithelial resistance of about 300 $\Omega.cm^2$, greater than that of the choroid plexus (reviewed in ref. 6).

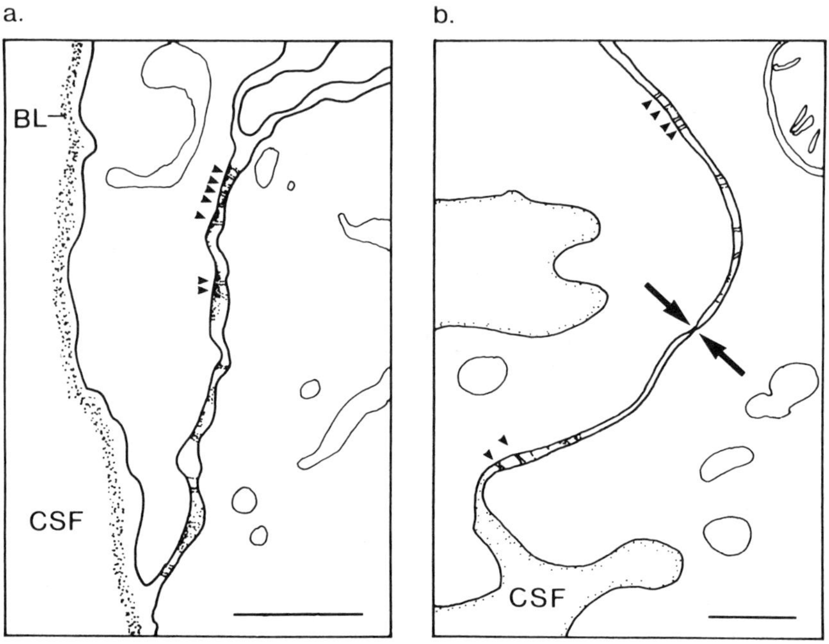

FIGURE 11. Junctions in the developing mammalian brain; drawings based on electron micrographs from K. Møllgård, with permission. **(a)** A 20-day-old pig embryo, superficial germinal matrix of neural tube. Junctions between cells of the undifferentiated neuroepithelium, in the layer that will subsequently form the glial "*membrana limitans*" beneath the pia, facing the subarachnoid cerebrospinal fluid (CSF). BL, pial basal lamina. Note the periodicities or "linkers" (*arrowheads*) spanning the cleft. Equivalent structures are seen in the ependyma lining the ventricles, in the region where single-stranded "strap" junctions will develop (FIG. 11b). Scale bar = 0.5 μm. **(b)** A 19-day-old sheep embryo, neuroependyma lining the cerebral vesicle, containing CSF. The junctional zone exhibits narrow intercellular clefts that appear occluded at focal points (*arrows*), corresponding to the single-stranded "strap junction" seen in freeze-fracture replicas.[32] Faint periodicities (*arrowheads*) resembling the "linkers" of Figure 11a are seen in wider parts of the cleft where cell membranes show clear parallel orientation. Scale bar = 0.5 μm. (Based on FIG. 1B in ref. 32.)

From the preceding discussion the simplest explanation of these observations is that parts of the glia-ependyma retain the barrier-forming potential of the embryonic ependyma. This leads to an additional question: to what extent do parenchymal glia of the adult vertebrate nervous system retain barrier-forming potential? It is noticeable that tumors and grafts within brain tissue that have leaky vessels are not associated with widespread leakage of intravascular tracers into adjacent parenchyma,[34] suggesting that the boundary between the tumor/graft and normal brain is relatively impermeable. This is a zone occupied by prominent sheets of reactive glia. Although the barrier may not be as effective as the endothelial blood-brain barrier, the combination of overlapping glial processes and junctional zones appears to provide effective restriction at least to proteins. Glial ensheathment around synaptic zones would also be expected to result in effective segregation of transmitter pools. Mammalian glia appear to retain useful barrier-forming properties into the adult.

WHAT IS THE ADVANTAGE OF THE SHIFT FROM GLIAL TO ENDOTHELIAL BARRIER?

Studies on *Sepia* emphasize that the blood-brain barrier should be regarded as a multicellular interface, with endothelium, pericytes, and glia cooperating. At one extreme, as in *Sepia* microvessels, the barrier is formed by glia, and the endothelium/pericytes play a relatively minor role. At the other extreme, as in the mammal, the endothelium is the restrictive site, the phagocytic pericytes act as a second line of defense, and the glia help to induce and maintain the endothelial barrier. The shift in barrier site during both ontogeny and phylogeny leads to a useful division of labor. Thus, shift of the barrier function to the pericyte layer (*Sepia* arterioles) permits the glia to concentrate on interstitial regulation. Further shift of the barrier to the endothelium frees pericytes for phagocytic duties and leaves the pericytes and glial cells to monitor and control the perivascular space. The maintenance of the endothelial barrier and the specific properties of the glial end-feet appear to involve complex inductive interactions between endothelium, pericytes, and glia.

WHAT IS THE SOURCE OF BARRIER-INDUCING AGENTS?

Svendgaard *et al.*[35] showed that certain properties specific to brain capillary endothelium could be induced in blood vessels growing into brain tissue transplanted into the anterior chamber of the eye; conversely, vessels growing into non-brain tissue grafted into brain lost their barrier properties. These studies implicated some influence from brain tissue in inducing the barrier. Stewart and Wiley,[36] using quail-chick chimeras, confirmed that the grafts are vascularized from the host tissue. Janzer and Raff[37] showed that impermeability to albumin could be induced by astrocytes, in vessels invading astrocytic clusters, grown on the iris and on the chorioallantois of the chick. More recently, electron microscopic studies have shown that the complexity of endothelial tight junctions is increased by co-culture with astrocytic conditioned medium (CM) and that CM can "tighten" the transendothelial resistance from ~ 400 to 600 $\Omega.cm^2$ (see chapter by Rubin *et al.*). An interesting example of the opposite kind of induction, where features characteristic of perivascular end-feet are induced in astrocytes by coculture with endothelium,[38] indicates that inductive processes work in both directions.

The comparative survey of glial barriers raises some interesting questions concerning induction. If glia produce an inductive influence, why do glial barriers form only at the blood-nervous tissue interface and not throughout the glial population within the brain parenchyma? The distinguishing feature of the barrier layer, whether glial or endothelial, is that it faces different cellular and fluid environments on its two surfaces. If barrier induction required two factors, one from plasma and one from brain, then the barrier would form at the cell layer exposed on its opposite faces to these two different influences. The same result would be achieved if a plasma factor at one face was necessary to induce the full apical/basal polarity of the interface layer, and the layer experiencing the strongest polarizing influence became, as a result, receptive to a brain-produced factor. When the endothelium is discontinuous, as in *Sepia* microvessels, the more complete perivascular glial layer will experience the strongest inductive influence. Once the endothelium becomes reasonably confluent, cell-cell junctions will start to form, and these will restrict the penetration of any plasma influence to the underlying glial layer, which will therefore lose its strong polarity. This discussion shifts the focus away from a single inductive factor from astrocytes acting on endothelial cells (the usual model, based on mammalian and avian experiments) to a more complex interaction, where the way in which inductive factors interact with cell monolayers is determined by the extent to which the monolayer has developed strong apical/basal polarity.

ANSWERS :

1. The BBB is primarily for ionic homeostasis around synapses.
2. Glia can form an effective BBB, using tight or 'matrix' junctions.
3. Glial and endothelial barriers share some but not all transport properties.
4. In evolution, BBB was originally glial, later endothelial.
5. In development, BBB was originally glial, later endothelial.
6. Factors inducing BBB may come from both brain and blood.

FIGURE 12.

CONCLUSION

Study of animal groups with glial barriers is not a scientific eccentricity; it offers novel experimental approaches to key questions concerning the nature and physiology of the blood-brain barrier. It permits the phrasing and testing of hypotheses, in many cases with greater freedom than in studies confined to higher vertebrates. Answers to some of the questions posed in FIGURE 1 are given in FIGURE 12.

ACKNOWLEDGMENTS

I would like to thank my collaborators, especially Helen Cserr, Magnus Bundgaard, Nancy Lane, Arthur Butt, and Berislav Zloković, for their contributions to this work, and the Director of the Marine Biological Association Laboratory, Plymouth, for facilities and encouragement.

REFERENCES

1. DAVSON, H., K. WELCH & M. B. SEGAL. 1987. Physiology and Pathophysiology of Cerebrospinal Fluid. Churchill Livingstone. London.

2. BRIGHTMAN, M. W. 1989. The anatomic basis of the blood-brain barrier. *In* Implications of the Blood-brain Barrier and Its Manipulation. E. A. Neuwelt, ed. Vol. 1: 53–83. Plenum Publishing Co. New York.
3. ABBOTT, N. J., N. J. LANE & M. BUNDGAARD. 1986. The blood-brain interface in invertebrates. Ann. N. Y. Acad. Sci. **481:** 20–41.
4. BRIGHTMAN, M. W., T. S. REESE, Y. OLSSON & I. KLATZO. 1971. Morphological aspects of the blood-brain barrier to peroxidase in elasmobranchs. Progr. Neuropathol. **1:** 146–161.
5. BUNDGAARD, M. & H. F. CSERR. 1981. A glial blood-brain barrier in elasmobranchs. Brain Res. **226:** 61–73.
6. ABBOTT, N. J., M. BUNDGAARD & H. F. CSERR. 1986. Comparative physiology of the blood-brain barrier. *In* The Blood-brain Barrier in Health and Disease. A. J. Suckling, M. G. Rumsby & M. W. B. Bradbury, eds. Ellis Horwood. Chichester.
7. BRADBURY, M. W. B. 1989. Transport across the blood-brain barrier. *In* Implications of the Blood-Brain Barrier and Its Manipulation. E. A. Neuwelt, ed. Vol. 1: 119–136. Plenum Publishing Co. New York.
8. ABBOTT, N. J., B. V. ZLOKOVIĆ & A. M. BUTT. 1988. Methods for study of the blood-brain barrier in non-mammalian species. *In* Peptide and Amino Acid Transport Mechanisms in the Central Nervous System. L. Rakić, D. J. Begley, H. Davson & B. V. Zloković, eds. :289–303. Macmillan Press and Serbian Academy of Sciences & Arts. London and Belgrade.
9. BUNDGAARD, M. & N. J. ABBOTT. 1992. Fine structure of the blood-brain interface in the cuttlefish *Sepia officinalis* (Mollusca, Cephalopoda). J. Neurocytol., in press.
10. ABBOTT, N. J. & M. BUNDGAARD. 1992. Electron-dense tracer evidence for a blood-brain barrier in the cuttlefish *Sepia officinalis.* J. Neurocytol., in press.
11. ABBOTT, N. J., M. BUNDGAARD & H. F. CSERR. 1985a. Brain vascular volume, electrolytes and blood-brain interface in the cuttlefish, *Sepia officinalis.* J. Physiol. **368:** 197–212.
12. ABBOTT, N. J., M. BUNDGAARD & H. F. CSERR. 1985b. Tightness of the blood-brain barrier and evidence for brain interstitial fluid flow in the cuttlefish, *Sepia officinalis.* J. Physiol. **368:** 213–226.
13. ABBOTT, N. J. & M. BUNDGAARD. 1987. Microvessel surface area, density and dimensions in brain and muscle of the cephalopod *Sepia officinalis.* Proc. R. Soc. B. **230:** 459–482.
14. BOULDIN, T. W. & M. R. KRIGMAN. 1975. Differential permeability of cerebral capillary and choroid plexus to lanthanum ion. Brain Res. **99:** 444–448.
15. LANE, N. J. & N. J. ABBOTT. 1992. Freeze-fracture evidence for a novel restricting junction at the blood-brain barrier of the cuttlefish *Sepia officinalis.* J. Neurocytol., in press.
16. ABBOTT, N. J., N. J. LANE & M. BUNDGAARD. 1992. A fibre-matrix model for the restricting junction of the blood-brain barrier in a cephalopod mollusc: Implications for capillary and epithelial permeability. J. Neurocytol., in press.
17. BUNDGAARD, M., N. J. ABBOTT & N. J. LANE. 1992. A novel restricting junction forms the blood-brain barrier of a cephalopod. Submitted.
18. CRONE, C. 1965. Facilitated transfer of glucose from blood into brain. J. Physiol. **181:** 103–113.
19. DICK, A. P., S. J. HARIK, A. KLIP & D. M. WALKER. 1984. Identification and characterization of the glucose transporter of the blood-brain barrier by cytochalasin B binding and immunological reactivity. Proc. Natl. Acad. Sci. USA **81:** 7233–7237.
20. OLDENDORF, W. H. 1970. Measurement of brain uptake of radiolabeled substances using a tritiated water internal standard. Brain Res. **24:** 372–376.
21. ABBOTT, N. J., D. BLEAKMAN & S. SHORT. 1986. Glucose transport across the glial blood-brain barrier of the cuttlefish *Sepia.* J. Physiol. **377:** 104P.
22. ZLOKOVIĆ, B. V., N. J. ABBOTT, M. TAYLOR & J. HART. 1992. Hexose uptake by the glial blood-brain barrier of the dogfish *Scyliorhinus canicula.* Submitted.
23. ABBOTT, N. J., A. M. BUTT & W. WALLIS. 1986. The Na-K ATPase of the blood-brain barrier (BBB)—a microelectrode study. Ann. N. Y. Acad. Sci. **481:** 390–391.
24. BUTT, A. M., H. C. JONES & N. J. ABBOTT. 1990. Electrical resistance across the blood-brain barrier in anaesthetized rats: A developmental study. J. Physiol. **429:** 47–62.

25. ABBOTT, N. J. & A. M. BUTT. 1986. A microelectrode study of K^+ transport at the dogfish glial blood-brain barrier. J. Physiol. **374:** 39P.
26. BUTT, A. M. 1986. Comparative physiology of the blood-brain barrier in the frog, *Rana pipiens* and *Rana temporaria,* and the dogfish, *Scyliorhinus canicula.* PhD Thesis, London University.
27. ABBOTT, N. J., J. HART, L. ROGAČ, M. TAYLOR & B. V. ZLOKOVIĆ. 1988. Amino acid transport by the glial blood-brain barrier of the anaesthetized dogfish. J. Physiol. **407:** 25P.
28. ABBOTT, N. J. & J. HART. 1988. Glutamate and GABA transport by the glial blood-brain barrier of the dogfish. J. Physiol. **407:** 141P.
29. TASAKA, J., S. SAKAI, T. TOSAKA & I. YOSHIHAMA. 1989. Glial uptake system of GABA distinct from that of taurine in the bullfrog sympathetic ganglia. Neurochem. Res. **14:** 271–277.
30. CSERR, H. F. & M. BUNDGAARD. 1984. Blood-brain interfaces in vertebrates: A comparative approach. Am. J. Physiol. **246:** R277–288.
31. BALINSKY, B. I. 1960. An Introduction to Embryology. W. B. Saunders Co. Philadelphia.
32. MØLLGÅRD, K., Y. BALSLEV, B. LAURITZEN & N. R. SAUNDERS. 1987. Cell junctions and membrane specializations in the ventricular zone (germinal matrix) of the developing sheep brain: A CSF-brain barrier. J. Neurocytol. **16:** 433–444.
33. ABBOTT, N. J., M. BUNDGAARD, N. J. LANE & K. MØLLGÅRD. 1988. Parallels between junctions in invertebrate brain and embryonic mammalian brain. J. Physiol. **400:** 72P.
34. DEANE, B. R., J. GREENWOOD, P. L. LANTOS & O. E. PRATT. 1984. The vasculature of experimental brain tumours. Part 4. The quantification of vascular permeability. J. Neurol. Sci. **65:** 59–68.
35. SVENDGAARD, N. A., A. BJORKLUND, J. E. HARDEBO & U. STENEVI. 1975. Axonal degeneration associated with a defective blood-brain barrier in cerebral implants. Nature **255:** 334–336.
36. STEWART, P. A. & M. J. WILEY. 1981. Developing nervous tissue induces formation of blood-brain barrier characteristics in invading endothelial cells: A study using quail-chick transplantation chimeras. Dev. Biol. **84:** 183–192.
37. JANZER, R. C. & M. C. RAFF. 1987. Astrocytes induce blood-brain barrier properties in endothelial cells. Nature **325:** 253–257.
38. TAO-CHENG, J.-H., Z. NAGY & M. W. BRIGHTMAN. 1990. Astrocytic orthogonal arrays of intramembranous particle assemblies are modulated by brain endothelial cells *in vitro.* J. Neurocytol. **19:** 143–153.

Astrocytic Induction of Endothelial Tight Junctions

Z. NAGY[a] AND K. MARTINEZ[b]

Department of Psychiatry[a] and Anatomy[b]
Semmelweis Medical University
Budapest, Hungary

The endothelium of brain microvessels constitutes a highly specialized, metabolically active barrier responsible for the stable environment of brain tissue. The passage of different substances across the blood-brain barrier (BBB) into cerebral interstitial fluid is regulated by this endothelial monolayer, which possesses a number of structural and enzymatic properties distinct from those of noncerebral capillaries.[1-4]

Capillary endothelial cells in the brain are linked by tight junctions (TJs) that form continuous belts around contiguous endothelial cells. This elaborate, complex junctional network has been considered a major structural component of the BBB to macromolecules.[5] Morphometric analysis of fracture faces of the junctions between cerebral endothelial cells demonstrates a multistranded complex of strands or ridges comprising belt-like TJs where the endothelial cells are ensheathed by astrocytes. Simple, less elaborate TJs with focal, potentially leaky junctional regions are restricted to collecting veins and arteries[6] in regions of endothelium covered by smooth muscle cells.

The ability of brain endothelial cells to form a barrier depends on the environment in which they grow. If a developmental, postnatal increase in the ensheathment of endothelium by astroglia occurs, there appears to be a related maturation of the BBB.[7,8] This maturation coincides with the disappearance of vesicular transport.[9,10] Furthermore, the astrocytic processes that envelope the brain capillaries are the most likely candidates to alter the capillaries' passive permeability by modulating the formation of TJs between the endothelial cells. A close interaction between endothelium and astroglia has repeatedly been proposed[2,11] and demonstrated convincingly by permitting permeable vessels from nonneuronal tissue to grow into embryonic, avascular brain fragments *in vivo.* Having entered a piece of brain, the once permeable vessels became impermeable to circulating tracer.[12] The component of brain environment responsible for this permeability change is apparently astrocytic. When dissociated astrocytes are grafted to the anterior chamber of the eye, the permeable vessels of the iris become impermeable upon penetrating the astrocytic domain.[13] To further test this hypothesis, fetal cerebellar tissue was examined atopically, but *in vivo,* by inserting it into the anterior chamber of the eye. In this way, changes in permeability of the revascularizing, permeable iris vessels could be followed during continuing development of the brain fragment *in vivo.* The results of these preliminary experiments are discussed in terms of freeze fracture analysis of bovine brain endothelium maintained *in vitro* either alone or in cocultures with rat astrocytes.

BLOOD-BRAIN BARRIER IN FETAL CEREBELLAR TISSUE IMPLANTED *IN OCULO*

Pieces of developing cerebellar anlage from 15-day-old rat embryos were transplanted into the anterior eye chamber of 200-g female CFY rats according to the method of Hoffer *et al.*[14] Following 4 weeks of "culturing" the graft *in oculo,* horseradish peroxidase (HRP) solution (Sigma type VI), 20 mg per 100 g of body weight, was slowly injected intravenously. After 5–10 minutes of circulation time, the animals were anesthetized with barbiturate and perfused with aldehyde fixatives through the left ventricle of the heart. The grafts and irises of five animals were sectioned. In the 40–80-μm thick tissue slices HRP was visualized with DAB reaction. Some of the slices were dehydrated and viewed by light microscopy; other

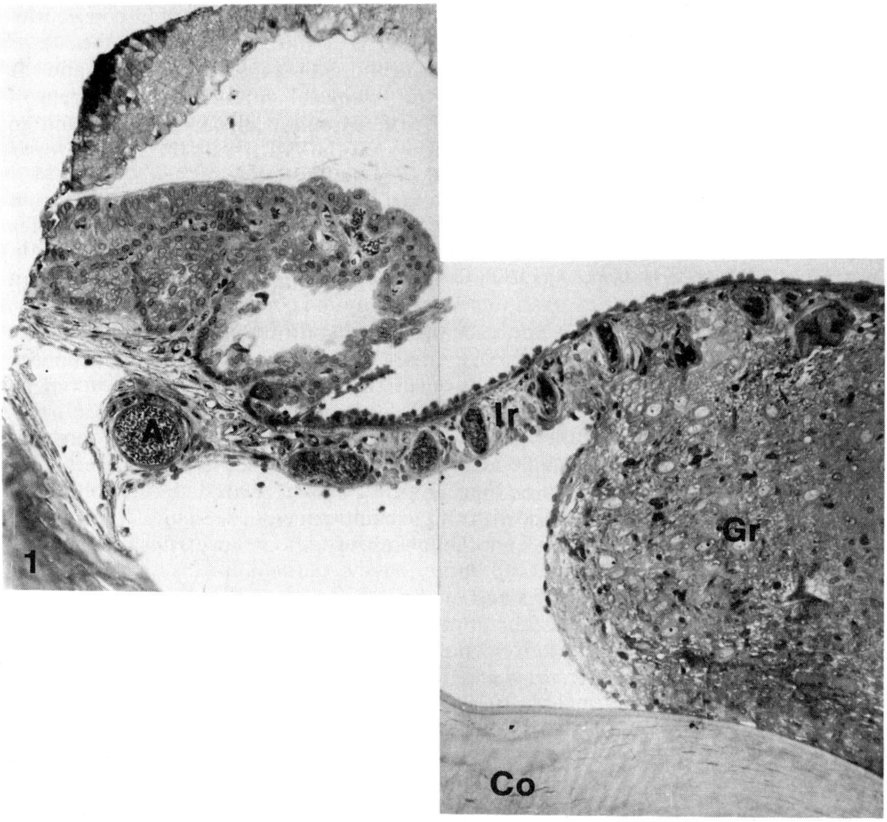

FIGURE 1. Cerebellar graft from a 15-day rat embryo implanted into the anterior eye chamber. Cerebellar tissue differentiated *in oculo* 4 weeks after implantation. Ir = iris; Co = cornea; Gr = graft; A = artery. Original magnification × 85; toluidine-blue staining.

sections were processed for ultrastructural examination. Astroglial cells were detected using a glial fibrillary acidic protein (GFAP) reaction.

The undifferentiated cerebellar tissue that was transplanted grew and differentiated within the eye to reach a final stage, similar to that of the normal adult trilaminar cerebellar cortex; however, the various areas of a given transplant seemed to mature at different times (FIG. 1). Maturation of cerebellar tissue, the types of cell contacts, and electrophysiology had been reported in detail.[15–18] After implantation, branches of microvessels arose from the host iris grown into the graft. Within a few days a fine-meshed capillary plexus developed (FIG. 2). The blood-cerebellar graft barrier to HRP was closed at the time of the experiment, whereas the iris microvasculature and the graft-nourishing capillary plexus were permeable to the tracer (FIG. 3).

The ultrastructure of the graft capillaries was similar to that observed *in vivo* in the brain. The endothelial cells were ensheathed with astroglial processes. These capillaries had very few pits or vesicles and had impermeable TJs to HRP between endothelial cells. In the nourishing capillary plexus around the cerebellar graft, patent interendothelial avenues were observed. The tracer permeated consecutive tight junctional compartments, and HRP-labeled pits were also detected (FIG. 4).

The striking contrast in the capillary permeability of the cerebellar graft and the host eye chamber tissue is a reflection of their structural and functional differences. Capillary endothelial cells in close contact with astroglial cells are impermeable to protein tracers such as HRP. It is interesting that the graft-nourishing microvessel branches became tight as they penetrated the cerebellar tissue.

ENDOTHELIAL-ASTROGLIAL INTERACTION *IN VITRO*

Primary cultures of bovine brain endothelium were prepared according to the method of Bowman *et al.*[19] From these primary cultures subcultures were prepared that were more homogeneous and highly enriched, with about 90% of cells taking up fluorochrome-labeled acetylated low-density lipoprotein, a marker of endothelial cells.[20] In primary astroglial cultures[21] the percentage of GFAP-positive cells was 60–85%, and in enriched secondary astroglial cultures[2] this percentage was 75–90%. Besides "solo" endothelial and astroglial cultures, a different coculture method was used: (1) dissociated endothelial cells from subcultures were seeded on a confluent astroglial bed; and (2) confluent endothelial subcultures on coverslips were placed facing a confluent astroglial bed.[2]

For control cocultures, conditioned medium from astroglial cultures, fibroblasts, from peripheral connective tissues, fibroblast-like cells from bovine trachea, and smooth muscle cells derived from bovine aorta were used in endothelial cultures. Further controls to test if astrocytes affect the junctions of noncerebral endothelium included endothelial cultures from bovine aorta, human umbilical veins, and bovine pulmonary artery.

To study the effect of brain microvessel endothelial cells on expression of astroglial membrane assemblies, different combinations of coculture were used. Primary astroglial cultures were cocultured with bovine pulmonary artery endothelial cells, bovine aorta endothelial cells, human umbilical vein endothelial cells, and fibroblast-like cells from bovine trachea.

The secondary astroglial cultures were cocultured with endothelial cells from bovine pulmonary artery, bovine aorta, bovine fibroblast-like cells from the trachea, rat fibroblasts from peripheral connective tissue, rat meningeal cells, rat cerebellar granular cells, and PC 12 cells. Detailed protocols of the experiments were published recently.[22]

The length and width of endothelial TJs and the assembly density of astroglial membranes were measured on freeze fracture electron micrographs.[22,23] At 6–10 days the TJs in solo endothelial cultures usually consisted of short 1–2-μm patches of junctional strands, occasionally with free ends, and enclosed many gap junctions (FIG. 5b). The frequency and extent of the TJs usually increased with time, but the

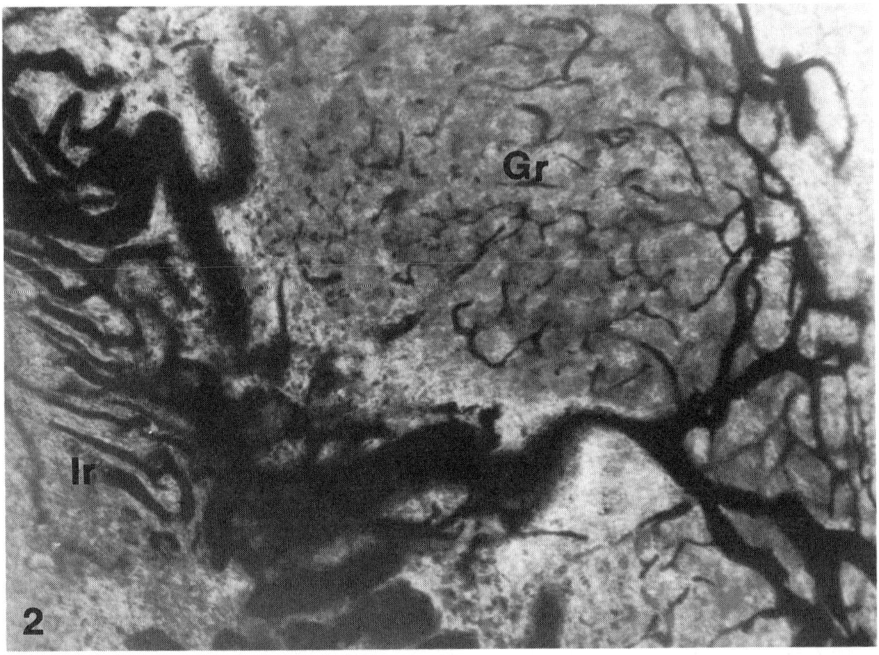

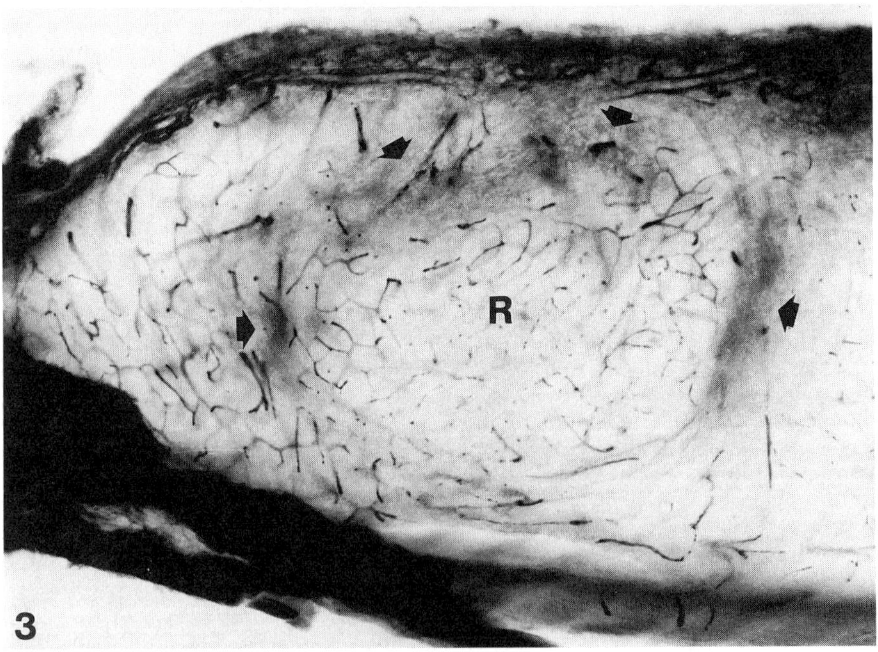

FIGURES 2 AND 3.

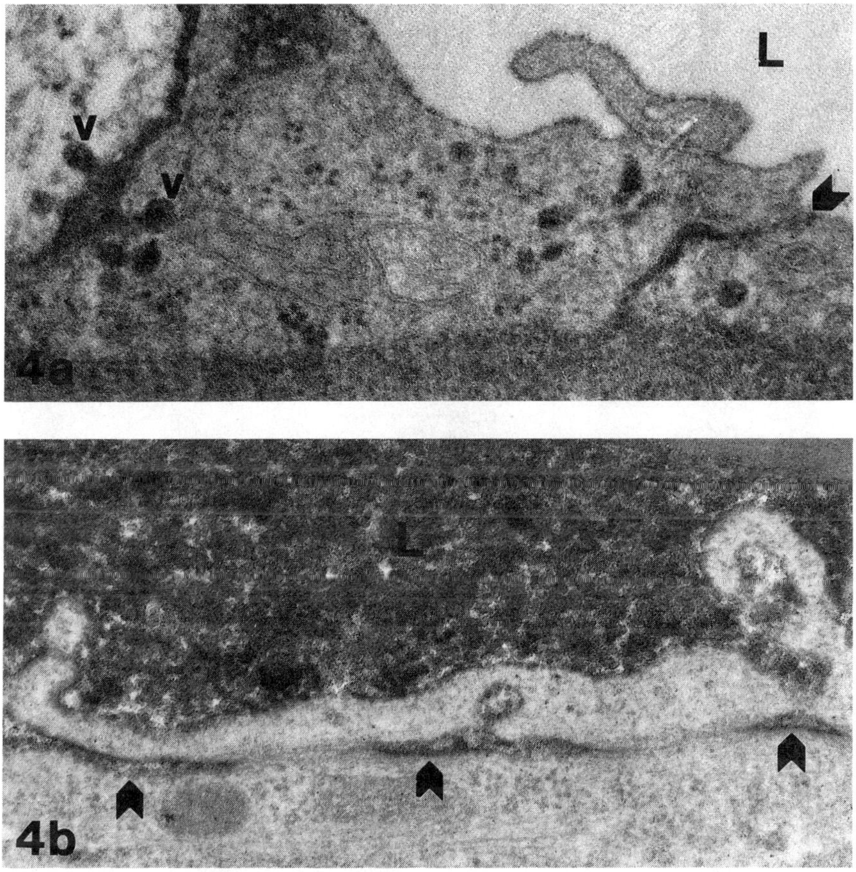

FIGURE 4. Electron micrographs of endothelial tight junctions (TJs) in selected areas where horseradish peroxidase (HRP) tracer penetrated the wall of the graft-nourishing microvessels. HRP reaction product is seen in compartments of TJs (*arrowheads*). L = lumen; En = endothelial cell; v = vesicle. Original magnification (**a**) × 52,000 and (**b**) × 60,000.

patches of TJs remained separate and never became belt-like as did TJs found *in vivo.*[6] The other difference from TJs *in vivo* was the prevalence of gap junctions among TJs *in vitro.* In mature brains, capillaries are devoid of gap junction.[6] In cocultures of endothelial and astroglial cells, the frequency as well as the length,

FIGURE 2. Microvasculature of host iris and branches of fine mesh of microvessel-network-covered cerebellar graft are outlined by horseradish peroxidase (HRP). Gr = graft; Ir = iris. Original magnification × 120; HRP-DAB reaction.

FIGURE 3. In the "rosette" of cerebellar graft, no horseradish peroxidase (HRP) tracer extravasation is seen. Microvessels of the iris and the segments of penetrating branches around the "rosette" are leaky to HRP tracer (*arrows*). R = rosette. Original magnification × 120; HRP-DAB reaction.

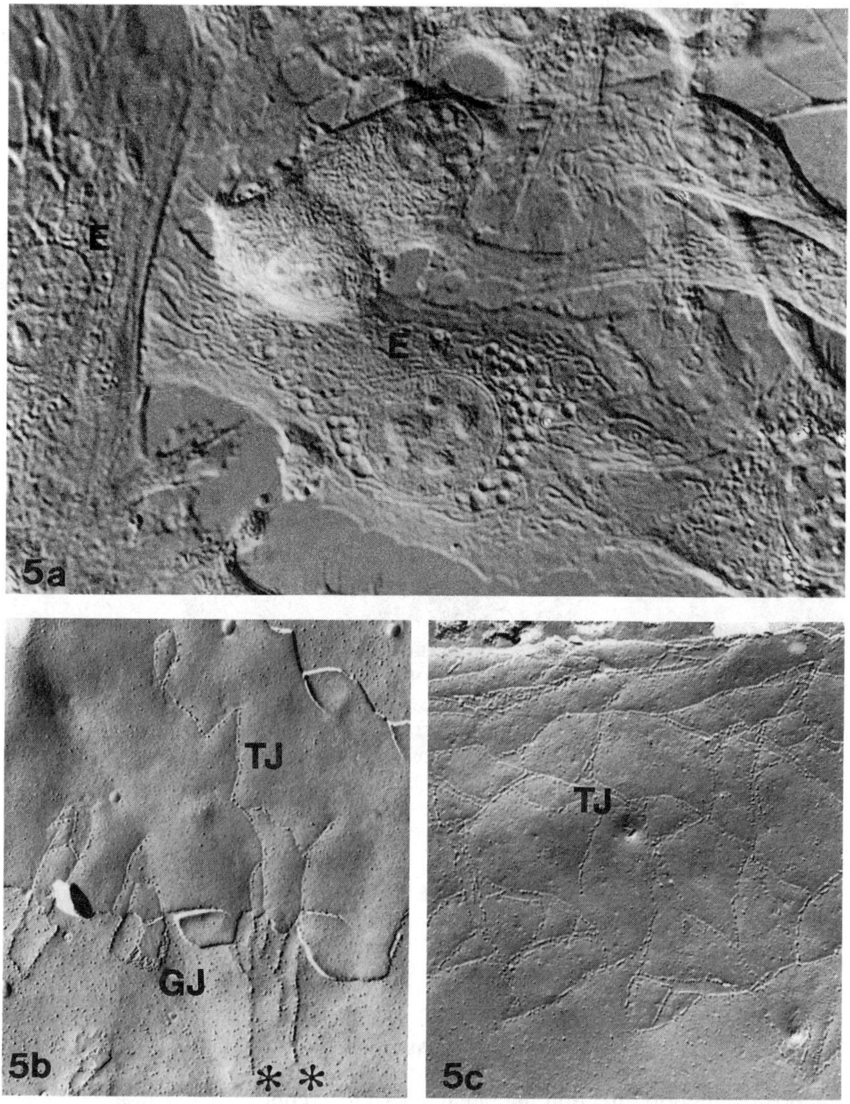

FIGURE 5. (**a**) Light micrograph of endothelial cells (**E**) from primary culture of bovine brain microvessel endothelium. Original magnification × 680. (**b**) Freeze-fracture morphology of TJs of endothelial cell from 6-day-old solo culture of bovine brain endothelium. TJs are short, fragmented, and enclose many gap junctions. TJ = tight junctional strands; GJ = gap junction; *stars* indicate free endings of strands. Original magnification × 40,000. (**c**) Same batch of endothelial cells, cocultured with rat astroglia for 6 days. TJs are longer, broader, and belt-like, similar to the TJs in brain capillary *in vivo.* Original magnification × 43,000.

width, and complexity of TJs increased significantly (FIG. 5c) compared to those in the sister solo endothelial cultures (FIG. 5b).

The enhancement of TJs was evident at 4 or 6 days depending on the coculture method used. In the former case, confluent endothelial cultures were placed facing the astroglial bed. Furthermore, the degree of enhancement increased with time. These augmented TJs formed continuous, complex, multistranded TJs similar to those found in brain capillaries *in vivo.*

Neither muscle cells nor fibroblasts affected the brain endothelial cultures. However, astroglial cells did not induce TJ formation when cocultured with noncerebral endothelial subcultures.

In solo primary astroglial cultures, assemblies were randomly distributed in fluctuating densities (e.g., 5–60 per μm^2 on day 10) (FIG. 6a), whereas in secondary cultures there generally were fewer assemblies (1–3 per μm^2). Assemblies were frequently scattered among the caveolae. In cocultures of astrocytes and brain endothelial cells, aggregations of assemblies were prominent in primary astroglial cultures (FIG. 6b), and an increase in assembly density was demonstrated in secondary cultures. Control, noncerebral endothelium used in these experiments did not induce assembly aggregation or density. However, meningeal cells and rat peripheral fibroblasts did increase the assembly density in astroglial membranes, whereas other control cell types did not.

DISCUSSION

Interaction of astrocytes and endothelial cells has been demonstrated in cerebellar tissue transplants *in vivo* and in endothelial-astroglial cocultures *in vitro.* The barrier properties of graft-host microvessels have been studied thoroughly in different transplantation models in which the brain was the host organ.[20,24–26] In the *in oculo* transplant experiment, the blood-brain barriers of the donor, 15-day, fetal cerebellar tissue were incomplete,[12,27,28] but after 4 weeks of implantation the blood-borne tracer HRP did not penetrate the endothelial barrier in the graft.

The nonbarrier microvessels originating from the iris became barrier capillaries as they penetrated the interior of the graft, where these capillaries were covered with astroglial foot processes.

Astroglial-endothelial interaction was further analyzed in a series of cocultures. The presence of astroglia largely restored the morphological properties of barrier TJ, that is, the formation of long, continuous, uninterrupted, and complex strands. The other effect of astroglia on endothelial TJs was the marked reduction in gap junctions. Presumably, astrocytes affect this junctional enhancement either by actual contact between the endothelium and astroglia or by a factor secreted by the astrocytes.[23,29]

Reciprocally, the function and significance of the assemblies of astroglial cells augmented by brain endothelial cells, meningeal cells, and even fibroblasts are unknown. It is important that *in vivo* the highest densities of assemblies were found in astrocytes facing endothelial or meningeal cells.[30] The difference in frequency and aggregation of assemblies between primary and secondary cultures of astrocytes may reflect a waning response by astrocytes to endothelial influences within the *in vitro* conditions of these experiments. The coculture technique as well as the *in oculo* approach may help delineate the interactions under which these mutual effects are exerted.

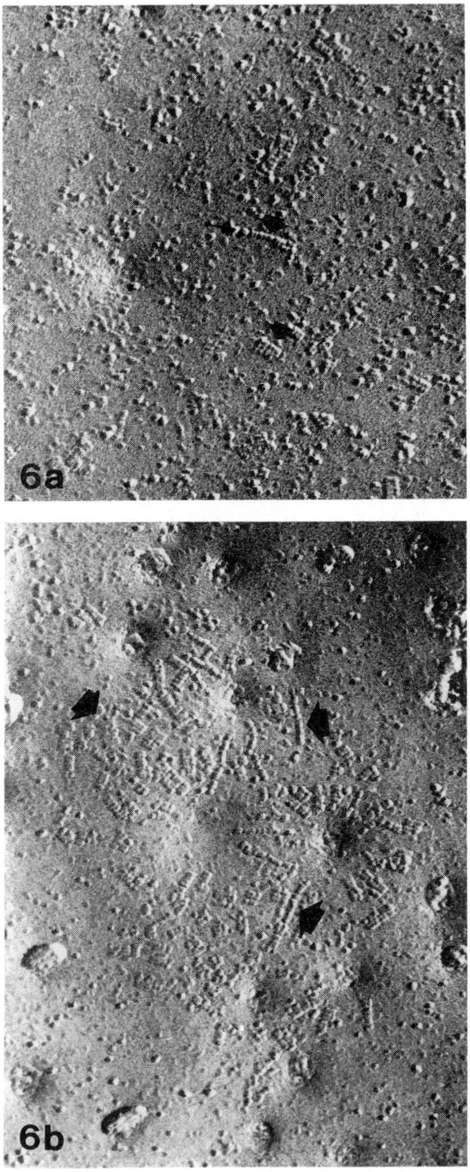

FIGURE 6. Primary astroglial cultures, assembly densities.

REFERENCES

1. Beck, D. W., R. L. Roberts & J. J. Olson. 1986. Glial cells influence membrane-associated enzyme activity at the blood-brain barrier. Brain Res. **381:** 131–137.
2. Debault, L. E. & P. A. Cancilla. 1980. Gamma-glutamyl transpeptidase in isolated brain endothelial cells: Induction by glia cells *in vitro.* Science **207:** 653–655.
3. Nagy, Z. 1990. Cerebral endothelium and the blood-brain barrier. *In* Physiology of the Blood-Brain Barrier. Eds B. B. Johansson & C. Owman, eds. :4–25. Elsevier Sci. Publ. Amsterdam.
4. Pardridge, W. M. 1983. Brain metabolism: A perspective from the blood-brain barrier. Physiol. Rev. **63:** 1481–1535.
5. Reese, T. S. & M. J. Karnovsky. 1967. Fine structural localisation of blood-brain barrier to exogenous peroxidase. J. Cell Biol. **34:** 207–217.
6. Nagy, Z., H. Peters & I. Huttner. 1984. Fracture faces of cell junctions in cerebral endothelium during normal and hyperosmotic conditions. Lab. Invest. **50:** 313–322.
7. Johanson, C. E. 1980. Permeability and vascularity of the developing brain: Cerebellum versus cerebral cortex. Brain Res. **190:** 3–16.
8. Luciano, D. S. 1968. Sodium movement across the blood-brain barrier in newborn and adult rats and its autoradiographic localization. Brain Res. **9:** 334–350.
9. Lossinsky, A. S., A. V. Vorbrodt & H. M. Wisniewski. 1984. Ultracytochemical study of transendothelial transport in cerebral micro-blood vessels during development (abstr.). J. Cell Biol. **99:** 283.
10. Stewart, P. A., E. M. Hayakawa & K. Hayakawa. 1985. Structural endothelial changes that underlie the maturation of the blood-brain barrier. Soc. Neurosci. Abstr. **II(I):** 674.
11. Bradbury, M. 1979. Why a blood-brain barrier? Trends Neurosci. **2:** 36–38.
12. Stewart, P. A. & M. J. Wiley. 1981. Developing nervous tissue induces formation of blood-brain barrier characteristics in indiving endothelial cells. Dev. Biol. **84:** 183–192.
13. Janzer, R. C. & M. C. Raff. 1987. Astrocytes induce blood-brain barrier properties in endothelial cells. Nature **325:** 253–257.
14. Hoffer, B., A. Sieger, T. Ljungberg & L. Olson. 1974. Electrophysiological and cytological studies of brain homografts in the anterior chamber of the eye: Maturation of the cerebellar cortex in oculo. Brain Res. **79:** 165–184.
15. Bragin, A. G. & O. S. Vinogradova. 1983. Comparison of neuronal activity in septal and hippocampal grafts developing in the anterior eye chamber. Dev. Brain Res. **10:** 279–286.
16. Dunnett, S. B., I. Q. Whishaw, T. S. Bunch & A. Fine. 1986. Acetylcholine-rich neuronal grafts in the forebrain of rats: Effects of environmental enrichment, neonatal noradrenaline depletion, host transplantation site and regional source of embryonic donor cells on graft size and acetylcholinesterase positive fibre outgrowth. Brain Res. **378:** 357–373.
17. Jensen, S., T. Sorensen, B. R. Finsen & J. Zimmer. 1989. Transplantation of immature murine brain tissue into the anterior eye-chamber of adult rats. Neurosci. Letts. **100:** 17–22.
18. Takacs, J., T. M. Nhon & J. Hamori. 1986. Electron microscopical study of synaptic glomeruli in cerebellum transplanted to the anterior eye chamber. Acta Biol. Hung. **37:** 259–276.
19. Bowman, P. D., S. R. Ennis, K. E. Rarey, A. L. Betz & G. W. Goldstein. 1983. Brain microvessel endothelial cells in tissue culture. A model for study of blood-brain barrier permeability. Ann. Neurol. **14:** 396–402.
20. Voyta, J. C., D. P. Via, C. E. Butterfield & B. R. Zetter. 1984. Identification and isolation of endothelial cells based on their increased uptake of acetylated-low density lipoprotein. J. Cell Biol. **99:** 2034–2040.
21. McCarthy, M. D. & J. de Vellis. 1980. Preparation of separate astroglial and oligodendroglial cell cultures from rat cerebral tissue. J. Cell Biol. **85:** 890–902.
22. Tao-Cheng, J.-H., Z. Nagy & M. W. Brightman. 1990. Astrocytic orthogonal arrays of intramembranous particle are modulated by brain endothelial cells *in vitro.* J. Neurocytol. **19:** 143–153.

23. TAO-CHENG, J.-H., Z. NAGY & M. W. BRIGHTMAN. 1987. Tight junctions of brain endothelium *in vitro* are enhanced by astroglia. J. Neurosci. **7**(10): 3293–3299.
24. ROSENSTEIN, J. M. & M. W. BRIGHTMAN. 1986. Alteration of the blood-brain barrier after transplantation of autonomic ganglia into the mammalian central nervous system. J. Comp. Neur. **250:** 339–351.
25. ROSENSTEIN, J. M. 1987. Adrenal medulla grafts produce blood-brain barrier dysfunction. Brain Res. **414:** 192–196.
26. SHUPECK, M., K. K. WARD, J. D. SCHMELZER & P. A. LOW. 1989. Comparison of nerve regeneration in vascularized and conventional grafts: Nerve electrophysiology, norepinephrine prostacyclin, malondialdehyde, and the blood-brain barrier. Brain Res. **493:** 225–230.
27. JOHANSON, C. E. 1989. Ontogeny and phylogeny of the blood-brain barrier. *In* Implications of the Blood-Brain Barrier and Its Manipulation. E. A. Neuwelt, ed. : 223–253. Plenum Med. Book Co. New York, London.
28. MOLLGARD, K. & N. R. SAUNDERS. 1975. Complex tight junctions of epithelial and endothelial cells in early foetal brain. J. Neurocytol. **4:** 453–468.
29. DEHOUCK, M. P., S. MERESSE, P. DELORME, J. C. FRICHART & R. CECCHELLI. 1990. An easier, reproducible, and mass-production method to study the blood-brain barrier *in vitro.* J. Neurochem. **54:** 1798– 1801.
30. ANDERS, J. J. & M. W. BRIGHTMAN. 1982. Particle assemblies in astrocytic plasma membranes are rearranged by various agents *in vitro* and cold injury *in vivo.* J. Neurocytol. **11:** 1009–1029.

Induction of Blood-Brain Barrier Endothelial Cell Differentiation[a]

WERNER RISAU

Max-Planck-Institut für Psychiatrie
Am Klopferspitz 18a
D-8033 Martinsried, FRG

The vascular system is a complex network of vessels connecting tissues and organs in the body. Endothelial cells form the inner lining of all blood vessels. These cells have common properties, such as a nonthrombogenic luminal surface, abluminal basement membrane, and von-Willebrand-Factor production. However, endothelial cells are functionally and morphologically heterogeneous in different tissues, which is important for tissue and organ function. On the basis of morphology they have been subdivided into three groups: continuous, fenestrated, and discontinuous capillaries.[1]

Capillaries in the central nervous system (CNS) that form the blood-brain barrier belong to the group of continuous capillaries. In contrast to those in lower animals (see preceding articles in this volume) the endothelial cells of brain capillaries form the blood-brain barrier in higher animals. This was established by Reese and Karnovsky[2] and Brightman and Reese[3] who showed that in rat brain capillaries the permeability barrier was located at the tight junctions between brain endothelial cells. (For review see refs. 4 and 5.) FIGURE 1 shows cross-sections of brain capillaries delineating the morphology and cellular constituents of the mammalian blood-brain barrier.

Three major functions are implicated in the term blood-brain barrier: (1) protection of the brain from the blood milieu as achieved by complex tight junctions between endothelial cells, which prevent the unspecific passage of molecules; (2) selective transport by specialized transport systems[4–6]; and (3) metabolism or modification of blood- or brain-borne substances.[7]

In certain areas of the brain, however, endothelial cells do not form tight junctions but are fenestrated and allow free exchange of molecules between the blood and adjacent neurons. Most of these areas are situated close to the ventricle and are therefore called circumventricular organs. They include the median eminence (hypothalamus), pituitary, choroid plexus, pineal gland, subfornical organ, organum vasculosum lamina terminalis, and area postrema. (For review see refs. 8 and 9.)

In these regions, direct exposure of certain neurons to the blood is of functional importance. For example, neuroendocrine hormones that are released from specialized nerve terminals can freely enter the blood and influence the activity of distant target organs (e.g., endocrine glands). Conversely, these neurons "sense" the level of circulating hormones, resulting in an adjustment of their activity by feedback control. To seal these areas from adjacent nervous tissue and cerebrospinal fluid, barriers of a different structure have developed. Here specialized ependymal and glial cells (in some organs called tanycytes) build up a barrier.

[a] This research is supported by the Max-Planck Gesellschaft and the Deutsche Forschungsgemeinschaft.

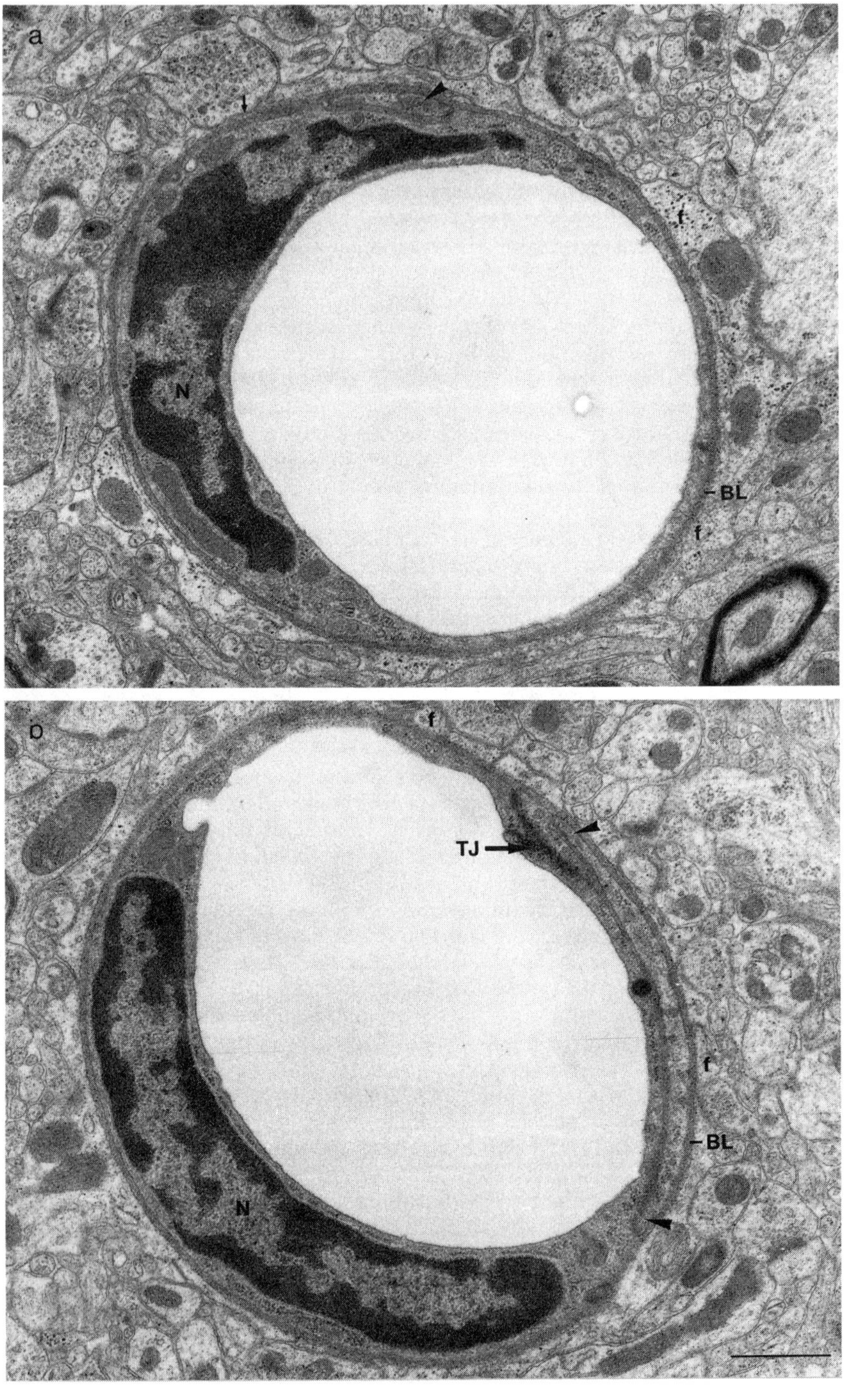

FIGURE 1.

This review focuses on the induction of the endothelial blood-brain barrier. To understand the cellular and molecular mechanisms involved in the establishment of a functional blood-brain barrier it is necessary to investigate the origin and differentiation of the cells concerned.

DEVELOPMENT OF THE BRAIN VASCULAR SYSTEM

A primary vascular plexus surrounds the neural tube very early in embryonic neural development (rat, day 10; chick, day 3). This vascular plexus is derived from migratory angioblasts, the precursors of embryonic blood vessels, that have invaded the head region.[10] From the perineural vascular plexus capillary sprouts invade the neuroectoderm. The endothelial cells first locally degrade the perineural basement membrane and then migrate deep into the neuroectoderm and branch in the subependymal layer. This pattern and subsequent secondary and tertiary invasions are highly reproducible.[10,11]

Endothelial cell proliferation is maximal in the chick brain around embryonic day 14, in the rat at postnatal days 6–8.[12] It then ceases, and the turnover is very low in the adult brain. The formation of new capillaries from preexisting blood vessels (i.e., angiogenesis) in the brain seems to be regulated by soluble factors produced by the embryonic brain. An endothelial cell growth and angiogenic factor has been isolated from embryonic chick brain and found to be identical to fibroblast growth factor.[13,14] It remains to be seen whether this factor is responsible, either alone or in combination with other factors, for the regulation of angiogenesis in the brain.

There are two possible ways in which endothelial cells could differentiate into a blood-brain barrier: (1) Predetermined neuroectoderm-derived endothelial precursor cells (angioblasts) could differentiate into the blood-brain barrier endothelium, or (2) endothelial cells from outside the CNS could invade the neuroectoderm and differentiate in response to the neural environment. The morphological studies just mentioned indicate that endothelial cells of brain capillaries are not derived from the neuroectoderm. Furthermore, embryonic brain tissue transplanted into ectopic sites of host embryos of a different species has been shown to induce angiogenesis from the host. The host origin of endothelial cells in the grafted tissue was demonstrated by the difference in nuclear morphology between chick and quail cells[15] and by the use of species-specific monoclonal antibodies.[16] Moreover, grafted neural tissue induced blood-brain barrier characteristics in host-derived endothelial cells.

DEVELOPMENT OF THE BLOOD-BRAIN BARRIER

Permeability of the blood-brain barrier is usually determined using inert tracers (i.e., those not taken up by brain endothelial cell carriers or receptors) such as horseradish peroxidase. Analyzing the permeability of CNS capillaries for this

←

FIGURE 1. Cross-sections of brain capillaries. A section of a seamless capillary (**a**) and that of a tight junction (TJ)-containing endothelial cell (**b**) are shown. Astrocytic foot processes (f) are closely attached to the abluminal basal lamina (BL) of the endothelial cells. The *arrow* points to an extremely fine astrocytic process between the endothelial cell basal lamina and a synaptic terminal. Pericytic processes embedded in the vascular basal lamina are indicated by *arrowheads*. N = nucleus of endothelial cell. Bar = 1 μm.

molecule during embryonic chick development showed that the capillaries gradually became impermeable around day 13.[17,18] Similar data have been presented for mouse embryos.[19] It should be emphasized that not all capillaries became impermeable at the same time. Rather, the actual time depended on the anatomical location (the barrier in the spinal cord forms earlier than does that in the telencephalon), the nature of the tracer, and the exact technique used.[17] Some controversy exists about the permeability of fetal barriers,[20-22] but irrespective of some variations in the exact timing of blood-brain barrier formation, it is clear that all capillaries present in the brain before day 13 in the chick and day 15 in the rat lack a mature barrier. Thus, differentiation of the blood-brain barrier seems to be independent of endothelial cell invasion and proliferation because rat brain endothelial cells *in vivo* commence barrier formation around embryonic day 16, while they are still proliferating.

DIFFERENTIATED PROPERTIES OF BRAIN ENDOTHELIAL CELLS *IN VIVO*

The transition from a permeable capillary to a capillary with features of a blood-brain barrier is characterized by several changes in endothelial cell morphology, biochemistry, and function that make these endothelial cells distinct from other endothelium in the body. These unique features allow precise control over the substances that leave or enter the brain.

The complex tight junctions between endothelial cells are primarily responsible for barrier function, and unlike simple tight junctions they provide a high electrical resistance (up to 2,000 $\Omega \cdot cm^2$; ref. 23). This complex network of continuous and branching tight junctions is only found in the capillaries and postcapillary venules of the brain cortex.[24] Furthermore, little bulk flow of molecules occurs through the cells because there are few pinocytotic vesicles and fenestrae. This physiological barrier protects the brain from neurotoxic substances present in the blood, such as glutamate and glycine which are neurotransmitters with extracellular levels in the CNS more than 1000-fold lower than those in the blood, or high potassium levels.

Unrestricted diffusion through the blood-brain barrier is only possible for substances that can cross biologic membranes by virtue of their lipophilic character, even though mechanisms may exist that actively export lipophilic molecules as soon as they have entered brain endothelium. For example, the P-glycoprotein is a transmembrane glycoprotein that actively exports lipophilic molecules out of cells. It confers insensitivity for drugs used for cancer chemotherapy (multi-drug-resistance) to cells that express this transporter and was shown recently to be expressed in the blood-brain barrier.[25]

Metabolism or modification of substances is yet another mechanism by which brain endothelial cells maintain homeostasis of the neuronal microenvironment. For example, blood-derived DOPA is metabolized by brain endothelial cells.[7]

Specific carrier systems that allow control over the exchange of substances between blood and nervous system are present in brain endothelial cells. As shown schematically in FIGURE 2, facilitated diffusion (e.g., glucose transporter) and active (ATP-dependent) transport mechanisms exist in the blood-brain barrier. (For a review of transport at the blood-brain barrier see refs. 4, 6, and 26.) It is still controversial if transcytosis via clathrin-coated vesicles occurs in brain endothelial cells.[26] The density of pinocytotic vesicles is remarkably low in brain endothelial cells, but clathrin-coated vesicles can be found. Evidence that the transferrin receptor and low density lipoprotein receptor may act as transcytosing receptors at the blood-brain barrier has been presented,[27,28] but direct evidence is still lacking. An endothelial model system equivalent to the MDCK kidney epithelial system in which transcytosis by the poly-Ig and Fc receptors has been demonstrated[29] would be very valuable.

Several proteins have been localized to brain capillaries *in vivo* (e.g., transferrin receptor,[30] gamma-glutamyl-transpeptidase,[31] alkaline phosphatase,[32] and butyrylcho-

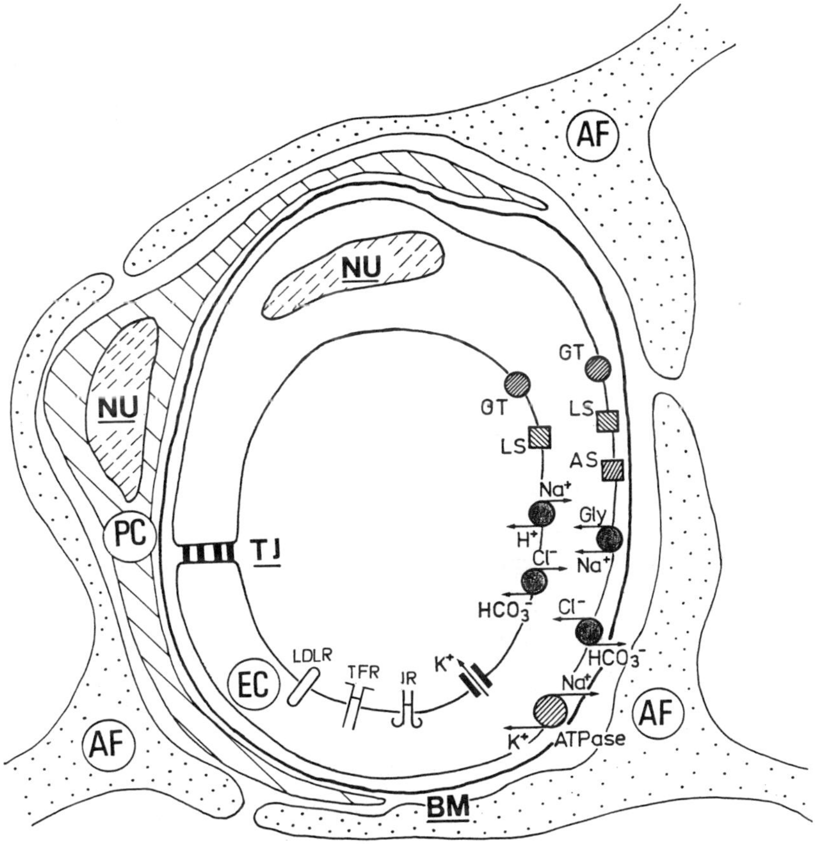

FIGURE 2. Schematic representation of a cross-section of a brain capillary. AF = astrocytic foot process; AS = A-system of amino acid transport (alanine-preferring); BM = basement membrane; EC = endothelial cell; GT = glucose transporter; IR = insulin receptor; LDLR = low density lipoprotein receptor; LS = L-system of amino acid transport (leucine-preferring); NU = nucleus; PC = pericyte; TFR = transferrin receptor; TJ = tight junctions. Only some receptors are shown, and the identity and distribution of some transporters have not yet been conclusively demonstrated.

linesterase.[33] By analogy to other cell types that express these proteins, they have been implicated in transport or metabolic processes at the blood-brain barrier. However, direct evidence for such a role is lacking. Furthermore, proteins have been identified in brain capillaries using monoclonal antibodies.[16,34–36] Some of these proteins may be blood-brain barrier specific in the sense that they are present at a much reduced or undetectable level in permeable capillaries of the body. It is unlikely that there are blood-brain barrier constituents that are only present in brain endothelial cells and in no other cell type. On the other hand, proteins present in other endothelial cells seem to be absent from brain endothelium. For example, the acetylated low density lipoprotein (acLDL) receptor,[37] thrombomodulin,[38] and the OX43 antigen[39] seem to be down-regulated in brain endothelial cells.

During embryonic and postnatal development, interendothelial junctional changes and the expression in brain endothelial cells of potential transport proteins and antigens have been correlated with the development of the permeability barrier.[16,19,40,41] This suggests a coordinated induction of complex tight junctions[41] and transporters.

PERICYTES AND THE BLOOD-BRAIN BARRIER

The barrier represents a complex cellular system that consists of endothelial cells, pericytes, perivascular microglia, astrocytes, and basal laminae (FIG. 1). Although the endothelial cell layer forms the barrier proper, the interaction of all cells may be necessary for induction or maintenance of a functional blood-brain barrier. Pericytes are adventitial (also called mural) contractile cells that in the capillary are the counterpart of large vessel smooth muscle cells.[42] They are embedded in the abluminal basal lamina of endothelial cells and extend long processes around the capillary tubes. The function of these cells is unclear, but evidence exists that they are important for the growth control of endothelial cells, the integrity of capillaries, and transport and possibly phagocytic processes.[42,43]

Pericytes seem to cover neural capillaries more than other capillaries.[44] Furthermore, pericytes in the brain seem to express certain proteins that are not present in pericytes in other organs,[45] suggesting that pericytes may have specific functions in brain capillaries.

Pericytes are closely associated with endothelial cells already at very early stages of development.[44] It is unclear whether brain pericytes differentiate from neuroectodermal, meningeal, or endothelial cells. In other tissues, pericytes, like smooth muscle cells, have been shown to differentiate from perivascular fibroblasts (graft cells in chick-quail chimeras[46,47]).

ASTROCYTES INDUCE THE BLOOD-BRAIN BARRIER *IN VIVO*

Using chick-quail transplantation experiments Stewart and Wiley[15] were the first to unambiguously demonstrate that blood-brain barrier characteristics can be induced in endothelial cells that had invaded brain transplants. Using the chick-specific monoclonal HT7 antibody we also observed the expression of this blood-brain barrier-specific antigen in chick endothelial cells that had invaded embryonic mouse brain transplanted to the chick chorioallantoic membrane.[16] Conversely, brain capillaries became permeable after invasion of somite transplants.[15] These results indicate that organ-specific characteristics of endothelial cells may be induced and maintained by the local environment.

A cross-section of a brain capillary (FIG. 1) shows that astrocytic endfeet are in close proximity to the endothelial cell plasma membrane and are separated only by the basal lamina. Certainly, in many sections pericytes and microglial cells are found between astrocytes and endothelial cells, but it is probable that each capillary endothelial cell encounters many direct contacts with the astrocytic processes. The differentiation of type-1 astrocytes (identified by glial fibrillary acidic protein [GFAP]) and the formation of the astroglial perivascular sheet occur at the same time as the permeability barrier develops.[48,49] This indicates that astrocytes are involved in blood-brain barrier formation.

Janzer and Raff[50] have provided direct evidence that purified type-1 astrocytes (when transplanted into the rat anterior eye chamber or the chick chorioallantoic membrane) induce a permeability barrier in invading endothelial cells. The fact that in many brain tumors morphological irregularities of the perivascular ensheathment (enlarged perivascular space, gaps in the basal lamina, and unusual or lacking glial investment) correlate with a breakdown of the blood-brain barrier (for a review see ref. 51) suggests that astrocytes are also necessary for the maintenance of the blood-brain barrier. In addition, it supports the concept that endothelial cell differentiation can be modulated by organ and tissue environments.

A peculiar membrane specialization is found at the gliovascular interface in astrocytic endfeet, the so-called orthogonal arrays of particles (OAP) or assemblies. The function of the OAP is unknown. As both the OAP and potassium conductance are distributed over the glial surface in a similar pattern, particularly in astrocytes and retinal Müller cells, these data support the idea that OAPs correspond to symmetric high conductance potassium channels.[52–54] Potassium homeostasis is crucial for brain function and is an important aspect of the blood-brain barrier. Interestingly, OAPs appear during ontogenesis when the blood brain barrier matures, suggesting that astrocyte differentiation, defined by general markers such as GFAP, and specialized properties of astrocytes such as the polarized distribution of OAP in their plasma membrane, are coupled with blood-brain barrier differentiation.

BLOOD-BRAIN BARRIER DIFFERENTIATION *IN VITRO*

Although methods to isolate brain capillaries and culture the derived cells have been available for more than 10 years,[55] it is still not a simple task to obtain pure brain endothelial cell cultures. Two basic methods have been used to isolate capillaries from cerebral cortex; one involves mechanical homogenization,[55–58] the other collagenase digestion.[59–61] The mechanical methods result in many fewer living cells compared to the enzymatic methods.[59,60] However, good endothelial cell cultures have been obtained using either method. Inasmuch as brain capillaries still consist of endothelial cells, pericytes, perivascular microglia, and astrocytes, all these cell types are present in isolated capillaries. Density gradients have been used to isolate endothelial cells from capillary fragments. From most of the recent studies it appears that pericytes represent the major contaminating cell type in derived cultures.[57,61–63] *In vitro* studies depend on the availability of a large number of pure endothelial cells in culture. As pericytes inhibit endothelial cell growth and disrupt endothelial monolayers, methods have to be devised to get rid of these cells. Two techniques have been successfully used to clean pericytes from primary endothelial cell cultures. One method employs the cloning of endothelial cells derived from bovine brain,[56,58,64] the other takes advantage of the fact that pericytes, but not endothelial cells, from rat brain express the Thy 1.1 antigen on their surface and can be killed by antibody-mediated complement lysis.[61]

These cultures offered the possibility of examining the maintenance of the differentiated state of brain endothelial cells which is characterized by the presence of complex tight junctions, low permeability, specific carrier systems, and expression of antigens. The aim is to maintain blood-brain barrier functions in cultured endothelial cells. This would provide the opportunity to investigate blood-brain barrier biology, physiology, and pathology *in vitro.*

Circumstantial evidence indicates that endothelial cells lose blood-brain barrier characteristics soon after *in vitro* culture possibly because of the withdrawal of astrocytes or astrocyte-derived factors. Reliably pure cultures of astrocytes from neonatal rat brain can be prepared, and these cells have been shown to induce a permeability barrier in endothelial cells *in vivo.*[50] Attempts have been made to reinduce blood-brain barrier properties in endothelial cells by coculturing them with astrocytes or glioma cells.[65,66] In such experiments, astrocytes were found to enhance the frequency of blood-brain barrier-like tight junctions in cultured endothelium and OAP in astrocytes. However, it was difficult to determine the level of induction obtained in these cocultures, and it was controversial whether cell-cell interactions or secreted products or both are needed for the observed effects. We recently observed that astrocytes cultured on one side of collagen filters induced or maintained a high complexity of tight junctions in primary brain endothelial cells cultured on the other side of the filter. Because these filters did not allow the penetration of astrocytic foot processes, we concluded that soluble products were responsible for this effect (J. Neuhaus, H. Wolburg, and W.R., unpublished experiments; see article in this volume).

A strong criterion for tight monolayers is electrical resistance and a permeability barrier for low molecular weight compounds such as sucrose or inulin. Rutten *et al.*[56] were the first to report high electrical resistance of endothelial cells *in vitro* over many population doublings. Similar results were published recently by Dehouck *et al.*[58] who also showed a permeability barrier for sucrose. This is consistent with our results. Therefore, some barrier characteristics seem to be maintained by cultured endothelial cells, although the complexity of tight junctions and the electrical resistance were not maintained at a similar level as that *in vivo.*

The question was whether other properties such as expression of specific proteins and transporters were maintained.[67,68] Several proteins were found to be present in brain capillaries but absent from other capillaries in the body. Consequently, they were proposed to be specific markers of the blood-brain barrier. Gamma-glutamyltranspeptidase (GGT) has been most widely used as a marker for brain capillaries and differentiated properties of brain endothelial cells.[31,60,69] However, our results showed that in primary cultures the majority of GGT activity is associated with pericytes, not endothelial cells.[64] Evidence indicates that GGT can be induced in brain endothelial cells *in vitro,*[66,70] most convincingly recently by Dehouck *et al.*[58] showing induction in cloned brain endothelial cells by coculturing with astrocytes. It has to be kept in mind, however, that GGT in brain capillaries is not expressed in all species possessing an endothelial barrier[19,71] and is also found in permeable capillaries.[72] The unexpected presence of GGT in brain pericytes supports the idea[45] that not only endothelial cells but also pericytes may have organ-specific characteristics. Although the function of GGT in brain capillaries remains unclear, our results suggest that the presence of this enzyme in a different cell type, the pericyte, necessitates a reconsideration of its significance for the blood-brain barrier.

ZO-1 is a monoclonal antibody against a 225-kD protein associated with the cytoplasmic portion of tight junctions.[73] The protein is present in mouse brain capillaries *in vivo* (FIG. 3a). In cultured mouse brain endothelial cells the protein was present on the plasma membrane (FIG. 3b and c) and was still expressed in cells after three passages. Recently, Stevenson *et al.*[74] showed that the amount of ZO-1 protein did not correlate with junctional complexity and electrical resistance in MDCK cells. They suggested that the phosphorylation state of this protein might be important for

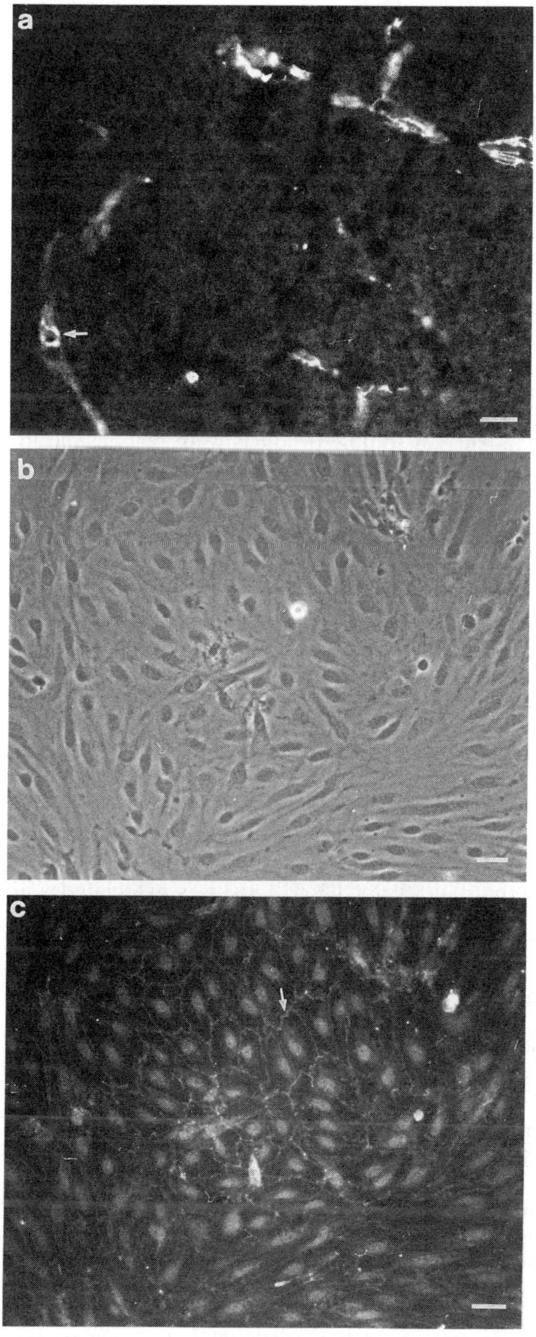

FIGURE 3. ZO-1 antigen is present in capillaries of the mouse brain. (**a**) Frozen section. In cultured mouse brain endothelial cells the protein is present at the cellular borders between cells (*arrow* in **c**; indirect immunofluorescence; **b,** phase contrast of the same field; 5-day-old primary culture). The nuclear fluorescence is an artifact due to a cross-reaction with the secondary goat anti-rat antibodies. Bars represent 50 μm.

the determination of the "tightness" of junctions. Thus, the regulation of protein phosphorylation in brain endothelial cells may be of major importance for the regulation of blood-brain barrier properties. These observations are consistent with our results using brain endothelial cells. We conclude that the mere presence of the ZO-1 protein defined by the monoclonal antibodies is not diagnostic for blood-brain barrier properties of endothelial cells.

The HT7 protein is a cell surface immunoglobulin-like glycoprotein and a marker for blood-brain barrier endothelium[16,75,76] (see also Seulberger *et al.,* this volume). It is expressed on the surface of chick brain endothelial cells *in vivo*[16] and *in vitro* (FIG. 4a; see also ref. 75). If these cells were cultured for a longer period of time and eventually formed a confluent monolayer, the protein remained present on those cells (FIG. 4b). Recently, the OX-47[77] and gp42[78] proteins have been identified as homologous proteins in rat and mouse, respectively. We found that the OX-47 protein was present in rat brain endothelial cells throughout the culture period in primary and passaged cells, suggesting that this protein is not down-regulated *in vitro* which is consistent with our results using HT7 antibodies in the chick (HT7 is a species-specific antibody). Compelling evidence indicates that the expression of this protein is modulated *in vivo* and *in vitro.* It is present on erythroblasts but not erythrocytes, and it was induced in endothelial cells that invaded a brain transplant.[16] In addition, the 1W5 protein[79] which is identical to the HT7 protein[75] is induced in neurons *in vitro,*[79] and the OX-47 protein is expressed by activated but not resting T lymphocytes.[77] These results raise the possibility that if down-regulation of this protein by brain endothelial cells occurs during the time in culture, this may be masked by its simultaneous induction by the culture conditions. Thus, at this point it is impossible to distinguish between maintenance of protein expression due to maintenance of differentiated characteristics of brain endothelial cells and induction due to culture conditions. Studies are in progress to address this issue.

The best understood carrier system at the blood-brain barrier is the glucose transporter.[80–82] This erythrocyte-type, insulin-independent glucose transporter (Glut-1[83]) is enriched in brain capillaries and probably regulates the influx of glucose into the brain. In contrast to GGT, the glucose transporter is a marker of blood-brain barrier endothelial cells.[76,82,84] Using immunocytochemistry we found that the transporter was present on the cell surface of endothelial cells in primary cultures but was subsequently dramatically down-regulated on further culture and proliferation of endothelial cells. Coculture of brain endothelial cells with astrocytes, glioma cells, and astrocyte-conditioned medium or varying the glucose concentration in the medium did not reinduce the transporter in endothelial cells.[85] These results show that endothelial cells freshly isolated from the brain express the glucose transporter on the cell surface, which confirms the notion that the glucose transporter is synthesized by brain endothelial cells and may be crucial for brain metabolism. Glucose uptake studies have shown that isolated brain microvessels are endowed with a high density of glucose transporter protein. However, it has not yet been possible to model transport of glucose through a blood-brain barrier-like endothelial monolayer. Our results suggest that expression of the glucose transporter by brain endothelial cells may be regulated by the brain environment in a complex manner dependent on blood-brain barrier-inducing factors (probably produced by astrocytes) and metabolic gradients between plasma and brain.

Whatever the factors are, our results suggest that the specialized properties of blood-brain barrier endothelial cells (tight junction complexity and expression of enzymes and transporters) are differentially regulated by the brain microenvironment. From a developmental point of view, we suggest that a primary event is the induction of complex tight junctions in endothelial cells that have invaded the

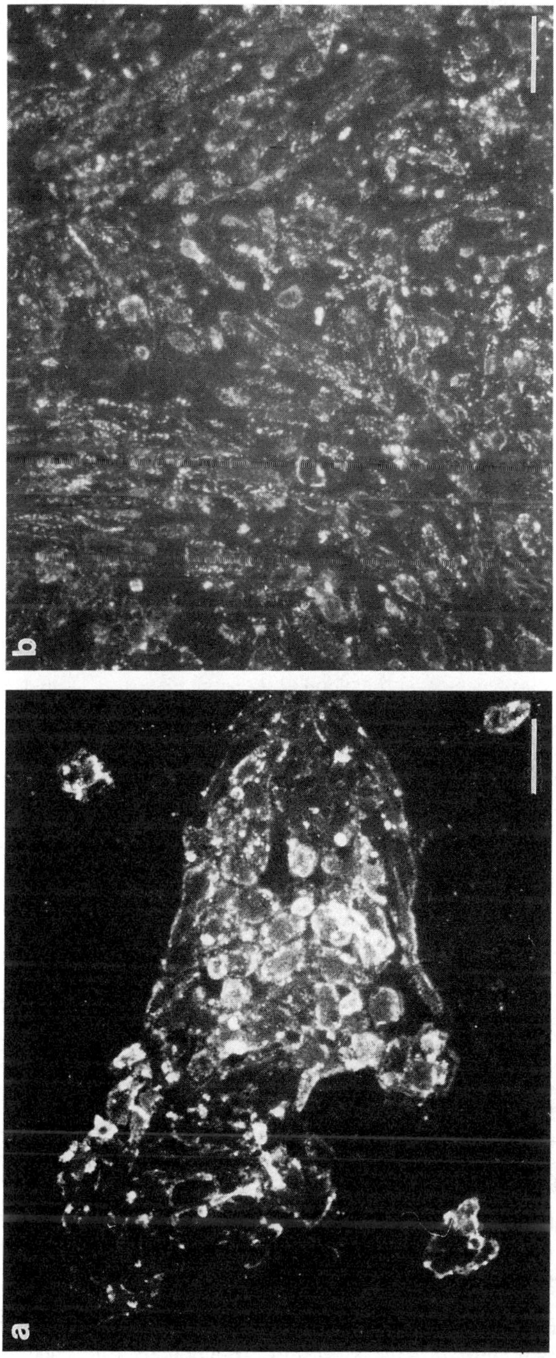

FIGURE 4. Indirect immunofluorescence of chick brain endothelial cells (**a,** 2-day-old; **b,** 9-day-old primary culture) using HT7 monoclonal antibodies. The protein is present on the surface of the cells. Bars represent 50 μm.

brain[5,41] and that subsequent expressions of enzymes and transporters are secondary events regulated by the physiological requirements of the neuronal compartment.

ACKNOWLEDGMENTS

I thank Ursula Albrecht, Andrea Dingler, Mücella Öcalan, Harald Seulberger, and Carolin Unger for their contributions to our blood-brain barrier projects, Hartwig Wolburg for the pictures shown in FIGURE 1 and for discussions, and Bruce Stevenson for ZO-1 monoclonal antibodies.

REFERENCES

1. MAJNO, G. 1965. Ultrastructure of the vascular membrane. *In* Handbook of Physiology, Circulation III. W. F. Hamilton & P. Dow, eds. :2293–2375. Washington. American Physiology Society.
2. REESE, T. S. & M. J. KARNOVSKY. 1967. Fine structural localization of a blood-brain barrier to exogenous peroxidase. J. Cell Biol. **34:** 207–217.
3. BRIGHTMAN, M. W. & T. S. REESE. 1969. Junctions between intimately apposed cell membranes in the vertebrate brain. J. Cell Biol. **40:** 648–677.
4. GOLDSTEIN, G. W. & A. L. BETZ. 1986. The blood-brain barrier. Sci. Am. **255:** 70–79.
5. RISAU, W. & H. WOLBURG. 1990. Development of the blood-brain barrier. Trends Neurosci. **13:** 174–178.
6. BETZ, A. L. & G. W. GOLDSTEIN. 1986. Specialized properties and solute transport in brain capillaries. Ann. Rev. Physiol. **48:** 241–250.
7. HARDEBO, J. E. & C. OWMAN. 1980. Barrier mechanisms for neurotransmitter monoamines and their precursors at the blood-brain barrier interface. Ann. Neurol. **8:** 1–11.
8. LEONHARDT, H. Ependym und circumventriculäre Organe. 1980. *In* Handbuch der mikroskopischen Anatomie des Menschen. A. Oksche & L. Vollrath, eds. :177–666. Springer. Berlin, Heidelberg, New York.
9. BOUCHAUD, C. & O. BOSLER. 1986. The circumventricular organs of the mammalian brain with special reference to monoaminergic innervation. Int. Rev. Cytol. **105:** 283–327.
10. BÄR, T. 1980. The vascular system of the cerebral cortex. Adv. Anat. Embryol. Cell Biol. **59:** 1–62.
11. FEENEY, J. F. JR. & R. L. WATTERSON. 1946. The development of the vascular pattern within the walls of the central nervous system of the chick embryo. J. Morphol. **78:** 231–303.
12. ROBERTSON, P. L., M. DU BOIS, P. D. BOWMAN & G. W. GOLDSTEIN. 1985. Angiogenesis in developing rat brain: An *in vivo* and *in vitro* study. Dev. Brain Res. **23:** 219–223.
13. RISAU, W. 1986. Developing brain produces an angiogenesis factor. Proc. Natl. Acad. Sci. USA **83:** 3855–3859.
14. RISAU, W., P. GAUTSCHI-SOVA & P. BÖHLEN. 1988. Endothelial cell growth factors in embryonic and adult chick brain are related to human acidic fibroblast growth factors. EMBO J. **7:** 959–962.
15. STEWART, P. A. & M. J. WILEY. 1981. Developing nervous tissue induces formation of blood-brain barrier characteristics in invading endothelial cells: A study using quail-chick transplantation chimeras. Dev. Biol. **84:** 183–192.
16. RISAU, W., R. HALLMANN, U. ALBRECHT & S. HENKE-FAHLE. 1986. Brain induces the expression of an early cell surface marker for blood-brain barrier specific endothelium. EMBO J. **5:** 3179–3183.
17. WAKAI, S. & N. HIROKAWA. 1978. Development of the blood-brain barrier to horseradish peroxidase in the chick embryo. Cell Tissue Res. **195:** 195–203.
18. DELORME, P., J. GAYET & G. GRIGNON. 1970. Ultrastructural study on transcapillary exchanges in the developing telencephalon of the chicken. Brain Res. **22:** 269–283.
19. RISAU, W., R. HALLMANN & U. ALBRECHT. 1986. Differentiation-dependent expression of

protein in brain endothelium during development of the blood-brain barrier. Dev. Biol. **117:** 537–545.
20. MØLLGÅRD, K. & N. R. SAUNDERS. 1986. The development of the human blood-brain and blood-CSF barriers. Neuropath. Appl. Neurobiol. **12:** 337–358.
21. SAUNDERS, N. R., K. M. DZIEGIELEWSKA & K. MOLLGARD. 1991. Letter to the editor. Trends Neurosci. **14:** 14.
22. RISAU, W. & H. WOLBURG. 1991. Reply. Trends Neurosci. **14:** 15.
23. CRONE, C. & S.-P. OLESEN. 1982. Electrical resistance of brain microvascular endothelium. Brain Res. **241:** 49–55.
24. NAGY, Z., H. PETERS & I. HÜTTNER. 1984. Fracture faces of cell junctions in cerebral endothelium during normal and hyperosmotic conditions. Lab. Invest. **50:** 313–322.
25. CORDON-CARDO, C., J. P. O'BRIEN, D. CASALS, L. RITTMAN-GRAUER, J. L. BIEDLER, M. R. MELAMED & J. R. BERTINO. 1989. Multidrug-resistance gene (P-glycoprotein) is expressed by endothelial cells at blood-brain barrier sites. Proc. Natl. Acad. Sci. USA **86:** 695–698.
26. BROADWELL, R. D. 1989. Transcytosis of macromolecules through the blood-brain barrier: A cell biological perspective and critical appraisal. Acta Neuropathol. **79:** 117–128.
27. DUFFY, K. R. & W. M. PARDRIDGE. 1987. Blood-brain barrier transcytosis of insulin in developing rabbits. Brain Res. **420:** 32–38.
28. MERESSE, S., C. DELBART, J. C. FRUCHART & R. CECCHELLI. 1989. Low-density lipoprotein receptor on endothelium of brain capillaries. J. Neurochem. **53:** 340–345.
29. MOSTOV, K. E. & N. E. SIMISTER. 1985. Transcytosis. Cell **43:** 389–390.
30. JEFFERIES, W. A., M. R. BRANDON, S. V. HUNT, A. F. WILLIAMS, K. C. GATTER & D. Y. MASON. 1984. Transferrin receptor on endothelium of brain capillaries. Nature **312:** 162–163.
31. ORLOWSKI, M., G. SESSA & J. P. GREEN. 1974. Gamma-glutamyl transpeptidase in brain capillaries: Possible site of a blood-brain barrier for amino acids. Science **184:** 66–68.
32. LEDUC, E. H. & G. B. WISLOCKI. 1952. The histochemical localization of acid and alkaline phosphatases, non-specific esterase and succinic dehydrogenase in the structures comprising the hematoencephalic barrier of the rat. J. Comp. Neurol. **97:** 242–279.
33. JOÓ, F. & B. CSILLIK. 1966. Topographic correlation between the hemato-encephalic barrier and the cholinesterase activity of brain capillaries. Exp. Brain Res. **1:** 147–151.
34. MICHALAK, T., F. P. WHITE, A. L. GARD & G. R. DUTTON. 1986. A monoclonal antibody to the endothelium of rat brain microvessels. Brain Res. **379:** 320–328.
35. STERNBERGER, N. H. & L. A. STERNBERGER. 1987. Blood-brain barrier protein recognized by monoclonal antibody. Proc. Natl. Acad. Sci. USA **84:** 8169–8173.
36. PARDRIDGE, W. M., J. YANG, J. EISENBERG & L. J. MIETUS. 1986. Antibodies to blood-brain barrier bind selectively to brain capillary endothelial lateral membranes and to a 46K protein. J. Cereb. Blood Flow Metab. **6:** 203–211.
37. GAFFNEY, J., D. WEST, F. ARNOLD, A. SATTAR & S. KUMAR. 1985. Differences in the uptake of modified low density lipopoteins by tissue cultured endothelial cells. J. Cell Sci. **79:** 317–325.
38. ISHII, H., H. H. SALEM, C. E. BELL, E. A. LAPOSATA & P. W. MAJERUS. 1986. Thrombomodulin, an endothelial anticoagulant protein, is absent from the human brain. Blood **67:** 362–365.
39. ROBINSON, A. P., T. M. WHITE & D. W. MASON. 1986. MRC OX-43: A monoclonal antibody which reacts with all vascular endothelium in the rat except that of brain capillaries. Immunology **57:** 231–237.
40. VORBRODT, A. W., A. S. LOSSINSKY & H. M. WISNIEWSKI. 1986. Localization of alkaline phosphatase activity in endothelial of developing and mature mouse blood-brain barrier. Dev. Neurosci. **8:** 1–13.
41. STEWART, P. A. & E. M. HAYAKAWA. 1987. Interendothelial junctional changes underlie the developmental 'tightening' of the blood-brain barrier. Dev. Brain Res. **32:** 271–281.
42. SIMS, D. E. 1986. The Pericyte—A review. Tissue & Cell **18:** 153–174.
43. FARRELL, C. R., P. A. STEWART, C. L. FARRELL & R. F. DEL MAESTRO. 1987. Pericytes in human cerebral microvasculature. Anat. Rec. **218:** 466–469.
44. SIMIONESCU, M., N. GHINEA, A. FIXMAN, M. LASSER, L. KUKES, N. SIMIONESCU & G. E.

PALADE. 1988. The cerebral microvasculature of the rat: Structure and luminal surface properties during early development. J. Submicrosc. Cytol. **20:** 243–261.

45. KRAUSE, D., B. VATTER & R. DERMIETZEL. 1988. Immunochemical and immunocytochemical characterization of a novel monoclonal antibody. Cell Tissue Res. **252:** 543–555.
46. JOTEREAU, F. & N. M. LE DOUARIN. 1978. The developmental relationship between osteocytes and osteoclasts: A study using the quail-chick nuclear marker in endochondral ossification. Dev. Biol. **63:** 253–265.
47. PARDANAUD, L., F. YASSINE & F. DIETERLENLIEVRE. 1989. Relationship between vasculogenesis, angiogenesis and hematopoiesis during avian ontogeny. Development **105:** 473–485.
48. PHELPS, C. H. 1972. The development of glio-vascular relationships in the rat spinal cord. Z. Zellforsch. **128:** 555–563.
49. SENJO, M., T. ISHIBASHI, T. TERASHIMA & Y. INOUE. 1986. Correlation between astrogliogenesis and blood-brain barrier formation: Immunocytochemical demonstration by using astroglia-specific enzyme glutathione S-transferase. Neurosci. Lett. **66:** 39–42.
50. JANZER, R. C. & M. C. RAFF. 1987. Astrocytes induce blood-brain barrier properties in endothelial cells. Nature **325:** 253–257.
51. GREIG, N. H. 1988. *In* Implications of the Blood-Brain Barrier and Its Manipulation. E. A. Neuwelt, ed. :77– 106. Plenum Medical Book Company. New York.
52. NEWMAN, E. A. 1985. J. Neurosci. **5:** 2225–2239.
53. NILIUS, B. & A. REICHENBACH. 1988. Pfluegers Arch. **411:** 654–660.
54. WOLBURG, H. & K. BERG. 1988. Glia **1:** 246–252.
55. JOÓ, F. 1985. The blood-brain barrier *in vitro:* Ten years of research on microvessels isolated from the brain. Neurochem. Int. **7:** 1–25.
56. RUTTEN, M. J., R. L. HOOVER & M. J. KARNOVSKY. 1987. Electrical resistance and macromolecular permeability of brain endothelial monolayer cultures. Brain Res. **425:** 301–310.
57. CARSON, M. P. & C. C. HAUDENSCHILD. 1986. Microvascular endothelium and pericytes: High yield, low passage cultures. In Vitro Cell Dev. Biol. **22:** 344–354.
58. DEHOUCK, M. P., S. MERESSE, P. DELORME, J. C. FRUCHART & R. CECCHELLI. 1990. An easier, reproducible, and mass-production method to study the blood-brain-barrier *in vitro.* J. Neurochem. **54:** 1798–1801.
59. WILLIAMS, S. K., J. F. GILLIS, M. A. MATTHEWS, R. C. WAGNER & M. W. BITENSKY. 1980. Isolation and characterization of brain endothelial cells: Morphology and enzyme activity. J. Neurochem. **35:** 374–381.
60. MISCHECK, U., J. MEYER & H.-J. GALLA. 1989. Characterization of gamma-glutamyl transpeptidase activity of cultured endothelial cells from porcine brain capillaries. Cell Tissue Res. **256:** 221–226.
61. RISAU, W., B. ENGELHARDT & H. WEKERLE. 1990. Immune function of the blood-brain barrier: Incomplete presentation of protein (auto-)antigens by rat brain microvascular endothelium *in vitro.* J. Cell Biol. **110:** 1757–1766.
62. RUPNICK, M. A., A. CAREY & S. K. WILLIAMS. 1988. Phenotypic diversity in cultured cerebral microvascular endothelial cells. In Vitro Cell Dev. Biol. **24:** 435–444.
63. MERESSE, S., M. P. DEHOUCK, P. DELORME, M. BENSAID, J. P. TAUBER, C. DELBART, J. C. FRUCHART & R. CECCHELLI. 1989. Bovine brain endothelial-cells express tight junctions and monoamine-oxidase activity in long-term culture. J. Neurochem. **53:** 1363–1371.
64. RISAU, W., A. DINGLER, U. ALBRECHT, M.-P. DEHOUCK & R. CECCHELLI. 1991. Blood-brain barrier pericytes are the main source of gamma-glutamyltranspeptidase activity in brain capillaries. J. Neurochem. Submitted.
65. TAO-CHENG, J.-H. & M. W. BRIGHTMAN. 1988. Development of membrane interactions between brain endothelial cells and astrocytes *in vitro.* Int. J. Dev. Neurosci. **6:** 25–37.
66. DEBAULT, L. E. & P. A. CANCILLA. 1980. Gamma-glutamyl transpeptidase in isolated brain endothelial cells: Induction by glial cells *in vitro.* Science **207:** 653–655.
67. SMITH, Q. R., S. MOMMA, M. AOYAGI & S. I. RAPOPORT. 1987. Kinetics of neutral amino acid transport across the blood-brain barrier. J. Neurochem. **49:** 1651–1658.
68. KALARIA, R. N. & S. I. HARIK. 1988. Adenosine receptors and the nucleoside transporter in human brain vasculature. J. Cereb. Blood Flow Metab. **8:** 32–39.

69. MEYER, J., U. MISCHECK, M. VEYHL, K. HENZEL & H. J. GALLA. 1990. Blood-brain-barrier characteristic enzymatic-properties in cultured brain capillary endothelial-cells. Brain Res. **514:** 305–309.
70. MAXWELL, K., J. A. BERLINER & P. A. CANCILLA. 1987. Induction of gamma-GTP in cultured cerebral endothelial cells by a product released by astrocytes. Brain Res. **410:** 309–314.
71. STEWART, P. A. 1980. Histochemical absence of gamma-glutamyl transpeptidase in chick brain capillary endothelium. Exp. Neurol. **67:** 442–446.
72. ALBERT, Z., M. ORLOWSKI, Z. RZUCIDLO & J. ORLOWSKA. 1986. Studies on gamma-glutamyl transpeptidase activity and its histochemical localization in the central nervous system of man and different animal species. Acta Histochem. **25:** 312–320.
73. STEVENSON, B. R., J. D. SILICIANO, M. S. MOOSEKER & D. A. GOODENOUGH. 1986. Identification of ZO-1: A high molecular weight polypeptide associated with the tight junction (zonula occludens) in a variety of epithelia. J. Cell Biol. **103:** 755–766.
74. STEVENSON, B. R., J. M. ANDERSON, D. A. GOODENOUGH & M. S. MOOSEKER. 1988. Tight junction structure and ZO-1 content are identical in two strains of Madin-Darby canine kidney cells which differ in transepithelial resistance. J. Cell Biol. **107:** 2401–2408.
75. SEULBERGER, H., F. LOTTSPEICH & W. RISAU. 1990. The inducible blood-brain barrier specific molecule HT7 is a novel immunoglobulin-like cell surface glycoprotein. EMBO J. **9:** 2151–2158.
76. ALBRECHT, U., H. SEULBERGER, H. SCHWARZ & W. RISAU. 1990. Correlation of blood-brain barrier function and HT7-protein distribution in chick brain circumventricular organs. Brain Res. **535:** 49–61.
77. FOSSUM, S., S. MALLETT & A. N. BARCLAY. 1991. The MRC OX-47 antigen of activated lymphocytes is a member of the immunoglobulin superfamily with an unusual transmembrane sequence. Eur. J. Immunol. **21:** 671–679.
78. ALTRUDA, F., P. CERVELLA, M. L. GAETA, A. DANIELE, F. GIANCOTTI, G. TARONE, G. STEFANUTO & L. SILENGO. 1989. Cloning of cDNA for a novel mouse membrane glycoprotein (gp42): Shared identity to histocompatibility antigens, immunoglobulins and neural-cell adhesion molecules. Gene **85:** 445–452.
79. SCHLOSSHAUER, B. & K. H. HERZOG. 1990. Neurothelin: An inducible cell-surface glycoprotein of blood-brain-barrier specific endothelial-cells and distinct neurons. J. Cell Biol. **110:** 1261–1274.
80. DICK, A. P. K., S. I. HARIK, A. KLIP & D. M. WALKER. 1984. Identification and characterization of the glucose transporter of the blood-brain barrier by cytochalasin B binding and immunological reactivity. Proc. Natl. Acad. Sci. USA **81:** 7233–7237.
81. FLIER, J. S., M. MUECKLER, A. L. MCCALL & H. F. LODSIH. 1986. Distribution of glucose transporter messenger RNA transcripts in tissues of rat and man. J. Clin. Invest. **79:** 657–661.
82. KALARIA, R. N., S. A. GRAVINA, J. W. SCHMIDLEY, G. PERRY & S. I. HARIK. 1988. The glucose transporter of the human brain and blood-brain barrier. Ann. Neurol. **24:** 757–764.
83. GOULD, G. W. & G. I. BELL. 1990. Facilitative glucose transporters: An expanding family. Trends Biol. Sci. **15:** 18–23.
84. HARIK, S. I., R. N. KALARIA, P. M. WHITNEY, L. ANDERSSON, P. LUNDAHL, S. R. LEDBETTER & G. PERRY. 1990. Glucose transporters are abundant in cells with occluding junctions at the blood eye barriers. Proc. Natl. Acad. Sci. **87:** 4261–4264.
85. DINGLER, A. 1989. Untersuchungen von Blut-Hirn Schranken Proteinen an kultivierten Gehirnendothelzellen. University of Tubingen. Diploma Thesis.

Differentiation of Brain Endothelial Cells in Cell Culture

L. L. RUBIN, K. BARBU, F. BARD, C. CANNON,
D. E. HALL, H. HORNER, M. JANATPOUR, C. LIAW,
K. MANNING, J. MORALES, S. PORTER, L. TANNER,
K. TOMASELLI, AND T. YEDNOCK

Athena Neurosciences
800-F Gateway Blvd.
South San Francisco, California 94080

Although most neurobiologists and neurologists are familiar with the concept of the blood-brain barrier (BBB), very few know what it is and how it forms. Nonetheless, it seems that the BBB will occupy a prominent position in neural science in the next few years for several reasons. First, new treatments for a variety of neurodegenerative and other neurological disorders undoubtedly will become available. Such treatments seem likely to involve the use of drugs that will not enter the brain in therapeutically effective levels when given peripherally. The only alternative is to infuse the drugs directly into cerebral ventricles, a procedure that is less than optimal for chronic therapy. The second reason is that there are situations in which the normal BBB is compromised for extended periods of time (such as in vasogenic brain edema) or in which undesired trafficking across the BBB occurs (such as the entry into the brain of monocytes carrying the HIV virus). In such cases, the ability to decrease the entry of potentially harmful substances would be desirable. Finally, there is clear and fundamental interest in trying to understand how endothelial cells (ECs) that make up brain capillaries and are responsible for the BBB become different from those that make up peripheral capillaries.

Experiments over the last 20 years have established that two presumably separate properties of brain capillary ECs combine to form a barrier that is poorly permeable to molecules that are neither sufficiently hydrophobic nor transported by specific membrane proteins.[1] First, between the brain ECs are tight junctions with high electrical resistance ($>1{,}000$ ohm·cm^2). These junctions severely limit molecular passage between neighboring ECs (*paracellular transport*). In addition, the junctions are probably important in allowing independent regulation of the ionic composition of brain extracellular fluid during varying states of neuronal activity. The second important property of brain ECs is that the amount of fluid-phase endocytosis in those cells is limited as compared to that of peripheral ECs, at least when assayed with the use of nonspecific tracer molecules such as horseradish peroxidase. This serves to limit the amount of *transcellular transport.*

Chick-quail transplantation experiments carried out by Stewart and Wiley[2] suggested strongly that tissue environment influences the permeability properties of ECs. Therefore, peripheral ECs that vascularize brain tissue acquire at least some of the normal properties of brain ECs (including the ability to exclude tracer dyes such as Evan's blue, a cationic dye that binds to bovine serum albumin). Furthermore, brain ECs that vascularize peripheral tissue become more like peripheral ECs in that they are permeable to such dyes.

These studies were extended by Janzer and Raff[3] who attempted to identify the source of the factors in brain that cause EC differentiation. They cultured type-1 astrocytes from neonatal rat and transplanted these cells grown on nitrocellulose filters either into rat eyes or onto the chorioallantoic membrane of chick embryos. They found that in either case the astrocytes were vascularized by ECs that acquired the ability to exclude Evan's blue dye. This suggests that the type-1 astrocyte can induce at least partial differentiation of ECs. This is sensible in that these astrocytes project to capillaries in normal brain.

It has been clear for some time that further evaluation of the effects of astrocytes on EC differentiation would be facilitated by having a cell culture system in which these cells could be studied. Several investigators have reported on various features of ECs cocultured with astrocytes or grown in the presence of various cell-derived conditioned media.[4–8] However, there have been relatively few reports of cell culture systems in which ECs were shown to have properties corresponding to those seen *in vivo,* including the presence of high resistance tight junctions, exclusion of small hydrophilic molecules, and low rates of fluid-phase endocytosis.

Our goal was to derive a cell culture system in which monolayers of ECs with these properties could be grown on permeable supports separating two chambers. Such a system could then be used in a simple way to screen various therapeutic compounds for their potential to pass from blood to brain. We also hoped that the information we obtained while trying to construct such a cell culture model would be useful in discovering techniques for enhancing the penetration of these compounds into the brain. In this article, we describe the development of a cell culture model of the BBB and its use in the design of drug delivery systems.

METHODS

Bovine brain ECs were isolated by modifications of standard protocols[5] and plated on collagen-fibronectin–coated tissue culture flasks. After approximately 7 days, the cells were trypsinized off the flasks, plated on Costar polycarbonate Transwells coated with collagen and fibronectin. Cells were grown in Eagle's minimal essential medium with either 5% fetal calf serum or plasma-derived horse serum. In a limited number of experiments, ECs were isolated from human brain biopsy samples that were transported directly from the operating room to our laboratory.

Rat brain type-1 astrocytes were prepared from 1-day rat cortex, essentially as described by Lillien *et al.*[9] Conditioned medium (CM) was collected by feeding confluent cultures with fresh medium and collecting it after 48 hours. The CM was diluted with an equal volume of fresh medium before use.

All other procedures were fairly standard. Resistance measurements were made by passing current and measuring voltage across EC monolayers using a device manufactured by WPI Instruments. Transport studies were performed by adding different radioactively labeled compounds to the top chamber (corresponding to the blood or apical side of the cells) and measuring their rate of appearance in the bottom chamber.

RESULTS

Cell Morphology

Prior to our experiments, an extensive analysis of MDCK cells, an epithelial cell line that forms high resistance tight junctions when cultured on permeable supports, had been made.[10] These studies had demonstrated that MDCK cells elaborate a junctional apparatus (including, for the present purposes, the true *zonula occludens* and the *zonula adhaerens*) that contains a variety of components including ZO-1 and cingulin—cytoplasmic proteins associated with the tight junction—filamentous actin, and the cell adhesion molecule E-cadherin.[11–13] All of these proteins distribute themselves in a belt-like fashion around the cell borders.

We anticipated that brain ECs that formed high resistance tight junctions might adopt a similar morphology. Instead, we found that cells grown according to standard protocols formed multiple layers lacking a precise localization of junctional proteins. This was particularly noticeable with filamentous actin. Such cultures also had low electrical resistance, varying from 10–40 ohm·cm^2, which is not much different from that seen with peripheral ECs.

We found that growing bovine or human brain ECs in a combination of astrocyte CM and in the presence of agents that elevate their levels of cyclic AMP caused a profound morphological change in the cells, producing single monolayers with relatively high electrical resistance (>500 ohm·cm^2). The distribution of tight-junction proteins in these treated cells was also much more similar to that found in MDCK cells.

Effect of Astrocyte CM

Cells grown in astrocyte CM from their time of initial plating had, on average, a two- to fourfold higher resistance than did cells grown in medium lacking astrocyte CM, achieving, on occasion, resistances of more than 150 ohm·cm^2. Also, the cells tended to maintain a single monolayer for a much longer period of time. Other types of cell-derived CM were less effective. Fibroblast CM, for example, did not promote the formation of single monolayers of higher resistance. Cells grown in astrocyte CM and then switched into CM prepared from fibroblasts, U2-51 astrocytoma or C6 glioma cells, also demonstrated lower resistance. If cells were not plated directly into astrocyte CM, but were supplemented with it at later times, the effect was also less pronounced.

Effect of Cyclic AMP Elevation

Although CM derived from astrocytes was important, it was not by itself sufficient to cause the cells to form extremely high resistance tight junctions. We found that agents that increase cyclic AMP levels in brain ECs caused a rapid and significant increase in monolayer resistance. Resistances reached over 1,000 ohm·cm^2 and were on average greater than 500 ohm·cm^2. Such agents include cyclic AMP analogs, cholera toxin and forskolin, which activate adenylate cyclase, and agonists that increase cyclic AMP by binding to receptors coupled to stimulatory G proteins. Increases in cyclic AMP levels were also associated with a redistribution of filamentous actin from a diffuse cytoplasmic (stress-filament) pattern to an exclusive localization around cell borders. Furthermore, these cells showed similar distributions of ZO-1 and an E-cadherin-like molecule that is present in brain ECs. Thus, when grown in a combination of astrocyte CM and enhanced cyclic AMP, brain ECs became much more like MDCK cells and more like brain capillary ECs.

Restricted Paracellular Transport in High Resistance Cultures

Traditionally, cultures of brain ECs have been much leakier than expected. That is, presumably because of low electrical resistance, the rate of transport of rather large hydrophilic tracer molecules across filters with ECs has been significant with respect to that occurring across cell-free filters. Our high resistance cultures are significantly less leaky, showing markedly less movement of molecules at least as small as sucrose. On the other hand, as expected, hydrophobic molecules of similar molecular weight pass through the monolayers rather readily. This type of molecular exclusion is important because it permits the study of specific transport processes that might otherwise be masked by a background of high leakage.

Fluid-Phase Endocytosis

As already mentioned, another well-established property of brain ECs is their relatively low rate of fluid-phase endocytosis. This process was assessed in brain and peripheral EC cultures by incubating the cells in a variety of fluorescent tracers, such as FITC-dextran. The rate of pinocytosis in brain cells was markedly lower than that in peripheral ECs. However, this low rate was seen even in cells grown in the absence of astrocyte CM and elevated cyclic AMP, and quantitative studies have indicated thus far that neither of these factors altered the rate. Thus, it appears as if brain ECs maintain this differentiated property for some time when placed in cell culture, whereas additional factors are necessary to induce the formation of high resistance tight junctions.

Design of Drug Delivery Systems

There are essentially two ways of enhancing penetration of molecules into the central nervous system. The first is to stimulate transcellular transport (either by receptor-mediated transcytosis or by fluid-phase endocytosis), and the second is to stimulate paracellular transport by reversibly modulating the permeability of EC tight junctions. In fact, there is some *in vivo* experimental and clinical experience using this latter technique. Intracarotid infusion of hyperosmotic mannitol seems to reversibly open tight junctions for several hours,[14] enhancing brain uptake of antitumor agents, for example.

We have been interested in developing techniques for discretely modulating the permeability of brain EC tight junctions. On the basis of previous studies with MDCK cells and our work with cultured brain ECs, we felt that there would be two general approaches. First, we expected that brain ECs would have cell adhesion molecules that would mediate the binding of cells to one another and trigger the formation of tight junctions. Second, it appeared that intracellular events would occur involving changes in cyclic AMP levels that would modulate the permeability of tight junctions. We have derived reagents that block cadherin binding and have investigated compounds that alter cyclic AMP levels in brain ECs. We have shown that both types of manipulation appear to alter tight junction permeability reversibly.

Cadherins and Tight Junctions

In cell adhesion molecule studies, we identified three EC cadherins. Two of them, one homologous to N-cadherin and the other to P-cadherin, are present in all ECs, including those from brain.[15] In addition, a cadherin somewhat homologous to E-cadherin is present in brain, but not in peripheral ECs, at least in mouse and rat. We have synthesized peptides with amino acid sequences derived from one particular portion of the extracellular domain of the cadherins, referred to as the HAV region.[16] Some, but not all, of these peptides were effective in reversibly opening tight junctions between brain ECs. We are currently synthesizing additional peptides and are also in the process of testing their ability to deliver CNS-active drugs in animals.

Second Messengers and Tight Junctions

The experiments described here suggested that elevated cyclic AMP levels are associated with an increase in tight junction resistance. We also found that the effect of increased cyclic AMP could be reversed rather readily. Once the agent used to stimulate cyclic AMP synthesis was removed from the culture medium, the transmonolayer resistance fell rapidly to low levels. The most likely explanation for this is that cyclic AMP is involved in the phosphorylation of proteins involved in determining tight junction resistance. As cyclic AMP levels decline, a rapidly acting phosphatase dephosphorylates these proteins and the resistance falls. This further suggests that, under normal circumstances, an active kinase must be present to keep these proteins in the phosphorylated state. All of our work with cultured cells predicts that any agent that either causes a decrease in cyclic AMP levels in brain ECs or inhibits the activity of protein kinase A should enhance tight junction permeability and provide the basis for a drug delivery system.

We have tested this possibility in mice using two assay systems. The first is a measurement of the amount of tracer molecule (either ^{125}I-bovine serum albumin or ^{3}H-sucrose) that enters the brain during various experimental manipulations. The second is a behavioral test that depends on increasing the penetration of morphine or enkephalin into the brain to produce analgesia (as measured in a standard hot plate assay). Our work to date suggests that several different agents, administered peripherally, that either decrease cyclic AMP levels or inhibit protein kinase A have the effect of increasing brain uptake of tracer molecules and of providing markedly increased analgesia. We are currently pursuing these results with the intent of deriving an agent that is as selective for brain ECs and as efficacious as possible.

CONCLUSION

We have succeeded in establishing cultures of bovine (and human) brain ECs that maintain several important differentiated characteristics. By using a combination of astrocyte CM and increased cyclic AMP levels, these cells can be induced to grow as a monolayer with tight junctions of relatively high electrical resistance. That this treatment was successful with both bovine and human brain ECs suggests the generality of these influences. These cells synthesize and localize in an appropriate fashion several proteins known to be associated with tight junctions, at least in epithelial cells. Prominent among these is an E-cadherin-like molecule that is normally present in brain ECs and may be important in the initial formation of tight junctions. The leakage of hydrophilic molecules through these monolayers is fairly minimal. These cells also have a low rate of fluid-phase endocytosis when compared to that of peripheral ECs in culture. Finally, we made use of this information to explore different strategies for modulating tight junction permeability for the purpose of drug delivery. To date, we have found that the manipulation of EC cyclic AMP levels enhances brain uptake of tracer molecules and allows effective delivery of analgesics to the brain of unanesthetized mice.

REFERENCES

1. GOLDSTEIN, G. W. & A. L. BETZ. 1986. Blood vessels and the blood-brain barrier. *In* Diseases of the Nervous System. A. K. Asbury, G. M. McKhann & W. I. McDonald, eds. **1:** 172–184. W.B. Saunders Company. Philadelphia, PA.
2. STEWART, P. A. & M. J. WILEY. 1981. Developing nervous tissue induces formation of blood-brain barrier characteristics in invading endothelial cells: A study using quail-chick transplantation chimeras. Dev. Biol. **84:** 183–192.
3. JANZER, R. C. & M. C. RAFF. 1987. Astrocytes induce blood-brain barrier properties in endothelial cells. Nature. **325:** 253–257.
4. ARTHUR, F. E., R. R. SHIVERS & P. D. BOWMAN. 1987. Astrocyte-mediated induction of tight junctions in brain capillary endothelium: An efficient *in vitro* model. Dev. Brain Res. **36:** 155–159.
5. AUDUS, K. L. & R. T. BORCHARDT. 1986. Characterization of an *in vitro* blood-brain barrier model system for studying drug transport and metabolism. Pharm. Res. 3: 81–87.
6. DEHOUCK, M.-P., S. MERESSE, P. DELORME, J.-C. FRUCHART & R. CECCHELLI. 1990. An easier, reproducible, and mass-production method to study the blood-brain barrier *in vitro.* J. Neurochem. **54:** 1798–1801.
7. MAXWELL, K., J. A. BERLINER & P. A. CANCILLA. 1987. Induction of γ-glutamyl transpeptidase in cultured cerebral endothelial cells by a product released by astrocytes. Brain Res. **410:** 309–314.
8. TAO-CHENG, J.-H. & M. W. BRIGHTMAN. 1988. Development of membrane interactions between brain endothelial cells and astrocytes *in vitro.* Int. J. Dev. Neurosci. **6:** 25–37.
9. LILLIEN, L. E., M. SENDTNER, H. ROHRER, S. M. HUGHES & M. C. RAFF. 1988. Type-2 astrocyte development in rat brain cultures is initiated by a CNTF-like protein produced by type-1 astrocytes. Neuron **1:** 485–494.
10. GUMBINER, B. 1987. Structure, biochemistry, and assembly of epithelial tight junctions. Am. J. Physiol. **253:** C749–C758.
11. STEVENSON, B. R., J. D. SILICIANO, M. S. MOOSEKER & D. A. GOODENOUGH. 1986. Identification of ZO-1: A high molecular weight polypeptide associated with the tight junction (zonula occludens) in a variety of epithelia. J. Cell Biol. **103:** 755–766.
12. CITI, S., H. SABANAY, R. JAKES, B. GEIGER & J. KENDRICK-JONES. 1988. Cingulin, a new peripheral component of tight junctions. Nature **333:** 272–276.
13. GUMBINER, B., B. STEVENSON & A. GRIMALDI. 1988. The role of the cell adhesion molecule uvomorulin in the formation and maintenance of the epithelial junctional complex. J. Cell Biol. **107:** 1575–1587.
14. GREIG, N. H. 1988. Drug delivery to the brain by blood-brain barrier circumvention and drug modification. 1988. *In* Implications of the Blood-Brain Barrier and Its Manipulation. E. A. Neuwelt, ed. **1:** 311–367. Plenum Medical Book Company. New York, NY.
15. GUMBINER, B. 1988. Cadherins: A family of Ca^{2+} dependent adhesion molecules. TIBS **13:** 75–76.
16. LIAW, C. W., C. CANNON, M. D. POWER, P. K. KIBONEKA & L. L. RUBIN. 1990. Identification and cloning of two species of cadherins in bovine endothelial cells. EMBO J. **9:** 2701–2708.
17. NOSE, A., K. TSUJI & M. TAKEICHI. 1990. Localization of specificity determining sites in cadherin cell adhesion molecules. Cell **61:** 147–155.

DISCUSSION PAPER

Astrocytes, Cerebral Endothelium, and Cell Culture

The Pursuit of an *in Vitro* Blood-Brain Barrier

JOHN GREENWOOD

Department of Clinical Science
Institute of Ophthalmology
London EC1V 9AT, UK

A great deal of research currently investigating the properties of the blood-brain barrier (BBB) is concerned with the molecular and cellular events that are responsible for the control and pathophysiology of this important structure. One of the central aims of many groups working in this field, including those at this meeting, is to recreate a blood-brain barrier *in vitro* by culturing brain capillary endothelial (BCE) cells. This goal has been pursued rigorously because of the considerable experimental flexibility offered by such a system over conventional *in vivo* studies. However, fundamental problems are associated with attempting to reproduce *in vitro* a complex *in vivo* structure, and this has generated a number of criticisms of this technique. It now appears that although we have still not achieved a fully representative *in vitro* blood-brain barrier, we are not far short of doing so.

One of the most important developments has been the ability to coculture BCE cells in close proximity to cultured astrocytes, thus attempting to recreate the association found *in vivo*. Although it has been mooted for some time that astrocytes may be important in helping to induce barrier properties, only recently has there been substantial *in vivo* and *in vitro* demonstration of this phenomenon.[1–4] Although BCE cells cultured in the absence of astrocytes retain some of their *in vivo* characteristics,[5,6] they lose others such as gamma-glutamyl transpeptidase[7] and tight junctions with high electrical resistance. These problems in many cases are exacerbated by passaging the cells.

In research presented at this meeting the similarity between BCE cells *in vivo* and their *in vitro* counterparts was assessed using a variety of criteria. The effect of coculturing astrocytes with BCE cells was investigated[8–10] as was the effect of adding astrocyte-conditioned medium (ACM) to the culture medium.[11] Both qualitative and quantitative methods were employed to evaluate those properties of the *in vivo* BCE cells that either were retained or could be induced in cultured BCE cells. It is encouraging that many different and new criteria are now being used and that each group is attempting to fulfil as many of these as possible. Four main approaches were used to evaluate the *in vitro* BBB:

1. *Transmonolayer resistance:* By culturing BCE cells on a permeable membrane it is possible to measure the permeability of the monolayer to ions by recording the electrical resistance across the endothelial cell monolayer. This method provides a highly sensitive indication of the degree of complexity of tight junction (TJ) formation and integrity of the *in vitro* barrier. However, this technique requires complete confluency of the monolayer without breaks or contamination from other cell types. The resistance across the mammalian BBB found *in vivo*[12] is in excess of 1,000 ohms·cm^2. *In vitro* an electrical resistance of around 400 ohms·cm^2 has been reported[13,14] in the absence of cocultured astrocytes that, on coculturing, increases to

approximately 600 ohms·cm^2. Few groups have achieved such high values in the absence of astrocytes. Rubin *et al.*[11] obtained resistance of the order of 40 ohms·cm^2 which increased to around 600 ohms.cm^2 when BCE cells were cultured in the presence of ACM. However, they also reported that in addition to ACM, cyclic AMP (cAMP) had to be elevated, with compounds such as forskolin, for such an increase in resistance to be achieved. Despite previous reports that elevation of cAMP in BCE cells *in vivo* induces pinocytosis and vesicular transport,[15] this was not the case with the *in vitro* studies of Rubin *et al.*,[11] thus confirming an earlier *in vitro* study.[16]

2. *Transmonolayer permeability to nontransported molecules:* Monolayers of BCE cells grown on permeable membranes also allow for the measurement of vectorial transfer of small inert molecules, such as sucrose,[11,14,17] that are normally excluded from a cellular barrier. Despite recording a decrease in the degree of permeability of a BCE cell monolayer to these tracers when cocultured with astrocytes, the level of exclusion still does not completely approach that of the *in vivo* BBB.

3. *Tight-junction complexity:* The degree of TJ complexity can be assessed by freeze-fracture electron microscopy. Although it has been reported that passaged BCE cells retain relatively complex TJs,[7] most investigators could only demonstrate a dramatic increase in TJ complexity of cultured BCE cells following coculture with astrocytes.[8–10,17] It must be borne in mind, however, that a demonstration of increased TJ complexity, as revealed by freeze-fracture techniques, does not necessarily indicate the presence of a functional barrier. Indeed, these studies must be carried out in conjunction with functional testing, as with radiotracers or electrical resistance. The fact that monolayers expressing complex TJ still do not reproduce the degree of transcellular resistance found *in vivo* adequately demonstrates this point. A useful quantitative method for assessing TJ formation was described that enables a realistic estimation of the complexity of TJs from relatively small areas of freeze-fractured junctions to be made.[17] It was also pointed out that with freeze fracture it is important to study both the E and P faces of complementary replicas in order to gain a complete picture of the complexity of the junction.[18]

4. *Characteristic endothelial cell markers:* The uptake of acetylated LDL[9] and the immunocytochemical detection of characteristic endothelial epitopes such as the Factor VIII related antigen/Von Willebrands factor is still being widely used. Other characteristic markers now being used include an inducible BBB endothelium-specific membrane glycoprotein of unknown function, termed HT7.[19] One suggested function of this molecule is to transport substances across the BBB, perhaps in a similar manner to that of the transferrin receptor which is also a BBB endothelial-specific marker. Antibodies against the glucose transporter[8] and against epitopes expressed at the TJ region of the cell[11] have also been used. A different approach in assessing the status of the cultured BCE cells was to visualize the cellular distribution of actin which becomes concentrated at the periphery of the cell upon treatment with ACM.[11] It was pointed out that although gamma-glutamyl transpeptidase is still used as a marker for BCE cells and has been induced in cultured BCE cells,[14,20] its value as an *in vitro* marker was questioned, as pericytes will express this enzyme strongly in culture.[8] Furthermore, it was also suggested that BBB characteristic molecules, such as the glucose transporter, are only expressed in cultured BCE cells upon induction of TJs.[8]

Despite considerable advances in the development of BCE cell culture systems, conflicting findings still remain. One of the most puzzling contradictions is the inability of some groups to induce barrier characteristics in cultured BCE cells when ACM is administered to either the "luminal" or the "abluminal" side, or both, despite being able to do so by coculturing astrocytes on the other side of a permeable membrane. It is unlikely that this is due to direct cell-to-cell contact between astrocytes and endothelial cells in coculture, as the pore size of most of the membranes used is too small (below 0.4 μ), thus preventing cell processes from growing through the membrane. The only explanation that could be given is that the factor (or factors) released from astrocytes was a highly labile molecule that was only active within a small distance from the cell. This still does not explain the ability of some groups[11] to induce barrier characteristics in cultured BCE cells by ACM alone. Furthermore, ACM from cultured neoplastic astrocytes is capable of inducing TJs[11] despite reports that ACM from such cells can disrupt the BBB *in vivo.*[21]

Substantial differences in technique in the isolation and culture of BCE cells still remain, and these differences may help explain the anomalies between the different groups. The chief differences in the culture techniques were:

Animal species: Bovine brain is probably the most popular source of BCE cells because of the large amount of starting material available. However, many groups are also currently using rat[20,22,23] and chick brain[8] as the source.

Isolation method: It was stressed that the isolation method must ensure that the end product gives capillary fragments without larger vessel contamination. Generally there appears to be a convergence towards the technique of enzymatic digestion with separation and purification on pre-formed density gradients.

Astrocyte cocultures: A further source of variability is introduced with the astrocyte coculture, with both neonatal and neoplastic cells being utilized. Moreover, in many coculture studies the BCE cells and astrocytes are derived from different species.

Growth substrate: Another important difference is the substrate upon which the BCE cells are seeded and grown. Collagen and gelatin are most commonly used; however, other substrates have been used and these different substrates may, among other things, play a part in determining the cell morphology, that is, spindle or cobble-stone shaped cells.

Culture medium: Substantial variability is found in the culture medium and supplements used to grow BCE cells. Moreover, the medium may be altered during the growth cycle, most frequently after the initial plating out of the capillary fragments. It was generally agreed, however, that supplementation of the medium with plasma-derived serum was more desirable than was fetal calf serum which induces the growth of nonendothelial cell types. Other supplements used included endothelial cell growth supplement (ECGS), fibroblast growth factor (FGF), and forskolin. As with the substrate, the constituents of the culture medium may play an important part in determining the cell phenotype.[11,20]

Passaging: Some differences of opinion still remain as to the use of primary or passaged cells, but on the strength of the preceding reports it seems that passaged cells will retain their BBB characteristics providing the conditions are correct. This important development allows for the production of cultures of greater purity and for the cloning of separate cell populations. By cloning out single colonies of BCE cells and passaging them, monolayers of higher permeability than those derived from mixed colonies were achieved.[8] This begs the question as to the homogeneity of a monolayer derived from many capillary fragments. Finally, the split ratio used during passaging differs considerably.

It is hardly surprising, therefore, that until some degree of agreement in the culture conditions can be achieved, fundamental differences in results will remain between different groups. It must therefore remain an objective of all groups working within this field to gradually elucidate which aspects of the numerous techniques are critical and to incorporate these into a single, more uniform technique. This is especially important as there is a substantial growth of interest in this field of study.

It was universally accepted that in primary cultures the presence of pericytes was a major factor in determining the viability of BCE cells. The presence of pericytes in the culture inhibits BCE cell growth, a finding previously reported in endothelial cells cultured from bovine adrenal cortex.[24] However, it has also been reported that pericyte-conditioned medium will promote the *in vitro* growth of retinal endothelial cells that are in part responsible for the blood-retinal barrier.[25] A variety of techniques were used to remove pericytes from primary cultures including the use of antibody Thy 1 which recognizes pericytes and which allows for destruction of these cells by complement fixation.[26] Alternatively, endothelial cells were removed for passaging by light trypsinization which tended to leave behind most of the pericytes. Furthermore, pericytes appear to stick to uncoated plastic, whereas endothelial cells do not. In one of the studies in which pericytes were added to cultured BCE cells, they were found capable of altering the permeability of the monolayer, but not apparently by the induction of TJs.[8]

Astrocytes will readily coculture with BCE cells in the same plane, but irrespective of which cell type is seeded first, contact inhibition occurs between the cell types. This prevents close contact other than that between cells at the periphery of a growing plaque of cells. In one report, passaged BCE cells could be induced to form tubular structures *in vitro,*[10] and unlike a monolayer arrangement, astrocytes seeded onto these plates migrated to, and spread out along the endothelial tubes. Pericytes were also able to migrate and spread out along the tubes at an even greater rate than were astrocytes. As with other coculture systems, TJ complexity assessed by freeze-fracture techniques was greatly increased when astrocytes associated themselves with the endothelial tubular structures. It was believed that the tubes were forming by the folding over of cells from the monolayer which was reflected in the appearance of these structures when seen ultrastructurally. Although the walls of these tubes were one cell thick, they were often composed of a number of individual endothelial cells. Furthermore, tubes appeared to form in a random manner with some in very close proximity to each other, a characteristic not normally associated with angiogenesis. However, if true budding and capillary formation can be demonstrated, this model may have some important role to play in the field of angiogenesis.

From the comparative studies of Abbott[27] and Butt[28] it is clear that in addition to the accepted characteristics of BCE, the extracellular matrix may also play a part in restricting movement of substances into the brain. In addition, it was pointed out that at the glial BBB the cell itself presents an enzymic and metabolic barrier,[27] a phenomenon that was recently demonstrated for the endothelial BBB.[29,30]

Finally a word of caution. Although it has now been irrevocably demonstrated that astrocytes play some part in the induction and maintenance of the BBB, it is highly unlikely that this is the whole story. Indeed, a number of anomalies require further investigation. Firstly, pial vessels in the rat[12] and frog[31] do not have associated astrocytic end-feet but do possess TJs of high electrical resistance. Conversely, some of the circumventricular regions such as the area postrema have a perivascular astrocytic sheath surrounding the capillaries, but the vessels are leaky. This certainly points to the involvement of factors other than the close proximity of astrocytes. A further finding that leads one to consider other factors is that in the developing rat the BBB becomes "tight" within hours immediately prior to birth.[12] This rapid change coincides with the observed onset of ion regulation by the BBB.[32] It is questionable that such a rapid induction can be achieved *in vitro;* however, it would be of interest to ascertain the temporal sequence of TJ formation following the administration of ACM. Furthermore, astrocyte development lags behind that of the BBB by 7–10 days in that at birth, when the BBB tightens, there are few astrocytic end-feet associated with blood vessels.[33] Finally, as demonstrated by Abbott,[27] Lane,[18] and Butt,[28] in some species astrocytes are themselves capable of forming TJs and barrier properties. It remains to be seen what factors induce the barrier properties at the glial BBB as well as at other barrier sites such as the epithelial cells of the choroid plexus and the retinal pigment epithelial cells of the blood-retinal barrier.

REFERENCES

1. BECK, D. W., H. V. VINTERS, M. N. HART & P. A. CANCILLA. 1984. Glial cells influence polarity of the blood-brain barrier. J. Neuropathol. Exp. Neurol. **43:** 219–224.
2. JANZER, R. C. & M. C. RAFF. 1987. Astrocytes induce blood-brain barrier properties in endothelial cells. Nature **325:** 253–257.
3. ARTHUR, F. E., R. R. SHIVERS & P. D. BOWMAN. 1987. Astrocyte-mediated induction of tight junctions in brain capillary endothelium: An efficient *in vitro* model. Dev. Brain Res. **36:** 155–159.
4. TAO-CHENG, J. H., Z. NAGY & M. W. BRIGHTMAN. 1987. Tight junctions of brain endothelium *in vitro* are enhanced by astroglia. J. Neurosci. **7:** 3293–3299.
5. GOLDSTEIN, G. W., A. L. BETZ & P. D. BOWMAN. 1984. Use of isolated brain capillaries and cultured endothelial cells to study the blood-brain barrier. Fed. Proc. **43:** 191–195.
6. HUGHES, C. C. W. & P. L. LANTOS. 1986. Brain capillary endothelial cells *in vitro* lack surface IgG Fc receptors. Neurosci. Lett. **68:** 100–106.
7. MÉRESSE, S., M.-P. DEHOUCK, P. DELORME, M. BENSAÏD, J.-P. TAUBER, C. DELBART, J.-C. FRUCHART & R. CECCHELLI. 1989. Bovine brain endothelial cells express tight junctions and monoamine oxidase activity in long-term culture. J. Neurochem. **53:** 1363–1371.
8. RISAU, W. 1991. Induction of blood-brain barrier endothelial cell differentiation by astrocytes. Ann. N.Y. Acad. Sci., this volume.
9. NAGY, Z. & K. MARTINEZ. 1991. Astrocytic induction of endothelial tight junctions. Ann. N.Y. Acad. Sci., this volume.
10. MINAKAWA, T., J. BREADY, J. BERLINER, M. FISHER & P. CANCILLA. 1991. *In vitro* interaction of astrocytes and pericytes with capillary-like tubular structures of brain microvessel endothelium. 1991. Ann. N.Y. Acad. Sci., this volume.
11. RUBIN, L. L., K. BARBU, F. BARD, C. CANNON, D. E. HALL, H. HORNER, M. JANATPOUR, C. LIAW, K. MANNING, J. MORALES, S. PORTER, L. TANNER, K. TOMASELLI & T. YEDNOCK. 1991. Differentiation of brain endothelial cells in cell culture. Ann. N.Y. Acad. Sci., this volume.
12. BUTT, M. B., H. C. JONES & N. J. ABBOTT. 1990. Electrical resistance across the blood-brain barrier in anaesthetized rats: A developmental study. J. Physiol. **429:** 47–62.
13. RUTTEN, M. J., R. L. HOOVER & M. J. KARNOVSKY. 1987. Electrical resistance and macromolecular permeability of brain endothelial monolayer cultures. Brain Res. **425:** 301–310.
14. DEHOUCK, M.-P., S. MÉRESSE, P. DELORME, J.-C. FRUCHART & R. CECCHELLI. 1990. An easier, reproducible, and mass-production method to study the blood-brain barrier *in vitro*. J. Neurochem. **54:** 1798–1801.
15. JOO, F. 1987. A unifying concept on the pathogenesis of brain oedemas. Neuropathol. Appl. Neurobiol. **13:** 161–176.
16. KEMPSKI, O., A. VILLACARA, M. SPATZ, R. F. DODSON, C. CORN, N. MERKEL & J. BEMBRY. 1987. Cerebromicrovascular endothelial permeability. *In vitro* studies. Acta Neuropathol. (Berl.) **74:** 329–334.
17. NEUHAUS, J., W. RISAU & H. WOLBURG. 1991. Induction of blood-brain barrier characteristics in bovine brain endothelial cells by rat astroglial cells in transfilter coculture. Ann. N.Y. Acad. Sci., this volume.
18. LANE, N. J. 1991. Morphology of glial blood-brain barriers. Ann. N.Y. Acad. Sci., this volume.
19. SEULBERGER, H., C. UNGER, U. ALBRECHT & W. RISAU. 1991. The inducible blood-brain barrier endothelium specific antigen HT7—A novel immunoglobulin-like membrane glycoprotein. Ann. N.Y. Acad. Sci., this volume.

20. Bauer, H. C., U. Tontsch, A. Amberger & H. Bauer. 1990. Gamma-glutamyl-transpeptidase (GGTP) and Na^+K^+-ATPase activities in different subpopulations of cloned cerebral endothelial cells: Responses to glial stimulation. Biochem. Biophys. Res. Comm. **168:** 358–363.
21. Ohnishi, T., P. B. Sher, J. B. Posner & W. R. Shapiro. 1990. Capillary permeability factor secreted by malignant brain tumor. J. Neurosurg. **72:** 245–251.
22. Male, D., G. Pryce, C. Hughes & P. Lantos. 1990. Lymphocyte migration into brain modelled *in vitro:* Control by lymphocyte activation, cytokines, and antigen. Cell. Immunol. **127:** 1–11.
23. Abbott, N. J., C. C. W. Hughes, J. Greenwood, P. Orsmond, N. Ramlakhan & P. A. Revest. 1990. Primary culture of rat brain capillary endothelial cells for physiological studies. J. Physiol. **423:** 2P.
24. Orlidge, A. & P. A. D'Amore. 1987. Inhibition of capillary endothelial cell growth by pericytes and smooth muscle cells. J. Cell Biol. **105:** 1455–1462.
25. Wong, H. C., M. Boulton, J. Marshall & P. Clark. 1987. Growth of retinal capillary endothelia using pericyte conditioned medium. Invest. Ophthalmol. & Visual Sci. **28:** 1767–1775.
26. Risau, W., B. Engelhardt & H. Wekerle. 1990. Immune function of the blood-brain barrier: Incomplete presentation of protein (auto-)antigens by rat brain microvascular endothelium *in vitro.* J. Cell Biol. **110:** 1757–1766.
27. Abbott, N. J. 1991. Permeability and transport of glial blood-brain barriers. Ann. N.Y. Acad. Sci., this volume.
28. Butt, A. M. 1991. Modulation of a glial blood-brain barrier. Ann. N.Y. Acad. Sci., this volume.
29. Cordon-Cardo, C., J. P. O'Brien, D. Casals, L. Rittman-Grauer, J. L. Biedler, M. R. Melamed & J. R. Bertino. 1989. Multidrug-resistance gene (P-glycoprotein) is expressed by endothelial cells at blood-brain barrier sites. Proc. Natl. Acad. Sci. USA **86:** 695–698.
30. Ghersi-Egea, J. F., A. Minn & G. Siest. 1988. A new aspect of the protective functions of the blood-brain barrier: Activities of four drug-metabolizing enzymes in isolated rat brain microvessels. Life Sci. **42:** 2515–2523.
31. Crone, C. & S.-P. Olesen. 1982. Electrical resistance of brain microvascular endothelium. Brain Res. **241:** 49–55.
32. Jones, H. C. & R. F. Keep. 1987. The control of potassium concentration in the cerebrospinal fluid and brain interstitial fluid of developing rats. J. Physiol. **383:** 441–453.
33. Caley, D. W. & D. S. Maxwell. 1970. Development of the blood-vessels and extracellular spaces during postnatal maturation of the rat cerebral cortex. J. Comp. Neurol. **138:** 31–48.

INTRODUCTION

Glia As Targets for Neuroactive Substances

BRIAN PEARCE

Pharmacology Department
The School of Pharmacy
London WC1N 1AX, UK

The anatomical arrangement of glia around neuronal cell bodies, their fibers, and synapses is suggestive of complex interactions between these cells in the overall functioning of the nervous system. However, the nature of these interactions is still poorly understood. Do glia play, in the main, a relatively passive supporting role in which they supply nutrients to the neurons and remove excess potassium ions and neurotransmitters from the extracellular space, or are they more actively involved in the integrative properties of the nervous system by influencing the passage of information along nerve fibers and across synapses? A prerequisite for the latter role is that glia must be capable of receiving, interpreting, and responding to chemical signal molecules released from neurons; in other words, they, like neurons, must possess functional receptors for neurotransmitters.

The search for and characterization of glial receptors is a relatively modern area of the neurosciences. This is due in part to the technical difficulties involved in studying glia, particularly central nervous system glia, *in situ.* A major step forward was made in the mid-1960s with the development of methods for the preparation and maintenance of glia in culture. During the 1970s numerous studies on primary cultures of glia and cell lines of glial origin were performed and reviewed in some detail by Van Calker and Hamprecht.[1] Their review indicated the presence of receptors for catecholamines, some peptides, and prostaglandins on both central and peripheral nervous system glia. Moreover, many of these receptors could be considered functional in that they were found coupled to the cyclic AMP intracellular second messenger system.

The 1980s saw the further development of cell culture and cell identification methods in which cultures enriched in particular types of glia became available and techniques such as ligand binding and electrophysiological recording were being applied to them. By the mid- to late 1980s it was becoming clear that glia possessed an array of receptors, some coupled to cyclic AMP and others to inositol lipid second messenger systems, that they possessed both voltage- and ligand-operated ion channels, and that glia are heterogeneous in their expression of receptors, suggesting some degree of specialization. In addition, researchers were now beginning to ask questions about the physiological relevance of these receptors. Some answers, or at least clues, were forthcoming in that receptor activation could be linked to relevant responses such as cell growth and the synthesis and release of neuroactive substances.[2,3]

One criticism has always been that apart from work on the invertebrate nervous system where neuron-glia relations remain intact, much of what we know about glial receptors comes from studies on cultured cells. Although this can give us an indication of the potential of glia, how do their responses in a culture dish relate to their role in the intact nervous system? This question has yet to be fully resolved; however, evidence now beginning to emerge shows that what we know about cultured glia holds true in general terms for glia *in situ*.

At the time of this writing (1990) the situation is that apart from a very few exceptions, glia or particular types of glia possess the same types of receptors coupled to the same intracellular signaling pathways and ion channels as are present in neurons. Glia can therefore be considered to at least have the necessary machinery to interact with neurons via chemical signals. The task for the 1990s is to build upon our existing knowledge and to ask meaningful questions about the nature of neuron-glial communication and its role in nervous system functions. The contributions to this section of the meeting's proceedings point to some of the directions we should take in future research.

REFERENCES

1. VAN CALKER, D. & B. HAMPRECHT. 1980. Effects of neurohormones on glial cells. *In* Advances in Cellular Neurobiology, **1:** 31–67. S. Federoff & L. Hertz, eds. Academic Press. New York.
2. MURPHY, S. & B. PEARCE. 1987. Functional receptors for neurotransmitters on astroglial cells. Neuroscience **22:** 381–394.
3. KIMELBERG, H. K. (Ed.) 1988. Glial Cell Receptors. Raven Press. New York.

Mechanisms of Axon-Schwann Cell Signaling in the Squid Nerve Fiber[a]

PETER D. EVANS,[b] VINCENZINA REALE,[c] ROSA MARIA MERZON,[d] AND JORGE VILLEGAS[d]

[b]*AFRC Laboratory of Molecular Signalling*
Department of Zoology
University of Cambridge
Cambridge CB2 3EJ, UK

[c]*Department of Clinical Biochemistry*
University of Cambridge
New Addenbrooke's Hospital
Cambridge CB2 3QB, UK

[d]*Unidad de Neurociencias*
Instituto Internacional de Estudios Avanzados (IDEA)
Caracas, Venezuela

The importance of axonal-glial cell signaling in the nervous system is becoming increasingly apparent as glial cells are seen to possess a range of voltage-gated ion channels and an ever-increasing series of receptors that can be activated by a wide range of neurotransmitters. In addition, the ability of glial cells to synthesize and release molecules conventionally thought of as neurotransmitters suggests that chemical messengers may also be released from glial cells to mediate both glial-neuron and glial-glial signaling pathways.

To understand the physiological significance of axon-glial cell signaling it is essential to study systems in which the anatomical relations between the two cellular components are maintained as intact as possible. To date, this requirement has only been achieved in relatively few preparations including the optic nerve of the mudpuppy, *Necturus maculosus,*[1,2] the segmental ganglion of the leech, *Hirudo medicinalis,*[3,4] the perineurial glia forming the blood-brain barrier of the cockroach, *Periplaneta americana,*[5] and the giant axon-Schwann cell preparation of the tropical squid, *Sepioteuthis sepioidiae.*[6–9] The present review will concentrate on studies on the mechanisms of axon-Schwann cell signaling in the latter preparation. We will initially review the anatomy of the preparation and the recording techniques, and then outline the basic nature of the signaling mechanism in this system. We will then discuss how this system can be modulated under physiological conditions by both aminergic and peptidergic pathways.

[a]The work reported in this chapter was partially supported by CONICIT research grants S1–0808 and S1–1176, by British Council travel grants awarded to P. D. Evans, V. Reale, and J. Villegas for exchange visits between our laboratories, by grants from the Marshall and Orr Bequest of the Royal Society towards the cost of consumables for part of the work, and by the collaborative support of Fundación para la Ciencia "José Gregorio Hernández" (FUNDACIENCIA) and its Estación de Investigaciones Marinas de Mochima.

ANATOMY AND RECORDING TECHNIQUES

Long lengths of single giant axons (300–400 μm in diameter), surrounded by their glial sheaths, can be dissected out from the squid stellate nerves that innervate the muscles of the mantle and that are involved in generating the rapid escape responses of the animal. The sheaths consist of a single layer of tesselated adaxonal glial cells, or Schwann cells,[10,11] and these are the cells on which all of the work on axonal-glial relations in this preparation have been performed because of their intimate contact with the giant axon. This layer is surrounded by a layer of endoneurial cells that share many of the properties of the Schwann cells, but that have not yet been studied in detail. The Schwann cell layer is very thin, 0.9–6 μm, with an average thickness of 1.9 μm. However, because of the tesselated nature of the cells, electrophysiological recordings are most likely to be made from the narrow interdigitating fingers of glial cytoplasm, and it is very difficult to maintain such recordings for extended periods of time in experiments that require many changes of solution.

Initial electrophysiological recordings of Schwann cell resting membrane potential in this preparation revealed that they were very low, in the region of −40 mV.[12] However, simultaneous long-term recording of membrane potentials of the giant axon and of one of its surrounding Schwann cells revealed that activation of the giant axon at high frequencies induced prolonged hyperpolarization of the Schwann cell membrane due to increased potassium conductance. Such experiments are difficult to perform routinely, as already mentioned, but membrane potential values for populations of Schwann cells can easily be obtained by a transient sampling technique.[11] FIGURE 1 shows typical examples of Schwann cell membrane potentials obtained using this technique in a recent study on the effects of the neuropeptide Substance P in this preparation.[13] In FIGURE 1A(i) the arrow indicates the giant axon membrane potential. The electrode is then transiently pushed through the axolemma into one of the surrounding Schwann cells to register its membrane potential before being returned to the axon. FIGURE 1A(ii) shows that the membrane potential of the Schwann cell hyperpolarizes by 6–8 mV after exposure of the preparation to 10^{-7}M Substance P. (See discussion to follow.) In addition, FIGURE 1B(i) indicates another type of response obtained using this technique when the recording electrode completely passes through a Schwann cell process, whose membrane potential is recorded as a transient notch (arrowhead) on the potential profile recorded by the electrode. If samples of Schwann cell membrane potential recorded in this way are plotted against time (e.g., FIG. 1C), this technique can be used to demonstrate that a hyperpolarization of the membrane potentials of a population of Schwann cells occurs immediately after giant axon stimulation and that this potential change has the same magnitude and time course as does that obtained by prolonged recordings from single Schwann cells.[12]

CHOLINERGIC RECEPTOR SYSTEM

Early pharmacological studies on this neurally induced long-lasting hyperpolarization indicated that it is mediated via a class of nicotinic cholinergic receptors.[14–16] The responses are mimicked by the application of carbachol, a nonmetabolized analog of acetylcholine, and nicotine, but not by muscarine. In addition, it is potentiated by eserine and blocked by both curare and α-bungarotoxin (α-BTX). The nicotinic cholinergic receptors were localized to the membranes of the Schwann cell autoradiographically using α-BTX labeled with radioactive iodine.[17] Additional histochemical studies provided evidence for the presence and location of other elements of a cholinergic system in this preparation, including acetylcholinesterase[18] and choline acetyltransferase[19] activities. Surprisingly, gas chromatography-mass spectroscopy analysis revealed that acetylcholine was being synthesized, stored, and released from the Schwann cell itself.[20] This evidence suggested that the mechanism

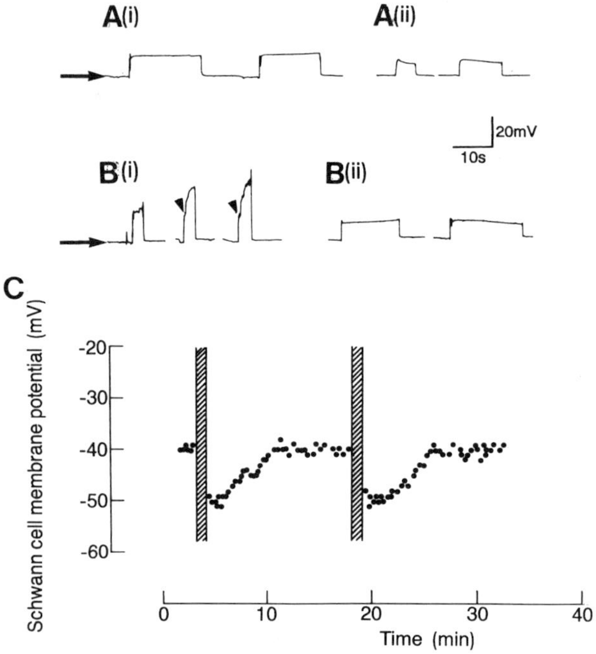

FIGURE 1. Examples of intracellular recordings of Schwann cell membrane potentials under control conditions (i) and after exposure to a 1-minute pulse of 10^{-7} M Substance P (ii) in two different nerve fibers (**A** and **B**). The *horizontal arrow* indicates the membrane potential of the giant axon. When the tip of the exploring microelectrode is passed through the axon to impale a Schwann cell, the potential difference trace rises. It drops back to the giant axon potential difference when the tip of the microelectrode is moved back into the axoplasm. In the last two traces in **B(i)** the Schwann cell membrane potential is seen as a transient notch (*arrowheads*) on the potential profile trace. (**C**) Hyperpolarization of the Schwann cell membrane induced by stimulation of the giant axon at 100 Hz for 1-minute periods (*vertical bars*). Each point represents the potential difference recorded in a different Schwann cell. Data are adapted from ref. 13.

of axon-Schwann cell signaling in the squid giant axon involves the release of an unknown agent from the axon upon stimulation. This agent initiates a change in the membrane properties of the Schwann cell by releasing acetylcholine from the Schwann cell which feeds back to activate nicotinic cholinergic receptors on the Schwann cell membrane. Such activation produces the long-term increase in potassium permeability underlying the acetylcholine-induced hyperpolarization of the Schwann cell membrane. Thus, the permeability properties of squid Schwann cells become more like those of other vertebrate[1] and invertebrate[3] glial cells for transient periods after giant axon activation.

The nicotinic cholinergic receptors on the Schwann cell are unusual in that they induce a long-lasting hyperpolarization which was recently shown to involve the second messenger cyclic AMP.[21] Evidence in support of the latter suggestion is that compounds that increase cyclic AMP levels by bypassing the receptor activation stage, such as phosphodiesterase inhibitors, forskolin, and cyclic AMP analogs, can all mimic the action of activation of the receptors by producing long-lasting hyperpolarization. In addition, the actions of carbachol can be potentiated in the presence of low doses of phosphodiesterase inhibitors and forskolin. Recently, the actions of forskolin in a number of preparations were demonstrated not to be due to adenylate cyclase activation.[22] However, the actions of forskolin on adenylate cyclase and on other sites of action can be differentiated by the use of the forskolin analog 1,9-dideoxyforskolin, which does not activate the adenylate cyclase system. We demonstrated recently that 1,9-dideoxyforskolin, at concentrations up to 2.5×10^{-5} M, neither induces hyperpolarization of the Schwann cell membrane nor potentiates the effect of a 1-minute pulse of 10^{-7} M carbachol in this preparation (Evans, Reale, Merzon, and Villegas, unpublished observations). Thus it is likely that the actions of forskolin on the Schwann cells are mediated by activation of the adenylate cyclase system. Direct receptor-mediated activation of adenylate cyclase would be the simplest explanation of the results just described, but they are also compatible with more complex modes of adenylate cyclase activation.[21] Recent evidence, however, suggests that these results may not be as unusual as first thought. Evidence is now accumulating for the existence of a wide range of distinct subtypes of nicotinic receptor[23,24] even in the squid nervous system.[25] In addition, nicotinic cholinergic receptors have been reported to increase cyclic AMP levels in chick embryo myocytes.[26] Furthermore, recent evidence shows that nicotinic cholinergic receptors can induce prolonged hyperpolarization of neurons in the rat dorsolateral septal nucleus,[27] which may also be mediated by a second messenger (J.P. Gallagher, personal communication).

MULTIPLE CLASSES OF GLUTAMATE RECEPTOR

One of the central questions in axon-Schwann cell signaling in the squid giant axon preparation that has long remained obscure is the nature of the signal released from the axon during stimulation to activate the Schwann cell cholinergic system. Early evidence suggested that this signal might be glutamate,[6] because the application of exogenous glutamate was able to induce hyperpolarization of the Schwann cell membrane through activation of the cholinergic system. Recently it was shown in two other species of squid, *Alloteuthis subulata* and *Loligo forbesi,* that Schwann cells of the giant axon possess glutamate receptors of the quisqualate/kainate subclass which can be activated by axonal stimulation.[28] However, there are a number of presumably species differences in the responses of the Schwann cells of the tropical squid *Sepioteuthis sepioidiae* with those reported for *Alloteuthis* and *Loligo.*[28] In view of this it would seem necessary to study each of the cholinergic, glutamatergic, aminergic, and peptidergic systems independently in these species because it may not be possible to extrapolate results on one system in one species to the same system in another species. Thus we have begun to characterize the nature of the glutamate receptors present on the Schwann cells of the tropical squid *Sepioteuthis sepioidiae* (Evans, Reale, Merzon, and Villegas, in preparation). FIGURE 2 shows that a

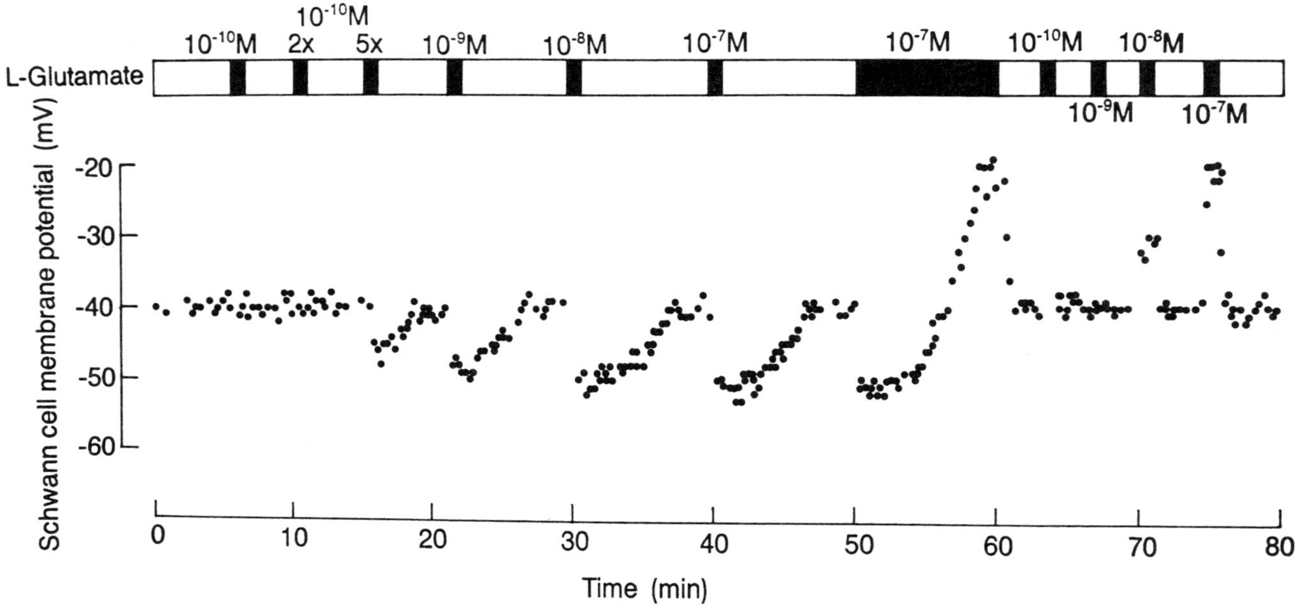

FIGURE 2. Effect of 1-minute pulses of various concentrations of L-glutamate and a 10-minute pulse of 10^{-7} M L-glutamate (*solid bars*) on the Schwann cell membrane potential. Each point represents the potential difference recorded in a different Schwann cell.

1-minute pulse of L-glutamate can produce a dose-dependent hyperpolarization of the Schwann cell membrane and that a prolonged pulse of L-glutamate gives a biphasic effect, with hyperpolarization followed by slow prolonged depolarization that lasts until the end of the glutamate pulse. In addition, after such a long glutamate pulse, the response to short glutamate pulses is reversed to give transient depolarizations. These gradually get smaller and reverse to give prolonged hyperpolarizations after about 20 minutes (data not shown). Previous work has indicated that the hyperpolarizations induced by L-glutamate are mediated by activation of the cholinergic system,[6–8] but the processes underlying the glutamate-induced depolarizations are unknown. We have now obtained evidence to suggest that both the slow and the rapid glutamate-induced depolarizations are mediated by the activation of glutamate receptors of the *N*-methyl-D-aspartate (NMDA) subclass of receptors. FIGURE 3 shows that the slow depolarizing effect of a long pulse of glutamate can be blocked by DL-amino-5-phosphonovaleric acid (APV), a specific NMDA type glutamate antagonist, with the slow hyperpolarization remaining untouched. Conversely, the slow hyperpolarization is blocked by DL-2-amino-4-phosphonobutyric acid (APB or AP4), a quisqualate/kainate type antagonist, whereas the slow depolarization is untouched by this compound. The conclusion we therefore draw is that in *Sepioteuthis* the glutamate-induced hyperpolarization is mediated by quisqualate/kainate subtype glutamate receptors, whereas the slow depolarization is due to the activation of NMDA type glutamate receptors. This finding in *Sepioteuthis* apparently is the first description of NMDA type glutamate receptors in any glial preparation. Furthermore, evidence for these conclusions comes from the fact that long pulses (10 minutes) of quisqualate or kainate up to concentrations of 10^{-5} M only produce hyperpolarizing responses of the Schwann cell membrane potential and short pulses of these compounds are unable to produce rapid depolarizations after a prolonged glutamate pulse. In contrast, long pulses of NMDA at concentrations between 10^{-9} M and 10^{-6} M produce exclusively a slow depolarization, whereas pulses of 10^{-5} M produce an initial hyperpolarization followed by a slow depolarization. In addition, after the initiation of a slow depolarization by either glutamate or NMDA, short 1-minute pulses of low doses of NMDA (around 10^{-9} M) can initiate rapid depolarizations. Pharmacological studies with a range of antagonists and agonists indicate pharmacological differences between the NMDA type glutamate receptors mediating the slow depolarization in response to a prolonged glutamate pulse and those mediating the rapid depolarizations found after the initiation of such a slow depolarization (Evans, Reale, Merzon, and Villegas, unpublished observations). Thus, in the tropical squid, *Sepioteuthis,* it appears that at least three distinct classes of glutamate receptor can be demonstrated. The quisqualate/kainate type receptors are activated after axonal stimulation to mediate a prolonged hyperpolarization of the Schwann cell membrane via activation of the cholinergic system. The two classes of NMDA receptor can also be activated after axonal stimulation (Evans, Reale, Merzon, and Villegas, unpublished observations), but the physiological implications of activation of the NMDA receptors are currently under investigation.

MODULATION BY OCTOPAMINE RECEPTORS

The extremely complex multistep interaction process between the squid giant axon and the Schwann cell presumably allows for amplification and possible spatial spread of the glial responses to low levels of a signal released from the axon. In addition, its multistep nature would also ideally suit it to be modulated by a number of different factors that could tailor the response of the Schwann cells to different

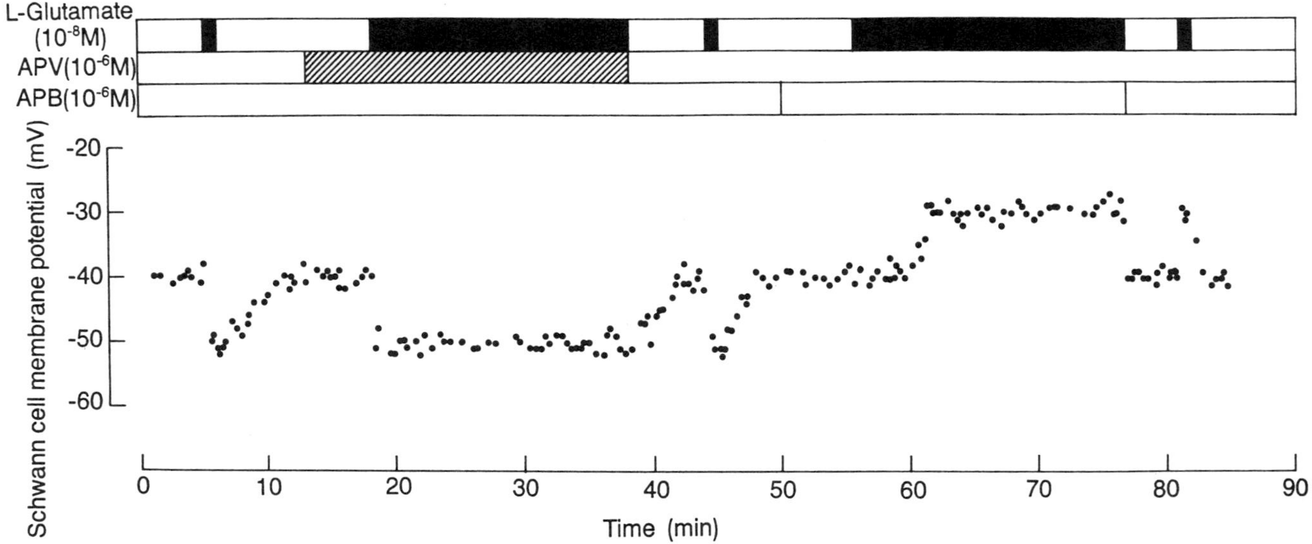

FIGURE 3. The biphasic response of the Schwann cell membrane potential to a prolonged pulse of L-glutamate (10^{-8} M) (*solid bar*) is converted into a rapidly induced sustained hyperpolarization in the presence of APV (10^{-6} M) (*hatched bar*) and into a slowly developing sustained depolarization in the presence of APB (10^{-6} M) (*open bar*). The hyperpolarizing response to a short pulse (1 minute) of L-glutamate reverses polarity after the induction of the maintained depolarization. Each point represents the potential difference recorded in a different Schwann cell.

behavioral situations in the intact animal. Thus we have investigated the modulatory influences of biogenic amines and peptides on this signaling pathway.

An initial survey of the effects of a range of biogenic amines on the Schwann cell membrane potential of the tropical squid (TABLE 1) revealed that it was selectively hyperpolarized by the phenolamine, octopamine, at concentrations above 10^{-8} M.[29] This response had many of the same pharmacological characteristics as octopamine receptor-mediated effects in other systems in terms of stereospecificity, amine specificity, and interactions with a range of agonists and antagonists. Evidence also suggested that it was again mediated by activation of adenylate cyclase. In addition, subthreshold doses of DL-octopamine (10^{-8} M), which alone produced no direct effect on the membrane potential of the Schwann cell, were able to potentiate the effects of submaximal doses of carbachol and of vasoactive intestinal peptide (VIP) (to be described). Because octopamine is only found at very low levels in the squid stellate nerve,[30] we have hypothesized that octopamine may be released into the squid hemolymph at times of stress, as in other invertebrates. This would mean that

TABLE 1. Actions of Amines on Schwann Cell Membrane Potential[a]

Amine	Hyperpolarizing Response ± SEM (mV)	n
D(−)-octopamine	12.3 ± 0.3	3
DL-octopamine	8.8 ± 0.3	17
Tyramine	8.6 ± 0.3	8
L(+)-octopamine	4.0 ± 0.1	3
L-adrenaline	3.2 ± 0.6	5
L-noradrenaline	2.8 ± 0.5	4
Phenylethanolamine	2.0, 3.0	2
5-Hydroxytryptamine	2.3 ± 0.3	3
DL-synephrine	0.8 ± 0.4	8
Dopamine	0	2
L-tyrosine	0	2

[a]The results are expressed as the mean ± standard error of the maximum hyperpolarization produced by the introduction of a 1-minute pulse of the amine into the superfusate at a concentration of 10^{-5} mol l^{-1}. n = number of observations for each amine on different nerve fibers. Results are taken from ref. 29.

hemolymph octopamine levels would be elevated at the same time as the squid would exhibit high levels of activity in the giant axons in connection with the escape response.

MODULATION BY NEUROPEPTIDE RECEPTORS

Glial cell responses in a wide range of preparations are also modulated by neuropeptides. However, the bulk of evidence for this has come from studies on the ability of neuropeptides to regulate cyclic AMP accumulation by clonal glial cell lines[31] and by primary cultures of glial cells.[32-35] The giant axon-Schwann cell preparation of the tropical squid provides an intact preparation in which anatomical relations are preserved in which to examine the effects of neuropeptides on axonal-Schwann cell signaling. A survey of the effects of a range of neuropeptides on Schwann cell membrane potential revealed that it was sensitive to a wide range of

TABLE 2. Hyperpolarizing Actions of a Range of Neuropeptides on Schwann-Cell Membrane Potential[a]

Neuropeptide	Hyperpolarizing Response ± SEM (mV)	n
	(a) VIP and Related Peptides	
VIP	10.9 ± 0.2	25
PHI	8.6 ± 0.9	4
$VIP_{10\text{-}28}$	7.0 ± 1.2	3
Secretin	0	4
Glucagon	0	4
	(b) Other Peptides	
Substance P	10.0 ± 0.3	10
Somatostatin	8.7 ± 0.5	5
APP	5.3 ± 0.5	4
[Met]-enkephalin	0	4

[a]The results are expressed as the mean ± SEM of mean of the maximum hyperpolarization produced by the introduction of a 1-minute pulse of the neuropeptide into the superfusate at a concentration of 10^{-7} M. n = the number of observations for each peptide. Results taken from ref. 37.

peptides from structurally diverse families[36] (TABLE 2). VIP is a very potent agent at inducing a long-lasting hyperpolarization of the Schwann cell membrane, with a threshold occurring between 10^{-10} and 10^{-9} M. This response is pharmacologically similar to VIP receptor-mediated effects in a range of other preparations because similar effects are produced by peptide histidine isoleucine (PHI) and by $VIP_{10\text{–}28}$, but not by glucagon and secretin. Further evidence for this point is that the actions of VIP-like peptides in this preparation can be competitively and reversibly blocked by VIP antagonists, such as (pCl-D-Phe,[6] Leu[17]) VIP and by (N-AcTyr,[1] D-Phe[7])-GRF(1–29) amide.[37] Additional blocking experiments reveal that the responses to VIP-like peptides are not mediated by interaction with either the cholinergic or the octopaminergic systems. In addition, they do not appear to be activated through a mechanism using adenylate cyclase, but the potentiation of the action of submaximal doses of VIP by lithium suggests involvement of the inositol phosphate pathway.

Long-lasting hyperpolarizations of the Schwann cell membrane potential can also be produced by a range of other neuropeptides, including Substance P, somatostatin, and avian pancreatic polypeptide (TABLE 2). [Met]-enkephalin does not produce any direct hyperpolarizations of the Schwann cell membrane at concentrations up to 10^{-7} M. However, at this concentration it blocks the hyperpolarizing actions of a range of agonists including VIP, octopamine, carbachol, Substance P, and somatostatin, as well as the hyperpolarization of the Schwann cell induced by stimulation of the giant axon for 1 minute at 100 Hz. At a concentration of 10^{-8} M the effect of [Met]-enkephalin can be partially antagonized by the presence of 10^{-5} M naloxone. The mode of action and physiological significance of the activation of the [Met]-enkephalin-activated receptors are unknown.

The actions of Substance P on the squid Schwann cell membrane potential were recently demonstrated to be mediated by receptors with a marked similarity to the SP-P or NK_1 subtype of Substance P receptor, where Substance P itself is the preferred agonist[13] (TABLE 3). A survey of the effectiveness of a range of Substance P fragments indicates that the direct hyperpolarizing effects of the Substance P molecule are mediated by peptides with an intact amidated C-terminal. However, a second subset of receptors, which can be activated by N-terminal fragments and analogs lacking an amidated C-terminal, are also present in this preparation. Interestingly, evidence was also obtained for the presence of a separate set of receptors activated by bradykinin, a neuropeptide with the same N-terminal pair of amino acids (Arg-Pro-) as Substance P. The non-subtype-specific Substance P antagonist D-Arg,[1] D-Trp,[7,9] Leu[11] Substance P (spantide) was a potent blocker of the effects of Substance P in this preparation, but it did not block the effects of VIP, carbachol, or octopamine.

The neuropeptides VIP and Substance P, as well as producing direct hyperpolarizations of the Schwann cell membrane potential, are also capable of modulating the responses of the Schwann cell to other agonists. Thus, subthreshold doses of VIP can potentiate the effects of both carbachol and octopamine, whereas subthreshold doses of Substance P can potentiate the effects of VIP, but not those of carbachol and octopamine.

The aforementioned effects of neuropeptides are interesting pharmacologically but raise the question of their physiological relevance. At least as far as the receptors activated by VIP on the squid Schwann cell are concerned, recent evidence suggests that they are likely to be the targets for an endogenously released VIP-like peptide in

TABLE 3. Hyperpolarizing Actions of Substance P and Structurally Related Tachykinins[a]

Neuropeptide	Hyperpolarizing Response (mV)	n
A. Substance P	11.1 ± 0.2	24
Neurokinin B (neuromedin K)	9.3 ± 0.5	4
Physalaemin	9.0 ± 0.7	4
Kassinin	8.3 ± 1.1	4
Elodoisin	8.0 ± 1.1	4
Neurokinin A (substance K)	7.5 ± 0.9	4
B. Substance P fragments		
Substance P (1–4)	0	4
Substance P (1–9)	0	4
Substance P free acid	0	4
Substance P methyl ester	0	4
Substance P	9.3 ± 0.3	27
Substance P (2–11)	5.4 ± 0.8	4
Substance P (3–11)	4.9 ± 0.3	4
Substance P (4–11)	6.0 ± 0.5	4
Substance P (5–11)	7.1 ± 0.7	5
Substance P (6–11)	8.0 ± 0.6	4
Substance P (7–11)	6.7 ± 0.5	4
Substance P (8–11)	8.3 ± 0.5	4
Substance P (9–11)	6.5 ± 0.5	8
C. Other peptides		
Substance P	12.9 ± 0.2	4
Bombesin	10.5 ± 0.2	4
Bradykinin	5.4 ± 0.2	8

[a]Results are expressed as the mean ± SEM of the hyperpolarizations produced by the introduction of a 1-minute pulse of the peptide into the superfusate at a concentration of 10^{-7} M. n = number of observations for each peptide. The result in **A, B,** and **C** were obtained with different batches of squid. Results are taken from ref. 13.

this preparation.[36,37] The hyperpolarization of the squid Schwann cell induced by firing the giant axon at 100Hz for 1 min is blocked in the presence of 10^{-8} M α-BTX. However, if the 100 Hz stimulation is given in the presence of 10^{-8} M α-BTX plus a subthreshold dose (10^{-8} M) of DL-octopamine, a further noncholinergic hyperpolarizing response can be revealed. The latter response can be blocked by 10^{-9} M (pCl-D-Phe,[6] Leu[17]) VIP, suggesting that activation of the giant axon in this preparation can cause the release of an endogenous VIP-like peptide. Because the aforementioned VIP-blocking agent also reduces the size of the hyperpolarization produced by 100 Hz stimulation in the absence of α-BTX, it suggests that the normal

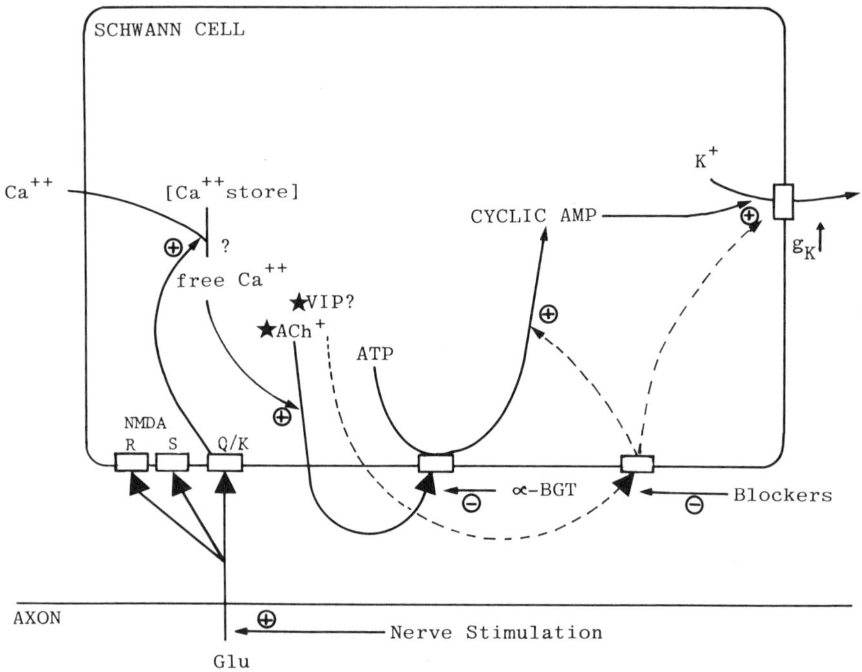

FIGURE 4. Diagrammatic representation of axon-Schwann cell signaling in the tropical squid, *Sepioteuthis sepioidiae.* NMDA = *N*-methyl-D-aspartate; R = rapid NMDA depolarizing receptor; S = slow NMDA depolarizing receptor; Q/K = quisqualate/kainate receptor; Glu = glutamate; + = potentiation; and − = inhibition.

long-lasting hyperpolarization of the Schwann cell membrane observed on stimulation of the giant axon is a composite response of a pure cholinergic hyperpolarization that is potentiated in amplitude and time course by the release of the VIP-like peptide. In addition, not enough VIP-like peptide is released under these conditions to give a direct hyperpolarization of the membrane in the presence of α-BTX to block the cholinergic response, and its presence can only be demonstrated if its actions are themselves potentiated by subthreshold doses of octopamine.

The source of the released endogenous VIP-like peptide has also been a matter of speculation, because the foregoing experiments do not distinguish between the release of peptide from the giant axon or the Schwann cell. However, recent experiments (Evans, Reale, Merzon, and Villegas, in preparation) suggest that the VIP-like peptide is likely to be co-released with acetylcholine from the Schwann cells inasmuch as the peptide can also be released by the application of short 1-minute pulses of glutamate to the preparation. In addition, conditions under which the release of acetylcholine from the Schwann cell is inactivated also block the release of the peptide, and the release of both substances recovers at the same rate. These results are consistent with the observations that VIP has been found in detectable concentrations in astrocytoma cell lines[38] and in nonneuronal fractions of brain tissue.[39,40]

CONCLUSIONS

In summary, our current understanding of the mechanisms underlying signaling between the squid giant axon and its surrounding Schwann cells reveals it to be a very complex process. FIGURE 4 shows a schematic representation of the main elements of this signaling process. Neuronal activation leads to the release of glutamate from the giant axon, which in the tropical squid can lead to the activation of three distinct classes of glutamate receptor. The physiological implications of stimulation of NMDA type receptors are not known, but activation of the quisqualate/kainate receptors leads to the corelease of acetylcholine and a VIP-like peptide from the Schwann cells. The exact response of the Schwann cell will likely depend on the relative amounts of acetylcholine and VIP-like peptide released under any particular pattern of physiological stimulation just as has been demonstrated for the co-release of these agents from neuronal systems.[41,42] Both the cholinergic and peptidergic components of the foregoing system can be modulated by the biogenic amine octopamine. If octopamine functions as a circulating neurohormone in the squid released under stressful circumstances, it would seem ideally situated to further modulate both components of the system under the exact conditions when they are most likely to be strongly activated, that is, under conditions favoring the initiation of an escape response. Most complex control mechanisms, in addition to having positive driving elements, also display negative control features to prevent overstimulation of the system. In interactions between the squid giant axon and its surrounding Schwann cells, either or both the enkephalin-activated system or the NMDA receptor-induced changes in the responsiveness of the system could be used to fulfill such a role. Thus, studies on the physiological significance of the latter two systems must have high priority in further studies on the mechanisms of signaling between the giant axon and Schwann cell of the tropical squid.

REFERENCES

1. ORKAND, R. K., J. G. NICHOLLS & S. W. KUFFLER. 1966. Effect of nerve impulses on the membrane potential of glial cells in central nervous system of amphibia. J. Neurophysiol. **29:** 788–806.
2. MARRENO, H., M. L. ASTION, J. A. COLES & R. K. ORKAND. 1989. Facilitation of voltage-gated ion channels in frog neuroglia by nerve impulses. Nature (Lond.) **339:** 378–380.
3. KUFFLER, S. W. & D. D. POTTER. 1964. Glia in the leech central nervous system: Physiological properties and neuron-glia relationship. J. Neurophysiol. **43:** 199–203.
4. WALZ, W. & W. R. SCHLUE. 1982. Ionic mechanisms of a hyperpolarizing 5-hydroxytryptamine effect on leech neuropile glial cells. Brain. Res. **250:** 111–121.

5. SCHOFIELD, P. K. & J. E. TREHERNE. 1985. Octopamine reduces potassium permeability of the glia that form the insect blood-brain barrier. Brain Res. **360:** 344–348.
6. VILLEGAS, J. 1978. Cholinergic properties of satellite cells in the peripheral nervous system. *In* Dynamic Properties of Glia Cells. E. Schoffeniels *et al.,* eds.: 207–215. Pergamon Press. Oxford.
7. VILLEGAS, J. 1981. Axon/Schwann cell relationships in the giant nerve fibre of the squid. J. Exp. Biol. **95:** 135–151.
8. VILLEGAS, J. 1984. Axon-Schwann cell relationship. Curr. Top. Membr. Transport. **22:** 547–571.
9. VILLEGAS, J., P. D. EVANS & V. REALE. 1988. Electrophysiology of Schwann cell receptors. *In* Glial Cell Receptors. H. K. Kimelberg, ed.: 141–157. Raven Press. New York, NY.
10. VILLEGAS, G. M. & R. VILLEGAS. 1968. Ultrastructural studies of the squid nerve fibers. J. Gen. Physiol. **51:** 44s–60s.
11. VILLEGAS, R., L. VILLEGAS, M. GIMÉNEZ & G. M. VILLEGAS. 1963. Schwann cell and axon electrical potential differences. Squid nerve structure and excitable membrane location. J. Gen. Physiol. **46:** 1047–1064.
12. VILLEGAS, J. 1972. Axon-Schwann cell interaction in the squid nerve fibre. J. Physiol. (Lond.) **225:** 275–296.
13. EVANS, P. D., V. REALE, R. M. MERZON & J. VILLEGAS. 1990. Substance P modulation of the membrane potential of the Schwann cell of the squid giant nerve fibre. Glia **3:** 393–404.
14. VILLEGAS, J. 1973. Effects of tubocurarine and eserine on the axon-Schwann cell relationship in the squid nerve fibre. J. Physiol. (Lond.) **232:** 193–208.
15. VILLEGAS, J. 1974. Effects of acetylcholine and carbamylcholine on the axon and Schwann cell electrical potentials in the squid nerve fibre. J. Physiol. (Lond.) **242:** 647–659.
16. VILLEGAS, J. 1975. Characterization of acetylcholine receptors in the Schwann cell membrane of the squid nerve fibre. J. Physiol. (Lond.) **249:** 679–689.
17. RAWLINS, F. A. & J. VILLEGAS. 1978. Autoradiographic localization of acetylcholine receptors in the Schwann cell membrane of the squid nerve fibre. J. Cell Biol. **77:** 371–376.
18. VILLEGAS, G. M. & J. VILLEGAS. 1974. Acetylcholinesterase localization in the giant nerve fibre of the squid. J. Ultrastruct. Res. **46:** 149–163.
19. HEUMANN, R., J. VILLEGAS & D. W. HERZFELD. 1981. Acetylcholine synthesis in the Schwann cell and axon in the giant nerve fiber of the squid. J. Neurochem. **36:** 765–768.
20. VILLEGAS, J. & D. J. JENDEN. 1979. Acetylcholine content of the Schwann cell and axon in the giant nerve fibre of the squid. J. Neurochem. **32:** 761–766.
21. EVANS, P. D., V. REALE & J. VILLEGAS. 1985. The role of cyclic nucleotides in modulation of the membrane potential of the Schwann cell of squid giant nerve fibre. J. Physiol. (Lond.) **363:** 151–167.
22. LAURENZA, A., E. MCHUGH SUTKOWSKI & K. B. SEAMON. 1989. Forskolin: A specific stimulator of adenylyl cyclase or a diterpene with multiple sites of action? Trends Pharmacol. Sci. **10:** 442–447.
23. CHIAPPINELLI, V. A. 1985. Actions of snake venom toxins on neuronal nicotinic receptors and other neuronal receptors. Pharmacol. Ther. **31:** 1–32.
24. BOULTER, J., J. CONNOLLY, E. DENERIS, D. GOLDMAN, S. HEINEMANN & J. PATRICK. 1987. Functional expression of two neuronal nicotinic acetylcholine receptors from cDNA clones identifies a gene family. Proc. Natl. Acad. Sci. **84:** 7763–7767.
25. CHEN, S.-J., R. SPATHIS & J. SCHMIDT. 1988. Binding sites for [^{3}H]-acetylcholine and ^{125}I-α-bungarotoxin in the optic ganglion of *Loligo pealii.* Comp. Biochem. Physiol. **90C:** 317–323.
26. ZANI, B. M., F. GRASSI, M. MOLINANO, L. MONACO & F. EUSEBI. 1986. Cyclic AMP regulates the life time of acetylcholine-activated channels in cultured myotubes. Biochem. Biophys. Res. Comm. **140:** 243–249.
27. WONG, L. A. & J. P. GALLAGHER. 1989. A direct nicotinic receptor-mediated inhibition recorded intracellularly *in vitro.* Nature (Lond.) **341:** 439–442.
28. LIEBERMAN, E. M., N. J. ABBOTT & S. HASSAN. 1989. Evidence that glutamate mediates axon-to-Schwann cell signalling in the squid. Glia **2:** 94–102.

29. REALE, V., P. D. EVANS & J. VILLEGAS. 1986. Octopaminergic modulation of the membrane potential of the Schwann cell of the squid giant nerve fibre. J. Exp. Biol. **121:** 421–443.
30. JUORIO, A. V. & P. B. MOLINOFF. 1974. The normal occurrence of octopamine in neural tissues of the *Octopus* and other cephalopods. J. Neurochem. **22:** 271–280.
31. VAN CALKER, D. & B. HAMPRECHT. 1980. Effects of neurohormones on glial cells. Adv. Cell. Neurobiol. **1:** 31–67.
32. ROUGON, G., M. NOBLE & A. W. MUDGE. 1983. Neuropeptides modulate the β-adrenergic response of purified astrocytes *in vitro.* Nature (Lond.) **305:** 715–717.
33. EVANS, T., K. D. MCCARTHY & T. K. HARDEN. 1984. Regulation of cyclic AMP accumulation by peptide hormone receptors in immunocytochemically defined astroglial cells. J. Neurochem. **43:** 131–138.
34. CHNEIWEISS, H., J. GLOWINSKI & J. PRÉMONT. 1984. Vasoactive intestinal polypeptide receptors linked to an adenylate cyclase, and their relationship with biogenic amine- and somatostatin-sensitive adenylate cyclases on central neuronal and glial cells in primary culture. J. Neurochem. **44:** 779–786.
35. KOH, S.-W. M., A. KYRITSIS & G. J. CHADER. 1984. Interaction of neuropeptides and cultured glial (Müller) cells of the chick retina: Elevation of intracellular cyclic AMP by vasoactive intestinal peptide and glucagon. J. Neurochem. **43:** 199–203.
36. EVANS, P. D., V. REALE & J. VILLEGAS. 1986. Peptidergic modulation of the membrane potential of the Schwann cell of the squid giant nerve fibre. J. Physiol. (Lond.) **379:** 61–82.
37. EVANS, P. D. & J. VILLEGAS. 1988. The action of vasoactive intestinal peptide antagonists on peptidergic modulation of the squid Schwann cell. J. Exp. Biol. **138:** 259–269.
38. SAID, S. I. & R. N. ROSENBERG. 1976. Vasoactive intestinal polypeptide: Abundant immunoreactivity in neural cell lines and normal nervous tissue. Science **192:** 907–908.
39. BESSON, J., W. ROTSZTEJN, M. LABURTHE, J. EPELBAUM, A. BEAUDET, C. KORDON & G. ROSSELIN. 1979. Vasoactive intestinal peptide (VIP): Brain distribution, subcellular localization and effect of deafferentation of the hypothalamus, in male rats. Brain Res. **165:** 79–86.
40. ROSTÈNE, W. H. 1984. Neurobiological and neuroendocrine functions of the vasoactive intestinal peptide (VIP). Prog. Neurobiol. **22:** 103–129.
41. LUNDBERG, J. M. 1981. Evidence for the coexistence of vasoactive intestinal polypeptide (VIP) and acetylcholine in neurons of cat exocrine glands. Acta Physiol. Scand. **112** (suppl 496): 1–57.
42. LUNDBERG, J. M., A. ÄNGGÅRD, J. FAHRENKRUG, G. LUNDGREN & B. HOLMSTEDT. 1982. Co-release of VIP and acetylcholine in relation to blood flow and salivary secretion in cat mandibular salivary gland. Acta Physiol. Scand. **115:** 525–528.

Role of Glutamate in Axon-Schwann Cell Signaling in the Squid[a,b]

EDWARD M. LIEBERMAN

Department of Physiology
School of Medicine
East Carolina University
Greenville, North Carolina 27858

Glial (Schwann) cells ensheathing axons historically have been considered to form a *passive* barrier, restricting the movement of substances between the periaxonal and bulk extracellular fluid spaces. This view originated with the observations of Frankenhaeuser and Hodgkin[1] that trains of action potentials in axons caused an accumulation of K^+ in the periaxonal space. The degree of accumulation was monitored by the magnitude of the transient decrease in the action potential undershoot and depolarization of the resting membrane potential. The potential changes were graded, depending on frequency and duration of the action potential trains with a time of recovery to prestimulation control values of tens of milliseconds. These properties were suggestive of a highly restrictive diffusion pathway away from the axon through glial intercellular spaces.

These observations were confirmed in studies by Adelman and colleagues[2,3] of the development and time of disappearance of voltage-clamp-generated K^+ tail currents and of the anatomical basis for series electrical properties of the squid giant axon. The direct effect of neuronal excitation on glial cells in the central nervous systems (CNSs) of both invertebrate[4,5] and vertebrate[6,7] animals demonstrated that K^+ that accumulated in the perineural space caused adjacent glial cells to depolarize in a Nernstian manner. These results added to the evidence that the glial barrier was primarily passive in its handling of excess periaxonal/perineural K^+.

As a result of these early investigations, models of periaxonal K^+ regulation have emphasized passive mechanisms whereby K^+ released during excitation serves as the primary, if not sole, signal for its own regulation. Two mechanisms, in addition to axonal reuptake processes, were proposed for periaxonal K^+ regulation, that is, *diffusion*[1,3,5] through glial intercellular spaces driven by a difference in concentration between the periaxonal and bulk extracellular spaces and *spatial buffering,*[7] the transcellular movement of K^+ driven by electrical gradients generated by asymmetric accumulation of K^+ across the glial cell layer.

The view that significant K^+ accumulation was an inevitable consequence of action potential generation led to the proposition that accumulation of periaxonal (perineural) K^+ and its effect on nerve excitability were a property of signal processing in the CNS of some animals.[8,9]

SCHWANN AND SCHWANN-LIKE CELLS OF SQUID AND CRAYFISH

The impetus for a reexamination of the excitation-induced periaxonal K^+ accumulation was the finding that accumulation of K^+ in the periaxonal space of squid axons

[a]This work was supported in part by grants from the Army Research Office (DAAG86-K-0023) and the Burroughs-Wellcome Fund.

[b]Original experimental data presented in this review are currently in press in Neuroscience.[26]

during excitation is not inevitable[10] and that in crayfish medial giant axons significant accumulation of K^+ rarely occurs under normal circumstances, regardless of the frequency and duration of action potential trains.[11] In the crayfish[12,13] 95% of the K^+ released into the periaxonal space during a voltage clamp pulse was cleared before the end of the pulse, and the remaining 5% cleared within 5–10 ms of the end of the pulse. In contrast to earlier investigations,[3,12] it also was demonstrated that K^+ is cleared from the periaxonal space of carefully isolated squid giant axons with a similarly rapid time course.[13]

Beginning with studies of the Schwann cell of the tropical squid *Sepioteuthis* in the late 1960s, Villegas[14] and colleagues dispelled the widely held view that the squid axon Schwann sheath was physiologically passive. The Schwann cells of both squid[14] and crayfish[24] are low potential cells (−40 mV) which respond to axonal stimulation with up to a 15 mV hyperpolarization, graded proportionately to the frequency and duration of stimulation.[15] This is in contrast to CNS glia of both invertebrates and vertebrates that have high resting potentials (up to −90 mV) and depolarize in response to excitation-induced periaxonal K^+ accumulation. The Schwann cell hyperpolarization is nicotinic/cholinergic receptor mediated[14] and associated with a cAMP second messenger system.[16] It also has been shown that the Schwann cell, not the axon, is the source of the ACh required for the hyperpolarization to occur and that K^+ alone is not a sufficient signal to induce the hyperpolarization response.[14] These results suggest that an additional axon-to-Schwann cell communication link must be operating.

GLUTAMATE AS THE MEDIATOR OF AXON-TO-SCHWANN CELL SIGNALING

During the 1970s it became increasingly apparent that glutamate was a likely transmitter at the neuromuscular junction of a variety of marine and aquatic invertebrate animals and a possible transmitter at the giant synapse of the squid.[17,18] In addition, glutamate is found in high concentration (25 mM) in the axoplasm of the squid giant axon.[19] It is not surprising then that glutamate would be investigated as a candidate for the axon-to-Schwann signal. Support for this hypothesis was provided by Villegas,[20] showing that externally applied glutamate (10^{-7} M) initiated the hyperpolarizing sequence in the squid Schwann cell.

Recent work from this laboratory[15] has shown that glutamate or a glutamate-like substance is released from the axon during excitation to activate a quisqualate/kainate receptor on the Schwann cell. Within seconds of the start of a train of action potentials in the squid giant axon, associated Schwann cells depolarize 8–10 mV. The depolarization is rapidly replaced with a long-lasting hyperpolarization which has been described by Villegas[14] as the nicotinic/cholinergic/cAMP-mediated response.

We found that the stimulation or glutamate-induced sequence of depolarization followed by hyperpolarization was blocked by 10^{-5} M 2-amino-4-phosphonobytyric acid (2-APB), whereas aspartate (10^{-7} M) which also caused a hyperpolarization was not blocked by 2-APB. Although an aspartate receptor seems to be present on the Schwann cell surface, it is not the excitation-released agonist responsible for the Schwann cell depolarization/hyperpolarization event. In tests with other glutamate agonists and antagonists, quisqualate (10^{-5} M) and kainate (10^{-4} M) also caused a hyperpolarization that was blocked by 2-APB or 10^{-4} M L-glutamic acid diethylester (GDEE). The glutamate receptor agonists *n*-methyl-D-aspartic acid (NMDA; 10^{-4} M), ibotenate (10^{-5} M), and α-amino-3-hydroxy-5-methyl-isoxazole (AMPA; 10^{-4} M) had no effect. Thus, the results support the conclusion that the glutamate receptor involved in axon-to-Schwann cell signaling is of a quisqualate/kainate type. On the basis of the pharmacology of glutamate activation of the depolarization/hyperpolarization sequence we tentatively suggested that the depolarization caused by glutamate is a factor in the release of Schwann cell ACh.[15]

Although the ionic basis for both the depolarization/hyperpolarization sequence is not completely understood at this time, it has been proposed that changes in Schwann cell membrane potential contribute to regulation of the K^+ concentration in the periaxonal microenvironment.[21]

STUDIES IN PROGRESS

Ionic Basis for the Glutamate-Induced Depolarization of the Schwann Cell and Its Relation to the Hyperpolarizing Response

Our most recent work[22] on the role of glutamate as a mediator of axon-to-Schwann cell communication dealt with the nature of the glutamate-initiated steps leading to the release of ACh by the Schwann cell. Although the depolarization precedes the hyperpolarizing response, it is not clear whether the former is a prerequisite for the latter.

The methodology for these studies has previously been published in detail[16,23,24] but will be briefly described here. Schwann cells associated with giant axons of the small squid, *Alloteuthis* or *Loliguncula,* and the large squid, *Loligo,* were used in our studies. Membrane potentials of many Schwann cells in the same axon during an experimental manipulation were sampled to give both a time course and an average population response. Standard 2.5 M KCl-filled glass micropipette electrodes and high impedance amplifier recording techniques were used. Axons and Schwann cells were studied during axonal stimulation (100 Hz/15 seconds), application of ACh or glutamate receptor agonists or antagonists, and ion substitutions. It should be noted that axon electrical properties were not sensitive to any of the glutamatergic or cholinergic agonists or antagonists used in these studies.

The resting membrane potential of a Schwann cell associated with the squid giant axon is approximately −40 mV (FIG. 1, TABLE 1). Axon stimulation at 100 Hz for 15 seconds causes Schwann cells to initially depolarize before hyperpolarizing to approximately −50 to −55 mV. The depolarization rapidly develops within seconds of the start of stimulation and is replaced by a hyperpolarization that peaks in 3–4 minutes before returning to its control level of −40 mV in 6–8 minutes. In the presence of *d*-tubocurarine (dTC; 10^{-8} M) used to block the development of hyperpolarization, stimulation causes a transient 8–10 mV depolarization lasting approximately 4–6 minutes. Glutamate (10^{-7} M) application causes a qualitatively similar sequence of potential changes (FIG. 2). It does appear, though, that the step(s) associated with the hyperpolarizing segment of the response desensitize with prolonged (10–12 minutes) glutamate treatment (FIG. 3). The quisqualate/kainate receptor antagonist 2-APB blocks both the stimulation and the glutamate-induced potential changes. It is especially significant that in the presence of 2-APB, the Schwann cell can be made to hyperpolarize with direct application of cholinergic agonists (but not stimulation), indicating that depolarization and hyperpolarization are pharmacologically independent steps in the glutamate-initiated sequence. Although depolarization can be isolated from hyperpolarization, that is, with dTC pretreatment the depolarization can occur in the absence of hyperpolarization, it is not apparent that the inverse is also true.

The ionic basis for the glutamate-induced depolarization of the Schwann cell was

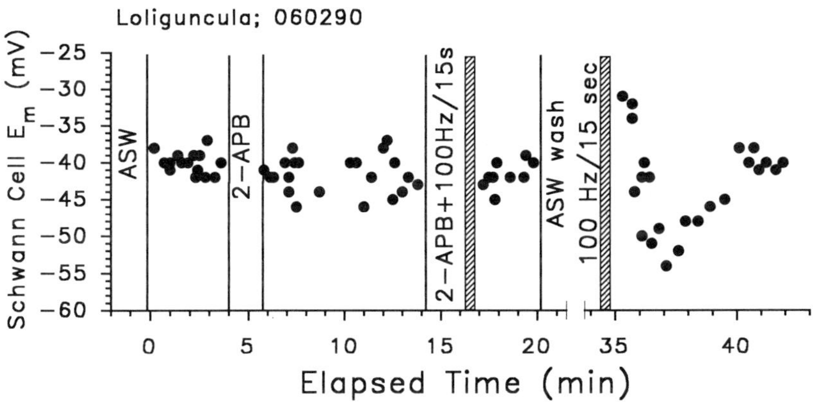

FIGURE 1. Effect of 100 Hz/15 second stimulation on the membrane potential of Schwann cells associated with the giant axon of the small squid *Loliguncula brevis.* Reading the panels from *left* to *right,* the resting membrane potential of Schwann cells in artificial seawater (ASW) averages −40.0 ± 0.5 mV (frame 1). 2-APB (10^{-4} M), a glutamate receptor antagonist, has no effect on the Schwann cell membrane potential (frame 2). Stimulation of the axon in the presence of 2-APB prevents any response from the Schwann cell (frame 3). Stimulation of the axon following the washout of 2-APB causes the Schwann cells to depolarize rapidly to approximately −32 mV. The depolarization is rapidly replaced by a transient hyperpolarization, peaking at −55 mV. The entire sequence of events lasts approximately 6–7 minutes.

studied by pharmacological and ion substitution procedures. Removal of extracellular Na^+ (replaced with Tris) or application of tetrodotoxin (TTX) in the presence of Na^+ and dTC blocked the depolarizing response to bath-applied glutamate (FIGS. 3 and 4).

In the absence of dTC, low external Na^+ or TTX treatment did not prevent the

TABLE 1. Effects of Ions on the Membrane Potential of Schwann Cells of Squid Giant Axons under a Variety of Agonist and Antagonist Conditions[a]

	Artificial Seawater (ASW)	*d*-Tubocurarine (dTC) (10^{-7} M)	Stimulation (100 Hz/15 s)	Stimulation/ dTC	Glutamate (10^{-7} M)	Glutamate/ dTC
ASW	−40.5 ± 0.1	−40.5 ± 0.2	−52.3 ± 0.5	−32.9 ± 0.5	−52.9 ± 0.9	−33.2 ± 0.2
(*n*)	(24)	(21)	(8)	(7)	(2)	(13)
0 Ca^{2+}[b]	−40.6 ± 0.4	. . .	−48.8 ± 1.0	. . .	−47.7	. . .
(*n*)	(5)		(6)		(1)	
0 Na^+	−41.2 ± 0.2	−40.4 ± 0.3	. . .	. . .	−51.5 ± 0.2	−41.5 ± 0.7
(*n*)	(4)	(9)			(2)	(6)
0 Cl^-	−49.9 ± 0.8	−49.2 ± 0.8	−43.1 ± 0.3	. . .	−41.6 ± 0.5	−44.0
(*n*)	(4)	(4)	(3)		(2)	(1)
TTX	−41.1 ± 0.1	−40.2 ± 0.3	. . .	. . .	−50.4 ± 0.7	−39.0 ± 0.9
(*n*)	(4)	(6)			(2)	(7)

[a]Data compiled from experiments on 29 axons; 2 or more axons were used for any one treatment condition.

[b]Also includes data from two experiments in which 75 mM Mg^{2+} was added to the nominally 0 Ca^{2+}-ASW.

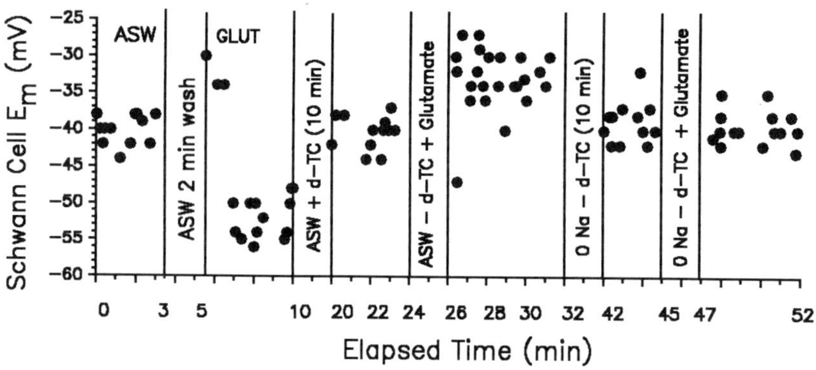

FIGURE 2. Effect of glutamate on the membrane potential of Schwann cells. Glutamate causes a transient depolarization/hyperpolarization sequence much as that seen with stimulation. If the Schwann cells are pretreated with 10^{-8} M dTC (frame 3), the addition of glutamate causes only a depolarization (frame 4). Treatment of Schwann cells in Na^+-free + dTC ASW prevents the Schwann cell from responding to glutamate (frame 6).

development of hyperpolarization with glutamate application even though the depolarization did not occur (FIGS. 3 and 4). It is clear from these results that the glutamate-induced depolarization is due to an increase in Na^+ influx, but most importantly that the depolarization is not a prerequisite for the development of hyperpolarization. Because it is also clear that glutamate-receptor activation is still a requirement for hyperpolarization, another receptor-activated step must be involved.

Quisqualate/kainate receptor activation has been associated with an increase in inositol phospholipid turnover and an increase in intracellular Ca^{2+} in a number of cell types including glia.[25] We therefore examined the possibility that glutamate, in addition to initiating Na^+ influx, causes activation of Ca^{2+} influx and/or release of Ca^{2+} from intracellular stores. An increase in intracellular Ca^{2+} may then act as it does in a variety of typical synaptic transmitter-releasing processes to release ACh and activate the hyperpolarization step in the Schwann cell. A decrease in extracellular Ca^{2+} to nominally zero, coupled with an increase in Mg^{2+} (75 mM), did not block the stimulation or glutamate-induced hyperpolarization, but low Ca^{2+}/high Mg^{2+} conditions did cause a slight decrease ($\sim$4 mV) in hyperpolarization (TABLE 1).

To determine if glutamate receptor activation caused an increase in intracellular Ca^{2+}, the axon and Schwann cells were loaded with the intracellular Ca^{2+} buffer Bapta/AM (30 μM). Under these circumstances glutamate or 100 Hz/15 second stimulation caused only a depolarization of the Schwann cell. These results implicate an intracellular release of Ca^{2+} by glutamate receptor activation and suggest the involvement of an inositol phosphate/Ca^{2+} triggering event for activation of the cholinergic step in the system.

Further studies of this mechanism are under way. Preliminary results are consistent with the existence of an IP_3 mechanism in the axon-Schwann cell preparation. [^{3}H]myoinositol incorporation into membranes of the axon-Schwann cell system is high (TABLE 2). Treatment of the preparation with glutamate tends to increase the incorporation of [^{3}H]myoinositol into the membrane fraction of the axon-Schwann cell preparation (25%), whereas 2-APB slightly decreases the membrane content of [^{3}H]myoinositol (15%). Particularly interesting is that, in combination, glutamate and 2-APB reduce the [^{3}H]myoinositol content of the membrane to 46% of its control value (statistically different from control at $p < 0.05$). Further work in this area is needed to quantitate these findings and to partition the IP_3 system between the Schwann cell and axon. Nonetheless, the results suggest that this approach may provide significant new information on mechanisms involved in the signaling system between axons and their associated Schwann cells.

Ionic Basis for the Cholinergically Induced Hyperpolarization of the Schwann Cell and its Relation to Periaxonal K^+ Regulation

Previous studies of axon-to-Schwann cell signaling in the crayfish giant axon indicate that cholinergically induced hyperpolarization of the Schwann cell is due to a decrease in an outwardly directed high Cl^- flux present in the resting condition.[24]

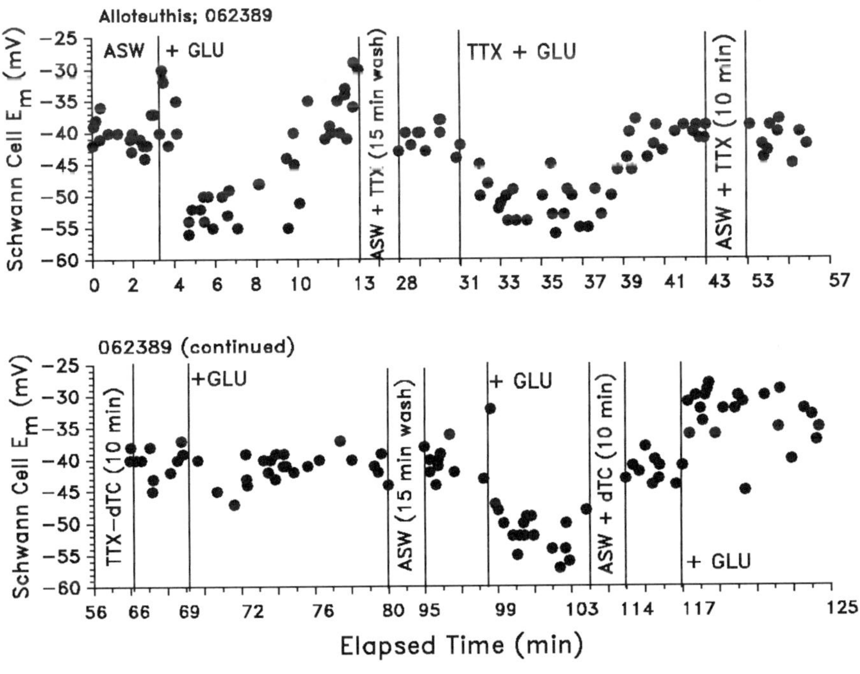

FIGURE 3. Effect of Na^+ and TTX on the depolarizing response to glutamate in dTC. Frame 2 (*upper panel*) demonstrates that prolonged treatment of Schwann cells with glutamate causes a desensitization to occur. The initial depolarization/hyperpolarization sequence occurs, but in 6–7 minutes the Schwann membrane potential reduces to the glutamate-induced depolarization level. Pretreatment of the Schwann cells with TTX (*panel 3*) or Na^+-free solution prevents the depolarization portion of the biphasic response but has no effect on the hyperpolarization induced by glutamate or stimulation (frame 4). Note also in frame 4 that prolonged glutamate treatment causes a desensitization that brings the membrane back to the normal resting membrane potential, suggesting that the TTX-sensitive Na^+ channel has been blocked. In the first two frames of the *lower panel* it is demonstrated that the combination of TTX and dTC renders glutamate ineffective in causing changes in Schwann cell potential. The actions of TTX and dTC are reversible as shown in the remainder of the frames.

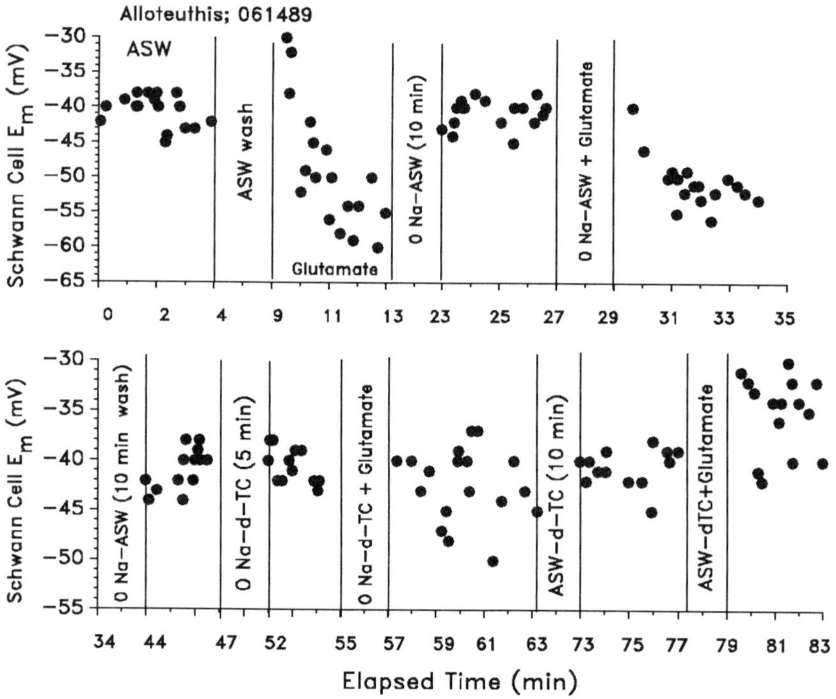

FIGURE 4. Effects of Na⁺-free solution on the glutamate-induced membrane potential changes of Schwann cells. Glutamate causes the typical depolarization/hyperpolarization sequence of Schwann cell membrane potential (frame 2, *upper panel*). In Na⁺-free conditions the glutamate causes a hyperpolarization only (frame 4). In the *lower panel* of experiments the combination of dTC and Na⁺-free conditions prevents all effects of glutamate. Returning the preparation to a Na⁺-containing solution containing dTC allows the depolarization to occur but blocks the cholinergically induced hyperpolarization.

This property of the Schwann cell is expressed as a long-lasting hyperpolarization when it is exposed to a Cl^--free solution. Schwann cells associated with the squid giant axon respond to Cl^--free solutions much as those of the crayfish. They hyperpolarize in Cl^--free solution to the extent expected following stimulation, glutamate, or ACh application in Cl^--containing solution (10–15 mV) (TABLE 1).

TABLE 2. Incorporation of [^{3}H]Myoinositol into the Membrane Fraction of Axons and Schwann Cells of the Squid

	Artificial Seawater	Glutamate (GLU)	2-Amino-4-Phosphonobutyric Acid (2-APB)	2-APB + GLU
Counts/g wt wgt[a] (CPM)	456,877	458,587	202,085	190,509
% Control (n = 7) ± SEM	100	123 ± 21	86 ± 25	46 ± 10

[a]Data from a single experiment to illustrate the quantity of incorporated myoinositol and the nature of myoinositol turnover in membranes of an axon-Schwann cell tissue.

Stimulation of an axon bathed in a Cl^--depleted solution causes the Schwann cell to transiently depolarize (8 mV) from its hyperpolarized level (TABLE 1). This is the same magnitude of depolarization expected with stimulation or glutamate treatment of a dTC-treated Schwann cell in Cl^--containing solution where its normal resting potential is −40 mV (TABLE 1). The results strongly suggest that in the squid, as in the crayfish, the cholinergically induced hyperpolarization is due to a decrease in outwardly directed Cl^- flux.

Modulation of the Cl^- conductance of the Schwann cell is an essential property for its participation in regulation of the periaxonal ionic microenvironment.[21,24] Hyperpolarization of the Schwann cell develops because the selectivity of the membrane for K^+ increases as the outwardly directed Cl^- conductance decreases. In these conditions the membrane potential is driven toward the K^+ equilibrium potential, and the K^+ electrochemical gradient is reduced, thereby decreasing the efflux of glial K^+ into the periaxonal space. There are two immediate consequences of this event. Firstly, the reduced glial K^+ efflux increases the apparent "dilution volume" for K^+, leaving the axon during stimulation. That is, axonal K^+ efflux can be increased up to an amount equal to the reduction in glial K^+ efflux without changing periaxonal K^+ concentration. The second consequence of a decreased Cl^- conductance-induced hyperpolarization is the reduced need for the glial cell to transport K^+ (at an energy cost) for its own intracellular ion concentration maintenance. Nonetheless, K^+ entering the periaxonal space from the stimulated axon keeps the Na^+/K^+ pump of both the axon and glial cells activated to at least normal resting levels. Thus, with no greater energy output than that expended under normal resting conditions the hyperpolarized glial cell can take up extra K^+ released by the axon during excitation.

This mechanism is best suited for K^+ regulation during prolonged periods of excitation when *diffusion* and *spatial buffering* mechanisms for clearance of excess K^+ are saturated.

FUTURE WORK

Although the studies described here provide a basis for understanding the relation between the axon during excitation and its surrounding Schwann sheath as it relates to ionic homeostasis of the periaxonal microenvironment, several questions remain to be answered.

The glutamate-induced Na^+ influx and depolarization, although not necessary for the Schwann cell hyperpolarization, are a simultaneous event to the proposed activation of the IP_3 second messenger system. Although its role can only be guessed, it might be speculated that the influx of Na^+ serves to "prime" the glial Na^+-pump mechanism to insure an activated Na^+/K^+ transporter required for the clearance of excess K^+ while helping to maintain Na^+ in the periaxonal space lost to the axon during action potential generation.

A second area of importance for further study is the mechanism of glutamate release and its quantitative coupling to K^+ efflux from the axon during excitation. It must be kept in mind that glutamate appears to be released from sites along the axon membrane where synaptic structures are not found and vesicular structures are only occasionally observed. It can be speculated that its release into the periaxonal space during excitation occurs through regulation (possibly by extracellular K^+) of a glutamate leak-pump mechanism. For example, extracellular glutamate could increase by a K^+-sensitive reversal of a glutamate uptake mechanism known to exist in excitable cells; through a decrease in uptake in the presence of a constant leak down its electrochemical gradient; or through an increase in its leak unbalanced by an increase in its uptake. In addition to its role in ion homeostasis of the periaxonal/perineural space, the release of glutamate along axons has significant implications for excitatory neuromodulatory activity in the CNS of vertebrates.

CONCLUSION

Recent work from this laboratory has provided evidence for the hypothesis that glutamate, released in a graded manner by the axon during excitation, is the mediator of axon-to-Schwann cell signaling and is the first step in the cascade leading to Schwann cell hyperpolarization. Although the mechanism has not been fully described, evidence suggests that activation of a quisqualate/kainate glutamate receptor results in a sequence of (1) an intracellular release of Ca^{2+}, (2) a Ca^{2+}-induced release of ACh from the Schwann cell, (3) activation of the Schwann cell cholinergic/cAMP system, and (4) Schwann cell hyperpolarization. Schwann cell hyperpolarization has been suggested to decrease the energy required to reabsorb excess K^+ from the periaxonal space and may be an important mechanism for long-term periaxonal ionic homeostasis and preservation of neural excitability.

ACKNOWLEDGMENTS

I would like to thank the directors and staff of the Marine Biological Association laboratory, Plymouth, England, and of the Duke University Marine Laboratory, Beaufort, North Carolina, for facilities.

REFERENCES

1. FRANKENHAEUSER, B. & A. L. HODGKIN. 1956. The after-effects of impulses in the giant nerve fibers of *Loligo*. J. Physiol. (Lond.) **131:** 341–376.
2. ADELMAN, W. J., JR., J. MOSES & R. V. RICE. 1977. An anatomical basis for the resistance and capacitance in series with the excitable membrane of the squid giant axon. J. Neurocytol. **6:** 621–646.
3. ADELMAN, W. J., JR. & Y. PALTI. 1972. The role of periaxonal and perineural spaces in modifying ionic flow across neural membranes. Curr. Top. Mem. Trans. **3:** 199–235.
4. KUFFLER, S. W. & D. D. POTTER. 1964. Glia in the leech central nervous system: Physiological properties and neuron-glia relationship. J. Neurophysiol. **27:** 290–319.
5. NICHOLLS, J. G. & S. W. KUFFLER. 1964. Extracellular space as a pathway for exchange between blood and neurons in the central nervous system of the leech: Ionic composition of glial cells and neurons. J. Neurophysiol. **27:** 645–671.
6. KUFFLER, S. W., J. G. NICHOLLS & R. K. ORKAND. 1966. Physiological properties of glial cells in the central nervous system of amphibia. J. Neurophysiol. **29:** 768–787.
7. ORKAND, R. K., J. G. NICHOLLS & S. W. KUFFLER. 1966. Effect of nerve impulses on the membrane potential of glial cells in the central nervous system of amphibia. J. Neurophysiol. **27:** 788–806.
8. BAYLOR, D. A. & J. G. NICHOLLS. 1968. Changes in extracellular potassium concentration produced by neuronal activity in the central nervous system of the leech. J. Physiol. (Lond.) **203:** 555–569.
9. BAYLOR, D. A. & J. G. NICHOLLS. 1968. After-effects of nerve impulses on signalling in the central nervous system of the leech. J. Physiol. (Lond.) **203:** 571–589.
10. ABBOTT, N. J., E. M. LIEBERMAN, Y. PICHON, S. HASSAN & Y. LARMET. 1988. Periaxonal K^+ regulation in the small squid *Alloteuthis*. Biophys. J. **53:** 275–279.

11. HASSAN, S. & E. M. LIEBERMAN. 1988. Studies of axon-glial cell interactions and periaxonal K^+ homeostasis. II. The effect of axonal stimulation, cholinergic agents and transport inhibitors on the resistance in series with the axon membrane. Neuroscience **25:** 961–969.
12. SHRAGER, P., J. C. STARKUS, M.-V. C. LO & C. PERACCHIA. 1983. The periaxonal space of crayfish giant axons. J. Gen. Physiol. **82:** 221–244.
13. PICHON, Y., N. J. ABBOTT, E. M. LIEBERMAN & Y. LARMET. 1987. Potassium homeostasis in the nervous system of cephalopods and crustacea. J. Physiol., Paris **82:** 346–356.
14. VILLEGAS, J. 1984. Axon-Schwann cell relationship. Curr. Top. Membr. Transp. **22:** 547–571.
15. LIEBERMAN, E. M., N. J. ABBOTT & S. HASSAN. 1989. Evidence that glutamate mediates axon-to-Schwann cell signaling in the squid. Glia **2:** 94–102.
16. EVANS, P. D., V. REALE & J. VILLEGAS. 1985. The role of cyclic nucleotides in modulation of the membrane potential of the Schwann cell of squid giant nerve fibre. J. Physiol. (Lond.) **363:** 151–167.
17. KRAVITZ, E. A., C. R. SLATER, K. TAKAHASHI, M. D. BOWNDS & R. M. GROSSFELD. 1970. Excitatory transmission in invertebrates—Glutamate as a potential neuromuscular transmitter compound. *In* Excitatory Synaptic Mechanisms. P. Andersen & J. K. S. Jansen, eds. :85–93. Universitets Forlaget Publisher. Uppsala, Sweden.
18. ADAMS, D. J. & J. I. GILLESPIE. 1988. The actions of L-glutamate at the postsynaptic membrane of the squid giant synapse. J. Exp. Biol. **140:** 535–548.
19. DEFFNER, G. J. 1961. The dialyzable free organic constituents of squid blood: A comparison with nerve axoplasm. Biochim. Biophys. Acta **47:** 378–388.
20. VILLEGAS, J. 1978. Cholinergic properties of satellite cells in the peripheral nervous system. *In* Dynamic Properties of Glial Cells. E. Schoffeniels, G. Franck, L. Hertz & D. B. Tower, eds. :207–215. Pergamon Press Ltd. Oxford.
21. LIEBERMAN, E. M. & S. HASSAN. 1988. Studies of axon-glial cell interactions and periaxonal K^+ homeostasis. III. The effect of anisosmotic media and potassium on the relationship between the resistance in series with the axon membrane and glial cell volume. Neuroscience **25:** 971–981.
22. LIEBERMAN, E. M. & E. SANZENBACHER. 1989. Possible mechanisms for glutamate mediation of axon-to-Schwann cell interaction in isolated giant axon of the squid. J. Physiol. (Lond.) **423:** 90P.
23. LIEBERMAN, E. M., J. VILLEGAS & G. VILLEGAS. 1981. The nature of the membrane potential of glial cells associated with the medial giant axon of the crayfish. Neuroscience **6:** 261–271.
24. BRUNDER, D. G. & E. M. LIEBERMAN. 1988. Studies of axon-glial cell interactions and periaxonal K^+ homeostasis. I. The influence of Na^+, K^+, Cl^- and cholinergic agents on the membrane potential of the adaxonal glia of the crayfish medial giant axon. Neuroscience **25:** 951–959.
25. PEARCE, B., J. ALBRECHT, C. MORROW & S. MURPHY. 1986. Astrocyte glutamate receptor activation promotes inositol phospholipid turnover and calcium flux. Neurosci. Lett. **72:** 335–340.
26. LIEBERMAN, E. M. & E. SANZENBACHER. 1991. Mechanisms of glutamate activation of axon-to-Schwann cell signaling in the squid. Neuroscience. In press.

Glutamate-Receptor Channels in Mammalian Glial Cells[a]

STUART G. CULL-CANDY AND DAVID J. A. WYLLIE

Department of Pharmacology
University College London
London WC1E 6BT, UK

Glutamate and aspartate are thought to be widespread fast excitatory transmitters throughout the mammalian central nervous system (CNS). At least three subtypes of glutamate receptors are known to occur in central neurons; these can be selectively activated by the glutamate analogs *N*-methyl-D-aspartate (NMDA), (S)-α-amino-3-hydroxy-5-methyl-4-isoxazole propionate (AMPA), and kainate.[1,2] Of these three receptor subtypes, the NMDA receptor and AMPA receptor can also be relatively selectively activated by aspartate and quisqualate, respectively, although in some cells quisqualate will also activate a type of glutamate receptor ("metabotropic" receptor) that is linked to a second messenger system.[3]

Patch-clamp methods applied to whole cells or outside-out patches[4] have given basic information about "fast" receptor channels (i.e., receptors directly linked to ion channels); second messenger responses are often lost, along with the cell contents, in the dialyzed whole cell or isolated patches.[5] These techniques have revealed various novel features of the different subtypes of neuronal glutamate channels.[6-10] We have applied these techniques to glutamate receptors in mammalian central glial cells. Initially patch-clamp recording was used to determine whether one type of central glial cell, the type-2 astrocyte, possessed glutamate receptors[11,12] as had previously been proposed from neurochemical studies.[13]

We have now extended these studies to address two specific questions.[14,15] Firstly, which types of central glial cells (and progenitor cells) possess glutamate receptors? It was hoped that this would give clues to the functional significance of these receptors. Secondly, do astrocytic glutamate receptors resemble neuronal type glutamate receptors in their basic properties, or are they a distinctly different population of receptor channels? It was hoped that this might also reveal some ideas about the function of such receptors, as well as give fundamental information about nonneuronal glutamate receptor properties in the CNS.[16]

CELL CULTURE METHODS AND SOLUTIONS

Cerebellar Neurons

Large cerebellar neurons were obtained from 17–19-day-old embryonic rats and were grown in tissue culture using methods previously described.[17-19] Briefly, cells were grown on glass coverslips coated with poly-L-lysine or collagen in Dulbecco's modified Eagle's medium (MEM) with 10% fetal calf serum and 10% horse serum at 37°C in an atmosphere of 95% air/5% CO_2. To reduce the number of glial cells and

[a]This work is supported by the Wellcome Trust. D.J.A.W. acknowledges grateful receipt of a Wellcome Trust Studentship.

granule neurons, cultures were treated with the antimitotic agent cytosine arabinoside (10 μM) 4 days after being plated. Fetal calf serum was not included in the culture medium after treatment with the antimitotic. These cultures contained Purkinje neurons and were also expected to contain Golgi cells and some deep cerebellar nuclei neurons. Cell soma diameters were approximately 15–30 μm.

Cerebellar granule neurons were grown in separate cultures, as previously described.[10] These were obtained from 2–14-day-old rat pups and were also grown in Dulbecco's MEM, with 10% fetal calf serum. Cultures possessed a high density of granule cells that could be readily identified on the basis of size (5–8 μm), morphology, and electrophysiological properties.

Cerebellar Glial Cells

Type-2 astrocytes were obtained from 7-day-old rat pups and isolated as previously described.[13] Cerebella were aseptically removed, and the tissue was chopped finely in two directions (at right angles to each other) with a McIllwain tissue chopper. Cells were dissociated in trypsin and DNase, plated on poly-L-lysine-coated coverslips, and cultured in Eagle's basal medium supplemented with 10% fetal calf serum, 2 mM glutamine, and 100 μg/ml gentamycin. Astrocytes, with an antigenic phenotype similar to that of type-1 astrocytes from the optic nerve,[20] were also present in our cultures. (We will refer to these as "type-1-like" astrocytes). The type-1-like astrocytes quickly formed a confluent monolayer; electrophysiological recordings were therefore made from type-2 astrocytes within 3–5 days of plating, when individual cells could still be readily distinguished. Cerebellar O-2A progenitor cells and oligodendrocytes were isolated from 1-day-old rat pups and cultured as described for type-2 astrocytes, but were grown in a culture medium that did not contain any fetal calf serum.[14]

During our electrophysiological experiments the different types of glial cells were identified morphologically: type-2 astrocytes were stellate shaped, with round, phase-bright cell bodies (~15–20 μm in diameter) and possessed processes up to 100 μm in length; type-1-like astrocytes had a flattened polygonal appearance and were phase dark; and 0-2A progenitor cells had a bipolar or multipolar morphology; oligodendrocytes tended to have larger cell bodies than did type-2 astrocytes and possessed a less extensive but more intricate network of processes. The morphological criteria used in our electrophysiological experiments were confirmed by labeling cells from the same cultures with antibodies to glial fibrillary acidic protein (GFAP), galactocerebroside (Gal-C, characteristic of oligodendrocytes), and the surface ganglioside GD-3 (antibody LB-1). Type-2 astrocytes were $GFAP^+/LB\text{-}1^+$; type-1-like astrocytes were $GFAP^+/LB\text{-}1^-$; O-2A progenitors were $GFAP^-/LB\text{-}1^-$; and oligodendrocytes were $GFAP^-/Gal\text{-}C^+$.

Patch pipettes contained (in mM): CsCl, 140 (or CsF, 110 and CsCl, 30); NaCl, 4; EGTA, 5; K-HEPES, 10; and $CaCl_2$, 0.5; pH = 7.2. In some experiments on type-1-like astrocytes patch pipettes contained (in mM): KCl, 140; $MgCl_2$, 2; EGTA, 0.5; K-HEPES, 0.5; and ATP (Mg), 0.5; pH = 7.2. External recording medium contained (in mM): NaCl, 150; KCl, 2.8; Na-HEPES, 10; and $CaCl_2$, 1.0; pH = 7.2. All external solutions were nominally magnesium free. Cells were viewed with water immersion phase contrast or differential interference contrast optics (total magnification 640). Electrophysiological experiments were all done at 20–24°C.

GLUTAMATE RECEPTOR CHANNELS IN CEREBELLAR NEURONS

A characteristic feature of glutamate receptor channels, recorded in outside-out patches from central neurons, is the presence of multiple conductance levels.[21,22] NMDA, quisqualate, and kainate all produce channels that open to several discrete conductance levels. The directly measured conductance amplitudes of these open states are similar for all three agonists and fall into at least five main groups with mean values of approximately 48, 38, 28, 18, and 8 pS. Examples of these multiple conductance levels produced by NMDA (10 μM), quisqualate (10 μM), and kainate (10 μM) in large cerebellar neurons are shown in FIGURE 1A, B, and C. The predominant conductance levels are indicated by dotted lines. These conductance values were estimated from the slopes of the relations between patch-potential and single-channel current amplitudes (I/V plots), examples of which are shown in the graph in FIGURE 1. These current-voltage plots show data gathered from four different patches; the single-channel currents represented in the graph were activated by NMDA, aspartate, glutamate, and quisqualate. The I/V plots were usually linear between +50 and −100 mV, with a mean reversal potential of approximately −4 mV as reported for various types of mammalian neurons.

The maximum conductance (48 pS) level activated in large cerebellar cells corresponds to the roughly 50 pS events previously reported for NMDA- and glutamate-activated channels in embryonic mouse midbrain neurons[6] and cerebellar granule neurons.[7] Furthermore, multiple conductance levels, qualitatively similar to those present in large cerebellar neurons, have also been described in hippocampal neurons,[22] cerebellar granule neurons,[10] and mouse midbrain neurons.[8,9]

From the single-channel recordings in FIGURE 1A, B, and C it is clear that none of these glutamate agonists is specific for any one particular conductance level, although the different agonists show some "preference" for particular levels. The preference of the agonists for the >30 pS openings shows the sequence NMDA > quisqualate > kainate. Indeed, more than 80% of the openings produced by NMDA and the putative neurotransmitters glutamate and aspartate were to the >30-pS levels. However, it is worth noting that in some large cerebellar cell patches the preponderance of openings produced by NMDA, glutamate, or aspartate was to the 38 rather than the 48 pS level.

The presence of direct transitions between particular conductance levels, during an opening, can be taken as reasonable evidence in support of the idea that the levels connected by a transition originate from the same channel. As is apparent from FIGURE 1A, B, and C, transitions between the 48 and 38 pS level and between the 38 and 18 pS level are commonly observed. Furthermore, apparent transitions also occurred between the 8 pS level and higher levels, but these transitions were more difficult to identify with confidence. From the experiments on large cerebellar cells, it is, clear however, that direct transitions can commonly occur between events *above* and *below* 30 pS in these cells.

A somewhat different picture has however emerged from the experiments on cerebellar granule neurons, although here glutamate agonists also produce multiple conductance openings that have at least five levels virtually identical in amplitude to those present in large cerebellar neurons.[10] The main differences are as follows. Firstly, in granule cells glutamate, aspartate, and NMDA open predominantly 50 pS levels, with relatively few openings to the lower levels. Secondly, and perhaps more importantly, direct transitions have been observed between 50, 40, and 30 pS levels (equivalent to 48, 38, and 28 pS in large cerebellar cells), indicating that these are all multiple states of the same channel (NMDA channel). However, few if any transi-

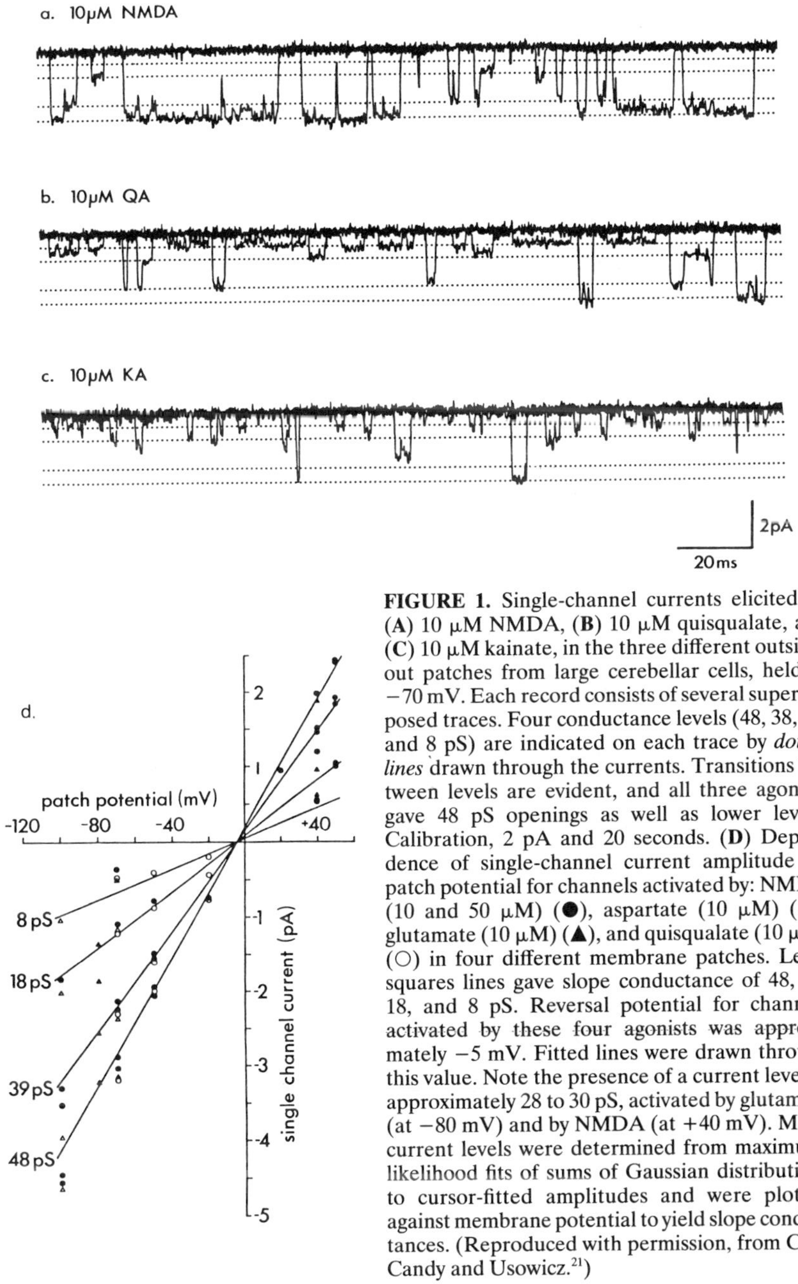

FIGURE 1. Single-channel currents elicited by (**A**) 10 μM NMDA, (**B**) 10 μM quisqualate, and (**C**) 10 μM kainate, in the three different outside-out patches from large cerebellar cells, held at −70 mV. Each record consists of several superimposed traces. Four conductance levels (48, 38, 18, and 8 pS) are indicated on each trace by *dotted lines* drawn through the currents. Transitions between levels are evident, and all three agonists gave 48 pS openings as well as lower levels. Calibration, 2 pA and 20 seconds. (**D**) Dependence of single-channel current amplitude on patch potential for channels activated by: NMDA (10 and 50 μM) (●), aspartate (10 μM) (△), glutamate (10 μM) (▲), and quisqualate (10 μM) (○) in four different membrane patches. Least squares lines gave slope conductance of 48, 39, 18, and 8 pS. Reversal potential for channels activated by these four agonists was approximately −5 mV. Fitted lines were drawn through this value. Note the presence of a current level of approximately 28 to 30 pS, activated by glutamate (at −80 mV) and by NMDA (at +40 mV). Mean current levels were determined from maximum-likelihood fits of sums of Gaussian distributions to cursor-fitted amplitudes and were plotted against membrane potential to yield slope conductances. (Reproduced with permission, from Cull-Candy and Usowicz.[21])

tions occurred between events of less than 30 pS (18 and 8 pS) and events of more than 30 pS (50 and 40 pS). In other words, in granule cells there is no reason to assume that events above and below 30 pS can originate from the same channel. It therefore seems likely that in these cells the 50 and 40 pS levels are associated *only* with NMDA receptors. Similarly, experiments of Llano and colleagues[23] have indicated that 50-pS conductance levels are absent from Purkinje neurons in organotypic cultures that lack NMDA receptors. However, it has proved surprisingly difficult to decide how universal this is. One reason that this is an important point is the implicit assumption that the 50 pS conductance confers novel features on the NMDA receptor, although presumably any 50 pS openings that may be associated with non-NMDA receptors need not possess these properties. We will be returning to this question later in the astrocyte experiments.

GLUTAMATE RECEPTORS IN ASTROCYTES

The ability of glutamate to depolarize glial cells[24,25] suggested the possibility that nonneuronal cells may possess glutamate receptors in the CNS. Given that glutamate would also be expected to depolarize glial cells by activating the Na^+-dependent electrogenic uptake mechanism,[26] this interpretation remained open to question. Furthermore, whole-cell recording from salamander retinal Müller cells[27] and rat type-1-like cerebellar astrocytes[10,14] indicated that glutamate could indeed produce a large electrogenic uptake current in these cells. Recent electrophysiological experiments have shown that glutamate receptors can occur in unidentified types of cortical astrocytes,[28] and from neurochemical studies it has been proposed that glutamate receptors may occur in type-2 astrocytes.[13] We therefore examined the responses of the various types of mammalian glial cells to excitatory amino acids.

Which Types of Glial Cell Possess Glutamate Receptors?

Type-2 Astrocytes

Patch-clamp recording has directly demonstrated that glutamate can evoke whole-cell and single-channel currents in type-2 astrocytes.[11,12] As shown in FIGURE 2, application of 30 μM glutamate, 30 μM kainate, or 30 μM quisqualate to type-2 astrocytes under whole-cell voltage clamp produced an inward current (in cells held at negative potentials). This current was accompanied by a marked increase in noise. Also, the quisqualate responses in astrocytes underwent dramatic desensitization (FIG. 2) as previously described in neurons, whereas the kainate responses exhibited little apparent desensitization, even after several minutes of exposure. The whole-cell current-voltage relationship (FIG. 3) indicated that the reversal potential was close to 0 mV. The presence of a noise increase, a reversal potential near 0 mV, and the fact that the currents were activated by glutamate and the agonists quisqualate, AMPA, and kainate (which are generally not considered to be taken up by the electrogenic mechanism) all support the idea that these responses resulted from the opening of ion channels.

Type-2 astrocytes did not respond to NMDA (or the selective NMDA-receptor agonist aspartate; see FIG. 5) regardless of the presence of glycine or the absence of Mg^{2+} [6,29] or the proton concentration.[30] All of these factors are known to influence

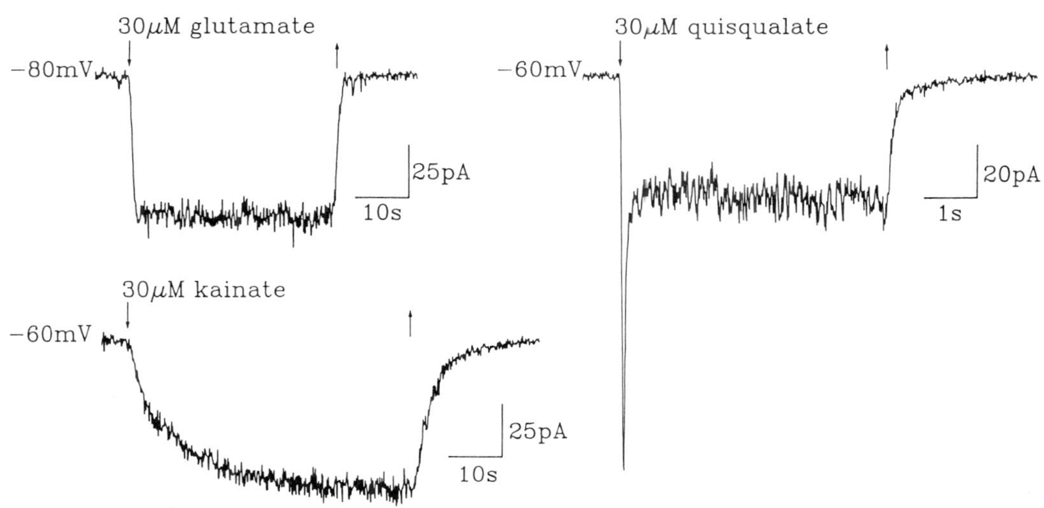

FIGURE 2. Whole-cell responses in type-2 astrocytes (three different cells) exposed to glutamate (30 μM, holding potential −80 mV), kainate (30 μM, −60 mV), and quisqualate (30 μM, 60 mV). *Arrows* indicate agonist application. Note the presence of a noise increase during the responses (particularly noticeable in the quisqualate response) and the marked desensitization of the quisqualate response. The slow time course of the kainate response in this example may reflect the slow application of drug. Calibration, 25 pA and 10 seconds (left hand records); 20 pA and 1 second (right hand record).

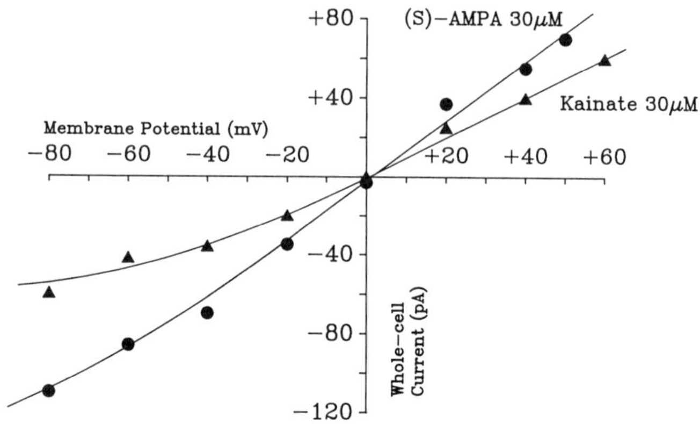

FIGURE 3. Current-voltage relationship for whole-cell currents evoked by 30 μM (S)-AMPA (●) or 30 μM kainate (▲) in a type-2 astrocyte. Currents reversed direction near 0 mV holding potential and showed some outward rectification.

NMDA-receptor responses in neurons. We were therefore unable to find any evidence of NMDA receptors in glial cells.

Ionophoretic mapping of glutamate receptors in type-2 astrocytes: As indirect evidence suggests that type-2 astrocytes may be one of the cells that send processes to the nodes of Ranvier *in vivo,*[31] and as these cells are found in white matter where neurotransmission does not occur, it has been suggested that such receptors may be involved in neuronal-glial signaling at the node.[11] We have therefore used ionophoretic mapping with quisqualate- and kainate-filled high resistance micropipettes (> 100 MΩ) to determine whether type-2 astrocytes possess glutamate receptors on their processes.

FIGURE 4 is an example of such an experiment. The drawing (FIG. 4A) depicts part of the cell body and one of the main processes of a cell that was mapped with a kainate-filled (kainate, 100 mM) pipette placed at various sites (labeled a–f in FIG. 4B) on, or near, the type-2 astrocyte. Brief pulses (100 ms, 100 nA) of kainate produced responses on the cell soma (position a) and on the processes (b, c, e, and f). As expected, moving the pipette laterally away from the process (position d) resulted in a rapid loss of the response, indicating that the response was not caused by kainate diffusing onto the soma. In this and other cells, responses could be obtained along the entire length of the processes. It would be difficult to obtain an accurate estimate of the density of the receptors in the processes, and we have not attempted to do so. However, given that reasonably large responses were obtained in regions that were electrically far from the recording electrode (e.g., positions e and f), it seems likely that the receptor density in the processes approached that in the cell body.

0-2A Progenitor Cells

If glutamate receptors in type-2 astrocytes have a functional significance *in vivo,* it seemed possible that they may also occur in the migratory 0-2A progenitor cells that give rise to the type-2 astrocyte and that might perhaps make initial contact with the axon by "detecting" the release of glutamate.[14,15]

Quisqualate (30 μM) and kainate (30 μM) both produced whole-cell current changes in cerebellar 0-2A progenitors. The responses reversed direction near 0 mV membrane potential and were accompanied by a clear noise increase. In common with type-2 astrocytes the O-2A progenitor cells did not respond to NMDA (30 μM). Interestingly, 0-2A progenitors from rat optic nerve (kindly supplied by Laura Lillien, grown according to the methods of Lillien and Raff[32]) also responded to kainate and quisqualate but not to NMDA, indicating that glutamate receptors in type-2 lineage cells are not confined to the cerebellum. Furthermore, glutamate receptors have recently also been observed in acutely dissociated 0-2A cells from optic nerve,[33] lending further support to the idea that these receptors occur *in vivo* and not solely in cultured astrocytes.

Oligodendrocytes Lack Glutamate Receptors

A majority of 0-2A cells differentiated into oligodendrocytes when cells were cultured in a low serum medium, as previously described.[34] However, these cells

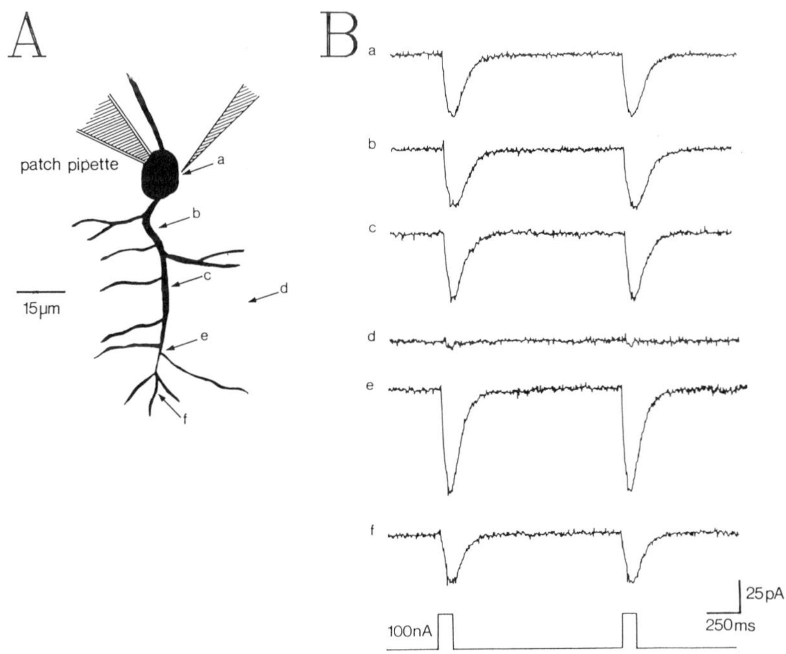

FIGURE 4. (**A**) Drawing of part of a type-2 astrocyte examined under whole-cell clamp. Receptor distribution was mapped with a kainate-filled ionophoretic micropipette positioned near the cell soma (a) or along the length of the main processes (at b, c, e, and f). (**B**) Ionophoretic application of kainate (pulse width 100 ms, amplitude 100 nA) evoked responses in the cell soma and along the entire length of the process. Note the large response obtained at e; the response at f approximately 60 μM from the soma was similar to the response at the tip of the process (not shown). The response was lost when the pipette was moved away from the cell (at d). Holding potential, −60 mV; calibration, 25 pA and 250 ms. (Reproduced, with permission, from Wyllie *et al.*[14])

produced no detectable receptor or uptake current response to glutamate, quisqualate, or kainate (30 μM).

Type-1-Like Astrocytes Possess Glutamate Uptake Currents

Astrocytes with an antigenic phenotype similar to that described for type-1 astrocytes from the optic nerve[20] were present in our cerebellar cultures. We were therefore interested in determining if these type-1-like astrocytes also possessed "fast" glutamate-receptor channels.

Our results indicate that in short-term cultures (<4 days *in vitro*) these cells lacked any detectable glutamate-receptor responses; hence, application of kainate, quisqualate, or NMDA did not evoke a whole-cell current change. As shown in FIGURE 5, when aspartate (or glutamate) was applied to the type-1-like astrocyte, it produced a large inward current that differed from glutamate responses in type-2 astrocytes because it was not accompanied by an increase in the noise level of the current trace, and it did not reverse direction at positive potentials (even at potentials as positive as +80 mV). Type-2 astrocytes did not respond to L-aspartate (FIG. 5). Taken together, these results indicated that aspartate was activating an electrogenic uptake carrier in type-1-like astrocytes (see also refs. 10 and 27), but not in type-2 astrocytes. We have confirmed that this current was indeed due to activation of an uptake current by replacing the external sodium in the bathing solution with lithium. From radiotracer experiments the uptake of glutamate and aspartate is known to depend on extracellular sodium ions.[35] In our experiments substitution of external sodium by lithium drastically reduced the glutamate-evoked currents in type-1-like astrocytes.

Interestingly, in older cultures (>7 days in culture) cells assumed to be type-1-like astrocytes responded to kainate and quisqualate, indicating the presence of glutamate receptors in these older cells. At this stage in culture it is not possible to identify the different types of glial cells by their morphology, as the cells become confluent and adopt a flattened appearance. However, it is likely that a majority of cells in these cultures are type-1-like astrocytes.[14]

ASTROCYTIC GLUTAMATE CHANNELS RESEMBLE "NEURONAL TYPE" CHANNELS

We have examined some of the biophysical and pharmacological properties of astrocytic glutamate receptors. In particular we wished to determine whether these receptors represented further subtypes of central glutamate receptors or whether they resembled glutamate receptor types present in neurons. Apart from the intrinsic interest of nonneuronal forms of central receptors, it was hoped that information about the functional importance of these receptors might be gained from an understanding of their properties. Furthermore, some of the recent molecular biology of glutamate receptors appears to relate as much to astrocytic as to neuronal glutamate receptors, making it useful to know more about their characteristics. For example, some studies on glutamate receptors expressed in oocytes have used mRNA derived from whole brain (see, for example, refs. 36 and 37). It seems likely that a proportion of the message may come from astrocytes as well as from neurons, so the non-NMDA receptors characterized may represent a mixture of astrocytic and neuronal receptors. This point is emphasized by the recent report that a cloned and sequenced kainate receptor appears to be localized in the Bergmann glia of the cerebellum.[38,39]

Multiple-Conductance Glutamate Channels in Type-2 Astrocytes

As shown in FIGURE 6a, glutamate (30 μM) applied to outside-out patches of cerebellar type-2 astrocytes invariably gave discrete single channel currents with clear multiple-conductance openings. The currents were annulled at 0 mV membrane potential and reversed direction at positive potentials; the multiple conductances were linearly dependent on patch potential between −90 and +20 mV, but there was some evidence that the I/V relationship deviated from linearity at more

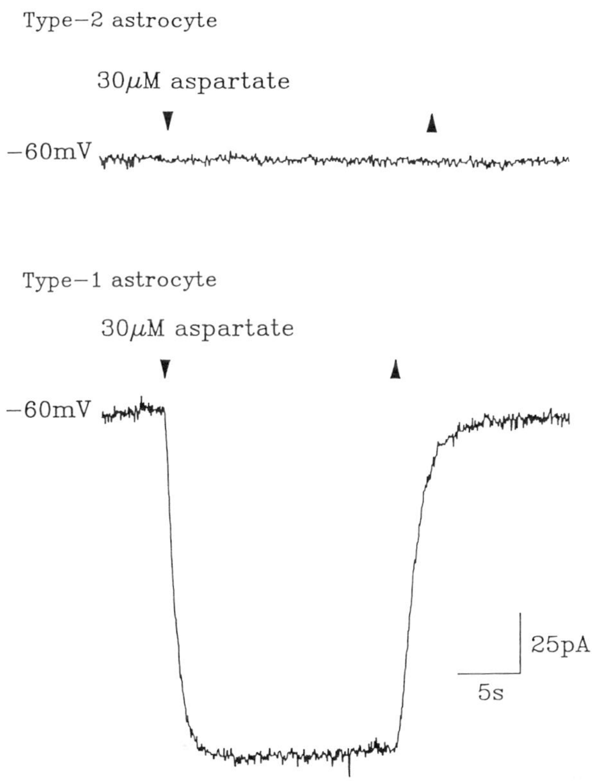

FIGURE 5. Application of aspartate (30 μM) to a type-2 astrocyte (*upper trace*; holding potential −60 mV) did not produce any current change (as expected if these cells lack NMDA receptors or an uptake current). Perfusion of aspartate onto a type-1-like astrocyte on the same coverslip (*lower trace*; −60 mV) resulted in a large inward current with no accompanying noise increase, indicating that the aspartate current was generated by an electrogenic uptake mechanism. Calibration, 25 pA and 5 seconds. (Reproduced, with permission, from Wyllie *et al.*[14])

positive potentials (as indicated by the dotted lines in FIG. 6b). The single-channel currents in the astrocytes appeared somewhat "noisier" than did channels in neurons. Nevertheless, the maximum level observed was close to 48 pS, as seen in central neurons, and the lower multiple conductance levels also resembled the lower conductances previously seen in neurons. Thus, there were at least five levels (FIG. 6

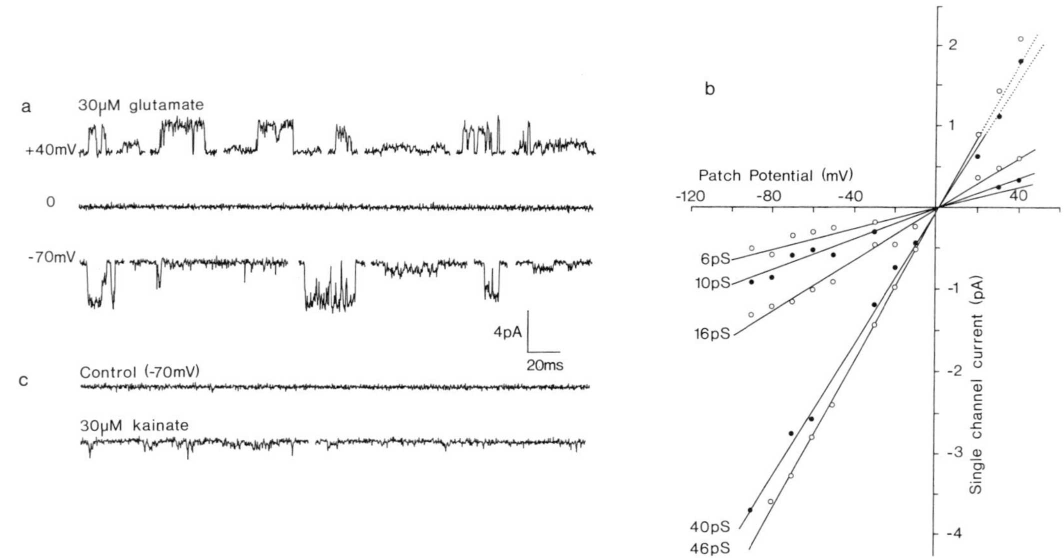

FIGURE 6. Single-channel currents produced by glutamate and kainate in an outside-out patch of a cerebellar type-2 astrocyte. (**a**) Multiple-amplitude currents produced by 30 μM glutamate are inward at −70 mV and outward at +40 mV; the reversal potential is near 0 mV. (**b**) Relationship between multiple current amplitudes and patch potential for channels activated by glutamate in **a.** The mean reversal potential is +1.5 mV, and each line has been drawn through this value with the slope calculated from least squares fitting. The points, at +40 mV, belonging to the two largest levels (*dotted lines*), deviate from the fitted lines which have slope conductances of 46, 40, 16, 10, and 6 pS. For outward currents at +40 mV, the two largest conductance levels are 54 and 47 pS. The linear relation in the potential range −90 to +20 mV, however, indicates that the slight outward rectification of whole-cell currents is not due to non-ohmic behavior of single-channel conductance. (**c**) *Upper trace:* Control noise level of the patch after washout of glutamate (−70 mV). *Lower trace:* Single-channel currents produced by 30 μM kainate in the same patch cannot be individually resolved, but appear as a noise increase. Calibration for **a** and **c:** 3 pA and 15 ms. Currents filtered at 1.56 kHz, −3 dB. (Reproduced, with permission, from Usowicz *et al.*[11])

and ref. 11). The 10 and 6 pS levels likely represent further subdivisions of the 8 pS level seen in neurons, and in common with neurons some 30 pS openings occur relatively infrequently. Openings of similar multiple levels were activated by quisqualate. Although 48 pS events were clearly discerned, it is worth noting that about 80% of openings to glutamate and quisqualate were to levels less than 30 pS.

FIGURE 7 shows typical recordings obtained from outside-out patches exposed to the three selective agonists quisqualate (30 μM), kainate (100 μM), and NMDA (100 μM). Quisqualate activated at least five conductance levels, indicated by dotted lines (48, 39, 15, 10, and 6 pS). We have not yet examined the types of transitions in detail; however, transitions were observed between levels above and below 30 pS. In some patches, kainate (30–100 μM) produced a noise increase, as expected if kainate was activating low conductance channels that were too small to be resolved directly. In other patches, such as the one depicted in FIGURE 7b, it gave some multiple conductance openings, with amplitudes similar to quisqualate openings of <30 pS, superimposed on a noise increase. This resembled the situation previously observed in cerebellar neurons. However, as in our whole-cell experiments no response to NMDA in outside-out patches was detectable from type-2 astrocytes (FIG. 7b, bottom trace).

Antagonism of Responses by CNQX

6-Cyano-7-nitroquinoxaline-2,3-dione (CNQX) acts as a competitive antagonist of non-NMDA receptor-mediated responses in neurons.[40] We were therefore interested to find out whether this antagonist also inhibited kainate, quisqualate, and AMPA responses in type-2 astrocytes. The whole-cell current responses (at either positive or negative holding potentials) to each of these agonists could be rapidly and reversibly antagonized by CNQX. Furthermore, inasmuch as kainate responses do not appreciably desensitize, it was possible to construct accurate inhibition curves and calculate the concentration at which CNQX half-inhibited a 100 μM kainate response (IC_{50}) in order to compare quantitatively its action on astrocytic and neuronal kainate receptors. CNQX produced half-inhibition (IC_{50}) of 100 μM kainate responses at a concentration of about 550 nM. This is slightly more potent than previously reported for neuronal membranes in the binding assay used by Honore and colleagues[40] (IC_{50} in binding study: 1.5 μM). However, the IC_{50} for CNQX inhibition of kainate responses in type-2 astrocytes was virtually identical to values that we previously obtained with cerebellar granule cells examined under whole-cell clamp (unpublished results).

Kainate-Quisqualate Interaction

A well-known phenomenon in neurons is the ability of quisqualate to apparently antagonize a kainate response.[41] Although the basis of this "interaction" is unclear, our experiments show that it also occurs in cerebellar type-2 astrocytes. Thus, 100 μM of kainate produced large inward currents (in cells held at negative holding potentials). If 30 μM of quisqualate was then applied together with kainate, a transient increase in the current amplitude was followed by a reduction in the whole-cell current response. This reduction in kainate current was maintained as

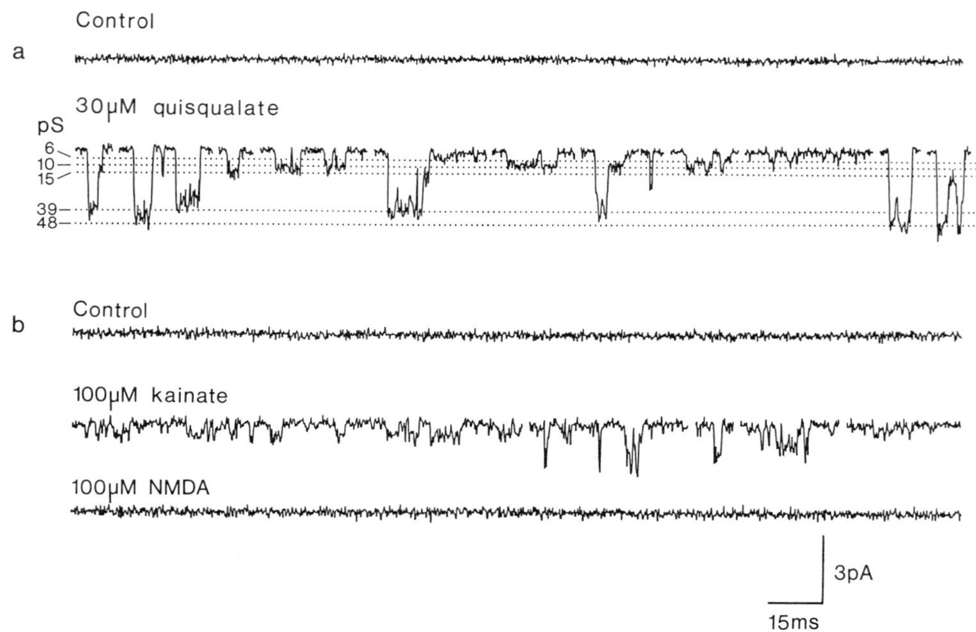

FIGURE 7. Single-channel currents produced in outside-out patches (−70 mV) of cerebellar type-2 astrocytes by the selective glutamate-receptor agonists quisqualate, kainate, and NMDA. (**a**) *Upper trace:* Control noise level. *Lower trace:* Currents produced by 30 μM quisqualate show multiple current amplitudes (indicated by *dotted lines*). These correspond to slope conductances of 48, 39, 15, 10, and 6 pS (interpolated reversal potential +1.9 mV). Values for the largest level were increased at potentials more positive than +20 mV (not shown), as for glutamate channels in FIGURE 6b. It is not clear whether the quisqualate currents reflect activation of several channels or one channel, but direct transitions between levels were infrequent. However, transitions were observed between levels above and below 30 pS, which is inconsistent with the possibility that levels above and below 30 pS arose from separate channels. (**b**) *Upper trace:* Control noise level. *Middle trace:* Kainate (100 μM) produced openings with directly resolvable multiple levels, and concomitantly activated conductances that are too small to be resolved individually, but appear as a noise increase. *Lower trace:* The same patch gave no currents in response to 100 μM of NMDA. Slope conductance estimates for the discernible levels were 19, 12, and 8 pS, with a reversal potential of −2 mV. (Reproduced, with permission, from Usowicz *et al.*[11])

long as quisqualate was present. Furthermore, the simultaneous application of kainate and quisqualate produced a clear noise increase in the current trace compared with the noise level in the presence of kainate alone. This phenomenon of quisqualate "antagonism" of the kainate response was seen in type-2 astrocytes held at positive or negative potentials, indicating that the effect was voltage independent (i.e., not due to channel block). Quisqualate produced a stronger inhibition of currents evoked by 100 μM kainate than by 30 μM kainate; this was previously taken to suggest that the antagonism exerted by quisqualate on kainate responses is uncompetitive[41]; however, in spinal cord neurons an increase in the concentration of kainate overcame the blocking actions of quisqualate, suggesting that the antagonism may be competitive in nature.[42] We have not attempted to examine in detail the mechanism by which quisqualate inhibits the kainate responses in astrocytes. Nevertheless, the presence of this interaction in these cells further supports the idea that glutamate receptors in astrocytes show similarities in their pharmacological properties to those found in neurons.

CONCLUSIONS

The existence of "fast" glutamate-activated ion channels in mammalian central glial cells *in vitro* indicates that glutamate receptors are not solely confined to neurons. It should also be noted that in this article we have considered only the properties of "fast" glutamate ion channels. However, good evidence also exists that a "metabotropic" glutamate receptor, possibly linked to phospholipase C, exists in astrocytes.[43,44] Many other second messenger-linked receptors have also been reported in astrocytes. (For a review see ref. 45.)

The receptors in type-2 astrocytes show some clear similarities to those found in central neurons. For example, the existence of multiple conductance levels, the block of quisqualate- and kainate-activated currents by CNQX, and the inhibition of kainate responses by quisqualate have all previously been described for glutamate receptors in mammalian neurons.[21,22,40,41] Interestingly, type-2 astrocytes show neuronal type characteristics in certain of their other membrane properties such as their antigenic phenotype[20]; in addition, they possess fast voltage-activated Na^+ channels that resemble neuronal type channels in their kinetic properties.[46]

Experiments on glutamate receptors in type-2 astrocytes have relevance to some general questions concerning central glutamate receptors. For example, in large cerebellar neurons that possess NMDA, quisqualate, and kainate receptors, the application of quisqualate frequently opened channels with conductances of roughly 50 pS. The possibility clearly exists that this is due to quisqualate activation of the NMDA receptor. It was therefore of particular interest that quisqualate produces 50 pS openings in astrocytes although these cells lack NMDA receptors. On the other hand, Purkinje neurons from organotypic cultures[23] that lack NMDA receptors failed to give 50 pS openings in response to quisqualate. Similarly, in granule cells the 40–50 pS events rarely, if ever, open to levels less than 30 pS and appear to be associated only with the NMDA receptor.[10] It seemed possible that glutamate receptors in astrocytes may differ from neuronal types of glutamate receptors; however, most of our data indicate some remarkable similarities. As the non-NMDA receptors in astrocytes give rise to 50 pS openings that are not blocked by Mg^{2+},[11] it would now be of interest to investigate if 50 pS openings to quisqualate or AMPA in granule cells and large cerebellar neurons are also insensitive to block by Mg^{2+}.

What could be the possible functions of glutamate receptors in type-2 astrocytes? In culture these receptors are found along the entire length of the cell processes, including their extremities. Recent studies suggest, from indirect evidence, that type-2 astrocytes may be one of the cells that extend "end-feet" to the nodes of Ranvier *in vivo.*[31] These glutamate receptors may therefore be appropriately placed to "detect" any axonal release of glutamate. Such a nonsynaptic release of neurotransmitter has previously been suggested to occur in the giant axon of the squid,[47] where glutamate is thought to mediate axon-to-Schwann cell signaling and in peripheral and central nerve trunks of the bullfrog.[48] It was recently suggested that vertebrate axons release factors that alter the properties of glial ion channels. Thus, in frog optic nerve the current voltage relationship of sodium currents in glial cells was shifted in a hyperpolarizing direction during axon stimulation.[49] This effect was thought to result from the release of a mediator (glutamate?) from the axon. However, a change in the ionic environment around the glial cell was not ruled out.

If glutamate receptors occur in type-2 astrocyte processes *in vivo,* then activation of such receptors could alter the concentration of sodium and potassium locally at the node of Ranvier, if the cell processes closely approach the node. Regardless of the precise location of type-2 astrocytes in white matter, their receptors would be expected to be activated by a generalized increase in glutamate concentration, which in turn may alter the excitability of the axon. Indeed, in kainic acid-lesioned rat hippocampus, kainate caused an efflux of potassium from astrocytes. Such potassium efflux could be responsible for the late slow potential recorded in some neurons.[50]

As well as possessing glutamate receptors, type-2 astrocytes have voltage-activated Ca^{2+} and Na^{+} channels.[46] Glutamate-induced depolarization of the astrocytes could therefore activate such channels to cause an alteration in the intracellular Ca^{2+} concentration or perhaps the release of neurotransmitter from type-2 astrocytes. Indeed, excitatory amino acids were previously shown to produce the release of GABA from type-2 astrocytes in culture.[13] However, the question of whether this can occur *in vivo* remains to be answered. Inasmuch as O-2A progenitor cells also possess glutamate receptors, the receptors could be involved in some way in the migration of the O-2A cell. For example, their activation could cause Ca^{2+} entry and process outgrowth. It would seem, however, that differentiation of the O-2A progenitor into oligodendrocytes (rather than type-2 astrocytes) is accompanied by a loss of expression of glutamate receptors in these particular cells.

Our results indicate that astrocytic glutamate receptors are very similar to the glutamate receptors present in central neurons. With recent advances in molecular biology it should soon be possible to decide not only whether these astrocytic and neuronal glutamate receptors possess similar biophysical and pharmacological properties, but also whether they share structural homology in their amino acid sequences.

REFERENCES

1. MAYER, M. L. & G. L. WESTBROOK. 1987. The physiology of excitatory amino acids in the vertebrate central nervous system. Progr. Neurobiol. **28:** 197–276.
2. WATKINS, J. C. & H. J. OLVERMAN. 1987. Agonists and antagonists for excitatory amino acid receptors. Trends Neurosci. **10:** 265–272.
3. SUGIYAMA, H., I. ITO & C. HIRONO. 1987. A new type of glutamate receptor linked to inositol phospholipid metabolism. Nature **325:** 531–533.
4. HAMILL, O. P., A. MARTY, E. NEHER, B. SAKMANN & F. J. SIGWORTH. 1981. Improved patch-clamp techniques for high-resolution current recording from cells and cell-free membrane patches. Pflugers Arch. **391:** 85–100.
5. NEHER, E. 1988. The use of the patch clamp technique to study second messenger related cellular events. Neuroscience **26:** 727–734.

6. NOWAK, L., P. BREGSTOVSKI, P. ASCHER, A. HERBET & A. PROCHIANTZ. 1984. Magnesium gates glutamate-activated channels in mouse central neurons. Nature **307:** 462–465.
7. CULL-CANDY, S. G. & D. C. OGDEN. 1985. Ion channels activated by L-glutamate and GABA in cultured cerebellar neurons of the rat. Proc. Roy. Soc. B **224:** 367–373.
8. ASCHER, P. & L. NOWAK. 1988a. Quisqualate and kainate activated channels in mouse central neurons in culture. J. Physiol. **399:** 227–245.
9. ASCHER, P. & L. NOWAK. 1988b. The role of divalent cations in the *N*-methyl-D-aspartate responses of mouse central neurons in culture. J. Physiol. **399:** 247–266.
10. CULL-CANDY, S. G., J. R. HOWE & D. C. OGDEN. 1988. Noise and single channels activated by excitatory amino acids in rat cerebellar granule neurons. J. Physiol. **400:** 189–222.
11. USOWICZ, M. M., V. GALLO & S. G. CULL-CANDY. 1989. Multiple conductance channels in type-2 cerebellar astrocytes activated by excitatory amino acids. Nature **339:** 380–383.
12. CULL-CANDY, S. G., V. GALLO & M. M. USOWICZ. 1989. Ion channel activated by quisqualate and kainate in cultured cerebellar type-2 astrocytes of the rat. J. Physiol. **415:** 45P.
13. GALLO, V., C. GIOVANINNI, R. SUERGIU & G. LEVI. 1989. Expression of excitatory amino acid receptors by cerebellar cells of the type-2 astrocyte cell lineage. J. Neurochem. **52:** 1–9.
14. WYLLIE, D. J. A., A. MATHIE, C. J. SYMONDS & S. G. CULL CANDY. 1991. Activation of glutamate-receptors and glutamate-uptake in identified macroglial cells in rat cerebellar cultures. J. Physiol. **432:** 235–258.
15. CULL-CANDY, S. G., A. MATHIE, C. J. SYMONDS & D. J. A. WYLLIE. 1989. Distribution of quisqualate and kainate receptors in rat type-2 astrocytes and their progenitor cells in culture. J. Physiol. **418:** 195P.
16. WYLLIE, D. J. A. & S. G. CULL-CANDY. 1990. Glutamate channels in rat cultured type-2 cerebellar astrocytes resemble neuronal-type glutamate channels. J. Physiol. **429:** 46P.
17. MOONEN, G., E. A. NEALE, R. L. MACDONALD, W. GIBBS & P. G. NELSON. 1982. Cerebellar macroneurons in microexplant cell culture. Methodology, basic electrophysiology and morphology after horseradish peroxidase injection. Dev. Brain Res. **5:** 59–73.
18. GROUL, D. L. 1983. Cultured cerebellar neurons: Endogenous and exogenous components of Purkinje cell activity and membrane response to putative transmitters. Brain Res. **263:** 223–241.
19. CULL-CANDY, S. G. & M. M. USOWICZ. 1989. On the multiple conductance single channels activated by excitatory amino acids in large cerebellar neurons of the rat. J. Physiol. **415:** 555–582.
20. RAFF, M. C., E. R. ABNEY, J. COHEN, R. LINDSAY & M. NOBLE. 1983a. Two types of astrocyte in cultures of developing white matter: Differences in morphology, surface gangliosides and growth characteristics. J. Neurosci. **3:** 1289–1300.
21. CULL-CANDY, S. G. & M. M. USOWICZ. 1987. Multiple-conductance channels activated by excitatory amino acids in cerebellar neurons. Nature **325:** 525–528.
22. JAHR, C. E. & C. F. STEVENS. 1987. Glutamate activates multiple single channel conductances in hippocampal neurons. Nature **325:** 522–525.
23. LLANO, I., A. MARTY, J. W. JOHNSON, P. ASCHER & B. GAHWILER. 1988. Patch clamp recordings of amino acid activated responses in "organotypic" slices in culture. Proc. Nat. Acad. Sci. USA **85:** 3221–3225.
24. BOWMAN, C. L. & H. K. KIMELBERG. 1984. Excitatory amino acids directly depolarize rat brain astrocytes in primary culture. Nature **311:** 656–659.
25. KETTENMANN, H., K. H. BACKUS & M. SCHACHNER. 1974. Aspartate, glutamate and γ-aminobutyric acid depolarize cultured astrocytes. Neurosci. Lett. **52:** 25–29.
26. BALCAR, V. J. & G. A. R. JOHNSTON. 1972. The structural specificity of the high affinity uptake of L -glutamate and L-aspartate in rat brain slices. J. Neurochem. **19:** 2657–2666.
27. BREW, H. & D. ATTWELL. 1987. Electrogenic glutamate uptake is a major current carrier in the membrane of axolotl retinal glial cells. Nature **327:** 707–709.
28. SONTHEIMER, H., H. KETTENMANN, K. H. BACKUS & M. SCHACHNER. 1988. Glutamate opens Na^+/K^+ channels in cultured astrocytes. Glia **1:** 328–336.

29. JOHNSON, J. W. & P. ASCHER. 1987. Glycine potentiates the NMDA response in cultured mouse brain neurons. Nature **325:** 529–531.
30. TRAYNELIS, S. F. & S. G. CULL-CANDY. 1990. Proton inhibition of *N*-methyl-D-aspartate receptors in cerebellar neurons. Nature **345:** 347–350.
31. ffRENCH-CONSTANT, C. & M. C. RAFF. 1986. The oligodendrocyte-type-2 astrocyte cell lineage is specialized for myelination. Nature **323:** 335–338.
32. LILLIEN, L. E. & M. C. RAFF. 1990. Analysis of the cell–cell interactions that control type-2 astrocyte development *in vitro.* Neuron **4:** 525–543.
33. BARRES, B. A., W. J. KOROSHETZ, K. J. SWARTZ, L. L. Y. CHUN & D. P. COREY. 1990. Ion channel expression by white matter glia: The O-2A glial progenitor cell. Neuron **4:** 507–524.
34. RAFF, M. C., R. H. MILLER & M. NOBLE. 1983b. A glial progenitor cell that develops *in vitro* into an astrocyte or an oligodendrocyte depending on the culture medium. Nature **303:** 390–396.
35. PETERSON, N. A. & E. RAGHUPATHY. 1974. Selective effects of lithium on synaptosomal amino acid transport systems. Biochem. Pharmacol. **23:** 2491–2494.
36. GUNDERSEN, C. B., R. MILEDI & I. PARKER. 1984. Glutamate and kainate receptors induced by rat brain messenger RNA in Xenopus oocytes. Proc. Roy. Soc. **B221:** 127–143.
37. VERDOORN, T. A., N. W. KLECKNER & R. DINGLEDINE. 1988. Rat brain *N*-methyl-D-aspartate receptors expressed in Xenopus oocytes. Science **238:** 1114–1116.
38. SOMOGYI, P., N. ESHHAR, V. I. TEICHBERG & J. B. D. ROBERTS. 1990. Subcellular localization of a putative kainate receptor in Bergmann glial cells using a monoclonal antibody in the chick and fish cerebellar cortex. Neuroscience **35:** 9–30.
39. GREGOR, P., I. MANO, I. MAOZ, M. MCKEOWN & V. I. TEICHBERG. 1989. Molecular structure of the chick cerebellar kainate binding subunit of a putative glutamate receptor. Nature **342:** 689–692.
40. HONORE, T., S. N. DAVIES, J. DREJER, E. J. FLETCHER, P. JACOBSEN, D. LODGE & F. E. NEILSEN. 1988. Quinoxalinediones: Potent competitive non-NMDA glutamate receptor antagonists. Science **241:** 701–703.
41. ISHIDA, A. T. & J. NEYTON. 1985. Quisqualate and L-glutamate inhibit retinal horizontal cell responses to kainate. Proc. Nat. Acad. Sci. USA **82:** 1837–1841.
42. O'BRIEN, R. J. & G. D. FISCHBACH. 1986. Characterization of excitatory amino acid receptors expressed by embryonic chick motoneurons *in vitro.* J. Neurosci. **6:** 3275–3283.
43. CORNELL-BELL, A. H., S. M. FINKBEINER, M. S. COOPER & S. J. SMITH. 1990. Glutamate induces calcium waves in cultured astrocytes: Long range signalling. Science **247:** 470–473.
44. MCNAUGHTON, L. A., L. LAGNADO, M. SOCOLOVSKY, S. P. HUNT & P. A. MCNAUGHTON. 1990. Glutamate elevates free $[Ca]_i$ in type-1 astrocytes cultured from rat cerebral cortex. J. Physiol. **424:** 48P.
45. MURPHY, S. & B. PEARCE. 1987. Functional receptors for neurotransmitters on astroglial cells. Neuroscience **22:** 381–394.
46. BARRES, B. A., L. L. Y. CHUN & D. P. COREY. 1989. Glial and neuronal forms of the voltage-dependent sodium channel: Characterization and cell type distribution. Neuron **2:** 1375–1388.
47. LIEBERMAN, E. M., N. J. ABBOTT & S. HASSAN. 1989. Evidence that glutamate mediates axon to Schwann cell signaling in the squid. Glia **2:** 94–102.
48. WEINREICH, D. & R. HAMMERSCHLAG. 1975. Nerve impulse enhanced release of amino acids from non-synaptic regions of peripheral and central nerve trunks of bullfrog. Brain Res. **84:** 137–142.
49. MARRERO, H., M. L. ASTION, J. A. COLES & R. K. ORKAND. 1989. Facilitation of voltage gated ion channels in frog neuroglia by nerve impulses. Nature **339:** 378–380.
50. MACVICAR, B. A., K. BAKER & S. A. CRICHTON. 1988. Kainic acid evokes a potassium efflux from astrocytes. Neuroscience **25:** 721–725.

Receptor Activation and Its Biochemical Consequences in Astrocytes[a]

G. P. WILKIN,[b] D. R. MARRIOTT,[b] A. J. CHOLEWINSKI,[c]
J. N. WOOD,[c] G. W. TAYLOR,[d] G. J. STEPHENS,[b]
AND M. B. A. DJAMGOZ[b]

[b]*Departments of Biochemistry and Biology (M.B.A.D.)*
Imperial College of Science, Technology & Medicine
London, SW7 2AZ, UK

[c]*Sandoz Institute for Medical Research*
Gower Place
London, WC1E 6BN, UK

[d]*Department of Clinical Pharmacology*
Royal Postgraduate Medical School
Ducane Road
London W12 OHS, UK

Our ability to study astrocytes has greatly been facilitated by the use of tissue culture techniques. It is now possible to obtain highly enriched (>95%) cultures of astrocytes from many parts of the central nervous system (CNS). In the main, these cultures have been derived from developing animals as it is not possible to obtain intact cells capable of growth in culture from adult animals unless the CNS is first damaged to generate gliosis.[1,2] However, the removal of astrocytes from the environment in which they would normally mature, together with the potential for them to become reactive following CNS trauma, makes it possible that such cells in culture are not fully representative of differentiated astrocytes in the undamaged nervous system. With respect to the expression of astrocyte cell surface receptors we believe that it is prudent to consider the wide range of receptors demonstrated thus far in culture as potential of *in vivo* expression until the presence of such receptors is confirmed in the intact CNS. The heterogeneity of cells *in vivo,* however, often makes the study of astrocyte receptors alone impossible, and it is necessary to take advantage of tissue culture to facilitate astrocyte separation from other cell types. The road forward now is to use information from both *in vivo* (or at least *in situ* in brain slices) and *in vitro* tissue culture studies to determine both the presence of receptors on astrocytes and the functions of these receptors.

EVIDENCE FOR SECOND MESSENGER SYSTEMS IN ASTROCYTES *IN VIVO*

With the above caution in mind it is nonetheless clear that cultured astrocytes are capable of expressing receptors for a wide range of classical neurotransmitters and neuropeptides (reviewed in refs. 3 and 4). Particular ligands are capable of activating

[a]The authors' work is supported in part by project grants from the Medical Research Council and the International Spinal Research Trust.

the synthesis of cAMP or cGMP or increasing the turnover of phosphatidylinositol (PI). Data on the synthesis of these second messengers elicited by a range of ligands in cultured astrocytes have been reviewed extensively elsewhere.[3,4] Here we review the evidence to date on the presence of second messengers, their synthetic enzymes, and other associated molecules in astrocytes *in vivo*.

Immunohistochemical studies on the localization of guanylate cyclase have proven inconsistent. Ariano *et al.*[5] reported that it is located in neurons and in the cerebellum, is confined primarily to Purkinje cells. Nakane and coworkers[6] showed weak staining of glia and granule cells in addition to strong Purkinje cell staining, whereas Zwiller *et al.*[7] reported strong staining in all neurons, astrocytes, and oligodendroglia in the cerebellum. We measured the activity of guanylate cyclase in a freshly isolated astrocyte-enriched cell fraction and in cultured astrocytes derived from postnatal rat cerebellum.[8] Both of the preparations exhibited enzyme activity. Indeed, detergent-treated cultured astrocytes showed activity several-fold greater than that of either cultured interneurons or an immature Purkinje cell-enriched fraction.

Good evidence for the synthesis of cGMP in astrocytes in brain slices was recently reported by de Vente *et al.*[9] Using an antiserum specific for cGMP they showed immunocytochemically that a subpopulation of glial fibrillary acidic protein (GFAP)-positive astrocytes responded to atrial natriuretic factor (ANF) with increased synthesis of cGMP. cGMP production can also be stimulated by ANF in cultured rodent astrocytes.[10,11] As far as we are aware, there are no reports in the literature on the immunocytochemical localization of adenyl cyclase. Ariano and Matus,[12] however, showed immunocytochemically that cAMP can be localized to astroglial processes in sections of rat brain.

Stimulation of receptors associated with PI turnover can lead to both activation of protein kinase C and increases in the concentrations of inositol phosphates. It is now apparent that protein kinase C comprises a family of related but distinct enzymes, α, β_1, β_{11}, δ, ϵ, and ζ.[13] Antibodies specific for individual subspecies of protein kinase C were used for immunohistochemical localization, although those tried so far revealed only neuronal compartments.[13] Mochly-Rosen and colleagues,[14] however, stained astroglia in sections with monoclonal antibodies against protein kinase C, but no data are yet available on the subspecies of the enzyme present. The other "arm" of the PI system concerns the inositol phosphates. Of particular interest is inositol, 1,4,5 trisphosphate (IP_3), which acts to increase cellular Ca^{2+} levels following its binding to certain intracellular receptors on membranes sequestering Ca^{2+}.[15] An IP_3 receptor protein has been purified[16] and antibodies raised against it.[17] An immunohistochemical study in rat cerebellum, however, revealed a localization within the Purkinje neurons, whereas the astroglia were apparently unlabeled.[17] This is a puzzle. Astrocytes *in vitro* were shown, for example, to respond to the excitatory amino acid agonist quisqualate with both increases in inositol phosphates and release of Ca^{2+} from internal stores.[18–21] It would be expected that an IP_3 receptor linked to Ca^{2+} release would be present in astrocytes. Clearly, experiments aimed at the unraveling of this puzzle are urgently required.

EVIDENCE FOR THE EXISTENCE OF MEMBRANE RECEPTORS ON ASTROCYTES *IN VIVO*

Using antibodies raised against the β_2-adrenergic receptor of the frog erythrocyte, Aoki *et al.*[22] localized it to both neurons and astrocytes in rat cortex and striatum. An interesting finding was that many stained astrocytic processes were found adjacent to similarly labeled neuronal dendrites, suggesting perhaps a functionally integrated response of glia and neurons to released norepinephrine.

Excitatory amino acids comprise the greater part of those neurotransmitters, causing postsynaptic excitation in the CNS. Receptors for these amino acids are at present divided into five classes, *N*-methyl-D-aspartate (NMDA), kainate, α-amino-3-hydroxy-5 methylisoxazole-4-proprionic acid (AMPA), 2-amino-4-phosphobutanoate (AP_4), and the metabotropic receptor.[23] Recently a putative kainate receptor protein ($M_r = 49{,}000$) was purified from chicken cerebellum.[24] Immunocytochemical studies utilizing a monoclonal antibody to this protein in both chick and fish cerebellar cortex revealed a clear-cut Bergmann glial localization.[25] Surprisingly, in this light of kainate's neurotoxic action, no neuronal labeling was seen. Somogyi and colleagues[25] speculate that excessive activation of glial cells might disrupt the normal function of glia, leading to neuronal damage.

Using an autoradiographic approach, Lightman *et al.*[26] localized opiate receptors to the GFAP-containing pituicytes of the rat pituitary gland. To rule out binding to the incoming vasopressin- and oxytocin-containing nerve terminals, the pituitary stalk was cut, with the consequent death and disappearance of these terminals. We confirmed these studies and showed that these opiate receptors were of the kappa subclass.[27] It was clear from both our studies and those of Lightman *et al.*[26] that the opiate receptors were present on pituicytes both before and after lesioning. After lesioning the amount of GFAP in the pituicytes as measured immunocytochemically rose dramatically, suggesting that these cells had undergone reactive gliosis. It would appear then that pituicytes are capable of expressing the kappa opiate receptor both in normal circumstances and after reactive gliosis. No clear-cut function has yet been demonstrated for these receptors, but it is possible that they play a role in a neuronal-glial relationship intimately concerned with hormone release (see ref. 28).

Burnard *et al.*[29] used kainate-lesioned hippocampal slices to study astroglia *in situ,* albeit that they were reactive. This model was also used in a study by MacVicar *et al.*[30] who demonstrated electrophysiologically the presence of GABA-activated chloride channels in such astrocytes. These authors proposed a role for these receptors and associated chloride channels in the control of extracellular chloride at times of high inhibitory activity. Still on the theme of GABA receptors, Wisden *et al.,*[31] using *in situ* hybridization histochemistry with a ^{32}P-labeled oligonucleotide probe, demonstrated the presence of the α_2 subunit of the $GABA_A$ receptor in the Bergmann glial layer of bovine cerebellum. Again the suggestion is that they may play a role in the control of Cl^- levels.

The $GABA_A$ receptor comprises various protein subunits that as a complex bind a number of ligands in addition to GABA itself.[32] One group of molecules with high affinity for the GABA receptor is the benzodiazepines, used therapeutically for their anxiolytic, sedative, anticonvulsant, and muscle-relaxant properties. In addition to these "central-type" benzodiazepine receptors which are generally accepted to be present exclusively on neurons, "peripheral-type" benzodiazepine sites are also located in the CNS. These are characterized by high-affinity binding of the benzodiazepine RO-5-4864. These sites appear to be located on glia. (For a review see ref. 33.) Autoradiographic studies showed binding over the choroid plexus, ependyma, pineal gland, and glial-rich parts of the olfactory bulb with additional diffuse staining over most other parts of the brain. Considerable increases in RO-5-4864 binding are seen over areas of gliosis following excitotoxic lesions of the CNS. RO-5-4864 increased the level of cGMP in the cerebellum and in this respect shares a common feature with ANF. In the CNS, ANF receptors are found in the same regions that show high densities of RO-5-4864 binding, and ANF is a potent displacer of diazepam from primary cultures of astrocytes. The possibility that ANF might be an endogenous ligand for the RO-5-4864 receptor is an intriguing one. The last receptor found on glia *in vivo* to be discussed here is that for Substance P (SP). Mantyh and colleagues[34] showed that ^{125}I-Bolton-Hunter Substance P (BHSP) binds to sections of lesioned rabbit optic nerve but not to normal unlesioned sections.[34] Other radiolabeled ligands including calcitonin gene-related peptide, cholecystokinin, galanin, glutamate, somatostatin, and vasoactive intestinal peptide (VIP) showed no binding. As the transected optic nerve comprises reactive astrocytes almost exclusively, the SP receptor apparently is synthesized by these cells following neuronal damage. Astrocytes in culture also expressed binding sites for ^{125}I-BHSP,[35–37] increased inositolphosphate turnover in the presence of nanomolar concentrations of SP,[37,38] and synthesized certain prostaglandins.[38,39] These phenomena will be discussed at greater length in the next section.

The key point we wish to make here about these *in vivo* studies is that once we have the assurance that the receptors are expressed *in vivo,* then functional studies on these receptors expressed *in vitro* become meaningful in the context of CNS function.

FUNCTIONAL STUDIES PERFORMED ON CULTURED ASTROCYTES

In this section we will summarize published data on the effects of various neuroactive agents on astrocytes downstream from the synthesis of second messengers and discuss recent investigations on the effects of two peptide receptors, SP and bradykinin (BK), on astrocytes in culture.

Vasoactive Intestinal Peptide (VIP)

VIP increased cAMP levels in cultured astrocytes from several species (reviewed in ref. 40). Studies in our laboratory revealed considerable differences in the responsiveness of astrocytes cultured from rat cortex, cerebellum, and spinal cord.[41] Cortical cells were strongly responsive in comparison with the other two regions. VIP concentrations are highest in the cortex and very low in the cerebellum and spinal cord.[42] It is possible that astrocytes in culture express densities of VIP receptors that reflect both *in vivo* VIP levels and their potential responsiveness to VIP-releasing neurons in the CNS. Interestingly, studies on the effects of VIP on cultured astrocytes revealed several functions for this peptide. First, it is a potent stimulator of glycogenolysis both in cerebral cortical slices and in astrocytes in culture.[43,44] Consideration of the distribution of VIP-containing neurons in cortex led Magistretti[45] to suggest that they are in a position to regulate glycogenolysis locally within restricted and radially oriented cortical domains. These domains contrast with the distribution of noradrenergic fibers that are orthogonal to the VIP-containing neurons. Magistretti[45] suggests that concomitant release of VIP and norepinephrine (through α_1 receptors) at the same points could lead to "cortical hot spots" where cAMP levels and rates of glycogenolysis possibly within local astrocytes would be greatly increased in preparation for local increases in neuronal activity.

Second, Brenneman and colleagues[46,47] showed that VIP is a peptide of importance in the survival of murine spinal cord neurons. Their evidence suggests that VIP causes the release of neurotrophic substances that increase the number of neurons surviving in dissociated cultures.[47] Third, the same investigators showed that treatment with VIP over a 5-day period increased astroglial mitosis.[47] Finally, Frohman *et al.*[48] showed that VIP is also able to significantly down-regulate the IFN-γ-induced Ia expression rat astrocytes.

Adrenergic Receptors

Stone and Ariano[49] recently reviewed the evidence for glia as targets for neuronally released noradrenaline in the CNS. There is a strong case in favor of the proposal that an important function of the noradrenergic system may be to regulate the metabolic and trophic actions of glia. The localization of β receptors on astrocytes *in vivo,* the presence of the bulk of brain glycogen in glia, and the influence of noradrenaline through both α and β receptors on cultured astrocyte metabolism of cAMP and glycogen strongly suggest that astroglia *in vivo* are also targets for neuronally released noradrenaline. The full consequences of glycogenolysis are unclear at present, but glucose and/or other metabolic substrates could possibly be made available to both astrocytes and local neurons (see ref. 50).

Just as VIP can cause glycogenolysis and also initiate the release of neuron survival-promoting activity, noradrenaline can activate glycogen breakdown and cause release of nerve growth factor.[51,52] Another parallel with VIP is that noradrenaline, possibly through β receptors, inhibits IFN-γ-induced Ia expression on cultured astrocytes.[53]

Apparently glutamate and GABA transport into rat cortical astrocyte cultures can be influenced through adrenoceptors. Hansson and Rönnbäck reported that a 15-second preincubation with the α_1 receptor-agonist phenylephrine significantly increased subsequent glutamate uptake, whereas GABA uptake was increased in the presence of the β-agonist isoproterenol.[54] Studies by Salm and colleagues[55] exploiting the use of Ca^{2+}-sensitive dyes showed that both α_1 and α_2 adrenergic receptors are able to mobilize calcium in cultured astrocytes.[55]

Finally β-adrenergic agonists such as isoproterenol stimulate [^{3}H]-taurine release from the glioma-derived LRM 55 cell line and primary astrocytes in culture.[56] Glial receptors and glial neurotransmitter release are the subject of another chapter in this volume by Gary Dutton and colleagues.

Serotonin Receptors

Just as adrenergic fibers spread throughout the CNS, deriving from a small center, the locus coeruleus, so serotonergic fibers in a similar fashion spread throughout the brain, their respective cell bodies being found in the brainstem raphe nuclei. High affinity serotonin receptors are found on cultured astrocytes, but data on the synthesis of second messengers associated with these receptors are conflicting, and a clear picture has not yet emerged (see ref. 57 for review). Like noradrenaline and VIP, however, serotonin affects glycogen metabolism.[44] A more recent and particularly interesting finding is the serotonin-linked release from astrocytes of a factor(s) that regulates the growth of serotonergic neurons.[58] Finally, serotonin has an inhibitory effect on the expression of GFAP in astrocytes cultured from neonatal rat brainstem.[59] Le Prince *et al.*[59] found that in the presence of serotonin, GFAP immunolabeling and mRNA decreased, and glial filaments localized to the periphery of the cytoplasm in contrast to controls in which staining was also seen in a perinuclear position. They suggested that the change in GFAP expression could be due to regulation of gene expression or the stability of the GFAP message.

Amino Acid Receptors

Several recent reports from different laboratories all emphasize the importance of glutamate in activating Ca^{2+} mobilization and influx into astrocytes.[19–21] Two types of receptor appear to be involved, one quisqualate-preferring, releasing Ca^{2+} from intracellular stores, the other kainate-preferring and allowing surface-membrane Ca^{2+} influx. The quisqualate-activated Ca^{2+} responses are oscillatory and can travel from cell to cell in cultured astrocytes. The importance of this phenomenon is not yet clear, but it has been suggested that it could underlie a signaling mechanism adapted to long-range transmission of information in an electrically coupled glial network.

Condorelli *et al.*[60] demonstrated that quisqualate can also act as an inhibitor of astrocyte division, whereas kainate, NMDA, and AMPA were ineffective in this process. This effect correlates with inositolphosphate hydrolysis which is increased in astrocytes in the presence of quisqualate, but not in the presence of kainate or AMPA.

Opioid Receptors

Met-enkephalin has also been shown to suppress astrocyte DNA synthesis in culture. Stiene-Martin and Hauser[61] reported that 1 μM met-enkephalin reduced [^{3}H]-thymidine incorporation into mouse cerebral cortical astrocytes. This effect was antagonized by naloxone. The subtype of receptor is unclear at present, but it may be the delta receptor, this being the one through which met-enkephalin would normally be expected to act.

Opiates also are reported to be involved in astrocyte glycogen metabolism. Pearce *et al.*[62] showed that although morphine augmented noradrenaline-induced increase in 2-[^{14}C]-deoxyglucose incorporation into glycogen, met-enkephalin antagonized it.

Receptors for Substance P and Bradykinin

One aspect of astrocyte receptors that we have been interested in is regional heterogeneity.[63] We might expect the profile of receptors on any one mature astrocyte *in vivo* to depend on or at least reflect its neuronal partners. The mature profile might be different from the one pertaining during development and might change again as a consequence of damage and gliosis. Comparative studies of cultured astrocytes obtained from neonatal rat brain revealed that cells from cortex, cerebellum, and spinal cord expressed different profiles of receptors.[41,64] One particularly interesting finding was that only spinal cord astrocytes responded to SP with increased PI turnover. In contrast, all three regions responded to bradykinin. This difference and the possible roles of SP and BK in inflammatory, pain, and immune responses[65–67] prompted us to investigate these two peptides further.

Substance P

SP and receptors for this peptide are found throughout the CNS. (For a review see ref. 68.) The strong case advanced for SP acting as a neurotransmitter at certain dorsal horn sensory afferents puts SP at the top of the list of potential peptide neurotransmitters. With the high concentrations of both the peptide and its receptors in the spinal cord (SC), our finding of an SP second messenger response in cultured rat SC astrocytes[64] alerted us to a potential neuron-glial interaction. The observation by Barber *et al.*[69] of SP-positive presynaptic boutons localized on astrocytes in spinal cord is particularly significant here.

Further studies in both our own laboratory[37] and that of Beaujouan and colleagues[36] in France demonstrated the presence of ^{125}I-Bolton Hunter SP binding on spinal cord astrocytes and confirmed the regional heterogeneity. SP-containing dorsal root ganglion neurons are believed to function both in conveying afferent

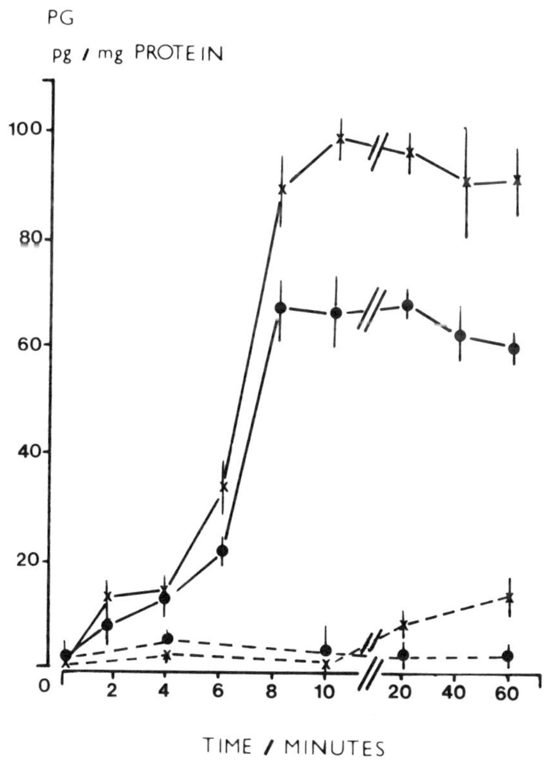

FIGURE 1. Time course of PGD_2 (X) and PGE_2 (●) release from spinal cord astrocytes exposed to Substance P (10^{-6} M). *Broken lines* show basal release. Each value represents the mean ± SEM of four dishes and is representative of two independent experiments. PG = prostaglandin. (Reproduced, with permission, from Marriott *et al.*[38])

somatosensory information to the spinal cord and in the efferent regulation of the peripheral tissues that they innervate. Thus, DRG neurons are implicated in afferent transmission of nociception in the CNS and in efferent regulation of inflammation. (See ref. 65.) A study by Mantyh and colleagues[34] now raises the possibility that SP may also be involved in the regulation of the glial response to injury in the CNS. They undertook an autoradiographic study using ^{125}I-BHSP binding to rabbit optic nerve sections obtained both before and after optic nerve section. Whereas no ^{125}I-BHSP binding sites were found in unlesioned tissue, the gliotic scar tissue exhibited a

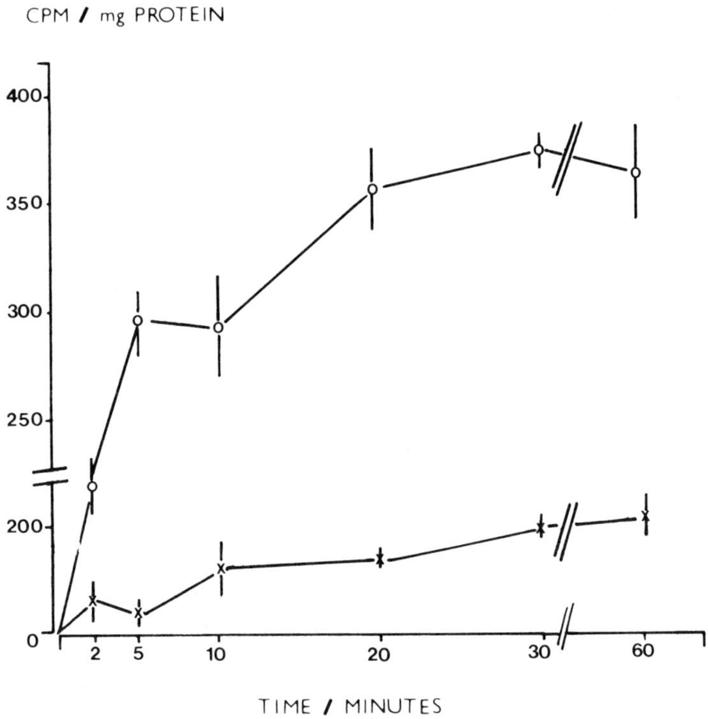

FIGURE 2. Time course of phosphoinositol accumulation in spinal cord astrocytes exposed to 10^{-6} M substance P (O). X shows basal turnover. Each value represents the mean ± SEM of four observations and is representative of five independent experiments. (Reproduced, with permission, from Marriott *et al.*[38])

density of binding sites comparable to the highest density observed in rabbit forebrain. We recently found SP-induced PI turnover in rat optic nerve astrocyte cultures (Marriott and Wilkin, unpublished observation). Clearly it is now important to determine if SP receptors are expressed in normal circumstances on astrocytes *in vivo* or if the expression of receptors on cultured astrocytes is due to a reactive gliotic response of the cells *in vitro*.

In either circumstance, undamaged or damaged CNS, what is the reason for the presence of SP receptors on astrocytes? A possibility that we considered in the light of work from other laboratories was prostaglandin release. SP, ATP, and interleukin 1 all stimulate the release of prostaglandins from rat cortical astrocytes.[70–72] We found that SP induces the release of prostaglandins from spinal cord astrocytes, but not from those derived from cortex or cerebellum.[38] A rapid release of PGD_2 and PGE_2 over the first 8 minutes of stimulation is paralleled by a slightly earlier increase of inositol phosphate synthesis (FIGS. 1 and 2). SP is effective at very low concentrations in both of these assays. The half-maximal response of phosphoinositol formation was approximately 10^{-10} M. The dose-response profiles for prostaglandin release show that PGD_2 was released in greater amounts than was PGE_2 (FIG. 3). The release of PGE_2 did not saturate within the concentrations of SP tested. As stated, cortical astrocytes did not respond to SP with either increased PI turnover or prostaglandin synthesis and release. Nevertheless, they were capable of prostaglandin release. Both calcium ionophore A23187 and 10% fetal calf serum elicited the release of both PGE_2 and PGD_2 from cortical astrocytes (FIG. 4).

Bradykinin Receptors

We found that a consistent response in astrocytes from several parts of the rat CNS has been BK-induced PI turnover.[64] This response is effected at nanomolar concentrations of BK, the ED_{50} being approximately 6 nM. BK is a nonapeptide that is released from precursors, the kininogens, by action of the enzyme kallikrein in response to tissue damage.[67] Two independent studies have shown that BK is present throughout the CNS.[73,74] The peptide acts locally in various tissues throughout the body and is normally present in extremely low concentrations in plasma and tissue because of its very rapid degradation (plasma half-life < 30 s). At least two BK receptors apparently exist, B_1 and B_2. B_2 receptors are thought to facilitate most of the actions of BK, mediation of pain and vasodilation, release of prostaglandins and histamine, stimulation of smooth muscle cells, and activation of sensory nerves and

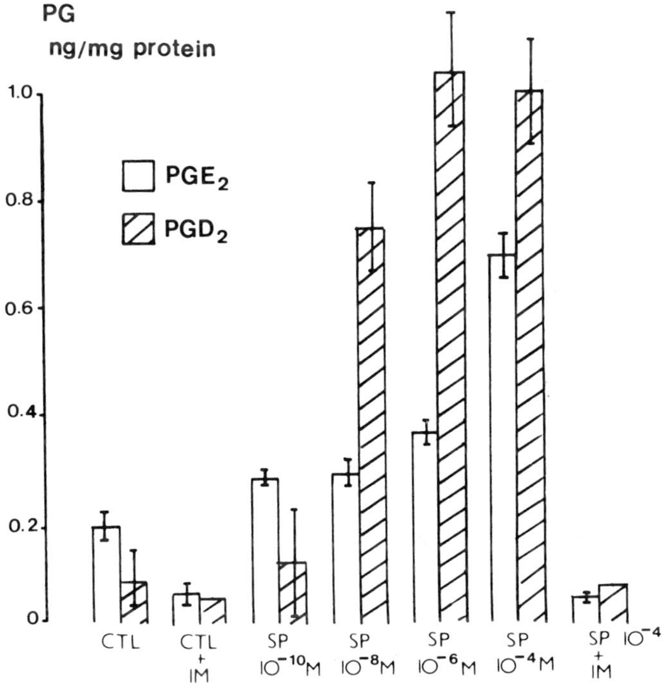

FIGURE 3. Dose dependence of PGD_2 and PGE_2 release from spinal cord astrocytes exposed to Substance P (60 min). Substance P (10^{-6} M) preferentially released PGD_2. Stimulated release of PGE_2 was not saturable within the range tested. Saturating levels of PGD_2 were obtained at 10^{-6} M Substance P. Each value represents the mean ± SEM of four dishes and is representative of two independent experiments. CTL = control; IM = indomethacin; PG = prostaglandin; SP = Substance P. (Reproduced, with permission, from Marriott *et al.*[38])

reflexes.[67] Receptor activation of BK leads to stimulation of the enzymes phospholipase C and A_2, with the subsequent synthesis of several second messengers, inositol trisphosphate, diacyclglycerol, and various cyclooxygenase and lipoxygenase products of arachidonate.[67]

As just indicated, astrocytes from various parts of the rat CNS respond to BK.[64] We therefore used cortical astrocytes as a model system to further study BK receptors. [^{3}H] BK binding to these astrocytes exhibited a time-dependent increase in specific binding at 4°C which reached a plateau by about 30 minutes. This binding

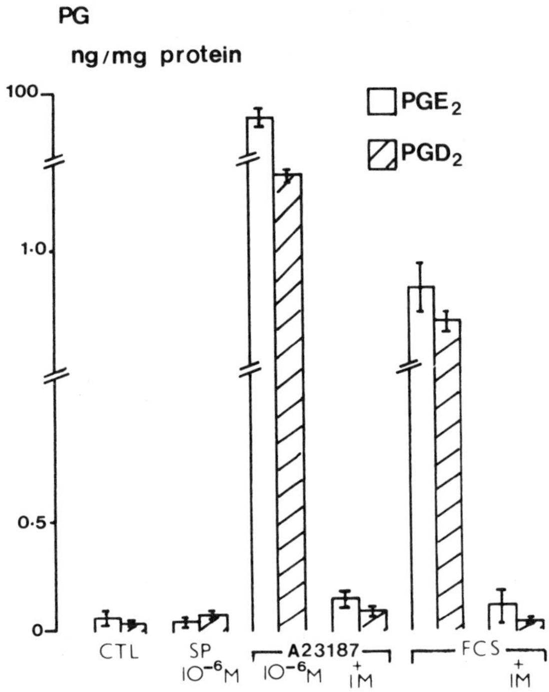

FIGURE 4. Effects of Substance P (10^{-6} M), calcium ionophore A23187 (10^{-6} M), and 10% FCS on release of PGD_2 and PGE_2 from cortical astrocytes. Calcium ionophore and FCS induced release of both prostaglandins. Neither prostaglandin was released by Substance P. Each value represents the mean ± SEM of four dishes and is representative of at least two independent experiments. CTL = control; FCS = fetal calf serum; PG = prostaglandin; SP = Substance P. (Reproduced, with permission, from Marriott *et al.*[38])

was saturable with increasing [^{3}H] BK concentrations, and Scatchard analysis of these data revealed a K_d of 9 nM and B_{max} of 77 fmol/mg protein. The receptor subtype was assessed by studying the ability of two specific BK antagonists to block the binding of 3 nM [^{3}H] BK. The two antagonists were Des-Arg^9[Leu^8]-BK, B_1 specific, and D-Arg [Hyp^3, D-Phe^7]-BK, B_2 specific. In competition, the B_2 antagonist inhibited [^{3}H] BK binding with an IC_{50} value of 50 nM, whereas the B_1 antagonist only inhibited 20% of the total specific binding at a concentration of 1 μM.

In this report we have not been concerned with electrophysiological evidence for receptor functioning in astrocytes; there is a substantial body of such evidence (see Cull-Candy in this volume). However, we would like to highlight some recent results that suggest that BK receptors on astrocytes are also electrophysiologically active.[75] In whole-cell patch-clamp recordings, application of micromolar BK to type I and II cortical astrocytes produced inward currents in a fraction of the cells, whereas kainate and glutamate were effective on almost every cell. In some 50% of astrocytes preloaded with fura-2, a rise in the intracellular Ca^{2+} signal did not appear to depend appreciably on the presence of extracellular Ca^{2+}, and in common with the membrane current response, it suffered significantly from desensitization. Another consequence of BK receptor activation is the release of prostaglandins from astrocytes (Marriott, Taylor, and Wilkin, unpublished results). The precise functional relation between a BK-induced rise in intracellular Ca^{2+}, the inward membrane current flow, and prostaglandin release is not yet clear however.

CONCLUSION

The accumulated knowledge on astrocyte receptors and their possible functions has grown rapidly over the last few years. *In vitro* systems have contributed greatly to this knowledge. However, we now need to understand the relation between astrocyte receptor expression and local neurotransmitter release *in vivo*. The next few years promise to be exciting as we discover the consequences of these varifl☆' ,systems between neurons and astrocytes.

REFERENCES

1. SINGH, V. K. & D. VAN ALSTYNE. 1978. Glial cells from normal adult rat brain established in continuous culture. Brain Res. **155:** 418–421.
2. VAN ALSTYNE, D., E. M. SMYRNIS & D. W. PATY. 1983. Differentiation of glioblasts from adult brain. Neurosci. Letts. **40:** 327–332.
3. MURPHY, S. & B. R. PEARCE. 1987. Functional receptors for neurotransmitters on astroglial cells—a commentary. Neuroscience **22:** 381–394.
4. KIMMELBERG, H. K. (ED.) 1988. Glial Cell Receptors. Raven Press. New York.
5. ARIANO, M., J. A. LEWICKI, H. J. BRANDWEIN & F. MURAD. 1982. Immunohistochemical localization of guanylate cyclase within neurons of rat brain. Proc. Natl. Acad. Sci. USA **79:** 1316–1320.
6. NAKANE, M., M. ICHIKAWA & T. DEGUCHI. 1983. Light and electron microscopic demonstration of guanylate cyclase in rat brain. Brain Res. **273:** 9–15.
7. ZWILLER, J., M. S. GHANDOUR, M. O. REVEL & P. BASSET. 1981. Immunohistochemical localization of guanylate cyclase in rat cerebellum. Neurosci. Letts. **23:** 31–36.
8. BUNN, S. J., J. GARTHWAITE & G. P. WILKIN. 1986. Guanylate cyclase activities in enriched preparations of neurones, astroglia and a synaptic complex isolated from rat cerebellum. Neurochem. Int. **8:** 179–185.
9. DE VENTE, J., C. G. MANSHANDEN, R. A. SIKKING, F. C. S. RAMAEKERS & H. W. M. STEINBUSCH. 1990. A functional parameter to study heterogeneity of glial cells in rat brain slices: Cyclic guanosine monophosphate production in atrial natriuretic factor (ANF)-responsive cells. Glia **3:** 43–54.
10. FRIEDL, A., C. HARMENING, F. SCHMELZ, B. SCHURICHT, M. SCHILLER & B. HAMPRECHT. 1989. Elevation by atrial natriuretic factors of cyclic GMP levels in astroglia-rich cultures from murine brain. J. Neurochem. **52:** 589–597.
11. SIMMONET, G., M. ALLARD, P. LEGENDRE, J. GABRION & J. D. VINCENT. 1989. Characteristics and specific localization of receptors for atrial natriuretic peptides at non-neuronal cells in cultured mouse spinal cord cells. Neuroscience **29:** 198–199.

12. ARIANO, M. A. & A. I. MATUS. 1981. Ultrastructural localization of cyclic GMP and cyclic AMP in rat striatum. J. Cell Biol. **91:** 287–292.
13. NISHIZUKA, Y. 1988. The molecular heterogeneity of protein kinase C and its implications for cellular regulation. Nature **334:** 661–665.
14. MOCHLY-ROSEN, D., A. I. BASBAUM & D. E. KOSHLAND, JR. 1987. Distinct cellular and regional localization of immunoreactive protein kinase C in rat brain. Proc. Natl. Acad. Sci. USA **84:** 4660–4664.
15. BERRIDGE, M. J. & R. F. IRVINE. 1989. Inositol phosphates and cell signalling. Nature **341:** 197–205.
16. SUPATTAPONE, S., P. F. WORLEY, J. N. BARABAN & S. H. SNYDER. 1988. Solubilization, purification and characterization of an inositol triphosphate receptor. J. Biol. Chem. **263:** 1530–1534.
17. ROSS, C. A., J. MELDOLESI, T. A. MILNER, T. SATOH, S. SUPATTAPONE & S. H. SNYDER. 1989. Inositol 1,4,5-triphosphate receptor localized to endoplasmic reticulum in cerebellar Purkinje neurons. Nature **339:** 468–470.
18. PEARCE, B., J. ALBRECHT, C. MORROW & S. MURPHY. 1986. Astrocyte glutamate receptor activation promotes inositol phospholipid turnover and calcium flux. Neurosci. Letts. **72:** 335–340.
19. CORNELL-BELL, A. H., S. M. FINKBEINER, M. S. COOPER & S. J. SMITH. 1990. Glutamate induces calcium waves in cultured astrocytes: Long-range glial signaling. Science **247:** 470–473.
20. JENSEN, A. M. & S. Y. CHIU. 1990. Fluorescence measurement of changes in intracellular calcium induced by excitatory amino acids in cultured cortical astrocytes. J. Neurosci. **10:** 1165–1175.
21. GLAUM, S. R., J. A. HOLZWARTH & R. J. MILLER. 1990. Glutamate receptors activate Ca^{2+} mobilization and Ca^{2+} influx into astrocytes. Proc. Natl. Acad. Sci USA **87:** 3454–3458.
22. AOKI, C., T. H. JONG & V. M. PICKEL. 1987. Ultrastructural localization of β-adrenergic receptor-like immunoreactivity in the cortex and neostriatum of rat brain. Brain Res. **437:** 264–282.
23. YOUNG, A. B. & G. E. FAGG. 1990. Excitatory amino acid receptors in the brain; membrane binding and receptor autoradiographic approaches. Trends Pharmacol. Sci. **11:** 126–133.
24. GREGOR, P., I. MANO, I. MAOZ, M. MCKEOWN & V. I. TEICHBERG. 1989. Molecular structure of the chick cerebellar kainate-binding subunit of a putative glutamate receptor. Nature **342:** 698–692.
25. SOMOGYI, P., N. ESHAR, V. I. TEICHBERG & J. D. B. ROBERTS. 1990. Subcellular localization of a putative kainate receptor in Bergmann glial cells using a monoclonal antibody in the chick and fish cerebellar cortex. Neuroscience **35:** 9–30.
26. LIGHTMAN, S. L., M. NINKOVIC, S. P. HUNT & L. L. IVERSEN. 1983. Evidence for opiate receptors on pituicytes. Nature **305:** 235–237.
27. BUNN, S. J., M. R. HANLEY & G. P. WILKIN. 1985. Evidence for a kappa-opioid receptor on pituitary astrocytes: An autoradiographic study. Neurosci. Letts. **55:** 317–323.
28. HATTON, G. I. 1990. Emerging concepts of the structure-function dynamics in adult brain: The hypothalamo-neurohypophysial system. Progr. Neurobiol. **34:** 437–470.
29. BURNARD, D. M., S. A. CRICHTON & B. A. MACVICAR. 1990. Electrophysiological properties of reactive glial cells in the kainate-lesioned hippocampal slice. Brain Res. **510:** 43–52.
30. MACVICAR, B. A., F. W. Y. TSE, S. A. CRICHTON & H. KETTENMAN. 1989. GABA-activates Cl^- channels in astrocytes of hippocampal slices. J. Neurosci. **9:** 3577–3583.
31. WISDEN, W., L. A. MCNAUGHTON, M. G. DARLISON, S. P. HUNT & E. A. BARNARD. 1989. Differential distribution of $GABA_A$ receptor mRNAs in bovine cerebellum: Localization of α_2 mRNA in Bergmann glia layer. Neurosci. Letts. **106:** 7–12.
32. STEPHENSON, F. A. 1988. Understanding the $GABA_A$ receptor: A chemically-gated ion channel. Biochem. J. **249:** 21–32.
33. HERTZ, L. & A. S. BENDER. 1988. Astrocytic benzodiazepine receptors: Their possible role in regulation of brain excitability. *In* Glial Cell Receptors. H. K. Kimelberg, ed. 159–181. Raven Press. New York.
34. MANTYH, P. W., D. J. JOHNSON, C. G. BOEHMER, M. D. CATTON, H. V. VINTERS, J. E. MAGGIO, H.-P. TOO & S. R. VIGNA. 1989. Substance P receptor binding sites are

expressed by glia *in vivo* after neuronal injury. Proc. Natl. Acad. Sci. USA **86:** 5193–5197.

35. TORRENS, Y., M. C. DAGUET DE MONTETY, M. EL ETR, J. C. BEAUJONAN & J. GLOWINSKI. 1989. Tachykinin receptors of the NK_1 type (substance P) coupled positively to phospholipase C on cortical astrocytes from the newborn mouse in primary culture. J. Neurochem. **52:** 1913–1918.
36. BEAUJOUAN, J. C., M. C. DAGUET DE MONTETY, Y. TORRENS, M. SAFFROY, M. DIETL & J. GLOWINSKI. 1990. Marked regional heterogeneity of ^{125}I-Bolton Hunter substance P binding and substance P induced activation of phospholipase C in astrocyte cultures from the embryonic or newborn rat. J. Neurochem. **54:** 669–675.
37. WILKIN, G. P., J. C. REID, D. R. MARRIOTT & A. J. CHOLEWINSKI. 1990. Peptide receptors on astrocytes: Characterization and regional heterogeneity. *In* Differentiation and Functions of Glial Cells. G. Levi, ed. : 265–274. Alan R. Liss. New York.
38. MARRIOTT, D. R., G. P. WILKIN & J. N. WOOD. 1991. Substance P induced release of prostaglandins from astrocytes: Regional specialization and correlation with phosphoinositol metabolism. J. Neurochem. **56:** 259–265.
39. HARTUNG, H.-P., K. HEININGER, B. SCHAFER & K. V. TOYKA. 1988. Substance P and astrocytes: Stimulation of the cyclooxygenase pathway of arachidonic acid metabolism. FASEB J. **2:** 48–51.
40. WILKIN, G. P. & A. CHOLEWINSKI. 1988. Peptide receptors on astrocytes. *In* Glial Cell Receptors. H. K. Kimelberg, ed. : 223–241. Raven Press. New York.
41. CHOLEWINSKI, A. J. & G. P. WILKIN. 1988. Astrocytes from forebrain cerebellum and spinal cord differ in their responses to vasoactive intestinal peptide. J. Neurochem. **51:** 1626–1633.
42. LOREN, I., P. C. EMSON, J. FAHRENKRUG, A. BJORKLUND, J. ALUMETS, R. HAKANSON & F. SUNDLER. 1979. Distribution of vasoactive intestinal polypeptide in the rat and mouse brain. Neuroscience **4:** 1953–1976.
43. MAGISTRETTI, P. J., J. H. MORRISON, W. J. SHOEMAKER, V. SAPIN & F. E. BLOOM. 1981. Vasoactive intestinal polypeptide induces glycogenolysis in mouse cortical slices: A possible regulatory mechanism for the local control of energy metabolism. Proc. Natl. Acad. Sci. USA **78:** 6535–6539.
44. MAGISTRETTI, P. J., M. MANTHORPE, F. E. BLOOM & S. VARON. 1983. Functional receptors for vasoactive intestinal polypeptide in cultured astroglia from neonatal rat brain. Regul. Pept. **6:** 71–80.
45. MAGISTRETTI, P. J. 1986. Intercellular communication mediated by VIP in the cerebral cortex. Peptides **7**(Suppl 1): 169–173.
46. BRENNEMAN, D. E. & G. A. FOSTER. 1987. Structural specificity of peptides influencing neuronal survival during development. Peptides **8:** 687–694.
47. BRENNEMAN, D. E., T. NICOL, D. WARREN & L. M. BOWERS. 1990. Vasoactive intestinal peptide: A neurotrophic releasing agent and an astroglial mitogen. J. Neurosci. Res. **25:** 386–394.
48. FROHMAN, E., T. FROHMAN, B. VAYUVEGULA, S. GUPTA & S. VAN DEN NOORT. VIP inhibits MHC expression on astrocytes. J. Neurol. Sci.
49. STONE, E. A. & M. A. ARIANO. 1989. Are glial cells targets of the central noradrenergic system? A review of the evidence. Brain Res. Rev. **14:** 297–309.
50. HERTZ, L. 1989. Is Alzheimer's disease an anterograde degeneration, originating in the brainstem and disrupting metabolic and functional interactions between neurons and glial cells? Brain Res. Rev. **14:** 335–353.
51. FURUKAWA, S., Y. FURUKAWA, E. SATOYOSHI & K. HAYASHI. 1987. Regulation of nerve growth factor synthesis/secretion by catecholamine in cultured mouse astroglial cells. Biochem. Biophy. Res. Commun. **147:** 1048–1054.
52. SCHWARTZ, J. P. 1988. Stimulation of nerve growth factor mRNA content in C6 glioma cells by a β-adrenergic receptor and by cyclic AMP. Glia **1:** 282–285.
53. FROHMAN, E. M., B. VAYUVEGULA, S. VAN DEN NOORT & S. GUPTA. 1988. Norepinephrine inhibits gamma-interferon-induced MHC class II (Ia) antigen expression on cultured brain astrocytes. J. Neuroimmunology **17:** 89–101.
54. HANSSON, E. & L. RONNBACK. 1989. Regulation of glutamate and GABA transport by adrenoceptors in primary astroglial cell cultures. Life Sci. **44:** 27–34.

55. SALM, A. K., L. LEREA, H. CASTROS & K. D. MCCARTHY. 1990. Distinct subsets of astroglia can be defined by their expression of neuroligand receptors that regulate intracellular calcium levels. *In* Differentiation and Functions of Glial Cells. G. Levi, ed.: 275–288. Alan R. Liss. New York.
56. MARTIN, D. L., W. SHAIN & V. MADELIAN. 1988. Receptor-mediated release of taurine from glial cells and signalling between neurons and glia. *In* Glial Cell Receptors. H. K. Kimelberg, ed.: 183–195. Raven Press. New York.
57. WHITAKER-AZMITIA, P. M. 1988. Astroglial serotonin receptors. *In* Glial Cell Receptors. H. K. Kimelberg, ed.: 107–120. Raven Press. New York.
58. WHITAKER-AZMITIA, P. M. & E. C. AZMITIA. 1989. Stimulation of astroglial serotonin receptors produces culture media which regulates growth of serotonergic neurons. Brain Res. **497:** 80–85.
59. LE PRINCE, G., M.-C. COPIN, H. HARDIN, M.-F. BELIN, J. P. BOUILLOUX & M. TARDY. 1990. Neuron-glia interactions: Effect of serotonin on the astroglial expression of GFAP and of its encoding message. Dev. Brain Res. **51:** 295–298.
60. CONDORELLI, D. F., F. INGRAO, G. MAGRI, V. BRUNO, F. NICOLETTI & R. AVOLA. 1989. Activation of excitatory amino acid receptors reduces thymidine incorporation and cell proliferation rate in primary cultures of astrocytes. Glia **2:** 67–69.
61. STIENE-MARTIN, A. & K. F. HAUSER. 1990. Opioid-dependent growth of glial cultures: Suppression of astrocyte DNA synthesis by metenkephalin. Life Sci. **46:** 91–98.
62. PEARCE, B., M. CAMBRAY-DEAKIN & S. MURPHY. 1985. Astrocyte opioid receptors: Activation modifies the noradrenaline-evoked increase in 2-[^{14}C]-deoxyglucose incorporation into glycogen. Neurosci. Letts. **55:** 157–160.
63. WILKIN, G. P., D. M. MARRIOTT & A. J. CHOLEWINSKI. 1990. Astrocyte heterogeneity. Trends Neurosci. **13:** 43–46.
64. CHOLEWINSKI, A. J., M. R. HANLEY & G. P. WILKIN. 1988. A phosphoinositide-linked peptide response in astrocytes: Evidence for regional heterogeneity. Neurochem. Res. **13:** 389–394.
65. MANTYH, P. W., M. D. CATTON, C. G. BOEHMER, M. L. WELTON, E. P. PASSARO, J. E. MAGGIO & S. R. VIGNA. 1989. Receptors for sensory neuropeptides in human inflammatory diseases: Implications for the effector role of sensory neurons. Peptides **10:** 627–645.
66. MCGILLIS, J. P., M. L. ORGANIST & D. G. PAYAN. 1987. Substance P and immunoregulation. FASEB J. **46:** 196–199.
67. TAYLOR, J. E., F. V. DEFENDIS & J. P. MOREAU. 1989. Bradykinin-antagonists: Therapeutic perspectives. Drug Dev. Res. **16:** 1–11.
68. NICOLL, R. A., C. SCHENKER & S. E. LEEMAN. 1980. Substance P as a transmitter candidate. Ann. Rev. Neurosci. **3:** 227–268.
69. BARBER, R. P., J. E. VAUGHN, J. R. SLEMMON, P. M. SALVATERRA, E. ROBERTS & S. E. LEEMAN. 1979. The origin, distribution and synaptic relationships of substance P axons in rat spinal cord. J. Comp. Neurol. **184:** 331–352.
70. HARTUNG, H. P., K. HEININGER, B. SCHAFER & K. V. TOYKA. 1988. Substance P and astrocytes: Stimulation of the cyclooxygenase pathway of arachidonic acid metabolism. FASEB J. **2:** 48–51.
71. PEARCE, B., S. MURPHY, J. JEREMY, C. MORROW & P. DANDONA. 1989. ATP-evoked Ca^{2+} mobilisation and prostanoid release from astrocytes: P_2-purinergic receptors linked to phosphoinositide hydrolysis. J. Neurochem. **52:** 971–977.
72. HARTUNG, H. P., B. SCHAFEL, K. HEININGER & K. V. TOYKA. 1989. Recombinant interleukin-1β stimulates eicosanoid production in rat primary culture astrocytes. Brain Res. **489:** 113–119.
73. PERRY, D. C. & S. H. SNYDER. 1984. Identification of bradykinin in mammalian brain. J. Neurochem. **43:** 1072–1080.
74. KARIYA, K., A. YAMAUCHI & T. SASAKI. 1985. Regional distribution and characterization of kinin in the CNS of the rat. J. Neurochem. **44:** 1892–1897.
75. STEPHENS, G. J., A. J. CHOLEWINSKI, M. B. A. DJAMGOZ & G. P. WILKIN. An electrophysiological study of excitatory amino acid and bradykinin receptors on cortical astrocyte subtypes in culture. In preparation.

Astrocyte Taurine[a]

GARY R. DUTTON,[b] MARYANN BARRY,[b]
MARTHA L. SIMMONS,[b] AND ROBERT A. PHILIBERT[c]

Departments of Pharmacology[b] and Psychiatry[c]
University of Iowa, College of Medicine
Iowa City, Iowa 52242

Astrocytes, the most abundant cells in the central nervous system (CNS), clearly play critical, active roles in controlling the development and maintenance of the brain under normal and adverse conditions. These glial elements mediate and directly contribute to regulating brain extracellular space, and important intercellular interactions with cells of both the vasculature and neuronal populations. Some of these functions include homeostatic mechanisms for the maintenance of balanced metabolic rates and ion concentrations and for the inactivation of numerous neuroactive substances. However, less well understood are the ways in which astrocytes may more directly control neuronal activity levels.

One substance that may participate in these events and whose investigation has recently experienced resurgence in excitable tissues is the sulfur-containing amino acid taurine (2-aminoethane sulfonic acid).[1] Taurine has long been known to be an essential requirement for normal brain development, to possess inhibitory neuromodulatory, Ca^{2+} modulatory, and anticonvulsant properties, and also to possibly act as an osmoregulator and an antioxidant in the CNS. This topic was recently the subject of an eloquent and comprehensive review by Huxtable.[1]

This chapter focuses on astrocyte taurine and how its presence in and release from these cells may subserve a variety of important CNS functions. Much information on this topic has been obtained through the employment of *in vitro* techniques using tissue slices or cultured cells, from measuring endogenous taurine or preloaded [^{3}H]taurine, or from *in vivo* intracranial microdialysis methods.[2] Work reported from this laboratory was performed largely on cultured cerebellar astrocytes, by measuring the release of endogenous taurine and other amino acids from these cells with HPLC.[3]

CNS taurine arises primarily from dietary sources and hepatic biosynthesis, whereas the brain's inherent synthetic capacity for this amino acid is low. Biochemical and immunohistochemical evidence shows that taurine and its synthetic enzymes, cysteine dioxygenase (which converts cysteine to cysteine sulfinate) and cysteine sulfinate decarboxylase (CSAD, which converts cysteine sulfinate to hypotaurine, which is then further oxidized to taurine), are present and unevenly distributed in brain. The taurine content of immature brain is high and decreases with age, whereas CSAD activity increases during this time. The exact cellular localization of both CSAD and taurine has not been unambiguously determined. These difficulties are, at least in part, due to problems with taurine and CSAD antibody specifities. For example, Tappaz *et al.*[4] maintain that two forms of CSAD exist, a CSAD II activity expressed by glutamate decarboxylase and a CSAD I activity identical to the liver enzyme, the latter of which is responsible for synthesizing taurine in the brain. These

[a]Work reported from this laboratory was supported by NIH grant NS20632. M.A.B. is a member of the Neuroscience Program, and M.L.S. is supported by the Life and Health Insurance Medical Research Fund.

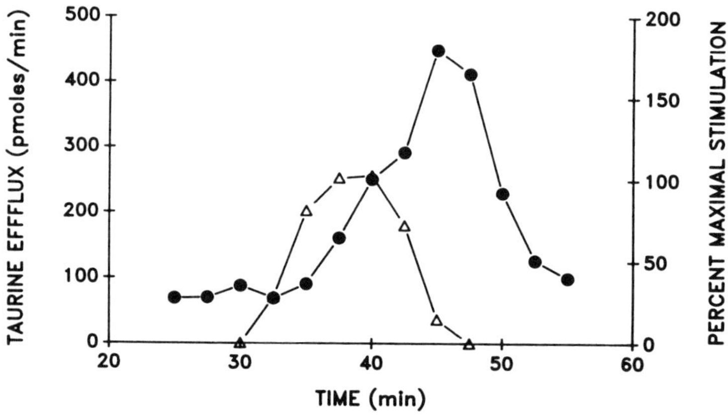

FIGURE 1. Time course of taurine efflux (●) during isoosmotic stimulation of cultured cerebellar astrocytes with 50 mM K^+ (△). Stimulation was for 10 minutes from the 30th to 40th minute.

authors have reported that CSAD I in the cerebellum is found exclusively in glial cells, whereas others[5] have evidence of activity largely in the Purkinje and other inhibitory neurons of this brain region. The overwhelming body of evidence points to neuronal, not glial, localization.[1] That is not to say, however, that taurine is present only in neurons. Taurine appears to be located in all cells of the CNS.

Taurine is a zwitterion at physiological pH, and its strongly acidic sulfonate group renders the molecule lipophobic and impermeant through bilayer membranes.[1] Taurine is relatively inert. Its metabolism occurs very slowly in the CNS, with a half-life of about 5 days.[6] A high-affinity uptake mechanism in both neurons and glia exists for taurine and other β-amino acids. This transport system is capable of maintaining a concentration gradient $[tau_i]/[tau_o]$ several thousand-fold in cultured astrocytes.[7] Taurine transport is also Na^+ dependent and sensitive to K^+, Cl^-, and pH changes.[8]

TAURINE RELEASE

K^+-Stimulated Efflux

Taurine in cultured astrocytes originates primarily from the serum in the culture medium. Under culture conditions used in this laboratory, the taurine concentration in the medium is about 25 μM. Assuming a cell volume of 5 μl/mg protein,[9] it is estimated that intracellular concentrations of approximately 55 mM may be attained in astrocytes maintained 17–19 days in culture. Astrocytes release taurine and, to a lesser extent, other amino acids in response to isosmotically elevating K^+.[9-11] This efflux is dose dependent, has a time course much slower than that of known neurotransmitter amino acids, and has been observed by others using brain slices.[12] A delay of about 10 minutes between maximal stimulation and maximal taurine release is always observed (FIG. 1). Thereafter, efflux returns to near basal levels, even in the presence of continued K^+ stimulation.[9] This nonneurotransmitter (nonvesicular?) type of delayed release is also seen for taurine and adenosine in cultured cerebellar neurons.[13]

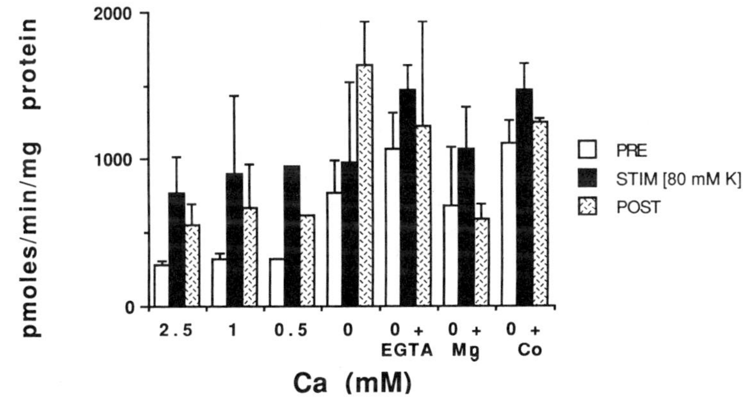

FIGURE 2. K^+-evoked (80 mM) cerebellar astrocyte taurine efflux with various concentrations of extracellular Ca^{2+}, and in the absence of added Ca^{2+} with added EGTA (1 mM), Mg^{2+} (10 mM), or Ca^{2+} (1 mM). PRE (□) = 10 minutes prestimulation; STIM (■) = 10-minute stimulation; POST (▨) = 10-minute poststimulation. $n = 3$.

Ion Dependencies

In earlier work, investigators noted that ions such as Na^+, Ca^{2+}, Cl^-, and Mg^{2+} influence the extent to which taurine is released from astrocytes.[14] We originally reported that isosmotically K^+-evoked taurine release was Ca^{2+} dependent. This stemmed from our observations that in the absence of added Ca^{2+} and 10 mM Mg^{2+} or 1 mM EGTA, no K^+-stimulated taurine release from these cells was observed over the range of 10–80 mM K^+, compared to standard conditions (2.5 mM Ca^{2+} and 1 mM Mg^{2+}).[10] In addition, with decreasing extracellular Ca^{2+} levels (0.1–0.0 mM), increased basal taurine levels eventually swamped 50 mM K^+-evoked release in the absence of added Ca^{2+}. In contrast to this observation, others have subsequently reported that the release of preloaded [^{3}H]taurine in the absence of added Ca^{2+} and 1 mM $CoCl_2$ is Ca^{2+} independent.[9] We now know that the presence of 10 mM Mg^{2+} in our system, which is slightly hyperosmotic (335 mosm), may have contributed to a hyperosmotic suppression of K^+-stimulated taurine efflux. However, in reassessing this issue we recently repeated these experiments, varying extracellular Ca^{2+} concentrations in the presence of isosmotically added 80 mM K^+. FIGURE 2 shows that reducing Ca^{2+} from 2.5 to 0.5 mM has no effect on either basal or 80 mM K^+-stimulated levels of taurine efflux. However, in the absence of added Ca^{2+}, basal levels rise about threefold. What we also see here, but had not seen in earlier experiments using 50 mM K^+, is an additional twofold Ca^{2+}-independent, K^+-evoked increase in taurine release above basal levels. These data suggest the possibility that some portion of 80 mM K^+-evoked taurine release may be independent of extracellular Ca^{2+}. On the other hand, EGTA (1 mM), Mg^{2+} (10 mM), or Co^{2+} (1 mM) added in the absence of Ca^{2+} produced no additional K^+-stimulated taurine release although reported elsewhere.[9] At the present time we cannot explain these discrepancies. Perhaps unidentified differences exist in the release properties of radiolabeled, preloaded taurine compared with endogenous taurine, as suggested by Huxtable.[1] In other work we showed that voltage-regulated Ca^{2+} channels of the N- and L-type do not appear to be involved in K^+-evoked taurine release, because dihydropyridine antagonists and ω-conotoxin were without effect.[15] Clearly the role of extracellular Ca^{2+} is a complicated issue, as recognized by other investigators.[1,8]

Another of our earlier investigations was designed to determine if K^+ channels play a role in K^+-evoked (50 mM) taurine release.[16] Cultured cerebellar astrocytes were challenged with Ba^{2+}, Cs^+, apamin, and tetramethylammonium. In all cases neither basal nor K^+-stimulated taurine efflux levels were altered, suggesting that K^+ channels were not involved in K^+ uptake and that this phenomenon probably occurs at least in part via passive Gibbs-Donnan equilibrium forces.

The role of Na^+ in K^+-evoked taurine release is not well understood. In our hands, complete, but not partial (50%), substitution of Na^+ by choline abolished K^+-induced taurine release from both cultured cerebellar[15] and hypothalamic[17] astrocytes. However, ouabain, an inhibitor of Na^+/K^+ ATPase, had no effect on K^+-evoked release.[16] In addition, the Na^+ ionophore, monensin (50 μM), had very little effect on basal taurine levels under standard conditions of Na^+ concentration.[15] In other systems, such as brain slices, retina, olfactory bulb, and cortex, taurine efflux appears to be stimulated by an increased ratio of intra- to extracellular Na^+.[1] Voltage-dependent Na^+ channels also do not seem to be involved in basal or K^+-evoked increases in taurine efflux levels, inasmuch as tetrodotoxin or veratridine were without effect.[15]

Cell Swelling

Isosmotically, but not hyperosmotically, increased KCl produces taurine efflux in a dose-dependent fashion.[9–11] This release is also Cl^- sensitive, because replacement by the impermeant anion, gluconate, in the presence of depolarizing K^+ levels, abolished taurine efflux.[9,18] Thus, it apparently is not K^+-induced membrane depolarization that causes taurine release, but rather cell swelling produced by K^+ uptake. The characteristically slow time course for taurine release is consistent with the occurrence of a postswelling regulatory volume decrease (RVD), as the cells regain their normal volume,[19] even in the presence of continued stimulation.[20]

We[10] and Pasantes-Morales and Schousboe[9] previously demonstrated that isosmotically elevated K^+ results mainly in the release of only endogenous and preloaded, radiolabeled taurine, respectively. Pasantes-Morales and Schousboe[9] speculate that when K^+ causes cell swelling, it cannot participate in the volume recovery phase. Therefore, under these conditions perhaps taurine, being the more abundant intracellular amino acid, acts as the major osmoeffector in regulating cellular volume decrease. In another study these investigators used hypoosmotic conditions to bring about astrocytic swelling and observed the release of endogenous and radiolabeled taurine and, to a much lesser extent, preloaded, labeled D-aspartate (a marker for glutamate and aspartate), GABA, and glycine.[20] Taurine release increased with increasing hypoosmolarity, with a time course related to the cell's volume regulatory phase. Kimelberg *et al.*[19] demonstrated comparable release of endogenous or preloaded, radiolabeled taurine and glutamate as well as labeled D-aspartate under hypotonic conditions in similar cultures of astrocytes. Results from this study suggested that release was not due in large part to the reversal of the Na^+-dependent uptake system, because isotonically elevated K^+ increased efflux to a lesser extent. Thus, we attempted to determine which endogenous amino acids are released by equiosmotically elevating K^+ (80 mM) under increasingly severe hypoosmotic (0 to −100 mosm) conditions, beginning from our standard conditions of 335 mosm. As shown in FIGURES 3 (serine and taurine) and 4 (aspartate and glutamate), while at 335 mosm elevated K^+ resulted in the efflux of only taurine, under conditions of

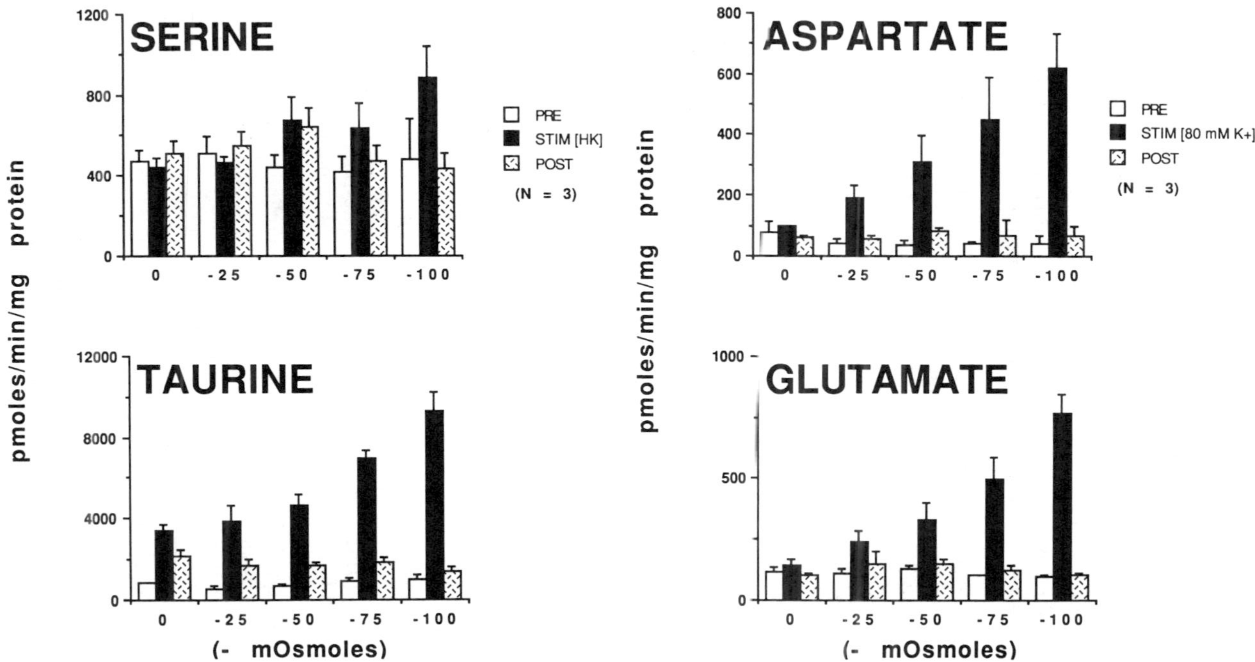

FIGURES 3 AND 4. Amino acid release from cultured cerebellar astrocytes exposed to 80 mM K^+ under conditions of decreasing osmolarity. Separate sets of cultures were washed with standard (335 mosm) perfusion buffer, or buffer of reduced osmolarity (decreased NaCl), three times for 10 minutes each, prior to a fourth 10-minute wash (□ = prestimulation). This was followed by a 10-minute stimulation in buffer of appropriate osmolarity, but containing 80 mM K^+ (■ = stimulation), and finally a 10-minute poststimulation (▨ = poststimulation). The 30-minute pre-washes were performed to allow completion of the regulatory volume decrease (RVD) and stable baseline following cell swelling produced by hyposmotic conditions.

decreasing osmolarity a dose-dependent increase occurred in the release of taurine, aspartate, and glutamate. However, this did not include serine (FIG. 3) or alanine and proline (data not shown). These results show that swelling produced by isotonically elevating K^+ results in a less robust release of amino acids compared to that produced by reducing osmolarity by an equivalent milliosmolar amount. Furthermore, taurine appears to be the most prominently released amino acid under both conditions, which is perhaps an indication of the importance of its role as an osmoregulatory solute. The differences in the levels of amino acids released, under these two conditions, also suggest that different efflux mechanisms might be involved.

In other experiments we showed that K^+-mediated taurine release is partially sensitive (~40% inhibited at 50 mM K^+) to furosemide (1 mM), a Cl^-/HCO_3^- anion exchange inhibitor, suggesting that an additional component of K^+-produced swelling involves a K^+/Cl^- cotransport system.[14] In any case, it is clear that taurine is released only when an increase in cell volume occurs and its release is prevented in the absence of swelling. However, some investigators believe that K^+-mediated astrocyte swelling may be even more complex. For example, it has been suggested that cell swelling may occur via a furosemide- and Cl^--sensitive cotransport mechanism after exposure of cells to lower (12 mM) extracellular increase in K^+. However, on exposure to higher (72 mM) K^+ levels, cell swelling may proceed by a furosemide-insensitive, Cl^--sensitive, and bumetanide-, SITS-, and Ba^{2+}-insensitive mechanism, which is largely mediated by passive Donnan forces.[9] This complex and as yet unresolved issue is the focus of much current research.[1] In any event, the cardinal signal for the release of taurine from astrocytes would appear to be an increase in cell volume,[21] and the mechanism(s) by which it occurs awaits further investigation.

Receptor-Mediated Release

Other substances such as glutamate,[22] adenosine,[23] kainate,[22] and weak organic acids[24] also produce astrocyte swelling and taurine efflux. However, here again more than one mechanism is apparently involved in this swelling.[1,14] Among the many substances, adenosine promotes taurine efflux from a rat spinal cord glioma cell line, LRM55, by what is thought to be an A_2 type adenosine receptor-mediated mechanism linked to cAMP elevation.[25] Therefore, we set out to determine if adenosine and its nucleotide analogs, adenosine mono-(AMP), di-(ADP)-, and tri-(ATP)phosphate, had similar effects on endogenous amino acid release from cultured cerebellar astrocytes. FIGURE 5 (taurine and proline) shows that taurine release is increased more than twofold by all four substances following a 20-minute stimulation. Proline, on the other hand, is released in a much more selective fashion quantitatively and preferentially by ADP and ATP. ADN- and AMP-evoked proline efflux moved only two- to threefold above basal levels. FIGURE 6 (serine and alanine) shows that all four compounds elicit the release of serine and alanine to almost equivalent, twofold increases above basal levels. Amino acid release in these experiments was selective, however, inasmuch as no aspartate or glutamate efflux was observed in these experiments (data not shown). In cultures of rat cortical astrocytes, Neary and coworkers[26] found that exposure to adenosine (and ADP and ATP) produced Ca^{2+} influx and a rise in intracellular Ca^{2+}, which appeared to be mediated through purinergic P_1 (adenosine) and P_2 (ADP and ATP) receptors. Thus, although we have demonstrated here that adenosine, AMP, ADP, and ATP are capable of stimulating taurine efflux and currently have no evidence as to whether they act through receptors coupled to second messengers (e.g., Ca^{2+}), the selective release of amino acids is perhaps reflective of some type of specific response taking place. However,

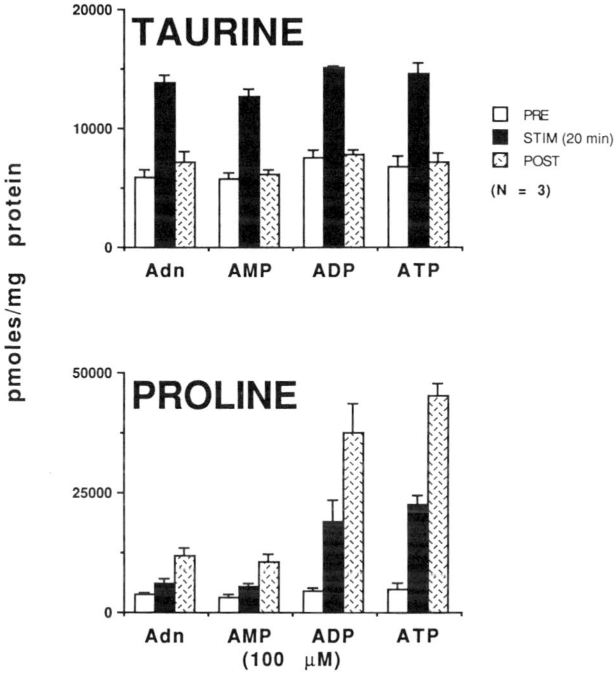

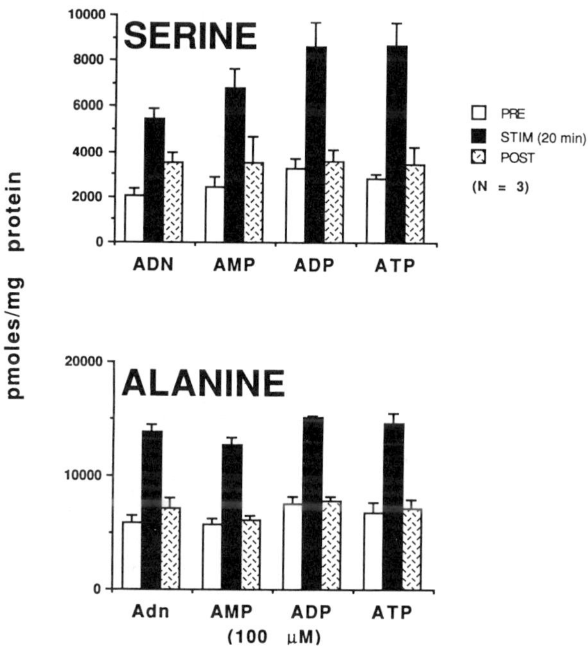

FIGURES 5 AND 6. Effects of adenosine and adenine nucleotides on amino acid release from cultured cerebellar astrocytes. Adn and ADN = adenosine; AMP = adenosine monophosphate; ADP = adenosine diphosphate; ATP = adenosine triphosphate; □ PRE = 20-minute prestimulation fraction; ■ STIM (20 min) = 20-minute stimulation fraction with 100 μM of the above (as indicated); ▨ POST = 20-minute poststimulation fraction. No aspartate or glutamate release was seen under these conditions.

why serine, alanine, and proline, but not glutamate and aspartate, are among the amino acids released is not clear.

In work performed largely with the glioma cell line LRM55, Shain, Martin, and co-workers[27,28] reported that taurine efflux is mediated through β-adrenergic receptors linked to cAMP. In addition, they have determined that such taurine release is not dependent on Ca^{2+} and is not coupled to membrane depolarization, and that taurine is not sequestered in a specific subcellular region within the cell.[8] They showed that the characteristics of taurine uptake and release reflect the activities of different transporter mechanisms and respond differently to osmolarity changes and anion transport blockers. Regarding taurine release, they speculated that a possible mechanism may involve the activation of tension-activated stretch channels,[11] but no direct evidence exists for this. We have performed our own sets of experiments to elicit taurine efflux through the use of various receptor agonists and second messenger analogs in an attempt to determine if receptor-mediated events can produce release. While both glutamate (FIG. 7) and kainate (FIG . 8) evoked taurine release

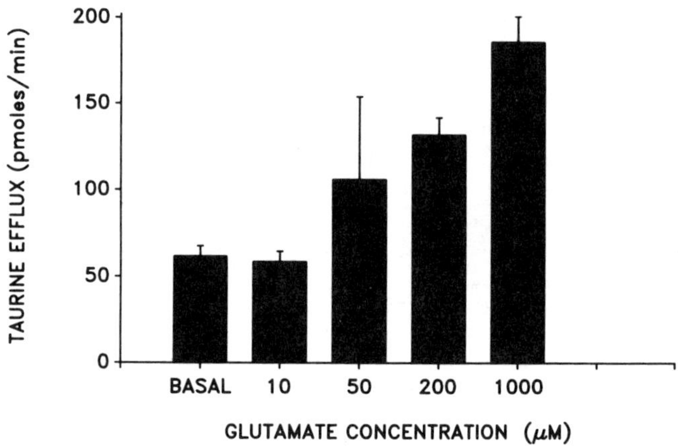

FIGURE 7. Dose response of glutamate-evoked (10-minute) stimulation of taurine efflux from cultured cerebellar astrocytes. $n = 3$. Basal = no added glutamate.

in a dose-dependent fashion, the kainate/quisqualate receptor antagonist CNQX inhibited kainate-mediated release (FIG. 8), but had no effect on that produced by glutamate (FIG. 9). Furthermore, the following compounds failed to alter basal taurine levels during a 10-minute stimulation: carbachol (100 μM), serotonin (200 nM), isoproterenol (200 nM), muscimol (100 μM), phorbol dibutyrate (1 μM), dibutyryl cyclic AMP (1 mM), and dibutyryl cyclic GMP (0.5 mM).[10] Thus, with the exception of kainate, we have been able to generate little evidence for direct neurotransmitter receptor-evoked taurine release in cultured cerebellar astrocytes.

Ca^{2+} Mobilization

To determine if a correlation exists between the mobilization of intracellular Ca^{2+} levels ($[Ca^{2+}]_i$) and evoked taurine efflux in astrocytes, in preliminary experiments we looked at the effects of kainate, glutamate, quisqualate, anisoosmotic

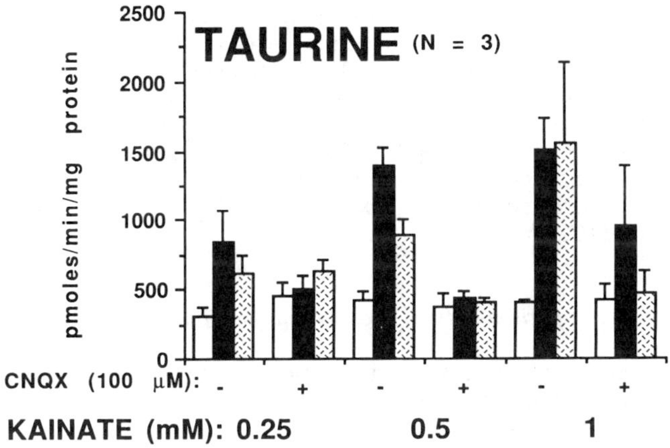

FIGURE 8. Kainate-evoked taurine efflux from cultured cerebellar astrocytes with and without CNQX (100 μM). □ = prestimulation (10 minutes); ■ = stimulation (10 minutes); ▨ = poststimulation

conditions, and elevated K^+ on fura-2/AM-loaded cells in the presence and absence of extracellular Ca^{2+} ($[Ca^{2+}]_o$). Overall, most of the data obtained are in agreement with the results of Enkvist *et al.*[29] Kainate and glutamate, each at 1 mM, and quisqualate (50 μM) produced elevated $[Ca^{2+}]_i$ of about 10–20 nM in the presence of $[Ca^{2+}]_o$ = 2.5 mM. However, kainate had no effect in the absence of $[Ca^{2+}]_o$ with 2 mM EGTA. Quisqualate, on the other hand, produced elevated $[Ca^{2+}]_i$ of approximately 10 nM. In anisoosmotic experiments using 308 mosm buffer as isotonic, we found that reductions of 45 and 75 mosm slightly (5–10 nM) increased $[Ca^{2+}]_i$ in the presence and absence of extracellular Ca^{2+}. However, using isosmotically elevated K^+ (40 mM), astrocytes responded by increases in $[Ca^{2+}]_i$ of 20–30 nM in the

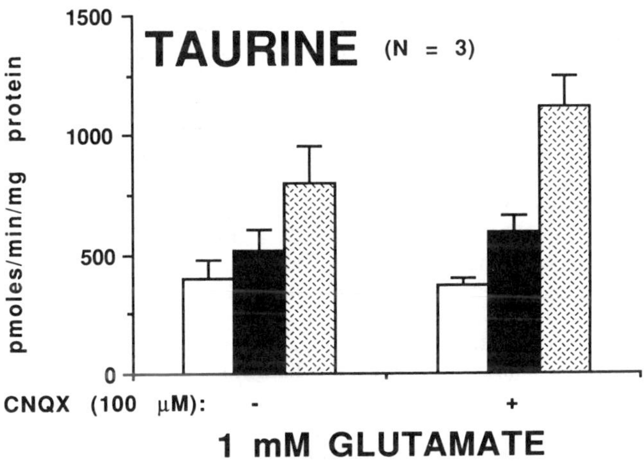

FIGURE 9. Glutamate-evoked taurine release from cultured cerebellar astrocytes with and without CNQX (100 μM). □ = prestimulation (10 minutes); ■ = stimulation; ▨ = poststimulation.

presence, but not absence, of $[Ca^{2+}]_o$. In experiments in which 40 mM K^+ was added hyperosmotically, $[Ca^{2+}]_i$ rose in both the presence and the absence of $[Ca^{2+}]_o$, albeit to a much smaller extent in the latter (5–10 nM) than in the former (10–40 nM) case. Osmotically shocking cells by 45 or 75 mosm produced only minimal $[Ca^{2+}]_i$ increases (~ 10 nM) when $[Ca^{2+}]_o$ was either present or absent. Overall these data show no clear correlation between $[Ca^{2+}]_i$ mobilization and conditions that produce cell swelling. These findings are in general agreement with those of Shain, *et al.* Martin,[30,31] who reported that cell swelling resulting in taurine efflux occurred independently of $[Ca^{2+}]_i$.

Comparison of Glial versus *Neuronal Taurine Release*

Cultured neurons also release taurine, which at first glance appears to be brought about by very different mechanisms than those in the case of astrocytes. For example, in previous work we[13,32] demonstrated that in cerebellar neuronal cultures, veratridine (50 μM) causes efflux which is blocked by tetrodotoxin (2 μM), whereas in astrocytes veratridine or tetrodotoxin alone do not affect basal or K^+-evoked taurine release levels, respectively. Also, the Ca^{2+}-channel agonist, BAY K 8644, augments K^+-evoked taurine release from neuronal cultures, which is sensitive to inhibition by the antagonist nifedipine. In contrast, neither of these dihydropyridines alters basal or K^+-stimulated taurine release in astrocytes.

These differences might be explained by suggesting that in cerebellar neuronal (largely glutamergic) cultures, it is largely the depolarization-evoked release of glutamate that remains at high concentrations in the extracellular space in the absence of astrocytes in the cultures. This effect may in part be responsible for a neuronal swelling-induced increase in taurine efflux. Thus, when glutamate release is prevented by the appropriate antagonists (e.g., tetrodotoxin vs. veratridine), so too is taurine release, because cell swelling is prevented. It is interesting to note that K^+-evoked release is not inhibited by tetrodotoxin in these neuronal cultures.[32]

Comparison of the similarities between astrocyte and neuronal taurine efflux also makes the case for similar release mechanisms for these two cell types even stronger. For example, decreasing extracellular Ca^{2+} over the range 1–0 mM results in increased basal release levels of taurine in both types of cultures, unless blocked by 10 mM Mg^{2+}. Removing Na^+ by substituting $choline^+$ in an isosmolar fashion essentially abolishes K^+-evoked (50 mM) taurine release from both neuronal and astrocyte cultures.[13,32] The time course of release from both astrocyte and neuronal cultures is also characterized by a delayed onset after stimulation and a slow rise to maximum release, which occurs about 10 minutes after maximum stimulation. Additionally, both neuronal and astrocytic kainic acid-evoked taurine release are sensitive to inhibition by the kainate/quisqualate antagonist CNQX. Finally, although at some variance with our previously reported results, other investigators have produced evidence that both neuronal and astrocytic K^+-evoked taurine release are Ca^{2+} independent[33] and that this K^+-induced taurine release is abolished by substitution of Cl^- by the impermeant anion gluconate.

It is currently not possible to decide whether taurine release from neurons and astrocytes occurs by similar or different mechanisms.

SUMMARY

The evidence presented, together with the lack of solid evidence for a specific receptor site, strongly suggests that taurine does not act as a traditional neurotransmitter in the CNS. In fact, the properties seen to be governing its efflux from both glial cells and neurons argue strongly in favor of a primary role in volume regulation. However, subsequent to its release into the extracellular space, it is possible that the inherent neuroactive properties (e.g., inhibitory neuromodulation and Ca^{2+}-level modulation) may be important at the synapse, at the cell plasma membrane, and intracellularly in further directing the level of neuronal activity. Whether or not the levels of released taurine are great enough to sustain these effects has still to be determined.

ACKNOWLEDGMENT

The authors would like to thank Kathryn Andrews for manuscript preparation.

REFERENCES

1. HUXTABLE, R. J. 1989. Taurine in the central nervous system and the mammalian actions of taurine. Progr. Neurobiol. **32:** 471–533.
2. PASANTES-MORALES, H., D. L. MARTIN, W. SHAIN & R. MARTIN DEL RIO (Eds.) 1990. Taurine: Functional Neurochemistry, Physiology, and Cardiology. Vol. 351. Wiley-Liss. New York.
3. ROGERS, K. L., R. A. PHILIBERT, A. J. ALLEN, J. MOLITOR, E. J. WILSON & G. R. DUTTON. 1987. HPLC analysis of putative amino acid neurotransmitters released from primary cerebellar cultures. J. Neurosci. Methods **22:** 173–179.
4. TAPPAZ, M., F. LEGAY, K. ALMARGHINI, S. HENRY & A. RÉMY. 1990. Cysteine sulfinic acid decarboxylases (CSD) in the brain. *In* Progress in Clinical and Biological Research. H. Pasantes-Morales, D. L. Martin, W. Shain & R. Martin del Rio, eds. Vol. 351: 53–68. Taurine: Functional Neurochemistry, Physiology, and Cardiology. Wiley-Liss. New York.
5. MAGNUSSON, K. R., J. E. MADL, J. R. CLEMENTS, J.-Y. WU, A. A. LARSON & A. J. BEITZ. 1988. Colocalization of taurine- and cysteine sulfinic acid decarboxylase-like immunoreactivity in the cerebellum of the rat with monoclonal antibodies against taurine. J. Neurosci. **8(12):** 4551–4564.
6. HUXTABLE, R. J. & S. E. LIPPINCOTT. 1982. Sources and turnover of taurine in newborn, weanling, and mature rats. Adv. Exp. Med. Biol. **139:** 23–46.
7. WALZ, W. & A. F. ALLEN. 1987. Evaluation of the osmoregulatory function of taurine in brain cells. Exp. Brain Res. **68:** 290–298.
8. SHAIN, W., V. MADELIAN, R. A. WANIEWSKI & D. L. MARTIN. 1990. Characteristics of taurine release from astroglial cells. *In* Progress in Clinical and Biological Research. H. Pasantes-Morales, D. L. Martin, W. Shain & R. Martin del Rio, eds. Vol. 351: 299–306. Taurine: Functional Neurochemistry, Physiology, and Cardiology. Wiley-Liss. New York.
9. PASANTES-MORALES, H. & A. SCHOUSBOE. 1989. Release of taurine from astrocytes during potassium-evoked swelling. Glia **2:** 45–50.
10. PHILIBERT, R. A., K. L. ROGERS, A. J. ALLEN & G. R. DUTTON. 1988. Dose-dependent, K^+-stimulated efflux of endogenous taurine from primary astrocyte cultures is Ca^{2+}-dependent. J. Neurochem. **51:** 122–126.
11. MARTIN, D. L., V. MADELIAN, B. SELIGMANN & W. SHAIN. 1990. The role of osmotic pressure and membrane potential in K^+-stimulated taurine release from cultured astrocytes and LRM55 cells. J. Neurosci. **10(2):** 571–577.
12. OJA, S. S. & P. KONTRO. 1987. Cation effects on taurine release from brain slices: Comparison to GABA. J. Neurosci. Res. **17:** 302–311.
13. PHILIBERT, R. A. & G. R. DUTTON. 1989. Dihydropyridines modulate K^+-evoked amino acid and adenosine release from cerebellar neuronal cultures. Neurosci. Letts. **102:** 97–102.
14. WALZ, W. 1989. Role of glial cells in the regulation of the brain ion microenvironment. Progr. Neurobiol. **33:** 309–333.
15. PHILIBERT, R. A., K. L. ROGERS & G. R. DUTTON. 1989. K^+-evoked taurine efflux from cerebellar astrocytes: On the roles of Ca^{2+} and Na^+. Neurochem. Res. **14(1):** 43–48.

16. Dutton, G. R. & R. A. Philibert. 1990. Taurine release from cultured astrocytes. *In* Differentiation and Functions of Glial Cells. G. Levi, ed. : 235–241. Alan R. Liss, Inc. New York.
17. Tigges, G. A., R. A. Philibert & G. R. Dutton. 1990. K^+- and temperature-evoked efflux from hypothalamic astrocytes. Neurosci. Letts. **119:** 23–26.
18. Martin, D. L., V. Madelian & W. Shain. 1990. Osmotic sensitivity of isoproterenol- and high-$[K^+]_o$-stimulated taurine release by cultured astroglia. *In* Progress in Clinical and Biological Research. H. Pasantes-Morales, D. L. Martin, W. Shain & R. Martin del Rio, eds. Vol. 351: 349–356. Taurine: Functional Neurochemistry, Physiology, and Cardiology. Wiley-Liss. New York.
19. Kimelberg, H. K., S. K. Goderie, S. Higman, S. Pang & R. A. Waniewski. 1990. Swelling-induced release of glutamate, aspartate, and taurine from astrocyte cultures. J. Neurosci. **10(5):** 1583–1591.
20. Pasantes-Morales, H. & A. Schousboe. 1988. Rapid communication, volume regulation in astrocytes: A role for taurine as an osmoeffector. J. Neurosci. Res. **20:** 505–509.
21. Pasantes-Morales, H. & R. Martin del Rio. 1990. Taurine and mechanisms of cell volume regulation. *In* Progress in Clinical and Biological Research. H. Pasantes-Morales, D. L. Martin, W. Shain & R. Martin del Rio, eds. Vol. 351: 317–328. Taurine: Functional Neurochemistry, Physiology, and Cardiology. Wiley-Liss. New York.
22. Kimelberg, H. K., S. Pang & D. H. Treble. 1989. Excitatory amino acid-stimulated uptake of $^{22}Na^+$ in primary astrocyte cultures. J. Neurosci. **9(4):** 1141–1149.
23. Bourke, R. S., J. B. Waldman, H. K. Kimelberg, K. D. Barron, B. D. San Filippo, A. J. Popp & L. R. Nelson. 1981. Adenosine-stimulated astroglial swelling in cat cerebral cortex *in vivo* with total inhibition by a non-diuretic acylaryloxyacid derivative. J. Neurosurg. **55:** 364–370.
24. Solis, J. M., A. S. Herranz, O. Herreras, N. Menéndez & R. Martin del Rio. 1990. J. Neurosci. Res. **26:** 159–167.
25. Madelian, V., S. Silliman & W. Shain. 1988. Adenosine stimulates cAMP-mediated taurine release from LRM55 glial cells. J. Neurosci. Res. **20:** 176–181.
26. Neary, J. T., C. van Breemen, E. Forster, L. O. B. Norenberg & M. D. Norenberg. 1988. ATP stimulates calcium influx in primary astrocyte cultures. Biochem. Biophys. Res. Commun. **157(1):** 1410–1416.
27. Madelian, V., D. L. Martin, R. Lepore, M. Perrone & W. Shain. 1985. β-receptor-stimulated and cyclic adenosine 3′,5′-monophosphate-mediated taurine release from LRM55 glial cells. J. Neurosci. **5(12):** 3154–3160.
28. Shain, W., V. Madelian & D. L. Martin. 1989. Inactivation of cyclic AMP-dependent taurine release from astroglia. J. Neurochem. **52:** 1455–1460.
29. Enkvist, K. M. O., I. Holopainen & K. E. O. Akerman. 1989. Glutamate receptor-linked changes in membrane potential and intracellular Ca^{2+} in primary rat astrocytes. Glia **2:** 397–402.
30. Martin, D. L., V. Madelian & W. Shain. 1989. Spontaneous and beta-adrenergic receptor-mediated taurine release from astroglial cells do not require extracellular calcium. J. Neurosci. Res. **23:** 191–197.
31. Shain, W., J. A. Connor, V. Madelian & D. L. Martin. 1989. Spontaneous and beta-adrenergic receptor-mediated taurine release from astroglial cells are independent of manipulations of intracellular calcium. J. Neurosci. **9(7):** 2306–2312.
32. Philibert, R. A., K. L. Rogers & G. R. Dutton. 1989. Stimulus-coupled taurine efflux from cerebellar neuronal cultures: On the roles of Ca^{2+} and Na^+. J. Neurosci. Res. **22:** 167–171.
33. Schousboe, A. & H. Pasantes-Morales. 1989. Potassium-stimulated release of [^{3}H]taurine from cultured GABAergic and glutamatergic neurons. J. Neurochem. **53:** 1309–1315.

DISCUSSION PAPER

Glial Receptors, Signaling, and Second Messengers

ANNE MUDGE

Department of Biology
Medawar Building
University College London
Gower Street
London WC1E 6BT, UK

The papers presented in this section clearly illustrate the potential for chemical signaling between neurons and glia in both vertebrate and invertebrate nervous systems. They also raise many intriguing questions about the consequences of this signaling as well as if or how receptors are matched to sites of transmitter and neuropeptide release.

The papers by Evans *et al.* and Lieberman have the advantage of studying invertebrate axons and glial cells when their anatomic relationships are intact. Figure 4 of Evans *et al.* illustrates the complexity of the signaling between the axon and the glial cell. What is the point of this interaction? Does it serve to link the supply of nutrients or trophic factors from glia to the metabolism of the axon, or is it a sophisticated way of maintaining the ionic balance of the fluid surrounding the cells during periods of intense nerve activity?

The same questions apply to the studies involving rodent astrocytes. Why do astrocytes need receptors for transmitters and neuropeptides? Wilkins *et al.* cite an example of a neuropeptide affecting release of a neurotrophic agent (see Brenreman *et al.*). Is this a general phenomenon? Assuming that release of a neurotrophic agent from glia would occur via the constitutive release pathway, is the synthesis of glial-derived trophic factors regulated by signaling molecules activating second-messenger systems? Likewise, does glycogen breakdown in response to elevating cAMP in astrocytes serve the energy requirements of the astrocytes or does it provide nutrients for the neurons? If the latter, what is the glycogen metabolite released by neurons? To release glucose, for example, the astrocyte would need to remove phosphate from glucose-6-phosphate: this conversion is usually made by liver cells but not other cells in the body. Are there special glycogen metabolic pathways in astrocytes?

Depolarization or osmotic changes cause the release of taurine from astrocytes as shown by Dutton *et al.* in this section. Perhaps glial cells serve as storage depots for many compounds that participate in the effective functioning of the nervous system.

Wilkin *et al.* also address the problem of why astrocytes derived from embryonic or newborn rats express so many receptors when grown *in vitro.* As they suggest, perhaps glial cells express a large repertoire of receptors during development and then some are selected by the availability of the appropriate ligand released by nearby neurons. Their work shows that there is some regional specificity as to neuropeptide receptors expressed by glia from different regions of the nervous system, which parallels the distribution of the ligand. How well are ligands and receptors matched and how is the matching achieved?

Origin, Growth Factor Responses, and Ultrastructural Characteristics of an Adult-Specific Glial Progenitor Cell

GUUS WOLSWIJK,[a] PETER M. G. MUNRO,[b]
PETER N. RIDDLE,[c] AND M. NOBLE[a]

[a]*Ludwig Institute for Cancer Research*
91 Riding House Street
London W1P 8BT, UK

[b]*Institute of Neurology*
Queen Square
London WC1N 3BG, UK

[c]*Imperial Cancer Research Fund*
Lincoln's Inn Fields
London WC2A 3PX, UK

Oligodendrocyte type-2 astrocyte (O-2A) progenitor cells isolated from the optic nerves of adult rats differ from their perinatal counterparts in morphology, antigenic phenotype, cell cycle time, rate of migration, and time course of differentiation.[1] Moreover, O-2A^{adult} progenitors divide in an asymmetrical manner and have the capacity for extensive self-renewal, that is, these cells have stem cell-like properties.[2] At least some of the differences appear to represent phenotypic specializations relevant to the differing roles of these cells in developing and adult animals.[1] The first O-2A^{adult} progenitors appear in the developing nerve approximately 7 days after birth, and these cells coexist with O-2A^{perinatal} progenitors for several weeks, after which the perinatal population disappears from the nerve.[1,3] In this paper, we examine the origin of the O-2A^{adult} progenitor cell, its response to platelet-derived growth factor (PDGF), and its ultrastructural characteristics.

RESULTS AND DISCUSSION

Using time-lapse microcinematography, we found that the most likely origin of O-2A^{adult} progenitors is a subpopulation of O-2A^{perinatal} progenitors.[2] Bipolar, rapidly dividing, and migrating O-2A progenitors were seen to give rise to unipolar cells that divided and migrated slowly. Thus, the population of O-2A^{perinatal} progenitors is a tripotential one, capable of giving rise to oligodendrocytes, type-2 astrocytes, and O-2A^{adult} progenitors. Moreover, a precursor population with properties appropriate for early development gives rise to a second population of precursor cells with properties suited to the needs of the mature organism.

O-2A^{perinatal} and O-2A^{adult} progenitors express their distinguishing properties when they are cultured on monolayers of type-1 astrocytes (the differentiated progeny of a separate glial lineage in the rat optic nerve) or in medium conditioned by type-1 astrocytes (Astro-CM). Moreover, cells with the properties of O-2A^{perinatal} and O-2A^{adult} progenitors can coexist in the same culture (for example, cultures of P21 optic nerve).[3] These observations suggest that the differences between O-2A^{perinatal} and

TABLE 1. Properties of O-2A^{perinatal} and O-2A^{adult} Progenitors Cultured in the Presence of PDGF

Characteristic	Source O-2A Progenitors: Perinatal Optic Nerves	Source O-2A Progenitors: Adult Optic Nerves
Morphology	Bipolar	Unipolar
Antigenic phenotype	$O4^-$ vimentin$^+$	$O4^+$ vimentin$^-$
Cell cycle time	20 ± 6 hours	59 ± 5 hours
Rate of migration	25 ± 5 μm/h	4 ± 1 μm/h

O-2A^{adult} progenitors are intrinsic to the cells themselves and are not the result of differences in the environment. Recent studies have shown that type-1 astrocytes secrete PDGF and that PDGF can mimic the effects of Astro-CM on O-2A^{perinatal} progenitors.[4] We have now found that PDGF is also able to mimic the effects of Astro-CM on O-2A^{adult} progenitors. O-2A^{adult} progenitors cultured in PDGF have a unipolar morphology, an $O4^+$vimentin$^-$ antigenic phenotype, a cell cycle time of approximately 59 hours, and a rate of migration of about 4.1 μm/hour (TABLE 1). Moreover, anti-PDGF antibodies are able to neutralize the majority of mitogenic activity in Astro-CM for O-2A^{adult} progenitors. Thus, the binding of type-1 astrocyte-derived PDGF to O-2A^{perinatal} and O-2A^{adult} progenitors elicits two entirely different programs of behavior in these two O-2A progenitor cell types.

It is very likely that O-2A^{adult} progenitors play a role in the generation of new oligodendrocytes after demyelinating damage to the central nervous system. Thus far, no markers have been identified that allow the identification of O-2A^{adult} progenitors in sections of adult brain. To be able to study the role of O-2A^{adult} progenitors *in vivo,* we have been trying to identify O-2A^{adult} progenitors in sections of adult optic nerve on the basis of their *in vitro* ultrastructural characteristics. O-2A^{adult} progenitors that have been cultured *in vitro* have characteristic ultrastructural features (FIG. 1), and cells with similar ultrastructural characteristics can be identified in sections of adult optic nerve (data not shown). The *in situ* identification of O-2A^{adult} progenitors in sections of adult brain may prove useful in studying the role of these cells in the repair of demyelinating damage.

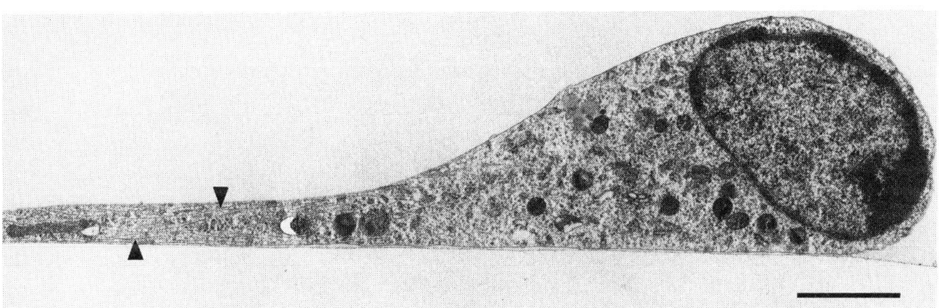

FIGURE 1. Vertical section through a cultured, resin-embedded O-2A^{adult} progenitor cell. Mitochondria and abundant free ribosomes are the principal cytoplasmic constituents of the cell body, with parallel bundles of microtubules predominating in the main process (*arrowheads*). The nucleus, which occupies the greater part of the cell body, is usually ringed with heterochromatin. Bar = 2 μm.

REFERENCES

1. WOLSWIJK, G. & M. NOBLE. 1989. Development **105:** 387–400.
2. WREN, D., G. WOLSWIJK & M. NOBLE. Submitted.
3. WOLSWIJK, G., P. N. RIDDLE & M. NOBLE. 1990. Development **109:** 691–698.
4. PRINGLE, N., E. J. COLLARINI, I. K. HART, M. C. RAFF & W. D. RICHARDSON. 1991. Ann. N.Y. Acad. Sci. This volume.

Studies Relating Differentiation to a Mechanism That Measures Time in O-2A Progenitor Cells

OLIVER BÖGLER AND MARK NOBLE

Ludwig Institute for Cancer Research
London W1P 8BT, UK

The scheduled appearance of oligodendrocytes in rat optic nerve can be recreated in culture by the addition of platelet-derived growth factor (PDGF).[1] In the presence of PDGF, oligodendrocyte type-2 astrocyte (O-2A) progenitor cells derived from perinatal optic nerve undergo a limited period of cell division before a cell-intrinsic mechanism causes them to differentiate into nondividing oligodendrocytes.[2]

Basic fibroblast growth factor (bFGF) is mitogenic for both immature oligodendrocytes and O-2A progenitor cells, but it does not prevent premature oligodendrocytic differentiation, suggesting that PDGF specifically enables O-2A progenitors to measure time.[3] In the presence of both PDGF and bFGF, O-2A progenitors divide indefinitely and do not differentiate into oligodendrocytes.[3] To understand further how cell division, differentiation, and measurement of time may be coordinated, we attempted to assess if the cell-intrinsic mechanism that measures time remains operative in conditions that prevent oligodendrocytic differentiation. We present evidence that in the presence of PDGF and bFGF, O-2A progenitor cells continue to measure time, as assessed by their behavior upon release from differentiation arrest.

RESULTS

Three lines of evidence suggest that O-2A progenitors continue to measure time in the presence of PDGF and bFGF. Firstly, populations of O-2A progenitors, pre-cultured in PDGF and bFGF for between 1 and 6 weeks and subsequently grown in conditions that promote timed differentiation, show an increased rate of differentiation that is greater with longer periods of pre-culture. In a typical experiment in which cells were analyzed 3 days after release from PDGF and bFGF, 17.7 ± 1.1% (mean ± SEM) of O-2A lineage cells were oligodendrocytes (galactocerebroside [GalC]-positive) in control cultures from postnatal day 7 (P7) optic nerve compared to 56.0 ± 6.7% and 91.4 ± 3.3% (mean ± SD) after 1 and 6 weeks of pre-culture in PDGF and bFGF, respectively. Accompanying the increased rate of differentiation was a smaller increase in O-2A lineage cell numbers, and the population of cells that had been pre-cultured in PDGF and bFGF for 6 weeks did not expand at all before differentiating (data not shown). Secondly, O-2A progenitors pre-cultured in PDGF and bFGF for 1 week gave rise to smaller clones of oligodendrocytes in single-cell culture, implying that they divided fewer times before differentiating; O-2A progenitors precultured in PDGF and bFGF for 4 weeks did not divide in single-cell microculture (FIG. 1). Lastly, in cultures derived from embryonic day 19 (E19) optic nerves that were exposed to PDGF and bFGF for 24 hours and subsequently to

PDGF only, the first oligodendrocytes appeared on the same schedule as in cultures exposed to PDGF alone throughout (data not shown).

DISCUSSION

Pre-culture in PDGF and bFGF causes O-2A progenitors to divide fewer times and to differentiate sooner, upon release from the differentiation arrest, suggesting that the cells continue to measure time. Furthermore, the observation that the longer the cells are maintained in PDGF and bFGF, the more rapidly they subsequently differentiate supports the idea that an increasing proportion of the cells remember that they have passed the point where they would ordinarily have differentiated. That bFGF does not preclude the measurement of time by these cells is supported by the fact that O-2A progenitors derived from E19 differentiate on schedule after a short exposure to PDGF and bFGF. We propose a model whereby cell division, differentiation, and the measurement of time are three independent cellular processes that

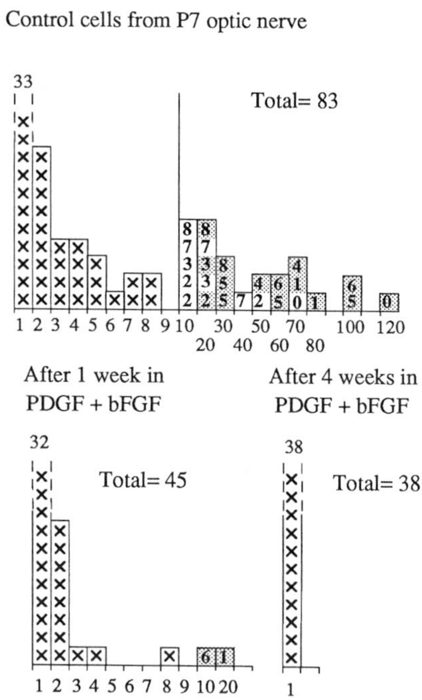

FIGURE 1. A stem and leaf plot showing the reduction in proliferative capacity before differentiation after pre-culture in PDGF and bFGF. O-2A progenitors from P7 optic nerve were pre-cultured in PDGF and bFGF for 1 or 4 weeks before analysis in single-cell microculture (as in ref. 2). Clones consisted entirely of $GalC^+$ oligodendrocytes or, if shaded, contained typically <35% $A2B5^+GalC^-$O-2A progenitors. Note that in the P7 control data set 4 clones consisted of 4 oligodendrocytes, 2 of 12, and 1 clone of mixed phenotype consisted of 74 cells.

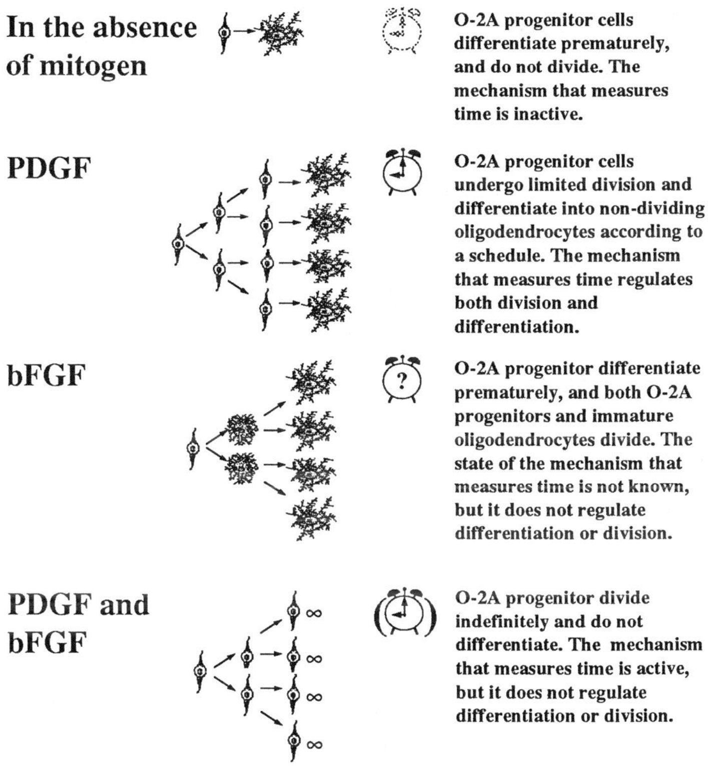

FIGURE 2. A model showing the role of the mechanism that measures time in the behavior of the O-2A progenitor cultured in various conditions. (O-2A progenitors are shown as bipolar and oligodendrocytes as multipolar cells.)

may be coordinately controlled to produce the complex repertoire of behavior of the O-2A progenitor (FIG. 2).

REFERENCES

1. RAFF, M. C., L. E. LILLIEN, W. D. RICHARDSON, J. F. BURNE & M. D. NOBLE. 1988. Nature **333:** 562–565.
2. TEMPLE, S. & M. C. RAFF. 1986. Cell **44:** 773–779.
3. BÖGLER, O., D. WREN, S. C. BARNETT, H. LAND & M. NOBLE. 1990. Proc. Natl. Acad. Sci. USA **87:** 6368–6372.

Expression of Myelin Glycolipids and Proteins in Mitotic and Postmitotic Murine Oligodendrocytes[a]

P. E. KNAPP AND R. P. SKOFF

Department of Anatomy and Cell Biology
Wayne State University
Detroit, Michigan 48201

Oligodendrocytes (OLs) express many different specific glycolipids and proteins during their differentiation. These include the glycolipids galactocerebroside (GC) and sulfatide as well as the proteins 2′,3′-cyclic nucleotide 3′-phosphohydrolase (CNPase) and proteolipid protein (PLP), among others. Previous studies showed that some of these markers are expressed earlier than others, but it is not clear which are expressed by dividing OLs (mitotic) and which are expressed after the final mitosis (postmitotic). We combined immunocytochemistry (ICC) for OL-specific antigens with either [^{3}H]-thymidine autoradiography or bromodeoxyuridine (BdU) ICC to determine the sequence of appearance of specific markers in mitotic and postmitotic OLs. Both thymidine and BdU are incorporated into DNA during the synthesis phase of the cell cycle, which in glia is approximately 8–10 hours long.[1] We used a monoclonal IgM antibody to sulfatide (007) that has been shown to be OL specific.[2] The monoclonal antibodies to GC[3,4] and BdU[5] and the rabbit antibodies to CNPase[3] and PLP[4] were previously described. IgG subclass specific secondary antibodies were used for double labeling with the GC and BdU antibodies.[5]

Mice 5–6 days postnatal were sacrificed from 1–120 hours after a single 1-hour pulse of thymidine. Vibratomed sections were immunostained, embedded in plastic, resectioned at 2 μm, and processed for autoradiography. Cultures derived from 2-day mouse cerebra[3] were pulsed with BdU for 1 hour and processed for ICC after 1–96 hours. The short exposures to thymidine (*in vivo*) or BdU (*in vitro*) ensure that cells labeled in the DNA synthesis phase will not move out of that phase and into mitosis during the pulse. One hour after a pulse of thymidine or BdU many dividing cells both *in situ* and *in vitro* are immunostained with antibody 007 (FIG. 1a and b). At 4 days *in vitro* only 1–2% of the sulfatide-positive population is BdU$^+$ after a 1-hour pulse. This figure seems low, but it is deceptive for two reasons. First, many sulfatide-positive cells will already be postmitotic before the BdU pulse. Also, assuming the synthesis phase to be 8–10 hours and the cell cycle to be 24 hours, only a third of the total cycling sulfatide-positive cells will be labeled during a 1-hour pulse. As OLs are found among the process-bearing cells in the top layer of the cultures, we asked a different question and found that about 12% of the BdU$^+$ process-bearing cells are sulfatide-positive after a 1-hour pulse. Because more than half the process-bearing cells express astrocyte markers, a large portion of the nonastrocytic, BdU$^+$ population is sulfatide positive. In certain areas of the cortex *in vivo,* the vast majority of thymidine-labeled cells are 007$^+$ at this age. Fewer thymidine or BdU$^+$ cells are immunostained for GC *in situ* or in culture. Specific glycolipids are therefore expressed by mitotic OLs, with sulfatide detected in many cells before GC. The

[a] This work was supported by NS 15338.

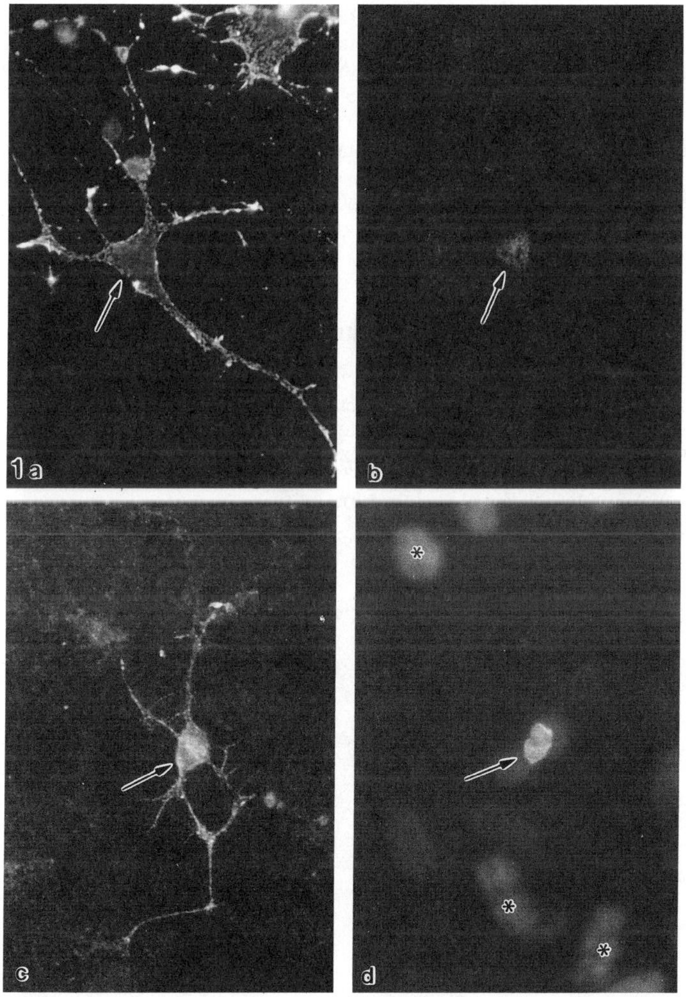

FIGURE 1. Primary cultures pulsed for 1 hour with BdU, then immunostained for BdU and either sulfatide (**a** and **b**) or PLP (**c** and **d**). In **a** and **b,** cultures were stained for sulfatide immediately after the 1-hour BdU pulse. A sulfatide-positive, process-bearing OL in **a** (*arrow*) has BdU immunostaining in its nucleus in **b.** As expected, the BdU⁺ nucleus is smaller than the cell body. In **c** and **d,** cultures were stained for PLP 96 hours after the BdU pulse. The PLP+, process-bearing OL in **c** (*arrow*) has BdU immunostaining in its nucleus in **d.** *Panel d* also shows that many nuclei in the bed layer astrocytes are BdU⁺ (*). These cells are in a different plane of focus compared to the OLs in the culture.

situation is very different for the OL-specific proteins that we examined. Neither CNPase nor PLP was detected in proliferating cells. CNPase was not seen until roughly 24 hours after the last mitosis either *in vivo* or *in vitro.* PLP appeared even later, by 72 hours *in vitro* (FIG. 1c and d) and by 96 hours *in vivo.*[4]

These results show that thymidine and BdU incorporation studies are useful for charting the subsequent development and differentiation of dividing cells. OL-specific glycolipids are generally expressed earlier in OL differentiation than are OL-specific proteins. The glycolipids are found in a percentage of dividing cells, whereas the proteins appear after cell division has been completed. It is clear that dividing cells can express markers of committed OLs, showing that commitment and proliferation are not mutually exclusive events in the OL lineage. After allowing for astrocytes, the percentage of BdU^+ process-bearing cells that immunostains for sulfatide is quite high. Thus, both *in vivo* and *in vitro,* committed OLs probably are a major source for generating new OLs.

REFERENCES

1. KORR, H., B. SCHULTZE & W. MAURER. 1975. J. Comp. Neurol. **160:** 477–490.
2. KNAPP, P. E. 1990. Trans. Am. Soc. Neurochem. **21:** 398.
3. KNAPP, P. E., R. P. SKOFF & T. J. SPRINKLE. 1988. J. Neurosci. Res. **21:** 249–259.
4. VERMEESCH, M. K., P. E. KNAPP, R. P. SKOFF, D. M. STUDZINSKI & J. A. BENJAMINS. 1990. Dev. Neurosci. **12:** 303–315.
5. KNAPP, P. E. 1991. J. Neurosci. Res. In press.

Protein Kinase C in Astrocytes

Phosphorylation and the Regulation of Cell Morphology[a]

PHILIP L. MOBLEY[b] AND BETH C. HARRISON[c]

Department of Pharmacology
University of Texas Health Science Center
San Antonio, Texas 78284

The regulation and maintenance of astrocytic morphology play an important role in the function of these cells. Shape changes occur during development, brain trauma, and neuronal degeneration. Additionally, the extension and retraction of astrocytic processes are thought to regulate neuron-neuron contact of the hypothalamic magnocellular neurosecretory cells. Therefore, an understanding of the processes that regulate astrocytic morphology is of considerable importance. A number of factors have been found that cause changes in astrocytic morphology. Rapid changes in the shape of cultured astrocytes can be induced by agents that mimic or promote the action of cyclic AMP.[1,2] The mechanism that mediates these changes appears to be distinct from that mediated by glial maturation factors that require several days to maximally alter astrocyte shape.[1]

Recent studies demonstrated that activators of protein kinase C (PK-C) also elicit changes in astrocytic morphology. Treatment of confluent cultures of rat brain astrocytes with 100 nM phorbol myristate acetate (PMA), an activator of PK-C, resulted in the conversion of flat, polygonal cells to process-bearing cells within 1–2 hours.[3,4] This effect could be blocked with colchicine, suggesting the involvement of cytoskeletal elements (FIG. 1). Activation of PK-C by PMA increased ^{32}P incorporation into a number of proteins including an acidic protein (pI 4.5) with a relative molecular weight (M_R) of 80,000, glial fibrillary acidic protein (GFAP), vimentin, an M_R 50,000 protein (pI 4.9), and two M_R 30,000 proteins (pI 5.5 and 5.7).[5] The effect of PMA on protein phosphorylation appeared to be mediated by the activation of PK-C, because no effect was observed in astrocytes depleted of PK-C.[5] Also, these PK-C depleted cells no longer exhibited a morphological response to PMA.

The shape changes observed with PMA are very similar to those induced by activators of cyclic AMP-dependent protein kinase (PK-A). Thus, these two kinase systems may converge at the level of protein phosphorylation to regulate astrocytic morphology. Treatment with 8-bromocyclic AMP (8BrcAMP) increased [^{32}P] incorporation into GFAP, vimentin, and both M_R 30,000 proteins. However, treatment with 8BrcAMP resulted in decreased [^{32}P] incorporation into the M_R 80,000 protein. Because of cytoskeletal involvement in the morphological effects of PMA, further studies were conducted to determine if phosphorylation elicited by treatment with PMA and 8BrcAMP occurred on the same tryptic fragments of GFAP and vimentin. Two-dimensional tryptic maps of GFAP and vimentin phosphorylated in intact cells

[a] This work was supported by NIH grant NS25766.

[b] To whom correspondence should be addressed.

[c] Current address: Department of Molecular and Cellular Biology, Pennsylvania State University, University Park, PA 16802.

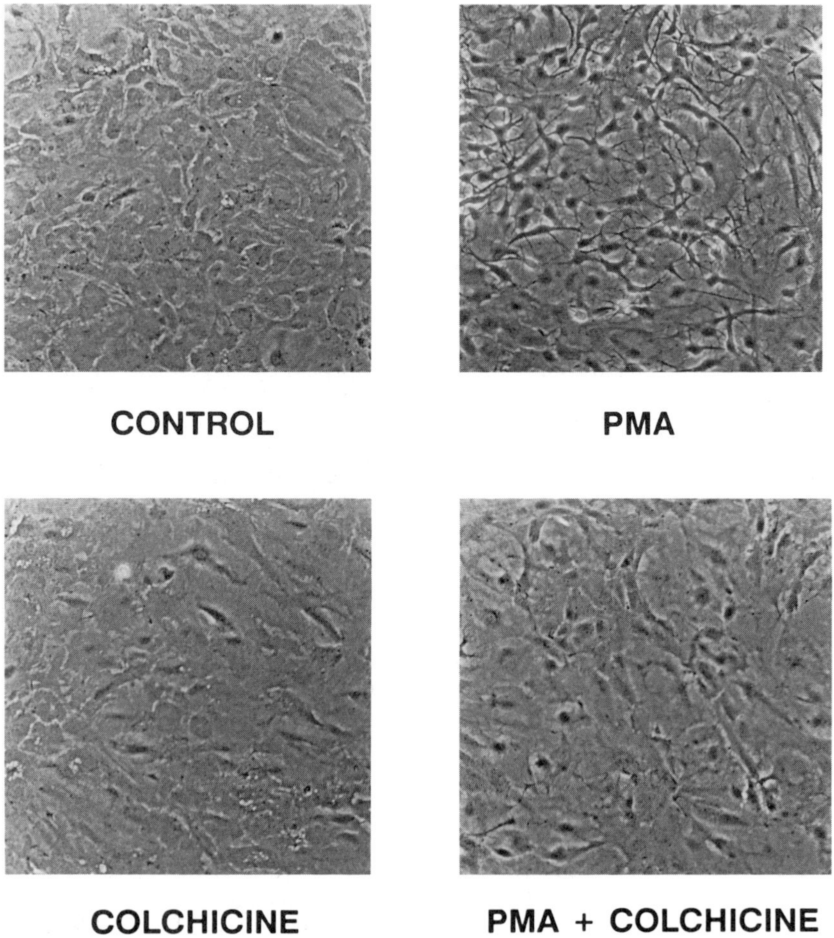

FIGURE 1. Effect of PMA and colchicine on astrocytic morphology. Confluent cultures of rat brain astrocytes were pretreated with 2 μM colchicine or left untreated, and then 1 hour later treated with PMA (300 nM final concentration) or its vehicle, dimethylsulfoxide (DMSO, 0.25 μl/ml). Phase contrast photographs were made 1 hour after treatment with PMA or its vehicle. Treatment was as follows: CONTROL = 1 hour of DMSO; PMA = 1 hour of PMA; COLCHICINE = 2 hours of colchicine, 1 hour of DMSO; PMA + COLCHICINE = 2 hours of colchicine, 1 hour of PMA.

by treatment with PMA or 8BrcAMP indicated that these two agents promoted phosphorylation at common sites on both proteins; however, PMA also promoted the phosphorylation of additional sites not affected by 8BrcAMP. Prominent [^{32}P] incorporation was observed in two tryptic fragments of vimentin (nos. 1 and 2 in FIG. 2) in response to treatment with 8BrcAMP, whereas three prominent phosphorylated fragments (nos. 1, 2, and 6 in FIG. 2) were observed following treatment with PMA. Prominent [^{32}P] incorporation was observed in one fragment (no. 1 in FIG. 2) of GFAP following treatment with 8BrcAMP and in two fragments (nos. 1 and 2 in FIG. 2) after treatment with PMA. Some [^{32}P] incorporation was also observed in two other fragments of GFAP (nos. 5 and 6 in FIG. 2) following treatment with PMA. Although a significant amount of [^{32}P] is incorporated in fragment no. 3 of GFAP under basal conditions, neither treatment increased incorporation in this fragment. Further studies indicated that treatment with either PMA or 8BrcAMP resulted in

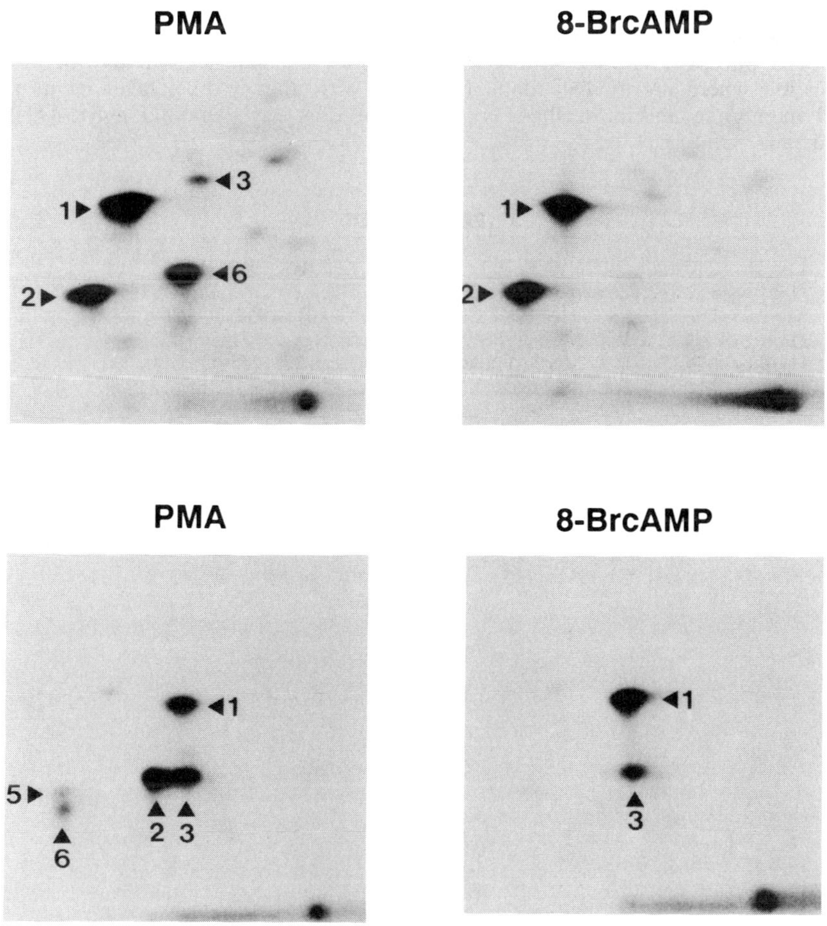

FIGURE 2. Two-dimensional tryptic maps of vimentin (*upper panels*) and GFAP (*lower panels*). Confluent cultures of rat brain astrocytes were prelabeled with [^{32}P]-orthophosphate for 2 hours and then exposed to either 300 nM PMA, 2 mM 8BrcAMP, or their vehicles for 15 minutes. Following the treatment period, the cells were washed and dissolved in 1% sodium dodecyl sulfate, then prepared for and subjected to two-dimensional gel electrophoresis. Spots corresponding to GFAP and vimentin were cut from the gels and subjected to trypsin digestion. The tryptic digests were spotted on cellulose plates and separated electrophoretically in one dimension and by ascending chromatography in the second dimension. Autoradiograms were then prepared from the dried plates.

phosphorylation of the *N*-terminal region of GFAP and vimentin, and only at serine residues. The phosphorylation response of these two agents apparently was mediated directly by PK-C and PK-A inasmuch as the same sites are phosphorylated using purified substrates and kinases.

The studies just described indicate that PK-C plays a role in the regulation of astrocytic morphology by cytoskeletal-dependent mechanisms. Of interest in this regard is the finding that PK-C phosphorylates the *N*-terminal region of both GFAP and vimentin. Several *in vitro* studies have indicated that phosphorylation of the *N*-terminal region of intermediate filament proteins inhibits their ability to exist in aggregates.[6] Although only a small proportion of the intermediate filament protein pool is phosphorylated by PMA in intact cells, phosphorylation may occur at locations where intermediate filaments interact with other cytoskeletal elements, the cell membrane, and intracellular organelles and thus play a key role in cytoskeletal rearrangement.

REFERENCES

1. Lim, R., K. Mitsunobu & W. K. P. Li. 1973. Exp. Cell Res. **79:** 243–246.
2. Narumi, S., H. K. Kimelberg & R. S. Bourke. 1978. J. Neurochem. **31:** 1479–1490.
3. Mobley, P. L., S. L. Scott & E. G. Cruz. 1986. Brain Res. **398:** 366–369.
4. Harrison, B. C. & P. L. Mobley. 1990. J. Neurosci. Res. **25:** 71–80.
5. Harrison, B. C. & P. L. Mobley. 1989. J. Neurochem. **53:** 1245–1251.
6. Inagaki, M., Y. Gonda, S. Ando, S. Kitamura, Y. Nishi & C. Sato. 1989. Cell Struct. Funct. **14:** 279–286.

Regulation of Glial Development by Cell Line Derived Factors

JOEL M. LEVINE[a] AND CHANGLIN DOU

Department of Neurobiology and Behavior
State University of New York at Stony Brook
Stony Brook, New York 11794

In dissociated tissue culture, oligodendrocytes and process-bearing or stellate astrocytes develop from a common bipotential precursor cell termed an O-2A progenitor cell.[1] The development and differentiation of these bipotential progenitor cells can be regulated, at least in part, by extrinsic factors. For example, platelet-derived growth factor (PDGF) and basic fibroblast growth factor (bFGF) are mitogens for O-2A progenitor cells. Although the development of oligodendrocytes in serum-free media is constitutive, the development of stellate astrocytes (type-2 astrocytes) is an induced event requiring the presence of fetal calf serum or factors extracted from optic nerve.[1] Recently, ciliary neurotrophic factor (CNTF) was shown to initiate astrocyte development, but the appearance of astrocytic properties induced by CNTF is transient.[2] Thus, O-2A progenitor cells are targets for multiple growth and differentiation-promoting factors and provide a useful bioassay for identifying such trophic activities.

We have begun to characterize trophic activities released into serum-free media by neuronal and glial clonal cell lines. These lines, which are derived from the central and peripheral nervous system, have lost the ability to control their growth and are likely to produce a variety of neurotrophic and gliotrophic activities.

To detect mitogens for O-2A progenitor cells, postnatal day 5 and 6 rat optic nerve cells were grown for 3 days in serum-free media conditioned (CM) by a variety of cell lines. The cultures were pulse-labeled with [^{3}H]-thymidine for 14–16 hours and immunofluorescently labeled with antibodies that identify O-2A progenitor cells (anti-NG2[3]) and oligodendrocytes (antigalactocerebroside). The percentage of labeled progenitor cells and the number of oligodendrocytes was determined for each CM used (TABLE 1). Most of the CMs tested contained mitogens for progenitor cells, and this mitogenic stimulation resulted in a low number of oligodendrocytes in the cultures. The same CMs stimulated ^{3}H-thymidine incorporation into quiescent NIH/3T3 cells, which respond to PDGF and bFGF, and CHO cells, which lack a PDGF receptor (TABLE 1). To determine if the cells produce PDGF or bFGF-like growth factors, we carried out Northern blotting of cellular RNA with cDNA probes for PDGF A chain, PDGF B chain, and bFGF. As shown in TABLE 2, most of the cell lines contained mRNAs that hybridized to these cDNA probes. Thus the ability of medium conditioned by these cell lines to stimulate progenitor cell division can be accounted for by the production of known mitogens for O-2A progenitor cells.

Of the cell lines tested, media conditioned by B92 cells[4] caused between 30 and 40% of the optic nerve progenitor cells to develop into type-2 astrocytes. This activity differed from that of CNTF in several regards. First, astrocyte induction by B92 CM is permanent, whereas CNTF effects are transient. Second, B92 CM is most effective when supplied to postnatal day 3 and 6 optic nerve cells, whereas CNTF is maximally

[a] To whom correspondence should be addressed.

TABLE 1. Effects of Cell Line Conditioned Media: Bioassays[a]

Condition	Percentage of Dividing O-2A Cells	Ratio (P/O)	3T3 Cells	CHO Cells
Defined media	6.4	0.07	...	...
PDGF (3 ng/ml)	25.0	0.87	3.7	1.0
bFGF (10 ng/ml)	49.7	7.5	3.3	1.1
Neuronal cell lines				
B103	52.2	0.44	13.6	3.1
B50	41.1	1.1	5.9	1.5
B35	45.7	0.82	4.6	1.9
N2A	27.6	0.33	1.4	5.8
PC12	11.1	1.28	1.9	1.8
Glial cell lines				
B82	52.8	0.83	8.7	2.3
B92	39.4	2.2	4.2	2.9
B49	40.6	0.88	7.1	3.3
C6	28.1	0.75	1.4	4.6
A172	40.8	0.53	2.3	1.2

[a]Postnatal day 5 and 6 rat optic nerve cells were grown for 3 days in either serum-free conditioned medium or defined medium supplemented with the indicated growth factors. After 3 days, the cells were pulse labeled for 14-16 hours with [^{3}H]-thymidine, fixed, and immunofluorescently labeled with antibodies that identify progenitor cells and oligodendrocytes. The percentage of dividing cells is a labeling index indicating the percentage of O-2A progenitor cells labeled with silver grains. The ratio of progenitor cells and oligodendrocytes is an indication of oligodendrocyte development in the cultures. For 3T3 and CHO cells, data shown is fold stimulation of [^{3}H]-thymidine incorporation into monolayers of quiescent cells compared to controls treated with serum-free defined media.

effective on embryonic or newborn optic nerve cells. B92 CM contains material that supports the growth of E8 chick ciliary ganglia neurons in short-term tissue culture; however, this activity was additive with both CNTF and bFGF. Preliminary results using a nitrocellulose Western blot assay developed by Carnow *et al.*[5] suggest that

TABLE 2. Northern Analysis: Summary[a]

	PDGF A[b]	PDGF B[c]	bFGF[d]
Neuronal lines			
B103	+	+	+
B50	+	+	Trace
B35	+	Trace	Trace
N2A	+	Not done	Trace
PC12	–	–	–
Glial lines			
B82	++	++	Not done
B92	+	+	+
B49	+	+	+
C6	+	Trace	+
A172	++	+	+

[a]Cellular RNA was separated on 1% agarose formaldehyde gels and electrophoretically transferred to nylon membranes. All blots were washed in 0.1 × SSC, 1% SDS at 60°C.
[b]cDNA probe was generated from plasmid PDGF-A 13.1.[6]
[c]cDNA probe was generated from plasmid PGGS 2.7 dBH.[7]
[d]cDNA and cRNA probes were generated from plasmid pRObFGF503.[8]

some of the activity that promotes the growth of ciliary neurons has a molecular weight of approximately 16,000 daltons.

These data demonstrate the utility of clonal cell lines as sources of neurotrophic and gliotrophic activities. Although the mitogenic activity contained within these CMs can be explained by the release of PDGF and bFGF-like substances, the ability of B92 CM to induce the appearance of type-2 astrocytes suggests that it contains a novel differentiation-inducing factor.

REFERENCES

1. RAFF, M. C. 1989. Glial cell diversification in the rat optic nerve. Science **243:** 1450–1455.
2. HUGHES, S. M., L. E. LILLIEN, M. C. RAFF, H. ROHRER & M. SENDTNER. 1988. Ciliary neurotrophic factor induces type-2 astrocyte differentiation in culture. Nature **335:** 70–73.
3. STALLCUP, W. B. & L. BEASLEY. 1987. Bipotential glial precursor cells of the optic nerve express the NG2 proteoglycan. J. Neurosci. **7:** 2737–2744.
4. SCHUBERT, D., S. HEINEMANN, W. CARLISLE, H. TARIKAS, B. KIMES, J. PATRICK, J. H. STEINBACH, W. CULP & B. L. BRANDT. 1974. Clonal cell lines from the rat central nervous system. Nature **249:** 224–227.
5. CARNOW, T. B., M. MANTHORPE, G. E. DAVIS & S. VARON. 1985. Localized survival of ciliary ganglionic neurons identifies neurotrophic factor bands on nitrocellulose blots. J. Neurosci. **5:** 1965–1971.
6. BETSHOLTZ, C., B. WESTERMARK, B. EK & C. H. HELDIN. 1986. DNA sequence and chromosomal localization of human platelet-derived growth factor A-chain and its expression in tumor cell lines. Nature **320:** 695–699.
7. IGARASHI, H., C. D. RAO, M. SIROFF, F. LEAL, K. C. ROBBINS & S. A. AARUNSON. 1987. Detection of PDGF-2 homodimers in human tumor cells. Oncogene **1:** 79–85.
8. SHIMASAKI, S., N. EMOTO, A. KOBA, M. MERCADO, F. SHIBATA, K. COOKSEY, A. BAIRD & N. LING. 1988. Complementary DNA cloning and sequencing of rat ovarian basic fibroblast growth factor and tissue distribution study of its mRNA. BBRC **157:** 256–263.

Differentiation of Aminergic Neurons *in Vitro* Is Not Paralleled by Glial Maturation[a]

CORDIAN BEYER,[b] JÜRGEN ENGELE,
CHRISTOPH PILGRIM, AND INGRID REISERT

Abteilung Anatomie und Zellbiologie
Universität Ulm
D-7900 Ulm, Germany

Our previous investigations on differentiation of aminergic neurons in cell cultures of fetal rat brain showed distinct differences in developmental schedules with respect to neuronal phenotype and brain region. Mature transmitter-related functions such as (re)uptake and stimulated release of transmitter developed in the following time sequence: serotoninergic neurons of the rhombencephalon,[1] dopaminergic neurons of the mesencephalon, and dopaminergic neurons of the diencephalon[2] (FIG. 1a and b). Even within one brain region, considerable differences can be observed; rhombencephalic noradrenergic neurons develop late in relation to other aminergic neurons[3] (data not shown).

Inasmuch as glial cells are believed to influence neuronal survival and differentiation, regionally diverse maturation of glia could be one of the factors responsible for region-specific neuronal differentiation. Maturation of glial cells was investigated in dissociated cell cultures of gestational day 14 rat brain containing both neurons and astroglia or their precursors.[4] Cultures were prepared from di-, mes-, and rhombencephalic tissue and raised with serum-supplemented modified Eagle's medium (MEM) or serum free medium. Development of glial-specific cytoskeletal proteins was determined by quantitative immunoblotting and counting of immunolabeled cells.

During cultivation, vimentin content increased in all brain regions examined in a similar way (FIG. 1c). In contrast, considerable regional differences in glial maturation were found in terms of GFAP content and numbers of GFAP-immunoreactive cells.[4] Both parameters increased rapidly in rhombencephalic cultures, followed by diencephalic cultures and mesencephalic cultures (cell numbers not shown).

The results thus demonstrate considerable heterogeneity of developmental schedules among astrocytes from different regions of the brain stem. Moreover, the temporal sequence of maturation among astrocytes of different origin does not correlate with the sequence in which different types of aminergic neurons develop that were cultured from these regions.

REFERENCES

1. REISERT, I., V. HAN, S. HARTWIG, G. AHNERT-HILGER & C. PILGRIM. 1989. Neuroscience **32:** 133–139.
2. ENGELE, J., C. PILGRIM, M. KIRSCH & I. REISERT. 1989. Brain Res. **483:** 98–109.
3. AHNERT-HILGER, G., J. ENGELE, I. REISERT & C. PILGRIM. 1986. Neuroscience **17:** 157–165.
4. BEYER, C., B. EPP, J. FASSBERG, I. REISERT & C. PILGRIM. 1990. Glia **3:** 55–64.

[a] This work was supported by Deutsche Forschungsgemeinschaft Re 413/2-2,3.

[b] Address for correspondence: Dr. Cordian Beyer, Abteilung Anatomie und Zellbiologie, Universität Ulm, Postfach 4066, D-7900 Ulm, Germany.

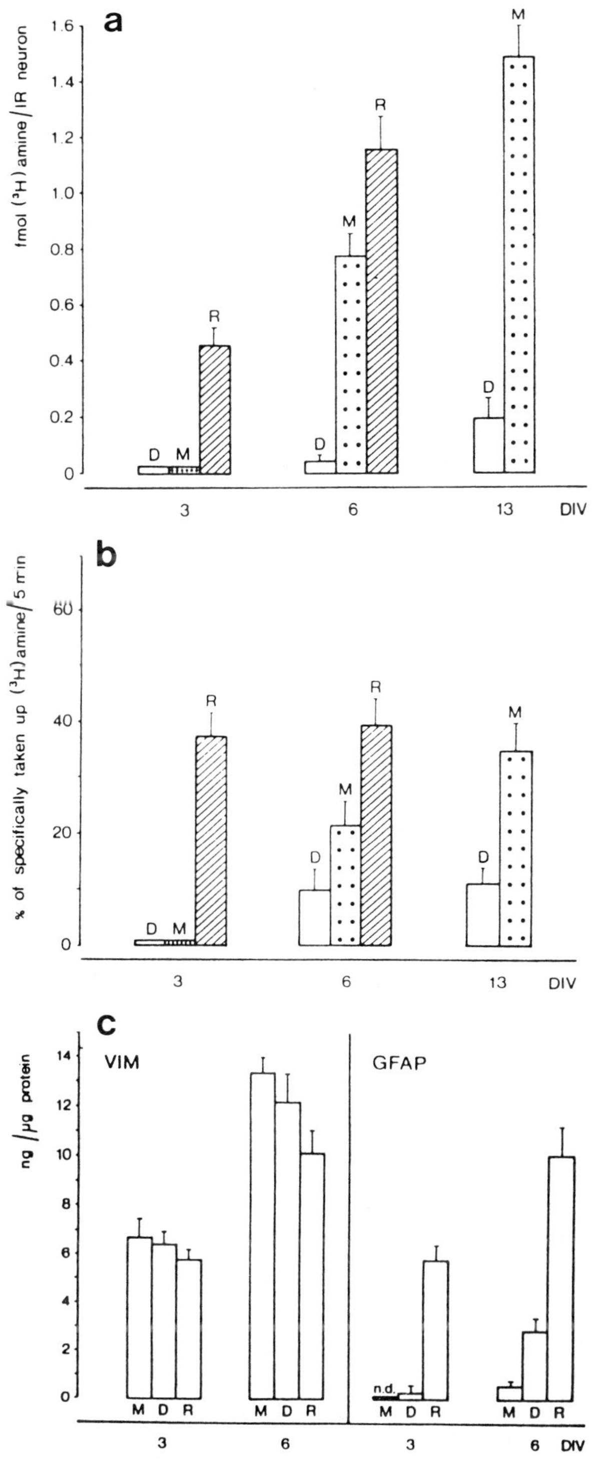

FIGURE 1. Uptake (**a**) and release (**b**) of (^{3}H)-labeled transmitter in di- (D), mes- (M), and rhombencephalic (R) cultures after 3, 6, and 13 days *in vitro* (DIV). Specific uptake of [^{3}H]dopamine (D,M) and [^{3}H]serotonin (R) was calculated by subtraction of nonspecific uptake as measured in the presence of fluoxeine (R) and benztropine (D,M) from total uptake and expressed per immunoreactive (IR) neuron (D,M = tyrosine hydroxylase; R = serotonin). Release was stimulated with 54 mM K^+ in the presence of 1.2 mM Ca^{2+}. Data for serotonin uptake and release at 13 DIV are not available. GFAP$^-$ and vimentin (VIM) were extracted and quantified by immunoblotting (c); nd = not detectable.

Astrocytes and O-2A Progenitors Migrate Toward Distinct Molecules in a Microchemotaxis Chamber

REGINA ARMSTRONG,[a,b] LIANA HARVATH,[c] AND
MONIQUE DUBOIS-DALCQ[a]

[a]*Laboratory of Viral and Molecular Pathogenesis
National Institutes of Neurological Disorders and Stroke
Bethesda, Maryland 20892*

[c]*Division of Blood and Blood Products
Center for Biologics Evaluation and Research
Food and Drug Administration
Bethesda, Maryland 20892*

In developing central nervous system (CNS) tissues, glial cells are generated in germinal zones near the ventricles.[1] Glial precursor cells must then migrate from these regions to their appropriate locations throughout the CNS. The signals that guide this glial cell migration are not yet understood. Therefore, we have used an *in vitro* assay to determine if various polypeptide growth factors and/or extracellular matrix molecules can stimulate directed glial cell migration. This study has focused on the migration of two major glial cell types, type-1 astrocytes and oligodendrocyte-type-2 astrocyte (O-2A) progenitors. Type-1 astocytes appear in the CNS before O-2A progenitors and may secrete factors that can influence the subsequent development of O-2A progenitors, which give rise to oligodendrocytes.[2]

Migration was evaluated with a microchemotaxis chamber in which cells in an upper compartment adhere to a polycarbonate filter and then migrate through 8-μm pores toward chemoattractants diffusing from a lower compartment.[3] This assay requires that a cell suspension added to the upper compartment is enriched for a particular cell type. Therefore, cell suspensions of purified type-1 astrocytes (96% expressed glial fibrillary acidic protein [GFAP]) and enriched O-2A progenitors (71% expressed A2B5 but not galactocerebroside or GFAP) were prepared from 2-day-old rat brains. As microglia are a likely contaminant of these preparations, we also prepared cell suspensions enriched in microglia (91% bound acetylated low density lipoprotein labeled with DiI) for comparison assays. These three types of cell suspension were then tested with the following potential chemoattractant molecules: fibronectin, laminin, complement-derived C5a, platelet-derived growth factor (PDGF), fibroblast growth factor (FGF), transforming growth factor β (TGFβ), and insulin-like growth factor I (IGF-I). Cells that migrated to the lower surface of the filter, in response to a chemoattractant, were visualized with a histological stain for cell counting or labeled with cell-type-specific markers for phenotype identification.

Our *in vitro* analysis revealed that specific molecules stimulated directed migration of each glial cell type (FIG. 1). Type-1 astrocytes moved toward laminin and C5a. In several experiments, the migrated cells on the lower surface of the filter were

[b]Address for correspondence: Dr. Regina Armstrong, Bldg. 36, Rm. 5D04, LVMP/NINDS, 9000 Rockville Pike, Bethesda, MD 20892.

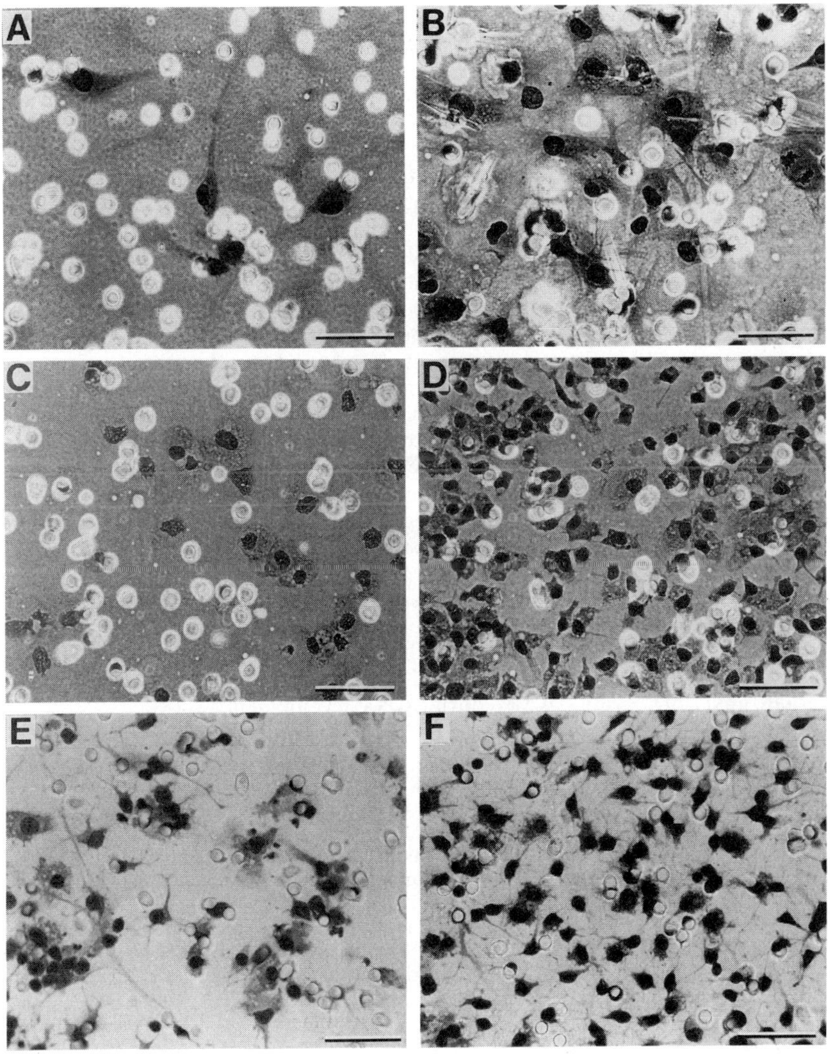

FIGURE 1. Migrated cells on the lower surface of each filter were fixed in methanol and stained with Diff-Quik histologic stain (similar to the Wright-Giemsa technique). The *panels on the left* show the minimal response observed when astrocytes (**A**), microglia (**C**), or O-2A progenitors (**E**) were placed in the upper compartment, whereas only Silberberg defined medium, without any test attractants, was present in the lower compartment. The number of migrated astrocytes increased when laminin (10 μg/ml) was added to the lower compartment (**B**). An increased number of microglia migrated in response to C5a (1:20 dilution of zymosan-activated rat serum) (**D**). Similarly, PDGF (10 ng/ml) stimulated increased migration of O-2A progenitor cells (**F**). Although each cell type exhibited a distinct morphology (compare **B, D,** and **F**), the phenotypes of migrated cells were confirmed in other experiments using cell-type-specific markers (see text).

identified as astrocytes that immunostained with antisera against GFAP. Cells in the microglia-enriched suspension also responded to C5a, as has been described previously.[4] These cells bound a marker for microglia lineage cells, low density lipoprotein labeled with DiI. O-2A progenitors migrated toward PDGF and were bipolar or multipolar cells that expressed cell surface gangliosides recognized by the A2B5 monoclonal antibody.

More detailed analysis showed that astrocyte migration toward laminin involved sites unique to laminin among the extracellular matrix components. Astrocyte migration was inhibited by the pentapeptide YIGSR-NH2, which corresponds to a sequence in the B1 chain of laminin and binds to the 67-kD laminin receptor of epithelial cells.[5] The pentapeptide YIGSK, which does not bind well to the laminin receptor, did not inhibit astrocyte migration toward laminin.

Although astrocyte migration occurred during a 4-hour incubation period, O-2A progenitor migration required an extended incubation period. O-2A progenitor migration toward PDGF was not apparent after 4 or 8 hours, but a vigorous response was present by 16 hours. The delayed migration does not appear to be due to a prerequisite of cell division before migration. Simultaneous analysis of migration and DNA synthesis, assessed by bromodeoxyuridine (BrdU) incorporation,[6] demonstrated that PDGF can indeed act as a chemoattractant and as a mitogen in the same O-2A progenitor cell. However, treatment with an antimitotic agent, cytosine arabinoside, markedly blocked O-2A progenitor incorporation of BrdU in response to PDGF in the lower well but only slightly impaired migration. Therefore, DNA replication is not a prerequisite for O-2A progenitor migration toward PDGF.

This *in vitro* study demonstrates that specific molecular signals can stimulate the directed migration of glial cells from developing CNS tissue. If gradients of relevant chemoattractants can be formed and maintained *in vivo* (by developmental regulation of synthesis, secretion, and/or binding), then these molecules may be able to mediate glial cell migration *in vivo* during CNS development.

REFERENCES

1. Paterson, J. A., A. Privat, E. A. Ling & C. P. Leblond. 1973. Investigation of glial cells in semithin sections III. Transformation of subependymal cells into glial cells, as shown by radioautography after 3H-thymidine injection into the lateral ventricle of the brain of young rats. J. Comp. Neurol. **149:** 83–102.
2. Raff, M. C. 1989. Glial cell diversification in the rat optic nerve. Science **243:** 1450–1455.
3. Harvath, L., W. Falk & E. J. Leonard. 1980. Rapid quantitation of neutrophil chemotaxis: Use of polyvinylpyrrolidone-free polycarbonate membrane in a multiwell assembly. J. Immunol. Meth. **37:** 39–45.
4. Yao, J., L. Harvath, D. L. Gilbert & C. A. Coulton. 1990. Chemotaxis by the CNS macrophage, the microglia. J. Neurosci. Res. **27:** 36–42.
5. Graf, J., R. C. Ogle, M. Sasaki, G. R. Martin, Y. Yamada & H. K. Kleinman. 1987. A pentapeptide from the laminin B1 chain mediates cell adhesion and binds the 67,000 laminin receptor. Biochemistry **26:** 6896–6900.
6. Gratzner, H. G. 1982. Monoclonal antibody to 5-bromo- and 5-iododeoxyuridine: a new reagent for detection of DNA replication. Science **218:** 474–475.

Synthesis and Action of Extracellular Proteases in Rat Hippocampus[a]

MYRNA A. R. DENT, DAVINA E. OWEN,
PETER J. ANDREWS, AND P. JOHN SEELEY

Norman and Sadie Lee Research Centre
Laboratory of Neurobiology
National Institute for Medical Research
Mill Hill
London NW7 1AA, UK

Extracellular proteases in general and the plasminogen activators (PAs) in particular are thought to be required for remodeling of tissue and for cell invasion. We are interested in the involvement of these materials in nervous system development and regeneration and are therefore studying their patterns of synthesis and the ability of the PA/plasmin system to alter cell interactions *in vitro*. Our interests are focused on the hippocampus. We measured protease activities in the hippocampus both in culture and *in vivo* using colorimetric and [^{125}I]fibrin-release techniques. Determinations were made under conditions of sample protein concentration and tissue extraction that optimized measurement of PA activity.

Protease activity was measured in buffer-perfused, dissected rat hippocampus over the entire range of ages examined from E16 to adult. Activity declined as development proceeded. The major part of the protease activity at pH 8.3 was plasminogen dependent (average 97%) and, of this, most of the activity was inhibited by amiloride (average 89%), indicating urokinase-type PA.

Proteases were also detected in neuronal cultures of dissociated E18 hippocampus maintained in defined medium. Activity was almost entirely plasminogen dependent (average 96%) and amiloride sensitive (average 81%). Most PA activity was released from the cells (average 86%), resulting in high levels in culture supernatants. The addition of purified plasmin or plasminogen to these cultures markedly changed the patterning of both perikarya and neurites. These reagents caused cleavage of cell-substratum contacts in preference to cell-cell contacts and thus resulted in cell clustering, increased fasciculation of neurites, and, in general, marked reduction in neurite length. Cell-substratum contacts differed in their plasmin sensitivity depending on both cell type (neuron versus glia) and the nature of the substratum (laminin versus polyethyleneimine).

The presence of PA activities *in vivo* and the influence of plasmin on cell patterning *in vitro* are consistent with roles for these reagents in the histogenesis of hippocampus.

[a]This work was supported by the British Neurological Research Trust, the International Spinal Research Trust, and the Medical Research Council.

Relation of Axons to Patterns of Glia in the Rat Barrel Field Cortex

LEI ZHANG AND NIGEL G. F. COOPER[a]

Department of Anatomy and Neurobiology
University of Tennessee
Memphis, Tennessee 38163

A transiently expressed boundary pattern of glia[1] and glial-associated molecules[1–5] is apparent in the barrel field cortex of early postnatal rodents. The early appearance of the glial-related boundaries lends support to the view[1–3] that the development of the somatotopic map is in part due to the differential expression of substrates, intrinsic to the central nervous system. To understand the mechanisms involved in the development of the barrel field cortex, and to better appreciate the interactions between extrinsic and intrinsic factors in this process, we examined the development of the sensory projections from thalamus to somatosensory cortex in normal and deafferent rats.

A crystal of the carbocyanine dye, DiI, was placed in the ventroposteromedial (VPM) and ventroposterolateral (VPL) nuclei of the thalamus of aldehyde-fixed brain dissected from postnatal day 1–7 (P1–P7; P1 represents the first 24 hours after birth) rat pups. The deafferent animals had a row of facial whiskers (c-row of mystacial vibrissae) removed by cauterization on P1. The distribution of thalamocortical afferents in cortical layer IV was examined in 40-μm vibratome sections with fluorescence microscopy, and the axonal patterns were compared with cresyl violet staining patterns and lectin binding patterns.

The data indicate that a pattern of discrete axonal arbors, characteristic of barrel glomeruli, is present in the posteromedial barrel subfield, which represents the mystacial vibrissae, by P2 (FIG. 1A). Axons are present in other parts of the prospective rattunculus, such as the lower lip and the forelimb (FIG. 1A); however, these sensory projections are not organized into barrel glomeruli at this age. This difference in degree of organization of thalamocortical axons most likely accounts for the lateral-to-medial gradient of development of lectin-delineated barrel boundaries observed previously.[4] The axonal arbors in the cortical representation of the c-row vibrissae are not perturbed in the P2 deafferent group (FIG. 1A) compared to normal pattern (FIG. 1B), but they are obviously altered in the P7 deafferent group (FIG. 1C). These data lead us to suggest that the effect of a lesion on the development of axonal arbors is not immediate, and that the effect is not simply one of arresting development. The results of this study suggest actual regression from an organized to a disorganized state. Thus, a pattern of sensory activity may be required to maintain the pattern of barrel glomeruli in the early stages of development.

There exist correlations between the axonal projection patterns and the neuronal and glial patterns. In the deafferent group, the affected c-row of vibrissae, is represented in the cortex by a disappearance of glycoconjugate-rich barrel boundaries (FIG. 2A) and a loss of neuron-rich barrel walls (FIG. 2B). Similar findings were shown for the vibrissal subfield of mouse cortex[3] and forelimb subfield of rat cortex.[4]

[a]PRESENT ADDRESS: Department of Anatomical Sciences and Neurobiology, University of Louisville, Louisville, Kentucky 40292.

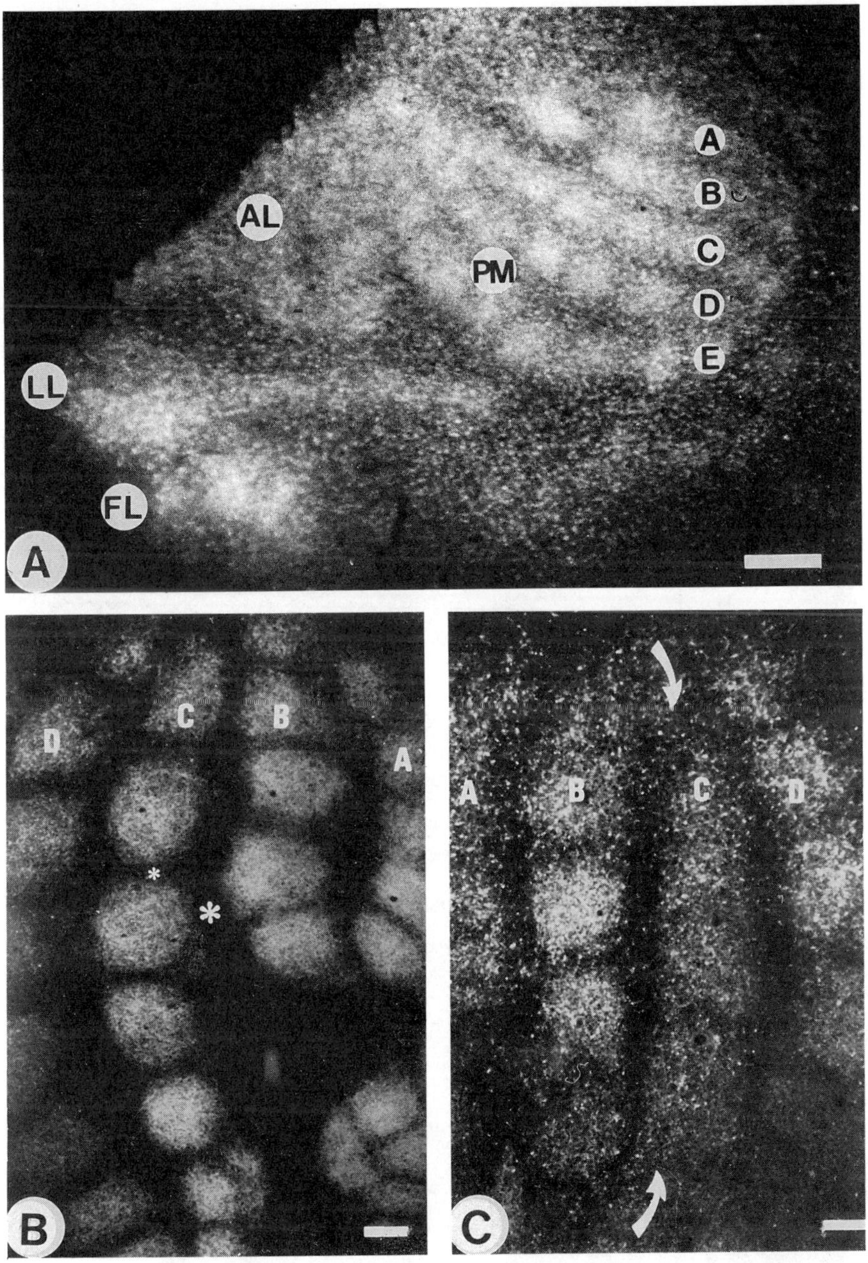

FIGURE 1. P2 and P7 rat somatosensory cortex. Use of DiI to demonstrate projections from VPM and VPL to cortical representation of whole body of rat in layer IV. The c-row of vibrissae was cauterized on P1. (**A**) P2 rat. Five individual rows (A, B, C, D, E) of barrel glomeruli and their individual axonal arbor patterns are clearly visible. Alterations to the axonal arbors are not evident in the c-row. The axonal projections to posterior medial (PM) and anterior lateral (AL) barrel subfield are better organized than are those to the lower lip (LL) and forelimb (FL) at this stage of development. Bar represents 680 μm. (**B** and **C**) Photomicrographs through the barrel field of the right and left hemispheres of the same P7 rat. (**B**) Right cerebral hemisphere. A, B, C, and D rows of barrel glomeruli are evident. Regions between rows (*large asterisk*) and between arcs (*small asterisk*) do not contain thalamocortical axons. Bar represents 350 μm. (**C**) Left cerebral hemisphere after cauterization of the c-row of vibrissae on the right face on P1. A, B, C, and D rows of barrel glomeruli are evident. Note the lack of barrel glomeruli in the c-row (*arrows*). This indicates that the loss of sensory input and/or activity results in the loss of aggregation of thalamocortical axons in layer IV. This alteration to the projection pattern takes more than 24 hours to develop. Bar represents 350 μm.

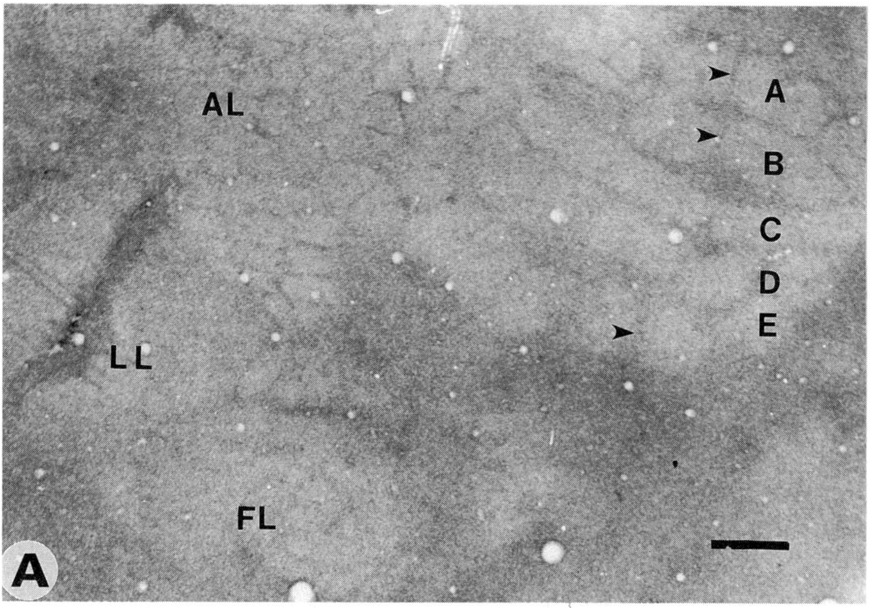

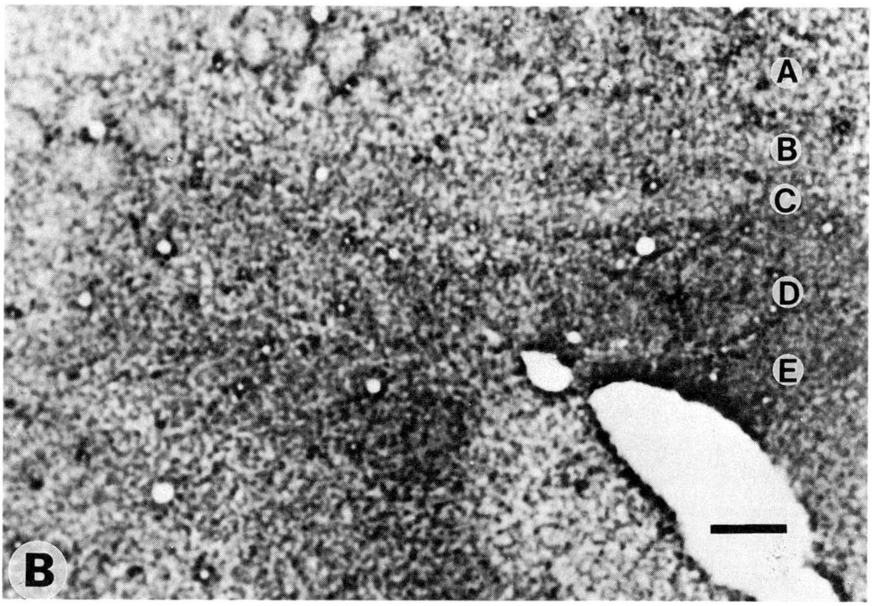

FIGURE 2. P6 rat somatosensory cortex. (**A**) In this experiment the c-row of mystacial vibrissae was cauterized on P1. The peroxidase-conjugated lectin, peanut agglutinin (PNA), was used in binding experiments that were performed on vibratome sections of cortex from P6 animals. The PNA receptors within the cortical c-row are altered. Some of the lectin-rich boundaries as seen in rows A, B, D, and E (*black arrowheads*) are not evident in the c-row. In this particular experiment, the c-row is narrower than the other four rows. Bar represents 480 μm. (**B**) Tangential section through layer IV of rat somatosensory cortex stained with cresyl violet. The c-row of vibrissae was cauterized on P1. The c-row of the posterior medial barrel subfield is altered such that barrel walls are not visible, and the c-row is narrower than the other four rows. Bar represents 400 μm.

It is to be noted here that the presence of the disorganized axons in the c-row representation is correlated with the altered pattern of boundaries and the altered pattern neuronal cell sorting in these animals. Presumably, the loss of a pattern of peripheral stimulation is related to the loss of the pattern of arborization of the thalamocortical axons. Nevertheless, the continued presence of the thalamocortical axons leads to the down-regulation of glial-related gene products, seen here as a lack of lectin binding sites within the c-row representation (FIG. 2A). The altered glial pattern is translated into an altered neuronal pattern as can be seen in the cresyl-violet-stained sections (FIG. 2B).

We believe that the significance of these data is that immature glia and their gene products serve as a substrate for the expression of peripheral influences on the development of cortical cytoarchitecture. The use of time per se, in any equation that is representative of the sensitive period, can most likely be replaced by the responsible factor, glia.

REFERENCES

1. COOPER, N. G. F. & D. A. STEINDLER. 1986. J. Comp. Neurol. **249:** 157–169.
2. COOPER, N. G. F. & D. A. STEINDLER. 1986. Brain Res. **380:** 341–348.
3. COOPER, N. G. F. & D. A. STEINDLER. 1989. Brain Res. **489:** 167–176.
4. MCCANDLISH, C., R. S. WATERS & N. G. F. COOPER. 1989. Exp. Brain Res. **77:** 425–431.
5. STEINDLER, D. A., N. G. F. COOPER, A. FAISSNER & M. SCHACHNER. 1989. Dev. Biol. **131:** 243–260.

Influence of Grafted Neurons on Glial Enzyme Expression[a]

MARILYN FISHER

Department of Anatomy & Cell Biology
University of Virginia Health Sciences Center
Charlottesville, Virginia 22908

Bergmann glia in cerebella of Lurcher (*Lc*) mutant mice show a striking metabolic response to the postnatal loss of Purkinje cells as evidenced by reduced expression of the enzymes glycerol-3-phosphate dehydrogenase (GPDH) and β-glucuronidase (β-GLU). Both of these enzymes normally undergo an age-related change of expression during the first postnatal weeks. GPDH activity rises dramatically, specifically in Bergmann glia and oligodendrocytes. β-GLU is expressed in most cell types in the cerebellum, but it undergoes a change in histochemical staining pattern during the first postnatal month. Early on, β-GLU activity in the Bergmann glia is rather diffusely spread throughout the cell body and processes. By 3 weeks of age, however, one sees a concentration of β-GLU reaction product into a dense supranuclear cap. In a wild-type mouse, once the mature staining patterns of these enzymes appear in Bergmann glia, they are maintained throughout the animal's life. In *Lc* mutants many Bergmann glia show the mature staining patterns of GPDH and β-GLU at around 20 days of age; however, both enzymes decrease rapidly as the Purkinje cell population disappears.[1,2] Studies with *Lc* chimeras, produced by embryo aggregation, showed that mutant Bergmann glia are competent to sustain apparently normal levels of enzyme expression as long as they lie within the domain of at least one surviving, wild-type Purkinje cell.[3] Thus, interaction between Bergmann glia and Purkinje cells is critical to the maintenance of mature levels of β-GLU and GPDH in Bergmann glia.

To determine if the reduced levels of GPDH and β-GLU found in Bergmann glia after Purkinje cell loss could be restored to normal levels, I made grafts of wild-type, fetal Purkinje cells into *Lc* mutant cerebella. The donor tissue is rhombic lip from 12-day fetuses of a mouse strain that does not express the adult isozyme of GPDH. *Lc* mutants from 21–56 days old were used as graft recipients. Tissue from the grafts was fixed 6–8 weeks after surgery and analyzed for GPDH and, in some cases, β-GLU expression.

Purkinje cell precursors can migrate into and differentiate within the molecular layer of hosts of all ages used. However, preliminary results indicate that there is a critical period in host age during which grafted Purkinje cells can influence GPDH and β-GLU expression. Bergmann glia near implanted Purkinje cells in hosts that were 21–35 days old at the time of grafting show apparently normal levels of GPDH and β-GLU expression. No host over 35 days old at the time of grafting showed high GPDH expression, but normal β-GLU staining was seen in a host that was 39 days old at the time of grafting. The results suggest that grafted Purkinje cell precursors differentiate to the point of being competent to influence GPDH and β-GLU expression in Bergmann glia, but can exert this influence only when they are introduced prior to the time when the glial cell has down-regulated expression in

[a] This work was supported by grant NS 25350 from the U.S. Public Health Service.

response to the loss of the native Purkinje cells. In other words, grafted Purkinje cells can sustain glial enzyme expression, but they apparently cannot reinduce it.

REFERENCES

1. FISHER, M. 1984. Proc. Natl. Acad. Sci. USA **81:** 4414–4418.
2. FISHER, M., P. TRIMMER & G. RUTHEL. 1991. Manuscript submitted.
3. FISHER, M. 1990. J. Neurogenet. **6:** 183–190.

Adhesion-Induced Differentiation of Oligodendrocytes Signals the Synthesis and Polarization of Heparan Sulfate Proteoglycans and Sulfated Glycoproteins[a]

S. SZUCHET[b,c] AND S. H. YIM[d]

[b]*Department of Neurology and The Brain Research Institute The University of Chicago Chicago, Illinois 60637*

[d]*The Myelin Section NINDS National Institute of Health Rockville, Maryland 20857*

An oligodendrocyte (OLG) enwraps 20–40 central nervous system internodes to deposit myelin; it also maintains myelin. Not all these internodes are ensheathed simultaneously. Therefore, OLGs must be able to spatially and temporally segregate components destined for myelin from those intended for housekeeping. To accomplish these tasks, OLGs are endowed with a polarized morphology; they also segregate the mRNAs coding for myelin proteins.[1,2] The signals that trigger the cytodifferentiation of OLGs are unknown. We have shown previously that adherence of OLGs to a substratum provides a signal for their biochemical and morphological differentiation.[3,4] In view of the evidence, from tissue culture studies, that basal lamina (BL) and extracellular matrix (ECM) are essential components in the signaling events that precede peripheral nervous system (PNS) myelination,[5] it was of interest to determine if ECM and BL elements are involved in the events that follow adhesion and if OLG polarization is one of these events. Here we report that OLGs synthesize heparan sulfate proteoglycans (HSPGs) and sulfated glycoproteins (Gps) upon substratum adhesion. These PGs and Gps are peripherally sited and spatially segregated, causing OLGs to become polarized. We postulate that: (a) these PGs and Gps are the determinants of OLG polarity; and (b) polarization is the first overt sign of OLG functional differentiation and precedes morphological differentiation.

Thirty-day-old OLG cultures were labeled with $[^{35}S]O_4^{2-}$ for 20 hours; heparin, salt, and/or detergent solutions were employed to sequentially extract labeled components. PGs were precipitated with cetylpyridinium chloride. The results indicated that ≈25% of PGs can be displaced by 100 μg/ml of heparin, that is, peripheral; operationally, the remainder are integral. To investigate *if* and *how* OLG peripheral PGs/Gps are modulated by substratum, we compared the expression and

[a]This work was supported by Grant PO1-NS 24575-03 from NIH.

[c]Address for correspondence: Dr. Sara Szuchet, The University of Chicago, Department of Neurology, Box 425, 5841 South Maryland Avenue, Chicago, Illinois 60637.

characteristics of PGs/Gps removed by heparin (100 μg/ml) from nonattached (B3.f) and attached (B3.fA) OLGs. B3.fA OLGs do not detach upon heparin treatment. Hence, to liberate components from the whole surface, B3.fA OLGs have to be exposed to heparin twice, first while OLGs are still attached (upper) and again after they have been detached (lower). This strategy provides a way of comparing components displayed on the upper and lower surfaces. In contrast, a single extraction suffices for B3.f OLGs because all the surface is available. Two sets of OLG cultures were set up that differed in adherence only (see legend to FIG. 1); cultures were labeled with $[^{35}S]O_4^{2-}$ for 20–24 hours. After isolating heparin-displaced components from B3.f and B3.fA OLGs, they were fractionated according to size by gel filtration chromatography on CL-6B Sepharose. The total amount of $[^{35}S]O_4^{2-}$-labeled material extracted with heparin from B3.f OLGs is small; it is composed of a high molecular weight class (F1) that elutes after the void volume (V_o) and a fraction (F2) that spreads throughout the included volume. The elution pattern of heparin-extracted components from the upper domain of B3.fA OLGs is dominated by a major, symmetrical peak, A2, that emerges in the included volume. Little material is eluted after V_o. In contrast, the elution profile obtained for material removed from the lower surface of B3.fA OLGs displays a major peak, B1, that appears after V_o. Not only is this pattern different from that obtained with B3.f OLGs, but it is also distinct from that derived from the B3.fA upper surface (data not shown). Thus, B3.fA OLGs have synthesized sulfated material that can be separated on the basis of size into two classes. This material is spatially segregated with the high molecular weight class concentrated on the lower cell surface, whereas the species with intermediate size is on the upper surface. We interpret these observations as indicating an adhesion-induced segregation of newly synthesized components with the creation of "sidedness." Similar studies performed on month-old cultures (and older) indicated that OLGs maintain their polarization.

It was of interest to establish if corresponding fractions from the upper and lower domains differed in composition or only in quantity. Heparin-removed fractions were exposed first to clostripain to facilitate subsequent analysis by gel electrophoresis. Each fraction was then divided into four aliquots and exposed to enzymes that digest different sugar chains (FIG. 1). Of the enzymes tested, heparitinase was the only one that degraded the heparin-released material (FIG. 1a and b). This finding establishes that all the heparin-displaceable PGs belong to the heparan sulfate class. The gel patterns in FIGURE 1a and b are distinct, indicating that PGs localized on the upper and lower OLG surfaces are not the same. There is a higher abundance or, alternatively, a higher extent of sulfation of PGs on the upper than on the lower domain (cf. FIG. 1a and b).

This study was performed with the bulk of material (i.e., unfractionated) that was removed from each of the OLG surfaces; the experiments were repeated using individual fractions that had been resolved by gel filtration. The major component of the lower surface, B1, had to be predigested with clostripain before being resolved by SDS-PAGE. The gel pattern obtained (not shown) resembles that shown in FIGURE 1b. Fractions A2 and B2 could be analyzed without predigestion with clostripain. Only heparitinase was used to degrade glycosaminoglycans. The results (FIG. 2a and b) confirmed the existence of differences in composition between the upper and lower surfaces and set on firm grounds the concept of an adhesion-induced segregation of peripheral PGs. FIGURE 2 reveals six bands ranging in $M_r \times 10^{-3}$ from 150 to 50 in A2 and one major band ($M_r \times 10^{-3} \approx 100$) in B2 that are not susceptible to heparitinase. We conjecture that these bands represent sulfated glycoproteins. Thus, heparin extracts both PGs and Gps. The use of clostripain in the analysis of fraction B1 precludes us from knowing if Gps are also present in this fraction.

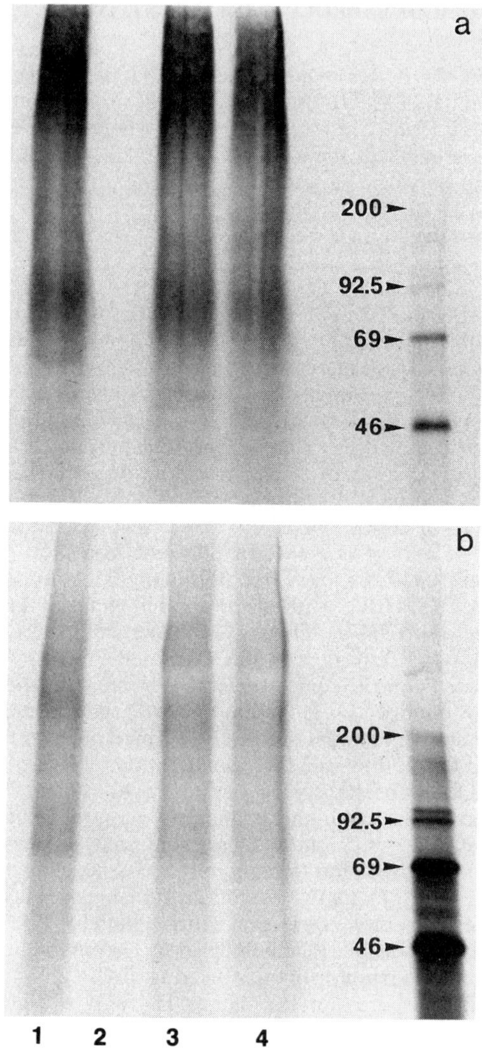

FIGURE 1. Electrophoretic patterns of heparin-extracted material from B3.fA OLGs after 24 hours of adherence on a polylysine substratum. Fluorograms. OLGs were isolated from 3–6-month-old lamb brains by our established procedure[6] and maintained in culture as described previously.[7] Freshly isolated OLGs were left in culture (nonattached) for 3 days. OLGs were harvested and washed thrice (to remove serum) with Dulbecco's modified Eagle's medium (DMEM), keeping the cells for a half hour at 37°C for each wash. OLGs were then divided into two aliquots: one was plated on plastic culture dishes where OLGs remain as floating clusters (B3.f; B3, for cells from band 3[5]; f, floating); the other was plated on polylysine-coated plates to which OLGs attach (B3.fA; A, adherent). Both cultures were labeled with 100 μCi/ml of carrier-free $[^{35}S]O_4^{2-}$ (1,142 Ci/mmol). From 20–24 hours later, medium was collected and cultures were washed with 20 mM Tris-buffered saline solution (TBS) thrice to remove excess label. Each culture was then exposed to a solution of 100 μg/ml of heparin in TBS plus a mixture of protease inhibitors [(PIs); 2 mM phenylmethylsulfonyl fluoride (PMSF), 10 mM N-ethylmaleimide, 10 mM EDTA, and 5 μg/ml of pepstatin] for 45 minutes at 37°C. B3.f OLGs were harvested and centrifuged for 5 minutes at 192 × g, and the supernatant containing extracted material was collected. Medium containing components

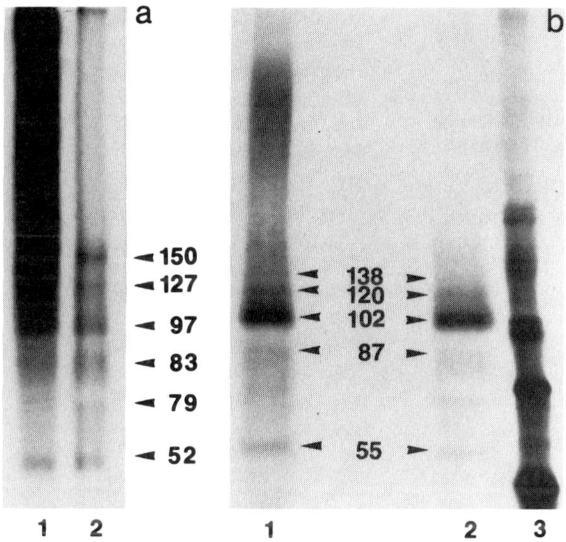

FIGURE 2. Comparison of the electrophoretic patterns obtained with fraction A2 with those obtained with fraction B2 (see text). Fluorograms. The experimental protocol was identical to that given in FIGURE 1. Fractions were pooled. Fractions A2 and B2 were aliquoted; one aliquot was left as the control, the other was treated with heparitinase as described in FIGURE 1. (**a**) Fraction A2. *Lane 1,* control; *lane 2,* treated with heparitinase. Note that the smear in *lane 1* disappeared after heparitinase treatment (*lane 2*), confirming that fraction A2 contains heparan sulfate proteoglycans. In addition, A2 contains a number of components whose molecular weight is not affected by enzyme treatment. We assume that these are sulfated glycoproteins. (**b**) Fraction B2. *Lane 1,* control; *lane 2,* after heparitinase treatment. A major component ($M_r \times 10^{-3} = 102$) and a couple of minor ones are not affected by the enzyme (cf. *lanes 1 and 2*), indicating that these are sulfated glycoproteins. Note that the patterns in **a** are different from those in **b**. Reproduced, with permission, from Yim and Szuchet (1991).

FIGURE 1 (*continued*) extracted from the upper surface of B3.fA OLGs was collected and saved; B3.fA OLGs were washed with TBS plus PIs to remove heparin, and cells were harvested using a specially designed Teflon wedge to detach them (examination by phase-contrast microscopy with trypan blue showed that OLGs sustain little to no damage with this treatment; OLGs cannot be detached with chelating agents). Detached B3.fA OLGs were treated with heparin again, this time to remove components on the lower surface (i.e., the one in contact with substratum). Cells were centrifuged and supernatant was saved. All supernatants were centrifuged at 10,000 × g for 30 minutes to eliminate cellular debris, dialyzed against distilled water, and concentrated by lyophilization. The bulk of material removed from the upper surface (FIG. 1a) was exposed to 5 U/ml of clostripain for 16 hours at 37°C to reduce the size of the molecules and thereby facilitate penetration in an acrylamide gel. Clostripain's action was arrested by boiling for 1 minute. The sample was then divided into four aliquots: one was left as the control; another was treated with 0.01 units of heparitinase at 43°C for 24 hours in the presence of 50 μg of unlabeled heparan sulfate; the third aliquot was exposed to 0.015 units of chondroitinase ABC at 37°C for 24 hours in the presence of 50 μg of unlabeled chondroitin sulfate; the fourth aliquot was treated with N-glycanase following the protocol recommended by the supplier (Genzyme Corporation). Each aliquot was placed in sample buffer and resolved with a 5–20% gradient SDS-PAGE. Material removed from the lower surface (FIG. 1b) was handled in an identical manner. *Lane 1,* control; *lane 2,* heparitinase-treated sample; *lane 3,* chondroitinase ABC; *lane 4, N*-glycanase; numbers and lane on the right are molecular weight standards. Note that heparitinase is the only enzyme that degrades the glycosaminoglycans, suggesting that most, if not all, of the proteoglycans are of the heparan sulfate class. Reproduced, with permission, from Yim and Szuchet (1991).

REFERENCES

1. COLMAN, D. R., G. KREIBICH, A. B. FREY & D. D. SABATINI. 1982. J. Cell Biol. **95:** 598–608.
2. TRAPP, B. D., T. MOENCH, M. PULLEY, E. BARBOSA, G. TENNEKOON & J. W. GRIFFIN. 1987. Proc. Natl. Acad. Sci. USA **84:** 7773–7777.
3. SZUCHET, S., P. E. POLAK & S. H. YIM. 1986. Dev. Neurosci. **8:** 208–221.
4. YIM, S. H., S. SZUCHET & P. E. POLAK. 1986. J. Biol. Chem. **261:** 11808–11815.
5. BUNGE, R. P., M. B. BUNGE & C. F. ELDRIDGE. 1986. Ann. Rev. Neurosci. **9:** 305–328.
6. SZUCHET, S., B. G. W. ARNASON & P. E. POLAK. 1980. J. Neurosci. Methods **3:** 7–19.
7. SZUCHET, S. & S. H. YIM. 1984. J. Neurosci. Res. **11:** 131–144.

The Axon May Control Schwann Cell Responses to Growth Factors

J. B. DAVIS[a] AND A. D. J. GOODEARL

Ludwig Institute for Cancer Research
London W1P 8BT, England

Schwann cells derived from neonatal rat sciatic nerve respond to few mitogens *in vitro.* Active mitogens include glial growth factor (GGF),[1] transforming growth factors (TGFs)-β1 and -β2,[2] and a number of membrane preparations derived from neuronal cells or cell lines.[3] Additionally, Schwann cells will respond to platelet-derived growth factors (PDGFs) and fibroblast growth factors (FGFs), but only when coadministered with an agent that raises intracellular levels of cAMP, such as forskolin, cholera toxin, or dibutyryl cAMP.[4] In the case of PDGF this cooperation occurs through the induction of growth factor receptor expression by forskolin.[5] As the pharmacological agents that have been used to elevate cAMP are not relevant to Schwann cell biology, we have sought a physiological equivalent and have investigated the role of axolemmal membranes as they are thought to raise cAMP levels in Schwann cells.[3]

Using an axolemmal membrane-enriched fraction (AEF) prepared from adult

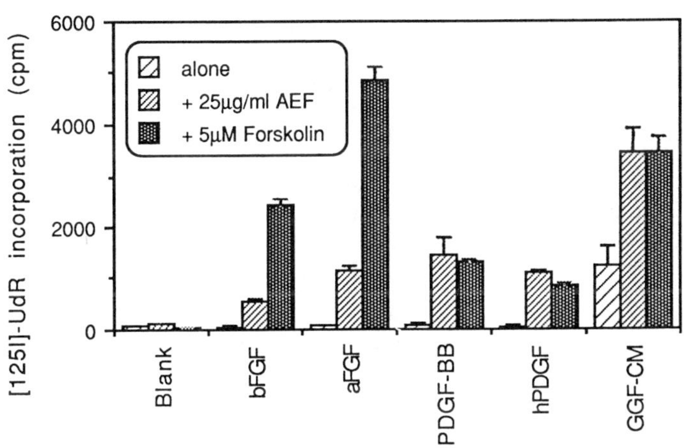

FIGURE 1. Cells were challenged for 48 hours with blank, 100 pM basic FGF, 10 nM acidic FGF, 50 ng/ml PDGF-BB, 3 nM human-platelet PDGF or 2.6 μg/ml partially purified GGF, either alone or in the presence of 25 μg/ml AEF or 5 μM forskolin. DNA synthesis was measured over the last 24 hours by measuring the incorporation of 5′-[^{125}I]-iodo-2′-deoxyuridine ([^{125}I]-UdR).

[a]Corresponding author: Dr. J. B. Davis, The Salk Institute, PO Box 85800, San Diego, CA 92186.

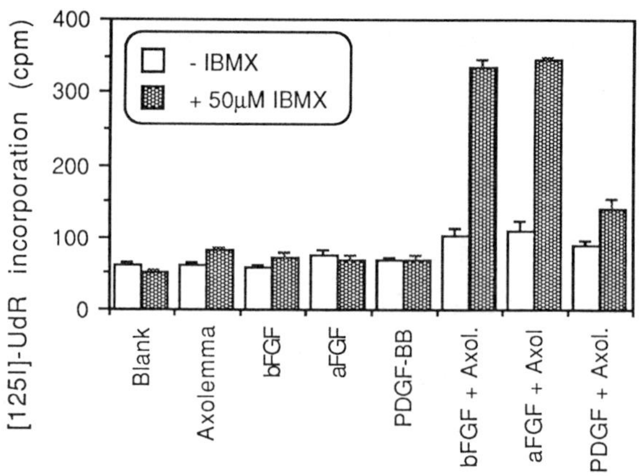

FIGURE 2. Schwann cells were challenged for 48 hours with blank, 12.5 μg/ml AEF (axolemma), 100 pM basic FGF, 10 nM acidic FGF, 50 ng/ml PDGF-BB, or combinations of these growth factors in the presence or absence of 50 μM IBMX. The degree of DNA synthesis over the last 24 hours was assessed by measuring the incorporation of [^{125}I]-UdR.

rat brain,[6] we show that AEF is capable of supporting a mitogenic response to FGFs and PDGFs and synergizes with GGF, in a similar manner to forskolin (FIG. 1). The AEF also has a mitogenic activity for Schwann cells in serum that is dependent upon platelet factors, like forskolin or cholera toxin,[4] and exerts its cooperative activity in a dose-dependent fashion. Isobutylmethylxanthine (IBMX) (50 μM) and 8-bromo-cyclicAMP (data not shown), which promote cAMP-mediated activities by inhibiting the breakdown of cAMP by its phosphodiesterase, enhance the activity of a submaximal dose of AEF (FIG. 2). This strongly suggests that stimulation with AEF leads to activation of adenylate cyclase, mimicking forskolin and cholera toxin which are specific activators of adenylate cyclase. However, we have not been able to measure a statistically significant increase in intracellular cAMP due to AEF, nor have we been able to do so at equivalent forskolin concentrations (0.01 μM), in keeping with results using other cell types. A commonly used concentration of forskolin (5 μM) results in a greater than 20-fold increase in cAMP. We believe that rather than discounting the role of cAMP, this suggests that the required changes in cAMP are subtle. It is possible that only very small increases in concentration are necessary or, perhaps more likely, production or distribution of cAMP is localized, leading to specific subcellular concentrations of cAMP that are not recorded by a crude whole cell measurement.

Thus, the axon may exert control over Schwann cell proliferation during development, regeneration, or aberrantly in tumorigenesis, by regulating the ability of Schwann cells to respond to growth factors. This control might be achieved using a cAMP-mediated regulation of receptor expression.

REFERENCES

1. LEMKE, G. & J. P. BROCKES. 1984. J. Neurosci. **4:** 75–83.
2. RIDLEY, A. J., J. B. DAVIS, P. STROOBANT & H. LAND. 1989. J. Cell Biol. **109:** 3419–3424.
3. RATNER, N., R. P. BUNGE & L. GLASER. 1986. Ann. N.Y. Acad. Sci. **486:** 170–181.
4. DAVIS, J. B. & P. S. STROOBANT. 1990. J. Cell Biol. **110:** 1353–1360.
5. WEINMASTER, G. & G. LEMKE. 1990. EMBO J. **9:** 915–920.
6. DE VRIES, G. H., J.-M. MATTHIEU, M. BENY, R. CHICHEPORTICHE, M. LAZDUNSKI & M. DOLIVO. 1978. Brain Res. **147:** 339–352.

Mitogenic Response and Phenotype of Short- and Long-Term Cultured Schwann Cells

Relationship to Intracellular cAMP Levels

H. J. S. STEWART, P. A. ECCLESTON, K. R. JESSEN, AND R. MIRSKY

Department of Anatomy
University College London
London WC1E 6BT, UK

Current opinion suggests that cAMP may play a fundamental role in Schwann cell biology. Elevation of intracellular cAMP levels induces Schwann cell differentiation[1,2] and acts as a Schwann cell mitogen in serum-containing medium.[3] cAMP-elevating agents are also able to synergize with platelet-derived growth factor (PDGF), acidic and basic fibroblast growth factors (FGFs), and glial growth factor (GGF) to induce Schwann cell mitosis in serum- or plasma-containing medium.[4]

In this study we used serum-free culture conditions to investigate the role played by cAMP in short-term (ST-Schwann cells) and long-term cultured Schwann cells (LT-Schwann cells). In tissue culture the proliferative capacity of Schwann cells changes with time. ST-Schwann cells (i.e., less than 2 weeks in culture), tested with an extensive list of potential mitogens, undergo mitosis only in response to an agent known to elevate intracellular cAMP in combination with either PDGF, GGF, or basic FGF (FIG. 1). However, after long-term culture (longer than 3 months) the Schwann cells spontaneously immortalize and are able to proliferate in serum-free medium. The response to mitogens is also altered because agents that elevate cAMP are no longer required for PDGF, GGF, and basic FGF to stimulate DNA synthesis. Any of these growth factors alone is sufficient to enhance LT-Schwann cell mitosis.

When the intracellular cAMP levels of these two Schwann cell types were compared, the basal cAMP level of the LT-Schwann cell was double that of the ST-Schwann cells (FIG. 2). This change in intracellular cAMP may account for the observed differences in mitogen requirements of ST-Schwann cells and LT-Schwann cells. In the LT-Schwann cell the already high cAMP level may bypass the requirement for a cAMP-elevating agent; thus, a single growth factor may stimulate DNA synthesis. The LT-Schwann cell was also found to have an altered molecular phenotype when compared to the ST-Schwann cell. The expression of the nerve growth factor receptor is down-regulated on LT-Schwann cells, whereas the antigens galactocerebroside C, 04, fibronectin, and the myelin protein P_0, all of which are absent on ST-Schwann cells, are present in variable or low levels on LT-Schwann cells. The expression of all these antigens has been shown to be governed by cAMP in the ST-Schwann cell.[1,5,6,7] Thus, the LT-Schwann cell expresses an altered phenotype and changes in mitogenic response that are consistent with an elevation in intracellular cAMP level. These findings support the hypothesis that in serum-free medium elevation of cAMP is necessary for Schwann cell DNA synthesis in response to growth factors.

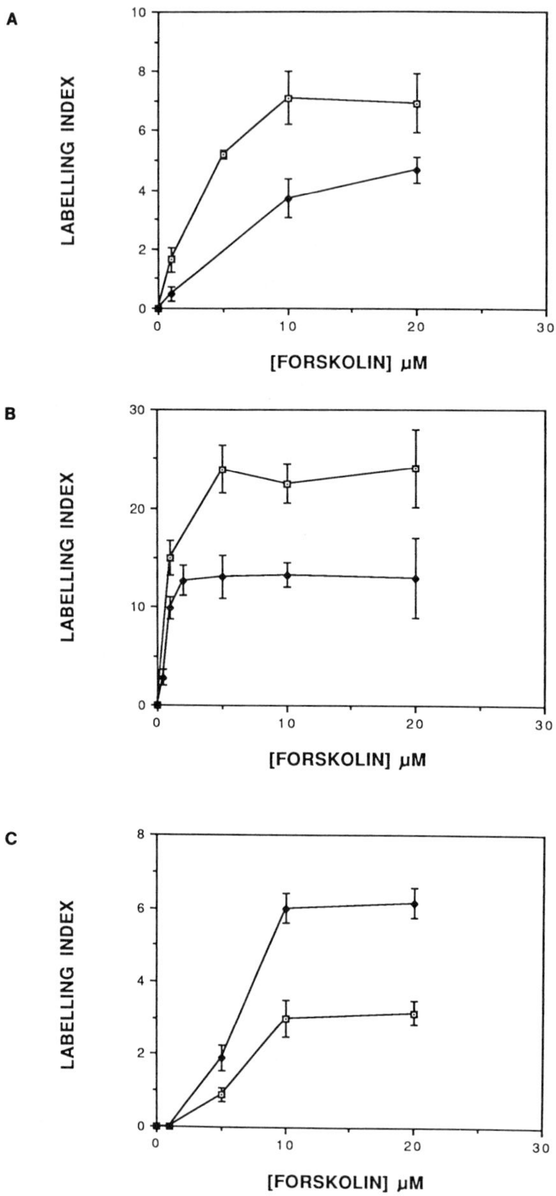

FIGURE 1. FGF, GGF, and PDGF each synergize with forskolin to cause DNA synthesis in Schwann cells. Short-term Schwann cell cultures were incubated in defined medium for 48 hours with various concentrations of forskolin together with (**A**) FGF either 100 ng/ml (⊡) or 50 ng/ml (◆), (**B**) GGF either 1.8 μg/ml (⊡) or 0.9 μg/ml (◆), and (**C**) PDGF either 5 ng/ml (⊡) or 2.5 ng/ml (◆). DNA synthesis was measured by incorporation of BrdU over the last 24 hours. Each experimental value represents the mean of at least three separate experiments.

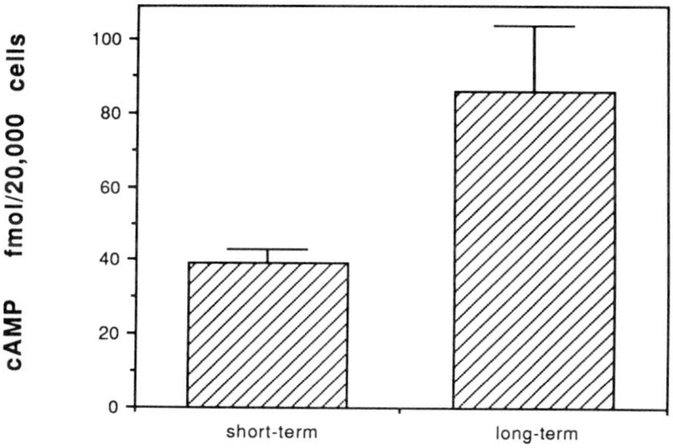

FIGURE 2. Intracellular cAMP is elevated in long-term cultured Schwann cells. cAMP levels in short- and long-term Schwann cells were measured using a radioimmunoassay kit.

REFERENCES

1. MORGAN, L., K. R. JESSEN & R. MIRSKY. 1990. J. Cell Biol. **112:** 457–467.
2. LEMKE, G. & M. CHAO. 1988. Development **102:** 203–215.
3. RAFF, M. C., A. HORNBY-SMITH & J. P. BROCKES. 1978. Nature **273:** 672–673.
4. DAVIS, J. B. & P. STROOBANT. 1990. J. Cell Biol. **110:** 1353–1360.
5. BARON VAN EVECOOREN, A., A. GANSÜLLER, M. GUMPEL, V. BAUMANN & H. K. KLEINMANN. 1986. Dev. Neurosci. **8:** 182–196.
6. MONUKO, K., G. SOBUE, V. R. REDDY, B. WURZER, H. KREIDER, P. HOTTA, A. H. BARON & D. PLEASURE. 1988. J. Neurosci. Res. **21:** 465–472.
7. SOBUE, G. & D. PLEASURE. 1984. Science **224:** 72–74.

Effect of EGF on DNA Labeling in Rat Cerebellar Immature Astrocytes Maintained Under Different Culture Conditions

Presence or Absence of Polylysine, Serum, or Both

R. AVOLA, G. MAGRÌ, F. INGRAO, L. INSIRELLO,
P. CARPANO, V. G. NICOLETTI, D. F. CONDORELLI,
N. RAGUSA, AND A. M. GIUFFRIDA STELLA

Institute of Biochemistry
Faculty of Medicine
University of Catania
Viale Andrea Doria 6
95125 Catania, Italy

The ability of epidermal growth factor (EGF) to stimulate cell growth was shown in a variety of cell types[1] including cells from the nervous system.[2] EGF increases cell proliferation, as shown by [methyl-^{3}H]thymidine incorporation into DNA of normal human glial cells (in serum-free medium supplemented with human serum albumin[3]) and also of cultured astrocytes from rat brain.[4,5]

The effect of EGF on the labeling of DNA, the various RNA species, cytoskeletal proteins, and histone and nonhistone chromosomal proteins in primary astroglial cell cultures from newborn rat cerebral hemispheres has previously been investigated.[5,6] Moreover, macromolecular synthesis and protein ADP ribosylation in fetal rat brain tissue slices after EGF treatment have also been studied.[7]

The aim of the present research was to investigate the effect of EGF on the DNA labeling of rat cerebellar immature astrocytes cultured in the presence or absence of polylysine in serum-free chemically defined medium (CDM) or in Basal Medium Eagle (BME) supplemented with 10% fetal calf serum (FCS).

Cerebellar astrocyte cultures were obtained by plating the cells on dishes precoated or uncoated with poly-L-lysine and maintained in BME supplemented with 10% FCS[8] or in CDM.[9] The results obtained indicated that continuous EGF treatment significantly stimulated [methyl-^{3}H]thymidine incorporation in immature astrocytes obtained from 1-day-old rat cerebella, seeded on dishes unprecoated with polylysine, and cultured in CDM from plating up to 11 days *in vitro* (DIV), when this increase was even more evident. In these serum-free culture conditions, cell growth and survival were markedly and significantly reduced in the absence of EGF. Twenty-four hours of EGF treatment in CDM significantly increased, at 2 and 3 DIV, cerebellar immature astrocyte DNA labeling, which was markedly reduced from 5 DIV onward (TABLE 1).

Furthermore, the continuous treatment with EGF of cerebellar astroglial cells, maintained in BME + 10% FCS, induced an effect similar to that found in cell cultures grown in CDM. In addition, EGF treatment for 24 hours in BME + FCS caused a significant increase in cerebellar astrocyte proliferation from 3 up to 9 DIV,

TABLE 1. Effect of EGF on DNA Labeling in Cerebellar Astrocytes Cultured in CDM or BME + FCS[a]

Culture Conditions	Days *in Vitro* (DIV)					
	2	3	5	7	9	11
A	9,501 ± 141	18,445 ± 1,091	51,379 ± 3,198	65,961 ± 2,475	74,987 ± 3,325	89,758 ± 5,984
B	▲ 482 ± 59	▲ 237 ± 21	▲ 247 ± 23	▲ 198 ± 35	▲ 117 ± 19	▲ 80 ± 3
C	●1,291 ± 168	▲ 2,150 ± 183	■ 850 ± 66	■ 2,874 ± 383	■ 1,466 ± 226	▲ 54,259 ± 4,694
D	■1,505 ± 198	▲ 960 ± 39	281 ± 28	299 ± 35	171 ± 39	● 196 ± 28
E		■31,453 ± 2,398	●72,271 ± 4,611	▲136,806 ± 1,310	■131,876 ± 2,703	▲150,872 ± 2,703
F	----	5,516 ± 906	7,946 ± 20	17,903 ± 172	27,338 ± 1,128	87,009 ± 2,494
G	----	● 7,930 ± 54	▲13,363 ± 51	● 24,802 ± 1,458	● 53,043 ± 3,426	87,973 ± 2,131

[a][Methyl-^{3}H]thymidine incorporation into astrocytes, obtained from 1-day-old rat cerebella, cultured on dishes unprecoated with polylysine and grown at different days (2, 3, 5, 7, 9, and 11 DIV) in serum-free CDM or in BME supplemented with 10% FCS. The results are expressed as DPM/dish ± SEM. The significance of the differences, analyzed by Student's t test, of B versus A, C versus B, D versus B, E versus A, and G versus F, is indicated as follows: ●$p < 0.05$; ■$p < 0.01$; ▲$p < 0.001$. A = BME + FCS; B = CDM; C = CDM + continuous EGF treatment; D = CDM + EGF treatment for 24 hours; E = BME + FCS + continuous EGF treatment; F = BME + FCS and then serum deprivation for 48 hours (controls for G); G = BME + FCS and then 24 hours of EGF treatment after 24 hours of serum deprivation.

whereas in CDM the mitogenic effect of EGF was markedly reduced from 5 DIV onward (TABLE 1).

Immunostaining analysis with specific antibodies against glial fibrillary acidic protein (GFAP) and vimentin demonstrated that these proliferating cerebellar cells, morphologically identified as immature astrocytes, express vimentin but not GFAP. Therefore, EGF treatment for 10 hours in primary rat astroglial cell cultures maintained in CDM containing insulin, transferrin, and BSA decreased GFAP mRNA levels, as shown by Northern blotting analysis, indicating a low degree of GFAP expression. GFAP mRNA levels decreased markedly when astroglial cell cultures were maintained in CDM 1–5 DIV, whereas serum readdition for 24 or 48 hours induced a high degree of GFAP expression.

REFERENCES

1. CARPENTER, G. & S. COHEN. 1979. Epidermal growth factor. Ann. Rev. Biochem. **48:** 193–216.
2. MORRISON, R. S., R. F. KEATING & J. R. MOSTAL. 1988. Basic fibroblast growth factor and epidermal growth factor exert differential trophic effects on CNS neurons. J. Neurosci. Res. **21:** 71–79.
3. WESTERMARK, B. 1976. Density dependent proliferation of human glial cells stimulated by epidermal growth factor. Biochem. Biophys. Res. Commun. **69:** 304–310.
4. SIMPSON, D. C., R. MORRISON, J. DE VELLIS & H. R. HERSCHMAN. 1982. Epidermal growth factor binding and mitogen activity on purified population of cells from the central nervous system. J. Neurosci. Res. **8:** 453–462.
5. AVOLA, R., D. F. CONDORELLI, S. SURRENTINO, L. TURPEENOJA, A. COSTA & A. M. GIUFFRIDA STELLA. 1988. Effect of epidermal growth factor and insulin on DNA, RNA and cytoskeletal proteins labeling in primary rat astroglial cell cultures. J. Neurosci. Res. **19:** 230–238.
6. AVOLA, R., D. F. CONDORELLI, L. TURPEENOJA, F. INGRAO, R. REALE, N. RAGUSA & A. M. GIUFFRIDA STELLA. 1988. Effect of epidermal growth factor on the labeling of the various RNA species and of nuclear proteins in primary rat astroglial cell cultures. J. Neurosci. Res. **20:** 54–63.
7. AVOLA, R., D. F. CONDORELLI, N. RAGUSA, F. INGRAO, G. MAGRI, V. NICOLETTI, L. INSIRELLO, S. REALE, A. COSTA & A. M. GIUFFRIDA STELLA. 1989. Effect of EGF on DNA, RNA synthesis and protein ADP-ribosylation in fetal rat brain tissue slices. J. Neurochem. **52** (Suppl.): S191C.
8. LEVI, G., F. ALOISI, M. T. CIOTTI & V. GALLO. 1984. Autoradiographic localization and depolarization-induced release of acidic amino acids in differentiating cerebellar granule cell cultures. Brain Res. **290:** 77–86.
9. BOTTENSTEIN, J. E. & G. M. SATO. 1979. Growth of rat neuroblastoma cell line in serum-free supplemented medium. Proc. Natl. Acad. Sci. USA **76:** 514–517.

A Molecular Approach to Myelination Using Recombinant Retroviruses[a]

GEOFFREY C. OWENS

Department of Anatomy and Neurobiology
Washington University School of Medicine
St. Louis, Missouri 63110

A peripheral myelin sheath originates from the plasma membrane of the Schwann cell that initially surrounds the axon as a unique 1:1 association is established between a Schwann cell and a length of axon destined to be myelinated.[1,2] This initial step must involve specific recognition and adhesive interactions between the Schwann cell and axonal surfaces. Based on its structure,[3-5] periaxonal location in mature myelin,[6] neurite binding activity,[7,8] and early expression during myelination,[9,10] myelin-associated glycoprotein (MAG) is considered to play a key role in the interaction between the axon and myelinating glial cell. To directly test the function of MAG, recombinant retroviruses were constructed containing MAG sequences (L-MAG isoform[5]) in the sense or antisense orientation (FIG. 1). Stably infected populations of primary Schwann cells were established, harboring either the sense or the antisense provirus, and the capacity of these genetically modified Schwann cells to differentiate was tested by coculturing them with purified dorsal root ganglion neurons.

Schwann cells infected with the sense virus constitutively expressed MAG without the normal requirement for neuronal contact to induce expression[10] and were unimpaired in their ability to form normal compact myelin (FIG. 2a). On the other hand, a proportion of the Schwann cells infected with the antisense virus did not express MAG and failed to form myelin (FIG. 2b). The inhibited cells expressed galactocerebroside, a myelin component induced by neuronal interaction, and assembled a basal lamina, but did not establish 1:1 associations with larger axons. In short-term coculture with sensory neurons, Schwann cells expressing recombinant L-MAG ensheathed larger axons after only 4 days provided that an exogenous basal lamina was supplied (FIG. 2c). Similar activity was rarely displayed by control

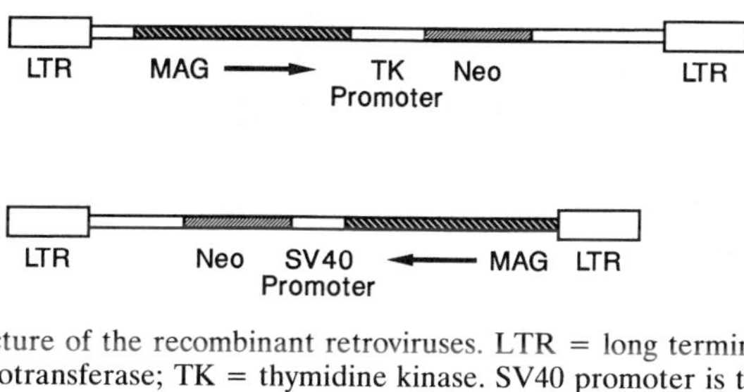

FIGURE 1. Structure of the recombinant retroviruses. LTR = long terminal repeat; Neo = neomycin phosphotransferase; TK = thymidine kinase. SV40 promoter is the promoter from the early region of SV40. The retroviral vector used to construct the sense virus was pMV7,[11] and for the antisense virus the vector was LNL6.[12]

[a] This work was funded by a grant from the National Multiple Sclerosis Society.

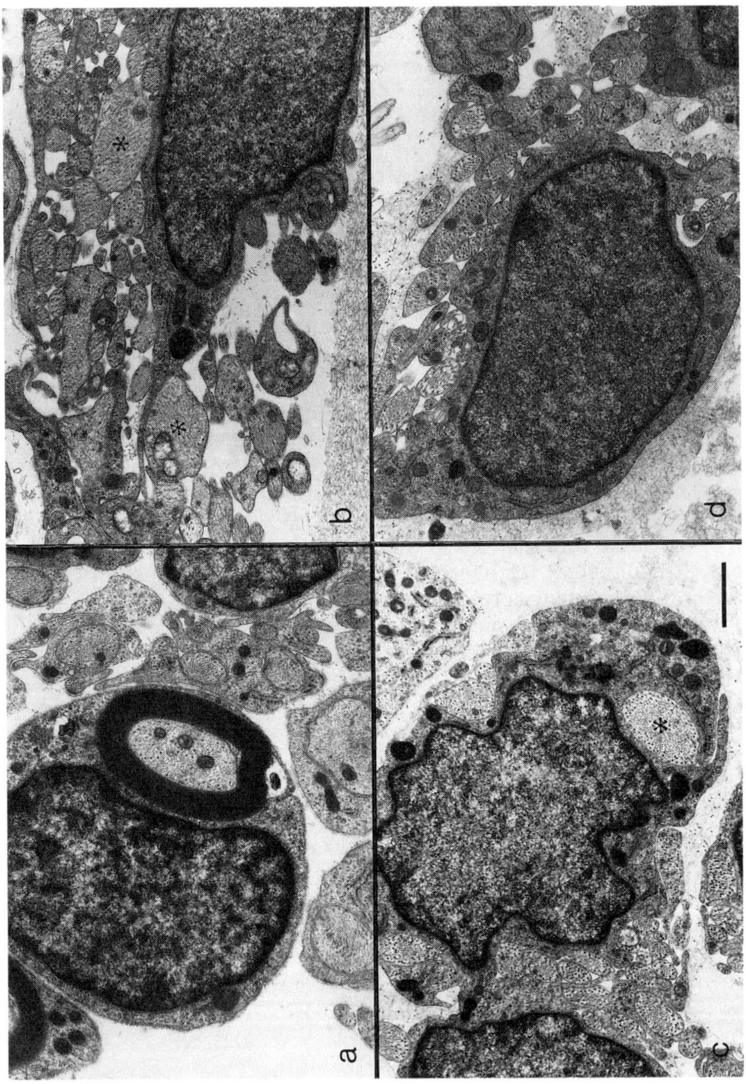

FIGURE 2. Coculture of genetically modified Schwann cells with sensory neurons. **a** and **b** are taken from neuron:Schwann cell cultures maintained in myelin promoting media[13] for 21 days. In **a** a Schwann cell infected with the sense virus has formed normal compact myelin, whereas in **b** a Schwann cell infected with the antisense virus has failed to form myelin despite prolonged interaction with neurons. The inhibited Schwann cell has not segregated and ensheathed larger axons (*asterisks*). Schwann cells infected with the sense virus and uninfected Schwann cells were cocultured with neurons for 4 days in the presence of an exogenous basal lamina matrix. In **c** an infected Schwann cell has segregated a large axon (*asterisk*) and surrounded it with a [illegible] In contrast to this, uninfected Schwann cells showed only limited interaction with axons (**d**). Bar = 1 μm.

Schwann cells, correlating with the low level of MAG induction (1–2% of cells) due to the limited time of interaction with neurons (FIG. 2d). Taken together, these data strongly support the idea that MAG is the critical Schwann cell component up-regulated by neuronal interaction that promotes the initial investment of axons destined to be myelinated.

ACKNOWLEDGMENTS

I would like to thank Dr. Jim Salzer for the MAG cDNA sense construct, Carol Boyd for excellent technical assistance, and Dr. Richard Bunge for advice and initial support.

REFERENCES

1. PETERS, A. & A. MUIR. 1959. Q. J. Exp. Physiol. **44:** 117–130.
2. WEBSTER, H. DEF., J. R. MARTIN & M. F. O'CONNELL. 1973. Dev. Biol. **32:** 401–416.
3. ARQUINT, M., J. RODER, L. S. CHIA, J. DOWN, Q. WILKINSON, H. HAYLEY, P. BRAUN & R. DUNN. 1987. Proc. Natl. Acad. Sci. USA **84:** 600–604.
4. LAI, C., M. A. BROW, K.-A. NAVE, A. B. NORANHA, R. H. QUARLES, F. E. BLOOM, R. J. MILNER & J. G. SUTCLIFFE. 1987. Proc. Natl. Acad. Sci. USA **84:** 4337–4341.
5. SALZER, J. L., W. P. HOLMES & D. R. COLMAN. 1987. J. Cell Biol. **104:** 957–965.
6. TRAPP, B. D. & R. H. QUARLES. 1982. J. Cell Biol. **92:** 877–882.
7. POLTORAK, M., R. SADOUL, G. KEILHAVER, C. LANDA, T. FAHRIG & M. SCHACHNER. 1987. J. Cell Biol. **105:** 1893–1899.
8. JOHNSON, P. W., W. ABRAMOW-NEWERLY, B. SEILHEIMER, R. SADOUL, M. B. TROPAK, M. ARQUINT, R. J. DUNN, M. SCHACHNER & J. C. RODER. 1989. Neuron **3:** 377–385.
9. MARTINI, R. & M. SCHACHNER. 1986. J. Cell Biol. **103:** 2439–2448.
10. OWENS, G. C. & R. P. BUNGE. 1989. Glia **2:** 119–128.
11. KIRSCHMEIER, P. T., G. M. HOUSEY, M. D. JOHNSON, A. S. PERKINS & I. B. WEINSTEIN. 1988. DNA **7:** 219–225.
12. BENDER, M., T. D. PALMER, R. E. GELINAS & A. D. MILLER. 1987. J. Virol. **61:** 1639–1646.
13. ELDRIDGE, C. F., M. B. BUNGE, R. P. BUNGE & P. M. WOOD. 1987. J. Cell Biol. **105:** 1023–1034.

Defining Critical Regions in the Promoter for Myelin Basic Protein Gene Transcription[a]

ARUNA ASIPU AND G. ERIC BLAIR[b]

Department of Biochemistry and Molecular Biology
University of Leeds
Leeds LS2 9JT, UK

The myelin basic proteins (MBPs) are a group of structurally related polypeptides that are derived by alternative splicing of a pre-mRNA transcribed from a 33-kb gene.[1,2] The synthesis of various isoforms has been shown to be developmentally regulated. An understanding of the molecular mechanisms that regulate MBP gene expression may provide an insight into the molecular basis of multiple sclerosis (MS) and other neurological disorders associated with demyelination. Here we describe studies to delineate regulatory elements responsible for MBP gene expression in different types of mammalian cells, by cloning the 5′-flanking region of the MBP gene in vectors designed to express bacterial chloramphenicol transacetylase (CAT).

Glial (C6 glioma, 33B oligodendrocytoma) and nonglial (HeLa, 3YI fibroblasts) cell lines were cultured in Dulbecco's modified Eagle's medium supplemented with 10% fetal calf serum (DMEM + 10% FCS). Primary cell cultures of oligodendrocytes were prepared from neonatal Wistar rat cerebral hemispheres.[3,4] Mixed brain cell cultures were grown for 7–8 days in DMEM + 10% FCS. Oligodendrocyte progenitor cells were then shaken off at 250 rpm for 18–24 hours and filtered through 70 μm followed by 20 μm Nytex filters (R. Reynolds, personal communication). The enriched cultures were either seeded on poly-L-lysine-coated coverslips for immunofluorescence or 60-mm dishes for transfection. Immunofluorescent antibody staining using monoclonal antibodies against galactocerebroside (GC) or MBP revealed that on the 21st day of culture (following birth), 60–70% of cells expressed GC and MBP, indicative of oligodendrocyte cell type. Cells were transfected with MBP-cat plasmids at this time.

The plasmid pEx-1, containing the first exon and intron of the MBP gene with its 5′-flanking sequence,[5] was digested with *Sac*I at position +225 (relative to the transcriptional start site) followed by *Bal*31 treatment. *Bam*H1 linkers were added, and the plasmids were re-ligated and transformed into *Escherichia coli.* Clones were picked and deletions determined by DNA sequencing. Fragments resulting from digestion with *Hind*III and *Bam*H1 (−1318/+70; −1318/−600; TABLE 1) were subcloned into a promoter test vector[6] containing a CAT reporter gene (pBLcat3). A partial DNA sequence of the approximate 1.3-kb 5′ region of the MBP gene was also obtained. Further deletions in the 5′-flanking region were created by restriction endonuclease digestion at unique sites in the plasmid pMBPcat3 (−1318/+70).

Glial and nonglial cell lines and primary oligodendrocytes were transfected with pMBPcat3 (−1318/+70) using the calcium phosphate method. CAT expression

[a]This work was supported by the Multiple Sclerosis Society of Great Britain and Northern Ireland.

[b]To whom correspondence should be addressed.

TABLE 1. Relative CAT Activities of MBPcat3 Constructs with Various Deletions[a]

Construct		% CAT Activity	
		C6 Cells	33B Cells
MBPcat3	−1318/+70	100	100
MBPA1cat3	−1318/−714	0.5	0.25
MBPA2cat3	−1318/−600	70	60
MBPB1cat3	−655/+70	240	270
MBPB2cat3	−396/+70	280	350
MBPB3cat3	−105/+70	0.25	0.5
MBPB4cat3	−53/+70	520	100
MBP0cat3	+40/+70	0.3	0.3

[a]CAT activity was determined as % acetylated chloramphenicol formed. The level of CAT activity from MBPcat3 (−1318/+70) was assigned a value of 100%.

vectors containing the Rous sarcoma virus (RSVcat) or JC virus (JCVcat) promoters were used as control expression plasmids. The JCVcat vector is particularly interesting because JC virus replicates selectively in glial cells, and its early promoter/enhancer shows specificity for expression in this cell type.[7] Forty-eight hours after transfection, cell lysates were prepared and CAT activity was assayed.[8] Expression of CAT driven by the entire MBP promoter (−1318/+70) was much greater in glial (C6 and 33B) cells than in nonglial (Hela, 3Y1) cells (FIG. 1). The MBP promoter was approximately three times as active in glial cells as was the JCV promoter. Relative to the constitutively expressed RSV promoter, expression of the MBP promoter was greatest in oligodendrocytes. These results show that the approximate 1.3-kb region upstream of the transcriptional initiation site confers gial cell-type expression of MBP.

To define more precisely the sequences responsible for glial cell-specific expression, plasmids containing deletions in the 5′-flanking region of the MBP gene were transfected into C6 or 33B cells (TABLE 1). The −53/+70 construct gave maximal activity, suggesting that a TATA sequence (located at −34) could alone be sufficient for promoter function. The −105/+70 construct gave very low activity which was restored in the −396/+70 construct, suggesting that the sequence between −53 and

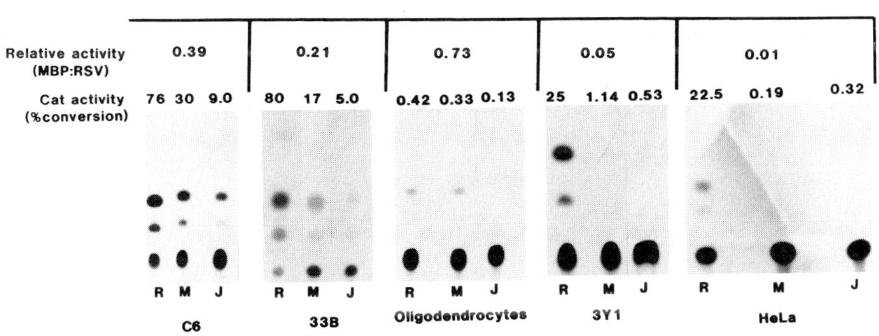

FIGURE 1. Cell-type expression of MBPcat3. Glial (C6, 33B, and primary oligodendrocytes) and nonglial (3Y1, HeLa) cells were transfected with different promoter-cat constructs, RSVcat (R), MBPcat 3 (M), and JCVcat (J). Relative activity refers to the ratio of CAT activity driven by the MBP promoter to that of the RSV promoter.

−105 had a negative effect on promoter function. Sequences between −1318 and −714 also had a negative regulatory effect. A large decrease in CAT activity was found when sequences between −714 and −105 were removed from the CAT constructs, indicating that this region has enhancer function. Interestingly, the sequence between −714 and −600 had appreciable promoter activity, and a possible TATA-like sequence could be identified around −611 to −615 bp.

This work has shown that the 1.3-kb upstream of the major start site for MBP transcription is necessary and sufficient for glial cell-type-specific transcription. Both positive and negative regulatory elements could be identified within this 1.3-kb region. A recent study[9] has described an approach to MBP expression similar to ours, but these authors used a neuronal-glial hybrid cell line, NG108-15, as the recipients in DNA transfection. Considerable agreement exists between the results presented here and those of Miura *et al.*,[9] the notable exceptions being that we find a strong negative regulatory element between −105 and −53 and a strong positive element between −53 and +70. Future work will be directed towards identification and characterization of *trans*-acting factors that enhance MBP expression in glial cells and repress expression in nonglial cells.

ACKNOWLEDGMENTS

We thank Dr. L. Hood for plasmid pEx-1 and Dr. R. Reynolds for advice on culturing oligodendrocytes.

REFERENCES

1. de Ferra, F., H. Engh, L. Hudson, J. Kamholz, C. Puckett, S. Molineaux & R. A. Lazzarini. 1985. Cell **43:** 721–727.
2. Newman, S. L., K. Kitamura & A. T. Campagnoni. 1987. Proc. Natl. Acad. Sci. USA **84:** 886–890.
3. Dubois-Dalcq, M., T. Behar, L. Hudson & R. A. Lazzarini. 1986. J. Cell Biol. **102:** 384–392.
4. Raff, M. C., R. H. Miller & M. Noble. 1983. Nature **303:** 390–396.
5. Takahashi, N., A. Roach, D. B. Teplow, S. B. Prusiner & L. Hood. 1985. Cell **42:** 138–148.
6. Luckow, B. & G. Schutz. 1987. Nucleic Acids Res. **15:** 5490–5493.
7. Kenny, S., V. Natarajan, D. Strike, G. Khoury & N. P. Salzman. 1984. Science **226:** 1337–1339.
8. Gorman, C. M., L. F. Moffat & B. H. Howard. 1982. Mol. Cell. Biol. **2:** 1044–1051.
9. Miura, M., T. Tamura, A. Aoyama & K. Mikoshiba. 1989. Gene **75:** 31–38.

Myelin Sheath Maintenance in the Absence of Axons

G. J. KIDD,[a] J. W. HEATH,[b] B. D. TRAPP,[a] G. J. LITTLE,[b]
AND P. R. DUNKLEY[b]

[a]*Meyer 6-181*
Johns Hopkins University
Baltimore, Maryland 21205

[b]*Faculty Medicine*
University of Newcastle
NSW 2308, Australia

Myelin breakdown in Wallerian degeneration has been attributed to loss of physical contact between the Schwann cell periaxonal membrane and the axon. However, studies of double myelination have established that some myelinating Schwann cells that become completely displaced from the axon continue to maintain an intact myelin sheath in the absence of axonal contact.[1,2]

The present study investigated whether myelin maintenance by these displaced outer Schwann cells might be regulated by diffusible axonal factors. Axonal degeneration was induced in doubly myelinated fibers of the superior cervical ganglion (SCG) either through surgical transection (rats and mice) or by guanethidine-induced (50 mg/kg day) degeneration of adrenergic neurons in rats. Responses of outer Schwann cells were determined by electron microscopy of serially sectioned SCG tissue. Internodes in contact with degenerating axons became disrupted, and myelin splitting, ovoid formation, and phagocytosis proceeded rapidly (1–5 days after surgery, 1–10 days of guanethidine) in both incompletely displaced outer internodes[2] and singly myelinated fibers. In contrast, completely displaced outer internodes retained the usual ordered arrangement of compact myelin lamellae at 16 weeks after surgery and after 6 weeks of guanethidine treatment, and surrounded inner bands of Büngner and numerous collagen fibrils. We conclude that myelin maintenance by Schwann cells does not require axonal trophic factors. Furthermore, these studies suggest that myelin breakdown is initiated by a positive signal that is presented only to Schwann cells in contact with degenerating axons.

REFERENCES

1. Kidd, G. J. & J. W. Heath. 1988. J. Neurocytol. **17:** 245–261.
2. Ibid. pp 263–276.

Differential Regulation of Nerve Growth Factor and Brain-Derived Neurotrophic Factor Expression in the Peripheral Nervous System

ICHIRO MATSUOKA, MICHAEL MEYER, M. HOFER, AND HANS THOENEN

Department of Neurochemistry
Max-Planck-Institute for Psychiatry
8033 Planegg-Martinsried, FRG

Whereas intact sciatic nerves of adult rats synthesize very low levels of nerve growth factor (NGF) *in vivo,* a massive increase of NGF synthesis is observed after a nerve lesion.[1] Application of interleukin 1 (IL-1) increases the level of NGF mRNA in cultured sciatic nerve explants as well as in fibroblasts isolated from the nerve.[2,3] However, the mechanisms regulating NGF synthesis in Schwann cells, the major nonneuronal cell type of the nerve, have not been clarified.

We found that forskolin (FK), a reversible activator of adenylate cyclase, rapidly increased the level of NGF mRNA without affecting its stability in highly enriched cultures of Schwann cells, whereas IL-1 and several peptide growth factors known to increase the NGF mRNA level in fibroblasts and astrocytes[4] failed to do so in Schwann cells. Forskolin was also effective in nerve explant cultures. The effects of FK were mimicked by cAMP analogs, but not by a cGMP analog or dideoxy-forskolin, a forskolin derivative that does not activate adenylate cyclase. Transforming growth factor-β1 decreased the NGF mRNA levels in Schwann cells, in contrast to its effects on astrocytes.[5] A Ca^{2+} ionophore and a phorbol ester potentiated the effect of FK on the NGF mRNA level. Pretreatment of Schwann cells with H-8, an inhibitor of cyclic nucleotide-dependent protein kinases, reduced both basal and induced NGF mRNA levels, suggesting an essential role of cAMP-dependent protein kinase for NGF mRNA regulation in Schwann cells.

Brain-derived neurotrophic factor (BDNF) is a recently cloned member of the family of NGF-like neurotrophic proteins,[6] which is mainly synthesized in the central nervous system.[6,7] So far no nonneuronal cell type capable of BDNF synthesis has been identified. We found that BDNF is expressed at very low levels in the rat sciatic nerve and at much higher levels in rat Schwann cell cultures (FIG. 1). In contrast to its effects on NGF mRNA, expression of BDNF in sciatic nerve explants is not changed by administration of IL-1 (FIG. 2). Opposite to its effects on NGF mRNA, forskolin decreased the basal as well as the ionomycin-induced levels of BDNF mRNA in cultured Schwann cells. Details of the mechanisms of BDNF regulation *in vivo* and *in vitro* remain to be investigated. It is not clear why both proteins are differentially regulated.

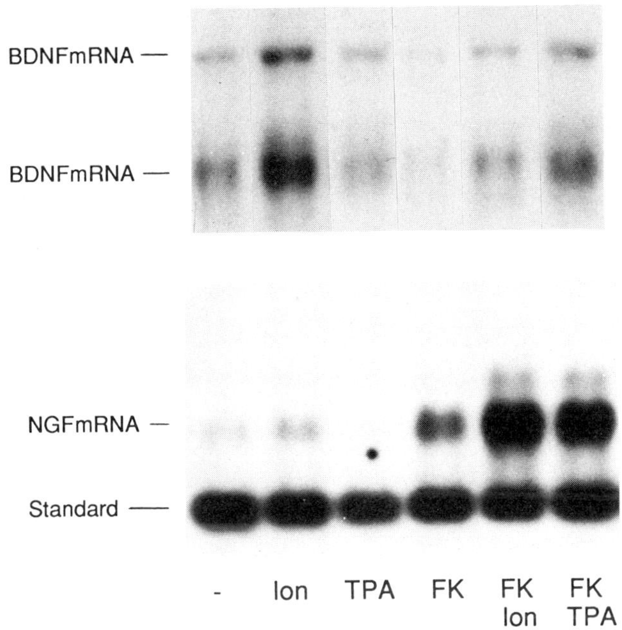

FIGURE 1. Northern blot analysis of NGF and BDNF mRNAs in cultured rat sciatic Schwann cells. Cells were incubated with 1 μg/ml ionomycin (Ion), 0.1 μg/ml TPA, and 20 μM forskolin (FK) for 3 hours.

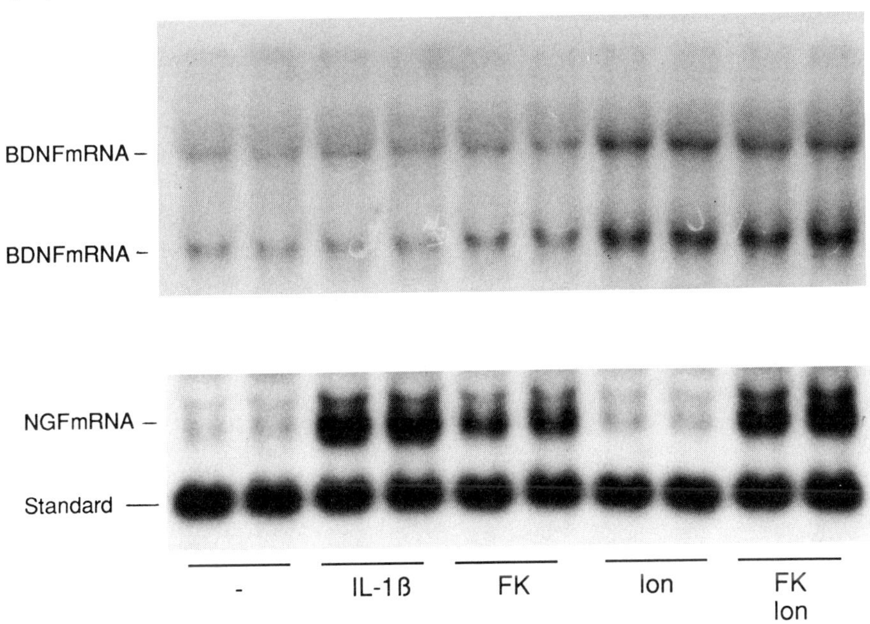

FIGURE 2. Northern blot analysis of NGF and BDNF RNAs in sciatic nerve explants. Sciatic nerve segments prepared from newborn rats (3 days in culture) were incubated with 60 U/ml IL-1β, 20 μM forskolin (FK), and 1 μg/ml ionomycin (Ion) for 3 hours. Nerve explants prepared from adult animals gave similar results.

REFERENCES

1. HEUMANN, R., S. KORSCHING, C. BANDTLOW & H. THOENEN. 1987. J. Cell. Biol. **104:** 1623–1631.
2. LINDHOLM, D., R. HEUMANN, M. MEYER & H. THOENEN. 1987. Nature **330:** 658–659.
3. LINDHOLM, D., R. HEUMANN, B. HENGERER & H. THOENEN. 1988. J. Biol. Chem. **263:** 16348–16351.
4. SPRANGER, M., D. LINDHOLM, C. BANDTLOW, R. HEUMANN, H. GNAHN, M. NÄHER-NOE & H. THOENEN. 1990. Eur. J. Neurosci. **2:** 69–76.
5. LINDHOLM, D., B. HENGERER, F. ZAFRA & H. THOENEN. 1990. Neuro. Rep. **1:** 9–12.
6. LEIBROCK, J., F. LOTTSPEICH, A. HOHN, M. HOFER, B. HENGERER, P. MASIAKOWSKI, H. THOENEN & Y.-A. BARDE. 1989. Nature **341:** 149–152.
7. HOFER, M., S. R. PAGLIUSI, A. HOHN, J. LEIBROCK & Y.-A. BARDE. 1990. EMBO J. **9:** 2459–2464.

Adhesion of Primary Schwann Cells to HNK-1 Reactive Glycosphingolipids

Cellular Specificity

LEILA K. NEEDHAM AND RONALD L. SCHNAAR

Departments of Neuroscience and Pharmacology
The Johns Hopkins University School of Medicine
Baltimore, Maryland 21205

A role for an unusual class of acidic glycolipids, sulfoglucuronyl-glycosphingolipids, in the maintenance of peripheral nervous system (PNS) myelin has been suggested by the finding that some patients with plasma cell dyscrasias and demyelinating

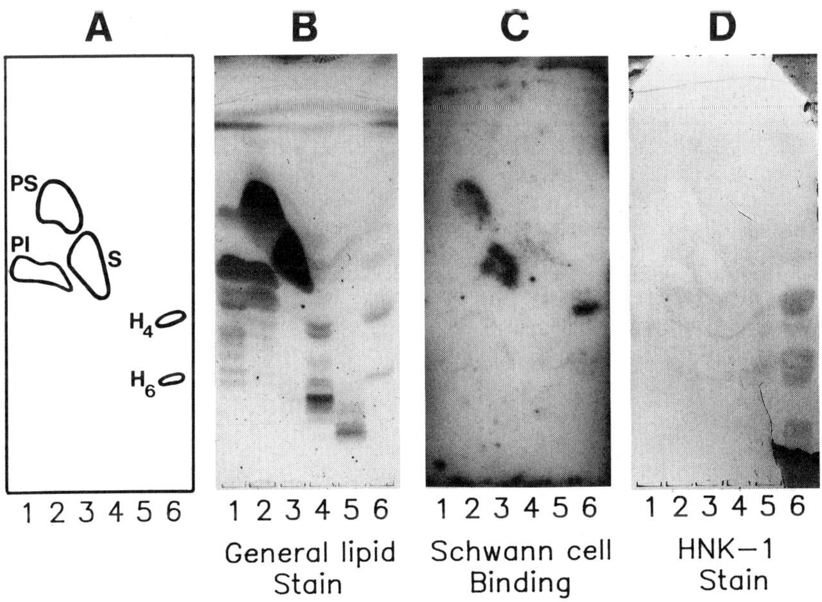

FIGURE 1. The binding of chick Schwann cells to sciatic nerve polar lipids fractionated by DEAE-Sepharose chromatography was examined. Aliquots equivalent to 100 mg wet weight of tissue of each of six fractions isolated by step-gradient elution of lipids from DEAE-Sepharose were applied to TLC plates and developed in chloroform:methanol:0.25% KCl (60:35:8). Resolved lipids were primarily: (*lane 1*) monosialogangliosides; (*lane 2*) phospholipids; (*lane 3*) sulfatide; (*lane 4*) disialogangliosides; (*lane 5*) trisialogangliosides; and (*lane 6*) HNK-1 reactive lipids. (*Panel A*) A key indicating the migration positions of the major HNK-1 reactive lipids (sulfate-3-glucuronyl-neolacto*tetra*osylceramide [H_4] and sulfate-3-glucuronyl-neolacto*hexa*osylceramide [H_6]), sulfatide (S), phosphatidylserine (PS), and phosphatidylinositol (PI); (*panel B*) plate treated with a general lipid stain (cupric sulfate char); (*panels C and D*) autoradiogram of radiolabeled chick Schwann cell binding (C) and the HNK-1 immunostain of the same plate (D).

peripheral neuropathy produce monoclonal IgM antibodies directed against these lipids.[1] These lipids, found in PNS myelin and axolemma,[2] are also recognized by the monoclonal antibody HNK-1. HNK-1 also recognizes carbohydrate epitopes found on several cell adhesion proteins. Recent studies suggest that these epitopes, on proteins or lipids, may act as adhesive ligands in cell-cell interactions.[3] We tested the ability of HNK-1 reactive glycolipids to support Schwann cell adhesion using a technique that measures binding of intact cells to lipids resolved on thin layer chromatography (TLC) plates.[4] After various lipid fractions were resolved by TLC, metabolically radiolabeled cells were incubated on the TLC plate surface. Nonadher-

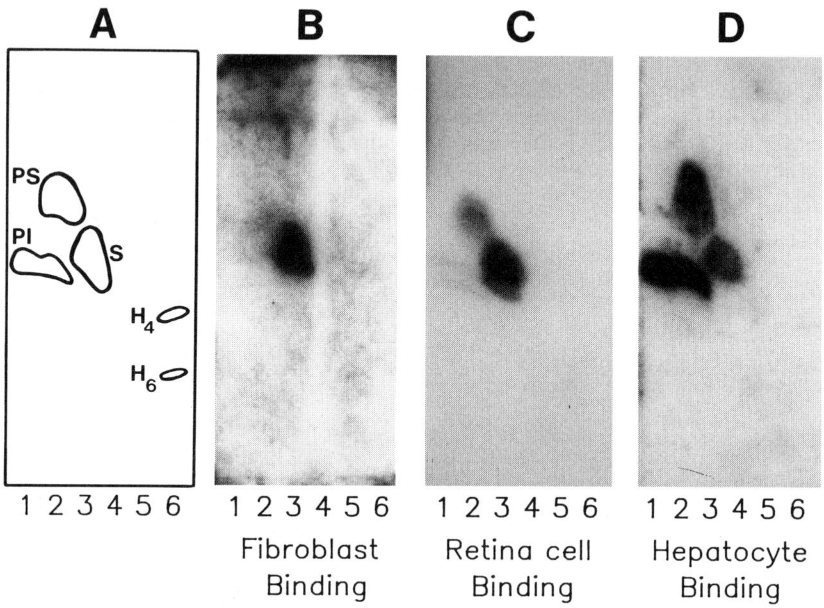

FIGURE 2. The binding of chick fibroblasts, neural retinal cells, and hepatocytes to sciatic nerve polar lipids fractionated by DEAE-Sepharose chromatography was examined. TLC plates were spotted and developed as described in the legend to FIGURE 1. (*Panel A*) A key indicating the migration positions of the major HNK-1 reactive lipids (sulfate-3-glucuronyl-neolacto*tetra*osylceramide [H_4] and sulfate-3-glucuronyl-neolacto*hexa*osylceramide [H_6]), sulfatide (S), phosphatidylserine (PS), and phosphatidylinositol (PI); (*panels B–D*) autoradiograms resulting from the binding of radiolabeled chick fibroblasts (B), neural retina cells (C), and hepatocytes (D).

ent cells were removed by centrifugation, and adherent cells were fixed in place. The plate was then subjected to autoradiography (to locate adherent cells) and immunostained with HNK-1. HNK-1 reactive lipids were partially purified from a crude acidic glycolipid fraction[5,6] of dog sciatic nerve by step-gradient elution from DEAE-Sepharose.[7] We previously reported[8] that when aliquots of the DEAE-Sepharose chromatography fractions were resolved by TLC, rat Schwann cells adhered to positions on the TLC plates corresponding to the migration of the major HNK-1 reactive lipid, sulfate-3-glucuronyl-neolactotetraosylceramide. As little as 450 pmol of this HNK-1 reactive lipid supported significant cell adhesion. Most other sciatic nerve lipids, including complex neutral lipids and gangliosides, did not support any cell adhesion. High concentrations of phosphatidylserine and sulfatide (5–7 nmol) did support rat Schwann cell adhesion, but they were less effective than were HNK-1 reactive lipids. We now report that cells isolated from chicken embryo sciatic nerve showed a pattern of binding to lipids nearly identical to that of rat Schwann cells (FIG. 1). The cellular specificity of recognition of the HNK-1 reactive glycolipids was then examined. Embryonic chicken fibroblasts, neural retinal cells, and hepatocytes failed to bind to HNK-1 reactive lipids, but they did bind to sulfatide, phosphatidylserine, and phophatidylinositol (FIG. 2). These data suggest that the HNK-1 reactive lipids may be structurally specific Schwann cell adhesion molecules.

REFERENCES

1. QUARLES, R. H., A. A. ILYAS & H. J. WILLISON. 1986. Chem. Phys. Lipids **42:** 235–248.
2. KOHRIYAMA, T., S. KUSUNOKI, T. ARIGA, J. E. YOSHINO, G. H. DEVRIES, N. LATOV & R. K. YU. 1987. J. Neurochem. **48:** 1516–1522.
3. KÜNEMUND, V., F. B. JUNGALWALA, G. FISCHER, D. K. H. CHOU, G. KEILHAUER & M. SCHACHNER. 1988. J. Cell Biol. **106:** 213–223.
4. SWANK-HILL, P., L. K. NEEDHAM & R. L. SCHNAAR. 1987. Anal. Biochem. **163:** 27–35.
5. SVENNERHOLM, L. & P. FREDMAN. 1980. BBA **617:** 97–109.
6. LADISCH, S. & B. GILLARD. 1985. Anal. Biochem. **146:** 220–231.
7. MAGNANI, J. L., S. L. SPITALNIK & V. GINSBURG. 1987. Meth. Enzymol. **138:** 195–207.
8. NEEDHAM, L. K. & R. L. SCHNAAR. 1990. Ann. N.Y. Acad. Sci. **605:** 416–419.

Expression of the 14E Antigen in Glial Cells in Human Normal and Pathological Central Nervous System Tissues[a]

J. NEWCOMBE,[b] A. BRAR, H. LI, AND M. L. CUZNER

The Multiple Sclerosis Society Laboratory
Institute of Neurology
London WC1N 1PJ, England

In tissue sections of normal adult human nervous system the monoclonal antibody 14E immunostains oligodendrocyte and Schwann cell perikarya, but not myelin. However, in actively demyelinating multiple sclerosis (MS) plaques the 14E antibody visualizes hypertrophic glial cells in addition to oligodendrocytes, which are increased in number. The IgM antibody was produced using oligodendrocytes isolated from normal adult brain as the immunogen.[1,2] The 14E antigen and its cellular distribution in normal and pathological central nervous system (CNS) tissues have been further investigated in this study.

Cryostat sections of adult human normal CNS from 15 control cases and of plaques and normal appearing CNS samples from 25 cases of MS were immunoperoxidase stained with 14E and glial cell-specific antibodies. In normal cerebellum 14E immunostained the nuclei but not the processes of Bergmann glia, a class of radial astrocytes. Only the processes of these cells were stained with antibodies directed against glial fibrillary acidic protein (GFAP), glutamine synthetase, or vimentin, whereas both the cell body and the processes of Bergmann glia were immunostained with an antibody against S-100. In blood vessel walls in normal CNS tissues, the nuclear membrane of columnar cells identified immunocytochemically as smooth muscle cells was immunostained by 14E, but in some active plaques both the cytoplasm and the nuclei of these cells were 14E positive.

The 14E antibody visualized oligodendrocyte nuclei and the nuclei, perikarya cytoplasm, and processes of scattered small clusters of hypertrophic glial cells in active MS plaques. Throughout these lesions a network of astrocyte processes was seen with GFAP (FIG. 1), vimentin, glutamine synthetase, S-100, and lipocortin I immunostaining. In a small number of active plaques, occasional groups of hypertrophic glia were visualized by these antibodies and by 14E. The morphology of 14E-positive hypertrophic cells was often bizarre, and two to six nuclei were seen in some cells, but similar perikarya were rarely visualized in plaques with GFAP immunostaining. The cells were not stained with the human macrophage and microglial marker EBM/11 and so differ from multinucleate giant cells observed in CNS tissue in HIV infection. Very dense networks of postreactive astrocytic processes were observed in chronic plaques, but only the nuclei of small numbers of glial

[a]This study was supported by the Multiple Sclerosis Society of Great Britain and Northern Ireland.

[b]Address for correspondence: Dr. J. Newcombe, The Multiple Sclerosis Society Laboratory, Department of Neurochemistry, Institute of Neurology, 1 Wakefield Street, London WC1N 1PJ, England.

and smooth muscle cells were stained with 14E. Therefore, the 14E-positive hypertrophic glial cells appear to be a subpopulation of reactive astrocytes.

Nuclei were isolated from white matter and active plaques, and membrane and soluble fractions of whole tissues and nuclei were prepared by centrifugation. Proteins in these fractions were examined by polyacrylamide gel electrophoresis and immunoblotting. The 14E antibody stained a single protein band of MW 23 kD in

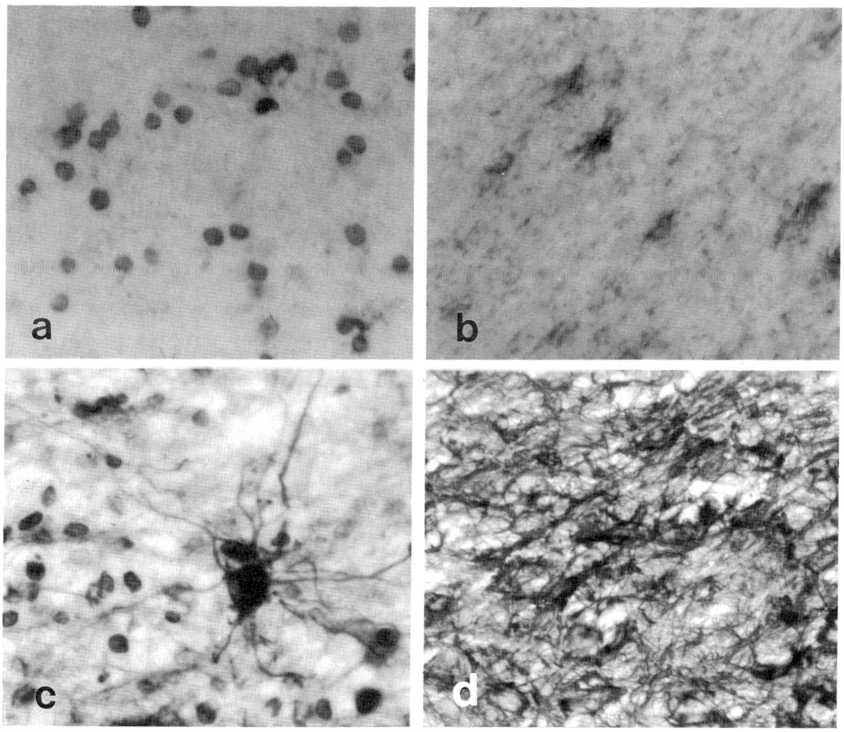

FIGURE 1. Cryostat sections (10 μm) fixed in acetone. (**a** and **b**) Normal control adult human brain ventricular white matter. (**a**) 14E immunostaining of oligodendrocyte nuclei; (**b**) monoclonal anti-GFAP antibody staining of astrocytes. (**c** and **d**) An actively demyelinating ventricular plaque from a case of MS. (**c**) 14E immunostaining of a hypertrophic glial cell and oligodendrocyte nuclei; (**d**) GFAP immunostaining of astrocyte processes in the same area of an adjacent section. Immunoperoxidase staining with an avidin-biotin complex system (Vector Laboratories), using diaminobenzidine tetrahydrochloride with nickel chloride (0.04%, w/v). No nuclear counterstaining; original magnification ×440.

these fractions with or without mercaptoethanol treatment, indicating that the 23-kD protein is not a subunit. In normal white matter, levels of 14E protein were much lower in the soluble fraction than in the membrane fraction, but the concentration was increased in both fractions of active plaques. The elevated expression of membrane fraction 14E in active plaques appears to reflect the presence of increased numbers of oligodendrocyte nuclei. The 14E antigen is currently being characterized by amino acid sequencing in order to determine the significance of soluble fraction 14E in a subpopulation of reactive astrocytes.

REFERENCES

1. NEWCOMBE, J. & M. L. CUZNER. 1988. J. Neuroimmunol. **19:** 11–20.
2. NEWCOMBE, J., B. N. HARDING & M. L. CUZNER. 1990. *In* Myelination and Dysmyelination. Ann. N.Y. Acad. Sci. **605:** 406–408.

Isolation and Characterization of the Myelin-Like Membranes Ensheathing Giant Axons in the Earthworm Nerve Cord[a]

BETTY I. ROOTS, BEATRICE CARDONE, AND PEDRO PEREYRA

Department of Zoology
University of Toronto
Toronto, M5S 1A1 Canada

The ventral nerve cord of the earthworm, *Lumbricus terrestris,* contains three dorsal giant axons that are ensheathed by multilamellar myelin-like membranes. The morphologic characteristics of the sheath have been described by Roots and Lane.[1] Initial characterization of isolated membrane-containing fractions by SDS-PAGE and immunoblotting techniques showed that the composition of the membranes is quite different from that of vertebrate myelin. The protein pattern is relatively simple

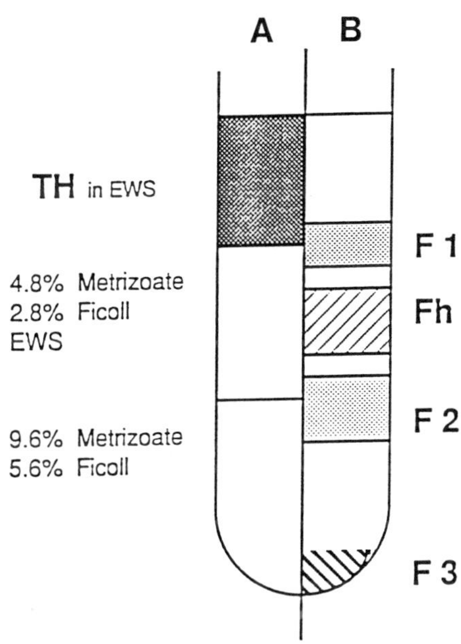

FIGURE 1. Separation of earthworm homogenate into fractions: **(A)** before centrifugation; **(B)** after centrifugation resulting in F_1 and F_2 fractions, F_h fraction (hemoglobin), and F_3 fraction (pellet).

[a]This work was supported by Natural Sciences and Engineering Research Council of Canada, Grant #A6052 to B.I.R.

with 80 and 42 kD proteins predominating. No positive results were obtained with antibodies against various vertebrate myelin proteins such as myelin basic protein (MBP), myelin-associated glycoprotein (MAG), proteolipid protein (PLP), and 2′3′-cyclic nucleotide phosphohydrolase (CNP).[2]

We have modified our earlier methods[2,3] to obtain two subfractions, F_1 and F_2, enriched in the myelin-like membranes. Ten to twenty earthworm nerve cords were

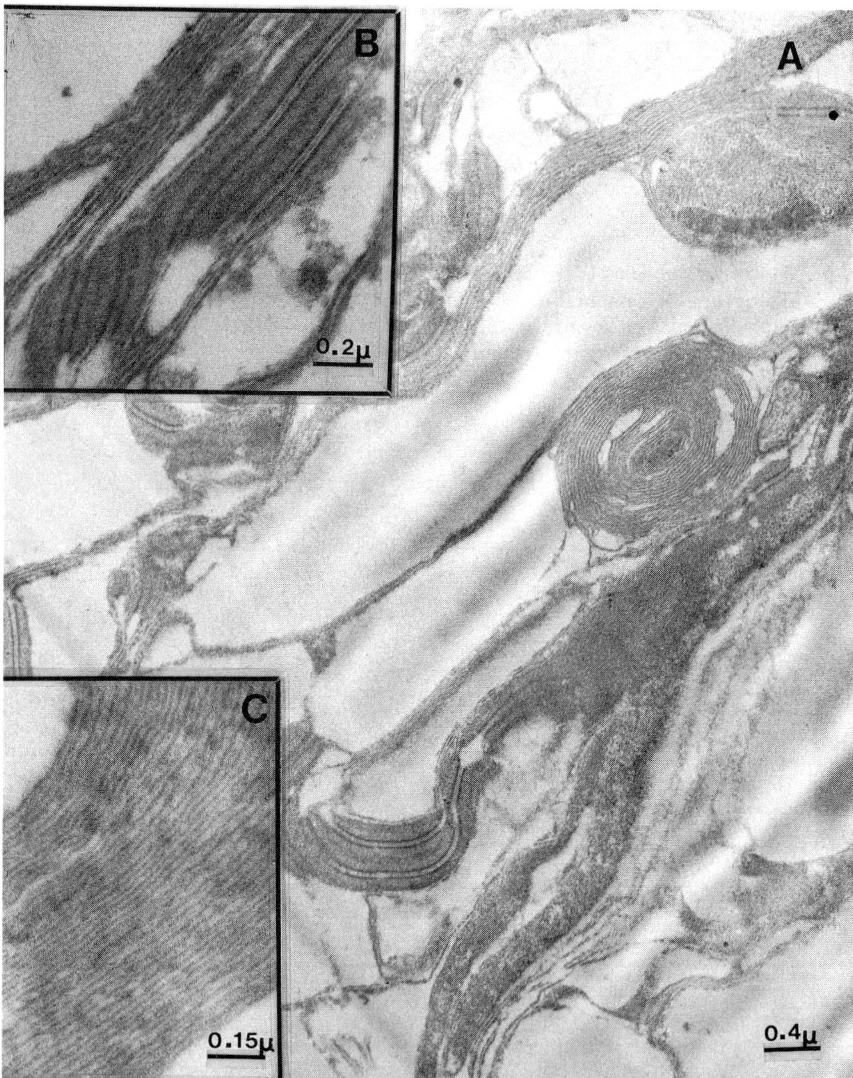

FIGURE 2. Electron micrograph of the F_1 fraction. **(A)** Low magnification micrograph showing membrane spirals. **(B)** High magnification view of desmosome-like complex. **(C)** Myelin membranes at high magnification.

excised, cut into 2 mm pieces, washed in earthworm saline solution (EWS [in mM]: NaCl 22.5, Na_2SO_4 32.5, KCl 2.5, $MgCl_2$ 0.5, $CaCl_2$ 1, Tris-HCL 5, and glucose 5; pH 7.4), then incubated for 30 minutes at 5°C in EWS containing peptidase inhibitors and heparin. The pieces were washed in EWS, then hand homogenized using a microteflon pestle. The homogenate (TH) was layered onto a two-step metrizoate/Ficoll gradient (Lymphoprep, Nycomed Cederlane) and separated using a TL-100 Ultracentrifuge (45,000 rpm, 30 minutes, 5°C, TLS-55 rotor, $4 \times 10^6 g$ min). F_1 and F_2 fractions were collected and pelleted (FIG. 1). The combined membrane-containing subfractions recovered from 20 nerve cords contained 1–2 mg protein. Aliquots of F_1 and F_2 fractions were prepared for electron microscopy.[4] The fractions contained multilamellar membrane spirals and associated desmosomal complexes (FIG. 2). The estimated membrane content observed in F_1 was approximately 50%, and in F_2 25–30%. The fractions also contained small and large vesicles, granular deposits, and other glial cell and axoplasmic inclusions.

We have prepared polyclonal (and are preparing monoclonal) antibodies to the F_1 and F_2 fractions. Fractions were lyophilized and reconstituted in a minimal amount of dH_2O. Approximately 100–200 μg of combined protein was used per immunization of BALB/c mice, 1:1 with Freund's complete adjuvant (first immunization only) and 1:1 with Freund's incomplete adjuvant (subsequent immunizations) every 14 days for 8 weeks. Mice were sacrificed, and serum (polyclonal) was collected by cardiac puncture. Spleens of immunized mice were fused with P3X63Ag8 murine myeloma cells for the future production of monoclonal antibodies, as per the methods of Galfre *et al.*[5]

Immunoblotting of SDS-PAGE gels indicates that these antibodies cross-react predominantly with proteins at molecular weights of 40–42 kD. Using these antibodies to label antigens in frozen earthworm nerve cord cross-sections, we were able to localize these proteins using immunofluorescence. Full details of immunoblotting and immunofluorescence techniques were given by Cardone and Roots.[4] The immunofluorescence studies indicate that these antibodies cross-react with the tissue that surrounds the giant axons, but do not show cross-reactivity with the axon contents. However, cross-reactivity is not confined to the sheath. Studies at the ultrastructural level and with monoclonal antibodies will elucidate the distribution of the proteins more precisely. We plan to use these antibodies to act as probes in our future studies of the phylogenetic distribution of these sheath component proteins and of the possible relation between these myelin-like membranes and vertebrate myelin.

REFERENCES

1. ROOTS, B. I. & N. J. LANE. 1983. Tiss. Cell. **15:** 695–709.
2. PEREYRA, P. M. & B. I. ROOTS. 1988a. Neurochem. Res. **13:** 893–901.
3. PEREYRA, P. M. & B. I. ROOTS. 1988b. Soc. Neurosci. Abstr. **14:** 584.
4. CARDONE, B. & B. I. ROOTS. 1990. Glia **3:** 180–192.
5. GALFRE, G., S. C. HOWE, C. MILSTEIN, G. W. BUTCHER & J. C. HOWARD. 1977. Nature **266:** 550–552.

A Growth Cone Collapsing Activity in Chicken Gray Matter

ROGER J. KEYNES, ALAN R. JOHNSON,
CAROLINE J. PICART, OLGA M. DUNIN-BORKOWSKI,
AND GEOFFREY M. W. COOK

Department of Anatomy
University of Cambridge
Cambridge CB2 3DY, UK

The development of a segmented pattern of peripheral spinal nerves provides a simple model system for the analysis of factors influencing axonal growth. In higher vertebrate embryos, outgrowing spinal axons are confined to the anterior half of the sclerotome of each successive somite.[1] Working on the hypothesis that axons are repelled from the posterior half sclerotome, we recently isolated a glycoprotein fraction from chick somites that causes collapse of dorsal root ganglion growth cones *in vitro.*[2] Two components (M_r 48 K and 55 K) of this fraction are present in the posterior, but not the anterior, half of the sclerotome, and bind the lectin peanut agglutinin (PNA). When immobilized on Sepharose 4B, both PNA and polyclonal antibodies raised against these components can be used to eliminate the collapse activity. The restriction of these components to the posterior half of the sclerotome indicates that the collapse detected *in vitro* is likely to be mirrored *in vivo* by an active repulsion between growth cones and posterior half sclerotome cells.[2]

We now find that the antibodies also cross-react with a component of M_r 48 K in the adult chicken forebrain. Moreover, when fragments of embryonic chick forebrain or retina are grafted in place in a 2-day embryo spinal cord, outgrowing central nervous system (CNS) axons also grow exclusively through the anterior half of the sclerotome. Finally, detergent-derived (2% CHAPS in phosphate-buffered saline solution) extracts of chicken gray matter contain potent collapsing activity for both CNS (retina) and peripheral nervous system (dorsal root ganglion) growth cones *in vitro.* So far, we have been unable to detect collapse activity in white matter extracts using the same assay procedures.

As for the somite-derived collapse activity, the activity in the gray matter can be eliminated using immobilized PNA. SDS-PAGE of the PNA binding fraction of gray matter detergent extracts also reveals two major components at 48 K and 55 K. We suggest, therefore, that brain-derived activity is likely to be similar or identical to that present in the embryonic mesoderm and that its presence could contribute to the failure of effective regeneration after injury to the adult CNS.

REFERENCES

1. KEYNES, R. J. & C. D. STERN. 1984. Segmentation in the vertebrate nervous system. Nature **310:** 786–789.
2. DAVIES, J. A., G. M. W. COOK, C. D. STERN & R. J. KEYNES. 1990. Isolation from chick somites of a glycoprotein fraction that causes collapse of dorsal root ganglion growth cones. Neuron **4:** 11–20.

Transplantation of Normal and Genetically Engineered Glia into Areas of Demyelination[a]

A. J. CRANG, R. J. M. FRANKLIN, AND W. F. BLAKEMORE

Department of Clinical Veterinary Medicine
University of Cambridge
Cambridge CB3 OES, UK

J. TROTTER AND M. SCHACHNER

Department of Neurobiology
University of Heidelberg
D6900 Heidelberg, Federal Republic of Germany

S. C. BARNETT AND M. NOBLE

Ludwig Institute for Cancer Research
London W1P 8BT, UK

Injection of the gliotoxic agent ethidium bromide (EB) into spinal white matter produces a central nervous system (CNS) lesion in which it is possible to investigate the ability of transplanted glial cells to reconstruct a glial environment around demyelinated axons. In the X-irradiated EB lesion there is no inherent repair, and all axons remain demyelinated for the lifetime of the experiment[1]; thus, any repair seen after transplantation of glial cells into X-irradiated lesions must be effected by the transplanted cells or their progeny. Using this lesion we showed that reconstruction of an intact glial environment is an interactive process involving both astrocytes and oligodendrocytes.[2] We also demonstrated that provided an appropriate immunosuppressive regimen is employed, it is possible to repair EB lesions in adult rats with perinatal mouse glial cells.[3] The X-irradiated EB lesion therefore provides a unique *in vivo* environment in which to examine the potential of transplanted xenogenic and allogenic genetically engineered glial cells to effect repair of a demyelinating lesion.

We employed two different *in vitro* protocols to establish populations of glial progenitor cells for transplantation: (1) retroviral infection of embryonic mouse glial cells with a temperature-sensitive (*ts*) oncogene, and (2) the growth factor-mediated continuous expansion of perinatal optic nerve glial progenitor cells.

Infection of mixed glial cell cultures prepared from E14 mice with a *ts*-SV40T oncogene/neomycin-resistance gene construct[4] generates immortalized populations of developmental-stage-specific glial cells when maintained in tissue culture at the permissive temperature (33°C) in medium containing the antibiotic G418. We have transplanted mixed populations, 04+ve immunoselected subpopulations and 04+ve clones of *ts*-immortalized glial cells into X-irradiated ethidium bromide lesions in rat

[a] This work was supported by The Multiple Sclerosis Society of Great Britain and Northern Ireland, the Wellcome Trust, the Deutsche Forschungs Gemeinschaft (SFB 317), and the Ludwig Institute for Cancer Research.

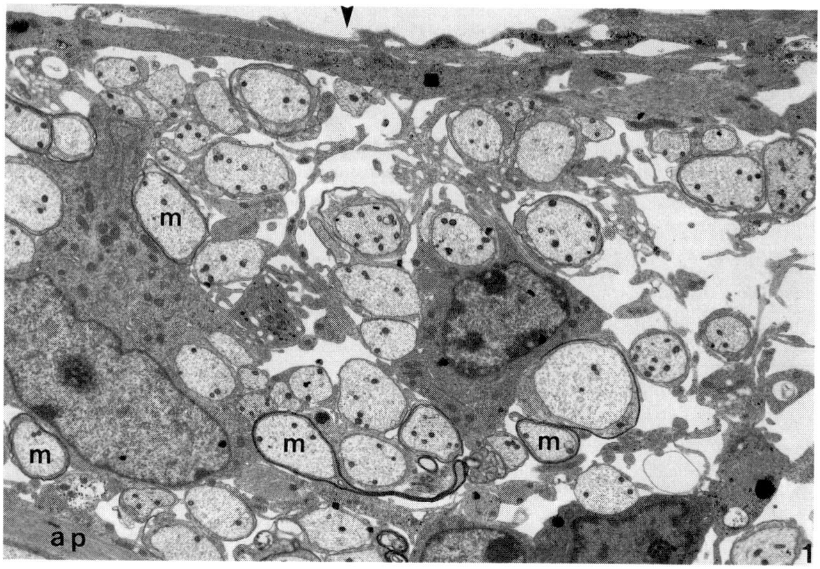

FIGURE 1. Remyelination in X-irradiated EB lesion after transplantation of a *ts*-immortalized mouse glial cell line. m = remyelinated axons; ap = astrocyte process; *arrow* indicates astrocyte processes arranged to form a tissue boundary. Lead citrate/uranyl acetate × 4000.

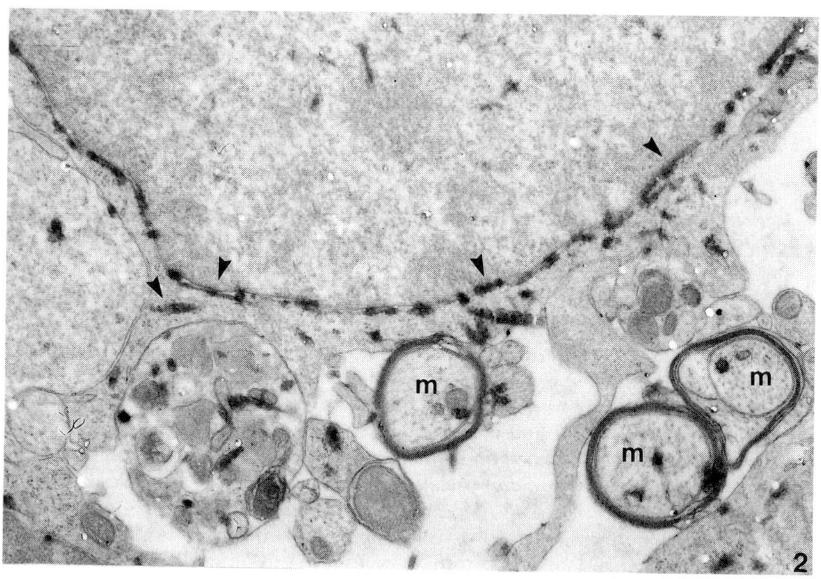

FIGURE 2. Remyelination in X-irradiated EB lesion after transplantation of clonal β-galactosidase transfected glial progenitors. m = remyelinated axons; *arrows* indicate electron-dense, rod-shaped, β-galactosidase reaction product in perinuclear and cytoplasmic locations. Unstained; original magnification × 17,000.

spinal cord. Our observations indicate that *ts*-immortalized immature glial cells, at body temperature, can differentiate to produce astrocytes and myelin-forming oligodendrocytes when injected into demyelinating lesions (FIG. 1).

Growth of glial cells isolated from P7 rat optic nerves in the presence of platelet-derived growth factor (PDGF) and basic fibroblast growth factor (bFGF) establishes cultures of dividing glial progenitor cells[5] that have been maintained in continuous culture for up to 14 months. To trace the progeny of these progenitors, we used lipofectin-mediated transfection to introduce the genes for β-galactosidase and neomycin resistance into O-2A progenitors in the presence of PDGF and bFGF. Clonal cultures were obtained by selection with the G418. Following transplantation of clonal cultures of O-2A progenitors into X-irradiated EB lesions, myelin-forming oligodendrocytes with perinuclear and cytoplasmic β-galactosidase reaction product were found within the lesioned tissue (FIG. 2).

Taken together, these two sets of experiments, using different protocols to establish clonal glial cell lines, demonstrate the feasibility of employing genetic engineering techniques to generate glial cells of specific phenotype that can be transplanted into the damaged CNS.

REFERENCES

1. BLAKEMORE, W. F. & R. C. PATTERSON. 1978. Suppression of remyelination in the CNS by X irradiation. Acta Neuropathol. **42:** 105–113.
2. BLAKEMORE, W. F. & A. J. CRANG. 1989. The relationship between type-1 astrocytes, Schwann cells and oligodendrocytes following transplantation of glial cell cultures into demyelinating lesions in the adult rat spinal cord. J Neurocytol. **18:** 519–528.
3. CRANG, A. J. & W. F. BLAKEMORE. 1991. Remyelination of demyelinated rat axons by transplanted mouse oligodendrocytes. Glia **4:** 305–313.
4. JAT, P. S. & P. A. SHARP. 1989. Cell lines established by a temperature-sensitive Simian virus 40 large T-antigen gene are growth restricted at the nonpermissive temperature. Molec. Cell. Biol. **9:** 1672–1681.
5. BÖGLER, O., D. WREN, S. C. BARNETT, H. LAND & M. NOBLE. 1990. Cooperation between two growth factors promotes extended self-renewal and inhibits differentiation of oligodendrocyte-type-2 astrocyte (O-2A) progenitor cells. Proc. Natl. Acad. Sci. USA **87:** 6368–6372.

Components of Astrocytic Extracellular Matrix Are Regulated by Contact with Axons[a]

MARCH D. ARD[b] AND ANDREAS FAISSNER[c]

[b]*Department of Anatomy*
University of Mississippi Medical Center
2500 N. State St.
Jackson, Mississippi 39216

[c]*Department of Neurobiology*
University of Heidelberg
Im Neuenheimer Feld 364
D-6900 Heidelberg, Federal Republic of Germany

During embryonic development *in vivo* and in immature stages of growth *in vitro,* astrocytes assemble an extracellular matrix (ECM) containing molecules known to promote neurite growth, such as laminin and fibronectin. However, as differentiation progresses the ECM diminishes. (For references on ECM of astrocytes, see Ard and Bunge.[1]) This function of astrocytes parallels events in the peripheral nervous system in which expression of the cell adhesion molecules N-CAM and L1 on Schwann cells[2] and N-CAM on muscle cells[3] is diminished by contact with axons. Denervation of Schwann cells or muscle cells restores a high level of expression of these molecules. Because the ability to reexpress adhesion molecules, including some ECM molecules, may be important for nervous system regeneration, we tested the regulation of ECM in cultures of astrocytes and neurons.

Astrocytes dissected from newborn rat cerebral cortex were cultured in flasks for 2–4 weeks in DMEM plus 10% fetal bovine serum, then subcultured on collagen-coated coverslips with 20% fetal bovine serum. Explants of embryonic day 15 rat retina were the source of axonal outgrowth. Some explants grew for 7 days; others were excised after 4–5 days and their axons left to degenerate for 2–3 days. At 7 days all cultures were immunostained. Polyclonal antibodies were used to recognize J1/tenascin,[4,5] laminin (Telios), and fibronectin (Telios). The chondroitin sulfate proteoglycan and 575 antigen were identified by monoclonals.[6]

The ECM secreted by the astrocytes cultured alone (FIG. 1a) reacted with antibodies to J1/tenascin, laminin, fibronectin, a chondroitin sulfate proteoglycan (CSPG), and an as yet uncharacterized antigen, 575. Immunostaining of J1/tenascin (FIG. 2a), laminin, and fibronectin occurred in a punctate pattern on cell surfaces and in fibrils accumulated between cells. Antibody to CSPG labeled rare cells, always in a punctate pattern. 575 labeled most but not all astrocytes; its labeling was abundant and closely followed the shapes of cells and processes.

Axons from retinal explants were located on the astrocyte layers by immunostaining phosphorylated neurofilaments (FIG. 1b; antibody from Sternberger-Meyer). Double labeling of axons and ECM components showed that in areas where axons

[a]This work was supported by NIH grant 2 S07 RR05386.

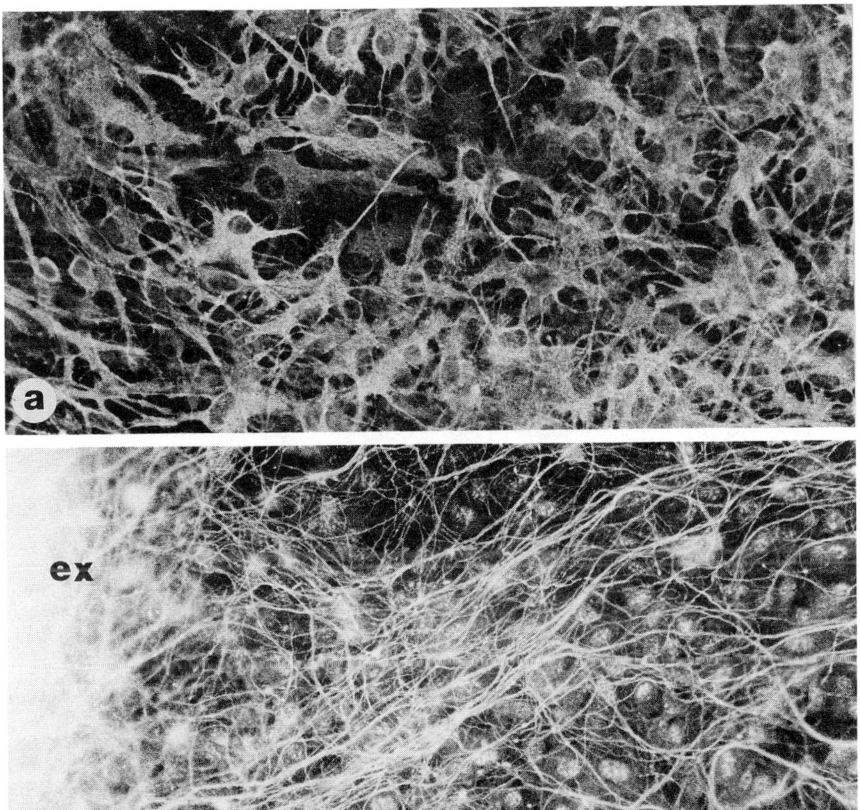

FIGURE 1. (**a**) Astrocytes in culture were labeled with antibodies to glial filaments to ascertain cell type and to visualize cell shape. Nuclei appear as negative images within the fluorescent cytoplasm. (**b**) Neurites grew from the explant at *left* (ex) over a layer of astrocytes. Both axons and astrocytic nuclei are visible because the antineurofilament antibody also labels a nuclear protein in cultured cells. (**c**) At left, wrinkled edges of collagen (nonspecifically fluorescent) mark the spot where a razor cut was made to remove the retinal explant from this culture. Antineurofilament labeling reveals only a few beaded, probably degenerating axons (*arrows*) remaining on the bed of astrocytes (with immunolabeled nuclei). Magnification × 280.

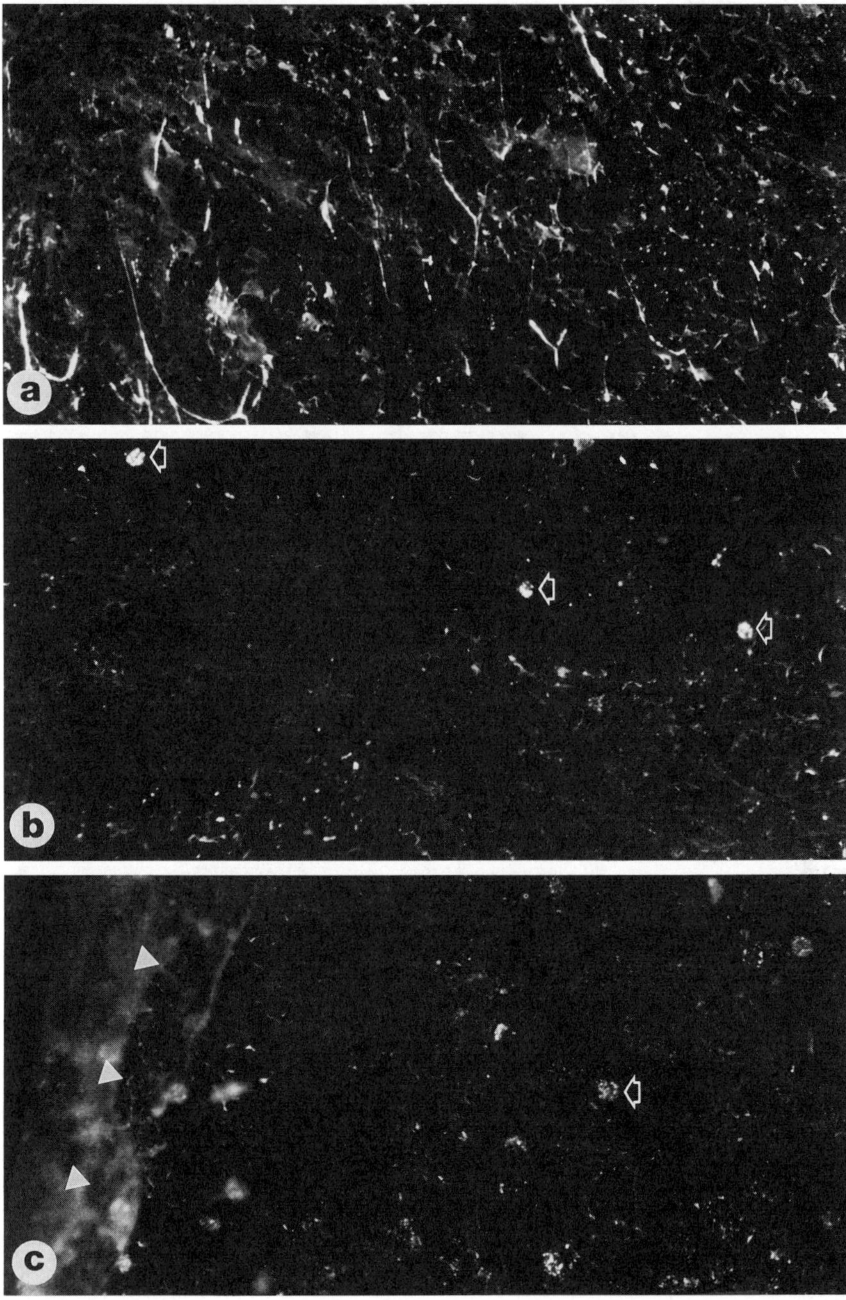

FIGURE 2. Cultures similar to those shown in Figure 1 were immunostained before fixation with antibodies to J1/tenascin. (**a**) Astrocytes cultured alone were labeled extracellularly, indicating the presence of J1/tenascin in the extracellular matrix. The label appears as both puncta and long fibrils. (**b**) In an area where neurites covered the astrocyte layer, immunostaining for J1/tenascin was greatly reduced. In addition to sites of specific J1/tenascin labeling,

FIGURE 2 (*continued*) three macrophages (*arrows*) from the retinal explant have bound the fluorescent antibody. (**c**) After excision of the explant and degeneration of the axons, immunostaining for J1/tenascin remains low and does not return to the level seen on uninnervated astrocytes (**a**). *Arrowheads* mark wrinkled collagen where the excision was made. *Arrow* marks one of several labeled macrophages present. Magnification × 280.

contacted astrocytes there was a reduction in immunostaining for J1/tenascin (FIG. 2b), laminin, and fibronectin. In contrast, CSPG and 575 labeling were unchanged.

Degeneration of axons in contact with astrocytes (FIG. 1c) did not alter the pattern of immunostaining with any of these antibodies (FIG. 2c) compared to that seen with axons intact. No up-regulation of ECM in response to denervation was seen. The lack of such a response may contribute to the failure of central nervous system regeneration.

REFERENCES

1. ARD, M. D. & R. P. BUNGE. 1988. J. Neurosci. **8:** 2844–2858.
2. SEILHEIMER, B., E. PERSOHN & M. SCHACHNER. 1989. J. Cell Biol. **108:** 1909–1915.
3. COVAULT, J. & J. R. SANES. 1985. Proc. Natl. Acad. Sci. USA **82:** 4544–4548.
4. KRUSE, J., G. KEILHAUER, A. FAISSNER, R. TIMPL & M. SCHACHNER. 1985. Nature **316:** 146–148.
5. FAISSNER, A., J. KRUSE, R. CHIQUET-EHRISMANN & E. MACKIE. 1988. Differentiation **37:** 104–114.
6. FAISSNER, A. 1988. Soc. Neurosci. Abstr. **14:** 920.

Effects of X-Irradiation on Proximal Nerve Stump Reaction after Nerve Crush and Axonal Regeneration

M. BRESJANAC,[a] J. SKETELJ,[a] AND D. LINDHOLM[b]

[a] *Institute of Pathophysiology*
Zaloška 4
61105 Ljubljana, Yugoslavia

[b] *Max-Planck-Institute for Psychiatry*
Martinsried, Germany

It was recently shown that following nerve crush, sensory and motor axons elongate well in the absence of nonneuronal cells in the distal stump[1] and away from their denervated targets.[2] A possible explanation might be the availability of a proximal source of trophic support. Following crush, a "proximal stump reaction" occurs: cells proliferate[3] and augment production of trophic substances.[4] We investigated the possibility that the proximal stump reaction significantly enhances axonal growth. Sensory and motor axonal elongation through acellular distal nerve stump was assessed, after X-irradiation had been used[5] to suppress proximal nonneuronal cell response to the injury.

MATERIALS AND METHODS

1. Rat sciatic nerve was crushed; a distal segment was frozen-thawed, cut, and deflected.[2] The control animals received only this treatment.
2. X-irradiation (2 × 30 Gy; 10 Gy/min; 100% dose was achieved 3 mm below the skin surface) was applied to the hind limb of the experimental animals. The first dose of irradiation was applied half an hour before surgery, and the second 30 Gy were given 48 hours later.
3. Total cell counts of proximal stump transverse sections and quantitative determinations of NGF and NGFrec mRNAs[6] in the stump were done to assess the effects of X-irradiation.
4. Axon elongation was followed by nerve-pinch test (sensory) and choline acetyltransferase (ChAT) activity distribution (motor axons).[1]

RESULTS

Total cell counts performed on transverse sections of proximal nerve stumps disclosed overall reduced cell numbers in irradiated versus control nerves at 5 days postoperative (1,238 ± 353 *vs.* 2,539 ± 394; $n = 4$; $p < 0.005$). The average total number of cells in a normal rat nerve transverse section was 807 ± 95 ($n = 4$).

Levels of NGF and NGFrec mRNAs in the proximal stumps were determined to see how the irradiation had affected the functional state of Schwann cells. The

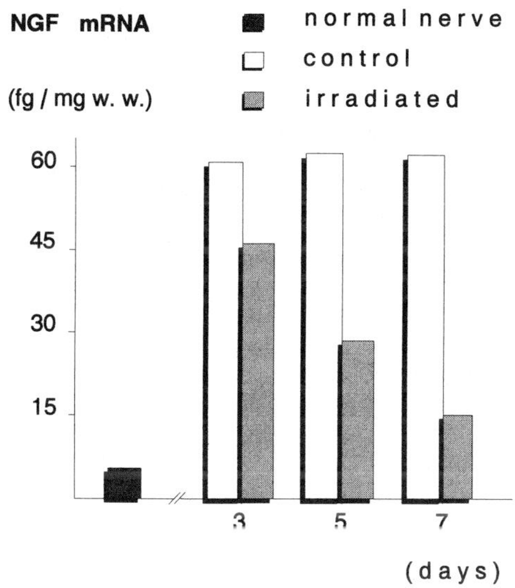

FIGURE 1. Levels of NGF mRNA in the proximal nerve stumps during the first week after treatment. Baseline level in an uninjured rat sciatic nerve is given for comparison.

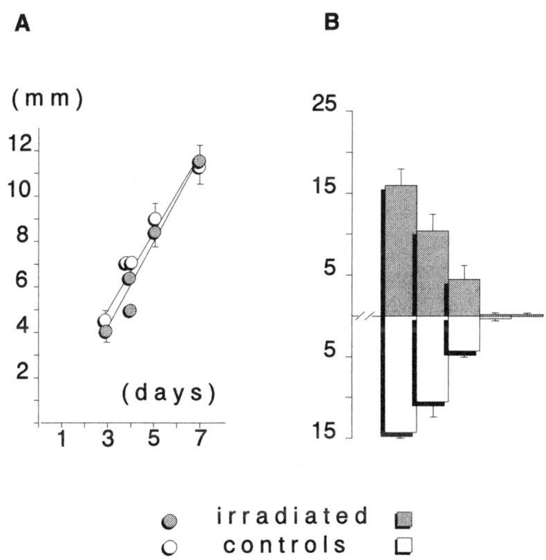

FIGURE 2. Assessment of axonal regeneration. (**A**) Results of the nerve-pinch test. Regression lines were fitted to the two groups of obtained data, using the least square method: irradiated nerves ($r = 0.97$); control nerves ($r = 0.96$). Means and SE ($n = 6$, except at 4 days postoperative) are shown. (**B**) ChAT activity in the consecutive 3-mm nerve pieces distal to the crush site at 5 days postoperative, indicating the presence of motor axons advancing through the acellular distal nerve stumps. Means and SE are shown for each group ($n = 5$) of the corresponding nerve pieces.

results of the quantitative Northern blotting procedure for NGF mRNA are shown in FIGURE 1. A similar decrease in time was found in NGFrec mRNA levels (data not shown).

FIGURE 2A shows the elongation of sensory axons after the described treatment. No significant differences were noted between the pinch test results of the two groups of animals ($p > 0.8$). FIGURE 2B shows the distance reached by motor axons at 5 days postoperative in the distal nerve stumps of the animals of both groups, as evidenced by the presence of ChAT.

CONCLUSIONS

The applied X-irradiation considerably reduced, rather than completely eliminated, the proximal stump reaction to the nerve crush. However, the reduction of levels of NGF and NGFrec mRNAs implicates that the Schwann cell functional status was notably compromised by the irradiation. This, in turn, did not significantly affect either the rate of sensory axon elongation or the motor axon advancement through the acellular distal nerve stump.

REFERENCES

1. SKETELJ, J., M. BRESJANAC & M. POPOVIC. 1989. J. Neurosci. Res. **24:** 153–162.
2. BRESJANAC, M. & J. SKETELJ. 1989. J. Neurosci. Res. **24:** 501–507.
3. FRIEDE, R. & M. A. JOHNSTONE. 1967. Acta Neuropathol. **7:** 218–231.
4. HEUMANN, R., S. KORSCHING, C. BANDTLOW & H. THOENEN. 1987. J. Cell Biol. **104:** 1623–1631.
5. SCARAVILLI, F., S. LOVE & R. MYERS. 1986. J. Neurocytol. **15:** 439–449.
6. HEUMANN, R., D. LINDHOLM, C. BANDTLOW, M. MEYER, M. J. RADEKE, T. P. MISKO, E. SHOOTER & H. THOENEN. 1987. Proc. Natl. Acad. Sci. USA **84:** 8735–8739.

Glial Cells Transplanted to the Rat Optic Tract

Influence on the Regrowth of Retinal Axons

A. R. HARVEY, M. CHEN, AND S. E. DYSON

Department of Anatomy and Human Biology
The University of Western Australia
Nedlands, Perth, Western Australia 6009

Considerable evidence from both *in vivo* and *in vitro* studies now indicates that Schwann cells and immature astrocytes can support the survival of retinal ganglion cells and promote the outgrowth of their axonal processes. This influence on regenerative growth is due to both diffusible and contact-mediated factors.

In the present study, in an attempt to promote the regrowth of retinal axons in the rat optic tract, pieces of nitrocellulose paper (Millicell-HA; pore size 0.45 μm) were coated with purified Schwann cells (FIG. 1A and C) or astrocytes (FIG. 1B). The papers were then placed in the lesioned left optic tract, between the caudal dLGN and rostral superior colliculus (SC), of anesthetized 13–18-day-old Wistar rats (FIG. 1D). Schwann cells were obtained from neonatal dorsal root ganglia or sciatic nerves; astrocytes were derived from the cerebral cortex or SC of neonatal rats. Uncoated and poly-L-lysine (PLL)-coated papers were also implanted as controls. Schwann cell-coated implants ($15–40 \times 10^3$ cells per implant) were placed with the cells lying on either the dorsal or the ventral surface (inverted implants).

Five to seven weeks later, retinal projections were traced by injecting host eyes with wheat germ agglutinin-conjugated horseradish peroxidase. After 24 hours animals were anesthetized and perfused; frozen or vibratome sections (40–50 μm) were processed for tetramethylbenzidine (TMB) histochemistry. Some sections were osmicated and prepared for electron microscopic examination.

Implants were found attached to the caudal dLGN in 45 of the 86 rats studied. Of these 45, sparse axonal growth was seen on the rostral end of 1 uncoated implant, on 2 of the PLL-coated papers, and on 4 of the 10 astrocyte-coated implants. The densest retinal axon regrowth was found on the implants initially coated with Schwann cells. In 15 of the 30 implants that were attached to the dLGN, retinal axons regrew between 50 and 1,120 μm (mean 500 μm). Regrowth of axons on either the dorsal or ventral side of the implants was correlated with the initial placement of the Schwann cells, facing either up or down in the lesion site. Examples of growth on the dorsal surface are shown in FIGURE 1E–G; growth on the ventral surface of inverted implants is shown in FIGURE 1H and I.

An example of the cellular coating that formed on the Schwann cell-coated side of the implant is shown in FIGURE 2A. There was extensive overgrowth by host glia, and the tissue became vascularized. Many of the axons were centrally myelinated, and oligodendrocytes could be found associated with the regrown axons (FIG. 2C). However, a number of the myelin sheaths had a characteristic basal lamina coat (FIG. 2B), showing that at least some transplanted Schwann cells survived in the host brain.

The data suggest that factors, expressed particularly by Schwann cells, can enhance and perhaps direct retinal axon growth in the lesioned rat optic tract. In an attempt to induce further and more sustained regrowth, Schwann cells are currently

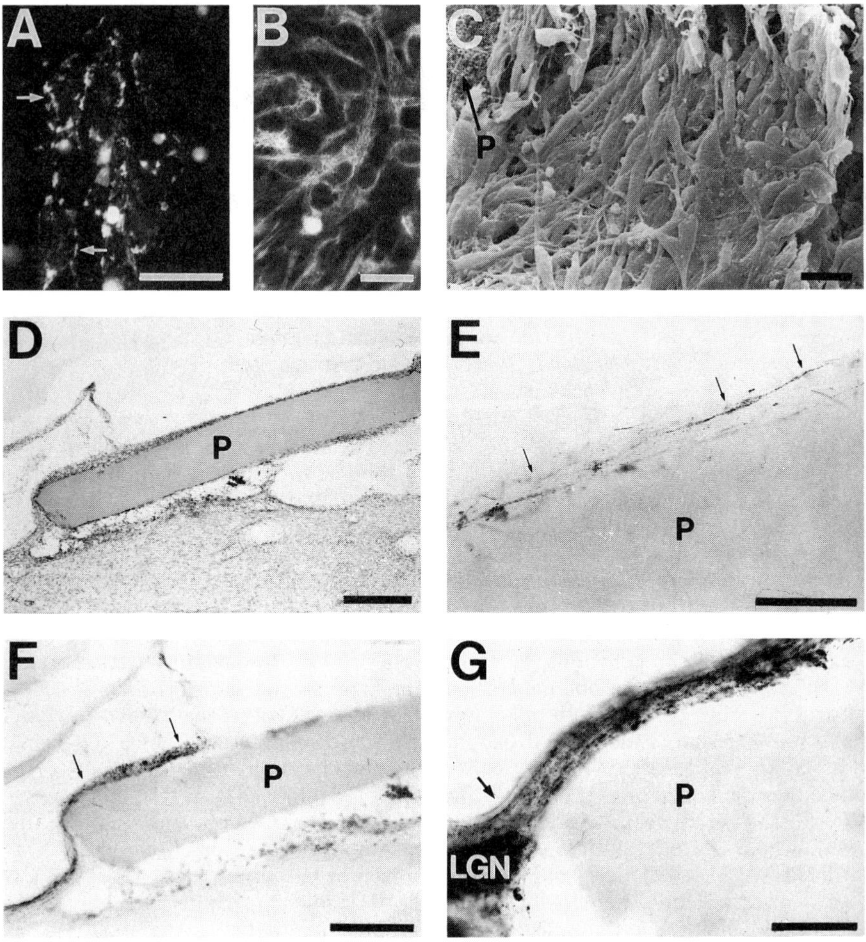

FIGURE 1. **(A)** FITC immunofluorescence showing cultured Schwann cells reactive for nerve growth factor receptor (*arrows*). **(B)** FITC immunofluorescence showing cultured astrocytes reactive for glial fibrillary acidic protein. **(C)** Scanning electron micrograph showing Schwann cells on nitrocellulose paper, prior to implantation into the host optic tract. A piece of the underlying paper (P) can be seen top left (*arrow*). **(D)** Nissl-stained sagittal section (rostral is to the left) of an implant (P) that had been coated on the dorsal surface with Schwann cells. Note the presence of cells on *both* sides of the paper. **(E–I)** TMB-reacted sagittal sections (rostral is to the left), showing HRP-labeled retinal axons (*arrows*) regrowing on the surface of implants

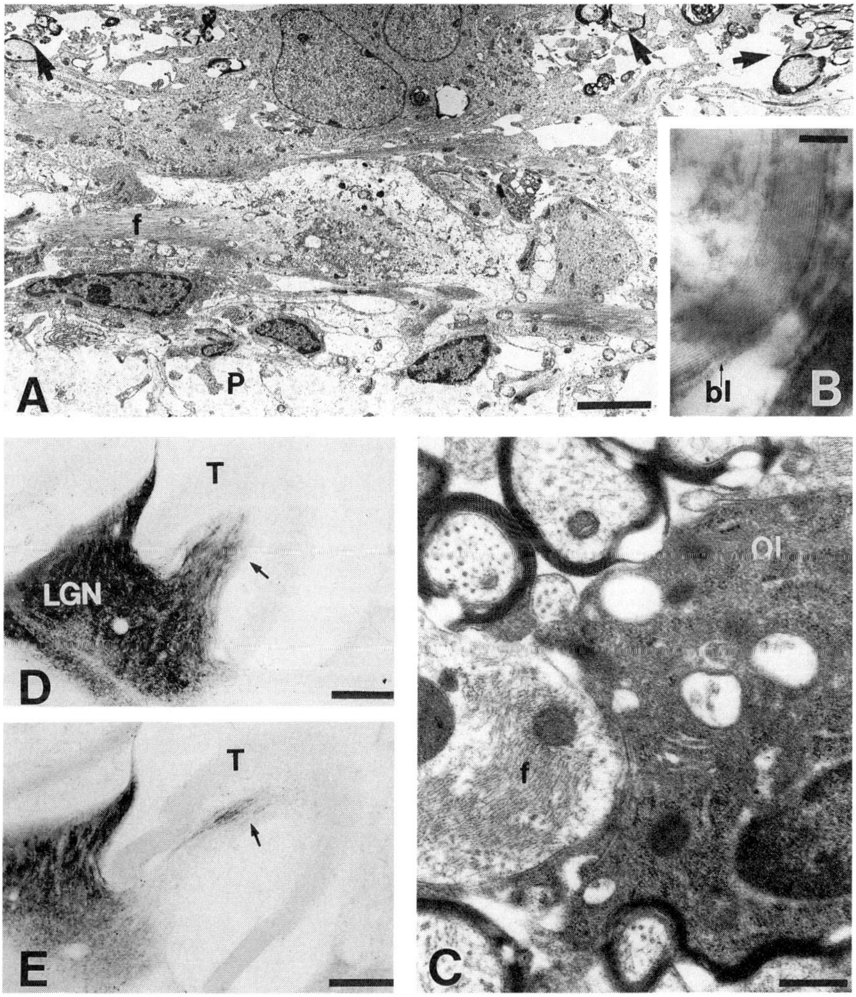

FIGURE 2. **(A)** Example of the tissue that forms on the dorsal surface of papers (P) initially coated with Schwann cells. Note clusters of myelinated axons (*arrows*) and different types of glia. The glial cells next to the paper may be the original implanted Schwann cells or invading microglia. Astrocytic processes containing filamentous bundles (f) are common. **(B)** Schwann-cell myelination occurs on some axons that have characteristic basal lamina coats (bl) and interlamellar distances typical of peripheral myelin. **(C)** Most of the myelination associated with the implanted papers is centrally derived, and oligodendrocytes (Ol) are seen in the tissue above the paper. **(D–E)** A nitrocellulose tube (T) filled with Schwann cells and implanted into the caudal part of the dorsal lateral geniculate nucleus (dLGN). Sagittal section; rostral is to the left. Note large numbers of retinal axons (*arrows*) growing within the lumen of the tube. Scale bars: **A** = 10 μm; **B** = 0.2 μm; **C** = 0.5 μm; **D, E** = 250 μm.

FIGURE 1. (*continued*) (P) that had been coated with Schwann cells. **(E–G)** Regrowth on the *dorsal* surface of implants where the Schwann cells were placed uppermost. **(H–I)** Regrowth on the *ventral* surface of inverted implants where the Schwann cell-bearing surface was placed downwards. LGN = dorsal lateral geniculate nucleus. Scale bars: **A** = 20 μm; **B, C** = 25 μm; **D** = 250 μm; **E, I** = 100 μm; **F, H** = 200 μm; **G** = 50 μm.

being transplanted within nitrocellulose tubes[1] rather than as two-dimensional arrays. An example of retinal axonal regeneration into the rostral end of one such tube is shown in FIGURE 2D and E.

REFERENCE

1. FAWCETT, J. W., E. HOUSDEN, L. S. THOMAS & R. L. MEYER. 1989. The growth of axons in three dimensional astrocyte cultures. Dev. Biol. **135:** 449–458.

Nerve Regeneration *in Vivo*

A Blind-Ended Tube Model for Studying Environmental Effects

ROBERT W. VAN BOVEN,[a] MARK R. GILBERT,[b] AND BEVERLY L. HARDING[b]

Departments of Neurological Surgery[a] and Neurology[b]
The Johns Hopkins University School of Medicine
Baltimore, Maryland 21205

Unlike the peripheral nervous system, neurons in the central nervous system (CNS) are unable to regenerate unless suitable conditions, such as a peripheral nerve graft, are provided.[1] One of the leading hypotheses is that CNS neurons are able to regenerate, but regrowth is inhibited by contact with glial cells.[2] We studied the effects of astrocytes on the regeneration of peripheral nerves in a model developed to selectively control the local cell population. A semipermeable, blind-ended tube (Colla-tec, Inc., Plainsboro, NJ) is used to transplant cultured cells to the site of nerve injury and regeneration. This model allows us to study the effects of specific cell interaction with elongating axons. Cellular influence of the distal nerve stump is eliminated. Furthermore, the permeability characteristics of the tube allow for sustained cell viability and ingrowth which does not occur in nonpermeable, blind-ended tubes.[3]

Cortical astrocyte cultures were prepared from newborn Sprague-Dawley rats exposed to dibutyryl-cAMP and suspended in a 1:1 mixture of cell culture medium and Matrigel. These cells (corresponding to 15–20 days postnatal age) were introduced into a 6 mm long, semipermeable, collagen tube (50 kD molecular weight cut-off).

Normal adult male Sprague-Dawley rats were anesthetized, their tibial nerves were transected 1 mm distal to the tibial-peroneal bifurcation, and the distal nerve branch was resected to the point of muscle penetration. The proximal stumps were secured 1 mm into tubes that contained either astrocytes in medium and Matrigel (experimental group) or cell-free medium and Matrigel (control group).

Two weeks later, thin sections were prepared for light microscopic (stained with toluidine blue) and transmission electron microscopic analysis of the nerve specimens. In the control group, myelinated and unmyelinated axons extended fully to the distal end of the tube. Numerous terminal endings were swollen with the accumulation of cytoskeletal elements. In contrast, the experimental group showed axons with apparent blocked regeneration at 2–3 mm into the tube. No signs of axonal engorgement were evident.

These findings suggest that this semipermeable tube model allows for regeneration in the absence of a distal nerve stump. Moreover, this *in vivo* model may be useful for studying the effects of glial-neuronal interaction on nerve regeneration.

REFERENCES

1. DAVID, S. & A. J. AGUAYO. 1981. Science **214:** 931–933.
2. LUIZZI, F. J. & R. J. LASEK. 1987. Science **237:** 642–645.
3. SHINE, H. D., P. G. HARCOURT & R. L. SIDMAN. 1985. J. Neurosci. Res. **14:** 393–401.

Induction of Blood-Brain Barrier Characteristics in Bovine Brain Endothelial Cells by Rat Astroglial Cells in Transfilter Coculture[a]

JOCHEN NEUHAUS,[b] WERNER RISAU,[c]
AND HARTWIG WOLBURG[b]

[b]*Institute of Pathology*
University of Tübingen
Liebermeisterstraße 8
D-7400 Tübingen, Germany

[c]*Max-Planck-Institute for Psychiatry*
Department of Neurochemistry
Am Klopferspitz 18A
D-8033 Martinsried, Germany

Astrocytes are believed to induce blood-brain barrier characteristics in endothelial cells of brain capillaries. The mechanism by which this induction works is unknown. *In vitro* studies coculturing astroglial and endothelial cells show that endothelial tight junctions (TJ) increase in the presence of astroglial cells.[1] Arthur *et al.*[2] showed that an astroglia-conditioned medium is sufficient to induce endothelial TJ formation. Dehouck *et al.*[3] observed in a transfilter coculture system the maintenance of endothelial markers and of blood-brain barrier characteristics. We used a similar *in vitro* model of the blood-brain barrier in which endothelial and astroglial cells are cultured separately on a filter. In contrast to Méresse *et al.*[4] we show a clear dependence of the complexity of endothelial TJ formation on the presence of astroglial cells.

Bovine brain endothelial cells (BBE) were cocultured with rat astroglial cells (RA). The cells were seeded separately on both faces of Cellagen™ membranes, avoiding direct contact between BBE and RA. Controls were BBE cultured without RA. Permeability of the BBE monolayer was measured and correlated with the complexity of TJ as demonstrated in freeze-fracture replicas (FIGS. 1 and 2). Replicas of BBE membranes were analyzed quantitatively by computer-supported morphometry to describe gradual changes in TJ complexity. The complexity index (CI) was determined by counting both the ramification points (r) of the network of TJ strands and the meshes (m) formed by TJ strands, and by measuring the total length (l) of all TJ strands. CI was calculated as $CI = 100\ r/l \cdot 100\ m/l$. In the presence of RA an increase in CI correlated with a reduced penetration of [^{14}C]-sucrose. Additionally, RA showed an increase in the density of orthogonal arrays of particles (OAP) when cocultured with BBE. Because the cellagen filter prevented any contact between endothelial and astroglial cells, the increase in TJ complexity in BBE as well as the increased OAP formation in RA did not require cell-cell contacts between RA and BBE. Therefore, soluble factors released by the cells seem to play a

[a]This work was supported by the Deutsche Forschungsgemeinschaft.

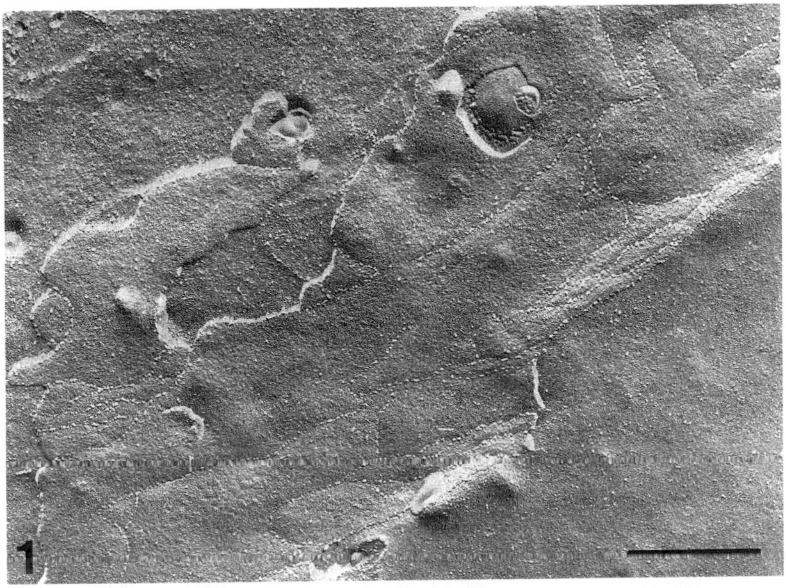

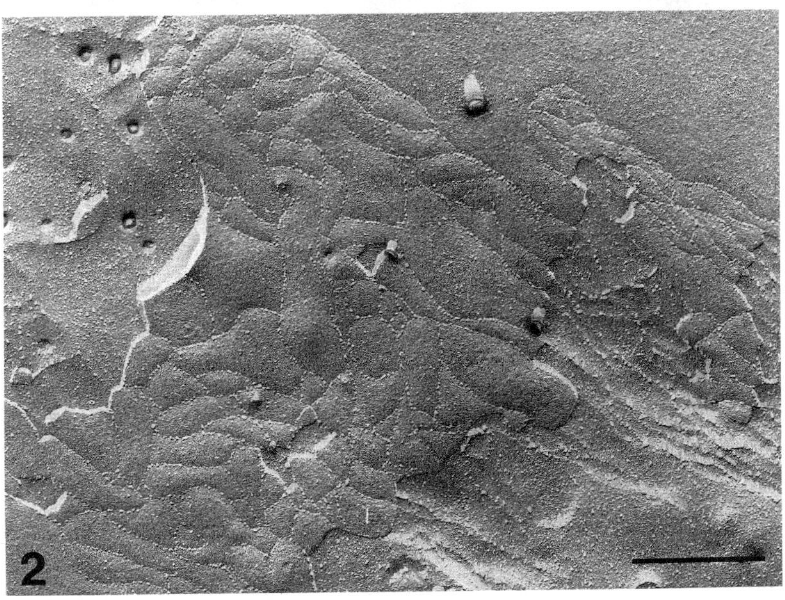

FIGURE 1. Freeze-fracture replica of cultured bovine brain endothelial cells. The tight junction complexity is low. Bar = 0.5 μm.

FIGURE 2. Freeze-fracture replica of bovine brain endothelial cells cocultured with rat astrocytes. Note the very complex tight junctions. Bar = 0.5 μm.

major role in the induction and maintenance of blood-brain barrier characteristics in brain endothelial cells and astrocytes.

REFERENCES

1. TAO-CHENG, J.-H., Z. NAGY & M. W. BRIGHTMAN. 1987. Tight junctions of brain endothelium *in vitro* are enhanced by astroglia. J. Neurosci. **7:** 3293–3299.
2. ARTHUR, F. E., R. R. SHIVERS & P. D. BOWMAN. 1987. Astrocyte-mediated induction of tight junctions in brain capillary endothelium: An efficient *in vitro* model. Dev. Brain Res. **36:** 155–159.
3. DEHOUCK, M.-P., S. MÉRESSE, P. DELORME, J.-C. FRUCHART & R. CECCHELLI. 1990. An easier reproducible and mass-production method to study the blood-brain barrier *in vitro*. J. Neurochem. **54:** 1798–1801.
4. MÉRESSE, S., M.-P. DEHOUCK, P. DELORME, M. BENSAID, J.-P. TAUBER, C. DELBART, J.-C. FRUCHART & R. CECCHELLI. 1989. Bovine brain endothelial cells express tight junctions and monoamine oxidase activity in long-term culture. J. Neurochem. **53:** 1363–1371.

Molecular Mechanisms Leading to Lesion-Induced Increases in Nerve Growth Factor Synthesis[a]

ROLF HEUMANN,[b,c] BASTIAN HENGERER,[b]
MICHAEL BROWN,[d] AND HUGH PERRY[d]

[b]*Max-Planck-Institute for Psychiatry*
8033 Martinsried, Germany

[d]*Department of Experimental Psychology*
South Parks Road
Oxford, UK

The rate of nerve growth factor (NGF) synthesis is very low in the nonneuronal cells surrounding the axons of the intact rat sciatic nerve. After nerve lesion, a biphasic increase occurs in levels of NGF mRNA in the distal nerve segment and in the tip of the proximal stump.[1] The initial increase in NGF mRNA (first phase) is preceded by an almost immediate elevation in the levels of the mRNAs coding for the nuclear protooncogenes c-fos and c-jun. The *trans*-activating effect of FOS on the NGF promoter was investigated by transfecting fibroblasts with an expression vector that carried a heavy metal inducible fos gene. On induction of exogenous FOS, the endogenous c-jun mRNA levels increased almost concomitantly, followed by a delayed increase in NGF mRNA levels. This change in NGF mRNA resulted from stimulation of NGF promoter activity as shown by cotransfection of an NGF promoter/CAT reporter construct with the fos expression vector. DNAse I footprint analysis and promoter activity studies, using specifically mutated constructs, show that an AP-1 consensus sequence located in the first intron of the NGF gene is essential for the FOS-mediated induction of the NGF promoter. These results identify the gene coding for the neurotrophic factor NGF as a novel target for FOS.

In contrast to the transient initial increase of NGF mRNA, the second persistent phase of increase in NGF synthesis is mediated, at least *in vitro,* by macrophages that synthesize interleukin-1. *In vivo* this persistent elevation of NGF synthesis is preceded by a similar increase in c-jun mRNA. However, assays of the transcriptional activities of various NGF promoter/CAT reporter constructs indicate that different mechanisms regulate the two phases of increase in NGF mRNA levels after lesion.

The role of macrophages was substantiated by experiments performed in mutant mice (strain C57/Ola). In this strain the lesion-induced invasion of macrophages and the onset of Wallerian degeneration are delayed.[2] Consequently, the second phase of increase in NGF mRNA is markedly attenuated. Moreover, in the absence of macrophages *in vivo,* the lesion-induced NGF receptor reexpression by Schwann cells is suppressed. These results suggest that macrophages not only are responsible

[a]Paper presented at conference but not submitted for publication. Abstract included at editor's request.

[c]Present Address: Molecular Neurobiochemistry, Ruhr University, 4630 Bochum, Germany.

for the degradation of severed axons and of myelin debris but also direct the synthesis of NGF and its receptor *in vivo.* The importance of these results in understanding the mechanisms of degeneration and regeneration of sensory neurons has been shown recently.[3]

REFERENCES

1. HEUMANN, R., S. KORSCHING, C. BANDTLOW & H. THOENEN. 1987. J. Cell Biol. **104:** 1623–1631.
2. LUNN *et al.* 1989. Eur. J. Neurosci **1:** 27–33.
3. BROWN, M. C., V. H. PERRY, E. R. LUNN, S. GORDON & R. HEUMANN. 1991. Neuron **6:** 359–370.

Anion Conductance Blocked by Divalent Cations in Cultured Rat Astrocytes[a]

T. JALONEN,[b,c] V. VARGA,[b] K. HARTIKAINEN,[b]
R. JANÁKY,[b] AND S. S. OJA[b]

[b]*Department of Biomedical Sciences*
University of Tampere
SF-33101 Tampere, Finland

[c]*The Nobel Institute for Neurophysiology*
Karolinska Institutet
S-10401 Stockholm, Sweden

Astroglial cells possess a variety of ion currents similar to those found in neurons, such as transmitter amino acid-sensitive ion currents[1,2] and potential-dependent potassium, sodium, and calcium currents.[3-5] Among others, voltage-gated chloride conductance has been demonstrated.[6] More recently, single-channel recordings have revealed a high-conductance anion channel with a potential-dependent channel opening probability and an increased occurrence following application of hypotonic solutions,[7] suggesting possible involvement of the anion currents in volume regulation. However, the properties and modulation of these anion channels are still largely unknown. Here we present indirect biochemical and direct electrophysiological evidence of their modulation by the divalent cations calcium, zinc, and cadmium.

Primary cultures of rat cerebral astrocytes were prepared as previously described.[3] About 90% of the cultured cells were positively stained for glial fibrillary acidic protein (GFAP). Single-channel recordings[8] were performed with the EPC-7 amplifier (List-Medical, FRG) from cells cultured 3–14 weeks. Data was stored on a Racal Store 4DS recorder (Racal, UK) and analyzed by hand or by pCLAMP software (Axon Instruments Inc., USA). The electrodes had a resistance of 3–10 MΩ. All measurements were made at room temperature (22°C). The astrocyte membrane preparation and L-glutamate binding assays were performed as described[9]; astrocytes from 4-week-old primary cultures were collected, and after hypoosmotic treatment the membrane fraction was sedimented by centrifugation (50,000 *g*, 30 minutes) and kept frozen at −20°C. For the binding experiments the samples were thawed and washed four times with an excess of 50 mM Tris-acetate buffer, pH 7.2.

The blocking effect of divalent cations was demonstrated by adding 0.5–20 mM Zn^{2+} and Cd^{2+} into the bath. The channel activity of the high-conductance anion channel disappeared at positive membrane potentials but continued as flickering channel openings at negative potentials (FIG. 1). In crude astrocyte membrane preparations, glutamate binding and sequestration were enhanced by $CaCl_2$ in a concentration-dependent manner, activation being near 300% at a 5 mM $CaCl_2$ concentration. This glutamate binding was inhibited by 0.5 mM Zn^{2+} and Cd^{2+} (FIG. 2). Inhibition by cadmium exceeded that of zinc. The inhibitory effect of both cations was more pronounced at higher $CaCl_2$ concentrations.

[a]Financial support was given by the Emil Aaltonen Foundation and Swedish Medical Research Council.

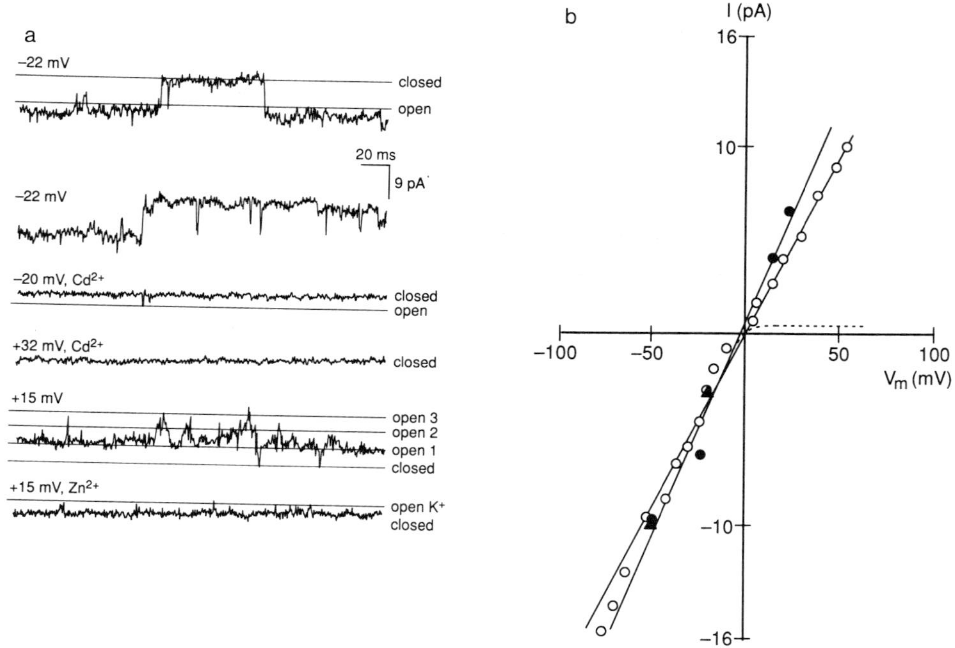

FIGURE 1. Single-channel activity and current-potential relation of the anion channel. **(a)** Samples of channel activity before and after the addition of 4 mM Cd^{2+} and 1 mM Zn^{2+} into the bath (the cytoplasmic side of the excised membrane patch). Both cadmium and zinc totally abolished the anion current at positive membrane potentials. Potassium channels of a smaller conductance were occasionally seen. **(b)** The current-potential relationship drawn according to the largest amplitude of the channel openings at each potential. Measurements were made from inside-out membrane patches with symmetrical extracellular solutions (146.6 NaCl, 5 KCl, 1.2 $MgCl_2$, 1 $CaCl_2$, 0.44 KH_2PO_4, 20-*N*-tris(hydroxymethyl)-methyl-2-amino-ethanesulfonic acid (TES), pH 7.4) (○), and intracellular solution (3 NaCl, 152.6 KCl, 1.2 $MgCl_2$, 0.44 KH_2PO_4, 20 TES, 1 [ethylene-*bis*(oxyethylenenitrilo)]-tetraacetic acid (EGTA), pH 7.2) in the bath and extracellular solution in the pipette (●). When 0.6 mM Zn^{2+} was added to the bath, the single-channel currents were still detected at negative (▲) but not at positive membrane potentials. The single-channel conductances range between 200 and 300 pS. The reversal potentials are as expected for anions.

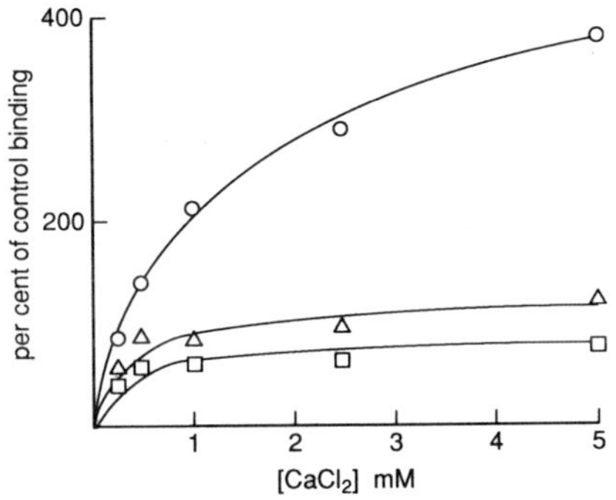

FIGURE 2. Inhibition of $CaCl_2$-activated L-[^{3}H]glutamate binding to astrocyte membrane fraction by $ZnCl_2$ and $CdCl_2$. Binding was determined in 50 mM Tris-acetate, pH 7.2, in the presence of increasing concentrations of $CaCl_2$ (○), $CaCl_2$ +0.5 mM $ZnCl_2$ (△), and $CaCl_2$ + 0.5 mM $CdCl_2$ (□). The incubation was carried out at 37°C for 30 minutes with 20 nM of [^{3}H]glutamate and 100–120 μg protein and terminated by centrifugation. Results are means of two experiments with triplicate samples.

Glutamate is known to cause cell swelling, depolarization, and calcium influx.[10–12] Swelling has been shown to cause cell depolarization[13] and depolarization to cause both Ca^{2+} and Cl^- channels to open.[5,14] Decreased Ca^{2+} concentrations can increase the astrocyte cell volume.[15] Our findings suggest a regulatory effect of divalent cations on high-conductance anion channel activity and a possible interaction with glutamate in astrocyte cell volume regulation. The exact mechanism of this interaction needs further study.

ACKNOWLEDGMENTS

We thank Mrs. Oili Pääkkönen and Mr. Ari-Pekka Viirtelä for skillful technical assistance.

REFERENCES

1. Kettenmann, H., K. H. Backus & M. Schachner. 1987. Brain Res. **404:** 1–9.
2. Sontheimer, H., H. Kettenmann, K. H. Backus & M. Schachner. 1988. Glia **1:** 328–336.
3. Jalonen, T. & I. Holopainen. 1989. Brain Res. **484:** 177–183.
4. Barres, B. A., L. L. Y. Chun & D. P. Corey. 1989. Neuron **2:** 1375–1388.
5. MacVicar, B. A. 1984. Science **226:** 1345–1347.
6. Gray, P. T. A. & J. M. Ritchie. 1986. Proc. Roy. Soc. Lond. B **228:** 267–288.
7. Jalonen, T., S. Johansson, I. Holopainen, S. S. Oja & P. Århem. 1989. Acta Physiol. Scand. **136:** 611–612.
8. Hamill, O. P., A. Marty, E. Neher, B. Sakmann & F. J. Sigworth. 1981. Pflügers Arch. **391:** 85–100.
9. Bridges, R. J., M. Nieto-Sampedro, M. Kadri & C. W. Cotman. 1987. J. Neurochem. **48:** 1709–1715.
10. Chan, P. H., L. Chu & S. Chen. 1990. J. Neurosci. Res. **25:** 87–93.
11. Bowman, C. L. & H. K. Kimelberg. 1984. Nature **311:** 656–659.
12. Pearce, B., J. Albrecht, C. Morrow & S. Murphy. 1986. Neurosci. Lett. **72:** 335–340.
13. Kimelberg, H. K. & E. O'Connor. 1988. Glia **1:** 219–224.
14. Bevan, S., S. Y. Chiu, P. T. A. Gray & J. M. Ritchie. 1985. Proc. Roy. Soc. Lond. B **225:** 299–313.
15. Olson, J. E., D. Fleischhacker, W. B. Murray & D. Holtzman. 1990. Glia **3:** 405–412.

Ultrastructure and Voltage-Dependent Sodium Currents at the Glia Limitans of the Frog Optic Nerve before and after the Axons Degenerate[a]

PAULA M. ORKAND,[b,c] ROSA BLANCO,[b,c]
HECTOR MARRERO,[b,d] AND RICHARD K. ORKAND[b,d,e]

[b]*Institute of Neurobiology*
San Juan, Puerto Rico 00901

Departments of Anatomy[c] *and Physiology*[d]
University of Puerto Rico Medical Sciences Campus
San Juan, Puerto Rico 00936

The surface of the frog optic nerve consists of astrocytic processes separated by narrow intercellular clefts. Beneath this glia limitans lie bundles of unmyelinated axons as well as a few myelinated axons and their associated oligodendroglia (FIG. 1A). With the loose patch-clamp technique,[1] voltage-dependent inward currents are recorded from the astrocyte surface (FIG. 2A).[2,3] The inward currents are blocked (>90%) by 1 μM tetrodotoxin and disappear when external Na is replaced by tetramethylammonium (TMA).

Sixteen weeks after removal of the retina all the axons have disappeared, leaving a "glial" nerve with a diameter 20–40% of normal. The glia limitans becomes highly folded, and the glial cells contain lipid droplets in addition to their normal inclusions (FIG. 1B). Voltage-dependent inward currents, although reduced in amplitude, can still be recorded from the glial surface (FIG. 2B). The time to peak current is longer than that under normal conditions, and they are little affected by the addition of 1 μM tetrodotoxin. However, when the Na is removed and replaced by TMA, the inward currents disappear.

REFERENCES

1. STUHMER, W., W. M. ROBERTS & W. ALMERS. 1983. The loose patch clamp. *In* Single-Channel Recording. B. Sakmann & E. Neher, eds.: 123–132. Plenum Press, New York.

[a]This work was supported by NIH grants NS-07464, NS-08513, and NSF RII8705802.
[e]Address for correspondence: Institute of Neurobiology, 201 Blvd. del Valle, San Juan, Puerto Rico 00901.

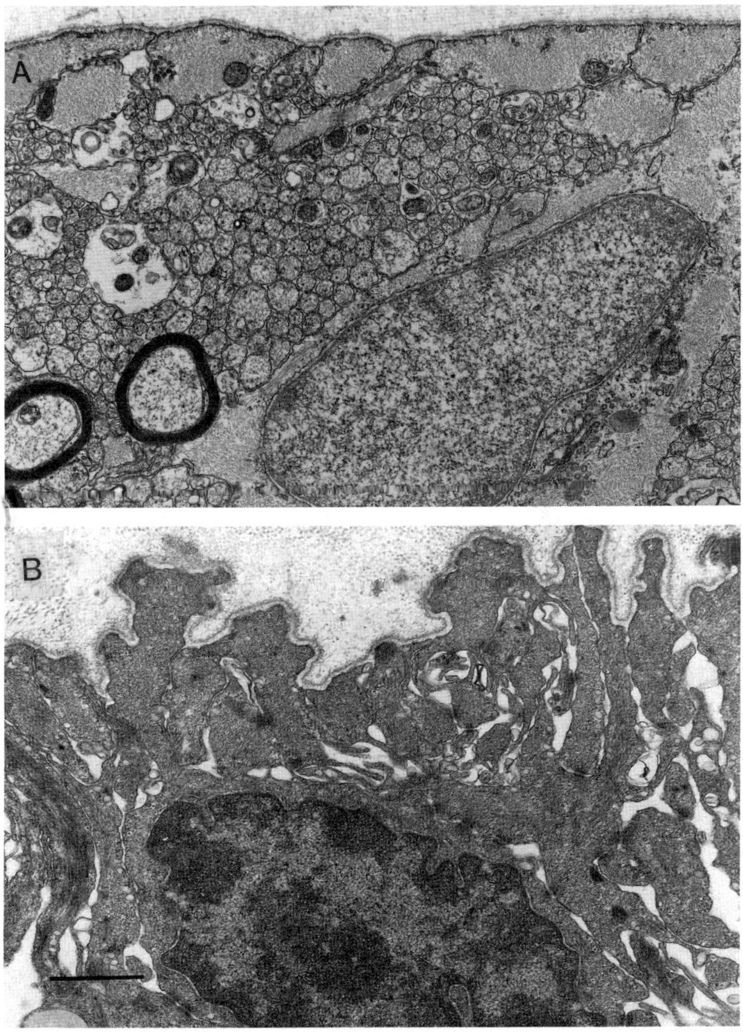

FIGURE 1. Electron micrographs of cross-sections of frog optic nerve. **(A)** Normal nerve. This electron micrograph shows part of the surface. The nerve consists of unmyelinated and myelinated axons (30:1) and sheets of filament-laden astrocyte cytoplasm that come together at the surface to form the glia limitans, leaving patent clefts between their processes. **(B)** Frog optic nerve 16 weeks after removal of the retina. The diameter of the nerves is now reduced 50–80%. They are made up of glial cells and their myriad processes separated by areas of extracellular space greater than that in normal nerves, but firmly attached to neighboring processes by numerous desmosomes. The surface of the nerve is made up of a highly folded astrocytic glia limitans. There is little debris of axonal or myelin degeneration except for occasional lipid droplets. Axons, unmyelinated and myelinated, have completely disappeared. Bar = 1 μm.

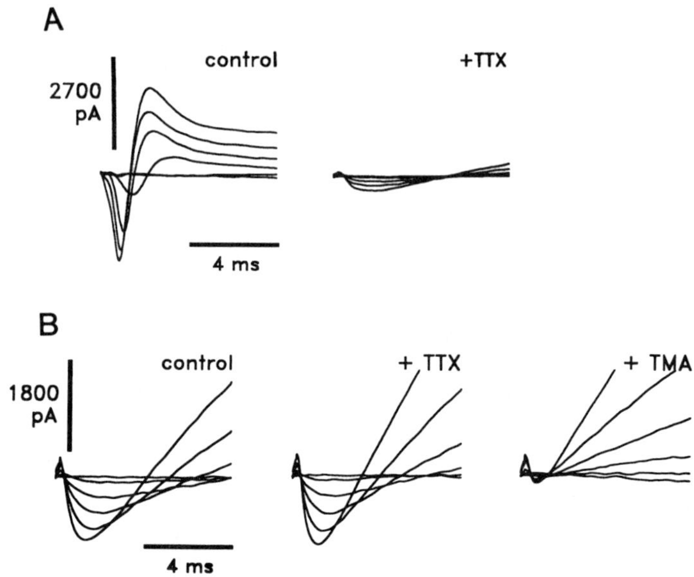

FIGURE 2. Current responses recorded with the loose-patch clamp technique from the surface of the frog optic nerve in response to a series of voltage steps in a normal nerve (**A**) and 16 weeks after removal of the retina (**B**). Recording electrodes (10–20 μm, 0.1–0.3 Mohm) were filled with Ringer's solution and pressed on the surface of the nerve until a seal formed with a resistance equal to that of the pipette. Symmetrical voltage steps were applied, and the resulting current flow was measured and corrected for leakage and capacitative artifacts (Stuhmer Electronik). Each trace is an average of three repetitions. (**A**) Responses from normal nerve to six voltage steps of 25 mV from 25–150 mV before and after the addition of 1 μM TTX. The TTX blocked most of the inward current reversibly. (**B**) Responses from a nerve containing only glial cells 16 weeks after removal of the retina. Six voltage steps of 30 mV from 30–180 mV. The inward currents are unaffected by the addition of 1 μM TTX (*middle*) and disappear after extracellular Na is replaced by TMA (*right side*).

2. Ortiz, S., O. Rodriguez, P. M. Orkand, R. K. Orkand & H. Marrero. 1988. Voltage-gated currents recorded from the surface of the frog optic nerve. PRHSJ **7:** 141–143.
3. Marrero, H., M. L. Astion, J. A. Coles & R. K. Orkand. 1989. Facilitation of voltage-gated ion channels in frog neuorglia by nerve impulses. Nature **339:** 378–380.

Accumulation of Intracellular Bicarbonate Accounts for the Missing Anion during Potassium-Evoked Swelling of Cortical Type-1-Like Astrocytes[a]

WOLFGANG WALZ

Department of Physiology
College of Medicine
University of Saskatchewan
Saskatoon, Saskatchewan S7N 0W0, Canada

During ischemia, hypoxia, trauma, and seizures glial cells show signs of swelling (cytotoxic swelling). This swelling is preceded by elevated external K^+ concentrations

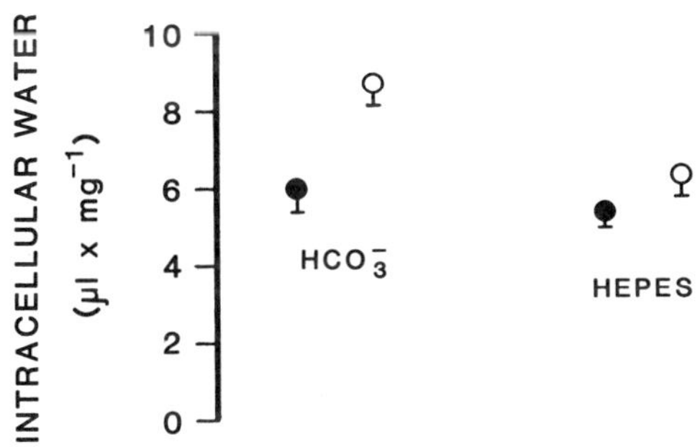

FIGURE 1. Intracellular water content in microliters per milligram of protein 5 minutes after exposure to 3 (*filled circle*) or 60 mM K^+ (*open circles*). Experiments were conducted in normal HCO_3^-/CO_2-buffered saline solution and in nominal HCO_3^--free saline solution with HEPES as buffer ($n = 5 \pm$ SEM). The water space was calculated using a urea washing procedure that is based on the finding that urea penetrates cell membranes but is not metabolized. It was found that in astrocytes urea reaches equilibrium distribution across the cell membrane within 2 minutes, and it is assumed that this means it reaches the same concentration in the cellular water as in the extracellular saline solution. The cells were exposed for 5 minutes to [^{14}C] urea (specific activity 60 kbq/ml) and then washed with sucrose solution (300 mM, 1°C). Each wash lasted about 3–4 seconds, and an efflux curve of the experimental data of urea removed was constructed. Two efflux rate constants were calculated using the peeling technique. The faster one represents the removal of extracellular urea, the slower one the efflux of intracellular urea. That rate constant was used to calculate the urea content per milligram of cell protein before the start of the washing procedure. The average urea space of astrocytes was found to be 6.4 ± 0.4 μl × mg^{-1} of protein.

[a]The author is the recipient of a Scientist Award of the Medical Research Council of Canada. Operating funds were provided by the MRC and the Saskatchewan Heart and Stroke Foundation.

of up to 80 mM. In primary cultures of astrocytes as well as *in vivo,* such swelling can be evoked by exposure to high K^+ alone. In solutions with increased K^+ the Na^+ ion concentration was decreased in a 1:1 molar fashion, leaving ionic strength and osmolarity constant. We use mouse astrocytes in primary cultures, which are identified as cortical type-1-like astrocytes by being GFAP positive and A2B5

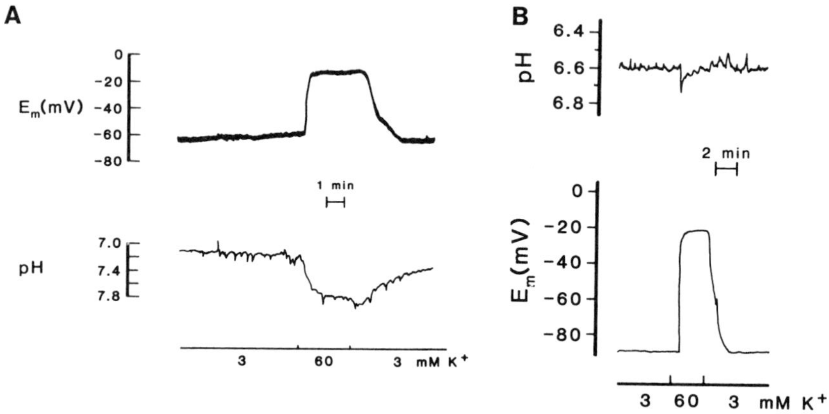

FIGURE 2. Intracellular recordings of the membrane potential and intracellular pH of cultured astrocytes with a two-channel ion-sensitive microelectrode. During the recordings the K^+ concentration of saline solution was increased for 2 minutes from 3 to 60 mM. (**A**) Typical recording of an astrocyte in HCO_3^-/CO_2 buffered saline solution. (**B**) Recording of an astrocyte in nominal HCO_3^--free HEPES buffered saline solution. Figure-eight borosilicate glass capillaries with internal filament (Clark Medical Instruments; 2 × 1.5 mm diameter) were twisted around each other under heat by 180 degrees in a microelectrode puller (Narishige PE-2). The slightly pre-pulled electrodes were allowed to cool and were then pulled to microelectrodes with a relatively short taper. This short taper together with the twisting made the electrodes more stable and less prone to vibrations during impalement. When filled with 0.5 M KCl, each barrel of an electrode had a resistance of 80–120 MΩ, and the tip diameter of both barrels together was about 0.3 μm. The tip of the prospective active barrel was filled with a 3% or 4% solution of tributylchlorosilane (Fluka) in carbon tetrachloride, and any air bubbles were removed with a fine soldering iron. The electrodes were then immediately placed on a hot plate (475°C) for 5 minutes. After cooling, the reference barrel was filled with 1 M lithium acetate containing 10 mM LiCl. The small amount of Fluka proton cocktail IIA was then introduced into the tip of the active barrel, and the rest of the barrel was filled with a backfill solution containing 100 mM NaCl and 100 mM sodium citrate, pH 6.0. The two barrels of an electrode were kept electrically isolated by injecting a short column of mineral oil into the backend of each barrel. The electrodes were then calibrated. They were recalibrated after each experiment. Because of their small tip size, electrodes often had a slope less than the theoretical 58 mV per pH unit, and only electrodes with a slope of 40 mV or more were used for experiments. An Axoprobe 1A amplifier was used, and signals were displayed on a Gould 2200 pen recorder and stored on a VCR recorder.

negative. At first the cells were incubated for 1 hour in normal salt solution. The cells were then exposed to 60 mM K^+ for 2–5 minutes. Radiotracer analysis[1] using ^{42}K, ^{36}Cl, and radioactive urea for volume measurements (FIG. 1) revealed an increase in the KCl and water content, all occurring within the first minute after the K^+ increase. Within this 1-minute period, the cell water content increased by 40% (FIG. 1), and the K^+ concentration calculated from water and ion content increased from 88 to 128 mM by 40 mM, whereas the Cl^- concentration increased by only 23 mM. We assumed that passive KCl fluxes via K^+ and voltage-dependent anion channels[2] are responsible for this increase. However, there is an anion deficit of roughly 20 mM that we could not account for. We therefore measured the intracellular pH of these astrocytes with microelectrodes (FIG. 2) during this K^+-evoked swelling and found an increase in the pH from 7.05 to 7.6. The intracellular HCO_3 concentration was calculated from:

$$[HCO_3^-]_i = 10^{(pH_i - pH_o)} \times [HCO_3^-]_o \qquad (1)$$

and was found to increase from 9 to 31 mM during the swelling. Thus its increase by 22 mM accounts for roughly the estimated anion deficit. Further experiments revealed that HCO_3^- is necessary for significant swelling to occur during the high K^+ exposure. The cells were incubated for 1 hour in a nominal HCO_3^-/CO_2-free HEPES-buffered saline solution. If then K^+ was increased to 60 mM, no significant swelling was observed (FIG. 1) and the pH did not increase (FIG. 2). Thus, HCO_3^- accumulation is an essential part of the K^+-evoked swelling response. Its presence in saline solution is necessary for swelling and the pH increase to occur. The mechanism of the response is not clear yet. Quite conceivably it could involve a carbonic anhydrase-mediated production of HCO_3^- or carrier-mediated transport into the cells via Na^+-HCO_3^- cotransport.[3] HCO_3^- influx through anion channels is possible, but because there is always an outward driving force for HCO_3^- existing at 3 as well as 60 mM K^+, such a passive flux via channels would have to work through a reduction of the outward driving force for HCO_3^- in swollen cells, which in turn would increase the internal HCO_3^- concentration. Similar HCO_3^--dependent astrocytic swelling is known from several different preparations and was assumed to involve no internal HCO_3^- accumulation but NaCl accumulation through stimulation of Na^+-H^+ and Cl^--HCO_3^- exchange. The evidence presented here is not compatible with NaCl accumulation. However, inasmuch as cytotoxic swelling *in vivo* is a more complex issue than is K^+-evoked swelling in cell culture and involves glutamate and acidosis, NaCl accumulation and HCO_3^- accumulation together could well play a role in these processes.

REFERENCES

1. WALZ, W. & S. MUKERJI. 1988. Exp. Neurol. **99:** 17–29.
2. WALZ, W. 1989. Progr. Neurobiol. **33:** 309–333.
3. DEITMER, J. W. & W. R. SCHLUE. 1989. J. Physiol. **411:** 179–194.

Cell Coupling Is Restricted to Subpopulations of Astrocytes Cultured from Rat Hippocampus and Optic Nerve

H. SONTHEIMER, J. E. MINTURN, B. R. RANSOM,
J. A. BLACK, A. H. CORNELL-BELL, AND S. G. WAXMAN

Department of Neurology
Yale University School of Medicine
New Haven, Connecticut 06510

PVA/EPVA Neuroscience Research Center
VA Medical Center
West Haven, Connecticut 06516

Astrocytes are thought to form a network of cells connected to each other by gap junctions.[1] These connections permit cells to share a common intracellular environment, because gap junctions allow the diffusion of small organic and inorganic molecules including second messengers.[2] Intracellular diffusion through gap junctions may facilitate the redistribution of excess extracellular K^+ to provide for K^+ homeostasis,[3] a process termed spatial buffering.[4] Coupling between astrocytes has been demonstrated in cortical slices,[5] in cortical[6] and cerebellar[7] cultures, and in the *in situ* optic nerve.[8] In optic nerve cultures two types of astrocytes, which differ in a number of properties including response to and release of growth factors, antigen profile,[9] and ion-channel pattern,[10] can be distinguished on the basis of binding of monoclonal A2B5 antibody.[11]

Using whole-cell voltage-clamp recordings we compared astrocytes from rat hippocampus and rat optic nerve (RON) in terms of coupling to adjacent cells and Na^+ current expression. Cell coupling was assessed by monitoring the spread of lucifer yellow (LY) into neighboring cells. In both hippocampal astrocytes and RON

FIGURE 1. Coupling of hippocampal astrocytes as revealed by dye injections and evaluation of current transients. Cells were filled with LY, and phase-contrast (**a** and **c**) and fluorescence photomicrographs (**b** and **d**) are shown. Filling and recordings of current transients were achieved using the whole-cell patch-clamp technique. Patch-clamp electrodes (input resistance 3–5 MΩ) were filled with a solution containing (in mM): *N*-methyl-D-glucamine (*N*mDg) 120, TEA-Cl 20, $MgCl_2$ 1, $CaCl_2$ 0.2, EGTA 10, HEPES (sodium salt) 10, and 0.2% Lucifer yellow (dilithium salt, Sigma). pH was adjusted to 7.4 using HCl. (**a, b**) Example of a cell extensively coupled to adjacent cells. Almost all cells in the field were dye coupled to the injected cell (original magnification ×200). (**c, d**) Example of an astrocyte coupled to about five adjacent cells. Cell processes of the injected cells were brightly stained. Neighboring cells showed clear staining of their somata, but there was little staining of cell processes. (**e**) Dye-coupled astrocytes consistently showed current transients that decayed with a double exponential time course (control). The two exponential functions with time constances of 1.1 and 12 msec, respectively, were superimposed on the original current trace. Acute exposure to octanol (10 μM, 5 minutes), suppressed the slower time constant but left the fast time constant virtually unaltered. Currents were larger before application of octanol (control) because of the increase in input resistance in the presence of octanol.

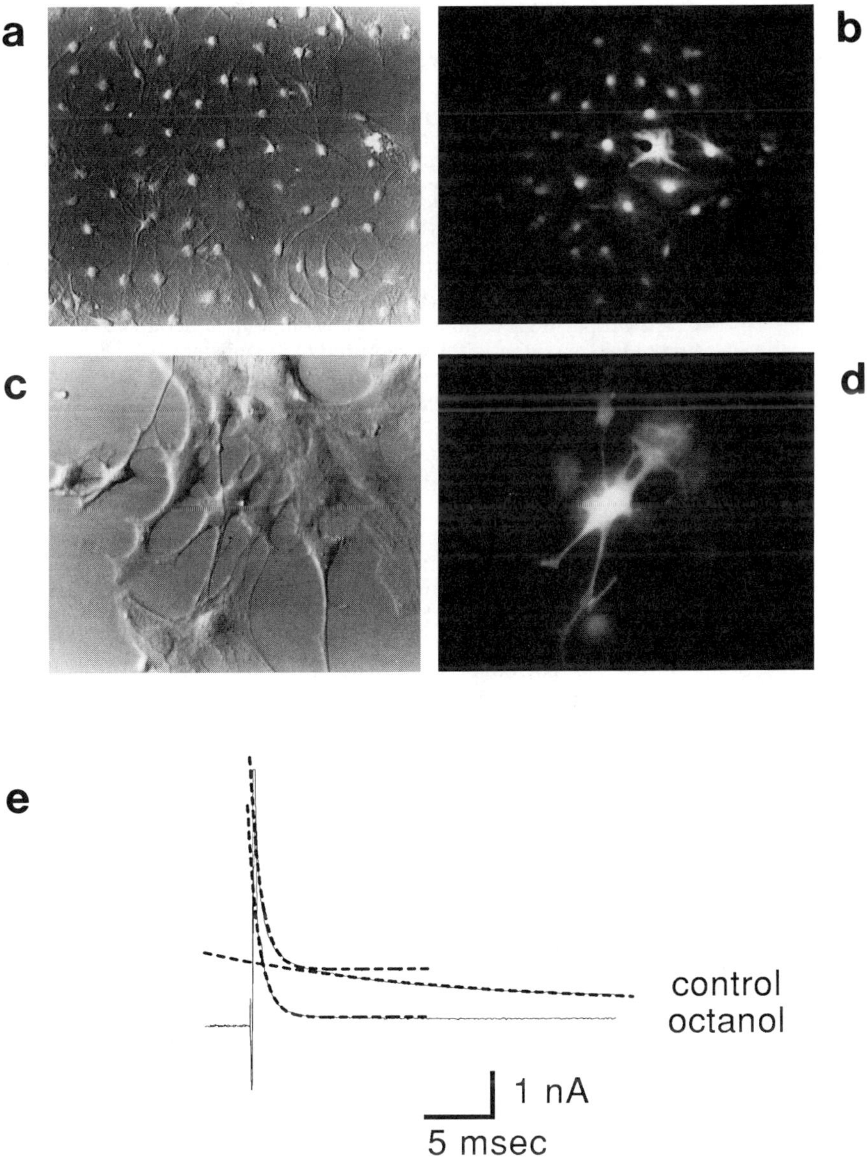

FIGURE 1.

astrocytes, only a small percentage of cells exhibited dye coupling within 1–5 days *in vitro* (DIV) (26% and 24%), respectively, but this increased to 46% in hippocampal astrocytes and 73% in RON astrocytes after 20 DIV. The degree of cell coupling varied greatly between cells studied. In most instances, groups of 3–10 cells were dye coupled (FIG. 1c and d), but more than 50 cells could be coupled (FIG. 1a and b). In hippocampal cultures astrocytes were morphologically heterogeneous and consisted of flat, fibroblast-like cells and stellate, process-bearing cells. Hippocampal astrocytes were all glial fibrillary acid protein positive ($GFAP^+$), but did not stain with A2B5. In optic nerve cultures, flat astrocytes were $A2B5^-$ (FIG. 2a and b), whereas stellate, process-bearing astrocytes were $A2B5^+$. In hippocampal cultures coupling was observed between both types of cell morphologies (FIG. 1c and d). In RON cultures, dye coupling was only observed between $A2B5^-$ astrocytes (FIG. 2a and b), whereas $A2B5^+$ astrocytes were always noncoupled (FIG. 2c, d, and e).

We consistently observed differences in the electrical response between coupled and noncoupled cells. Current transients in response to de- or hyperpolarizing voltage steps decayed with a single exponential time course in noncoupled cells, but in contrast, the sum of two exponentials was required to fit transients of dye-coupled cells (FIG. 1e). The correlation of dye coupling and double exponential current decay provides an indicator to assess astrocyte coupling. In hippocampal astrocytes dye coupling was abolished, and the slower time-constant was suppressed by acute exposure to octanol (10 μM; FIG. 1e), known to inhibit gap-junctional conductance,[12] leaving the fast time-constant virtually unaltered. RON astrocytes could not be acutely uncoupled with octanol (up to 100 μM), and dye coupling was still visible in cells exposed to octanol for up to 60 minutes.

Na^+ currents could be recorded in a subpopulation of astrocytes in both culture systems. In hippocampal astrocytes two types of Na^+ currents were observed that differed in their steady-state inactivation (h_∞) curve. At 1–5 DIV, h_∞ curves showed a midpoint close to −60 mV, whereas h_∞ midpoints were always close to −85 mV at 6–20 DIV. Similarly, two Na^+ current characteristics were observed in cultured RON astrocytes, but cells with both characteristics coexisted in cultures at 1–15 DIV. The more hyperpolarized h_∞ curves (midpoint close to −85 mV) were observed in $A2B5^-$ astrocytes, whereas $A2B5^+$ astrocytes expressed the more depolarized form (−60 mV[13]). Interestingly, Na^+ currents in hippocampal astrocytes were only recorded in noncoupled cells; in RON, Na^+ currents were observed in both coupled and noncoupled astrocytes. In both culture systems, coupled and noncoupled astrocytes

FIGURE 2. Dye coupling of RON astrocytes as revealed by dye injections. Lucifer yellow was injected into two morphologically distinct types of astrocytes with either stellate **(c, e)** or flat, fibroblast-like **(a)** appearance. Dye spread into adjacent cells was observed only in fibroblast-like astrocytes **(a).** In stellate astrocytes, the dye remained exclusively in the injected cell **(c, e).** Cells were identified according to their expression of A2B5 antigen **(b, d).** After immunostaining for A2B5, fluorescent photomicrographs **(b, d)** show the same microscopic field illustrated in **a** and **c.** These revealed that the fibroblast-like cells were $A2B5^-$ astrocytes, whereas the stellate appearing cells were $A2B5^+$ astrocytes. **(a, b)** Two adjacent $A2B5^-$ astrocytes. The injected $A2B5^-$ astrocyte showed dye coupling to an adjacent $A2B5^-$ astrocyte. **(c, d)** Three adjacent $A2B5^+$ astrocytes. The injected $A2B5^+$ astrocyte did not exhibit dye coupling to any other cells. **(e)** Time course of dye coupling in $A2B5^-$ and $A2B5^+$ astrocytes. The percentage of astrocytes of both antigenic phenotypes was plotted as a function of time in culture. Cells from three time groups were pooled: 1–6, 7–12, and 13–20 days *in vitro* (DIV). The percentage of $A2B5^-$ astrocytes exhibiting dye coupling increased with time in culture from 24%, between days 1 and 6, to 73% after 13–20 DIV. In contrast, $A2B5^+$ astrocytes did not exhibit dye coupling at any time in culture up to 20 DIV.

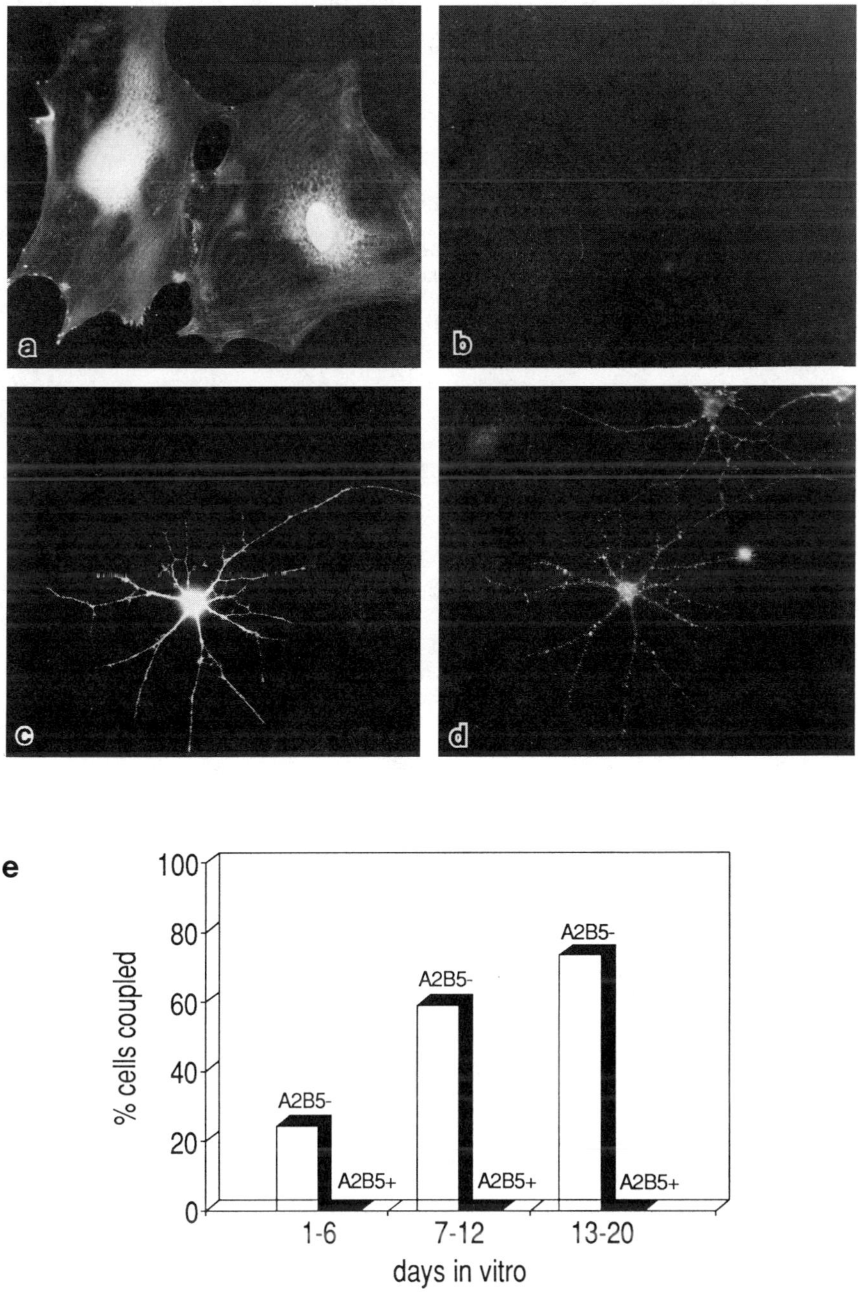
a
b
c
d
e
100
80
60
40
20
0
% cells coupled
A2B5-
A2B5+
A2B5-
A2B5+
A2B5-
A2B5+
1-6
7-12
13-20
days in vitro

FIGURE 2.

coexisted at all ages studied, suggesting that their coupling state may reflect functional specializations. It is likely that coupled cells participate in spatial buffering. Noncoupled cells may serve different roles and may be involved in structural or regenerative functions, such as those after CNS injury.

REFERENCES

1. Mugnaini, E. 1986. Astrocytes. S. Fedoroff & A. Vernadakis, eds.: 329–371. Academic Press. Orlando.
2. Loewenstein, W. R. 1981. Physiol. Rev. **61:** 829–913.
3. Gardner-Medwin, A. R. 1985. J. Physiol. (Lond) **365:** 52P.
4. Orkand, R. K. 1977. Glial cells. Handbook of Physiology (Am. Physiol. Soc.), **1:** 855–875.
5. Gutnick, M. J., B. W. Connors & B. R. Ransom. 1981. Brain Res. **213:** 486–492.
6. Kettenmann, H. & B. R. Ransom. 1988. Glia **1:** 64–73.
7. Fischer, G. & H. Kettenmann. 1985. Exp. Cell Res. **159:** 273–279.
8. Butt, A. M. & B. R. Ransom. 1989. Glia **2:** 470–475.
9. Raff, M. C. 1989. Science **243:** 1450–1455.
10. Barres, B. A., L. L. Y. Chun & D. P. Corey. 1990. Ann. Rev. Neurosci. **13:** 441–474.
11. Raff, M. C., E. R. Abney, J. Cohen, R. Lindsay & M. Noble. 1983. J. Neurosci. **3:** 1289–1300.
12. Spray, D. C., J. C. Saez, D. Brosius, M. V. L. Bennett & E. L. Hertzberg. 1986. Proc. Natl. Acad. Sci. USA **83:** 5494–5497.
13. Barres, B. A., L. L. Y. Chun & D. P. Corey. 1989. Neuron **2:** 1375–1388.

Endothelin Increases Rubidium Uptake through Calcium-Activated Potassium Channels in C6 Glioma Cells

SURACHAI SUPATTAPONE AND
CHRISTOPHER C. ASHLEY

University Laboratory of Physiology
Oxford OX1 3PT, UK

Astrocytes are thought to redistribute excess extracellular potassium from the brain through spatial buffering.[1,2] Unlike neurons, which have large sodium and chloride conductances, astrocytes closely obey the Nernst equation for potassium. In this way, they act like potassium "electrodes" within the brain. Because astrocytes are electrically coupled, as a network they are able to remove potassium from areas of elevated potassium, such as areas near active neurons, and transport the ions transcellularly to areas with lower potassium levels, such as the glia limitans, the junction between brain astrocytes and blood capillary endothelial cells. We[3] previously showed that the hormone endothelin, which is released by endothelial cells near astrocytes at the blood-brain barrier, can increase the cytosolic free calcium level in astrocytes. We now demonstrate that endothelin increases $^{86}Rb^+$ (a radioactive substitute for potassium) uptake in the C6 glioma cell line and that this uptake probably occurs through calcium-activated potassium channels.

C6 cells were grown to a confluent monolayer on 26 × 33 mm polystyrene eight-well multiplates in Dulbecco's modified essential medium with 10% fetal calf serum. Cells were preincubated for 10 minutes at 36°C in (mM): NaCl 150, KCl 5, $CaCl_2$ 2, $MgSO_4$ 0.4, glucose 25, and HEPES-Tris 25, pH 7.4 (flux buffer), followed by incubation in 1 ml of flux buffer with 0.25 mCi/ml $^{86}Rb^+$ and agonists. After incubation between 0 and 30 minutes, cells were quickly rinsed three times with 2 ml of ice-cold phosphate-buffered saline solution. After the last wash, the cells were scraped off in 1 ml of 5% trichloroacetic acid followed by low-speed centrifugation. The radioactivity of the supernatant was determined by Cerenkov radiation. The sediment was dissolved in 1 M NaOH and protein concentration determined by absorbance at 280 nm.

Under these conditions, basal rubidium influx was found to be 30 nmol/min per milligram of protein, and influx under all conditions was linear to 6 minutes and saturated by 30 minutes. The addition of 1 μM ionomycin, a calcium ionophore, increased the influx to 60 nmol/min per milligram of protein, suggesting the presence of a calcium-activated potassium channel. The addition of a maximal dose of 50 nM rat endothelin increased the rubidium influx to 48 nmol/min per milligram of protein. The half-maximal dose of endothelin required to increase rubidium influx was 15 nM. The effects of both ionomycin and endothelin could be blocked by removal of calcium from the flux buffer or by the addition of 5 μg/ml charybdotoxin, a specific inhibitor of the maxi-K type calcium-activated potassium channel. The effect of rat endothelin could be mimicked by porcine endothelin and sarafotoxin S6b, but not sarafotoxin S6c.

In conclusion, these studies show that endothelin activates a charybdotoxin-sensitive, calcium-activated potassium channel in C6 glioma cells. Thus, endothelin may be an important regulator of glial spatial buffering of potassium in the brain. It is most likely that these endothelin-evoked potassium channels are involved in clearing elevated potassium from astrocytes into the bloodstream, rather than clearing elevated potassium at the synapse. The most likely candidate for that role would be a voltage-independent, nonrectifying potassium channel.

REFERENCES

1. ORKLAND, R. K., J. G. NICHOLLS & S. W. KUFFLER. 1966. J. Neurophysiol. **29:** 788–806.
2. GARDNER-MEDWIN, A. R. 1983. J. Physiol. **335:** 353–374.
3. SUPATTAPONE, S., A. W. M. SIMPSON & C. C. ASHLEY. 1989. Biochem. Biophys. Res. Comm. **165:** 1115–1122.

Single Ion Channel Currents from Vesicles in Teased Rat Spinal Roots[a]

S. QUASTHOFF AND P. GRAFE

Department of Physiology
University of Munich
D-8000 München 2, FRG

Myelinated axons in the peripheral nervous system are an anatomical structure with intensive glial-neuronal interactions. One of the scientific tools for the study of such interactions is the electrophysiological recording. Patch-clamp techniques have revealed single ion channel currents in the membrane of Schwann cells[1] as well as axons.[2] However, because of the lack of an adequate method, electrophysiological properties of the *nonsomatic* Schwann cell membrane have not yet been explored. We are reporting on a technique that allows recordings of single channel currents from Schwann cell membrane in internodes of adult rat spinal root axons.

Isolated rat spinal roots were split mechanically into several bundles of myelinated axons. Within about 30 minutes after this procedure small bleb-like vesicles occurred at the edges of the fiber bundles (FIG. 1). Microscopic examination with high resolution optics indicated that the Schwann cell membrane was the most likely origin of the vesicles.

Standard patch-clamp recordings were performed with a high rate of success on the vesicles in the inside-out configuration. Currents were regularly observed from single ion channels with conductances that depended on the ionic composition of the solution used. Pipette and bath solutions were identical. *Solution A* (in mmol/L): K^+ methylsulfate 120, $MgCl_2$ 3, EGTA 1, and HEPES 5; pH was 7.2. Single ion channels with a mean conductance of 209 ± 6 pS ($n = 9$) and a linear current-voltage relationship in the range −50 to +50 mV were observed (FIG. 2). Currents from up to five such channels were recorded from a single patch. In the voltage range of ±25 mV the channel spent most of its time in the open state. Channel openings were hundreds of milliseconds long, separated only by brief closed intervals of a few milliseconds. Beyond this potential range, strong voltage-dependent inactivation was observed. *Solution B* (in mmol/L): NaCl 118, Na^+ gluconate 20, KCl 3, $MgCl_2$ 1, $CaCl_2$ 1.5, EGTA 2, and HEPES 6; pH was 7.2. Mean single channel conductance in this solution was 280 pS ($n = 3$). The other electrophysiological parameters remained unaltered.

The characteristics of the single channel currents observed in the present study resemble the properties of a high conductance anion channel described by Gray *et al.*[3] in Schwann cells of 1–2-day-old rats. The physiological significance of these Cl^-

[a]This work was supported by the Deutsche Forschungsgemeinschaft (SFB 220) and the Wilhelm Sander Stiftung.

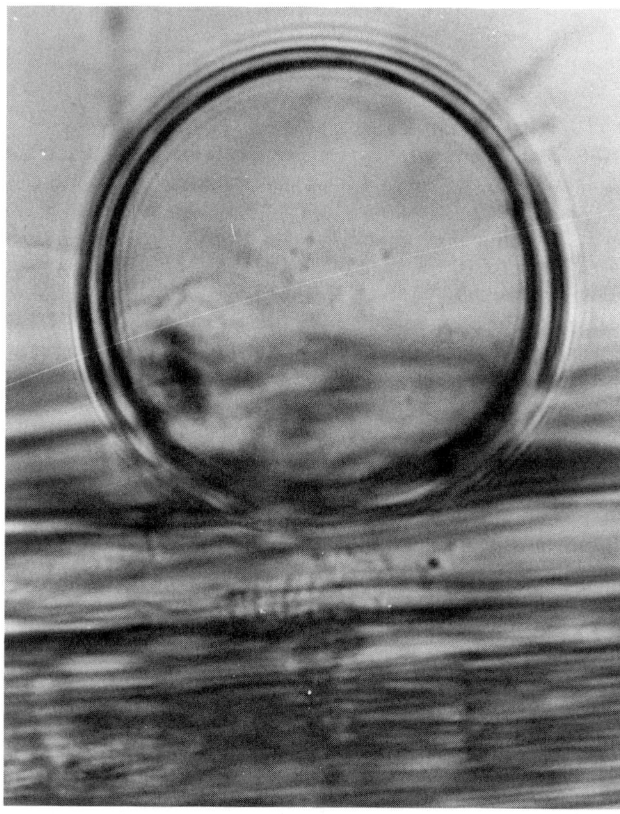

FIGURE 1. Isolated rat spinal root. Appearance of a small bleb-like vesicle at the edge of a fiber bundle with high resolution optics (original magnification ×1,000). For details of the preparation see text.

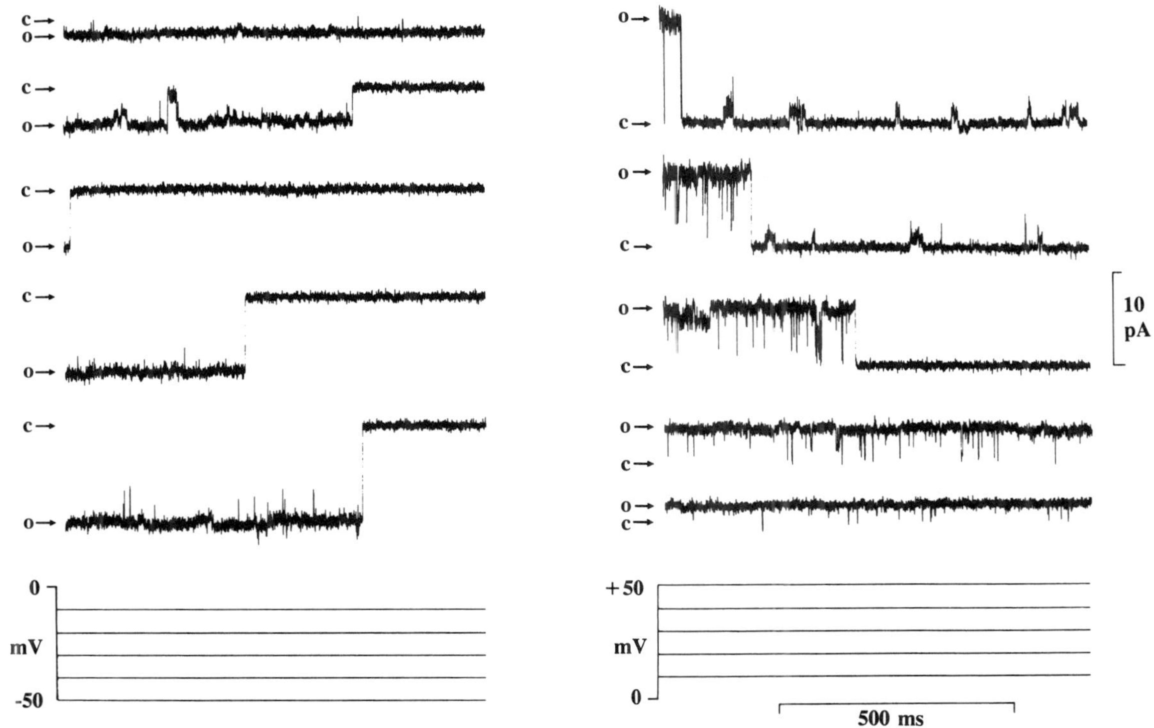

FIGURE 2. Single-channel currents from a Schwann cell vesicle (patch-clamp recording in the inside-out configuration). Pipette and bath solutions were identical (in mmol/L): K^+ methylsulfate 120, $MgCl_2$ 3, EGTA 1, and HEPES 5; pH was 7.2, room temperature. The onset of the current traces is 20 ms after the beginning of a voltage command. (For pulse protocol, see lowermost panel.) Voltage indicates pipette potential.

channels is unclear. However, it is very possible that they contribute to K^+ uptake into Schwann cells by promoting passive KCl influx.[4]

REFERENCES

1. BARRES, B. A., L. L. Y. CHUN & D. P. COREY. 1990. Ann. Rev. Neurosci. **13:** 441–474.
2. JONAS, P., M. E. BRÄU, M. HERMSTEINER & W. VOGEL. 1989. Proc. Natl. Acad. Sci. USA **86:** 7238–7242.
3. GRAY, P. T. A., S. BEVAN & J. M. RITCHIE. 1984. Proc. R. Soc. (Lond.) Ser. B. **221:** 395–409.
4. GRAFE, P. & K. BALLANYI. 1987. Can. J. Physiol. Pharmacol. **65:** 1038–1042.

Properties of Ion Channels in Cultured Adult Human Schwann Cells

JAMES McLARNON[a] AND SEUNG KIM[b]

[a]*Department of Pharmacology and Therapeutics and*
[b]*Division of Neurology*
Faculty of Medicine
University of British Columbia
Vancouver, BC, V6T 1Z3, Canada

Considerable data are available to suggest that glial cells in the central nervous system (CNS) subserve a role in the regulation of extracellular K^+. At present, however, evidence for a similar function for the peripheral counterparts of oligodendrocytes, Schwann cells, is lacking. Although ion channels relatively specific for anions,[1] sodium,[2] and outward potassium[3] currents have been observed in Schwann cells, no detailed report on channels passing inward K^+ current has appeared; such channels would be integral to cellular function as part of a regulatory system for extracellular K^+. In this work we used single-channel patch-clamp recordings to determine properties of ion channels in cultured adult human Schwann cells and which of these properties may be relevant to K^+ regulation.

The methods used to prepare the cultured Schwann cells from adult human trigeminal ganglia have been described.[4] Immunofluorescence staining by rabbit laminin antibody was used to identify the Schwann cells that were studied 12–30 days after isolation.

In most experiments the cell-attached patch-clamp configuration was used with the bath solution containing (in mM): NaCl 140, KCl 5, $CaCl_2$ 0.2, and HEPES 10; pH was 7.3. The pipette solution contained (in mM): KCl 140, NaCl 5, $CaCl_2$ 0.2, and HEPES 10; pH was 7.3. In some experiments the inside-out patch-clamp mode was used to record the unitary currents. In separate experiments the resting membrane potential was measured (-48 ± 6 mV, $n = 7$), and a value of -50 mV is assumed for the resting potential in this work. (Therefore, a driving force of -50 mV is associated with a pipette potential of 0 mV.)

Cell-attached recordings showed the presence of two distinct ion channels in the cultured adult human Schwann cells. At the cell resting potential a channel with a current amplitude near 1 pA was recorded, current amplitude increasing with increased patch hyperpolarization (FIG. 1a, b, and c). The current-voltage relation for the channel is shown in FIGURE 1d; similar data were recorded from six additional patches, and the slope conductance of the channel was 17 pS with zero-current potential near 0 mV. Excision of the patch to an inside-out mode did not significantly alter either the slope of the I/V relation or the zero-current potential. These data are consistent with identification of a channel passing inward K^+ current and inconsistent with significant permeability for other ions. In the cell-attached mode the mean open time was near 20 ms at the cell resting potential and was decreased to 3 ms at -120 mV. This channel may have some function in the regulation of extracellular K^+; for large depolarizations of Schwann cells (for example, when adjacent neurons were very active) the driving force for inward K^+ current would be small; therefore, substantive inward K^+ flow would require long mean open times for the channel. In this regard an inward rectifier K^+ channel with a similar, but stronger, voltage dependence of mean open time has been documented in adult human oligodendrocytes.[5]

In some patches ($n = 5$) depolarizing and hyperpolarizing voltage steps from resting potential caused activation of a large conductance channel (FIG. 2a and b). The current-voltage relation is shown in FIGURE 2c, and the conductance of the channel was 350 pS with current reversal near the cell resting potential. Excision of the patch to an inside-out mode had no obvious effect on channel kinetics, conductance, or zero-current potential. These data are thus consistent with activation of an anion channel in the Schwann cell membrane, because for the on-cell patches the Cl^- is in approximate equilibrium across the patch, and for the excised patches the Cl^- concentrations are equal on each side of the patch. For both on-cell and excised patch recordings the anion channel exhibited rapid activation (channel opening occurring during the onset of the step) and inactivation within 500 ms after the onset of the step.

The presence, in cultured adult human Schwann cells, of ion channels that allow movement of inward K^+ and Cl^- suggest that these cells could be involved in the

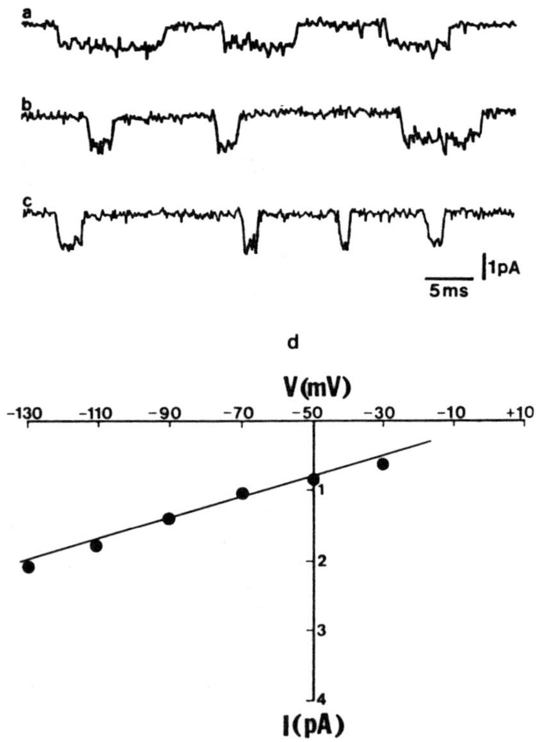

FIGURE 1. Unitary currents (downward openings from the baseline) in cultured adult human Schwann cells. The potentials were: **(a)** −50 mV, **(b)** −70 mV, and **(c)** −90 mV; the low-pass filter was set at 2 kHz. **(d)** Current-voltage relation for the patch recordings shown above.

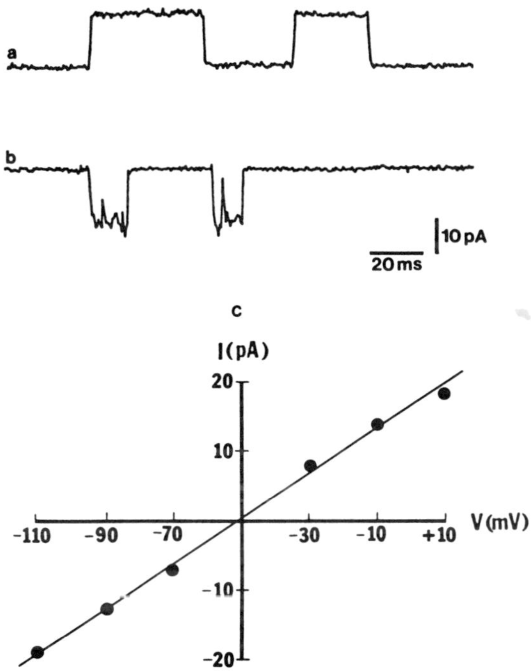

FIGURE 2. Unitary currents through a large conductance channel in human Schwann cells. The patch potential in **a** was −10 mV and in **b,** −90 mV; in both cases the cell resting potential was assumed to be −50 mV. Channel activation occurred during the applied depolarizing or hyperpolarizing step. The records show two openings for each potential, and after the second closure no further openings were recorded despite maintained step commands; in most cases the channels evinced only a single opening prior to inactivation. **c,** Current-voltage relation for the large conductance channel.

regulation of extracellular K^+. As pointed out previously,[6–8] the cellular entry of K^+ accompanied by Cl^- would presumably allow for storage of K^+ near entry sites and thus be readily available for release. Such an arrangement has some utility compared with spatial buffering[9] of K^+ in which, in the absence of Cl^- entry, inward K^+ would be shuttled to more distal points. In some instances, however, as for Müller cells, spatial buffering would serve as a useful homeostatic process.[10] Further studies on Schwann cells will be useful in the elaboration of the single-channel properties of the anion and cation channels to better understand the roles of the channels in supporting the regulation of extracellular K^+ and perhaps also in involving myelination of peripheral nerve.

REFERENCES

1. Gray, P. T. A., S. Bevan & J. M. Ritchie. 1984. Proc. R. Soc. Lond. B **221:** 395–409.
2. Chiu, S. Y., P. Shrager & J. M. Ritchie. 1984. Nature **311:** 156–157.
3. Shrager, P., S. Y. Chiu & J. M. Ritchie. 1985. Proc. Natl. Acad. Sci. USA **82:** 948–952.

4. MORETTO, G., S. U. KIM, D. H. SHIN, D. E. PLEASURE & N. RIZZURO. 1984. Acta Neuropathol. (Berl.) **64:** 15–21.
5. MCLARNON, J. G. & S. U. KIM. 1989. Neurosci. Lett. **101:** 107–112.
6. BEVAN, S., S. Y. CHIU, P. T. A. GRAY & J. M. RITCHIE. 1985. Proc. R. Soc. Lond. **225:** 299–313.
7. KETTENMANN, H. 1987. Can. J. Physiol. Pharmacol. **65:** 1033–1037.
8. BARRES, B. A., L. Y. CHUN & D. P. COREY. 1988. Glia **1:** 10–30.
9. ORKAND, R. K., J. G. NICHOLLS & S. W. KUFFLER. 1966. J. Neurophysiol. **29:** 788–806.
10. NEWMAN, E. A. 1985. J. Neurosci. **5:** 2225–2239.

Electrophysiological Properties of Squid Giant Axon Schwann Cells

Relevance to K⁺ Clearance[a]

N. JOAN ABBOTT,[b,c] Y. PICHON,[d] E. R. BROWN,[e]
I. INOUE,[f] F. KUKITA,[g] PATRICIA A. REVEST,[c]
AND I. C. H. SMITH[c]

[c] *Physiology Group*
Biomedical Sciences
King's College
London WC2R 2LS, UK

[d] *Laboratoire de Neurobiologie Cellulaire et Moléculaire*
CNRS
Gif-sur-Yvette, F-91198, France

[e] *Marine Biological Association Laboratory*
Plymouth PL1 2PB, UK

[f] *Institute for Enzyme Research*
Tokushima University
770 Japan

[g] *National Institute for Physiological Research*
Okasaki, 444 Japan

The giant axon of the squid is surrounded by a closely apposed layer of flattened Schwann cells (glia) about 1–2 μm thick. In early studies on the giant axon of the squid *Loligo forbesi*, Frankenhaeuser and Hodgkin[1] showed an exponential decline of undershoots during a burst of action potentials, evidence of potassium accumulation just outside the axon membrane. The data could be fitted by a model in which K^+ accumulated in a space of width $\theta \approx 30$ nm, from which it was cleared via a pathway of permeability $P = 6 \times 10^{-5}$ cm.s^{-1}. The calculated values of θ and P agreed with estimates of, respectively, the width of the periaxonal space, and dimensions of the Schwann cell intercellular clefts, suggesting that K^+ was cleared via diffusion through the extracellular spaces of the Schwann cell layer. Later current-clamp and electron microscopic (EM) studies were compatible with this model.[2]

Recent work on the small squid *Alloteuthis* has shown that if axons are left intact in the mantle muscle, or carefully dissected to preserve intact Schwann cells, much less K^+ accumulation is seen.[3,4] Voltage clamp studies of fresh *Loligo* axons kept at <4°C also showed lack of K^+ accumulation.[5] Moreover, high quality EM of well-fixed Schwann cells revealed a system of transcellular tubules (the glial tubular system [GTS]) that proved to be the major extracellular diffusion route.[6,7] However, even allowing for the contribution of the GTS, the extracellular space could not account

[a] This work was supported by the Royal Society, Wellcome Trust, and SERC.

[b] Address for correspondence: Dr. N. J. Abbott, Physiology Group, Biomedical Sciences Division, King's College, Strand, London WC2R 2LS.

for all the K^+ clearance observed.[7] A possible explanation was that clearance into the Schwann cell contributed to the apparently high θ and P.

Recent experiments have concentrated on the physiological properties of the Schwann cells that would enable them to clear K^+. Microelectrode impalements in Schwann cells have been held for periods of 30 seconds to 140 minutes in isolated axons and in axons left undisturbed attached to a block of mantle muscle.[8,9] Schwann cell resting potentials (E_m) recorded with 2.5 M KCl-filled microelectrodes in both *Alloteuthis* and *Loligo* were in the range of -25 to -80 mV, with a median of ≈ -40 mV. Approximately -40 mV apparently represents the real resting membrane potential of healthy Schwann cells; in stable preparations, sampling from individual Schwann cells along the axon with sharp electrodes shows potentials around -40 mV, and cells with this resting potential respond to electrical stimulation and application of carbachol and glutamate with a characteristic hyperpolarization.[10] The -40 mV E_m is compatible with the membrane having approximately equal K^+ and Cl^- permeability at rest, as suggested by ion substitution experiments.[11] Impalements with blunt electrodes generally gave low potentials (~ -25 mV), or showed cells that started at -25 to -40 mV and then hyperpolarized. The percentage of hyperpolarized cells increased with time after dissection. Preparations treated with Ca^{2+}-free solution during dissection or impalement showed fewer high membrane potential cells. It was concluded that the hyperpolarized state is caused by spontaneous or electrode-induced Ca^{2+} entry.[9] Some cells showed oscillations of potential (generally hyperpolarizing), particularly following axonal stimulation, that were reduced in amplitude by removal of external Ca^{2+}.[8] Both the gradual hyperpolarization and the oscillations could be explained by the presence of Ca^{2+}-activated K^+ channels. Pharmacological agents and natural axonal stimulation may activate the Schwann cell hyperpolarization *in situ,* using the same mechanism.

Cells of the split-open giant axon preparation of *Loligo* injected iontophoretically with Lucifer Yellow (LY) and horseradish peroxidase (HRP) appeared elongated (600×20 μm) and aligned along the long axis of the giant axon.[9] Light microscopy and EM of the HRP-filled cells confirmed that they were within the Schwann cell layer. Although the cell width agrees with previous EM calculations, the length of the fine processes is about ten times greater than previously estimated.

Dye (LY) and electrical coupling was observed between cells, with some evidence for greater coupling at more hyperpolarized potentials.[9] Taken together with the observations of E_m in resting and hyperpolarized cells, these results suggest that Schwann cells in the resting state may take up K^+ mainly as KCl, whereas those exposed to high K^+ from the axonal side during periods of stimulation-evoked hyperpolarization could remove K^+ via a "K^+ spatial buffer" mechanism, making use of the elongated cell processes to redistribute K^+.

Experiments were designed to separate the mechanisms for K^+ removal via Schwann cells. Mild K^+ accumulation was induced in healthy preparations at 10–14°C by repetitive stimulation (40–60 Hz) for 5–10 seconds. The action potential undershoots declined along a complex time course which could be fitted by the sum of three exponentials, each permitting calculation of a notional θ and P.[12] Raising or lowering the osmolarity of the artificial sea water (ASW) affected the dimensions of the tubular (GTS) route in the direction expected, leaving the mesaxonal clefts relatively unaffected.[13] Ouabain (1 mM, inhibitor of Na^+-K^+-ATPase) affected both the fast phase (probably by causing cell swelling) and the slow phases of K^+ accumulation. Bumetanide (150 μm, inhibitor of Na^+-K^+-2Cl co-transport) affected the medium and slow phases of K^+ accumulation.[12] The membranes lining the GTS are the most likely site for the ion channels and carriers responsible for K^+ clearance, as they account for a high proportion of the membrane surface area. Mathematical modeling shows that the GTS is an effective way of removing K^+ from the axon surface while keeping it close enough for effective recycling.

It is concluded that the morphology and passive membrane properties of the Schwann cells and their carrier-mediated transport systems, and the permeability changes underlying the hyperpolarization induced by axonal stimulation, all contribute to the mechanisms by which Schwann cells regulate the K^+ liberated into the periaxonal space during axonal activity.

REFERENCES

1. FRANKENHAEUSER, B. & A. L. HODGKIN. 1956. After-effects of impulses in the giant nerve fibres of *Loligo.* J. Physiol. (Lond.) **131:** 341–376.
2. ADELMAN, W. J., J. MOSES & R. V. RICE. 1977. An anatomical basis for the resistance and capacitance in series with the excitable membrane of the squid giant axon. J. Neurocytol. **6:** 621–646.
3. ABBOTT, N. J., E. M. LIEBERMAN, Y. PICHON & Y. LARMET. 1988. Periaxonal K^+ regulation in the small squid *Alloteuthis;* studies on isolated and *in situ* axons. Biophys. J. **53:** 275–279.
4. ASTION, M. L., J. A. COLES, R. K. ORKAND & N. J. ABBOTT. 1988. K^+ accumulation in the space between giant axon and Schwann cell in the squid *Alloteuthis.* Biophys. J. **53:** 281–285.
5. PICHON, Y., N. J. ABBOTT, E. M. LIEBERMAN & Y. LARMET. 1987. Potassium homeostasis in the nervous system of cephalopods and crustaceans. J. Physiol. (Paris) **82:** 346–356.
6. ZWAHLEN, M. J., C. SANDRI & N. G. GREEFF. 1988. Transglial pathway of diffusion in the Schwann sheath of the squid giant axon. J. Neurocytol. **17:** 145–159.
7. BROWN, E. R. 1991. Axon-Schwann cell interaction in the squid. PhD Thesis, London University.
8. ABBOTT, N. J., Y. PICHON & E. R. BROWN. 1990. Long-term microelectrode recording from Schwann cells surrounding the isolated giant axon of the squid; steady and oscillating potentials. J. Physiol. (Lond.) **430:** 128P.
9. BROWN, E. R., Q. BONE, N. J. ABBOTT & K. P. RYAN. 1991. Morphology and electrical properties of Schwann cells around the giant axon of the squids *Loligo forbesi* and *Loligo vulgaris.* Proc. Roy. Soc. B **243:** 255–262.
10. BROWN, E. R. & N. J. ABBOTT. 1991. Membrane potential in squid Schwann cells: Stable impalements in the split open giant axon preparation of *Loligo.* J. Physiol. (Lond.), in press.
11. LIEBERMAN, E. M. & E. SANZENBACHER. 1989. Possible mechanisms for glutamate mediation of axon-to-Schwann cell interaction in the isolated giant axon of the squid. J. Physiol. (Lond.) **423:** 90 P.
12. ABBOTT, N. J., Y. PICHON, I. INOUE, E. R. BROWN, E. M. LIEBERMAN, P. A. REVEST & I. C. H. SMITH. 1989. Role of Schwann cells in periaxonal ion regulation in the squid. Acta Physiol. Scand. **136** (Suppl. 582): 42.
13. ABBOTT, N. J. & E. R. BROWN. 1989. Effects of osmotic manipulation on extracellular routes for ion regulation in the squid Schwann cell. J. Physiol. (Lond.) **418:** 16 P.

Electrogenic Glutamate Uptake

Basic Properties, Modulation, and Failure in Anoxia[a,b]

DAVID ATTWELL, BORIS BARBOUR, HELEN BREW, MONIQUE SARANTIS, AND MAREK SZATKOWSKI

Department of Physiology
University College London
London, WC1E 6BT, UK

Glutamate uptake into glial cells and neurons keeps the extracellular glutamate concentration below neurotoxic levels and helps to terminate the synaptic action of neurotransmitter glutamate. We have monitored the activity of the glutamate uptake carrier in glial cells from the salamander and rabbit retina, using the whole-cell patch-clamp technique to measure the membrane current associated with glutamate uptake.

Applying glutamate to isolated glial cells evokes an inward membrane current that is abolished when the extracellular sodium or intracellular potassium are removed.[1-3] This current is also evoked by glutamate analogues with the pharmacology of glutamate uptake, and we attribute it to the activation of an electrogenic glutamate uptake carrier that cotransports sodium ions into the cell and countertransports potassium ions out of the cell.

Glutamate uptake is modulated by several factors. Membrane depolarization, raised extracellular potassium concentration, and extracellular arachidonic acid all reduce the glutamate uptake current.[1-5] In anoxia, when the extracellular potassium level rises, leading to cell depolarization, and when arachidonic acid is known to be released, the resulting inhibition of uptake will contribute to raising the extracellular glutamate concentration, possibly to neurotoxic levels.[6,7]

REFERENCES

1. Brew, H. & D. Attwell. 1987. Nature **327:** 707–709.
2. Barbour, B., H. Brew & D. Attwell. 1988. Nature **335:** 433–435.
3. Sarantis, M. & D. Attwell. 1990. Brain Res. **516:** 322–325.
4. Barbour, B., M. Szatkowski, N. Ingledew & D. Attwell. 1989. Nature **342:** 918–920.
5. Barbour, B., H. Brew & D. Attwell. 1990. J. Physiol. **400:** 169–193.
6. Nicholls, D. & D. Attwell. 1990. Trends Pharmacol. Sci. **11:** 462–468.
7. Szatkowski, M., B. Barbour & D. Attwell. 1990. Nature **348:** 443–446.

[a] Paper presented at conference, but not submitted for publication. Abstract included at editor's request.

[b] This work was supported by the Wellcome Trust, M.R.C. and S.E.R.C.

The Inducible Blood-Brain Barrier Endothelium-Specific Antigen HT7, A Novel Immunoglobulin-Like Membrane Glycoprotein

HARALD SEULBERGER, CAROLIN UNGER, URSULA ALBRECHT, AND WERNER RISAU

Max-Planck-Institut für Psychiatrie
Am Klopferspitz 18a
D-8033 Martinsried, Germany

Homeostasis of the neural microenvironment is important for neuronal function and integrity. Therefore, the neuronal tissue must be protected from the changing metabolite and ion concentrations in the blood. In higher vertebrates the endothelial cells of brain capillaries form a selective barrier between the blood and the neuronal cells. Complex tight junctions and a low number of pinocytotic vesicles are the structural basis of this blood-brain barrier. Specific transport of nutrients, metabolites, and ions across this barrier is mediated by carriers, ion channels, and receptors located at the cell surface of blood-brain barrier endothelium.[1]

We have characterized a monoclonal antibody (HT7) that recognizes an antigen on the chick blood-brain barrier endothelium.[2] The first expression of the HT7 antigen in brain endothelium at day 11 of chick embryonic development correlates well with the development of other properties of blood-brain barrier capillaries, like the impermeability to tracers injected into the blood stream, and the induction of other marker proteins in brain capillaries (alkaline phosphatase, day 12; transferrin receptor, day 11).[3] Endothelial cells of the brain circumventricular organs (choroid plexus, area postrema, pituitary gland, pineal gland, median eminence, and subfornical organ) that lack a functional blood-brain barrier do not express the antigen.[4] It is also not detectable on other endothelial cells in the body (kidney, skin, muscle, and lung), but it can be induced in endothelial cells of the chick chorioallantoic membrane by mouse brain tissue.[2] The expression of the antigen by other cell types, which form tissue barriers, like the pigment epithelial cells of the retina (blood-eye barrier) and the epithelial cells of the choroid plexus (blood-cerebrospinal fluid barrier) and the epithelial cells of the kidney and of the liver bile ducts, indicates that the antigen may be involved in adhesion, cell surface recognition, or transport processes. This assumption is supported by the presence of the antigen on the luminal surface of brain endothelium that was demonstrated by the binding of intravenously injected monoclonal antibodies to brain capillaries.[2] Furthermore, the presence of the antigen at the basolateral surface and in cytoplasmic compartments of the blood-brain barrier endothelium could be demonstrated using immunoelectron microscopy.[4] The antigen was also found in retinal neurons and erythroblasts, but not in mature erythrocytes.

As the HT7 monoclonal antibody does not recognize the denatured antigen on Western immunoblots, we raised an antiserum against the immunopurified antigen. With this antiserum in Western immunoblots, the antigen from brain capillaries, kidney tissue, and an erythroblast cell line was identified as a highly glycosylated protein of an apparent molecular weight of 43–52 kD in a reducing SDS polyacrylamide gel (FIG. 1). The apparent molecular weight of the protein from the erythroblasts is higher (48–52 kD) than that from brain and kidney (43–48 kD). We isolated the protein from plasma membrane protein preparations of different sources (brain, kidney, and erythroblasts) by an HT7 antibody affinity column and reversed phase HPLC.[5] Irrespective of the source, the NH_2-terminal sequence of the protein was identical. Deglycosylation of the protein by endoglycosidase F reduced the molecular weight to 32–34 kD (FIG. 1). Inhibition of N-glycosylation by culturing the erythroblast cell line in the presence of 1 μg/ml tunicamycin resulted in a reduction of the apparent molecular weight to about 26 kD (FIG. 1). On the basis of amino acid sequence information, we designed degenerate oligonucleotides for the polymerase chain reaction and amplified a cDNA fragment. This probe was used to clone a 2-kbp cDNA encoding the HT7 protein when expressed in COS-1 cells. Screening of databases and alignments to published protein sequences revealed that the HT7 protein (deduced molecular weight, 26,893 Da) was a novel member of the immunoglobulin superfamily with two immunoglobulin-like domains in the extracellular part of the molecule (186 amino acids). Comparison of the immunoglobulin domains with

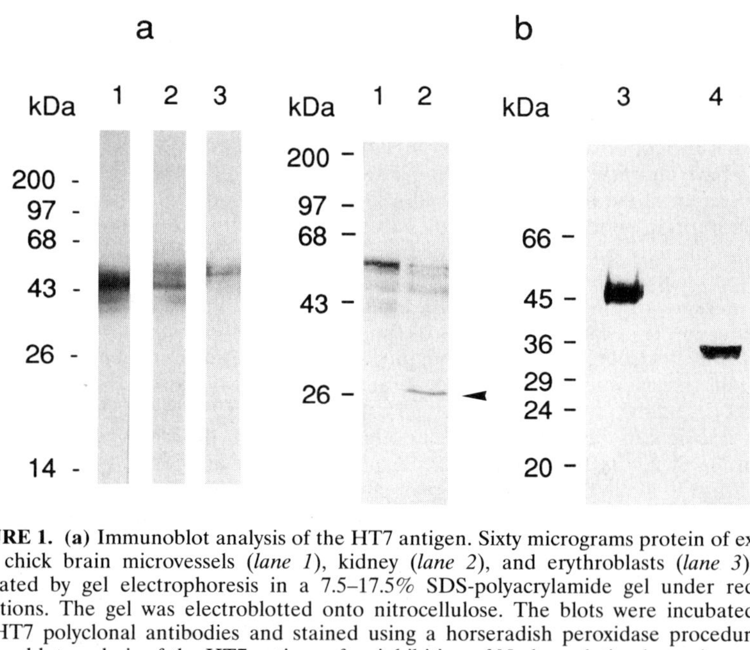

FIGURE 1. **(a)** Immunoblot analysis of the HT7 antigen. Sixty micrograms protein of extracts from chick brain microvessels (*lane 1*), kidney (*lane 2*), and erythroblasts (*lane 3*) were separated by gel electrophoresis in a 7.5–17.5% SDS-polyacrylamide gel under reducing conditions. The gel was electroblotted onto nitrocellulose. The blots were incubated with anti-HT7 polyclonal antibodies and stained using a horseradish peroxidase procedure. **(b)** Immunoblot analysis of the HT7 antigen after inhibition of N-glycosylation by tunicamycin or deglycosylation of the purified HT7 protein by incubation with endoglycosidase F. Sixty micrograms protein of extracts from erythroblasts cultured in the absence (*lane 1*) or presence (*lane 2*) of 1 μg/ml tunicamycin were loaded on a 12.5% SDS polyacrylamide gel and treated as described in **a.** The inhibition of N-glycosylation results in a molecular weight of 26 kD (*arrowhead*). Treatment of purified HT7 protein with endoglycosidase F for 24 hours at 37°C resulted in a reduction of the apparent molecular weight from 43–48 kD (*lane 3*) to about 34 kD (*lane 4*). A 12.5% SDS-gel was used and treated as described in **a.**

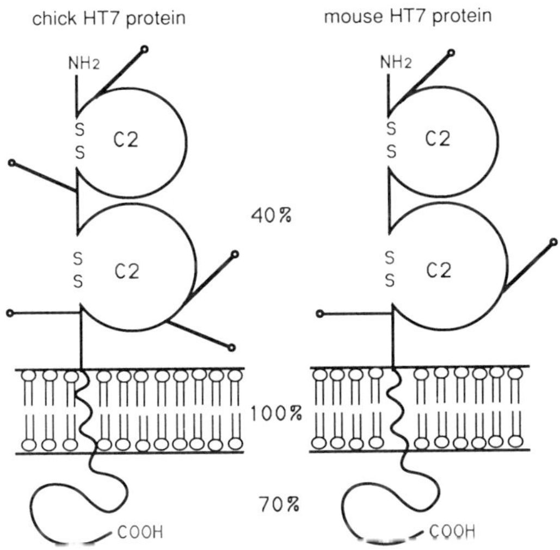

FIGURE 2. Schematic model of the possible structure of the chick and mouse HT7 protein. The circles held together by disulfide bonds represent possible immunoglobulin-like domains. Putative N-glycosylation sites are marked with *small circles.* The homology in the amino acid sequences of the extracellular, transmembrane, and intracellular domains between the chick and the mouse proteins is indicated by percent identity.

those of other members of the immunoglobulin superfamily indicated that they are of the C2 type according to the criteria of Williams and Barclay.[6] In addition, the HT7 protein has a hydrophobic signal sequence (26 amino acids), one membrane-spanning domain (21 amino acids), and a cytoplasmatic tail (39 amino acids). The protein does not show a high homology to any other protein. The high degree of glycosylation is reflected by five N-glycosylation consensus sequences. By low stringency hybridization using the chick cDNA probe we recently cloned a homologous murine cDNA. The identity between the chick and the mouse protein is 100% in the transmembrane region, 70% in the cytoplasmatic tail, and 40% between the two immunoglobulin-like domains (FIG. 2). The high degree of conservation in the transmembrane and cytoplasmic domain suggests that these structural domains play an important functional role. As other members of the immunoglobulin superfamily are either adhesion molecules, receptors for growth factors, or receptors involved in endocytosis or transcytosis, we are currently investigating if the HT7 protein has any of these functions.

In conclusion, the HT7 antigen was identified as a novel constituent of blood-brain barrier endothelium. Therefore, it may be an important tool for the investigation of the mechanism and regulation of blood-brain barrier formation *in vivo* and *in vitro* and the loss of this regulation under pathologic conditions such as brain tumors. The expression of the protein in polarized cells forming tissue barriers and the structure of this novel member of the immunoglobulin superfamily indicate that it may have an important function in cell surface recognition, cell adhesion, or transcellular transport. In addition, it may be a useful receptor for drug targeting to the brain. This is often desirable considering the impermeability of the blood-brain barrier for hydrophilic molecules.

REFERENCES

1. Risau, W. & H. Wolburg. 1990. Trends in Neurosci. **13:** 174–178.
2. Risau, W., R. Hallmann, U. Albrecht & S. Henke-Fahle. 1986. EMBO J. **5:** 3179–3183.
3. Risau, W., R. Hallmann & U. Albrecht. 1986. Dev. Biol. **117:** 537–545.
4. Albrecht, U., H. Seulberger, H. Schwarz & W. Risau. 1990. Brain Res. **535:** 49–61.
5. Seulberger, H., F. Lottspeich & W. Risau. 1990. EMBO J. **9:** 2151–2158.
6. Williams, A. F. & A. N. Barclay. 1987. Ann. Rev. Immunol. **6:** 381–405.

Blood-Brain Barrier Function in Central Demyelinating Lesions Repaired by Schwann Cell Remyelination

P. A. FELTS,[a] K. J. SMITH,[a,b] AND E. TILT[a]

[a] *Department of Anatomy and Neurobiology*
Eastern Virginia Medical School
Norfolk, Virginia 23501

[b] *Departments of Neurology and Anatomy*
United Medical and Dental Schools
Guy's Campus
London, SE1 9RT, UK

A suggested therapy for multiple sclerosis (MS) is the induction of remyelination of the central demyelinated axons by Schwann cells (SC). In support of such therapy it was recently shown that such remyelinated axons are able to conduct action potentials securely.[1] However, regions remyelinated by SC typically contain few, if any, astrocytes, namely, cells that appear capable of inducing blood-brain barrier (BBB) function in endothelial cells.[2,3] The absence of astrocytes, therefore, raises the question of whether a competent BBB would be formed at the capillaries within the lesion. Because the loss of the BBB in MS lesions may be an important element in the pathogenesis of the disorder,[4] we assessed the integrity of the BBB in regions of remyelination by SC.

Demyelinating lesions were induced under anesthesia (pentobarbital and halothane) in the dorsal columns of 48 adult male Sprague-Dawley rats by the intraspinal injection of ethidium bromide (EBr; 1 μl of 0.5 mg/ml), either alone or in conjunction with local beta irradiation (40 Gy). After 1–54 weeks the animals were reanesthetized, and horseradish peroxidase (HRP) (150–250 mg/kg body weight, Sigma Type II) was administered intravenously, after premedication with diphenhydramine hydrochloride. After 15 minutes the vasculature was flushed, and the animals were fixed by perfusion with aldehydes. Adjacent transverse vibratome sections through the lesion were processed for (1) extravasated HRP, (2) the presence of astrocytes by glial fibrillary acidic protein (GFAP) immunohistochemistry, and (3) light and electron microscopy.

The lesions were approximately 4–5 mm in length, affected at least 70% of the ipsilateral dorsal column, and were dominated by segmental demyelination from 1–3 weeks. Remyelination by SC became increasingly widespread thereafter, and although most lesions were completely remyelinated by approximately 12 weeks, small pockets of demyelination persisted in some lesions even at 54 weeks. There was an approximately threefold increase in the number of capillaries within the lesion. Extravasated HRP was present in all lesions that contained demyelinated axons (irrespective of their age), but was generally absent once repair by SC remyelination was complete. $GFAP^+$ cells were absent in regions remyelinated by SC, indicating a lack of astrocytes. The results show that the barrier to HRP can be restored when remyelination by SC is complete despite the formation of new capillaries within the lesion. Restoration of a competent BBB may not be achieved if demyelinated fibers persist.

REFERENCES

1. FELTS, P. A. & K. J. SMITH. Conduction properties of central nerve fibers remyelinated by Schwann cells. Brain Res., in press.
2. JANZER, R. C. & M. C. RAFF. 1987. Astrocytes induce blood-brain barrier properties in endothelial cells. Nature **325:** 253–257.
3. TAO-CHENG, J. H., Z. NAGY & M. W. BRIGHTMAN. 1987. Tight junctions of brain endothelium *in vitro* are enhanced by astroglia. J. Neurosci. **7:** 3293–3299.
4. POSER, C. M. 1986. Pathogenesis of multiple sclerosis. A critical reappraisal. Acta Neuropathol. (Berl.) **71:** 1–10.

Development of Nonneural Elements in the Central Nervous System of *Drosophila*

J. S. EDWARDS,[a,b] L. S. SWALES,[c] AND C. M. BATE[d]

[a] *Zoology Department*
Cambridge University
Cambridge CB2 3EJ, UK

[c] *AFRC Laboratory of Molecular Signalling*
Zoology Department
Cambridge University
Cambridge CB2 3EJ, UK

[d] *Zoology Department*
Cambridge University
Cambridge CB2 3EJ, UK

The development, diversity, and functional relations of the nonneural elements in the insect central nervous system (CNS) are only partially understood. Neurons and some or all glial categories are derived from neuroectoderm. The tracheal supply, with its tracheolar ramifications that resemble glia, is invaginated from the epidermis. A final category of cells, the sheath cells associated with the extracellular neural lamella that surrounds the CNS, has been subject to debate. It is shown here that these cells are of mesodermal origin and thus fundamentally distinct from other components of the CNS.

Both the neural lamella and associated sheath cells are absent in *Drosophila* embryos which are homozygous for a loss-of-function mutation in the gene *twist* (twi^{1d96}/twi^{1d96}). Twist function is required for gastrulation, and in its absence mesodermal cells fail to segregate; all structures of mesodermal origin are consequently absent.[1] The condition is lethal, but the CNS develops and undergoes shortening more or less as in the wild-type. Transverse sections of the thoracic region of the CNS taken from 20-hour old embryos lack the neural lamella and associated sheath cells found in the wild-type (FIG. 1). The use of ionic lanthanum as an electron-opaque marker of ion penetration into the CNS shows that blood-brain barrier glial cells are intact and functional. The barrier glial cells are thus shown to be independent of the overlying sheath cells in their differentiation of normal physiologic functions. Other glial elements in the cortical layer of neuron cell bodies and in the neuropile appear to be normal in *twist* embryos.

A second mutation used in this study, *Delta,* has a dominant adult phenotype affecting wing shape and a recessive lethal phenotype in the embryo in which neurogenesis is disrupted. In homozygous *Delta* embryos excess neuroblasts develop at the expense of ventral epidermis.[2] In such embryos all elements of the CNS develop, but sheath and barrier glial cells fail to establish continuous layers. The CNS is thus permeable to lanthanum ions. The inability of the inverting cell layers to envelope the CNS appears to be due to the excess volume of supernumerary ganglion cells.

[b] On leave from Zoology Department NJ15, University of Washington, Seattle, Washington 98195.

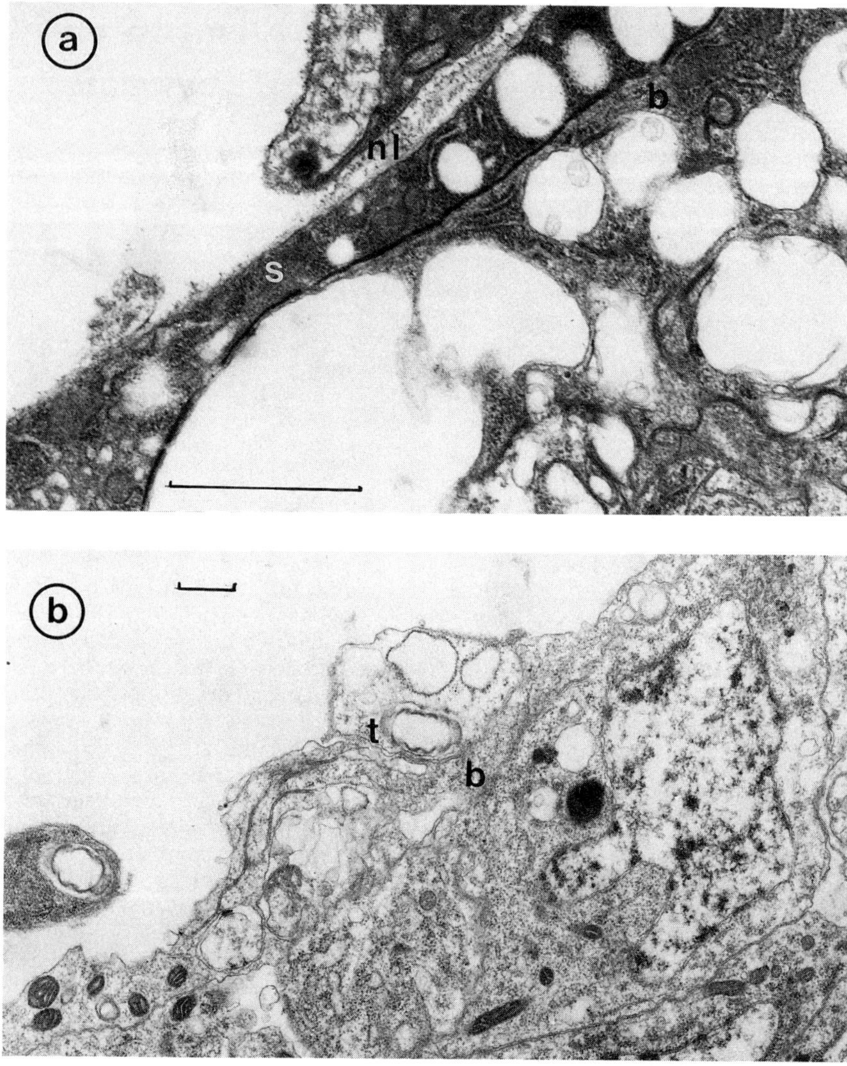

FIGURE 1. (**a**) Electron micrograph, transverse section wild-type embryonic *Drosophila* CNS after 1-hour immersion in 0.1 mM lanthanum in *Drosophila* saline solution. A dark sheath cell (s) situated diagonally has begun to secrete the neural lamella (nl). Lanthanum forms the electron-dense layer situated beneath the sheath cell and limited by the outer surface of the barrier glial cell (b). Scale bar = 1 μm. (**b**) Electron micrograph, transverse section twist embryonic *Drosophila* CNS. Sheath cells are absent. Lanthanum deposits lie below a tracheal cell (t), on the external surface of a barrier glial cell (b). Scale bar = 1 μm.

REFERENCES

1. Nusslein-Volhard, C. E., E. Wieschaus & H. Kluding. 1984. Roux's Arch. Dev. Biol. **193:** 267–282.
2. Lehmann, R., F. Jimenez, U. Dietrich & J. A. Campos-Ortega. 1983. Roux's Arch. Dev. Biol. **192:** 62–74.

Inhibition of Potassium-Activated Release of D-Aspartate from Astrocytes by an Anion Transport Inhibitor[a]

H. K. KIMELBERG AND S. GODERIE

Division of Neurosurgery, A-60
Albany Medical College
Albany, New York 12208

Astrocytic swelling is an early response in experimental and human head injury as well as in other pathologic states such as ischemia. In previous studies we observed that the astrocytic swelling otherwise associated with a traumatic and hypoxic head injury model[1] was prevented by treatment with an anion transport inhibitor (L644,711) given after trauma but before the imposed period of hypoxia; this led to an improved outcome from the combined injury.[2,3] In *in vitro* studies using primary astrocytic cultures we observed that rapid swelling of the cells by exposure to hypotonic medium results in their regulating their volumes back towards normal levels, a process known as regulatory volume decrease or RVD. During this process the cells rapidly lose taurine and also amino acids such as glutamate and aspartate. Interestingly, this swelling-induced release of preloaded, radiolabeled taurine, glutamate, and aspartate was inhibited by L644,711 and other anion transport inhibitors (SITS and furosemide).[4] In the present study in which a perfusion technique was used, we show that exposure of cultured astrocytes to elevated K^+ in the medium (50 mM) also caused the increased release of previously accumulated [^{3}H] D-aspartate which was blocked by L644,711.

METHODS

Primary astrocyte cultures were grown from the midbrain region of 1-day-old Sprague-Dawley rats[4] on cell support film (Bellco Glass Inc.) for 3–4 weeks by anchoring pieces of the film with autoclaved vacuum grease to the bottom of individual wells in six-well plastic trays. The cell suspensions were then seeded onto the film at 2×10^5 cells per well in 2.5 ml of MEM + 10% HS. The cells were loaded by adding 20 μCi [^{3}H] D-aspartate in a final concentration of 10^{-4} M D-aspartate in growth medium and letting them incubate overnight (approximately 16 hours). The cell support film was then transferred to a 60-mm petri dish and washed once in 5 ml of a HEPES-buffered medium of the following composition (in mmol/L): NaCl, 122; KCl, 3.3; $CaCl_2$, 1.3; $MgSO_4$, 0.4; KH_2PO_4, 1.2; HEPES, 25, adjusted with NaOH to obtain a pH of 7.4 M, and glucose 10. In HCO_3^--buffered medium 25 mM HEPES was replaced with 25 mM HCO_3^- and the solution bubbled with a 5% CO_2/95% air mixture. The cell support film was then inserted into a cut 1-ml tuberculin syringe and perfused at a rate of approximately 1 ml/min. One-milliliter aliquots were

[a]This work was supported by grant NS 23750 from the National Institute of Nervous Disorders and Stroke.

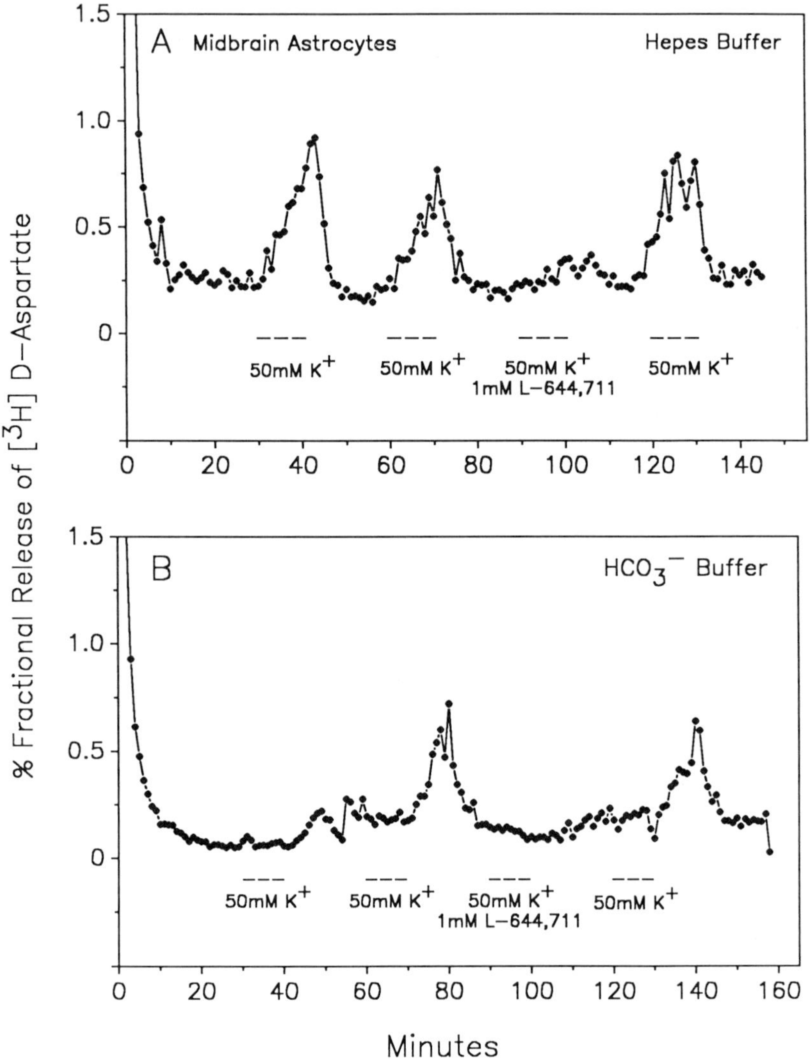

FIGURE 1. Release of previously accumulated [^{3}H] D-aspartate from astrocytes prepared from the midbrain region of 1-day-old Sprague-Dawley rats. Release was measured at 37°C using a constant perfusion system in which control HEPES (**A**) or bicarbonate (**B**)-buffered medium was changed, as indicated, to medium containing 50 mM K^+ by isotonic replacement of Na^+ or 50 mM K^+ plus the anion transport inhibitor L-644,711, at 1 mM. Results are expressed as a percentage of the radioactivity present at the beginning of each 1-minute collection period obtained by sequentially adding back all the radioactivity released. See Methods section for further experimental details.

collected, 5-ml scintillation fluid was added to each tube, and the radioactivity was counted. The syringe was kept at 37°C using a heating coil.

RESULTS AND DISCUSSION

As shown in FIGURE 1A, after an initial brief washout a stable baseline efflux of 0.2–0.3% was obtained by perfusion of the cells with normal HEPES buffer. When the perfusion medium was changed to one in which the K^+ concentration was raised to 50 mM by replacement of Na^+, a gradual increase in the rate of efflux increasing to 0.9% after a 10-minute exposure was seen. This rapidly reversed when the perfusion medium was changed back to the control medium. After 20 minutes a second exposure to 50 mM K^+ resulted in a similar increase, but in the third exposure, when 1 mM L-644,711 was also present, no increase in efflux was seen. Subsequent exposure to 50 mM K^+ gave a normal control result. In FIGURE 1B a similar result was seen in HCO_3^--containing medium, but the responses to 50 mM K^+ were less consistent and seemed to be delayed.

These data show that exposure of primary astrocytic cultures to medium containing elevated K^+ leads to increased recovery of preloaded [^{3}H] D-aspartate in the perfusate. These results are consistent with swelling of primary astrocytic cultures due to exposure to high K^+ causing increased release of [^{3}H] D-aspartate, inasmuch as previous results showed that swelling of primary astrocytic cultures in hypotonic medium caused marked release of previously accumulated [^{3}H] D-aspartate and that this release was also inhibited by L-644,711 (see introduction). Additional experiments that would distinguish between K^+-induced depolarization or swelling, or both, would involve raising K^+ to 50 mM but reducing Cl^- to maintain a constant $K^+ \times Cl^-$ product; this has been shown to eliminate the K^+-stimulated release of [^{3}H] taurine from cultured astrocytes.[5] We also need to show that swelling from exposure to either hypotonic medium or high potassium under these conditions does not cause release of cells or lysis, but inhibition by L-644,711 suggests that a specific transport process is indeed involved. Extracellular K^+ in ischemia, head injury, and other pathologic states is known to increase up to 80 mM.[6] Under these conditions, release of excitatory amino acids such as glutamate or aspartate from astrocytes might contribute to the excitotoxicity, and inhibition of such release by L-644,711 might contribute to its beneficial effect in head injury.[2,3] We have also shown that early administration of L-644,711 eliminates the astrocytic swelling otherwise observed after head injury.[1] Elimination of swelling would also eliminate the swelling-induced release of amino acids, but inhibition of such release should also reduce RVD after exposure to hypotonic medium. We have found that this actually occurs in astrocytic cultures (unpublished observations). Inhibition of astrocytic swelling *in situ* by L-644,711 could be due to blockade of $Cl^-HCO_3^-$ exchange or diffusion.[2,3] We therefore need to distinguish between inhibition of K^+-induced glutamate or aspartate release due to inhibition of K^+-induced swelling by L-644,711 or a direct effect of L-644,711 on the aspartate amino-acid release process itself.

REFERENCES

1. BARRON, K. D., M. P. DENTINGER, H. K. KIMELBERG, L. R. NELSON, R. S. BOURKE, S. KEEGAN, R. MANKES, & E. J. CRAGOE, JR. 1988. Ultrastructural features of a brain injury model in cat. I. Vascular and neuroglial changes and the prevention of astroglial swelling

by a fluorenyl (aryloxy) alkanoic acid derivative (L-644,711). Acta Neuropathol. (Berl.) **75:** 295–307.

2. CRAGOE, E. J. JR., O. W. WOLTERSDORF, JR., N. P. GOULD, A. M. PIETRUSZKIEWICZ, C. ZIEGLER, Y. SAKURAI, G. E. STOKKER, P. S. ANDERSON, R. S. BOURKE, H. K. KIMELBERG, L. R. NELSON, K. D. BARRON, J. E. ROSE, D. SZAROWSKI, A. J. POPP & J. P. WALDMAN. 1986. Agents for the treatment of brain edema. 2. [(2,3,9,9a-tetrahydro-3-oxo-substituted-1H-fluoren-7-yl)oxy] alkanoic acids and some of their analogues. J. Med. Chem. **29:** 825–841.
3. KIMELBERG, H. K., E. J. CRAGOE, JR., L. R. NELSON, A. J. POPP, D. SAZROWSKI, J. W. ROSE, O. W. WOLTERSDORF, JR. & A. M. PIETRUSZKIEWICZ. 1987. Improved recovery from a traumatic-hypoxic brain injury in cats by intracisternal injection of an anion transport inhibitor. Central Nervous System Trauma. **4:** 3–14.
4. KIMELBERG, H. K., S. K. GODERIE, S. HIGMAN, S. PANG & R. A. WANIEWSKI. 1990. Swelling-induced release of glutamate, aspartate, and taurine from astrocyte cultures. J. Neurosci. **10**(5): 1583–1591.
5. PASANTES-MORALES, H. & A. SCHOUSBOE. 1989. Release of taurine from astrocytes during potassium-evoked swelling. Glia **2:** 45–50.
6. HANSEN, A. J. 1985. Effect of anoxia on ion distribution in the brain. Physiol. Rev. **65:** 101–148.

Peptidylglycine Amidating Monooxygenase (PAM), an Enzyme Required for Neuropeptide Biosynthesis, Is Present in Schwann Cells and Some Glia

C. HARKER RHODES,[a,b] RUTH H. ANGELETTI,[c] AND F. ARTHUR McMORRIS[d]

[a]*Dartmouth Medical School*
Hanover, New Hampshire 03756

[c]*Albert Einstein College of Medicine*
Bronx, New York 10461

[d]*Wistar Institute*
Philadelphia, Pennsylvania 19104

The biologically active form of about half the known neuroendocrine peptides is the peptide amide. All of these molecules apparently are synthesized from a glycine-extended precursor that is converted to the amidated peptide and glyoxal by the enzyme peptidylglycine amidating monooxygenase (PAM).[1] The fact that PAM is required for the biosynthesis of the amidated peptides means that the enzyme can be used as a histochemical or immunohistochemical marker for cells producing an amidated peptide even when the peptide itself has not been identified.

An antiserum was raised against a synthetic peptide corresponding to the NH_2-terminus of the enzyme and used to identify PAM-containing cells by immunohistochemistry.[2] The specificity of the antiserum was tested by using immobilized IgG purified from the serum as an immunoaffinity reagent to purify PAM from brain homogenate. PAGE analysis of the purified protein showed a prominent peak of the appropriate molecular weight for the mature form of PAM and several faint higher molecular weight bands that probably represent PAM precursors. The specificity of the antiserum in immunohistochemical experiments was demonstrated using immobilized peptide to eliminate staining by the antiserum.

In those immunohistochemical experiments PAM immunoreactivity was demonstrated in neurons known to produce an amidated peptide, in some neurons not previously known to produce an amidated peptide, and, surprisingly, in Schwann cells and some glia. This result suggests that these nonneuronal cells are peptidergic.

[b]Address for correspondence: C. H. Rhodes, Department of Pathology, Dartmouth Medical School, Hanover, NH 03756.

The presence of PAM in Schwann cells was confirmed by experiments in which amidating enzymatic activity was measured in peripheral nerve 10 days after nerve section. Those experiments showed that nearly half the amidating activity in sciatic nerve persisted after axonal degeneration, strongly suggesting that the PAM immunoreactivity observed in Schwann cells is due to *bona fide* PAM.

In CNS glia, there is no way to mechanically remove the adjacent neurons or cause them to degenerate *in vivo*; however, it is possible to grow primary glial cultures in which all of the neurons have died. With the use of the culture procedures described by McMorris *et al.*,[3,4] nearly pure astrocytic cultures and mixed glial cultures enriched for oligodendroglia were produced. The composition of the cultures was determined by staining them with antisera for the astrocyte marker glial fibrillary acidic protein, the neuronal marker neurofilament protein, the oligodendroglial marker cyclic nucleotide phosphohydrolase, and several markers for immature glia. Significant levels of amidating activity were detected in the cells and in the culture medium in both types of culture (TABLE 1).

The primary glial cultures were grown in the presence of serum that contains some amidating activity. To show that the glia were producing PAM and not simply concentrating it from the medium, metabolic labeling experiments were performed. Astrocyte and oligodendroglial cultures were labeled with [^{35}S]-methionine and the labeled proteins immunoprecipitated with the anti-PAM antiserum. The immunoprecipitated proteins were separated by PAGE and detected by autoradiography. As can be seen in FIGURE 1, these cultured glia not only contain PAM enzymatic activity, but also are producing a protein with PAM immunoreactivity and of the appropriate molecular weight for mature PAM.

These results suggest that Schwann cells and some glia are producing PAM and, by inference, also producing an amidated peptide. The biochemical identity of that peptide and its physiological role are unknown.

TABLE 1. Amidating Enzyme Activity in Cultured Astrocytes and Oligodendrocytes and in Their Cell Culture Medium[a]

Sample	Amidating Activity (arbitrary units)
Astrocyte cell pellet	47,500
Oligodendroglial pellet	42,900
Astrocyte medium after culture	206,000
Astrocyte medium before culture	102,000
Difference	104,000
Oligodendroglia medium after culture	305,000
Oligodendroglia medium before culture	19,500
Difference	285,500

[a]These measurements were made on cultures similar to those illustrated in FIGURE 1. There is amidating enzyme activity in serum. The astrocytes were grown in medium containing 10% serum; the oligodendroglia were grown in 1% serum. Amidating enzyme activity was measured in the cell culture medium before and after several days in culture, and the difference represents activity generated by the cells. The figures are for the total amount of enzyme activity present per culture dish and are in arbitrary units.

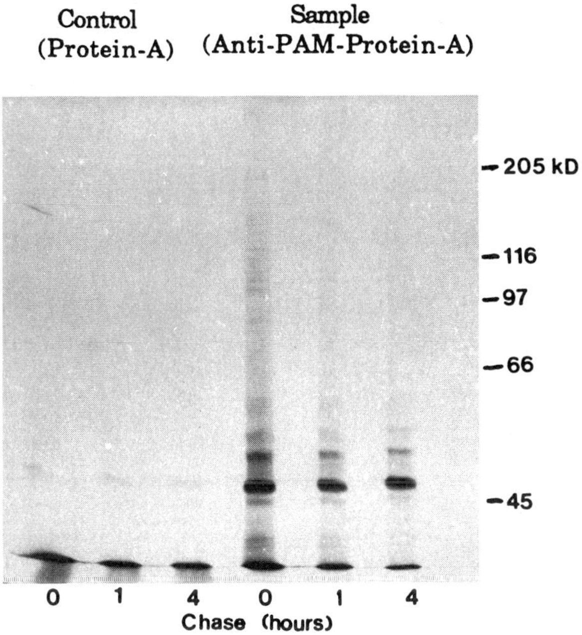

FIGURE 1. Autoradiogram of metabolically labeled protein from oligodendroglial cultures immunoprecipitated with anti-PAM antiserum. A mixed glial culture enriched for oligodendroglia and containing no neurons was labeled for 30 minutes with [^{35}S]-methionine, and after a 0-, 1-, or 4-hour chase period, the cells were homogenized in 1% NP-40. The samples were split in half, and PAM was purified by immunoprecipitation with anti-PAM IgG covalently bound to protein-A sepharose from one portion. The other portion was treated identically except that protein-A sepharose without anti-PAM IgG was used to control for nonspecifically adsorbed protein. The purified proteins were separated on a 7.5% polyacrylamide gel in the presence of SDS and the labeled proteins detected by autoradiography. The lanes on the left are the control samples and those on the right are the anti-PAM samples. This experiment demonstrates that these cultured glia are synthesizing protein that is bound by anti-PAM IgG and is of the appropriate molecular weight for *bona fide* PAM. Similar results were obtained with the astrocyte cultures.

REFERENCES

1. BRADBURY, A. F., M. D. A. FINNE & D. G. SMYTH. 1982. Mechanism of C-terminal amide formation by pituitary enzymes. Nature **298:** 686–688.
2. RHODES, C. H., R.-Y. XU & R. H. ANGELETTI. 1990. Peptidylglycine alpha-amidating monooxygenase (PAM) in Schwann cells and glia as well as neurons. J. Histochem. Cytochem. **38:** 1301–1311.
3. MCMORRIS, F. A., R. M. SMITH, S. DESALVO & R. W. FURLANETTO. 1986. Insulin-like growth factor I/somatomedin C; A potent inducer of oligodendrocyte development. Proc. Natl. Acad. Sci. USA. **83:** 822–28.
4. RIABLE, D. W. & F. A. MCMORRIS. 1989. Cyclic AMP regulates the rate of differentiation of oligodendrocytes without changing the lineage commitment of their progenitors. Dev. Biol. **133:** 437–446.

Fluoroacetate, a Selective Inhibitor of the Glia Tricarboxylic Acid Cycle, Attenuated the Release of Luteinizing Hormone-Releasing Hormone from the Hypothalamus of Ovariectomized Rats

T. J. WU,[a,b] N. H. MCARTHUR,[c] AND P. G. HARMS[a,c,d]

Departments of Animal Science[a] and Veterinary Anatomy[c]
Texas A&M University
College Station, Texas 77843

Recent morphological evidence suggests involvement of glia in luteinizing hormone-releasing hormone (LHRH) release from the hypothalamus.[1] The present study was conducted to further investigate the role of glia in LHRH release by assessing the effect of glia metabolism on LHRH release from hypothalamic explants in static culture. This was accomplished by selective inhibition of the tricarboxylic acid cycle (TCA) in glia by fluoroacetate (FA). Subsequent to selective uptake by glia, FA is converted into fluorocitrate which in turn blocks aconitase and thereby inhibits the glia TCA.[2]

MATERIALS AND METHODS

Adult female rats (225–250 g; Harlan Sprague-Dawley Inc.) were maintained in controlled environmental conditions of a 12L:12D cycle and temperature (24°C) and had access to rat chow and water *ad libitum.* Three weeks before tissue procurement, the rats were ovariectomized under ketamine (80 mg/kg body weight)/xylazine (12 mg/kg body weight) anesthesia. On the day of experimentation, the hypothalamus (including the preoptic area, mediobasal hypothalamus, and infundibulum) was procured and incubated (one hypothalamus per chamber) in 600 μl of Kreb's-ringer's bicarbonate medium (KRB; with 10 mM glucose, 0.1% bovine serum albumin, and 0.06 g/L bacitracin; pH 7.6 under 95% O_2 and 5% CO_2 for 3 hours at 37°C. At 30, 60, 90, 120, 150, and 180 minutes, 500-μl samples of medium were removed from each tube and replaced with fresh medium (FIG. 1). The hypothalamic explants ($n = 8$ per group) were randomly assigned to 0, 0.1, or 1.0 mM FA with KRB throughout the 180-minute period. From 120–150 minutes of incubation, the explants were subjected to depolarization with a modified KRB (K; containing 60 mM potassium ions instead of the 4.7 mM in KRB). The samples collected were centrifuged and then stored at −20°C until quantitation for LHRH concentration by radioimmunoassay.

[b]Tom Slick Senior Graduate Fellow, College of Agriculture, Texas A&M University, College Station, Texas.

[d]To whom correspondence should be addressed.

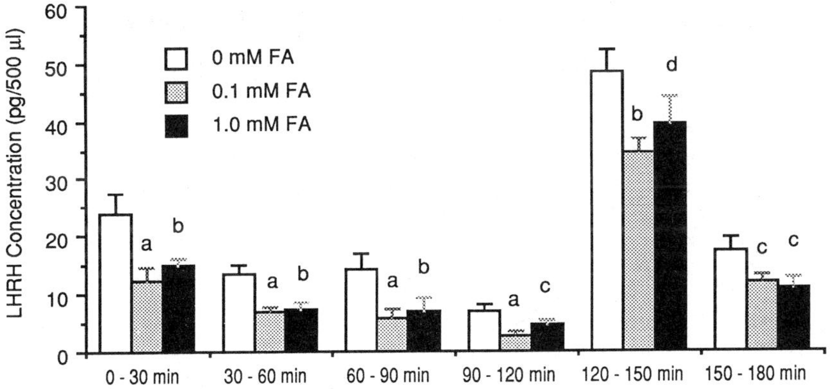

FIGURE 1. Mean concentration of LHRH released from hypothalamic explants procured from ovariectomized rats. Explants were incubated in a static culture system containing 600 μl of 0 mM, 0.1 mM, or 1.0 mM fluoroacetate (FA) in KRB during 0–30, 30–60, 60–90, 90–120, 120–150, and 150–180 minutes of incubation. Letters (a, b, c, and d) above the bars represent the levels of difference ($p < 0.005$, $p < 0.01$, $p < 0.05$, and $p < 0.1$, respectively) when compared with the 0 mM FA treatment.

RESULTS AND CONCLUSIONS

Basal LHRH release was attenuated when explants were exposed to either 0.1 mM ($p < 0.005$) or 1.0 mM ($p < 0.05$) FA. The effect of FA occurred within 30 minutes of incubation. Similarly, depolarization-induced LHRH release (120–150 min) was also attenuated in the presence of 0.1 mM ($p < 0.01$; 34.27 ± 2.41 pg) or 1.0 mM ($p < 0.10$; 39.37 ± 4.74 pg) FA compared to control (48.37 ± 4.07 pg). The present study demonstrated that alteration of glia metabolism influenced LHRH release. Hence, these results suggest that glia may be regulators, either directly or indirectly, of LHRH release from the hypothalamus.

REFERENCES

1. Lehman, M. N., F. J. Karsch, J. E. Robinson & A.-J. Silverman. 1988. J. Comp. Neurol. **273:** 447–458.
2. Szerb, J. C. & B. Issekutz. 1987. Brain Res. **410:** 116–120.

Capsaicin-Induced c-*fos* in Peripheral Nervous System Glial Cells

JANET WINTER AND CAROLINE EVISON

Sandoz Institute for Medical Research
5 Gower Place
London WCIE 6BN, UK

c-*fos* and other immediate early genes (IEGs) are transiently expressed in various cell types in response to growth factors, hormones, and the like. In the nervous system, IEGs may function as mediators of the long-term changes in neurons evoked by external stimuli, by turning on genes involved in plasticity, for example. Peripheral nerve stimulation causes c-*fos* induction in postsynaptic dorsal horn neurons, but not in the sensory neurons that convey the information.[1] In PC12 cells, the calcium influx triggered by depolarization seems to be important in inducing c-*fos* expression.[2] In sensory ganglia, the pain-producing excitotoxin, capsaicin, stimulates a subset of neurons with a resultant large influx of calcium.[3,4] We therefore thought that c-*fos* might be induced by capsaicin in this subset, and its presence would be a useful indicator of capsaicin sensitivity.

Adult rats were injected with 50 mg/kg of capsaicin (under halothane anesthesia) or vehicle alone (10% ethanol, 10% Tween 80, 80% phosphate-buffered saline solution [PBS]). Two hours later animals were sacrificed, and lumbar dorsal root ganglia (DRG) and superior cervical ganglia (SCG) were dissected, fixed in 4% paraformaldehyde, and cryoprotected in 20% sucrose in PBS. 10 μm cryostat sections were thaw-mounted onto gelatin-coated microscope slides and immunostained for the presence of c-*fos* with a rabbit anti-c-*fos* antibody (kindly donated by Dr. Gerard Evan). Amersham biotinylated anti-rabbit antibodies, followed by streptavidin fluorescein, were used to visualize the bound anti-c-*fos*. Very few DRG neuron nuclei (in sections from control or capsaicin-treated animals) were labeled with the c-*fos* antibody. However, many glial cell nuclei scattered throughout the DRG were labeled (FIG. 1) in capsaicin-treated but not in control rats.

When SCG sections were labeled, again neuronal staining with c-*fos* antibody was not evident, but in this sympathetic ganglion, which contains few or no capsaicin-sensitive neurons, labeled glial cell nuclei were also rare.

We interpret this example as one of neuron-glial interaction, with the capsaicin-sensitive neurons somehow activating nearby glial cells when capsaicin is administered to the animal. This activation involves c-*fos* expression. As SCG neurons are capsaicin insensitive, they do not activate local glial cells.

Additionally, sciatic nerves were dissected from animals that had been treated with vehicle or capsaicin 1 day previously and dissociated enzymatically with trypsin and collagenase and then mechanically through a Pasteur pipette (as described previously for DRG neurons[5]). Cells were plated into polyornithine and laminin-coated tissue culture dishes, grown for 1 day in F14 medium with 4% Ultraser-G (Gibco), and then fixed. Glial cells cultured from the sciatic nerves of capsaicin-treated animals behave differently than do those from control animals. They have obviously longer, more branching processes, which may be a consequence of c-*fos* induction turning on other genes.

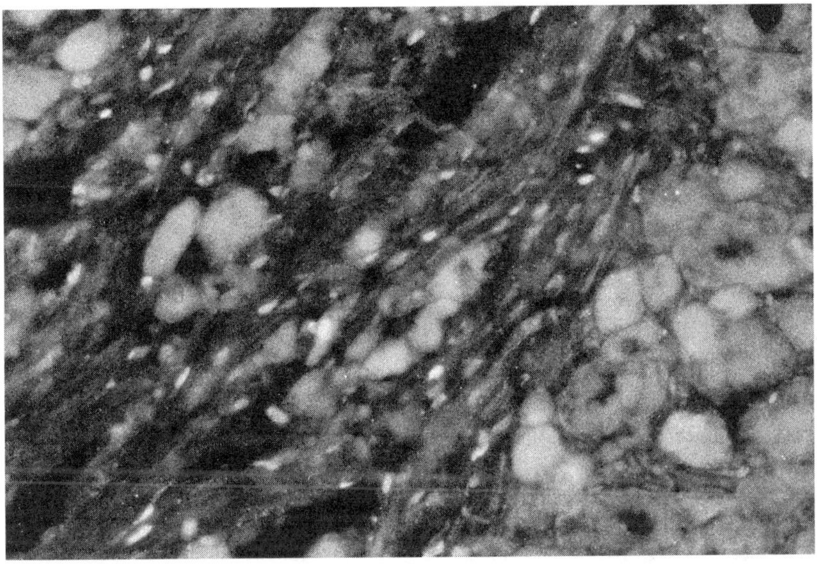

FIGURE 1. Cryostat section of a dorsal root ganglion taken from an adult rat 2 hours after systemic capsaicin treatment. The section was stained with an anti-c-*fos* antibody. Only glial cell nuclei stained, not nuclei in the larger, neuronal cell bodies.

We therefore think that some factor, such as a neuropeptide, is released from DRG neurons when stimulated by capsaicin. This factor then activates the neighboring glial cells, and c-*fos* is expressed. The increased ability of the glial cells to elaborate processes *in vitro* may be related to this activation or it may be a consequence of degeneration of the capsaicin-sensitive axons, thus destroying the normal neuron-glial interactions in the nerve.

REFERENCES

1. HUNT, S. P., A. PINI & G. EVAN. 1987. Nature **328:** 632–634.
2. SHENG, M., G. MCFADDEN & M. E. GREENBERG. 1990. Neuron **4:** 571–582.
3. WINTER, J., C. A. FORBES, J. STERNBERG & R. M. LINDSAY. 1988. Neuron **1:** 973–981.
4. WOOD, J. N., J. WINTER, I. F. JAMES, H. P. RANG, J. YEATS & S. BEVAN. 1988. J. Neurosci. **8:** 3208–3220.
5. LINDSAY, R. M. 1988. J. Neurosci. **8:** 2394–2405.

Neuron-Glia Interaction

Effect of Serotonin and DBcAMP on the Expression of GFAP and Its Encoding Message

M. TARDY,[a] G. LE PRINCE,[a] C. FAGES,[a] B. ROLLAND,[a]
J. NUNEZ,[a] AND M. F. BELIN[b]

[a]*Inserm U 282*
Hôpital Henri Mondor
94010 Creteil, France

[b]*Inserm U 171*
Laboratoire d'Anatomie Pathologie
Faculté de Médecine Alexis Carrel
Lyon, France

The close association between neurons and glia suggests that gene expression in these cell types is likely to be influenced by mutual interactions. Several lines of evidence indicate that glia influence the growth, migration, differentiation, and metabolism of neurons, but the effect of the neuronal environment on glial cells is far from being well understood.

Glial fibrillary acidic protein (GFAP) is a major component of the glial intermediate filaments found in astrocytes, and this protein is generally accepted as a specific marker of these cells. GFAP increases after any neuronal damage, and the protein is now employed as an index of astrogliosis. The significance of such an increase neither is understood nor has made clear the role of intermediate filaments and of GFAP in this cell type.

We therefore recently undertook a developmental study using a combination of immunoblotting and Northern blotting of the GFAP expression during postnatal development in the mouse brain and in astroglia-enriched primary cultures. We found some interesting regulations concerning the developmental pattern of GFAP. First, we observed a positive regulation between GFAP and its encoding message during the astroglial proliferation period.[1] Then, a discrepancy appeared between the GFAP level and its steady-state mRNA level during the later maturation stage. The GFAP level remained high, whereas its mRNA decreased. Run-on experiments performed on isolated nucleus showed that the high GFAP-mRNA transcription that occurred during the perinatal period decreased thereafter and was stabilized at a rather low rate. These data suggest that in addition to transcriptional control of GFAP synthesis, GFAP may be influenced by translational controls and/or GFAP-mRNA stabilization mechanisms.

Chemical signals that may represent the connection between neuronal and glial activity include neurotransmitter substances. These substances may be released into the extracellular environment, and astrocytes apparently not only recognize neuronal signals but also respond to them in different ways. Astroglial primary cultures exhibit a large variety of neurotransmitter receptor systems. Many of them modulate cAMP accumulation (catecholamines, adenosine, and prostaglandin E). Others, induce phosphatidylinositol (PI) breakdown (GLU, histamine, and serotonin). A number of astroglial functions are reported to be influenced by receptor stimulation, among them are activation of cAMP-dependent protein kinase and the subsequent increase in phosphorylation of some proteins including GFAP, but it is not yet clear what part phosphorylation of the protein may play.

In the present study, we investigated the effect of a cAMP derivative and of serotonin on the GFAP expression of astrocytes in primary cultures. dBcAMP has been used at 1 mM, a concentration that induces morphological changes even in the presence of fetal calf serum, transforming the flat polygonal, cerebral hemisphere, astroglial cells into differentiated process-bearing cells.

dBcAMP was added to the culture medium either for 48 hours or chronically for 10 days. The first condition corresponds to stabilized morphological changes[2] and the second one to a reactive stage of the cells.[3] In both conditions, GFAP immunolabeling increased with the observed shape changes. After the 48-hour contact, we were unable to observe any change in either the GFAP or the GFAP-mRNA level. The chronic contact, however, resulted in an increase in the protein content and its encoding message level.

When grown in the presence of 10^{-5} M serotonin for 48 hours, astrocytes obtained from the brainstem (an area rich in serotoninergic neurons) remained apparently unchanged morphologically. Immunolabeling, however, performed with an anti-GFAP antibody showed a modified GFAP distribution and a depleted immunolabeling. The perinuclear region was devoid of GFAP labeling, and GFAP labeled spots appeared scattered near the plasmic membrane. The level of both GFAP and GFAP-mRNA were decreased at the same rate.[4]

The mechanism of neuroglial interaction is a basic process that controls homeostasis in the central nervous system. Neurotransmitters such as serotonin or catecholamines may be involved in these processes at least in those astrocytes that possess receptors for such ligands. In response to both types of ligands, the immunochemical pattern of GFAP changes. In the cAMP-derivative system the reactive filaments appear well organized, radiating from the nucleus to the formed processes where they were strongly reactive. In the serotonin effect, the gliofilaments appear rather disorganized without any preferential organization at the periphery. These changes may be related to a more general process that involves different signals and may result in a "differentiation" process of astroglial cells. Increasing cAMP levels for a relatively short period results in a posttranslational modification of GFAP or for a longer time in a transcriptional effect related either to increased mRNA synthesis and/or to a modification of the message stability. In contrast, increasing PI breakdown induces a decrease in GFAP and GFAP-mRNA levels, indicating an inhibitory effect of serotonin on the expression of GFAP. The exact mechanism whereby the GFAP-mRNA levels are increased or decreased remains to be elucidated.

Finally, these data suggest that neuromediators may act as modulators of glia both by regulating their functional activity or/and by a trophic effect on the expression of cytoskeletal activity.

REFERENCES

1. Tardy, M., C. Fages, H. Riol, G. Le Prince, P. Rataboul, C. Charriere-Bertrand & J. Nunez. 1989. Developmental expression of the glial fibrillary acidic protein mRNA in the central nervous system and in cultured astrocytes. J. Neurochem. **52:** 162–167.

2. Bridoux, A. M., C. Fages, D. Couchie, J. Nunez & M. Tardy. 1986. Protein synthesis in astrocytes; spontaneous and cyclic AMP induced differentiation. Dev. Neurosci. **8:** 31–43.
3. Fedoroff, S., W. A. McAuley, J. D. Houle & R. M. Devon. 1984. Astrocyte cell lineage similarity of astrocytes that form in the presence of dBcAMP in cultures to reactive astrocytes *in vivo.* J. Neurosci. Res. **12:** 15–27.
4. Le Prince, G., M. C. Copin, H. Hardin, M. F. Belin, J. P. Bouilloux & M. Tardy. 1990. Neuron-glia interactions: Effect of serotonin on the astroglial expression of GFAP and of its encoding message. Dev. Brain Res. **51:** 295–298.

Closing Remarks

J. G. NICHOLLS

Department of Pharmacology
Biocenter, University of Basel
Klingelbergstrasse 70
CH-4056 Basel, Switzerland

It is plainly impossible to try to present a comprehensive and useful summary of a symposium on glial-neuronal interaction with so many excellent papers, new advances, and new insights. To pick out highlights would be invidious and merely a reflection of my own taste. Instead, it may perhaps be appropriate to compare the present state of our knowledge with that in the early 1960s when studies on the physiology of neuroglia began in Stephen Kuffler's laboratory at Harvard. His style of experimentation and rigorous thinking set the stage for the advances that were to come. Then, as now, open questions abounded. When areas in which new advances are likely to be made are considered, another name that inevitably appears is that of John Treherne. The last series of experiments he made, on microglia, provided completely new information about the functional role of these supporting cells that had hitherto been so neglected.

What then was the status of research on glial cells in 1960? Interest in glia was virtually restricted to neuroanatomists. They had a vast amount of information about the structure of glial cells and about their relation to neurons and blood vessels. In addition, the advent of electron microscopy had added support to the idea that glial cells were not merely a type of connective tissue but were metabolically active and likely to perform some essential role. Ideas about the possible functions of glia were: first, that oligodendrocytes were responsible for making myelin, as were Schwann cells; second, that glial cells produced scar tissue after injury to the CNS; and third, that astrocytes transported materials from capillaries in the brain to neurons. This concept had originally been proposed by Golgi and had gained unthinking acceptance. In retrospect, it seems a most complicated speculation, implying that molecules move through glial cells, across their membranes, in through cytoplasm and then out through other membranes to reach the neurons.

A major impetus for new thinking about glial cells and their possible functions was provided by the papers of Svaetichin and Galambos. Svaetichin, a pioneer of microelectrode technique working at IVIC in Caracas, recorded S potentials from horizontal cells in the retina; as he (erroneously) considered horizontal cells to be glial, he proposed a series of complex metabolic interactions between neurons and glia.[1] A fringe benefit of these studies was the interest inspired in Jorge Villegas who began his long series of studies on Schwann cells, so elegantly represented in this volume. The paper by Galambos (present at this meeting!) was catalytic. Published in the Proceedings of the National Academy of Sciences it proposed a contribution of glial cells to higher functions.[2] Less obvious or dramatic were the elegant observations made by Bernard Katz and Ricardo Miledi[3] who demonstrated unequivocally that Schwann cells came to occupy the vacant end-plate sites at denervated neuromuscular junctions and there secreted acetylcholine. (Even today this seems the most direct demonstration of secretion of a transmitter by a satellite cell that demonstrably interacts with postsynaptic receptors, giving rise to synaptic potentials.)

The initial impetus for Stephen Kuffler's studies on glial cells came from a marvelous course that was taught at Harvard in those years, called "Area III." Steve wanted to know how a course on the nervous system could be taught while omitting any mention of the cell type, glia, that was most prevalent in our brains. Textbooks of physiology made no mention of glial cells, except in relation to myelin formation. Even now with certain exceptions glial cells are barely mentioned in neurobiology texts. Kuffler and Potter set out to answer basic questions about the physiological properties of glial cells. Rather than trying to find out if they were involved in learning, memory, or higher functions, they set themselves a more modest goal: what were the resting potentials of glial cells; do glial cells give action potentials? How do they interact with neurons? Are they required for neurons to give action potentials? Do glial cells interact with each other? Are all glial cells similar in their membrane properties?

A major technical problem was how to make such measurements in glial cells with intracellular electrodes in the mammalian brain where they are heterogeneous and inextricably mixed up with the neurons. The approach that Kuffler used was to search for a simplified preparation in which such problems could be tackled directly. It is worth mentioning that Kuffler's previous work had involved recording from cat retina (in which he had for the first time made an opening into the mammalian visual system in terms of receptive field analysis), the crustacean stretch receptor (in which he provided the first full description of sensory receptor potentials and inhibitory mechanisms), and the unequivocal demonstration of gamma-aminobutyric acid as the inhibitory transmitter at crustacean neuromuscular junctions. A hallmark of Kuffler's approach was to select the right preparation to answer the questions that were being asked. David Potter in a library search came up with the leech as an ideal preparation. The advantages of the leech had already been recognized by Retzius, Cajal, Del Rio Hortega, and others. In the leech CNS the glial cells are huge and can be seen directly under the dissecting microscope. Unlike the arrangement in the mammalian central nervous system, where numerous glial cells surround one single neuron, in the leech a single large glial cell surrounds many cell bodies and another large glial cell surrounds axons or neuropil processes. Together, Kuffler and Potter showed directly that the resting potentials of glial cells were higher than those of neurons, that glial cells did not give impulses, that they were not necessary for neurons to give impulses, and that glial cells were linked by electrical junctions. It should be emphasized that these experiments were fiendishly hard to do. Microelectrodes were not particularly sharp at that time; I can remember making 100 electrodes to get 2 or 3 that would penetrate glial cells reliably. Moreover, no marking techniques were available.

By the time I came to work with Kuffler in 1962 the problems we were interested in concerned the ionic mechanisms responsible for producing resting potentials in glial cells and the pathways taken by ions and small molecules as they moved through the nervous system from blood to neurons and in the opposite direction. It became evident that the cytoplasm of glial cells contained a high concentration of potassium which dominated the membrane potential. Moreover, the narrow extracellular spaces seen in the brain by electron microscopists were open and constituted the pathway for molecules to diffuse through the nervous system. At a meeting Kuffler and I attended in the early 1960s there was serious debate about the dimensions of these spaces and whether everything flowed through glia. A somewhat pompous question posed to Kuffler was what he thought the exact dimensions of extracellular space in the brain were? He replied that he did not really care as long as the extracellular space was not 0% or 100%.

Shortly after these experiments when Dick Orkand joined us, we performed experiments on frogs and mud puppies to see if findings made in the leech had any relevance in higher animals. The effects of repetitive nerve stimulation in producing potassium accumulation were analyzed in the optic nerve of *Necturus.* Later Dennis and Gerschenfeld[4] made the first recordings in mammalian optic nerve. A particular source of pleasure for us was the elegant and careful experiments made by Bruce Ransom and Sidney Goldring[5] on the physiological properties of glial cells in the cat cortex that were so strikingly similar to those in the leech.

It is perhaps worthwhile to emphasize the usefulness of Kuffler's approach. A colleague asked me recently, after a talk I had given on regeneration in the leech, why anybody would work on an invertebrate other than *Aplysia.* Kuffler's work showed again and again how important insights obtained from different animals could be. Indeed, he pointed out that the work on the leech was not a roundabout way of studying glial cell properties; it was in fact a short cut to analyzing their behavior in the mammalian central nervous system. His experiments were always designed to provide direct answers to straightforward questions. In addition, he was restrained in his speculation. It was always clear to us that glial cells must produce spatial buffering of potassium. If potassium concentration was raised in one region and depolarized the glial cell, potassium would enter and leave elsewhere. What was not clear to us was how to make rigorous quantitative calculations for the changes in concentration produced by this mechanism without detailed knowledge of membrane properties in different regions of the glial cell as well as knowledge of its geometry. It seemed to us a quantitative rather than a qualitative problem. A further major contribution of Kuffler which was clearly reflected in his work on glia was the concept of "neurobiology" as a discipline. Before he founded the Department of Neurobiology at Harvard, the nervous system had been studied separately by anatomists, physiologists, and biochemists in separate departments who never or almost never spoke to each other.

Although so many major advances have been made since those early days, numerous open questions about glia remain. Many of these are referred to in the papers in this volume. One question of importance concerns the role of glia, if any, in synaptic function. Another concerns Golgi's old hypothesis of transfer of materials between capillaries and neurons in the brain. Somehow, an extraordinary observation by Kelly and Van Essen[6] is largely neglected nowadays. They showed that glial cells in the visual cortex became depolarized by precisely oriented optical stimuli. When a beam of light with the appropriate orientation activated neurons, the glial cells in that column became depolarized. Early on, Orkand, Kuffler, and I had thought that glial cells by their membrane potential and depolarization provided an index of the level of electrical activity in a particular region of the brain. Moreover, depolarization, because it was activated by potassium rather than transmitter, arose from firing in excitatory or inhibitory nerves, from incoming and outgoing fibers indiscriminantly. It was natural to wonder if the brain would make use of this information. It seems attractive to speculate that glial cells that have such an intimate relation to capillaries might perhaps serve to regulate blood flow through an active area. This seems to me more likely than the idea that they serve a nursing function for the neurons.

Another major area of future promise concerns the possible functions of microglia. In one of his last papers, John Treherne with Peter Smith and Elizabeth Howes showed that in the cockroach (again an invertebrate!) the blood-brain barrier breaks down after injury. In this animal the blood-brain barrier is made up of glial cells.[7] Hematocytes, resembling macrophages or microglia, invade the site of the lesion and there reconstitute the barrier, becoming transformed into glial cells. These experiments by Treherne and his colleagues point the way to new thinking about microglia, particularly in relation to wandering cells that enter the brain from the blood and can perform functions in relation to responses to injury or to foreign agents. In addition, the role of microglia in removing myelin after damage to the central nervous system has become an important new area of research.

Both Stephen Kuffler and John Treherne shared the characteristic of making experiments themselves, with their own hands, to find out how things work. Their interest was not in prizes, in breakthroughs, or in the making of major new hypotheses. In the long run, their more modest aims produced results that stood the test of time and served as stepping stones for the exciting panorama of new experiments made possible by the techniques available today. As Steve said to me years ago, late at night when we were performing experiments, "Remember, Johnny, *these* are the good old days!"

REFERENCES

1. Svaetichin, G., M. Langer, G. Mitarai, R. Fatehchand, E. Vallecalle & J. Villegas. 1961. Glial control of neuronal networks and receptors. *In* The Visual System: Neurophysiology and Psychophysics. R. Jung & H. Kornhuber, eds.: 445–463. Springer. Berlin.
2. Galambos, R. 1961. A glia-neuron theory of brain function. Proc. Natl. Acad. Sci. USA **47:** 129–136.
3. Katz, B. & R. Miledi. 1959. Spontaneous subthreshold activity at denervated amphibian end-plates. J. Physiol. **146:** 44–45.
4. Dennis, M. J. & H. M. Gerschenfeld. 1969. Some physiological properties of identified mammalian neuroglial cells. J. Physiol. **203:** 211–222.
5. Ransom, B. R. & S. Goldring. 1973. Ionic determinants of membrane potential of cells presumed to be glia in cerebral cortex of cat. J. Neurophysiol. **36:** 855–868.
6. Kelly, J. P. & D. C. Van Essen. 1974. Cell structure and function in the visual cortex of the cat. J. Physiol. **238:** 515–547.
7. Smith, P. J. S., E. A. Howes & J. E. Treherne. 1987. Mechanisms of glial regeneration in an insect central nervous system. J. Exp. Biol. **132:** 59–78.

Index of Contributors